S al

Fluid Mechanics with Engineering Applications

Ninth Edition

Joseph B. Franzini
Stanford University

E. John Finnemore
Santa Clara University

Prepared by
E. John Finnemore
Santa Clara University

WCB McGraw-Hill

Boston Burr Ridge, IL Dubuque, IA Madison, WI New York San Francisco St. Louis
Bangkok Bogotá Caracas Lisbon London Madrid
Mexico City Milan New Delhi Seoul Singapore Sydney Taipei Toronto

WCB/McGraw-Hill

A Division of The ***McGraw-Hill*** *Companies*

Solutions Manual to accompany
FLUID MECHANICS: WITH ENGINEERING APPLICATIONS

1 2 3 4 5 6 7 8 9 0 BKM/BKM 9 3 2 1 0 9 8

ISBN 0-07-021915-X

http://www.mhhe.com

CONTENTS

PREFACE

This Solutions Manual was prepared for instructors who use *Fluid Mechanics with Engineering Applications*, ninth edition, as a textbook. It is requested that recipients of this manual circulate it only to their teaching colleagues, for if it gets into the hands of students, its purpose is defeated.

Changes from the eighth edition.

The most notable changes in the ninth edition of the textbook which affect problems are:

- A new introductory chapter emphasizes and recommends approaches to problem solving;
- About 75% of the problems are new or changed from the previous edition;
- Quantities previously described as being in English units are now more correctly referred to as being in British Gravitational (BG) units;
- 'Dual' problems, posed in both English and SI units (in parentheses) in the previous edition, have been discontinued;
- A new category of problems called Exercises has been added. These are placed at the ends of many sections, and consist of simpler, straightforward drill problems on the current section, for which answers are provided in the back of the book;
- In the more advanced chapters emphasis has been placed on the use of programming techniques in problem solving. These techniques include mathematics software such as Mathcad, spreadsheets, and advanced programmable calculators, and use of these is included in worked examples and problems. Exercises and problems in which such computing aids could be helpful are indicated by a □ icon in the margin. The various programming techniques are discussed in new Appendix B of the textbook.

Important features of this Solutions Manual.

Problem statements have been added to the solutions manual, in italics, to avoid the need to correlate two different books to understand a problem. A few problem statements had to be corrected after publication of (earlier printings of) the textbook, and students will need to be informed of such changes; these are identified by an asterisk (*) in the "Special Features" column of the Problem Selection Guide (preceding each chapter's solutions), and likewise in the statements with the solutions.

The solutions to the exercises and problems are categorized by the sections of the text. For each section the problems follow the exercises. Many problems, of course, depend on more than one section; in such cases they are categorized under the most advanced section used in their solution. Some chapters have a few miscellaneous problems at the end.

References to the equations and figures used in the solutions are provided extensively; these clarify the solution procedure and serve to identify other sections of the book that such problems depend on.

Ways to help instructors select problems for assignments are provided in the following section, "Suggested Ways of Using this Solutions Manual in Selecting Problems for Assignment," and in the Problem Selection Guides preceding the solutions for each chapter.

The format, notations, and accuracy used in this solutions manual are described in the second following section.

Appendices A and B of this manual provide some solutions of Sample Problems (in the textbook) by Mathcad and by HP 48G calculator; copies could be used as course handouts. These useful equation solvers are discussed in the textbook in Secs. B.1 and B.3, and at various locations throughout, with examples particularly in Secs. 8.14, 8.24, and 10.3.

Appendix C of this manual lists the more important corrections to be made to the second printing[1] of the textbook (produced late in 1997). Because revision of this ninth edition was considerably more extensive than previous revisions, more errors crept in. About 100 pages of the first printing contained corrections which were made to the second printing. The majority of the corrections remaining to be made (Appendix C) relate to material after Chapter 8, and most affect answers to Exercises.

Acknowledgments. We are very grateful to the following for their assistance in the production of this manual: users of the previous manual who suggested improvements; students who helped uncover errors in the solutions or suggested improvements to the problems of the previous edition; engineer Richard R. Blackburn and students Luong D. Do, Georgi J. Hall, Hoang T. Nguyen, David C. Ohlson, Ajay Patel, Paul J. Reginato, and Tony Tran who helped revise the solutions; and students Jose A. Alvarez, Jerome A. Blaha, Christina Calderón, Andrew S. Duncan, Ralph E. Garcia, Jr., Michael A. Loomis. Victor O. Lopez, Stephen F. McMahon, and Steven M. Sidell, who helped record the solutions on word processors, and proof read and edit them; and especially Gulshan Finnemore, my wife, for her loving support and assistance of many kinds throughout.

Future Improvements. Although the solutions to all the problems in this ninth edition have been carefully checked, it is possible that a few errors remain. If errors are detected, it would be greatly appreciated if you would bring them to the attention of Professor E. John Finnemore, Department of Civil Engineering, Santa Clara University, Santa Clara, California 95053 or by e-mail to EFinnemore@SCU.Edu . Likewise, we would be pleased to receive your suggestions for improvements to the problems, the textbook, or to this solutions manual. Thank you.

E. John Finnemore

[1] See Appendix C of this manual for information on identification of the various printings.

SUGGESTED WAYS OF USING THIS SOLUTIONS MANUAL IN SELECTING PROBLEMS FOR ASSIGNMENT

First:

Refer to the Problem Selection Guide which precedes the solutions for each chapter. This categorizes the exercises and problems by text section, and gives convenient summary information on the solutions, such as their length, difficulty, and system(s) of units used. It also identifies multi-part problems, and special features required in their solution such as calculus, trial and error, or graphing. Under "Special features," a dagger (†) indicates answers are sensitive to values graphed or read from figures; a box (□, suggestive of the marginal icons) indicates that programmed computing aids could be of help.

Problems with multiple parts are identified in the Problem Selection Guide as some instructors may not wish to assign all parts for solution. The "Similar" column will help instructors avoid assigning two very similar problems (usually one is in BG units and one is in SI), and it could help find an alternative to a preferred problem.

Note that at the end of some chapters there are also some "miscellaneous" problems which are based on mixtures of material from various sections.

Second:

Refer to the exercise or problem statements and solutions themselves, which are categorized by text sections. There one may review the method of solution, and the extent of the solution's dependence on preceding sections and chapters (all "source" equations, figures, and tables needed in the solution are cited by number).

When assigning end-of-chapter problems note that answers to them are not given at the back of the book. One reason is that this would reduce the learning of trial and error procedures (more reasons are discussed near the end of Sec. 1.3 of the textbook). A possible compromise is to give students a *range* which includes the answer, e.g: 4 < Ans < 20 cfs.

FORMAT, NOTATIONS, AND ACCURACY
used in this solutions manual

Each solution is separated from its problem statement by a blank line, on which, in the left margin, is indicated the units used in the problem, as follows:

BG – British Gravitational (English)
SI – International System (metric)
B – Both (BG and SI)
N – None

Answers are indicated in the solutions by solid triangles, usually ◂ to the right.

The solutions have all been computed to calculator accuracy. Although results are reported to only three or four significant figures, as is usual and appropriate in engineering practice, it is important to note that in multi-step solutions such 'rounded' results from early steps are not used as input to subsequent steps. For then different methods of rounding could yield different answers. In all cases the full accuracy (eight significant figures) is retained and used throughout the calculations, regardless of how the results are reported.

Chapter 1

Introduction

PROBLEM SELECTION GUIDE

Sec	Exer/Prob	Units	Difficulty	Length	Parts	Similar	Special features
1.5	*Dimensions and Units*						
	X[1]1.5.1	N	Easy	Short	1		
	1.5.2	N	Easy	Short	1		
	1.5.3	N	Easy	Short	1		
	1.5.4	N	Easy	Medium	1		
	1.5.5	N	Easy	Short	1		
	1.5.6	B	Easy	Short	3		
	1.5.7	B	Easy	Short	3		Uses Table A.1
	1.5.8	B	Easy	Short	2		
	1.5.9	B	Easy	Short	2		

[1] For all Exercises (marked "X"), answers are given in Appendix D of the textbook.
X = Exercise, P = (end-of-chapter) Problem, S = Sample Problem.

Chapter 1
INTRODUCTION

Sec. 1.5: Dimensions and Units -- *Exercises*

1.5.1 N *Demonstrate that Eq. (6.5) is dimensionally homogeneous.*

Eq. 6.5: $\Sigma F = \dot{m}_2 V_2 - \dot{m}_1 V_1 = \rho_2 Q_2 V_2 - \rho_1 Q_1 V_1$

Term 1: Dimensions of $F = F = MLT^{-2}$

Terms 2−3: Dimensions of $\dot{m}V = (M/T)(L/T) = MLT^{-2}$

Terms 4−5: Dimensions of $\rho QV = (M/L^3)(L^3/T)(L/T) = MLT^{-2}$

All terms have the same dimensions, MLT^{-2} (or F) ◄

1.5.2 N *Demonstrate that Eq. (10.57) is dimensionally homogeneous. Note that C_d is dimensionless.*

Eq. 10.57: $Q = C_d A\sqrt{2gh}$

Term 1: Dimensions of $Q = L^3/T$

Term 2: Dimensions of $C_d A\sqrt{2gh} = (1)L^2\sqrt{(L/T^2)L} = L^3/T$

Both terms have the same dimensions, L^3/T ◄

1.5.3 N *Demonstrate that Eq. (11.2) is dimensionally homogeneous. Note that $\forall$ is a volume, h_L has the dimensions of length, and ν is kinematic viscosity.*

Eq. (11.2): $\nu = \dfrac{\pi D^4 h_L}{128 \forall L} gt$

Term 1: Dimensions of $\nu = L^2/T$

Term 2: Dimensions $= L^4(L)(LT^{-2})T/[(L^3)(L)] = L^2/T$

Both terms have the same dimensions, L^2/T ◄

1.5.4 N *Demonstrate that Eq. (12.4) is dimensionally homogeneous.*

Eq. 12.4: $-\dfrac{dp}{\gamma} - dz - d\dfrac{V^2}{2g} = \dfrac{2\tau_0\, ds}{\gamma r} + \dfrac{ds}{g}\dfrac{dV}{dt}$

Term 1: Dimensions of $dp/\gamma = (F/L^2)/(F/L^3) = L$

Term 2: Dimensions of $dz = L$

Term 3: Dimensions of $d(V^2/2g) = (L/T)^2/(L/T^2) = L$

Term 4: Dimensions of $2\tau_0 ds/(\gamma r) = (F/L^2)L/[(F/L^3)L] = L$

Term 5: Dimensions of $(ds/g)(dV/dt) = [L/(LT^{-2})][LT^{-1}/T] = L$

All terms have the same dimensions, L ◄

1.5.5 *Demonstrate that Eq. (13.45) is dimensionally homogeneous. Note that k is dimensionless.*
N

Eq. 13.45: $\dot{m}_{max} = \dfrac{A_2 p_0}{\sqrt{T_0}}\sqrt{\dfrac{k}{R}\left(\dfrac{2}{k+1}\right)^{(k+1)/(k-1)}}$

Term 1: Dimensions of $\dot{m}$ (or dm/dt) $= M/T$

Term 2: Dimensions of $\dfrac{Ap}{\sqrt{TR}} = \dfrac{L^2(FL^{-2})}{\sqrt{\theta(L^2T^{-2}\theta^{-1})}} = \dfrac{F}{LT^{-1}} = \dfrac{MLT^{-2}}{LT^{-1}} = M/T$

Both terms have the same dimensions, M/T ◄

1.5.6 *Using information from inside the cover of this book, determine the weight of a U.S. gallon of water in the following units: (a) pounds; (b) newtons; (c) dynes.*

B

Inside cover: 1 imperial gal = 1.201 US gal and weighs 10 lb, 1 lb = 4.448 N, 1 N = 10^5 dyne.

(*a*) (10 lb/imp gal)/(1.201 US gal/imp gal) = 8.33 lb/US gal ; 1 US gal weighs 8.33 lb ◄

(*b*) (8.33 lb)(4.448 N/lb) = 37.0 N ◄

(*c*) (37.0 N)(10^5 dyne/N) = 3.70×10^6 dyne ◄

1.5.7 *Using information from inside the cover of this book, determine the weight of one liter of water at 15°C in the following units: (a) pounds; (b) newtons; (c) dynes.*

B

Table A.1 at 15°C: γ = 9.798 kN/m³ = 9798 N/m³

Inside covers: 1 L = 10^{-3} m³, 1 N = 10^5 dyne, 1 N = 0.2248 lb

(*b*) (9798 N/m³)(10^{-3} m³/L) = 9.80 N/L ; 1 L weighs 9.80 N ◄

(*a*) (9.80 N)(0.2248 lb/N) = 2.20 lb ◄

(*c*) (9.80 N)(10^5 dyne/N) = 980×10^3 dyne ◄

1.5.8 *Using information from inside the cover of this book, convert 15 million U.S. gallons per day (mgd) into (a) BG and (b) SI units.*

B

Inside cover: 1 mgd = 1.547 cfs, 1 cfs = 0.0283 m³/s

(*a*) 15 mgd = (15 mgd)(1.547 cfs/mgd) = 23.2 cfs ◄

(*b*) 23.2 cfs = (23.2 cfs)(0.0283 m³/s)/(1 cfs) = 0.657 m³/s ◄

1.5.9 *Using information from inside the cover of this book, convert 75 km/h into (a) SI and (b) BG units.*
B

Inside back cover: 1 m/s = 3.28 ft/sec.

(*a*) 75 km/h = $\left(\dfrac{75\text{ km}}{\text{h}}\right)\left(\dfrac{1000\text{ m}}{1\text{ km}}\right)\left(\dfrac{1\text{ h}}{3600\text{ s}}\right)$ = 20.8 m/s ◄

(*b*) 20.8 m/s = (20.8 m/s)(3.28 ft/sec)/(1 m/s) = 68.4 ft/sec ◄

Chapter 2

Properties of Fluids

PROBLEM SELECTION GUIDE

Sec	Exer/Prob	Units	Difficulty	Length	Parts	Similar	Special features
2.3	*Density, Specific Weight, Specific Volume, and Specific Gravity*						
	X[1]2.3.1	BG	Easy	V Short	1	P2.2	
	2.3.2	BG	V Easy	V Short	1	2.3.3	
	2.3.3	SI	V Easy	V Short	1	2.3.2	
	2.3.4	B	V Easy	V Short	1		
	2.3.5	BG	V Easy	V Short	1	2.3.6	
	2.3.6	SI	V Easy	V Short	1	2.3.5	
	2.3.7	BG	V Easy	V Short	1	2.3.8	
	2.3.8	SI	V Easy	V Short	1	2.3.7	
	2.3.9	SI	Medium	Short	1		
	P 2.1	SI	V Easy	V Short	1		
	2.2	SI	Easy	V Short	1	X2.3.1	
	2.3	BG	Easy	Medium	3		
2.5	*Compressibility of Liquids*						
	X 2.5.1	BG	Easy	V Short	1		
	2.5.2	BG	Easy	Short	3		
	2.5.3	BG	V Easy	V Short	2		
	2.5.4	B	Easy	V Short	2		2-D interpolation; unit conversions
	2.5.5	BG	Easy	V Short	1	2.5.6	
	2.5.6	SI	Easy	V Short	1	2.5.5	
	2.5.7	BG	Easy	Short	1	2.5.8	
	2.5.8	SI	Easy	Short	1	2.5.7	
	P 2.4	SI	Medium	Medium	3		Interpolation
2.6	*Specific Weight of Liquids*						
	X 2.6.1	B	V Easy	V Short	1		†
	2.6.2	BG	Easy	Short	1		
	2.6.3	SI	Easy	Short	1		
	P 2.5	SI	Medium	Medium	1		
	2.6	BG	Medium	Short	1		Interpolation

/cont...

[1] For all Exercises (marked "X"), answers are given in Appendix D of the textbook.
† Answers are sensitive to values which are read from graphs.
X = Exercise, P = (end-of-chapter) Problem, S = Sample Problem.

Sec	Exer/Prob	Units	Difficulty	Length	Parts	Similar	Special features
2.7	***Property Relations for Perfect Gases***						
	X 2.7.1	BG	Easy	V Short	1		
	2.7.2	BG	Easy	Short	1		
	2.7.3	SI	Easy	V Short	1		
	2.7.4	BG	Easy	Short	1	2.7.5	
	2.7.5	SI	Easy	Short	1	2.7.4	
	2.7.6	BG	Easy	Short	3	2.7.7	
	2.7.7	SI	Easy	Short	3	2.7.6	
	2.7.8	SI	Medium	Medium	3		
	2.7.9	N	Medium	Short	1		Derivation
	P 2.7	SI	Easy	Short	2		
	2.8	SI	Easy	Short	3		
	2.9	BG	Medium	Medium	5		
2.8	***Compressibility of Gases***						
	X 2.8.1	BG	V Easy	V Short	1	2.8.2	
	2.8.2	SI	V Easy	V Short	1	2.8.1	
	2.8.3	BG	Medium	Short	1	2.8.4	
	2.8.4	SI	Medium	Short	1	2.8.3	
	2.8.5	SI	Easy	Short	2		
	P 2.10	BG	Easy	Short	2	2.11	
	2.11	SI	Easy	Short	2	2.10	
2.11	***Viscosity***						
	X 2.11.1	B	Easy	Short	1		Unit conversions
	2.11.2	B	Easy	Short	3		†
	2.11.3	B	V Easy	V Short	1		† Unit conversions (minor)
	2.11.4	BG	V Easy	V Short	1		†
	2.11.5	BG	V Easy	V Short	1		
	2.11.6	SI	Easy	V Short	1	2.11.7	
	2.11.7	BG	Easy	V Short	1	2.11.6	
	2.11.8	SI	Easy	Short	2		†
	2.11.9	B	Easy	Short	1		† Unit conversions
	2.11.10	BG	Medium	Short	1		
	2.11.11	BG	Medium	Short	1		†
	2.11.12	N	Easy	Short	1		Integration
	2.11.13	BG	Easy	Medium	1		Unit conversion (minor)
	P 2.12	SI	Medium	Short	1		Unit conversion (minor)
	2.13	SI	Medium	Short	1		
	2.14	BG	Hard	Medium	1		† Integration
	2.15	SI	Medium	Short	1		Integration
	2.16	N	Hard	Medium	1		Integration

/cont...

Sec	Exer/Prob	Units	Difficulty	Length	Parts	Similar	Special features
2.12	*Surface Tension*						
	X 2.12.1	SI	Easy	V Short	1	2.12.2	†
	2.12.2	BG	V Easy	V Short	1	2.12.1	†
	2.12.3	BG	Easy	Short	1		† Interpolation
	2.12.4	SI	Easy	V Short	1		
	2.12.5	BG	V Easy	V Short	1		
	P 2.17	BG	Easy	Short	1		
	2.18	SI	Easy	Short	1		
2.13	*Vapor Pressure of Liquids*						
	X 2.13.1	BG	Easy	Short	1		Interpolation
	2.13.2	SI	V Easy	V Short	1		
	P 2.19	BG	V Easy	V Short	1		

Chapter 2
PROPERTIES OF FLUIDS

Sec. 2.3: Density, Specific Weight, Specific Volume, and Specific Gravity -- *Exercises*

2.3.1 *If a certain gasoline weights 46 lb/ft^3, what are the values of its density, specific volume, and specific gravity relative to water at 60°F? Use Appendix A.*

BG

Eq 2.1: $\rho = 46/32.2 = 1.429$ slugs/ft^3 ◄

Eq 2.2: $\upsilon = 1/1.429 = 0.700$ ft^3/slug ◄

Table A.1: ρ_{water} at 60°F = 1.938 slugs/ft^3; $s = 1.429/1.938 = 0.737$ ◄

2.3.2 *A certain gas weighs 0.12 lb/ft^3 at a certain temperature and pressure. What are the values of its density, specific volume, and specific gravity relative to air weighing 0.075 lb/ft^3?*

BG

Eq. 2.1: $\rho = 0.12/32.2 = 0.003\,73$ slugs/ft^3 ◄

Eq. 2.2: $\upsilon = 1/0.003\,73 = 268$ ft^3/slug ◄

$s = 0.12/0.075 = 1.600$ ◄

2.3.3 *A certain gas weighs 18.0 N/m^3 at a certain temperature and pressure. What are the values of its density, specific volume, and specific gravity relative to air weighing 12.0 N/m^3?*

SI

Eq. 2.1: $\rho = 18/9.81 = 1.835$ kg/m^3 ◄

Eq. 2.2: $\upsilon = 1/1.835 = 0.545$ m^3/kg ◄

$s = 18/12 = 1.500$ ◄

2.3.4 *The specific weight of glycerin is 78.7 lb/ft^3. Compute its density and specific gravity. What is its specific weight in kN/m^3?*

B

Eq. 2.1: $\rho = 78.7/32.2 = 2.44$ slugs/ft^3 ◄

$s = 78.7/62.4 = 1.261$ ◄ so $\rho = 1.261$ Mg/m^3

Eq. 2.1: $\gamma = 9.81(1.261) = 12.37$ kN/m^3 ◄

2.3.5 *If the specific weight of a liquid is 54 lb/ft^3, what is its density?*

BG

Eq. 2.1: $\rho = 54/32.2 = 1.677$ slugs/ft^3 ◄

2.3.6 *If the specific weight of a liquid is 7.6 kN/m^3, what is its density?*

SI

Eq. 2.1: $\rho = 7600/9.81 = 775$ kg/m^3 ◄

2.3.7 *If the specific volume of a gas is 350 ft^3/slug, what is its specific weight in lb/ft^3?*

BG

Eqs. 2.1, 2.2: $$\gamma = \rho g = \frac{g}{\upsilon} = \frac{32.2\ \text{ft/sec}^2}{350\ \text{ft}^3\text{/slug}}\left(\frac{\text{lb}}{\text{slug}\cdot\text{ft/sec}^2}\right) = 0.0920\ \text{lb/ft}^3$$ ◄

2.3.8 *If the specific volume of a gas is 0.80 m³/kg, what is its specific weight in N/m³?*

SI

Eqs. 2.1, 2.2: $\gamma = \rho g = \frac{g}{v} = \frac{9.81\ \text{m/s}^2}{0.80\ \text{m}^3/\text{kg}}\left(\frac{\text{N}}{\text{kg}\cdot\text{m/s}^2}\right) = 12.26\ \text{N/m}^3$ ◄

2.3.9 *Initially when 1000.00 mL of water at 10°C are poured into a glass cylinder the height of the water column is 1000.0 mm. The water and its container are heated to 80°C. Assuming no evaporation, what then will be the depth of the water column if the coefficient of thermal expansion for the glass is 3.6×10⁻⁶ mm/mm per °C?*

SI

Table A.1: At 10°C, $\rho = 999.7\ \text{kg/m}^3$; at 80°C, $\rho = 971.8\ \text{kg/m}^3$.

Mass of water $= \rho V = \rho_{10} V_{10} = \rho_{80} V_{80}$

$(999.7\ \text{kg/m}^3)(1000.0\ \text{mL}) = (971.8\ \text{kg/m}^3) V_{80}$; so $V_{80} = 1028.71\ \text{mL} = 1028\ 710\ \text{mm}^3$

$A_{10} = V_{10}/h_{10} = 1000\,000.00\ \text{mm}^3/1000.0\ \text{mm} = 1000.0\ \text{mm}^2$

$A_{80} = A_{10}[1 + (80 - 10)3.6 \times 10^{-6}]^2 = 1000.5\ \text{mm}^2$

$h_{80} = V_{80}/A_{80} = 1028\,710/1000.5 = 1028.2\ \text{mm}$ ◄

Sec 2.3: Density, Specific Weight, Specific Volume, and Specific Gravity -- *Problems*

2.1 *If the specific weight of a gas is 16.40 N/m³, what is its specific volume in m³/kg?*

SI

Eq. 2.2: $v = \frac{1}{\rho} = \frac{g}{\gamma} = \frac{9.81\ \text{m/s}^2}{16.40(\text{kg}\cdot\text{m/s}^2)/\text{m}^3} = 0.598\ \text{m}^3/\text{kg}$ ◄

2.2 *If a certain liquid weighs 7600 N/m³, what are the values of its density, specific volume, and specific gravity relative to water at 15°C? Use Appendix A.*

SI

Eq. 2.1: $\rho = 7600/9.81 = 775\ \text{kg/m}^3$ ◄ Eq. 2.2: $v = 1/775 = 0.001\ 291\ \text{m}^3/\text{kg}$ ◄

Table A.1: ρ_{water} at 15°C $= 999.1\ \text{kg/m}^3$; $s = 775/999.1 = 0.775$ ◄

2.3 *Find the change in volume of 12.00 lb of water at ordinary atmospheric pressure for the following conditions: (a) reducing the temperature 40°F from 160°F to 120°F, (b) reducing the temperature 40°F from 120°F to 80°F, and (c) reducing the temperature 40°F from 80°F to 40°F. Calculate each and note the trend in the changes in volume.*

BG

$W = \gamma V$, so $V = W/\gamma = 12.00/\gamma$

T °F	γ (Table A.1) lb/ft³	$V = 12.00/\gamma$ ft³
160	61.00	0.196 72
120	61.71	0.194 46
80	62.22	0.192 86
40	62.43	0.192 22

(*a*) $\Delta V_{160-120} = 0.002\ 26\ \text{ft}^3$ decrease in volume ◄

(*b*) $\Delta V_{120-80} = 0.001\ 59\ \text{ft}^3$ decrease in volume ◄

(*c*) $\Delta V_{80-40} = 0.000\ 65\ \text{ft}^3$ decrease in volume ◄

For $\Delta T = 40°$ F, the volume change increases with temperature ◄

Sec. 2.5: Compressibility of Liquids -- *Exercises*

2.5.1 *Water in a hydraulic press is subjected to a pressure of 4500 psia at 68°F. If the initial pressure is 15 psia, approximately what will be the percentage decrease in specific volume? Use Table 2.1*

BG

From Table 2.1: $E_\upsilon \approx (320{,}000 + 348{,}000)/2 = 334{,}000$ psi

Eq. 2.3: $\Delta\upsilon/\upsilon_1 = -(4500 - 15)/365{,}000 = -0.013\,43$ or 1.34% decrease ◄

2.5.2 *At a depth of 5 miles in the ocean the pressure is 11,930 psi. Assume specific weight at the surface is 64.00 lb/ft³ and that the average volume modulus is 340,000 psi for that pressure range. (a) What will be the change in specific volume between that at the surface and at that depth? (b) What will be the specific volume at that depth? (c) What will be the specific weight at that depth?*

BG

(*a*) $\upsilon_1 = 1/\rho_1 = g/\gamma_1 = 32.2/64.00 = 0.503$ ft³/slug

Eq. 2.3: $\Delta\upsilon = -0.503(11{,}930 - 0)/340{,}000 = -0.017\,65$ ft³/slug ◄

(*b*) From Eq. 2.3: $\upsilon_2 = \upsilon_1 + \Delta\upsilon = 0.503 - 0.017\,65 = 0.485$ ft³/slug ◄

(*c*) $\gamma_2 = g/\upsilon_2 = 32.2/0.485 = 66.3$ lb/ft³ ◄

2.5.3 *(a) What is the percentage change in the specific volume in Exercise 2.5.2? (b) What is the percentage change in the specific weight in Exercise 2.5.2?*

BG

(*a*) $\Delta\upsilon/\upsilon_1 = 0.017\,65/0.503 = 3.51\%$ decrease ◄

(*b*) $\Delta\gamma/\gamma_1 = (66.3 - 64.00)/64.00 = 3.64\%$ increase ◄

2.5.4 *To two significant figures what is the bulk modulus of water in MN/m² at 50°C under a pressure of 30 MN/m²? Use Table 2.1.*

B

From inside cover: 50°C = 122°F; 30 MN/m² = 4351 psi

From Table 2.1, by linear interpolation in two directions: $E_\upsilon \approx 360{,}400$ psi

$360{,}400(6895/10^6) = 2485$ MN/m² ≈ 2500 MN/m² ◄

2.5.5 *At normal atmospheric conditions, approximately what pressure in psi must be applied to water to reduce its volume by 3% ?*

BG

Table 2.1: At normal atmospheric conditions, $E_\upsilon \approx 320{,}000$ psi

From Eq. 2.3*a*: $\Delta\upsilon/\upsilon_1 = -0.03 = -\Delta p/320{,}000$; $\Delta p = 9600$ psi ◄

2.5.6 *At normal atmospheric conditions, approximately what pressure in MPa must be applied to water to reduce its volume by 2% ?*

SI

Table 2.1: At normal atmospheric conditions, $E_\upsilon \approx 320{,}000$ psi

(320,000 psi)(6895) = 2.21×10^9 Pa = 2210 MPa

From Eq. 2.3*a*: $\Delta\upsilon/\upsilon_1 = -0.02 = -\Delta p/2210$; $\Delta p = 44.1$ MPa ◄

2.5.7 *Water is contained in a long rigid cylinder whose inside diameter is 0.650 in. A plunger is used to apply pressure to the water. The length of the column of water is 20.00 in. What will be the length of the column of water if a force of 500 lb is applied to the plunger. Assume no leakage and zero friction between the plunger and the cylinder.*

BG

$p_1 = 0;\ p_2 =$ Force/Area $= 500/\pi(0.325)^2 = 1507$ lb/in^2

Since the tube is rigid, from Eq. 2.3: $\Delta\upsilon/\upsilon = \Delta L/L = (L_2 - L_2)/L_1 = -(p_2 - p_1)/E_\upsilon$

So $L_2 - 20.00 = -20.00(1507 - 0)/320{,}000 = -0.0942$ in ; $L_2 = 20.00 - 0.0942 = 19.91$ in ◄

2.5.8 *A rigid cylinder, inside diameter 15 mm, contains a column of water 500 mm long. What will the column length be if a force of 2 kN is applied to its end by a frictionless plunger? Assume no leakage.*

SI

$p_1 = 0;\ p_2 =$ Force/Area $= 2\ \text{kN}/\pi(0.0075\ \text{m})^2 = 11\ 320\ \text{kN/m}^2 = 11.32$ MPa

$E_\upsilon \approx 320{,}000\ \text{psi}(6895) = 2.1 \times 10^9\ \text{Pa} = 2210$ MPa

Since the tube is rigid, from Eq. 2.3: $\Delta\upsilon/\upsilon = \Delta L/L = (L_2 - L_2)/L_1 = -(p_2 - p_1)/E_\upsilon$

so $L_2 - 0.5 = -(0.5\ \text{m})(11.32 - 0)/2210 = -0.002\ 56$ m

$L_2 = 0.5 - 0.002\ 56 = 0.497$ m or 497 mm ◄

Sec. 2.5: Compressibility of Liquids -- *Problems*

2.4 *Compare the change in volume of 2.00 m^3 of 80°C water for the following conditions: (a) a temperature decrease of 10°C from 80°C to 70°C with pressure constant, (b) a pressure increase of 10 MN/m^2 with temperature remaining constant at 80°C, and (c) a temperature increase from 80°C to 90°C combined with a pressure decrease of 20 MN/m^2.*

SI

(*a*) Table A.1: $\gamma_{80} = 9.530$ kN/m^3, $\gamma_{70} = 9.589$ kN/m^3

$W = \gamma V = \gamma_{80}V_{80} = \gamma_{70}V_{70}$ so $9.530(2) = 9.589V_{70}$, from which $V_{70} = 1.987\ 69$ m^3

$\Delta V_{80-70} = 2.000\ 00 - 1.987\ 69 = 0.012\ 31$ m^3 decrease in volume ◄

(*b*) 10 MN/m^2 $\equiv 10 \times 10^6/6894 = 1450$ psi ; 80°C $\equiv 32 + (9/5)80 = 176°$ F

From Table 2.1, by linear interpolation: $E_\upsilon = 326{,}000$ psi

$E_\upsilon \equiv 326{,}000(6895) = 2.24 \times 10^9$ Pa $= 2240$ MN/m^2

Letting the constant mass of water be m, then $m = \rho V$ and so $1/\rho = V/m$

But from Eq. 2.2: $\upsilon = 1/\rho$ so $\upsilon = V/m$

Using this relation to substitute for υ into Eq. 2.3, remembering that m is constant:

$(\upsilon_2 - \upsilon_1)/\upsilon_1 = (V_2 - V_1)/V_1 = -\Delta p/E_\upsilon$

So $\Delta V = -V\Delta p/E_\upsilon = -(2.00\ \text{m}^3)(10/2240) = -0.008\ 91$ m^3 ◄

The minus sign indicates a reduction in volume.

(*c*) Table A.1: $\gamma_{80} = 9.530$ kN/m^3, $\gamma_{90} = 9.466$ kN/m^3

Constant weight $= \gamma_{80}V_{80} = \gamma_{90}V_{90}$ so $9.530(2) = 9.466V_{90}$ from which $V_{90} = 2.013\ 52$ m^3

and due to temperature increase, $\Delta V_{80-90} = 0.013\ 52$ m^3 increase.

From Eq. 2.3: $\Delta V = -V_1\Delta p/E_\upsilon = -2(-20/2240) = +0.017\ 82$ m^3 (+ indicates a volume increase)

Summing the changes for both causes: $\Delta V = 0.013\ 52 + 0.017\ 82 = 0.0313$ m^3 increase ◄

Sec. 2.6: Specific Weight of Liquids -- *Exercises*

2.6.1 *Approximately what is the specific weight of water in kN/m³ at a temperature of 170°F under normal atmospheric conditions? What is the approximate specific weight at 170°F under a pressure of 2000 psia? Use Fig. 2.1.*

B

From Fig. 2.1: At 14.7 psi and 170°F, $\gamma \approx 9.57$ kN/m³ ◀

At 2000 psi and 170°F, $\gamma \approx 9.63$ kN/m³ ◀

2.6.2 *A vessel contains 2.0 ft³ of water at 40°F and atmospheric pressure. If the water is heated to 60°F what will be the percentage change in its volume? What weight of water must be removed to maintain the volume at its original value? Use Appendix A.*

BG

Table A.1: $\gamma_{40} = 62.43$ pcf, $\gamma_{60} = 62.37$ pcf

Weight of water $= \gamma\mathcal{V} = \gamma_{40}\mathcal{V}_{40} = \gamma_{60}\mathcal{V}_{60}$; $62.43(2) = 62.37\mathcal{V}_{60}$; $\mathcal{V}_{60} = 2.001\,924$ ft³

$\Delta\mathcal{V}/\mathcal{V}_{40} = 0.001\,924/2 = 0.0962\%$ increase ◀

Must remove (0.001 924 ft³)(62.37 lb/ft³) = 0.1200 lb ◀

2.6.3 *A cylindrical tank (diameter = 10.00 m and depth = 6.00 m) contains water at 20°C and is brimful. If the water is heated to 50°C, how much water will spill over the edge of the tank? Assume the tank does not expand with the change in temperature. Use Appendix A.*

SI

Table A.1: $\gamma_{20} = 9.789$ kN/m³, $\gamma_{50} = 9.689$ kN/m³

Tank vol $= \mathcal{V}_{20} = \pi 5^2(6) = 471.24$ m³; weight of water in tank $= 9.789(471.2) = 4613$ kN

$\mathcal{V}_{50} = 4613/9.689 = 476.10$ m³; $\Delta\mathcal{V} = 476.10 - 471.24 = 4.86$ m³ ◀

Sec. 2.6: Specific Weight of Liquids -- ***Problems***

2.5 *Repeat Exer. 2.6.3 for the case where the tank is made of a material which has a coefficient of thermal expansion of 4.8×10^{-6} mm/mm per °C.*

SI

Table A.1: $\gamma_{20} = 9.789$ kN/m³, $\gamma_{50} = 9.689$ kN/m³

At 20°C, $\mathcal{V}_{tank} = \pi 5^2(6) = 471.24$ m³; $\Delta T = 50 - 20 = 30°$ C

At 50°C, $\mathcal{V}_{tank} = 471.24[1 + (\Delta T)4.8 \times 10^{-6}]^3 = 471.24(1.000\ 432)$

$\Delta\mathcal{V}_{tank} = 471.24(0.000\ 432) = 0.204$ m³ increase

Weight of water $= \gamma\mathcal{V} = \gamma_{20}\mathcal{V}_{20} = \gamma_{50}\mathcal{V}_{50}$; $9.789(471.24) = 9.489\mathcal{V}_{50}$ from which $\mathcal{V}_{50} = 476.10$ m³

$\Delta\mathcal{V}_{water} = 476.10 - 471.24 = 4.86$ m³ increase ; spill $\Delta\mathcal{V} = 4.86 - 0.204 = 4.66$ m³ ◀

2.6 *A heavy closed steel chamber is filled with water at 50°F and at atmospheric pressure. If the temperature of the water and the chamber is raised to 90°F, what will be the new pressure of the water? The coefficient of thermal expansion of the steel is 6.5×10^{-6} in/in per °F; Assume the chamber is unaffected by the water pressure. Use Fig. 2.1.*

BG

$\mathcal{V}_{90} = \mathcal{V}_{50}[1 + (90 - 50)6.5 \times 10^{-6}]^3 = 1.000\,780\mathcal{V}_{50}$

Table A.1: $\gamma_{50} = 62.41$ lb/ft³ at p = atmos.

Wt of water $= \gamma\mathcal{V} = \gamma_{50}\mathcal{V}_{50} = \gamma_{90}\mathcal{V}_{90}$; $62.41\mathcal{V}_{50} = (1.000\,780\mathcal{V}_{50})\gamma_{90}$; $\gamma_{90} = 62.36$ pcf

Fig. 2.1, by interpolation: $p_{90} \approx 1240$ psia ◀

Sec. 2.7: Property Relations for Perfect Gases -- *Exercises*

2.7.1 *A hydrogen-filled balloon of the type used in cosmic-ray studies is to be expanded to its full size, which is a 100-ft-diameter sphere, without stress in the wall at an altitude of 150,000 ft. If the pressure and temperature at this altitude are 0.14 psia and −67°F respectively, find the volume of hydrogen at 14.7 psia and 60°F which should be added on the ground. Neglect the balloon's weight.*

BG

Eq. 2.4: $pv = RT$; $v = V/m = Vg/W$; $pVg/W = RT$; here pV/T = constant

$$\frac{14.7V_1}{460 + 60} = \frac{0.14 \times (4/3)\pi 50^3}{460 - 67}; \quad V_1 = 6600 \text{ ft}^3 \quad ◄$$

2.7.2 *If natural gas has a specific gravity of 0.6 relative to air at 14.7 psia and 68°F, what are its specific weight and specific volume at that same pressure and temperature. What is the value of R for the gas? Solve without using Table A.2.*

BG

Table A.5 for air: gas constant R = 1715 ft·lb/(slug·°R)

$$\text{Eq. 2.5:} \quad \gamma_{\text{air}} = \frac{gp}{RT} = \frac{32.2(14.7)144}{1715(460+68)} = 0.0753 \text{ lb/ft}^3$$

Sec. 2.3: $\gamma_{\text{ngas}} = s(\gamma_{\text{air}}) = 0.6(0.0753) = 0.0452 \text{ lb/ft}^3$ ◄

Eqs. 2.1 and 2.2: $v = 1/\rho = g/\gamma = 32.2/0.0452 = 713 \text{ ft}^3\text{/slug}$ ◄

Eq. 2.5: For a given T and p, gp/T is constant

$\therefore\ \gamma R = \gamma_{\text{air}}R_{\text{air}} = \gamma_{\text{ngas}}R_{\text{ngas}}$ and $R_{\text{ngas}} = R_{\text{air}}(\gamma_{\text{air}}/\gamma_{\text{ngas}}) = R_{\text{air}}/s_{\text{ngas}}$

or $R_{\text{ngas}} = 1715/0.6 = 2858$ ft·lb/(slug·°R) ◄

2.7.3 *A gas at 40°C under a pressure of 20 000 mb abs has a specific weight of 241 N/m³. What is the value of R for the gas? What gas might this be? Refer to Appendix A, Table A.5.*

SI

$$\text{Eq. 2.5:} \quad R = \frac{gp}{\gamma T} = \frac{9.81 \text{ m/s}^2(20\,000 \times 100 \text{ N/m}^2)}{241 \text{ N/m}^3(273+40)\text{K}} = 260 \text{ m}^2/(\text{s}^2\cdot\text{K}) \quad ◄$$

Table A.5: This gas with R = 260 might be oxygen ◄

2.7.4 *Calculate the density, specific weight and specific volume of air at 120°F and 50 psia.*

BG

$$\text{Eq. 2.5:} \quad \gamma = \frac{gp}{RT} = \frac{32.2(50 \times 144)}{1715(460+120)} = 0.233 \text{ lb/ft}^3 \quad ◄$$

Eq. 2.1: $\rho = \gamma/g = 0.233/32.2 = 0.007\,24 \text{ slugs/ft}^3$ ◄

Eq. 2.2: $v = 1/\rho = 138.2 \text{ ft}^3\text{/slug}$ ◄

2.7.5 *Calculate the density, specific weight and specific volume of air at 50°C and 3400 mb abs.*

SI

$$\text{Eq. 2.5:} \quad \gamma = \frac{gp}{RT} = \frac{9.81(3400 \times 100)}{287(273+50)} = 36.0 \text{ N/m}^3 \quad ◄$$

Eq. 2.1: $\rho = \gamma/g = 36.0/9.81 = 3.67 \text{ kg/m}^3$ ◄

Eq. 2.2: $v = 1/\rho = 0.273 \text{ m}^3\text{/kg}$ ◄

2.7.6 *(a) If water vapor in the atmosphere has a partial pressure of 0.50 psia and the temperature is 90°F, what is its specific weight? (b) If the barometer reads 14.50 psia, what is the partial pressure of the (dry) air, and what is its specific weight? (c) What is the specific weight of the atmosphere (air plus the water vapor present)?*

BG

(*a*) Table A.5 for water vapor: $R = 2760\ \text{ft}^2/(\text{sec}^2\cdot°\text{R})$

Eq. 2.5 for water vapor: $\gamma = \dfrac{32.2(0.50 \times 144)}{2760(460+90)} = 0.001\,527\ \text{lb/ft}^3$ ◄

(*b*) $p_{air} = 14.50 - 0.50 = 14.00$ psia ◄

Table A.5 for air: $R = 1715\ \text{ft}^2/(\text{sec}^2\cdot°\text{R})$

Eq. 2.5: Air, $\gamma = \dfrac{32.2(14.00 \times 144)}{1715(460+90)} = 0.0688\ \text{lb/ft}^3$ ◄

(*c*) Atmos, $\gamma = \gamma_{air} + \gamma_{wvapor} = 0.0688 + 0.001\,527 = 0.0703\ \text{lb/ft}^3$ ◄

2.7.7 *(a) If water vapor in the atmosphere has a partial pressure of 3500 Pa and the temperature is 30°C, what is its specific weight? (b) If the barometer reads 102 kPa abs, what is the partial pressure of the (dry) air, and what is its specific weight? (c) What is the specific weight of the atmosphere (air plus the water vapor present)?*

SI

(*a*) Table A.5 for water vapor: $R = 462\ \text{m}^2/(\text{s}^2\cdot\text{K})$

Eq. 2.5: $\gamma = \dfrac{9.81\ \text{m/s}^2 \times 3500\ \text{N/m}^2}{[462\ \text{m}^2/(\text{s}^2\cdot\text{K})](273+30)\text{K}} = 0.245\ \text{N/m}^3$ ◄

(*b*) $p_{air} = 102 - (3500/1000) = 98.5$ kPa abs ◄

Table A.5 for air: $R = 287\ \text{m}^2/(\text{s}^2\cdot\text{K})$

Eq. 2.5: $\gamma_{air} = \dfrac{gp}{RT} = \dfrac{(9.81\ \text{m/s}^2)(98\,500\ \text{N/m}^2)}{[287\ \text{m}^2/(\text{s}^2\cdot\text{K})](273+30)\text{K}} = 11.11\ \text{N/m}^3$ ◄

(*c*) $\gamma_{atmos} = \gamma_{air} + \gamma_{wvapor} = 11.11 + 0.245 = 11.36\ \text{N/m}^3$ ◄

2.7.8 *If an artificial atmosphere consists of 20% oxygen and 80% nitrogen by volume, at 101.32 kN/m² abs and 20°C, what are (a) the specific weight and partial pressure of the oxygen, (b) the specific weight and partial pressure of the nitrogen, and (c) the specific weight of the mixture?*

SI

Table A.5: R(oxygen) $= 260\ \text{m}^2/(\text{s}^2\cdot\text{K})$; R(nitrogen) $= 297\ \text{m}^2/(\text{s}^2\cdot\text{K})$

Eq. 2.5: 100% O_2: $\gamma = \dfrac{(9.81\ \text{m/s}^2)(101.32\ \text{kN/m}^2)}{[260\ \text{m}^2/(\text{s}^2\cdot\text{K})](273 + 20)\text{K}} = 13.05\ \text{N/m}^3$

100% N_2: $\gamma = \dfrac{(9.81\ \text{m/s}^2)(101.32\ \text{kN/m}^2)}{[297\ \text{m}^2/(\text{s}^2\cdot\text{K})](273 + 20)\text{K}} = 11.42\ \text{N/m}^3$

(*a*) Each m³ of mixture contains 0.2 m³ of O_2 and 0.8 m³ of N_2.

So for 20% O_2: $\gamma = 0.20(13.05) = 2.61\ \text{N/m}^3$ ◄

From Eq. 2.5 for 20% O_2: $p = \dfrac{\gamma RT}{g} = \dfrac{2.61(260)293}{9.81} = 20{,}300\ \text{Pa} = 20.3\ \text{kPa}$ ◄

/cont...

(*b*) For 80% N_2: $\gamma = 0.80(11.42) = 9.14\ \text{N/m}^3$ ◀

$$p = \frac{\gamma RT}{g} = \frac{9.14(297)293}{9.81} = 81{,}060\ \text{Pa} = 81.1\ \text{kPa} \quad ◀$$

(*c*) Mixture: $\gamma = 2.61 + 9.14 = 11.75\ \text{N/m}^3$ ◀

2.7.9 *Prove that Eq (2.7) follows from Eqs (2.4) and (2.6).*

N

Eq. 2.4: $v = RT/p$; Eq. 2.6: $p_1 v_1^n = p_2 v_2^n$

Eliminate v: $$p_1\left(\frac{R_1T_1}{p_1}\right)^n = p_2\left(\frac{R_2T_2}{p_2}\right)^n \text{ ; but } R_1 = R_2$$

Thus $$\frac{T_1^n}{p_1^{n-1}} = \frac{T_2^n}{p_2^{n-1}} \quad \text{or} \quad \left(\frac{T_2}{T_1}\right)^n = \left(\frac{p_2}{p_1}\right)^{n-1} \qquad \text{Finally} \quad \frac{T_2}{T_1} = \left(\frac{p_2}{p_1}\right)^{(n-1)/n} \quad \text{Q.E.D.} \quad ◀$$

Sec. 2.7: Property Relations for Perfect Gases -- *Problems*

2.7 *(a) Calculate the density, specific weight and specific volume of oxygen at 20°C and 50 kN/m² abs. (b) If the oxygen is enclosed in a rigid container of constant volume, what will be the pressure if the temperature is reduced to −100°C?*

SI

(*a*) Eq. 2.5: $$\gamma = \frac{gp}{RT} = \frac{(9.81\ \text{m/s}^2)(50\ \text{kN/m}^2)}{[260\ \text{m}^2/(\text{s}^2\cdot\text{K})](273 + 10)\text{K}} = 0.006\,44\ \text{kN/m}^3 = 6.44\ \text{N/m}^3 \quad ◀$$

Eq. 2.1: $\rho = \gamma/g = 6.44/9.81 = 0.656\ \text{kg/m}^3$ ◀

Eq. 2.2: $v = 1/\rho = 1/0.656 = 1.524\ \text{m}^3/\text{kg}$ ◀

(*b*) Eq. 2.4: $pv = RT$; $v = \text{const}$, $R = \text{const}$

$$\therefore \frac{p}{T} = \text{constant} = \frac{50}{273 + 20} = \frac{p_2}{273 - 100} \text{ ;} \quad p_2 = 29.5\ \text{kN/m}^2 \quad ◀$$

2.8 *If the specific weight of water vapor in the atmosphere is 0.1020 N/m³ and of the (dry) air is 11.62 N/m³ when the temperature is 20°C, (a) what are the partial pressures of the water vapor and the dry air in kPa, (b) what is the specific weight of the atmosphere (air and water vapor), and (c) what is the barometric pressure in kPa and millibars?*

SI

(*a*) Eq. 2.5: $\gamma_{\text{wvapor}} = gp/RT$; Table A.5 for water vapor: $R = 462\ \text{m}^2/\text{s}^2\cdot\text{K}$

$$\therefore p = \frac{\gamma RT}{g} = \frac{(0.1020\ \text{N/m}^3)(462\ \text{m}^2/\text{s}^2\cdot\text{K})(273 + 20)\text{K}}{9.81\ \text{m/s}^2} = 1407\ \text{Pa} = 1.407\ \text{kPa} \quad ◀$$

Eq. 2.5: $\gamma_{\text{air}} = gp/RT$; Table A.5 for air: $R = 287\ \text{m}^2/\text{s}^2\cdot\text{K}$

$$\therefore p = \frac{\gamma RT}{g} = \frac{(11.62\ \text{N/m}^3)(287\ \text{m}^2/\text{s}^2\cdot\text{K})(273 + 20)\text{K}}{9.81\ \text{m/s}^2} = 99\,600\ \text{Pa} = 99.6\ \text{kPa} \quad ◀$$

(*b*) $\gamma_{\text{atmos}} = \gamma_{\text{air}} + \gamma_{\text{wv}} = 11.62 + 0.102 = 11.72\ \text{N/m}^3$ ◀

(*c*) $p_{\text{atm}} = p_{\text{air}} + p_{\text{wv}} = 99.6\ \text{kPa} + 1.407\ \text{kPa} = 101.0\ \text{kPa}$ ◀

Inside back cover: 1 kPa = 10 mb. $\therefore p_{\text{atm}} = 101.0\ \text{kPa} = 1010\ \text{mb}$ ◀

2.9 *When the ambient air is at 70°F, 14.7 psia, and contains 21% oxygen by volume, 4.5 lb of air are pumped into a scuba tank, capacity 0.75 ft³. (a) What volume of ambient air was compressed? (b) When the filled tank has cooled to ambient conditions, what is the (gage) pressure of the air in the tank? (c) What is the partial pressure (psia) and specific weight of the ambient oxygen? (d) What weight of oxygen was put in the tank? (e) What is the partial pressure (psia) and specific weight of the oxygen in the tank?*

BG

(*a*) From Table A.2: $\gamma_1 = 0.074\,95\ \text{lb/ft}^3$. $\therefore\ V_{air} = \dfrac{W}{\gamma_1} = \dfrac{4.5\ \text{lb}}{0.074\,95\ \text{lb/ft}^3} = 60.0\ \text{ft}^3$ ◀

$$(b)\ \gamma_2 = \frac{W}{V} = \frac{4.5\ \text{lb}}{0.75\ \text{ft}^3} = 6.00\ \text{lb/ft}^3$$

From Section 2.7 $n = 1$ for isothermal conditions, so from Eq. 2.6:

$$\frac{p_1}{\rho_1} = \frac{p_2}{\rho_2};\quad \therefore\ p_2 = \left(\frac{\rho_2}{\rho_1}\right)p_1 = \left(\frac{\gamma_2}{\gamma_1}\right)p_1 = \left(\frac{6.00\ \text{lb/ft}^3}{0.074\,95\ \text{lb/ft}^3}\right)14.7\ \text{psia} = 1177\ \text{psia}$$

$p_2 = 1177 - 14.4 = 1162$ psig ◀

(*c*) Eq. 2.5 for 100% O_2: $\gamma = \dfrac{(32.2\ \text{ft/sec}^2)(14.7\ \text{lb/in}^2)(144\ \text{in}^2/\text{ft}^2)}{[1554\ \text{ft}^2/(\text{sec}^2\cdot{}^\circ\text{R})](460+70)^\circ\text{R}} = 0.0828\ \text{lb/ft}^3$

$\therefore$ for 21% O_2: $\gamma = 0.21(0.0828) = 0.017\,38\ \text{lb/ft}^3$ ◀

From Eq. 2.5 with 21% O_2: $p = \dfrac{\rho RT}{g} = \dfrac{(0.017\,38\ \text{lb/ft}^3)[1554\ \text{ft}^2/(\text{sec}^2\cdot{}^\circ\text{R})](530^\circ\text{R})}{32.2\ \text{ft/sec}^2}$

$p = 445$ psfa $= 3.09$ psia ◀

(*d*) $W_{O_2} = \gamma_1 V_{\text{air}} = (0.017\,38\ \text{lb/ft}^3)60\ \text{ft}^3 = 1.043\ \text{lb}$ ◀

$$(e)\ \gamma_2 = \frac{W}{V} = \frac{1.043\ \text{lb}}{0.75\ \text{ft}^3} = 1.391\ \text{lb/ft}^3$$ ◀

$$p = \frac{\gamma_2 RT}{g} = \frac{1.391(1554)530}{32.2} = 35{,}579\ \text{psfa} = 247\ \text{psia}$$ ◀

Sec. 2.8: Compressibility of Gases -- *Exercises*

2.8.1 *Methane at 21 psia is compressed isothermally, and nitrogen at 15 psia is compressed isentropically. What is the modulus of elasticity of each gas? Which is the more compressible?*

BG

Methane, isothermal: $n = 1$; $E_v = np = (1)21 = 21$ psi ◀

N_2, isentropic: $n = k = 1.40$; $E_v = 1.40(15) = 21.0$ psi ◀

E_v(methane) $= E_v(N_2)$, the compressibilities are equal ◀

2.8.2 *Methane at 120 kPa abs is compressed isothermally, and nitrogen at 90 kPa abs is compressed isentropically. What is the modulus of elasticity of each gas? Which is the more compressible?*

SI

Methane, isothermal: $n = 1$; $E_v = np = (1)120 = 120$ kPa ◀

N_2, isentropic: $n = k = 1.40$; $E_v = 1.40(90) = 126$ kPa ◀

E_v(methane) $< E_v(N_2)$, methane is more compressible ◀

2.8.3 *Helium at 25 psia and 65°F is isentropically compressed to one-fifth of its original volume. What is its final pressure?*

BG

Isentropic process: $n = k$; Table A.5 for helium: $R = 12{,}420$ ft·lb/(slug·°R), $k = 1.66$

Eq. 2.4: $v = \dfrac{RT}{p} = \dfrac{12{,}420(460+65)}{25(144)} = 1811$ ft³/slug

Eq. 2.6: $pv^{1.66} =$ constant ; $25(1811)^{1.66} = p_2(1811/5)^{1.66}$; $p_2 = 362$ psia ◄

2.8.4 *Helium at 165 kN/m² abs and 15°C is isentropically compressed to one-fifth of its original volume. What is its final pressure?*

SI

Isentropic process: n = k ; Table A.5 for helium: $R = 2077$ N·m/(kg·K), $k = 1.66$

Eq. 2.4: $v = \dfrac{RT}{p} = \dfrac{2077(273+15)}{165(1000)} = 3.63$ m³/kg

Eq. 2.6: $165(3.63)^{1.66} = p_2(3.63/5)^{1.66}$; $p_2 = 2390$ kPa abs ◄

2.8.5 *(a) If 10 m³ of nitrogen at 30°C and 120 kPa is permitted to expand isothermally to 30 m³, what is the resulting pressure? (b) What would the pressure and temperature have been if the process had been isentropic? The isentropic exponent k for nitrogen is 1.40.*

SI

(*a*) Isothermal: $p_1 v_1 = p_2 v_2 =$ constant ; 120(10/mass) = p_2(30/mass); $p_2 = 40$ kPa abs ◄

(*b*) Isentropic: $p_1 v_1^{1.40} = p_2 v_2^{1.40}$

$120(10/\text{mass})^{1.40} = p_2(30/\text{mass})^{1.40}$; $p_2 = 120(10/30)^{1.40} = 25.8$ kPa abs ◄

Eq. 2.7: $T_2 = (273+30)(25.8/120)^{0.40/1.40} = 195.3$ K $= -77.7$°C ◄

Sec. 2.8: Compressibility of Gases -- *Problems*

2.10 *(a) If 12 ft³ of carbon dioxide at 60°F and 15 psia is compressed isothermally to 2 ft³, what is the resulting pressure? (b) What would the pressure and temperature have been if the process had been isentropic? The isentropic exponent k for carbon dioxide is 1.28.*

BG

(*a*) Isothermal: $n = 1$; Eq. 2.6: $pv =$ constant; 15(12/mass) = p_2(2/mass); $p_2 = 90.0$ psia ◄

(*b*) Isentropic: $n = k = 1.28$; Eq. 2.6: $pv^n = pv^{1.28} =$ constant; $15(12/\text{mass})^{1.28} = p_2(2/\text{mass})^{1.28}$

$p_2 = 15(12/2)^{1.28} = 148.6$ psia ◄

Eq. 2.7: $T_2 = T_1(p_2/p_1)^{0.28/1.28} = (460 + 60)(148.6/15)^{0.219} = 859$°R $= 399$°F ◄

2.11 *(a) If 350 L of carbon dioxide at 20°C and 120 kN/m² abs is compressed isothermally to 50 L, what is the resulting pressure? (b) What would the pressure and temperature have been if the process had been isentropic? The isentropic exponent k for carbon dioxide is 1.28.*

SI

(*a*) Isothermal, Eq. 2.6: $pv =$ constant; 120(0.35/mass) = p_2(0.05/mass); $p_2 = 840$ kN/m² abs ◄

(*b*) Isentropic: $n = k = 1.28$; Eq. 2.6: $pv^n = pv^{1.28} =$ constant;

$120(0.35/\text{mass})^{1.28} = p_2(0.05/\text{mass})^{1.28}$; $p_2 = 120(0.35/0.05)^{1.28} = 1448$ kN/m² abs ◄

Eq. 2.7: $T_2 = (273 + 20)(1448/120)^{0.28/1.28} = 505$ K $= 232$°C ◄

Sec. 2.11: Viscosity -- ***Exercises***

2.11.1 *A liquid has an absolute viscosity of 3.5 × 10^{-4} lb·sec/ft^2. It weighs 58 lb/ft^3. What are its absolute and kinematic viscosities in SI units?*

B

$$\mu = 3.5\times10^{-4}\frac{\text{lb·sec}}{\text{ft}^2}\left(\frac{\text{ft}^2}{144\ \text{in}^2}\right)\frac{6.895\ \text{kN/m}^2}{\text{lb/in}^2} = 16.76\times10^{-6}\ \text{kN·s/m}^2$$

$$= 16.76\ \text{mN·s/m}^2 = 16.76\ \text{centipoise}\quad ◄$$

Eq. 2.11: $$\upsilon = \frac{\mu}{\rho} = \frac{16.76\ \text{mN·s/m}^2}{(58/62.4)\text{g/cm}^3} = \frac{16.76\ \text{m(kg·m/s}^2\text{)s/m}^2}{929\ \text{kg/m}^3}$$

$$= 18.03\times10^{-6}\ \text{m}^2\text{/s} = 0.1803\ \text{stokes}\quad ◄$$

2.11.2 *(a) What is the ratio of the viscosity of water at a temperature of 70°F to that of water at 200°F? (b) What is the ratio of the viscosity of the crude oil in Fig. 2.3 (s = 0.925) to that of the gasoline (s = 0.680), both being at a temperature of 60°F? (c) In cooling from 300 to 80°F, what is the ratio of the change of the viscosity of the SAE 30 Western oil to that of the SAE 30 Eastern oil?*

B

(*a*) Fig. 2.3: $1.0\times10^{-3}/(3.0\times10^{-4}) = 3.33$ ◄

(*b*) Fig. 2.3: $9.0\times10^{-2}/(3.1\times10^{-4}) = 290$ ◄

(*c*) Fig. 2.3: $$\frac{3.0\times10^{-1} - 3.2\times10^{-3}}{1.7\times10^{-1} - 3.4\times10^{-3}} = 1.782\quad ◄$$

2.11.3 *At 60°F what is the kinematic viscosity of the gasoline in Fig. 2.4, the specific gravity of which is 0.680? Give the answer in both BG and SI units.*

B

Fig. 2.4: 4.8×10^{-6} ft^2/sec ◄

Fig. 2.4: 4.5×10^{-7} m^2/s = 0.0045 cm^2/s = 0.45 centistokes ◄

2.11.4 *To what temperature must the fuel oil with the higher specific gravity in Fig. (2.4) be heated in order that its kinematic viscosity may be reduced to three times that of water at 40°F?*

BG

Table A.1 for water at 40° F: kinematic viscosity $\nu = 16.64\times10^{-6}$ ft^2/sec.

Fuel oil: $\nu = 3(16.64\times10^{-6}) = 4.99\times10^{-5}$ ft^2/sec.

Fig. 2.4 for fuel oil (s = 0.968) with $\nu = 4.99\times10^{-5}$ ft^2/sec: T = 375° F ◄

2.11.5 *Compare the ratio of the absolute viscosities of air and water at 70°F with the ratio of their kinematic viscosities at the same temperature and at 14.7 psia.*

BG

Table A.1 for water at 70°F and 14.7 psia: $\mu = 20.50\times10^{-6}$ lb·sec/ft^2, $\nu = 10.59\times10^{-6}$ ft^2/sec

Table A.2 for air at 70°F and 14.7 psia: $\mu = 0.382\times10^{-6}$ lb·sec/ft^2, $\nu = 0.164\times10^{-3}$ ft^2/sec

Absolute ratio = μ_{air}/μ_{water} = 0.382/20.50 = 1:53.7 ◄

Kinematic ratio = ν_{air}/ν_{water} = 164/10.59 = 15.5:1 ◄

2.11.6 *A flat plate 250 mm × 800 mm slides on oil (μ = 0.65 N·s/m²) over a large plane surface. What force is required to drag the plate at 1.5 m/s, if the separating oil film is 0.5 mm thick?*

SI

Eq. 2.9: $\tau = \mu \dfrac{dv}{dy} = 0.65 \dfrac{1.5}{0.000\ 54} = 1950\ \text{N/m}^2$

From Eq. 2.9: $F = \tau A = 1950(0.25 \times 0.8) = 390\ \text{N}$ ◄

2.11.7 *A flat plate 1 ft × 2 ft slides on oil (μ = 0.02 lb·sec/ft²) over a large plane surface. What force is required to drag the plate at 5 fps, if the separating oil film is 0.02 in thick?*

BG

Eq. 2.9: $\tau = \mu \dfrac{dV}{dy} = (0.02\ \text{lb}\cdot\text{sec/ft}^2) \dfrac{(5\ \text{ft/sec})(12\ \text{in/ft})}{0.02\ \text{in}} = 60\ \text{lb/ft}^2$

From Eq. 2.9: $F = \tau A = (60\ \text{lb/ft}^2)(1\ \text{ft} \times 2\ \text{ft}) = 120\ \text{lb}$ ◄

2.11.8 *A space 20 mm wide between two large plane surfaces is filled with SAE 30 Western lubricating oil at 30°C. (a) What force is required to drag a very thin plate of 0.4-m² area between the surfaces at a speed of 0.2 m/s if this plate is equally spaced between the two surfaces? (b) If it is at a distance of 5 mm from one surface?*

SI

Fig. 2.3 for SAE Western lubricating oil at 30°C: $\mu = 0.215\ \text{N}\cdot\text{s/m}^2$

(*a*) Eq. 2.9: $\tau = 0.215\left(\dfrac{0.2}{10/1000}\right) = 4.30\ \text{N/m}^2$; From Eq. 2.9: Force = 4.30(2)0.4 = 3.44 N ◄

(*b*) Eq. 2.9: $\tau_1 = 0.215\left(\dfrac{0.2}{5/1000}\right) = 8.60\ \text{N/m}^2$; $\tau_2 = 0.215\left(\dfrac{0.2}{15/1000}\right) = 2.87\ \text{N/m}^2$

From Eq. 2.9: $F_1 = \tau_1 A = 8.60(0.4) = 3.44\ \text{N}$; $F_2 = \tau_2 A = 2.87(0.4) = 1.147\ \text{N}$

$\therefore$ Force $= F_1 + F_2 = 4.59\ \text{N}$ ◄

2.11.9 *A journal bearing consists of an 80-mm shaft in an 80.4-mm sleeve 120 mm long, the clearance space (assumed to be uniform) being filled with SAE 30 Western lubricating oil at 40°C. Calculate the rate at which heat is generated at the bearing when the shaft turns at 150 rpm. Express the answer in kN·m/s, J/s, Btu/hr, ft·lb/sec, and hp.*

B

Fig. 2.3: $\mu = 1.1 \times 10^{-1} = 0.11\ \text{N}\cdot\text{s/m}^2$

Eq. 2.9: $\tau = \mu \dfrac{dv}{dy} = \mu \dfrac{r\omega}{\Delta D/2} = 0.11 \dfrac{40(150 \times 2\pi/60)}{0.2} = 346\ \text{N/m}^2$

Torque $= (\tau A)r = 346(\pi \times 0.08 \times 0.12)0.04 = 0.417\ \text{N}\cdot\text{m}$

Rate of heat generation $= T\omega = 0.417\left(\dfrac{2\pi \times 150}{60}\right) = 6.55\ \text{N}\cdot\text{m/s}$

= 6.55 J/s = 0.006 55 kN·m/s = 22.4 Btu/hr = 0.008 78 hp = 4.83 ft·lb/sec ◄◄

(using conversion factors from inside the back cover).

2.11.10 *A hydraulic lift of the type commonly used for greasing automobiles consists of a 10.000-in-diameter ram which slides in a 10.006-in-diameter cylinder, the annular space being filled with oil having a kinematic viscosity of 0.0035 ft²/sec and specific gravity of 0.85. If the rate of travel of the ram is 0.6 fps, find the frictional resistance when 10 ft of the ram is engaged in the cylinder.*

BG

$\gamma = 0.85 \times 62.4 = 53 \text{ lb/ft}^3, \ \rho = \dfrac{53}{32.2} = 1.647 \text{ slug/ft}^3$

Eq. 2.11: $\mu = \nu\rho = 0.0035(1.647) = 0.005\,77 \text{ lb·sec/ft}^2$

Eq. 2.9: $\tau = \mu\dfrac{dv}{dy} = 0.005\,77\left(\dfrac{0.6}{0.003/12}\right) = 13.84 \text{ lb/ft}^2$

From Eq. 2.9: Friction force $= \tau A = 13.84(10 \times \pi \times 10/12) = 362$ lb ◄

2.11.11 *A journal bearing consists of a 7.20-in shaft in a 7.21-in sleeve 8 in long, the clearance space (assumed to be uniform) being filled with SAE 30 Eastern lubricating oil at 100°F. Calculate the rate at which heat is generated at the bearing when the shaft turns at 80 rpm. Express the answer in Btu/hr.*

BG

Fig. 2.3: $\mu = 0.0018 \text{ lb·sec/ft}^2$; $u = \pi\left(\dfrac{7.2}{12}\right)\left(\dfrac{80}{60}\right) = 2.51$ fps

Eq. 2.9: $\tau = (0.0018)\dfrac{2.51}{0.005/12} = 10.86 \text{ lb/ft}^2$

From Eq. 2.9: Friction $= \tau A = 10.86\left(\dfrac{8}{12}\pi\dfrac{7.2}{12}\right) = 13.64$ lb

Per Sample Prob. 2.9: Energy loss rate $= T\omega = Fu = 13.64(2.51) = 34.3$ ft·lb/sec

Heat generation rate $= \dfrac{34.3 \text{ ft·lb}}{\text{sec}}\left(\dfrac{\text{Btu}}{778 \text{ ft·lb}}\right)\dfrac{3600 \text{ sec}}{\text{hr}} = 158.7$ Btu/hr ◄

2.11.12 *In using a rotating-cylinder viscometer, a bottom correction must be applied to account for the drag on the flat bottom of the inner cylinder. Calculate the theoretical amount of this torque correction, neglecting centrifugal effects, for a cylinder of diameter d, rotated at a constant angular velocity ω, in a liquid of viscosity μ, with a clearance Δh between the bottom of the inner cylinder and the floor of the outer one.*

N

Let r = variable radius. $dA = 2\pi r\,dr$, $\tau = \mu r\omega/\Delta h$

Torque $= \int r \times \tau dA = \dfrac{2\pi\mu\omega}{\Delta h}\int_0^{d/2} r^3 dr = \dfrac{\pi\mu\omega}{32\Delta h}d^4$ ◄

2.11.13 *Assuming a velocity distribution as shown in the diagram, which is a parabola having its vertex 12 in from the boundary, calculate the velocity gradients for y = 0, 3, 6, 9, and 12 in. Also calculate the shear stresses in lb/ft² at these points if the fluid viscosity is 500 cP.*

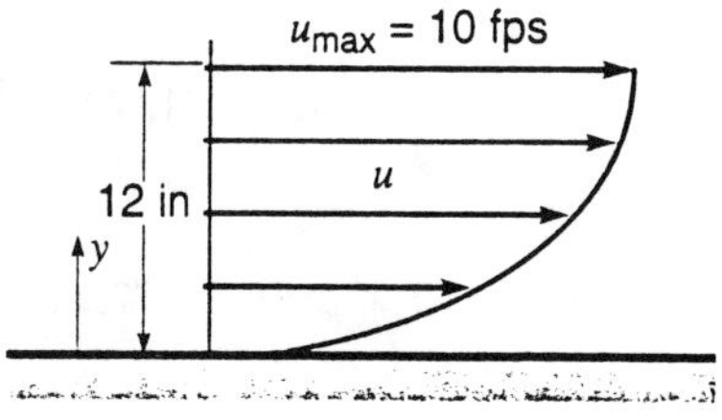

Figure X2.11.13

BG

Back cover: $\mu = 500 \text{ cP} = 5P = 0.5 \text{ N·s/m}^2$

$\mu = 0.5(0.020\,885) = 0.010\,44 \text{ lb·sec/ft}^2$. Parabola: $Y = aX^2$

For u in in/sec and y in inches: $120 - u = a(12 - y)^2$

$u = 0$ at $y = 0 \rightarrow a = 120/12^2 = 5/6$; $u = 120 - (5/6)(12 - y)^2$; $du/dy = (5/3)(12 - y)$

Eq. 2.9: $\tau = 0.010\,44\,du/dy$

y (in)	0	3	6	9	12	
du/dy (sec^{-1})	20	15	10	5	0	◄◄
τ (lb/ft²)	0.209	0.1566	0.1044	0.0522	0	◄◄

2.12 *The absolute viscosity of a certain gas is 0.0234 cP while its kinematic viscosity is 181 cSt, both measured at 1013 mb abs and 100°C. Calculate its approximate molecular weight, and suggest what gas it may be.*

SI

Eq. 2.11: $\rho = \dfrac{\mu}{\upsilon} = \dfrac{0.0234 \text{ cP}}{181 \text{ cSt}} = \dfrac{2.34 \times 10^{-5} \text{ N·s/m}^2}{181 \times 10^{-6} \text{ m}^2\text{/s}} = 0.1293 \text{ kg/m}^3$

From Eqs. 2.1 and 2.5: $R = \dfrac{p}{\rho T} = \dfrac{101.3 \text{ kN/m}^3}{(0.1298 \text{ kg/m}^3)(273+100)\text{K}} = 2100 \text{ N·m/(kg·K)}$

Sec. 2.7: $M = R_0/R = 8312/2100 = 3.96$ ◄ ; Table A.5: The gas may be helium ◄

2.13 *A hydraulic lift of the type commonly used for greasing automobiles consists of a 300.00-mm-diameter ram which slides in a 300.15-mm-diameter cylinder, the annular space being filled with oil having a kinematic viscosity of 0.0004 m²/s and specific gravity of 0.85. If the rate of travel of the ram is 0.2 m/s, find the frictional resistance when 2 m of the ram is engaged in the cylinder.*

SI

$\gamma = 0.85(9810) = 8340 \text{ N/m}^3, \quad \rho = 8340/9.81 = 850 \text{ kg/m}^3$

Eq. 2.11: $\mu = \upsilon\rho = 0.0004(850) = 0.340 \text{ N/m}^2$

Eq. 2.9: $\tau = \mu(du/dy) = 0.340 \times 0.2/(0.075/1000) = 907 \text{ N/m}^2$

From Eq. 2.9: Friction force $= \tau A = 907(2\pi 300/1000) = 1709$ N ◄

2.14 *Repeat Exer. 2.11.11 for the case where the sleeve has a diameter of 7.68 in. Compute as accurately as possible the velocity gradient in the fluid at the shaft and sleeve.*

Exer. 2.11.11: A journal bearing consists of a 7.20-in shaft in a sleeve 8 in long, the clearance space (assumed to be uniform) being filled with SAE 30 Eastern lubricating oil at 100°F. Calculate the rate at which heat is generated at the bearing when the shaft turns at 80 rpm. Express the answer in Btu/hr.

BG

Fig. 2.3: $\mu = 0.0018 \text{ lb·sec/ft}^2$; Fig. 2.7*a*: $u_1 = \pi(7.2/12)(80/60) = 2.51$ fps

7.20 in = 0.600 ft diameter = 0.300 ft radius ; 7.68 in = 0.640 ft diameter = 0.320 ft radius

$T = \tau A r = \tau[2\pi r(8/12)]r, \quad \text{i.e.,} \quad \tau = T/(4.19r^2) \text{ lb/ft}^2$

Eq. 2.9: $\tau = 0.0018(du/dy) = -0.0018(du/dr) = T/(4.19r^2)$

$$-du = (132.6T/r^2)dr \; ; \quad -\int_{u_1}^{0} du = 132.6T\int_{0.300}^{0.320} r^{-2}dr = 132.6T\left[\frac{r^{-1}}{-1}\right]_{0.300}^{0.320}$$

$$0 + 2.51 = 132.6T\left(-\frac{1}{0.32} + \frac{1}{0.30}\right) = 132.6T(0.208); \quad T = \frac{2.51}{27.6} = 0.0910 \text{ ft·lb}$$

$\tau_1 = 0.0910/(4.19r^2) = 0.241 \text{ lb/ft}^2 = 0.0018\,(du/dy)_1$; $(du/dy)_1 = 134.1$ fps/ft ◄

$(du/dy)_2 = 134.1(0.30/0.32)^2 = 117.8$ fps/ft ◄

Rate of energy loss $= T\omega = (0.0910)\,80(2\pi/60) = 0.762$ ft·lb/sec

Rate of heat generation $= 0.762(3600/778) = 3.53$ Btu/hr ◄

2.15 *A disk spins within an oil-filled enclosure, having 3-mm clearance from flat surfaces each side of the disk. The disk surface extends from radius 10 to 90 mm. What torque is required to drive the disk at 600 rpm if the oil's viscosity is 0.10 N·s/m²?*

SI

Let r = variable radius. $dA = 2\pi r\, dr,\ \tau = \mu r\omega/\Delta h$

$$\text{Torque} = 2\int r \times \tau dA = \frac{4\pi\mu\omega}{\Delta h}\int_{r_1}^{r_2} r^3 dr = \frac{4\pi\mu\omega}{\Delta h}\frac{(r_2^4 - r_1^4)}{4} = \frac{\pi\mu\omega}{\Delta h}(r_2^4 - r_1^4)$$

$$= \frac{\pi \times 0.1\frac{\text{N·s}}{\text{m}^2}\left(600 \times 2\pi\frac{\text{rev/min}}{60\ \text{s/min}}\right)\frac{(90^4 - 10^4)\text{mm}^4}{(10^3\ \text{mm/m})^4}}{\frac{3\ \text{mm}}{1000\ \text{mm/m}}} = 0.432\ \text{N·m} \quad ◀$$

2.16 *It is desired to apply the general case of Sample Prob. 2.9 to the extreme cases of a journal bearing (α = 0) and an end bearing (α = 90°). But when α = 0, r = tan α = 0, so T = 0; when α = 90°, contact area = ∞ due to b, so T = ∞. Therefore, devise an alternative general derivation which will also provide solutions to these two extreme cases.*

Sample Prob. 2.9: Oil of viscosity μ fills the gap of thickness Y. Obtain an expression for the torque T required to rotate the truncated cone at constant speed ω rad/sec.

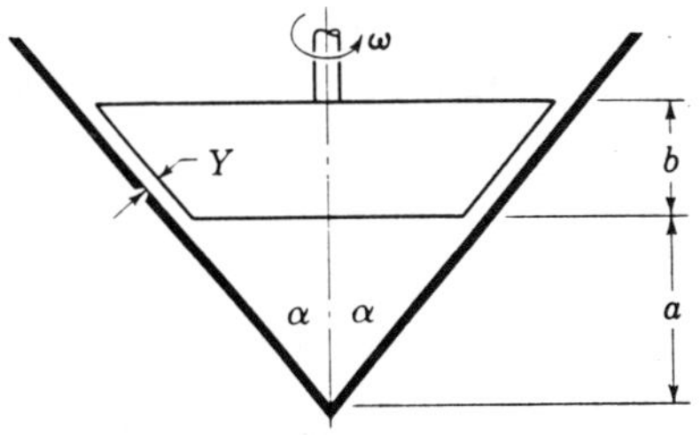

Figure S2.9

N

(*i*) Prevent r from going to zero by prescribing $r_{\text{min}} = R$ (see sketch);

(*ii*) Prevent contact area from becoming infinite by prescribing the sloping dimension c ($=b/\cos\alpha$).

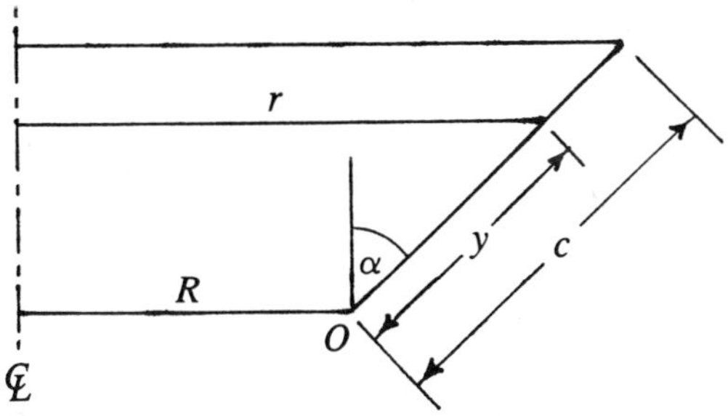

Define y to be the distance along the sloping surface to radius r, with y = 0 at point O (and dispense with a and b). Then $r = R + y\sin\alpha$

with (per Fig. S2.9) rotation velocity

$$U = \omega r = \omega(R + y\sin\alpha)$$

Eq. 2.9: $\tau = \mu\frac{du}{dy} = \mu\frac{U}{Y} = \frac{\mu\omega}{Y}(R + y\sin\alpha)$; $dA = 2\pi r dy = 2\pi(R + y\sin\alpha)dy$; $dF + \tau dA$

$dT = rdF = r\tau dA = (2\pi\mu\omega/Y)(R + y\sin\alpha)^3 dy$; $T = \int dT = (2\pi\mu\omega/Y)\int_0^c (R + y\sin\alpha)^3 dy$

Expanding and integrating $T = \frac{2\pi\mu\omega}{Y}\left(R^3c + \frac{3}{2}R^2c^2\sin\alpha + Rc^3\sin^2\alpha + \frac{c^4}{4}\sin^3\alpha\right)$ ◀

(*a*) Journal bearing: α = 0, sin α = 0, so $T = \frac{2\pi\mu\omega}{Y}R^3c$ ◀

(*b*) Flat end bearing: α = 90°, R = 0, so sin α = 1, let $D = 2c$,

then $T = \frac{2\pi\mu\omega}{Y}\frac{(D/2)^4}{4} = \frac{\pi\mu\omega}{32Y}D^4$ ◀ (This agrees with the solution to Exer. 2.11.12.)

Sec. 2.12: Surface Tension -- *Exercises*

2.12.1 *Distilled water at 20°C stands in a glass tube of 6.0-mm diameter at a height of 22.0 mm. What is the true static height?*

SI

20°C = 68°F; tube dia = 6.0 mm = 0.236 in. Fig. 2.8: Capillary rise ≈ 0.16 in = 4.06 mm

True static height ≈ 22.00 − 4.06 = 17.94 mm ◄

2.12.2 *Tap water at 68°F stands in a glass tube of 0.27-in diameter at a height of 6.00 in. What is the true static height?*

BG

Fig. 2.8: Capillary rise ≈ 0.070 in. True static height = 6.00 − 0.070, say 5.93 in ◄

2.12.3 *Use Eq. (2.12) to compute the capillary rise of water to be expected in an 0.30-in-diameter tube. Assume pure water at 68°F. Compare the result with Fig. 2.8.*

BG

Interpolating from Table A.1: σ = 0.004 99 lb/ft.

$$\text{Eq. 2.12:}\quad h = \frac{2\sigma}{\gamma r} = \frac{2(0.004\,99\ \text{lb/ft})(12\ \text{in/ft})}{(62.4\ \text{lb/ft}^3)(0.15/12\ \text{ft})} = 0.1535\ \text{in} \quad ◄$$

Fig. 2.8 shows a capillary rise of ≈ 0.127 in ◄

2.12.4 *Compute the capillary rise in mm of water at 5°C expected in a 1-mm-diameter tube.*

SI

From Table A.1 at 5°C: σ_{water} = 0.0749 N/m, γ = 9.807 kN/m³

$$\text{Eq. 2.12 with } \theta = 0:\quad h = \frac{2\sigma}{\gamma r} = \frac{2(0.0749\ \text{N/m})}{(9.807\ \text{N/m}^3)(0.0005\ \text{m})} = 0.0305\ \text{m} = 30.5\ \text{mm} \quad ◄$$

2.12.5 *Use Eq. (2.12) to compute the capillary depression of mercury (θ = 140°) to be expected in a 0.06-in-diameter tube. At 68°F the surface tension of mercury is 0.0318 lb/ft.*

BG

$$\text{Eq. 2.12:}\quad h = \frac{2\sigma\cos 140°}{\gamma r} = \frac{2(0.0318)0.766}{13.56\times 62.4(0.03/12)} = 0.0230\ \text{ft} = 0.276\ \text{in} \quad ◄$$

Sec. 2.12: Surface Tension -- *Problems*

2.17 *Water at 40°F stands in a glass tube of 0.05 in diameter at a height of 7.32 in. Compute the true static height.*

BG

From Table A.1 at 40°F: γ = 62.43 lb/ft³, σ = 0.005 14 lb/ft

$$\text{Eq. 2.12 with } \theta = 0°:\quad h = \frac{2\sigma}{\gamma r} = \frac{2(0.005\,14\ \text{lb/ft})}{62.4\ \text{lb/ft}^3(0.05/12\ \text{ft})} = 0.0395\ \text{ft} = 0.474\ \text{in}$$

True static height = 7.32 in − 0.474 in = 6.85 in ◄

2.18 *(a) Derive an expression for capillary rise (or depression) between two vertical parallel plates. (b) How much would you expect 15°C water to rise (in mm) if the clean glass plates are separated by 1.5 mm?*

SI

(*a*) $(\cos\theta)2\sigma Z = \gamma bhZ$, so $h = 2\sigma\cos\theta/(b\gamma)$ ◄

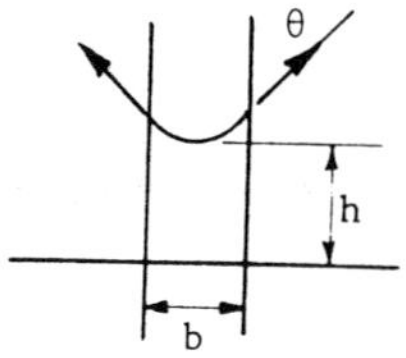

(*b*) For water, $\theta = 0°$

Table A.1 at 15°C: $\gamma = 9798\ \text{N/m}^3$, $\sigma = 0.0735\ \text{N/m}$

$$\therefore\ h = \frac{2(0.0735\ \text{N/m})\cos 0°}{(1.5/1000\ \text{m})(9798\ \text{N/m}^3)} = 0.010\,00\ \text{m} = 10.00\ \text{mm}$$ ◄

Sec. 2.13: Vapor Pressure of Liquids -- ***Exercises***

2.13.1 *At approximately what temperature will water boil in Mexico City (elevation 7400 ft)? Refer to Appendix A.*

BG

Table A.3, by interpolation: p_{at} = 11.21 psia at 7400 ft elevation

Table A.1, by interpolation: p_v water = 11.21 psia at about 198.6°F. ∴ Water boils at 198.6°F ◄

2.13.2 *At what pressure in millibars absolute will 50°C water boil?*

SI

Table A.1 at 50°C: p_v = 12.33 kN/m^2 abs; 1 millibar = 100 N/m^2

Will boil at 12.33 kN/m^2 abs = 123.3 millibars abs ◄

Sec. 2.13: Vapor Pressure of Liquids -- ***Problems***

2.19 *Water at 150°F is placed in a beaker within an airtight container. Air is gradually pumped out of the container. What reduction below standard atmospheric pressure of 14.7 psia must be achieved before the water boils?*

BG

Table A.1 at 150°F: p_v = 3.72 psia

14.7 − 3.72 = 10.98 psi; pressure must be reduced by 10.98 psi ◄

Chapter 3

Fluid Statics

PROBLEM SELECTION GUIDE

Sec	Exer/Prob	Units	Difficulty	Length	Parts	Similar	Special features
3.2	*Variation of Pressure in a Static Fluid*						
	X[1]3.2.1	BG	V Easy	V Short	1	3.2.2	
	3.2.2	SI	V Easy	V Short	1	3.2.1	
	3.2.3	SI	V Easy	V Short	1	3.2.4	Uses Sec. 2.3
	3.2.4	BG	V Easy	V Short	1	3.2.3	Uses Sec. 2.3
	P 3.1	BG	Hard	Medium	1		Differ'n, integr'n. Uses Secs 2.3 & 2.5
3.3	*Pressure Expressed in Height of Fluid*						
	X 3.3.1	SI	V Easy	V Short	1	3.3.2	
	3.3.2	BG	V Easy	V Short	1	3.3.1	
	3.3.3	BG	V Easy	V Short	1		
	3.3.4	BG	Easy	Short	1		Integration
	P 3.2	BG	Medium	Medium	1		
3.4	*Absolute and Gage Pressures*						
	X 3.4.1	B	Easy	Medium	1		Unit conversions
	3.4.2	SI	Easy	Short	1	3.4.3-4	
	3.4.3	BG	Easy	Short	1	3.4.2-4	
	3.4.4	SI	Easy	Short	1	3.4.2-3	
	3.4.5	BG	Easy	V Short	1	3.4.6	
	3.4.6	SI	Easy	V Short	1	3.4.5	
	3.4.7	BG	Easy	Short	1		
	P 3.3	BG	Medium	Medium	1	3.4	Uses $pv = RT$ (Sec. 2.7)
	3.4	SI	Medium	Medium	2	3.3	Uses $pv = RT$ (Sec. 2.7)
3.5	*Barometer*						
	X 3.5.1	BG	V Easy	V Short	1		
	3.5.2	SI	V Easy	V Short	1		

/cont...

[1] For all Exercises (marked "X"), answers are given in Appendix D of the textbook.
* Note a correction made to early printings of the textbook.
X = Exercise, P = (end-of-chapter) Problem, S = Sample Problem.

Sec	Exer/Prob	Units	Difficulty	Length	Parts	Similar	Special features
3.6	*Measurement of Pressure*						
	X 3.6.1	SI	V Easy	Short	1		Unit conversions
	3.6.2	BG	Easy	Short	1		
	3.6.3	BG	Easy	Short	1		
	3.6.4	BG	Easy	Short	2	3.6.5	*
	3.6.5	SI	Easy	Short	2	3.6.4	*
	3.6.6	BG	Easy	Short	1		
	3.6.7	BG	Easy	Short	1	3.6.8	Uses $pv = RT$ (Sec. 2.7)
	3.6.8	SI	Easy	Short	1	3.6.7	Uses $pv = RT$ (Sec. 2.7)
	3.6.9	BG	Easy	Short	1		
	3.6.10	SI	Easy	Medium	1		
	P 3.5	SI	Medium	Medium	1		
	3.6	N	Medium	Short	1		
	3.7	BG	Medium	Long	1		
3.8	*Center of Pressure*						
	X 3.8.1	N	Medium	Short	1		Optional integration
	3.8.2	N	Medium	Short	1		
	3.8.3	N	Easy	V Short	1		
	3.8.4	N	Easy	V Short	1		
	3.8.5	BG	Easy	Short	1		
	3.8.6	BG	Easy	Medium	3	3.8.7	
	3.8.7	SI	Easy	Medium	3	3.8.6	
	3.8.8	BG	Easy	Short	2		
	3.8.9	SI	Easy	Medium	1		
	3.8.10	BG	Easy	Short	1	3.8.11	
	3.8.11	SI	Easy	Short	1	3.8.10	
	3.8.12	N	Easy	Short	1		Proof
	3.8.13	N	Medium	Medium	2		Integration
	3.8.14	BG	Easy	Medium	1	3.8.15	
	3.8.15	SI	Easy	Medium	1	3.8.14	
	3.8.16	BG	Easy	Medium	2		
	3.8.17	SI	Easy	Short	1		
	3.8.18	BG	Easy	Medium	1		Moments
	3.8.19	SI	Easy	Short	2		Moments
	P 3.8	N	Medium	Medium	1		
	3.9	BG	Medium	Long	3		*
	3.10	SI	Medium	Medium	1		
	3.11	BG	Medium	Medium	2		
	3.12	BG	Medium	Medium	1		Moments
	3.13	BG	Medium	Medium	2		Moments
	3.14	BG	Medium	Short	1		Moments

/cont...

Sec	Exer/Prob	Units	Difficulty	Length	Parts	Similar	Special features
3.9	***Force on Curved Surface***						
	X 3.9.1	BG	V Easy	V Short	1		
	3.9.2	SI	Medium	Medium	3	3.9.3	
	3.9.3	BG	Medium	Medium	3	3.9.2	
	3.9.4	BG	Medium	Medium	1	3.9.5	
	3.9.5	SI	Medium	Medium	1	3.9.4	
	3.9.6	BG	Easy	Short	1	3.9.7	
	3.9.7	SI	Easy	Short	1	3.9.6	
	3.9.8	B	Easy	Short	1		Unit conversions
	3.9.9	BG	Easy	Medium	4		
	3.9.10	BG	Medium	Medium	1		
	P 3.15	BG	Medium	Medium	3		
	3.16	BG	Medium	Long	1		Requires planimetry; moments or cutout
	3.17	BG	Medium	Long	1		Moments or cutout
	3.18	SI	Medium	Medium	3		Integration; moments
3.10	***Buoyancy and Stability of Submerged and Floating Bodies***						
	X 3.10.1	BG	Easy	Short	1		
	3.10.2	SI	Easy	Medium	1		Differentiation
	3.10.3	BG	Easy	Short	1		Simultaneous equations
	3.10.4	BG	V Easy	V Short	1		
	3.10.5	SI	V Easy	V Short	1		
	3.10.6	BG	Easy	Short	1		Simultaneous equations
	3.10.7	SI	Easy	Medium	2		
	3.10.8	BG	Easy	Short	1		
	3.10.9	SI	Easy	Short	2		
	3.10.10	BG	Medium	Medium	3	3.10.11	Moments
	3.10.11	SI	Medium	Medium	3	3.10.10	Moments
	P 3.19	SI	Medium	Medium	1		
	3.20	BG	Medium	Medium	1	3.21	Uses $pv = RT$ (Sec. 2.7)
	3.21	BG	Medium	Medium	1	3.20	Uses $pv = RT$ (Sec. 2.7)
	3.22	BG	Medium	Medium	1		* Moments
	3.23	BG	Medium	Medium	2	3.24	Moments
	3.24	SI	Medium	Medium	2	3.23	Moments
	3.25	BG	Medium	Long	2		Moments
3.11	***Fluid Masses Subjected to Acceleration***						
	X 3.11.1	BG	Easy	Short	1	3.11.2	
	3.11.2	SI	Easy	Short	1	3.11.1	
	3.11.3	BG	Easy	Short	1	3.11.4	
	3.11.4	SI	Easy	Short	1	3.11.3	
	3.11.5	BG	Easy	Short	1	3.11.6	
	3.11.6	SI	Easy	Short	1	3.11.5	
	3.11.7	BG	Easy	Medium	1	3.11.8	Sketch
	3.11.8	SI	Easy	Medium	1	3.11.7	Sketch
	P 3.26	BG	Medium	Medium	1		
	3.27	SI	Medium	Medium	1	S3.10	

Chapter 3
FLUID STATICS

Sec 3.2: Variation of Pressure in a Static Fluid -- *Exercises*

3.2.1 *Neglecting the pressure upon the surface and the compressibility of water, what is the pressure in pounds per square inch at a depth of an ore deposit 12,000 ft below the surface of the ocean? The specific weight of ocean water under ordinary conditions is 64.0 lb/ft³.*

BG

Eq. 3.4: $p = \gamma h = 64.0(12{,}000)/144 = 5330$ psi ◄

3.2.2 *Neglecting the pressure upon the surface and the compressibility of water, what is the pressure in kPa at a depth of a wreck 4 km below the surface of the ocean? The specific weight of ocean water under ordinary conditions is 10.05 kN/m³.*

SI

Eq. 3.4: $p = \gamma h = 10.05(4000) = 40\ 200\ \text{kN/m}^2$ ◄

3.2.3 *A pressure gage at elevation 7.5 m on the side of an industrial tank containing a liquid reads 63.0 kN/m². Another gage at elevation 4.0 m reads 90.2 kN/m². Compute the specific weight and density of the liquid.*

SI

From Eq. 3.3: $\gamma = \Delta p/\Delta h = (90.2 - 63.0\ \text{kN/m}^2)/(7.5 - 4.0\ \text{m}) = 7.77\ \text{kN/m}^3$ ◄

Eq. 2.1: $\rho = \gamma/g = 7.77/9.81 = 0.792\ \text{g/cm}^3 = 792\ \text{kg/m}^3$ ◄

3.2.4 *A pressure gage at elevation 18.0 ft on the side of an industrial tank containing a liquid reads 11.4 psi. Another gauge at elevation 12.0 ft reads 13.7 psi. Compute the specific weight, density, and specific gravity of the liquid.*

BG

From Eq. 3.3: $\Delta p = \gamma(\Delta h)$; $(13.7 - 11.4)144 = \gamma(18 - 12)$; $\gamma = 55.2\ \text{lb/ft}^3$ ◄

Eq. 2.1: $\rho = \gamma/g = 55.2/32.2 = 1.714\ \text{slugs/ft}^3$ ◄ $s = 55.2/62.4 = 0.885$ ◄

Sec. 3.2: Variation of Pressure in a Static Fluid -- *Problem*

3.1 *Repeat Exer. 3.2.1, but consider the effects of compressibility (E_υ = 300,000 psi). Neglect changes in density caused by temperature variations. (Hint: As a starting point, express Eq. (2.3) in terms of γ and integrate to determine γ as a function of z.)*

Exer. 3.2.1: Find the pressure (psi) at a depth of 12,000 ft in the ocean (γ = 64.0 pcf at the surface). Neglect the pressure on the surface.

BG

From Eqs. 2.2 and 2.1: $\upsilon = 1/\rho = g/\gamma$

Differentiating, $\dfrac{d\upsilon}{d\gamma} = \dfrac{-g}{\gamma^2} = \dfrac{-\upsilon}{\gamma}$ or $\dfrac{d\upsilon}{\upsilon} = \dfrac{-d\gamma}{\gamma}$; Also, from Eq. 3.2: $dp = -\gamma\, dz$

$\therefore$ from Eq. 2.3*a*: $E_\upsilon = -\dfrac{dp}{d\upsilon/\upsilon} = -\dfrac{\gamma\, dz}{d\gamma/\gamma}$; $dz = -\dfrac{E_\upsilon\, d\gamma}{\gamma^2}$; integrate both sides

$z = \dfrac{E_\upsilon}{\gamma} + c$; at $z = 0$, $\gamma = 64$, so $c = -\dfrac{E_\upsilon}{64}$; $z = \dfrac{E_\upsilon}{\gamma} - \dfrac{E_\upsilon}{64}$; so $\gamma = \dfrac{64}{1 + \dfrac{64}{E_\upsilon} z}$

But also (Eq. 3.2): $\gamma = -dp/dz$ (Note: p increases as elevation z decreases.)

Equating γ expressions, separating variables, and integrating:

$$\int_p^0 dp = -64\int_{-12{,}000}^{0} \frac{dz}{1+(64/E_v)z} \; ; \; 0-p = -64\left(\frac{E_v}{64}\right)\left[\ln\left(1+\frac{64}{E_v}z\right)\right]_{-12{,}000}^{0}$$

$$p = (300{,}000 \text{ psi})\left[\ln 1 - \ln\left(1+\frac{(64 \text{ pcf})(-12{,}000 \text{ ft})}{(300{,}000 \text{ lb/in}^2)(144 \text{ in}^2/\text{ft}^2)}\right)\right] = 5380 \text{ psi} \quad ◄$$

Sec 3.3: Pressure Expressed in Height of Fluid -- *Exercises*

3.3.1 *An open tank contains 6.0 m of water covered with 2.5 m of oil (γ = 8.0 kN/m³). Find the gage pressure at the interface and at the bottom of the tank.*

SI

Eq. 3.4: $p = \gamma h = (8 \text{ kN/m}^3)2.5 \text{ m} = 20 \text{ kN/m}^2 = 20$ kPa at interface ◄

$p_b = 20 + (9.81)6 = 78.9 \text{ kN/m}^2 = 78.9$ kPa at tank bottom ◄

3.3.2 *An open tank contains 9 ft of water covered with 2.5 ft of oil (s = 0.86). Find the gage pressure at the interface between the liquids and at the bottom of the tank.*

BG

Eq. 3.4: $p = \gamma h = 0.86(62.4)2.5/144 = 0.932$ psi at interface ◄

$p_b = 0.932 + 62.4(10)/144 = 4.83$ psi at tank bottom ◄

3.3.3 *If air had a constant specific weight of 0.0765 lb/ft³ and were incompressible, what would be the height of air surrounding the earth to produce a pressure at the surface of 14.70 psia?*

BG

Eq. 3.5: $h = p/\gamma = 14.70(144)/0.0765 = 27{,}700$ ft ◄

3.3.4 *If the specific weight of a sludge can be expressed as γ = 64.0 + 0.2h, determine the pressure in psi at a depth of 12 ft below the surface. γ is in lb/ft³, and h in feet below the surface.*

BG

Eq. 3.2: $dp = \gamma dh = (64 + 0.2h)dh$; integrating both sides: $p = 64h + 0.1h^2$

For h = 12 ft: p = 64(12)/144 + 0.1(12)²/144 = 5.43 psi ◄

Sec. 3.3: Pressure Expressed in Height of Fluid -- *Problem*

3.2 *On a certain day the barometric pressure at sea level is 30.0 inHg and the temperature is 60°F. The pressure gage on an airplane flying overhead indicates that the atmospheric pressure at that point is 9.7 psia and that the temperature is 42°F. Calculate as accurately as you can the height of the airplane above sea level. Assume a linear decrease of temperature with elevation.*

BG

At sea level: p_o = 30.0 inHg(14.696 psia/29.92 inHg) = 14.74 psia

Per Sample Prob. 3.1(*d*) with linear lapse rate *b*: $T = a + bz$ and so

$$\frac{p_z}{p_o} = \left(\frac{a+bz}{a}\right)^{-g/Rb} = \left(\frac{T}{T_o}\right)^{-g/Rb} \; ; \; \frac{9.7}{14.74} = \left(\frac{42+460}{60+460}\right)^{-g/Rb} \; ; \; -g/Rb = \log 0.658/\log 0.965 = 11.87$$

Table A.5 for air: R = 1715 ft²/(sec²·°R). ∴ $-32.2/(1715b) = 11.87$; b = -0.001 582°F/ft

$T = a + bz$: $42 + 460 = 60 + 460 - 0.001582z$; z = 11,380 ft ◄

Sec 3.4: Absolute and Gage Pressures -- *Exercises*

3.4.1 *The absolute pressure on a gas is 38 psia and the atmospheric pressure is 925 mb abs. Find the gage pressure in psi, kPa, and mb.*

B

Inside back cover: p_{abs} = (38 psia)(0.001 kN/m^2)/(0.0001450 psi) = 262 kPa abs

p_{atm} = 925 mb abs = (925 mb abs)(0.1 kPa/mb) = 92.5 kPa abs

Eq. 3.7: Gage pressure = $p_{abs} - p_{atm}$ = 262 − 92.5 = 169.6 kPa ◄

Gage pressure = 169.6 kPa(mb/0.1 kPa) = 1696 mb ◄

Gage pressure = 169.6 kPa(0.0001450 lb/in^2)/(0.001 kPa) = 24.6 psi ◄

3.4.2 *If the atmospheric pressure is 860 mb abs and a gage attached to a tank reads 370 mmHg vacuum, what is the absolute pressure within the tank?*

SI

Eq. 3.4: $p_{gage} = \gamma h$ = (13.56 × 9.81 kN/m^3)(0.37m) = 49.2 kN/m^2 vac. = −49.2 kN/m^2

p_{atm} = (860 mb abs)(0.1 kPa/mb) = 86.0 kPa abs

Eq. 3.7: $p_{abs} = p_{atm} + p_{gage}$ = 86.0 − 49.2 = 36.8 kN/m^2 abs ◄

3.4.3 *If the atmospheric pressure is 14.10 psia and a gage attached to a tank reads 10.0 inHg vacuum, what is the absolute pressure within the tank?*

BG

$$\frac{\text{inHg (actual)}}{\text{psi (actual)}} = \frac{\text{inHg (standard)}}{\text{psi (standard)}} \; ; \quad \frac{-10.0}{x} = \frac{29.92}{14.696} \; ; \quad x = -4.91 \text{ psi pressure within tank}$$

Eq. 3.7: $p_{abs} = p_{atm} + p_{gage}$ = 14.10 − 4.91 = 9.19 psia ◄

3.4.4 *If the atmospheric pressure is 970 mb abs and a gage attached to a tank reads 220 mmHg vacuum, what is the absolute pressure within the tank?*

SI

$$\frac{-220 \text{ mm}}{x} = \frac{760 \text{ mmHg}}{101.325 \text{ kN/m}^2} \; ; \quad x = -29.3 \text{ kPa} = \text{pressure in tank}$$

p_{atm} = (970 mb)(0.1 kPa/mb) = 97.0 kPa

Eq. 3.7: $p_{atm} = p_{atm} + p_{gage}$ = 97.0 − 29.3 = 67.7 kPa abs ◄

3.4.5 *A gage is connected to a tank in which the pressure of the fluid is 45 psi above atmospheric. If the absolute pressure of the fluid remains unchanged but the gage is in a chamber where the air pressure is reduced to a vacuum of 28 inHg, what reading in psi will then be observed?*

BG

$$p = p_{fluid} + \text{inHg (actual)}\frac{\text{psi (standard)}}{\text{inHg (standard)}} = 45 + 28\frac{14.696}{29.92} = 58.8 \text{ psi}$$ ◄

3.4.6 *A gage is connected to a tank in which the pressure of the fluid is 315 kPa above atmospheric. If the absolute pressure of the fluid remains unchanged but the gage is in a chamber where the air pressure is reduced to a vacuum of 668 mmHg, what reading in kPa will then be observed?*

SI

$$p = p_{fluid} + \text{mmHg (actual)}\frac{\text{kPa (standard)}}{\text{mmHg (standard)}} = 315 \text{ kPa} + 668 \text{ mmHg}\frac{101.325 \text{ kPa}}{760 \text{ mmHg}} = 404 \text{ kPa}$$ ◄

3.4.7 *If the atmospheric pressure is 29.92 inHg, what will be the height of water in a water barometer if the temperature of the water is 70°F; 120°F? Be as precise as possible.*

BG

Standard atmosphere: 29.92 inHg = 14.696 psi

At 70°F, from Table A.1: p_{vapor} = 0.36 psia, γ = 62.30 lb/ft^3

Water barometer height = (14.696 − 0.36)144/62.30 = 33.14 ft ◄

At 120°F, from Table A.1: p_{vapor} = 1.69 psia, γ = 61.71 lb/ft^3

Water barometer height = (14.696 −1.69)144/61.71 = 30.35 ft ◄

Sec. 3.4: Absolute and Gage Pressures -- *Problems*

3.3 *The tire of an airplane is inflated at sea level to 65 psi. Assuming the tire does not expand, what is the pressure within the tire at elevation 35,000 ft? Assume standard atmosphere. Express the answer in psi and psia.*

BG

Let subscript 1 indicate sea level, subscript 2 indicate elevation 35,000 ft.

Eq. 3.7: $(p_1)_{abs} = p_{atm} + p_{gage}$ = 14.696 + 65 = 79.7 psia

Table A.3 for standard atmosphere: T_1 = 59°F = 519°R, T_2 = −65.6°F = 394.4°R

Eq. 2.4: $R = \dfrac{p_1 v_1}{T_1} = \dfrac{p_2 v_2}{T_2}$; $v_1 = v_2$; $\dfrac{79.7}{519} = \dfrac{p_2}{394}$; $(p_2)_{abs}$ = 60.6 psia ◄

From Table A.3: $(p_2)_{atm}$ = 3.47 psia

Eq. 3.7: $(p_2)_{gage} = (p_2)_{abs} - (p_2)_{atm}$ = 60.6 − 3.47 = 57.1 psi ◄

3.4 *The tire of an airplane is inflated at sea level to 300 kPa. Assuming the tire does not expand, what in the pressure within the tire at elevation (a) 10 000 m and (b) 20 000 m? Assume standard atmosphere. Express answers in both kPa and kPa abs.*

SI

At sea level (subscript 1):

Table A.3 for standard atmosphere: $(p_1)_{atm}$ = 101.325 kPa, T_1 = 15°C = 15 + 273 K = 288 K

Eq. 3.7: $(p_1)_{abs} = (p_1)_{atm} + (p_1)_{gage}$ = 101 + 380 = 481 kPa abs

(*a*) At 10,000 m (subscript 2):

Table A.3: $(p_2)_{atm}$ = 26.499 kPa, T_2 = −49.898°C = 273 − 49.9 K = 223.1 K

Eq. 2.4: $R = \dfrac{p_1 v_1}{T_1} = \dfrac{p_2 v_2}{T_2}$; $v_1 = v_2$; $\dfrac{481}{288} = \dfrac{p_2}{223.1}$; $(p_2)_{abs}$ = 373 kPa abs ◄

Eq. 3.7: $(p_2)_{gage} = (p_2)_{abs} - (p_2)_{atm}$ = 373 − 27 = 346 kPa ◄

(*b*) At 16,000 m (subscript 3):

Table A.3: $(p_3)_{atm}$ = 10.352 kPa, T_3 = −56.5°C = 273 − 56.5 K = 216.5 K

$\dfrac{481}{288} = \dfrac{p_3}{216.5}$; $(p_3)_{abs}$ = 362 kPa abs ◄ Eq. 3.7: $(p_3)_{gage}$ = 362 − 10.35 = 351 kPa ◄

The gage pressure in the tire will gradually decrease with elevation up to about 12,000 m. Above that elevation the gage pressure will increase because of the temperature variation in the atmosphere.

Sec 3.5: Barometer -- *Exercises*

3.5.1 *If the atmospheric pressure were equivalent to 33.1 ft of water, what would be the reading on a barometer containing an alcohol (s = 0.79) if the vapor pressure of the alcohol at the temperature of observation were 2.12 psia.*

BG

(33.1 ft of water)(14.696 psia/33.9 ft water) = 14.35 psia

From Eq. 3.8: $y = (p_{\text{atm}} - p_{\text{vapor}})/\gamma = (14.35 - 2.12)144/[(0.79)62.4] = 35.72$ ft ◄

3.5.2 *If the atmospheric pressure is 890 mb abs what would be the reading in meters of a barometer containing water at 70°C?*

SI

Eq. 3.5: $h = p/\gamma = (89 \text{ kN/m}^2)/9.8 \text{ kN/m}^3 = 9.08$ m

From Table A.1 at 70°C, $p_v/\gamma = 3.20$ m ; barometer reading = 9.08 − 3.20 = 5.88 m ◄

Sec 3.6: Measurement of Pressure -- *Exercises*

3.6.1 *In Sample Prob. 3.4 suppose the atmospheric pressure at B is 1035 mb abs. What would be the absolute pressure at A? Express in mb abs and in mHg.*
Sample Prob. 3.4: $(p_A)_{gage} = 112.0$ kPa.

SI

Given: $p_{\text{atm}} = 1035$ mb abs; Using inside cover: $(p_A)_{\text{gage}} = 112.0 \text{ kN/m}^2 = 1120$ mb

Eq. 3.7: $(p_A)_{\text{abs}} = p_{\text{atm}} + p_{\text{gage}} = 1035 + 1120 = 2155$ mb abs ◄

$(p_A/\gamma)_{\text{abs}} = (215.5 \text{ kN/m}^2)/(13.56 \times 9.81 \text{ kN/m}^3) = 1.620$ mHg abs ◄

3.6.2 *In the figure, originally there is a 6-in manometer reading. Atmospheric pressure is 14.70 psia. If the absolute pressure at A is doubled, what will be the manometer reading?*

BG

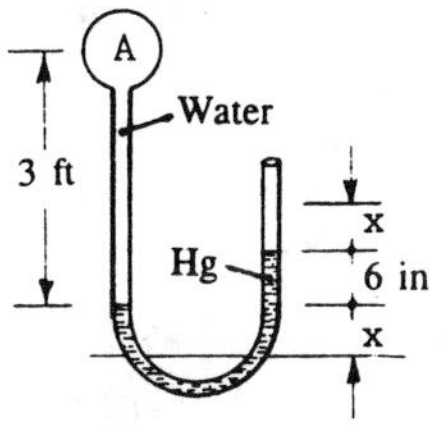

$p_{\text{atm}} = 14.70$ psia (given) = 33.91 ft of water

Eq. 3.9: $(p_A/\gamma)_{\text{abs}} = 33.91 + (6/12)(13.56) - 3 = 37.7$ ft

$2(37.7) = 75.4 = 33.91 + [(6 + 2x)/12]13.56 - [3 + (x/12)]$; $x = 17.32$ in.

Hence new manometer reading = 6 + 2(17.32) = 40.6 inches ◄

3.6.3 *A mercury manometer (Fig. 3.11) is connected to a pipeline carrying water at 150°F and located in a room where the temperature is also 150°F. If the elevation of point B is 10 ft above A and the mercury manometer reading is 48 in, what is the pressure in the pipe in psi? Repeat, assuming all temperatures are 68°F. Be as precise as possible, and note the effect of temperature. Note that at 150°F the specific gravity of mercury is 13.45.*

BG

Fig. 3.10: $p_A = \gamma_{\text{water}} z + \gamma_{\text{Hg}} y$

Table A.1 at 150°F: $\gamma_{\text{water}} = 61.20 \text{ lb/ft}^3$

$p_A = 61.20(10) + (62.4 \times 13.45)48/12 = 3969$ psf = 27.6 psi ◄

Table A.1 at 68°F: $\gamma_{\text{water}} = 62.31 \text{ lb/ft}^3$

$p_A = 62.30(10) + (62.4 \times 13.56)48/12 = 4008$ psf = 27.8 psi ◄

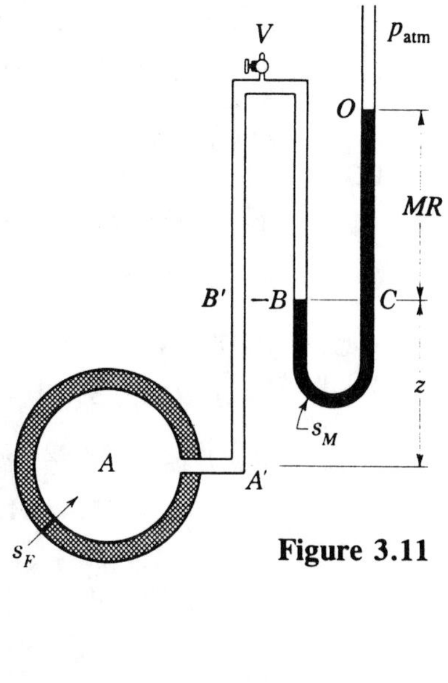

Figure 3.11

3.6.4 *(a) Two vessels are connected to a differential manometer using mercury (s = *13.56), the connecting tubing being filled with water. The higher-pressure vessel is 5 ft lower in elevation than the other. Room temperature prevails. If the mercury reading is 4.0 in, what is the pressure difference in feet of water, in psi? (b) If carbon tetrachloride (s = 1.59) were used instead of mercury, what would be the manometer reading for the same temperature difference? [*Note this correction to the textbook.]*

BG

(a) Let A be the lower vessel

Eq. 3.12: $p_A/\gamma - p_B/\gamma = (4/12)(13.56 - 1.0) + 5$

$p_A/\gamma - p_B/\gamma = 5 + 4.19 = 9.19$ ft of water ◀

$\Delta p = 9.19(14.696/33.91) = 3.98$ psi ◀

(b) Eq. 3.12: $p_A/\gamma - p_B/\gamma = 9.19$ ft of water $= 5.0 + (MR/12)(1.59 - 1.0)$

$MR = 4.19 \times 12/0.59 = 85.2$ in of CCl_4 ◀

3.6.5 *(a) Two vessels are connected to a differential manometer using mercury (s = *13.56), the connecting tubing being filled with water. The higher-pressure vessel is 1.5 m lower in elevation than the other. Room temperature prevails. If the mercury reading is 100 mm, what is the pressure difference in m of water, in kPa? (b) If carbon tetrachloride (s = 1.59) were used instead of mercury, what would be the manometer reading for the same temperature difference? [*Note this correction to the textbook.]*

SI

(a) Let A be the lower vessel

Eq. 3.12: $p_A/\gamma - p_B/\gamma = 1.5 + 0.10(13.56 - 1.0) = 1.5 + 1.256 = 2.76$ m ◀

$\Delta p = (2.76\ \text{m})(9.81\ \text{kN/m}^3) = 27.0$ kPa ◀

(b) Eq. 3.12: $(p_A/\gamma - p_B/\gamma) = 1.5 + MR(1.59 - 1.0)$

$\therefore\ 2.76 - 1.5 = MR(0.59)$; $MR = 2.13$ m of CCl_4 ◀

3.6.6 *Refer to the manometer of Fig. 3.14b. A and B are at the same elevation. Water is contained in A and rises in the tube to a level 68 in above A. Glycerin is contained in B. The inverted U-tube is filled with air at 15 psi and 70°F. Atmospheric pressure is 14.70 psia. Determine the difference in pressure between A and B if the manometer reading is 12 in. Express the answer in psi. What is the absolute pressure in B in inches of mercury, feet of glycerin?*

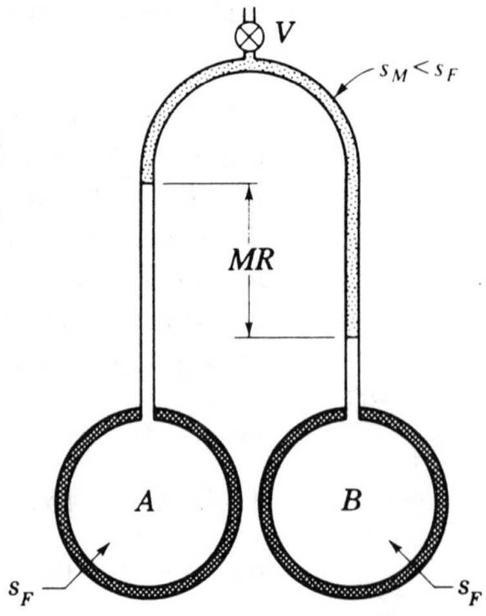

BG

$\gamma_{\text{air}} = 0$; Table A.4 for glycerin: $s = 1.26$

Eq. 3.12, modified: $p_A/\gamma - 68/12 + 1.26(68 - 12)/12 = p_B/\gamma$

$p_A/\gamma - p_B/\gamma = 5.67 - 5.88 = -0.213$ ft water

$14.70(0.213)/33.91 = 0.0925 = 0.09$ psi (B is at the higher pressure) ◀

$(p_B/\gamma)_{abs} = 29.92 + (29.92)15/14.70 + (68 - 12)1.26/13.56$

$= 65.7$ inHg ◀

$= (65.7 \times 13.56)/(1.26 \times 12) = 58.9$ ft glycerin ◀

3.6.7 *Gas confined in a rigid container exerts a pressure of 25 psi when its temperature is 40°F. What pressure would the gas exert if the temperature were raised to 165°F? Barometric pressure remains constant at 29.0 inHg.*

BG

$$p_{atm} = (\text{inHg})_{atm}\frac{\text{lb/in}^2\ (\text{standard})}{\text{inHg (standard)}} = (29.0)(14.70/29.92) = 14.25 \text{ psi}$$

Eq. 3.7: $p_{abs} = p_{atm} + p_{gage} = 14.25 + 25 = 39.2$ psia at 40°F

Eq. 2.4: $R = \frac{p\,v}{T} = \frac{39.2\,v}{460 + 40} = p'_{abs}\frac{v}{460 + 165}$; $p'_{abs} = 39.2\frac{625}{500} = 49.1$ psia

$p'_{gage} = 49.1 - 14.25 = 34.8$ psi ◀

3.6.8 *Gas confined in a rigid container exerts a pressure of 200 kPa when its temperature is 5°C. What pressure would the gas exert if the temperature were raised to 80°C? Barometric pressure remains constant at 29.0 inHg.*

SI

29 inHg = 737 mmHg ; p_{atm} = 737 mm(101.325 kPa/760 mm) = 98.2 kN/m^2

Eq. 3.7: $p_{abs} = p_{atm} + p_{gage} = 98.2 + 200 = 298.2$ kPa

Eq. 2.4: $R = \frac{p\,v}{T} = \frac{298.2\,v}{273 + 5} = p'_{abs}\frac{v}{273 + 80}$

$p'_{abs} = 378.7$ kN/m^2 abs ; $p'_{gage} = 378.7 - 98.2 = 280$ kPa ◀

3.6.9 *In the figure, atmospheric pressure is 14.70 psia; the gage reading at A is 3.5 psi; the vapor pressure of the alcohol is 1.4 psia. Compute x and y.*

BG

Working in terms of absolute pressure heads,

$$\frac{1.4(144)}{0.90(62.4)} + x = \frac{(14.70 + 3.5)144}{0.90(62.4)} \; ; \quad x = 43.1 \text{ ft} \quad ◀$$

$$0 + \frac{45(13.55)}{12(0.9)} - \left(\frac{45}{12} + y\right) = \frac{(14.70 + 3.5)144}{0.90(62.4)} \; ;$$

$y = 6.04$ feet ◀

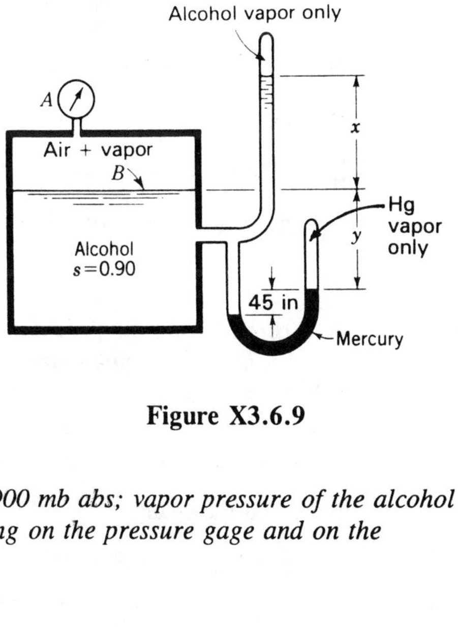

Figure X3.6.9

3.6.10 *In Fig. X3.6.9 assume the following: atmospheric pressure = 900 mb abs; vapor pressure of the alcohol = 130 mb abs; x = 3.50 m and y = 1.80 m. Compute the reading on the pressure gage and on the manometer.*

Fig. X3.6.9: See Solution 3.6.9 above.

SI

13 kPa + (0.90)9.81(3.5) = 43.9 kN/m^2 = $(p_B)_{abs}$

Eq. 3.7: $(p_A)_{gage} = (p_A)_{abs} - p_{atm} = 43.9 - 90 = -46.1$ kPa ◀

= 46.1 kPa vac (760 mmHg/101.325 kPa abs) = 346 mmHg vac ◀

$0 + MR\frac{13.56}{0.90} - MR - 1.8 = \left(\frac{p_B}{\gamma}\right)_{abs}$; however, $(p_B/\gamma)_{abs} = 43.9/(0.90 \times 9.81) = 4.97$ m

$(15.07)MR - MR = 1.8 + 4.97$ m ; $MR = 0.481$ m = 481 mm ◀

3.5 *What would be the manometer reading in Sample Prob. 3.4 if $p_B - p_A = 120$ kPa?*

Sample Prob. 3.4: Manometer fluid = Hg. $\gamma_A = 8.4$ kN/m^3, $\gamma_B = 12.4$ kN/m^3, $p_B = 207$ kPa, find p_A. Express all pressure heads in terms of liquid B.

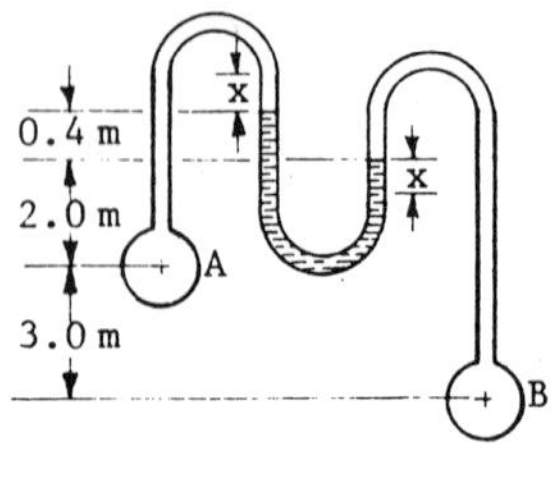

SI

$$\frac{p_B}{\gamma} - \frac{p_A}{\gamma} = \frac{120\ \text{kN/m}^2}{12.4\ \text{kN/m}^3} = 9.68\ \text{m}\ \ \text{(of liquid B)}$$

$$\frac{p_A}{\gamma} - (2.4 + x)\frac{8.4}{12.4} + (2x + 0.4)\frac{13.56(9.81)}{12.4} + (5 - x) = \frac{p_B}{\gamma}$$

$-1.626 - 0.677 + 21.46 + 4.29 + 5.0 - x = 9.68$; $x = 0.1017$ m

Manometer reading $= 0.4 + 2x = 0.603$ mHg ◀

3.6 *The diameter of tube C in Fig. 3.11 is d_1, and that of tube B is d_2. Let z_0 be the elevation of the mercury above A when both mercury columns are at the same level. R is the distance the right-hand column of mercury rises above z_0 when the fluid in A is under pressure. Let γ' be the specific weight of the mercury (or any other measuring fluid), while γ is the specific weight of the fluid in A and the connecting tubing. Prove that $p_A = \gamma z_0 + [\gamma' + (\gamma' - \gamma)(d_1/d_2)^2]R = M + NR$ where M and N are constants. It is seen that this equation involves only one variable, which is the reading R on the scale for column C. It also shows the significance of having d_2 large compared with d_1.*

Fig. 3.11: See Solution 3.6.3 above.

N

Let x = depression of Hg in tube B below equilibrium level z_0

From 0 to A: $0 + (R + x)\gamma' + (z_0 - x)\gamma = p_A$

Since Hg is incompressible, $R d_1^2(\pi/4) = x d_2^2(\pi/4)$; $x = R(d_1/d_2)^2$

$\therefore\ p_A = [R + R(d_1/d_2)^2]\gamma + z_0\gamma - \gamma R(d_1/d_2)^2 = \gamma z_0 + [\gamma' + (\gamma' - \gamma)(d_1/d_2)^2]R$ Q.E.D. ◀

3.7 *At a certain point the pressure in a pipeline containing gas ($\gamma = 0.05$ lb/ft^3) is 5.6 in of water. The gas is not flowing. All temperatures are 60°F. What is the pressure in inches of water at another point in the line whose elevation is 650 ft greater than the first point? Make and state clearly any necessary assumptions.*

BG

Assume $\gamma_{gas} = 0.05$ lb/ft^3 = constant

Note that the change in pressure in the atmosphere must be considered.

Assume $\gamma_{air} = 0.076$ lb/ft^3 = constant ; Let A be the lower point.

$(p_A/\gamma)_{abs} = (p_A/\gamma)_{atm} + 5.6/12$ ft of water (1)

$(p_B/\gamma)_{abs} = (p_B/\gamma)_{atm} + x/12$ ft of water (2)

$(p_A/\gamma)_{atm} - (p_B/\gamma)_{atm} = 650$ ft of air $= 650(0.076/62.4) = 0.792$ ft of water

$(p_A/\gamma)_{abs} - (p_B/\gamma)_{abs} = 650$ ft of gas $= 650(0.05/62.4) = 0.521$ ft of water

Subtract (2) from (1) and substitute the other relationships.

$(p_A/\gamma)_{abs} - (p_B/\gamma)_{abs} = (p_A/\gamma)_{atm} - (p_B/\gamma)_{atm} + (5.6/12) - (x/12)$

$0.521 = 0.792 + (5.6/12) - (x/12)$; $x = 8.85$ inches of water ◀

Sec 3.8: Center of Pressure -- *Exercises*

3.8.1 *If a triangle of height d and base b is vertical and submerged in liquid with its vertex at the liquid surface, derive an expression for the depth to its center of pressure.*

N

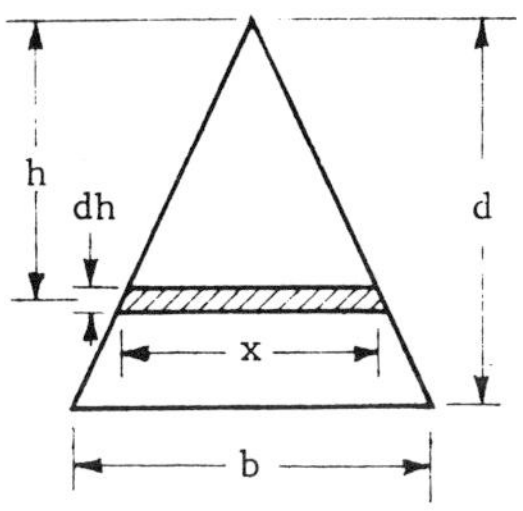

Eq. 3.14: $F = \int p\,dA = \int \gamma h\,dA = \int \gamma h x\,dh$

But $x = h(b/d)$ so $F = \gamma \dfrac{b}{d}\displaystyle\int_0^d h^2 dh$

As $h_p F = \int h\,dF = \gamma \dfrac{b}{d}\displaystyle\int_0^d h^3 dh$

$$h_p = \frac{\int_0^d h^3 dh}{\int_0^d h^2 dh} = \left.\frac{h^4/4}{h^3/3}\right|_0^d = \frac{3}{4}d \quad ◀$$

Alternatively, Eq. 3.18: $h_p = h_c + \dfrac{I_c}{h_c A} = \dfrac{2}{3}d + \dfrac{(1/36)bd^3}{(2/3)d(bd/2)} = \dfrac{2}{3}d + \dfrac{d}{12} = (3/4)d$ ◀

3.8.2 *Repeat Exer. 3.8.1 for the same triangle but with its vertex a distance a below the liquid surface.*

Exer. 3.8.1: The triangle shown in the figure with Solution 3.8.1 is vertical. Derive an expression for the depth to its center of pressure.

N

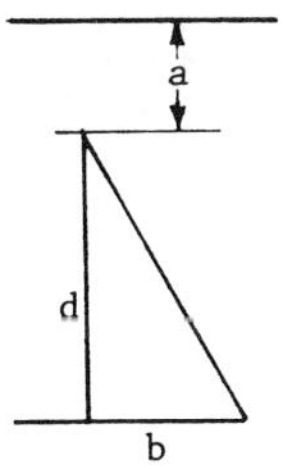

Eq. (3.18): $h_p = (a + 2d/3) + \dfrac{bd^3/36}{(a + 2d/3)(bd/2)} = (a + 2d/3) + \dfrac{d^2}{(a + 2d/3)\,18}$

$$= \frac{18(a^2 + 4ad/3 + 4d^2/9) + d^2}{18(a + 2d/3)} = \frac{6a^2 + 8ad + 3d^2}{6(a + 2d/3)} \quad ◀$$

3.8.3 *If a triangle of height d and base b is vertical and submerged in a liquid with its base at the liquid surface, derive an expression for the depth to its center of pressure.*

N

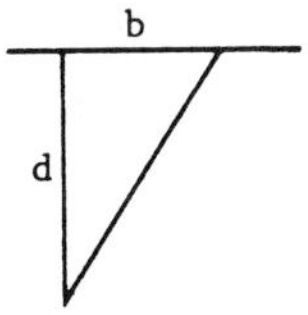

Eq. (3.18): $h_p = d/3 + \dfrac{bd^3/36}{(d/3)(bd/2)} = d/3 + d/6 = d/2$ ◀

3.8.4 *A circular area of diameter d is vertical and submerged in a liquid. Its upper edge is coincident with the liquid surface. Derive an expression for the depth to its center of pressure.*

N

Eq. (3.18): $h_p = d/2 + \dfrac{\pi d^4/64}{(d/2)\,\pi d^2/4} = d/2 + d/8 = 5d/8$ ◀

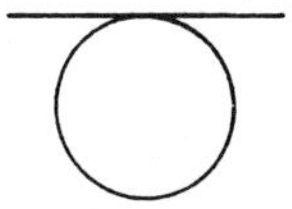

3.8.5 *Refer to Sample Problem 3.7. If the air pressure were 5 psi rather then 3 psi, compute the total force and determine the location of the center of pressure.*

Sample Prob. 3.7 (refer to it): At 3 psi, F_A = 5430 lb, ΣF = 7380 lb.

BG

$F_A = 5430(5/3) = 9050$ lb ; $\Sigma F = 7380 + (9050 - 5430) = 11{,}000$ lb ◀

$11{,}000(h_p) = 9050(3) + 675(2.28) + 941(3.85) + 333(4.18) = 33{,}700$

$h_p = 33{,}700/11{,}000 = 3.06$ ft ◀

3.8.6 *A rectangular plate submerged in water is 5 by 4 ft, the 5-ft side being horizontal and the 4-ft side being vertical. Determine the magnitude of the force on one side of the plate and the depth to its center of pressure if the top edge is (a) at the water surface; (b) 1 ft below the water surface; (c) 100 ft below the water surface.*

BG

Eq. 3.16: $F = \gamma h_c A = (62.4)h_c(4 \times 5) = 1248h_c$

(*a*) $h_c = 2$ ft ; $F = 1248(2) = 2496$ lb ◀

(*b*) $h_c = 3$ ft ; $F = 1248(3) = 3744$ lb ◀

(*c*) $h_c = 102$ ft ; $F = 1248(102) = 127{,}296$ lb ◀

Eq. 3.18: $h_p = y_c + \dfrac{I_c}{y_c A} = y_c + \dfrac{bh^3}{12 y_c bh} = y_c + \dfrac{h^2}{12 y_c}$

(*a*) $y_c = 2$ ft ; $h_p = 2 + \dfrac{4^2}{(12)2} = 2.67$ ft ◀

(*b*) $y_c = 3$ ft ; $h_p = 3 + \dfrac{4^2}{(12)3} = 3.44$ ft ◀

(*c*) $y_c = 102$ ft ; $h_p = 102 + \dfrac{4^2}{12(102)} = 102.0$ ft ◀

3.8.7 *Repeat Exer. 3.8.6 changing all dimensions from feet to meters, i.e., the plate is 5 m by 4 m.*
Exer. 3.8.6: A rectangular plate submerged in water measures 5 ft H × 4 ft V. Find F on one side and h_p if the top edge is (a) in the water surface, (b) 1 ft below it, (c) 100 ft below it.

SI

Eq. 3.16: $F = \gamma h_c A = 9.81$ kN/m^3(h_c)(5m ×4m) $= 196.2 h_c$

(*a*) $F = 196.2(2) = 392$ kN ◀ (*b*) $F = 196.2(3) = 589$ kN ◀

(*c*) $F = 196.2(102) = 20\,000$ kN ◀

Eq. 3.18: $h_p = y_c + \dfrac{I_c}{y_c A} = y_c + \dfrac{bh^3}{12 y_c bh} = y_c + \dfrac{h^2}{12 y_c}$

(*a*) $y_c = 2$ m ; $h_p = 2 + \dfrac{4^2}{(12)2} = 2.67$ m ◀ (*b*) $y_c = 3$ m ; $h_p = 3 + \dfrac{4^2}{(12)3} = 3.44$ m ◀

(*c*) $y_c = 102$ m ; $h_p = 102 + \dfrac{4^2}{12(102)} = 102.0$ m ◀

3.8.8 *A rectangular plate 5 ft by 4 ft is at an angle of 30° with the horizontal, and the 5-ft side is horizontal. Find the magnitude of the force on one side of the plate and the location of its center of pressure when the top edge is (a) at the water surface; (b) 1 ft below the water surface.*

BG

(*a*) $h_c = 0 + 2\sin 30° = 1.0$ ft ; Eq. 3.16: $F = \gamma h_c A = 62.4(1.00)(5 \times 4) = 1248$ lb ◀

$y_o = d/2 = 2$ ft ; Eq. 3.18: $y_p = y_c + \dfrac{bd^3/12}{y_c(bd)} = y_c + \dfrac{d^2}{12 y_c} = 2 + \dfrac{4^2}{12(2)} = 2.67$ ft ◀

(*b*) $h_c = 1 + 2\sin 30° = 2.0$ ft ; $F = 62.4(2.0)(5 \times 4) = 2500$ lb ◀

Per Fig. 3.16: $y_c = h_c/\sin 30° = 2/\sin 30° = 4$ ft

Eq. 3.18: $y_p = y_c + \dfrac{d^2}{12 y_c} = 4 + \dfrac{4^2}{12(4)} = 4.33$ ft ◀

3.8.9 *A rectangular area is 5 by 6 m, with the 5 m side horizontal. It is placed with its centroid 4 m below a water surface and rotated about a horizontal axis in the plane area and through its centroid. Find the magnitude of the force on one side and the distance between the center of pressure and the centroid of the plane when the angle θ = 90, 60, 30, 0°.*

SI

Eq. 3.16: $F = \gamma h_c A = (9.81)4(5\times6) = 1177$ kN for any angle θ ◄

From Eq. 3.18: $$y_p - y_c = \frac{I_c}{y_c A} = \frac{bd^3/12}{y_c(bd)} = \frac{d^2}{12y_c} = \frac{6^2}{12y_c} = \frac{3}{y_c} = \frac{3\sin\theta}{h_c} = \frac{3}{4}\sin\theta$$

From which

θ	$y_p - y_c$	
90°	0.750 m	◄
60°	0.650 m	◄
30°	0.375 m	◄
0°	0 m	◄

3.8.10 *A plane surface is circular and 5 ft in diameter. If it is vertical and the top edge is 2 ft below the water surface, find the magnitude of the force on one side and the depth to the center of pressure.*

BG

$h_c = 2 + d/2 = 4.5$ ft ; Eq. 3.16: $F = \gamma h_c A = 62.4(4.5)\pi(5)^2/4 = 5510$ lb ◄

From Eq. 3.18 and Table A.7: $$y_p - y_c = \frac{I_c}{y_c A} = \frac{\pi d^4}{64}\left(\frac{1}{h_c}\right)\frac{4}{\pi d^2} = \frac{d^2}{16h_c} = \frac{5^2}{16(4.5)} = 0.347 \text{ ft}$$

$h_p = 0.347 + h_c = 4.85$ ft ◄

3.8.11 *A plane surface is circular and 1.5 m in diameter. If it is vertical and the top edge is 1 m below the water surface, find the magnitude of the force on one side and the depth to the center of pressure.*

SI

$h_c = 1 + d/2 = 1.75$ m ; Eq. 3.16: $F = \gamma h_c A = 9.81(1.75)\pi(0.75)^2 = 30.3$ kN ◄

From Eq. 3.18 and Table A.7: $$y_p - y_c = \frac{I_c}{y_c A} = \frac{\pi d^4}{64}\left(\frac{1}{h_c}\right)\frac{4}{\pi d^2} = \frac{d^2}{16h_c} = \frac{1.5^2}{16(1.75)} = 0.0804 \text{ m}$$

$h_p = 0.0804 + 1.75 = 1.830$ m ◄

3.8.12 *Prove that for a plane area such that a straight line can be drawn through the midpoints of all horizontal elements, the center of pressure must lie on this line.*

N

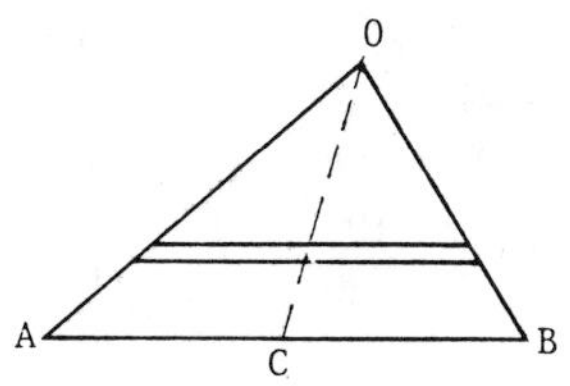

By inspection:

The straight line OC is drawn through the midpoints of all horizontal elements. Each element of hydrostatic force lies on this line. Hence, the resultant of all such elementary forces must also lie on this line. Thus the center of pressure lies on this line. Q.E.D. ◄

3.8.13 *A vertical right-triangle of height d and base b submerged in liquid has its vertex at the liquid surface. Find the distance from the vertical side to the center of pressure by (a) inspection; (b) calculus.*

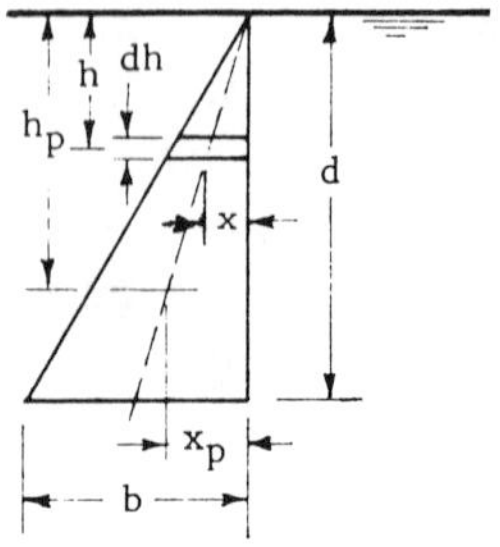

N

(*a*) By inspection: $h_p = (3/4)d$ from Exer. 3.8.1

Using Exer. 3.8.12: $x_p = (3/4)(b/2) = (3/8)b$ ◄

(*b*) By calculus: with $x = \frac{1}{2}\frac{hb}{d}$ (to center of each strip) and $dA = \frac{hb}{d}dh$

$$x_p = \frac{\int xh\,dA}{\int h\,dA} = \frac{\int_0^d \frac{hb}{2d}(h)\frac{hb}{d}\,dh}{\int_0^d (h)\frac{hb}{d}\,dh} = \frac{\frac{b^2}{2d^2}\int_0^d h^3\,dh}{\frac{b}{d}\int_0^d h^2\,dh} = \frac{(b)\frac{d^4}{4}}{(2d)\frac{d^3}{3}} = \frac{3}{8}b \quad ◄$$

3.8.14 *The right-triangular plate shown in Fig. X3.8.14 is submerged in a vertical plane with its base horizontal. Determine the depth and horizontal position of the center of pressure.*

BG

Eq. 3.18: $y_p = y_c + \frac{I_c}{y_c A}$; Table A.7 : $I_c = \frac{bh^3}{36}$

where : $A = \frac{4(6)}{2} = 12\text{ ft}^2$; $y_c = \frac{h}{3} + 1 = 3\text{ ft}$; $I_c = \frac{bh^3}{36} = 24\text{ ft}^4$

$\therefore\ y_p = y_c + \frac{I_c}{y_c A} = 3 + \frac{24}{3(12)} = 3.67\text{ ft}$ ◄

Below P: $y' = 1 + 6 - y_p = 3.33$ ft

To median, $x_p = (1/2)(3.33)(4/6) = 1.111$ ft ◄

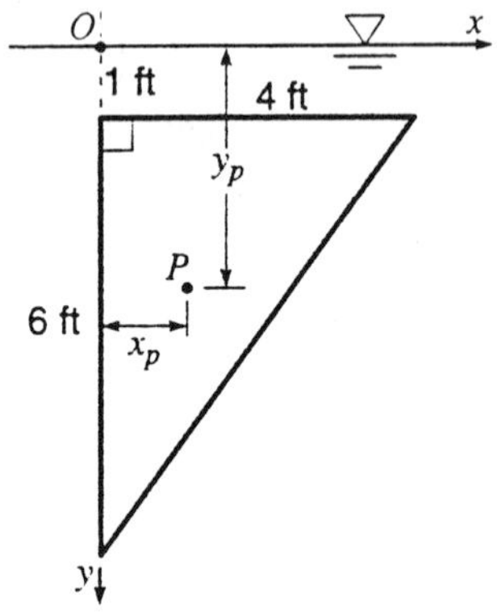

Figure X3.8.14

3.8.15 *Repeat Exer. 3.8.14, but with the base 1.2 m long and submerged 0.3 m, and the vertical edge is 1.8 m long.*

Exercise 3.8.14: For the submerged right triangular plate of Fig. X3.8.14 (above), find the depth and horizontal position of the center of pressure.

SI

Eq. 3.18: $y_p = y_c + \frac{I_c}{y_c A}$; Table A.7 : $I_c = \frac{bh^3}{36}$; $A = (1.2)1.8/2 = 1.080\text{ m}^2$

$y_c = \frac{h}{3} + 0.3 = 0.9\text{ m}$; $I_c = \frac{1.2(1.8)^3}{36} = 0.1944\text{ m}^4$; $\therefore\ y_p = 0.9 + \frac{0.1944}{0.9(1.080)} = 1.100\text{ m}$ ◄

Below P: $y' = 0.3 + 1.8 - y_p = 1.000$ m ; to median, $x_p = 0.5(1.000)1.2/1.8 = 0.333$ m ◄

3.8.16 *Figure X3.8.16 shows a cylindrical tank with 0.25-in-thick walls, containing water. What is the force on the bottom? What is the force on the annular surface MM? What is the weight of the water? Find the longitudinal (vertical) tensile stress in the sidewalls BB if (a) the tank is suspended from the top; (b) it is supported from the bottom. Neglect the weight of the tank.*

BG

Eq. 3.16: Force on bottom $F_B = \gamma h_c A = 62.4(2)\pi = 392$ lb ◄

Force on surface MM, $F_S = \gamma h_c A = 62.4(1)(\pi/4)(2^2 - 1^2) = 147.0$ lb ◄

Weight of water $= 62.4\,\pi[1^2 + (1/2)^2] = 245$ lb ◄

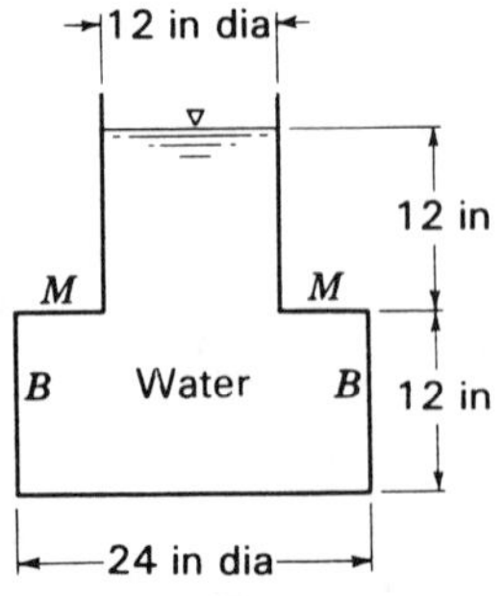

Figure X3.8.16

/cont...

(a) Suspended from top, $\sigma = \dfrac{F_B}{A} = \dfrac{392}{\pi(r_1^2 - r_2^2)} = \dfrac{392}{\pi[12^2 - (23.5/2)^2]}$

$= 392/18.65 = 21.0$ lb/in^2 ◀

(b) Supported from bottom, $\sigma = F_S/A = 147/18.65 = 7.88$ lb/in^2 ◀

3.8.17 *Find the magnitude and point of application of the force on the circular gate shown in Fig. X3.8.17.*

SI

Eq. 3.16: $F = \gamma h_c A = 9.81(2 + 0.8\sin 60°)\pi(0.8)^2 = 53.1$ kN ◀

$$\frac{I_c}{y_c A} = \frac{(\pi/64)(1.6)^4}{(2/\cos 30° + 0.8)\pi(0.8)^2} = 0.0515 \text{ m}$$

Eq. 3.18: $y_p = y_c + \dfrac{I_c}{y_c A}$ (slope distance)

$y_p = 2/\cos 30° + 0.8 + 0.0515 = 3.16$ m ◀

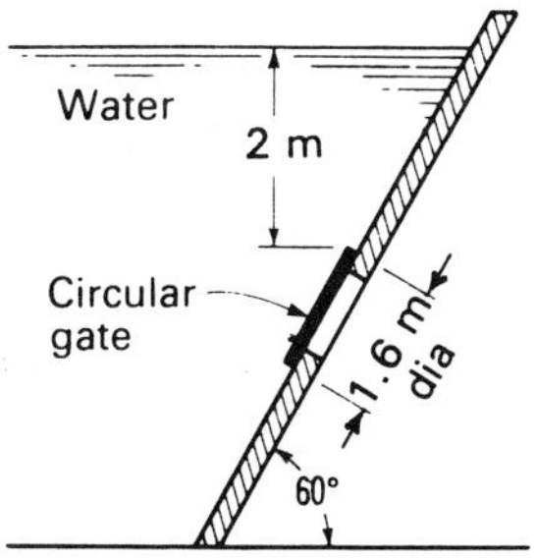

Figure X3.8.17

3.8.18 *The gate MN in Fig. X3.8.18 rotates about an axis through N. If the width perpendicular to the plane of the figure is 3 ft, what torque applied to the shaft through N is required to hold the gate closed?*

BG

Eq. 3.16: $F_L = \gamma h_c A = 62.4(3.4 + 1.7)(3.4 \times 3) = 3250$ lb

$$(y_p - y_c)_L = \frac{I_c}{y_c A} = \frac{bh^3/12}{y_c(bh)} = \frac{h^2}{12 y_c} = \frac{3.4^2}{12(3.4 + 1.7)} = 0.1889 \text{ ft}$$

$F_R = \gamma h_c A = (62.4)1.0(2 \times 3) = 374$ lb ; $(y_p)_2 = (1/3)3 = 1$ ft

$\Sigma M_N = 3250(6.8 - 5.1 - 0.1889) - 374(0.667) - \text{Torque} = 0$;

Torque $= 4660$ ft·lb ◀

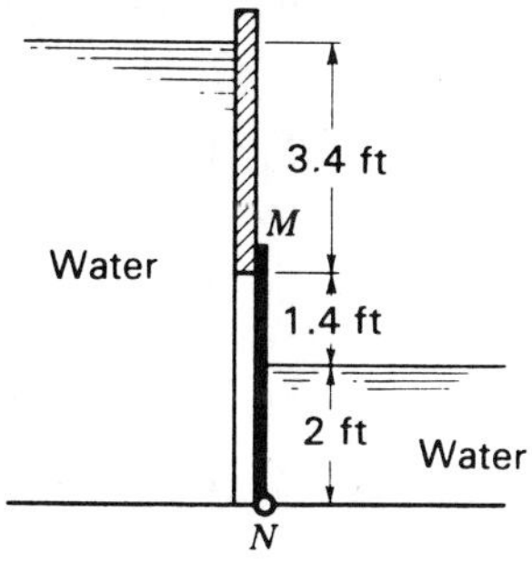

Figure X3.8.18

3.8.19 *In Fig. X3.8.19 the rectangular flashboard MN shown in cross-section is pivoted at B. (a) What must be the maximum height of B above N if the flashboard is on the verge of tipping when the water surface rises to M? (b) If the flashboard is pivoted at the location determined in (a) and the water surface is 1 m below M, what are the reactions at B and N per m length of board perpendicular to the figure?*

SI

(a) For critical stability, center of pressure must be at B.

Thus B will be $5.4/3 = 1.80$ m above N. ◀

(b) For 4.4 m depth of water, Eq. 3.16: $F = 9.81(4.4/2)4.4$

$= 95.0$ kN/m

$\Sigma M_B = 0 = 95.0(1.8 - 4.4) - 1.8 N_x$; $N_x = 17.59$ kN/m ◀

$\therefore\ B_x = 95.0 - 17.59 = 77.4$ kN/m ◀ $B_y = 0$ ◀

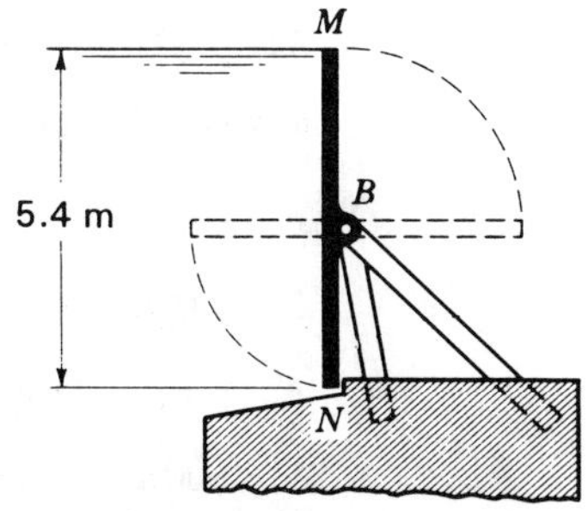

Figure X3.8.19

3.8 *A vertical semicircular area has its diameter in a liquid surface. Derive an expression for the depth to its center of pressure.*

N

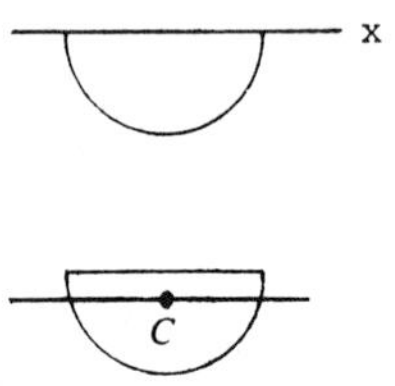

Eq. 3.18 and Table A.7: $h_p = \dfrac{4r}{3\pi} + \dfrac{I_c}{(4r/3\pi)(\pi r^2/2)} = \dfrac{4r}{3\pi} + \dfrac{I_c}{2r^3/3}$

$$I_x = (1/2)\frac{\pi d^4}{64} = (\pi/2)\frac{(2r)^4}{64} = (1/2)\frac{\pi r^4}{4} = \frac{\pi r^4}{8}$$

$$I_c = \frac{\pi r^4}{8} - (1/2)\pi r^2\left(\frac{4r}{3\pi}\right)^2 = \frac{\pi r^4}{8} - \frac{16 r^4}{18\pi} = r^4\left(\frac{\pi}{8} - \frac{8}{9\pi}\right) = r^4\left(\frac{9\pi^2 - 64}{72\pi}\right)$$

$$h_p = \frac{4r}{3\pi} + \frac{r^4(9\pi^2 - 64)/72\pi}{(2/3)r^3} = \frac{4r}{3\pi} + \frac{3}{2}r\left(\frac{25}{226}\right) = 0.424r + 0.1659r = 0.590r \quad ◀$$

3.9 **Repeat Exer. 3.8.6 for the case where a 2-ft-thick layer of oil (s = 0.8) is resting on the water, and replace "water surface" by "oil surface." [*Note this statement is a correction to the textbook.]*

Exer. 3.8.6: A rectangular plate, 5 ft H by 4 ft V, is submerged in water. Determine the magnitude of the force on one side of the plate and the location of its center of pressure if the top edge is (a) at the water surface; (b) 1 foot below the water surface; (c) 100 ft below the water surface.

BG

Eq. 3.18: $y_p = y_c + \dfrac{I_c}{y_c A} = y_c + \dfrac{bh^3/12}{y_c(bh)} = y_c + \dfrac{h^2}{12 y_c}$

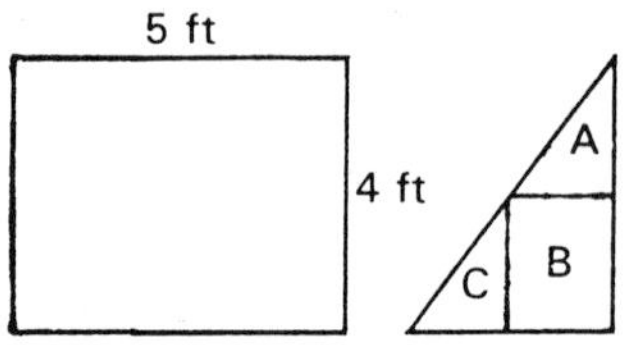

(*a*) $F_A = \gamma h_c A = 0.8(62.4)1(2\times5) = 499$ lb at $y_p = 1.333$ ft

$F_B = 0.8(62.4)2(2\times5) = 998$ lb at $y_p + 1 = 3$ ft

$F_C = 62.4(1/2\times2)(2\times5) = 624$ lb at $y_p + 2 = 3.33$ ft

$\Sigma F = 2120$ lb ◀

$2120 h_p = 499(1.333) + 998(3) + 624(3.33)$; $h_p = 2.71$ ft ◀ (See alternate below)

(*b*) $F_A = 0.8(62.4)1.5(1\times5) = 374$ lb at $y_p = 1.556$ ft

$F_B = 0.8(62.4)2(3\times5) = 1498$ lb at $y_p + 2 = 3.50$ ft

$F_C = 62.4(0.5\times3)(3\times5) = 1404$ lb at $y_p + 2 = 4.00$ ft ; $\Sigma F = 3280$ lb ◀

$3280 h_p = 374(1.556) + 1498(3.5) + 1404(4.00)$; $h_p = 3.49$ ft ◀

(*c*) $F_A = 0$

$F_B = 0.8(62.4)2(4\times5) = 1997$ lb at $y_p + 2 = 102$ ft

$F_C = 62.4(100)(4\times5) = 124{,}800$ lb at $y_p + 2 = 102.01$ ft ; $\Sigma F = 126{,}800$ lb ◀

$126{,}800 h_p = 1997(102) + 124{,}800(102.01)$; $h_p = 102.0$ ft ◀

Alternate solution (*a*):

$F_{oil} = 0.8(62.4)2(4\times5) = 1997$ lb at $y_p = 2.67$

$F_{water-oil} = 0.2(62.4)1(2\times5) = 125$ lb at $y_p + 2 = 3.33$ ft ; $\Sigma F = 2122$ ft ◀

$2122 h_p = 1997(2.67) + 125(3.33)$; $h_p = 2.71$ ft ◀

3.10 *The Utah-shaped plate shown in Fig. P3.10 is submerged in oil (s = 0.94) and lies in a vertical plane. Find the magnitude and location of the hydrostatic force acting on one side of the plate.*

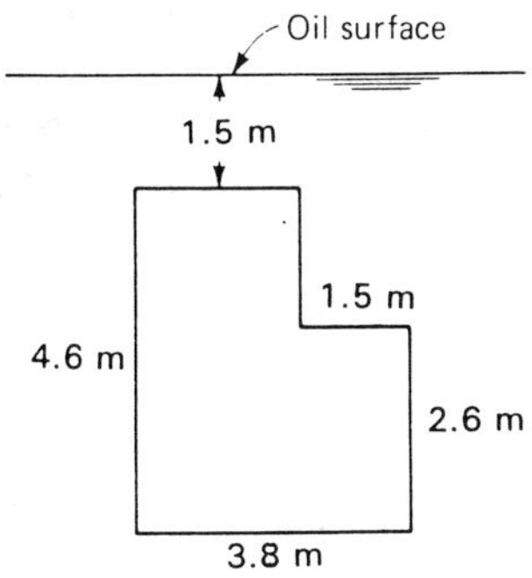

Figure P3.10

SI

Eq. 3.16: $F = \gamma h_c A$

$F_1 = (9.81 \times 0.94)(1.5 + 2.3)(2.3 \times 4.6) = 371 \text{ kN}$

Eq. 3.18: $h_p = y_c + \dfrac{I_c}{y_c A}$

$(h_p)_1 = 3.8 + \dfrac{(1/12)2.3(4.6)^3}{3.8(2.3 \times 4.6)} = 4.26 \text{ m}$

$F_2 = (9.81 \times 0.94)(1.5 + 2 + 1.3)(1.5 \times 2.6) = 173 \text{ kN}$

$(h_p)_2 = 4.8 + \dfrac{(1/12)1.5 \times 2.6^3}{4.8(1.5 \times 2.6)} = 4.92 \text{ m}$

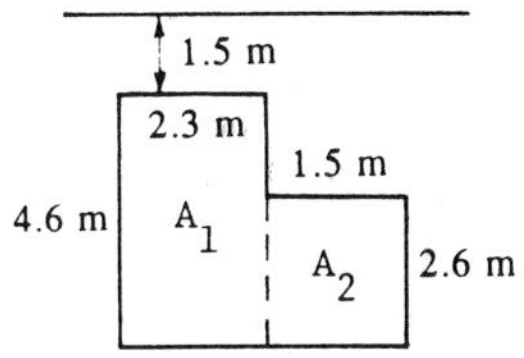

$F = F_1 + F_2 = 543 \text{ kN}$ ◄

$F(h_p) = F_1(h_p)_1 + F_2(h_p)_2$;

$h_p = (371 \times 4.26 + 173 \times 4.92)/543 = 4.47$ m below oil surface. ◄

3.11 *The common type of irrigation head gate shown in Fig. P3.11 is a plate that slides over the opening to a culvert. The coefficient of friction between the gate and its sliding ways is 0.6. Find the force required to slide open this 600-lb gate if it is set (a) vertically; (b) on a 2:1 slope (n = 2), as is common.*

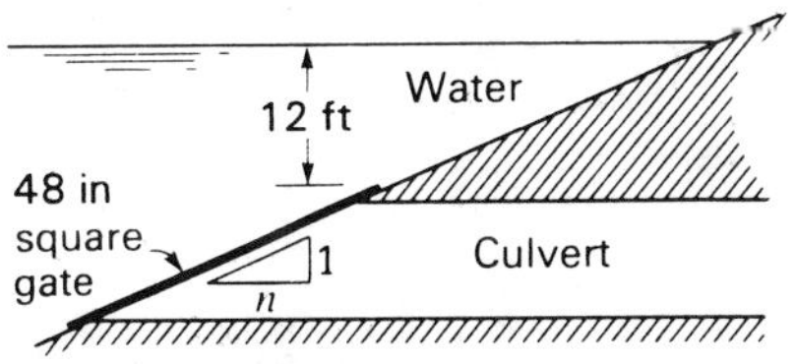

Figure P3.11

BG

(*a*) Gate vertical: Eq. 3.16: $h_c = 12 + 4/2 = 14$;

$F_x = \gamma h_c A = 62.4(14)4^2 = 13{,}980 \text{ lb}$

Let T = force parallel to gate required to open gate:

$\Sigma F_y = 0$; $T - 600 - 0.6(13{,}980) = 0$; $T = 8990 \text{ lb}$ ◄

(*b*) On 2:1 slope: Eq. 3.16: $h_c = [12 + (4/2)\sin\theta]$

$F = \gamma h_c A = 62.4[12 + (4/2)\sin\theta]4^2 = 12{,}870 \text{ lb}$

$\Sigma F_x = 0$; $N = F + W\cos\theta$

$\Sigma F_y = 0$; $T = \mu N + W\sin\theta = \mu(F + W\cos\theta) + W\sin\theta$

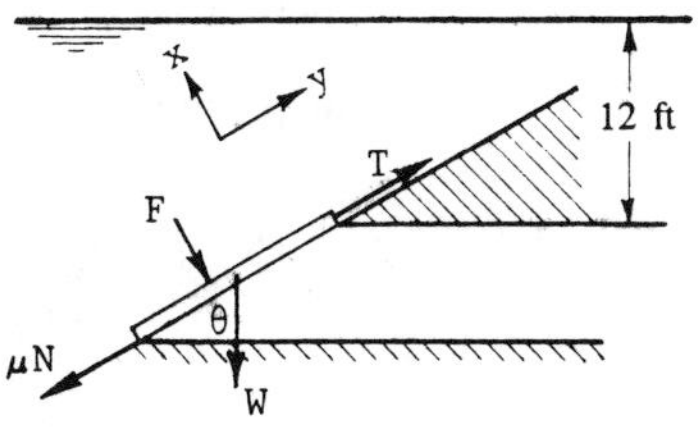

$T = 0.6\left(12{,}870 + 600\dfrac{2}{\sqrt{5}}\right) + \dfrac{600}{\sqrt{5}} = 8310 \text{ lb}$ ◄

3.12 *In the drainage of irrigated lands it is frequently desirable to install automatic flap gates to prevent a flood from backing up into the lateral drains from a river. Suppose a 50-in-square flap gate, weighing 1750 lb, is hinged 35 in above the center, as shown in Fig. P3.12, and the face is sloped 4° from the vertical. Find the depth to which water will rise behind the gate before it will open.*

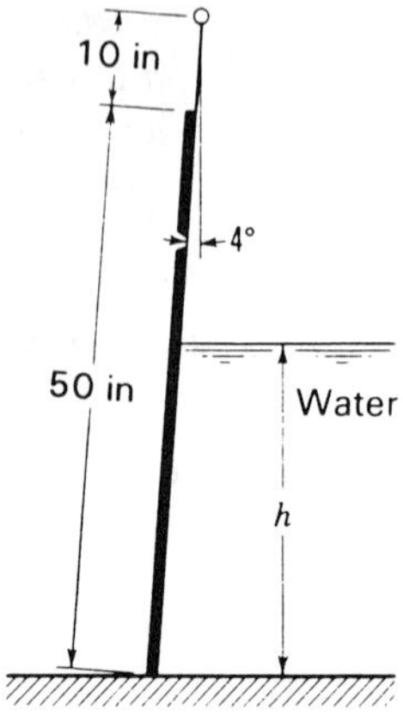

Figure P3.12

BG

Closing moment of gate about hinge = 1750(35/12)sin4° = 356 ft·lb

Find depth h, producing opening moment of 356 ft·lb

Eq. 3.16: $F = \gamma h_c A = 62.4\left(\dfrac{h}{2}\right)\dfrac{50h}{12\cos 4°} = 130.3h^2$

Moment arm $= \dfrac{50+10}{12} - \dfrac{h}{3\cos 4°} = 5.00 - 0.334h$

Then $130.3h^2(5.00 - 0.334h) = 356$; $h^3 - 14.96h^2 + 8.18 = 0$

This cubic equation may be solved by trial and error, etc, yielding $h = 0.759$ ft. ◄

3.13 *Find the minimum value of z for which the gate in Fig. P3.13 will rotate counterclockwise if the gate is (a) rectangular, 5 ft by 4 ft; (b) triangular, 4 ft base as axis, height 5 ft. Neglect friction in bearings.*

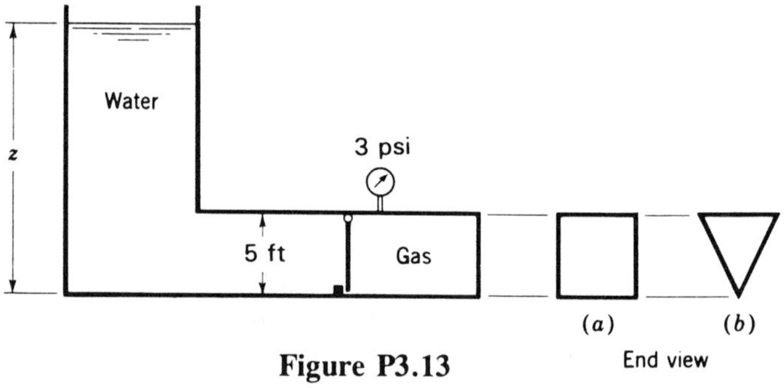

Figure P3.13

BG

(*a*) Closing moment due to gas:

(5 × 4)(3 × 144)2.5 = 21,600 ft·lb

Opening moment due to water:

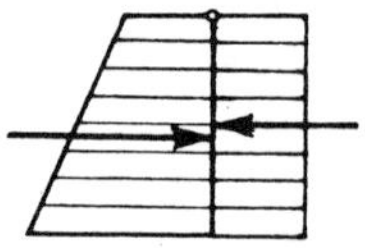

$$F_{\text{water}}[h_p - (z-5)],\ \text{where}\ h_p = (z-2.5) + \frac{I_c}{h_c A},\ I_c = \frac{bh^3}{12}$$

$$= 62.4(z-2.5)(5\times 4)\left[(z-2.5) + \frac{(1/12)4(5)^3}{(z-2.5)20} - (z-5)\right]$$

$$= 1248(z-2.5)\left[2.5 + \frac{25}{12(z-2.5)}\right] = 104(30z-50) = 3120z - 5200$$

Equate moments and solve for z: $3120z - 5200 = 21{,}600$; $z = 8.59$ ft ◄

(*b*) Moment due to water:

$$F_{\text{water}}[h_p - (z-5)],\ \text{where}\ h_p = \left(z - \frac{5\times 2}{3}\right) + \frac{I_c}{h_c A},\quad I_c = \frac{bh^3}{36}$$

$$= 62.4\left(z - \frac{10}{3}\right)\left(\frac{5\times 4}{2}\right)\left[\left(z - \frac{10}{3}\right) + \frac{(1/36)4(5)^3}{\left(z - \frac{10}{3}\right)10} - (z-5)\right]$$

$$= 624\left(z - \frac{10}{3}\right)\left[\frac{5}{3} + \frac{25}{18(z - 10/3)}\right] = 624(5/3)(z-2.5) = 1040z - 2600$$

Moment due to gas:

(3 × 144)(2 × 5)(5/3) = 7200 ft·lb ; equating moments, $z = 9.42$ ft ◄

3.14 *Referring to Fig. P3.14, what minimum value of b is necessary to keep the rectangular masonry wall from sliding if it weighs 160 lb/ft^3 and the coefficient of friction is 0.45? With this minimum b value, will it also be safe against overturning? Assume that water does not get underneath the block.*

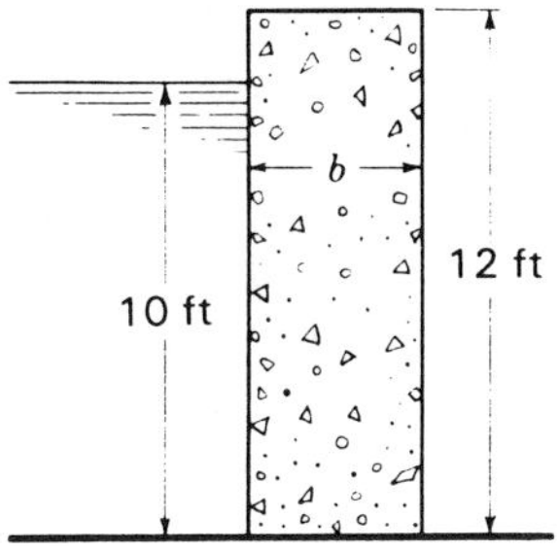

Figure P3.14

BG

$F_x = \gamma h_c A = 62.4(5)10 = 3120 \text{ lb/ft}$

$W = (160)12b = 1920b$; $R_V = W$

$\Sigma F_x = 3120 - 0.45(1920b) = 0$; $b = 3.61$ ft ◄

Consider moments about toe of F_x, W, R_V (distance x_t from toe):

For equilibrium, $\Sigma M_{\text{toe}} = 0 = 3120(10/3) - (1920)3.61(3.61/2) + (1920)3.61x_t$

$$x_t = \frac{12{,}520 - 10{,}400}{(1920)3.61} = 3.06 \text{ ft}$$

As $x_t > 0$, the dam <u>is</u> safe against overturning, assuming water does not get under the block. ◄

Sec 3.9: Force on a Curved Surface -- *Exercises*

3.9.1 *A vertical-thrust bearing for a large hydraulic gate is composed of a 10-in-radius bronze hemisphere mating into a steel hemispherical shell in the gate bottom. At what pressure must lubricant be supplied to the bearing so that a complete oil film is present if the vertical thrust on the bearing is 750,000 lb?*

BG

Projected area = $\pi r^2 = 100\pi \text{ in}^2$; $p = F/A = 750{,}000/(100\pi) = 2390$ psi ◄

3.9.2 *A tank with vertical ends contains water and is 6 m long normal to the plane of Fig. X3.9.2. The sketch shows a portion of its cross-section where MN is one-quarter of an ellipse with semiaxes b and d. If b = 2.5 m, d = 4 m, and a = 1.0 m, find, for the surface represented by MN, the magnitude and position of the line of action of (a) the horizontal component of force; (b) the vertical component of the force; (c) the resultant force and its direction with the horizontal.*

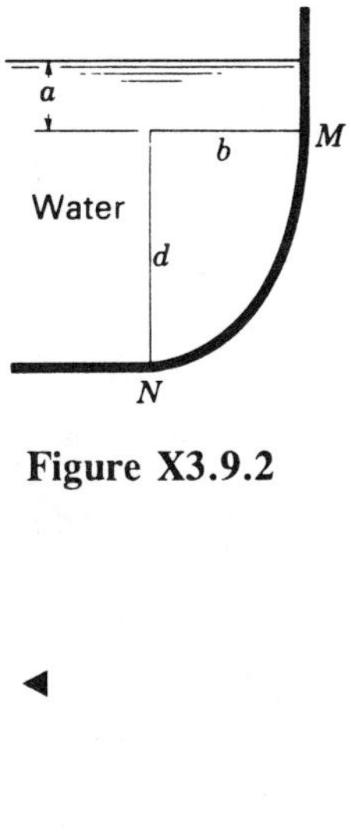

Figure X3.9.2

SI

(*a*) Eqs. 3.20, 3.16: $F_x = \gamma h_c A = (9.81)3.0(4 \times 6) = 706$ kN ◄ Acts at:

$$h_p = y_c + \frac{I_c}{y_c A} = y_c + \frac{h^2}{12y_c} = 3.0 + \frac{4^2}{12(3.0)} = 3.44 \text{ m below surface}$$ ◄

(*b*) Using Table A.7: Eq. 3.21: $F_y = W = 6\left(\frac{2.5\,\pi 4}{4} + 1.0 \times 2.5\right)9.81 = 609$ kN ◄

Acts at: $x_c = \dfrac{4b}{3\pi} = \dfrac{4(2.5\text{ m})}{3\pi} = 1.061$ m to right of N ◄

(*c*) $R = \sqrt{F_x^2 + F_y^2} = \sqrt{706^2 + 609^2} = 933$ kN ◄

Acts through intersection of F_x and F_y. ◄ $\theta = \tan^{-1}(F_y/F_x) = \tan^{-1}(609/706) = 40.8°$ ◄

3.9.3 *Find the answers called for in Exer. 3.9.2 if a = 2 ft, b = 6 ft, d = 9 ft, the tank is 12 ft long, and MN represents a parabola with vertex at N.*

Exer. 3.9.2: Refer to Fig. X3.9.2 with Solution 3.9.2 above. The tank has vertical ends and contains water. Find, for the surface represented by MN, the magnitude and position of the line of action of (a) the horizontal component of force; (b) the vertical component of the force; (c) the resultant force and its direction with the horizontal.

BG

(*a*) Eqs. 3.20, 3.16: $F_x = \gamma h_c A = (62.4)6.5(9 \times 12) = 43{,}800$ lb ◄

$$\text{Acts at: } h_p = y_c + \frac{I_c}{y_c A} = y_c + \frac{h^2}{12 y_c} = 6.5 + \frac{9^2}{12(6.5)} = 7.54 \text{ ft below surface} \quad ◄$$

(*b*) Using Table A.7: Eq. 3.21: $F_y = W = 12[(2/3)9 \times 6 + 2 \times 6]62.4 = 35{,}900$ lb ◄

Acts at: $x_c = (3/8)b = (3/8)6$ ft $= 2.25$ ft to right of N ◄

(*c*) $R = \sqrt{F_x^2 + F_y^2} = \sqrt{43{,}800^2 + 35{,}900^2} = 56{,}700$ lb ◄

Acts through intersection of F_x and F_y. ◄ $\theta = \tan^{-1}(F_y/F_x) = \tan^{-1}(35{,}900/43{,}800) = 39.4°$ ◄

3.9.4 *The cross section of a tank is as shown in Fig. X3.9.4. BC is a cylindrical surface. If the tank contains water to a depth of 10 ft, determine the magnitude and location of the horizontal- and vertical-force components on the wall ABC.*

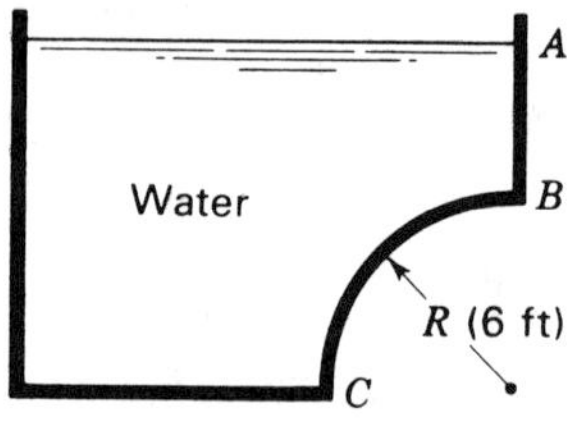

Figure X3.9.4

BG

Eqs. 3.20, 3.16: $F_x = \gamma h_c A = 62.4(1/2 \times 10)10 = 3120$ lb/ft ◄

F_x acts at: $y_p = (2/3)10 = 6.67$ ft below surface ◄

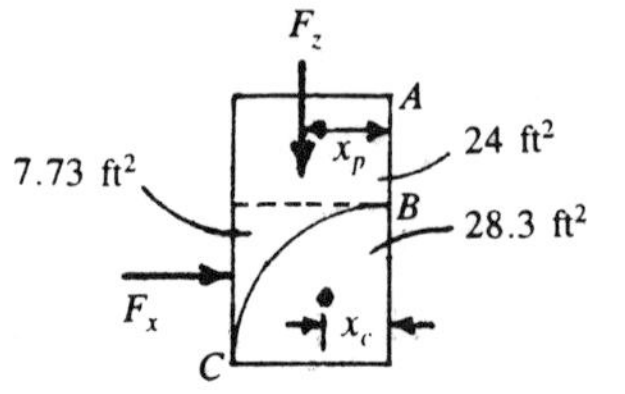

$(\pi/4)6^2 = 28.3$ ft² ; $10(6) - 28.3 = 31.7$ ft²

$x_c = 4r/(3\pi) = 2.55$ ft (Table A.7)

Eq. 3.21: $F_z = W = 62.4(31.7) = 1980$ lb/ft ◄

$x_p(31.7) + 2.55(28.3) = (10 \times 6)3$

F_z acts at: $x_p = 3.40$ ft to left of AB ◄

3.9.5 *Find the answers called for in Exer. 3.9.4 if the radius of BC is 2 m and the water depth is 3.5 m.*

Exer. 3.9.4: In Fig. X3.9.4 (above) BC is a cylindrical surface. Find the magnitude and location of the horizontal and vertical force components acting on tank wall ABC.

SI

Eqs. 3.20, 3.16: $F_x = \gamma h_c A = 9.81(1/2 \times 3.5)3.5 = 60.1$ kN/m ◄

F_x acts at: $y_p = (2/3)3.5 = 2.33$ m below surface

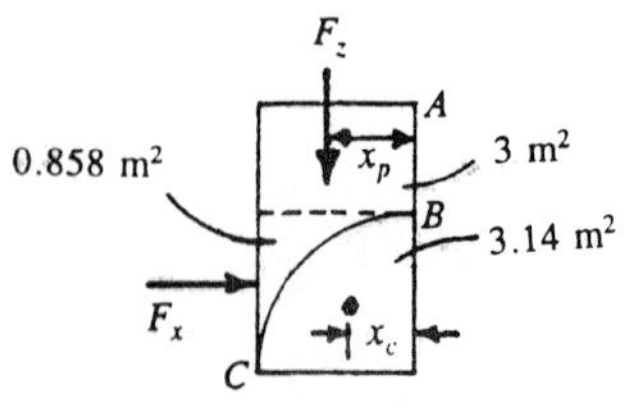

$(\pi/4)2^2 = \pi$ m² ; $3.5(2) - \pi = 3.86$ m²

Table A.7: $x_c = 4r/(3\pi) = 0.849$ m

Eq. 3.21: $F_z = W = 9.81(3.86) = 37.9$ kN/m ◄

$x_p(3.86) + 0.849(\pi) = (3.5 \times 2)1$

F_z acts at: $x_p = 1.123$ m to the left of AB ◄

3.9.6 *Repeat Exer. 3.9.4 where the tank is closed and contains gas at a pressure of 10 psi.*

Exer. 3.9.4: In Fig. X3.9.4 (above) BC is a cylindrical surface. Find the magnitude and location of the horizontal and vertical force components acting on tank wall ABC.

BG

Eq. 3.14: $F_x = pA = (10 \times 144)10 = 14{,}400$ lb/ft ◄ located 5 ft below top. ◄

$F_z = (10 \times 144)6 = 8640$ lb/ft ◄ located 3.0 ft to left of *AB* ◄

3.9.7 *Repeat Exer. 3.9.5 where the tank is closed and contains gas at a pressure of 75 kPa.*

Exer. 3.9.4: In Fig. X3.9.4 (above) BC is a cylindrical surface. Find the magnitude and location of the horizontal and vertical force components acting on tank wall ABC.

SI

$F_x = 75(3.5) = 263$ kN/m^2 ◄ at 1.75 m below top ◄

$F_z = 75(2) = 150$ kN/m^2 ◄ at 1.0 m to left of *AB* ◄

3.9.8 *A spherical steel tank of 18 m diameter contains gas under a pressure of 380 kN/m^2. The tank consists of two half-spheres joined together with a weld. What will be the tensile force across the weld in kN/m? If the steel is 22.0 mm thick, what is the tensile stress in the steel? Express in kPa and in psi. Neglect the effects of cross-bracing and stiffeners.*

B

Eq. 3.14: Force $= pA = (380\text{ kN/m}^2)(\pi 18^2/4)\text{ m}^2 = 96\,700$ kN

Force/length $= 110\,000/(\pi D) = 96\,700/(18\pi) = 1710$ kN/m ◄

$$\sigma = \frac{\text{Force/length}}{\text{thickness}} = \frac{1710\text{ kN/m}}{0.022\text{ m}} = 77\,700\text{ kPa}$$ ◄

From inside back cover: (77 700 kPa)(0.000 1450 psi/Pa)(1000 Pa/kPa) = 11,270 psi ◄

3.9.9 *The hemispherical body shown in Fig. X3.9.9 projects into a tank. Find the horizontal and vertical forces acting on the hemispherical projection for the following cases: (a) the tank is full of water with the free surface 5 ft above A; (b) the tank contains CCl_4 (s = 1.59) to the level of A overlain with water having its free surface 5 ft above A; (c) the tank is closed and contains only gas at a pressure of 6 psi; (d) the tank is closed and contains water to the level of A overlain with gas at a pressure of 2 psi. Assume the gas weighs 0.075 lb/ft^3.*

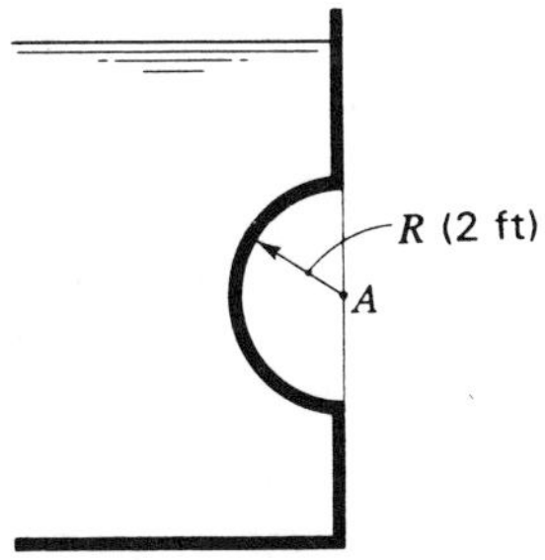

Figure X3.9.9

BG

Projected area on vertical plane $= \pi r^2 = \pi 2^2 = 12.57$ ft^2

Volume of hemisphere $= \left(\frac{1}{2}\right)\frac{4}{3}\pi r^3 = \left(\frac{1}{2}\right)\frac{4}{3}\pi 2^3 = 16.76$ ft^3

(*a*) Eq. 3.16: $F_x = \gamma h_c A = 62.4(5)(12.57) = 3920$ lb ◄ $F_z = 62.4(16.76) = 1046$ lb ◄

(*b*) $F_x = 3920 + (1.59 - 1.00)62.4\left(\frac{4 \times 2}{3\pi}\right)12.57 = 4120$ lb ◄

$F_z = \frac{1046}{2} + \frac{1046}{2}(1.59) = 1354$ lb ◄

(*c*) $F_x = 6(144)12.57 = 10{,}860$ lb ◄ $F_z = 0.075(1/2)(4/3)\pi 2^3 = 1.257$ lb ◄

(*d*) $F_x = 2(144)12.57 + 62.4\left(\frac{4 \times 2}{3\pi}\right)\frac{12.57}{2} = 3950$ lb ◄

$F_z = (1046/2) + (16.76/2)0.075 = 523$ lb ◄

3.9.10 *Determine the force required to hold the cone in the position shown in Fig. X3.9.10. Assume the cone is weightless.*

BG

Sketch shows vertical projection of 1/2 cone above opening.

p_{gas} = 0.6 psi − 0.8(62.4)6/144 = 0.6 psi − 2.08 psi = −1.48 psi

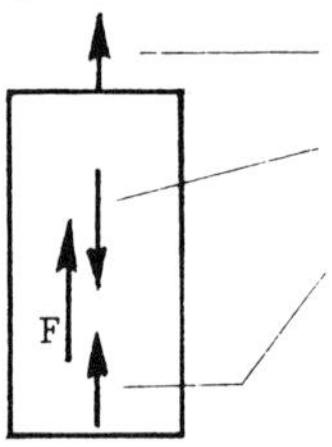

$1.48(144)\pi(0.938)^2 = 589$ lb

$62.4(0.8)\pi(0.938)^2 9.5 = 1310$ lb

$(\pi/3)0.938^2(3.5)62.4(0.8) = 160.9$ lb

$\Sigma F_z = 589 - 1310 + 160.9 + F = 0$

$F = 561$ lb ◀

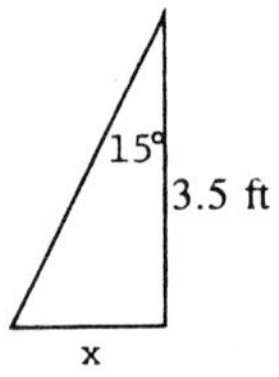

x = 3.5 tan 15°
= 0.938 ft

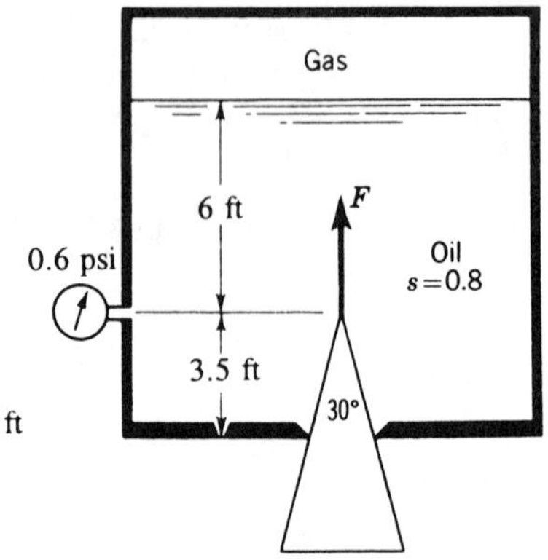

Figure X3.9.10

Sec. 3.9: Force on Curved Surface -- *Problems*

3.15 *(a) Find horizontal and vertical forces per foot of width on the Tainter gate shown in Fig. P3.15. (b) Locate the horizontal force and indicate the line of action of the vertical force without actually computing its location. (c) Locate the vertical force (hint: consider the resultant).*

BG

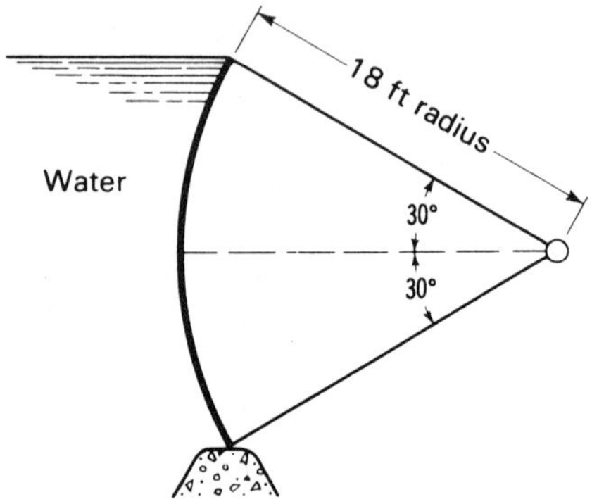

Figure P3.15

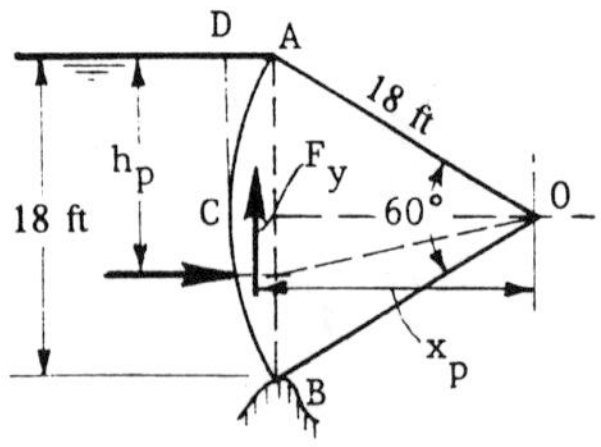

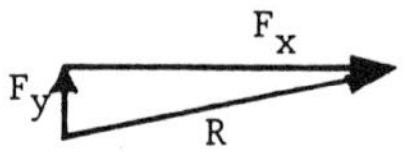

(*a*) Eq. 3.16: $F_x = \gamma h_c A = 62.4(18/2)18 = 10{,}110$ lb/ft ◀

Eq. 3.21: $F_y = W = \gamma$(Area *ABCD* − Area *ACD*) = 62.4(Area of segment *ABC*)

$$= 62.4\left[\pi(18)^2\frac{60}{360} - \frac{18\times 18\cos 30°}{2}\right] = 1831 \text{ lb/ft}$$ ◀

(*b*) $h_p = (2/3)18 = 12$ ft ◀ F_y acts through the centroid of segment *ABC* ◀

(*c*) Since all forces are normal to the circular arc, the resultant *R* must pass through *O*, and x_p may be determined from similarity of force and space triangles. Thus

$$\frac{x_p}{F_x} = \frac{(12 - 18/2)}{F_y}\ ;\quad x_p = 3\left(\frac{10{,}110}{1831}\right) = 16.56 \text{ ft}$$ ◀

3.16 *A tank has an irregular cross section as shown in Fig. P3.16. Determine as accurately as possible the magnitude and location of the horizontal- and vertical-force components on a one-foot length of the wall ABCD when the tank contains water to a depth of 8 ft. To determine areas, use a planimeter or count squares (1 ft grid); make a cardboard cutout, or take approximate moments of the squares, to locate the centroid.*

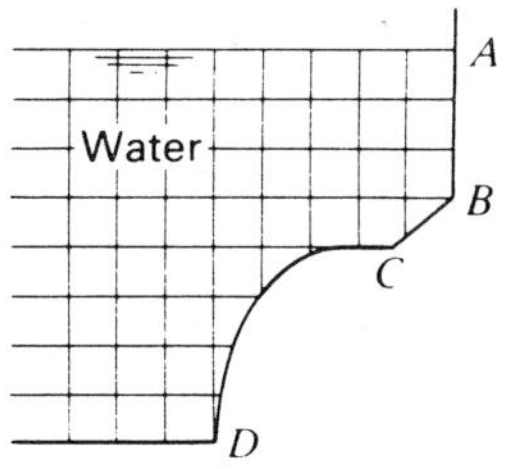

Figure P3.16

BG

Eqs. 3.20, 3.16: $F_x = \gamma h_c A = 62.4(1/2 \times 8)8 = 1997$ lb/ft ◀

F_x acts at: $y_p = (2/3)8 = 5.33$ ft below the surface ◀

Planimeter or count squares to determine water column area A_s above BCD.

$A_s = 21.6$ ft^2 (by calculation!). Eq. 3.21: $F_z = W = \gamma A_s = 62.4(21.6) = 1348$ lb/ft ◀

Make cardboard cutout or take moments of squares to locate centroid of A_s:
$x_p = 2.76$ ft (by calculation!) F_z acts about 2.76 ft to left of AB ◀

3.17 *Repeat Exer. 3.9.4 where the tank is closed and contains 4 ft of water overlain with a gas that is under a pressure of 0.8 psi.*

Exer. 3.9.4: In Fig. X3.9.4 (see Solution 3.9.4) BC is a cylindrical surface. Find the magnitude and location of the horizontal and vertical force components acting on tank wall ABC.

BG

Effect of gas pressure is transmitted through the water.

$F_x = F_{gas} + F_{water} = (0.8 \times 144)10 + 62.4(1/2 \times 4)4 = 1152 + 499 = 1651$ lb/ft ◀

$z_p = [1152(5) + 499(10 - 4/2)]/1651 = 5.91$ ft below top of tank ◀

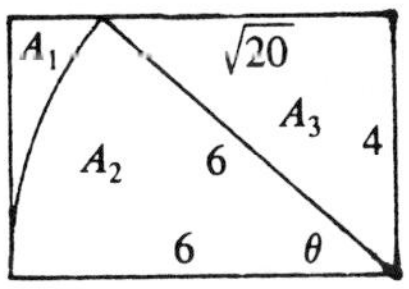

$\sin\theta = 4/6$; $\theta = 41.8°$

$A_1 + A_2 + A_3 = A$; $A_1 + \pi 6^2(41.8/360) + (1/2)4\sqrt{20} = 6(4)$

$A_1 = 1.921$ ft^2 = water area vertically above curved surface

$F_z = (0.8 \times 144)6 + 62.4(1.921) = 691 + 119.8 = 811$ lb/ft ◀

Take moments (or use cardboard cutouts) to locate centroid of F_z: $x_{c2} = 3.65$, $x_{c3} = 1.491$ ft from AB

$1.921x_{c1} + 13.14(3.65) + 8.94(1.491) = 24(3)$; $x_{c1} = 5.55$ ft from AB

$x_p = [(691)3 + (119.8)5.55]/811 = 3.38$ ft from AB ◀

3.18 *The cross-section of a gate is shown in Fig. P3.18. Its dimension normal to the plane of the paper is 8 m, and its shape is such that x = 0.2y². The gate is pivoted about O. Develop analytic expressions in terms of the water depth y upstream of the gate for the following: (a) horizontal force; (b) vertical force; (c) clockwise moment acting on the gate. Compute (a), (b), and (c) for the case where the water depth is 2.5 m.*

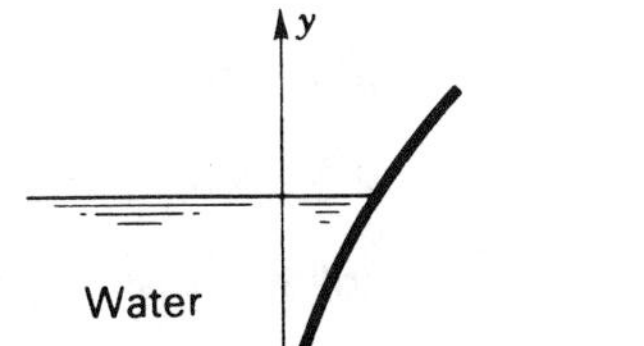

Figure P3.18

SI

(*a*) For 8-m width: $F_x = \gamma h_c A = \gamma(y/2)10y = 4\gamma y^2$ ◀

For $y = 2.5$ m, $F_x = 4(9.81)2.5^2 = 245$ kN ◀

(*b*) $F_z = W = \gamma(\text{side area})8 = 10\gamma\int_0^y x dy = 8\gamma\int_0^y 0.2y^2 dy = \gamma(0.2y^3/3)8 = 0.533\gamma y^3$ ◀

For y = 2.5 m, $F_z = 0.533(9.81)2.5^3 = 81.8$ kN ◀

(*c*) Clockwise $M = 4\gamma y^2(1/3)y + 8\gamma\int x(x/2)dy = (4/3)\gamma y^3 + 4\gamma(0.04y^5/5) = (4/3)\gamma y^3 + 0.0320\gamma y^5$ ◀

For y = 2.5 m, $M = (4/3)(9.81)2.5^3 + 0.0320(9.81)2.5^5 = 235$ kN·m clockwise ◀

Sec 3.10: Buoyancy and Stability of Submerged and Floating Bodies -- *Exercises*

3.10.1 *An iceberg in the ocean floats with one-eighth of its volume above the surface. What is its specific gravity relative to ocean water, which weight 64 lb/ft³. What portion of its volume would be above the surface if the ice were floating in pure water?*

BG

Let $\forall$ = total volume of iceberg ; $\forall_1 = (7/8)\forall$ = volume submerged.

$W = B$; $\gamma'\forall = (7/8)\gamma\forall$; $\therefore s' = (\gamma'/\gamma) = 7/8 = 0.875$ relative to ocean water ◄

and $\gamma' = (7/8)64 = 56.0$ lb/ft³, $s = 56.0/62.4 = 0.897$, and so

10.26% of its volume would be above the water surface in fresh water. ◄

3.10.2 *A hydrometer (Fig. 3.22a) consists of a 6-mm-diameter cylinder of length 180 mm attached to a 20-mm-diameter weighted sphere. The cylinder has a mass of 0.6 g and the mass of the sphere is 6.4 g. At what level will this device float in liquids having specific gravities 0.8, 1.0, and 1.2? Is the scale spacing on the cylindrical stem uniform? Why or why not?*

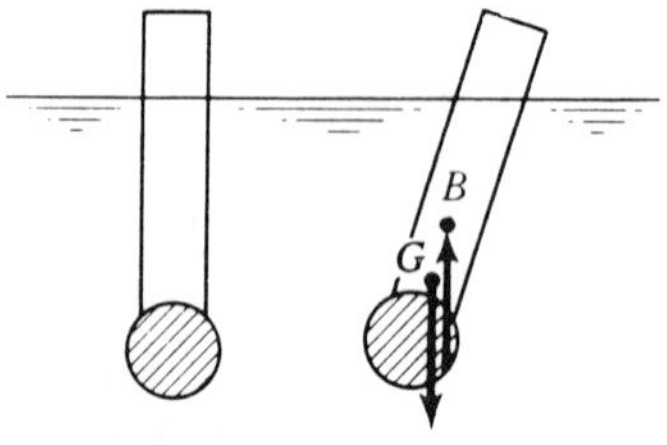

Figure 3.22(a)

SI

Vol of sphere = $(\pi/6)(20)^3 = 4190$ mm³

Volume of submerged cylinder = $\pi 3^2 y = 28.3y$ mm³, where y is the submerged length of the cylinder in mm.

Mass of hydrometer = 7 grams = $s(4190 + 28.3y)/1000$ grams ; $y = [(7000/s) - 4190]/28.3$ mm

If $s = 0.8$, $y = 161.3$ mm ◄

diff = 61.9 mm

If $s = 1.0$, $y = 99.4$ mm ◄

diff = 41.2 mm

If $s = 1.2$, $y = 58.2$ mm ◄

Scale spacing is not uniform because submergence is not directly proportional to specific gravity, as from the above equation: $y = 7000/(28.3s) - 4180/28.3$ and $dy/ds = -7000/(28.3s^2)$ ◄

3.10.3 *Determine the volume of an object that weighs 5.5 lb in water and 8 lb in oil (s = 0.82). What is the specific weight of the object?*

BG

$W - \forall(62.4) = 5.5$ (1)

$W - \forall(0.82 \times 62.4) = 8$ (2)

Solve (1) and (2) simultaneously to find $W = 19.39$ lb and $\forall = 0.223$ ft³ ◄

Specific weight = 19.39/0.223 = 87.1 lb/ft³ ◄

3.10.4 *A balloon weighs 160 lb and has a volume of 7200 ft³. It is filled with helium, which weighs 0.0112 lb/ft³ at the temperature and pressure of the air, which in turn weighs 0.0807 lb/ft³. What load will the balloon support, or what force in a cable would be required to keep it from rising?*

BG

(Buoyant force) (Wt of helium) (Balloon)

$\Sigma F_z = 0$; $7200(0.0807) - 7200(0.0112) - 160 - T = 0$

Required force in the cable, $T = 340$ lb ◄

3.10.5 *For the conditions shown in Figure X3.10.5, find the force F required to lift the concrete-block gate if the concrete weighs 23.6 kN/m³. Neglect friction.*

SI

$\Sigma F_z = 0$; $F + (\text{salt w.}) - (\text{fresh w.}) - (\text{conc. block}) = 0$

$F + 9.81(1.025)1.5\pi(0.3)^2 - 9.81(3)\pi 0.3^2 - 23.6(0.3)\pi 0.3^2 = 0$

$F + 4.26 - 8.32 - 2.00 = 0$; $F = 6.06$ kN ◄

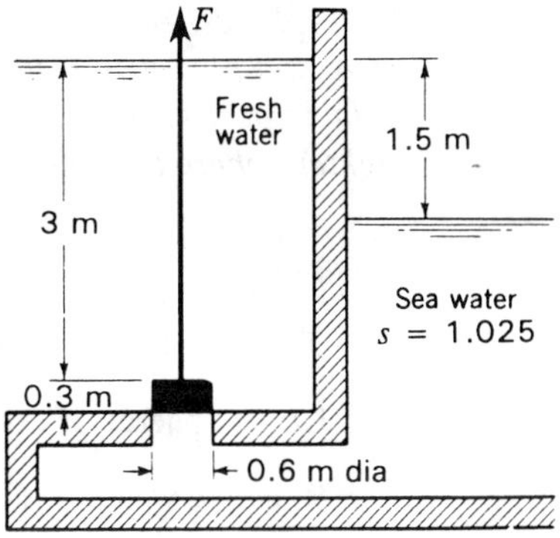

Figure X3.10.5

3.10.6 *An 8-in diameter solid cylinder of height 3 in weighing 3.4 lb is immersed in liquid (γ = 52 lb/ft³) contained in a tall, upright metal cylinder having a diameter of 9 in (Fig. X3.10.6). Before immersion, the liquid was 3 in deep. At what level will the solid cylinder float? Find the distance z between the bottoms of the two cylinders.*

BG

Volume A = Volume B ;

so $\pi 4^2 x = \pi(4.5^2 - 4^2)y$; $16x = 4.25y$ or $x = 0.266y$

F_B = Weight = 3.4 = $52\pi(4/12)^2(x+y)/12$ so $x + y = 2.25$ in

Solving, $y = 1.776$ in, $x = 0.472$ in, $\therefore$ $z = 3 - x = 2.53$ in ◄

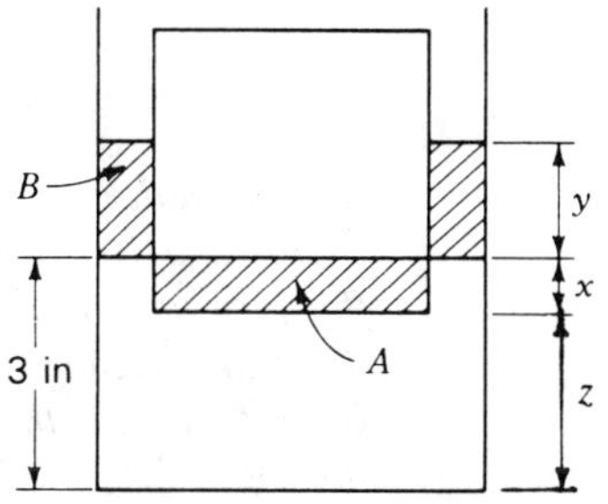

Figure X3.10.6

3.10.7 *A cylindrical bucket 250 mm in diameter and 400 mm high weighing 20.0 N contains oil (s = 0.80) to a depth of 180 mm. (a) When placed to float in water, what will be the immersion depth to the the bottom of the bucket? (b) What is the maximum volume of oil that the bucket can hold and still float?*

SI

(*a*) $\Sigma F_z = W_B + W_{oil} - \gamma_{water} h \pi r^2 = 0$; 0.020 kN $+ (0.8)9.81\pi(0.125)^2 0.18 - 9.81h\pi(0.125)^2 = 0$

$0.020 + 0.0693 - 0.482h = 0$; $h = 0.893/0.482 = 0.1855$ m $= 185.5$ mm ◄

(*b*) When immersed 400 mm, the oil depth h is given by

$0.020 + (0.8)9.81\pi(0.125)^2 h - 9.81(0.40)\pi(0.125)^2 = 0$

$0.020 + 0.385h - 0.1926 = 0$; $h = 0.1726/0.385 = 0.448$ m $= 448$ mm

448 mm > 400 mm, therefore the bucket will float when full of oil

Max vol $= \pi(0.125)^2 0.40 = 0.01963\,\text{m}^3 = 19.63$ L ◄

3.10.8 *A metal block 1.5 ft square and 1 ft deep is floated on a body of liquid which consists of a 10-in layer of water above a layer of mercury. The block weighs 120 lb/ft³. What is the position of the bottom of the block? If a downward vertical force of 600 lb is applied to the center of this block, what is the new position of the bottom of the block? Assume that the tank containing the fluid is of infinite dimensions.*

BG

(*a*) Block weighs $(1 \times 1.5^2)120 = 270$ lb

$270 = [(x/12)(13.6 \times 62.4) + (10/12)62.4]1.5^2$; $x = 0.962$ inches into the mercury ◄

(*b*) In this case the top of block is below the water surface.

Thus, $600 + 270 = \left[\frac{x}{12}(13.6 \times 62.4) + \frac{12 - x}{12}62.4\right]1.5^2$; $x = 4.95$ inches into the mercury ◄

3.10.9 *Two spheres, each 1.5 m in diameter, weigh 8 and 24 kN respectively. They are connected with a short rope and placed in water. (a) What is the tension in the rope and what portion of the lighter sphere's volume portrudes from the water? (b) What should be the weight of the heavier sphere in order for the lighter sphere to float halfway out of the water? Assume that the sphere volumes remain constant.*

SI

(*a*) Lower sphere: $F_B = \gamma(\text{vol. displaced}) = 9.81(\pi/6)1.5^3 = 17.34$ kN

$T = 24 - 17.34 = 6.66$ kN ◄

Upper sphere: $F_B = 8\text{ kN} + T = 14.66$ kN

% above water = (17.34 − 14.66)/17.34 = 15.41% of volume ◄

(*b*) Upper sphere: $F_B = (1/2)\gamma(\text{vol. sphere}) = (1/2)9.81(\pi/6)1.5^3 = 8.67$ kN

$W = B$; $W_L + 8 = 17.34 + 8.67$; $W_L = 18.00$ kN ◄

3.10.10 *A solid, half-cylinder-shaped log, 1.50 ft radius and 10 ft long, floats in water with the flat face up (Figure X3.10.10). (a) If the draft (immersion depth of the lowest point) is 0.90 ft, what is the uniform specific weight of the log? (b) The log tilts about its axis (zero net applied force) by less than 23°. Is it in stable equilibrium? Justify your answer with a sketch and logic. (c) If the log tilts by 20° (right side down; zero net applied force), what is the magnitude and sense of any moment that results?*

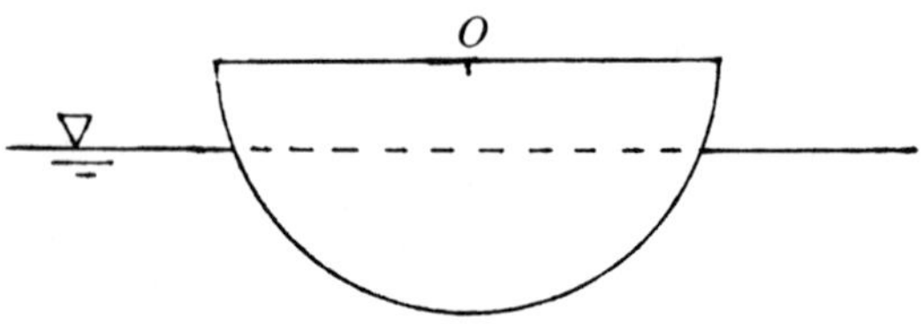

Figure X3.10.10

BG

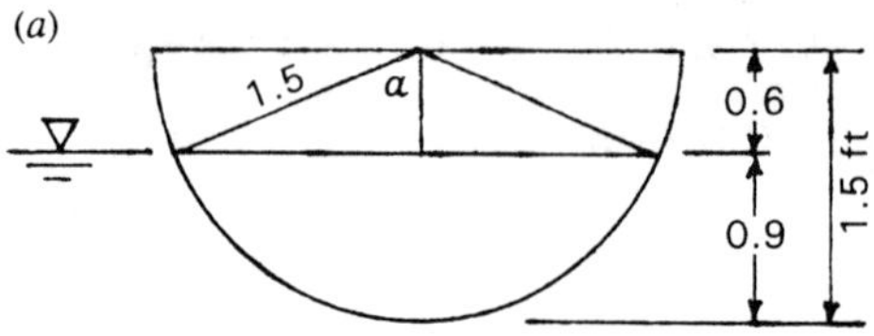

(*a*) $\gamma_s \pi r^2 L/2 = 62.4\left[\frac{2\alpha}{360}\pi r^2 - 2\left(\frac{1}{2}\right)0.6\sqrt{1.5^2 - 0.6^2}\right]L$

$\cos\alpha = 0.6/1.5 = 0.4$; $\alpha = 66.4°$

$\therefore \quad \gamma = 62.4(2.61 - 0.825)/3.53 = 31.5 \text{ lb/ft}^3$ ◄

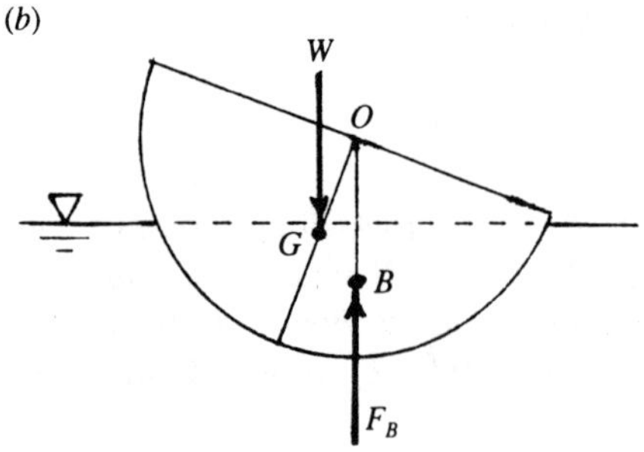

(*b*) When the body tilts per the sketch, *B* remains below axis *O*, but *G* moves to the left. So *W* and F_B create a righting moment. Therefore, yes the log is in stable equilibrium. ◄

(*c*) From sketch (*b*) above and Table A.7:

$OG = 4r/(3\pi) = 4(1.5)/(3\pi) = 0.637$ ft

Moment $= Wa = W(OG \sin 20°)$

$= 31.5(10\pi 1.5^2/2)(0.637 \sin 20°)$

$= 242$ ft·lb counterclockwise ◄

3.10.11 *A solid, half-cylinder-shaped log, 0.48 m in radius and 2.5 m long, floats in water with the flat face up (see Fig. X3.10.10). (a) If the draft (immersion depth of the lowest point) is 0.30 m, what is the uniform specific weight of the log? (b) The log tilts about its axis (zero net applied force) by less than 22°. Is it in stable equilibrium? Justify your answer with a sketch and logic. (c) If the log tilts by 18° (left side down; zero net applied force), what is the magnitude and sense of any moment that results?*

SI

(*a*)

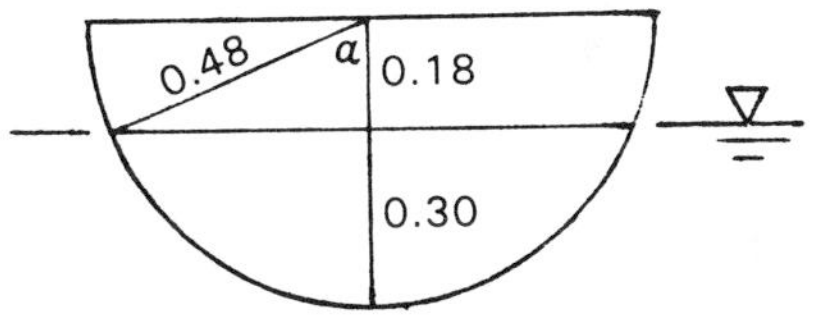

(*a*) $\gamma \pi r^2 L/2 =$

$$(9810\ \text{N/m}^3)\left[\frac{2\alpha}{360}\pi r^2 - 2\left(\frac{1}{2}\right)0.18\sqrt{0.48^2 - 0.18^2}\right]L$$

$\cos\alpha = 0.18/0.48 = 0.375, \quad \alpha = 68.0°$

$\therefore \quad \gamma = 9810(0.273 - 0.0801)/0.3619 = 5240\ \text{N/m}^3$ ◄

(*b*)

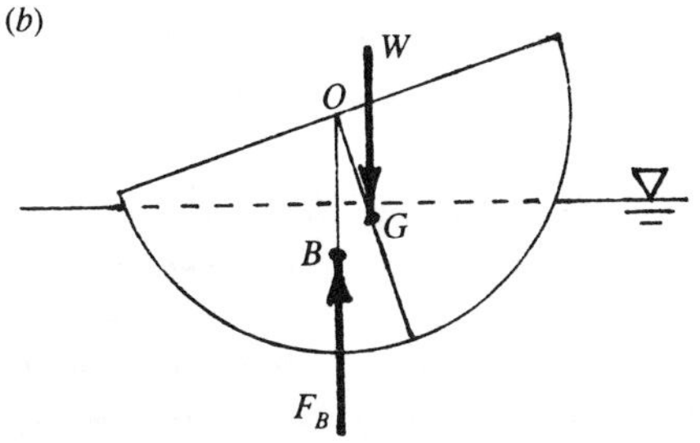

(*b*) As the body tilts as per sketch, B remains below axis O, but G moves to the right. So W and F_B create a righting moment. ∴ Yes, the log is in stable equilibrium. ◄

(*c*) From sketch (*b*) above and Table A.7:

$OG = 4r/(3\pi) = 0.204$ m

Moment $= Wa = W(OG \sin 18°)$

$= 5240(2.5\pi 0.48^2/2)(0.204 \sin 18°)$

$= 298$ N·m clockwise ◄

Sec. 3.10: Buoyancy and Stability of Submerged and Floating Bodies -- *Problems*

3.19 *Find the approximate value of the maximum specific gravity of liquid for which the device of Exer. 3.10.2 will be stable.*

Exer. 3.10.2: A hydrometer, Fig. 3.22a (see Solution 3.10.2), consists of a 6-mm-diameter cylinder of length 180 mm attached to a 20-mm-diameter weighted sphere. The cylinder has a mass of 0.6 g and the mass of the sphere is 6.4 g.

SI

Device is stable if center of buoyancy is above center of gravity.

Critical point of stability occurs when line of action of weight and buoyancy forces are coincident when device is tilted.

This occurs when center of gravity = center of buoyancy.

Use center of sphere as point of reference.

$$\bar{y}_m = \frac{6.4(0) + 0.6(10 + 90)}{6.4 + 0.6} = 8.57 \text{ mm above center of sphere}$$

Volume displaced by sphere = 4190 mm^3 , volume displaced by cylinder = 28.3y mm^3

$$\bar{y}_b = \bar{y}_g = 8.57 = \frac{4190(0) + 28.3y(10 + y/2)}{4190 + 28.3y}$$

From quadratic equation, $y = 49.0$ mm ; but mass of hydrometer = mass of fluid displaced, i.e.,

$7 = s(4190 + 28.3y)/1000$

Thus for $y = 49.0$ mm, the max measurable specific gravity, $s = 1.256$ ◄

3.20 *A 2.0-ft³ object weighing 650 lb is attached to a balloon of negligible weight and released in the ocean (γ = 64 lb/ft³). The balloon was originally inflated with 5.0 lb of air to a pressure of 20 psi. To what depth will the balloon sink? Assume that air temperature within the balloon stays constant at 50°F.*

BG

Find volume occupied by 5 lb of air. Table A.5 for air: $R = 1715\ \text{ft}^2/(\text{sec}^2\cdot{}^\circ R)$

Eq. 2.5: $\gamma = \dfrac{(14.7+20)144(32.2)}{1715(460+50)} = 0.1840$ pcf. Thus 5 lb of air occupies $5/0.1840 = 27.2\ \text{ft}^3 = \mathcal{V}_1$

At equilibrium depth: $\Sigma F_z = B - W = 0 = 64(2.0 + \mathcal{V}_2) - (650 + 5)$

so new volume of balloon, $\mathcal{V}_2 = 8.23\ \text{ft}^3$

From Eq. 2.4 with constant temperature: $pv = \text{const} = p\mathcal{V}/m$

and so for constant mass of air, $p\mathcal{V} = \text{const}$, i.e. $p_1\mathcal{V}_1 = p_2\mathcal{V}_2$

$(14.7 + 20)144(27.2) = (14.7 + p_2)144(8.23)$; $p_2 = 99.8$ psi

Depth $h = p_2/\gamma = (99.8)144/64 = 225$ ft ◄

3.21 *Work Prob. 3.20 with all data the same except assume the balloon was originally inflated with 5.0 lb of air to a pressure of 10 psi. In this latter case the balloon is more elastic because a lower pressure is obtained with the same amount of air.*

Prob. 3.20: A 2.0-ft³ object weighing 650 lb is attached to a balloon of negligible weight and released in the ocean (γ = 64 lb/ft³). To what depth will the balloon sink? Assume that air temperature within the balloon stays constant at 50°F.

BG

Find volume occupied by 5 lb of air. Table A.5 for air: $R = 1715\ \text{ft}^2/(\text{sec}^2\cdot{}^\circ R)$

Eq. 2.5: $\gamma = \dfrac{(14.7+10)144(32.2)}{1715(460+50)} = 0.1309$ pcf. This 5 lb of air occupies $5/0.1309 = 38.2\ \text{ft}^3 = \mathcal{V}_1$

At equilibrium depth: $\Sigma F_z = B - W = 0 = 64(2 + \mathcal{V}_2) - (650 + 5)$

so new volume of balloon, $\mathcal{V}_2 = 8.23\ \text{ft}^3$

From Eq. 2.4 with constant temperature: $pv = \text{const} = p\mathcal{V}/m$

and so for constant mass of air, $p\mathcal{V} = \text{const}$, i.e. $p_1\mathcal{V}_1 = p_2\mathcal{V}_2$

$(14.7 + 10)144(38.2) = (14.7 + p_2)144(8.23)$; $p_2 = 99.8$ psi

Depth $h = p_2/\gamma = (99.8)144/64 = 225$ ft ◄

3.22 *A wooden pole weighing 2 lb/ft has a cross-sectional area of *6.7 in² and is supported as shown in Fig. P3.22. The hinge is frictionless. Find θ. [*Note this correction to the textbook.]*

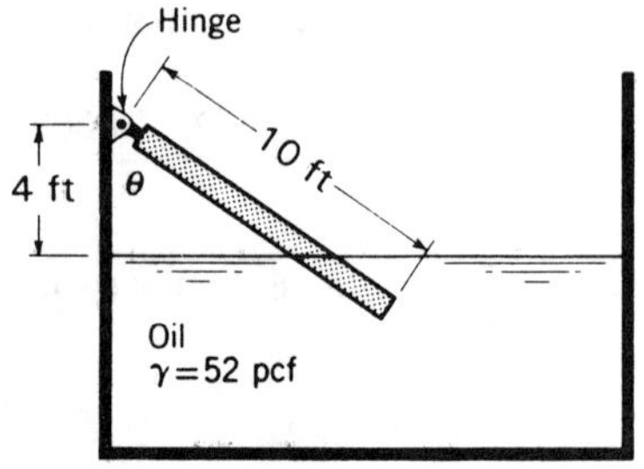

Figure P3.22

BG

$W = 2(10) = 20$ lb at $(1/2)(10)\sin\theta$ from hinge

$M = 100\sin\theta$ ft·lb clockwise

$F_B = \gamma_{\text{oil}}Ax = 52(6.7/144)x$ at $(10 - x/2)\sin\theta$

$M = 2.42x(10 - x/2)\sin\theta$ counterclockwise

Since $\Sigma M = 0$, we have $100 = 2.42x(10 - x/2) = 24.2x - 1.210x^2$

By quadratic equation, $x = 14.16$ ft (impossible) or 5.84 ft

$\cos\theta = \dfrac{4}{10-x} = \dfrac{4}{4.16} = 0.961$; $\theta = 16.12°$ ◄

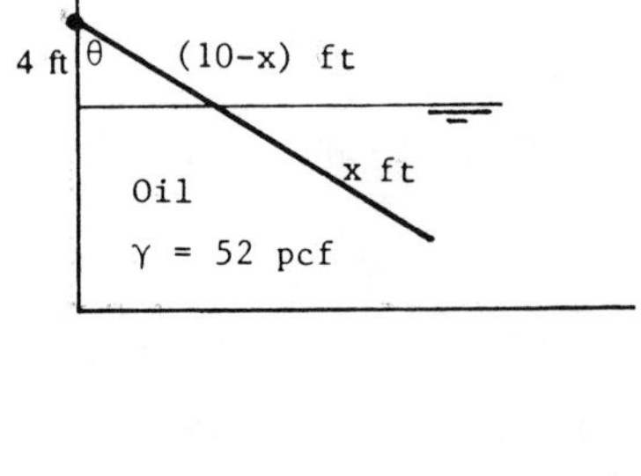

3.23 *A rectangular block of uniform material and length L = 3 ft, width b = 1 ft, and depth d = 0.20 ft, is floating in a liquid. It assumes the position shown in Fig. P3.23 when a uniform vertical load of 1.30 lb/ft is applied at P. (a) Find the weight of the block. (b) If the load is suddenly removed, what is the righting moment before the block starts to move? (Hint: Refer also to Fig. 3.19.)*

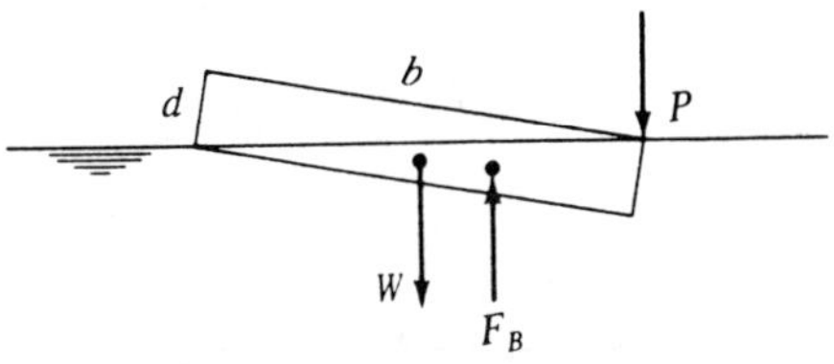

Figure P3.23

BG

$OP = \sqrt{1^2 + 0.20^2} = 1.020$ ft ; $BK = (2/3)d/2 = d/3$

$BK_x = BK\sin\alpha = (0.2/3)(0.2/1.020) = 0.01307$ ft

$GB_x = GK - BK_x = 1.020/6 - 0.01307 = a = 0.1569$ ft

$BP_x = KP + BK_x = 1.020/3 + 0.01307 = 0.353$ ft

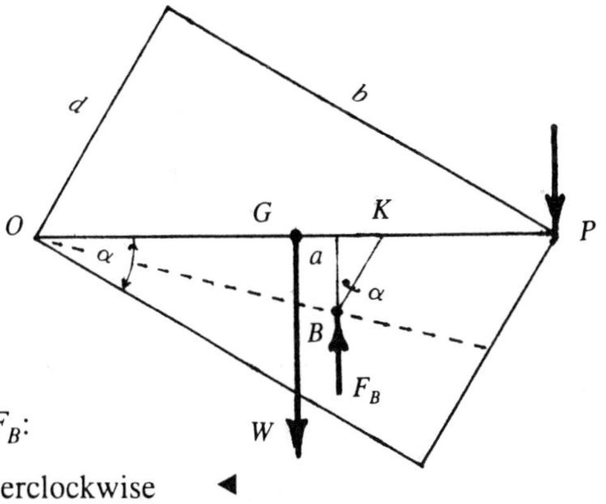

(*a*) Moments about *B*: $W(BG_x) = 1.3(3)BP_x$

$W(0.1569) = 3.9(0.353)$; $W = 8.77$ lb ◄

(*b*) ΣF_y: $F_B = W + 3.9$; $\therefore W < F_B$; Fig 3.19 for $W < F_B$:

Righting Moment $= Wa = 8.77(0.1569) = 1.377$ lb·ft counterclockwise ◄

3.24 *A rectangular block of uniform material and length L = 800 mm, width b = 250 mm, and depth d = 50 mm, is floating in a liquid. It assumes the position shown in Fig. P3.23 when a uniform vertical load of 20 N/m is applied at P. (a) Find the weight of the block. (b) If the load is suddenly removed, what is the righting moment before the block starts to move? (Hint: Refer also to Fig. 3.19.)*

Fig. 3.23: Refer to Solution 3.23.

SI

$OP = \sqrt{250^2 + 50^2} = 255$ mm ; $BK = (2/3)d/2 = d/3$; $BK_x = BK\sin\alpha = (50/3)(50/255) = 3.27$ mm

$GB_x = GK - BK_x = 255/6 - 3.27 = 39.2$ mm $= a$; $BP_x = KP + BK_x = 255/3 + 3.27 = 88.3$ mm

(*a*) Moments about *B*: $W(GB_x) = 20(0.8)BP_x$; $W(39.2) = 16(88.3)$; $W = 36.0$ N ◄

(*b*) ΣF_y: $F_B = W + 16N$, $\therefore W < F_B$; Fig 3.19 for $W < F_B$:

Righting Moment $= Wa = 36.0(0.0392 \text{ m}) = 1.412$ N·m counterclockwise ◄

3.25 *A solid block, 4 in wide by 4 in deep and 3 in high weighs 0.90 lb. It floats in liquid (γ = 55 lb/ft³) inside a cubic container of side 5 in. Before immersion the liquid was 2 in deep. (a) At what level will the block float? Find the distance z from the bottom of the block to the bottom of the container. (b) If the block is tilted by a couple (no net force) to an angle of 15° so that two sides remain vertical, what will be the righting moment in in·lb?*

BG

(*a*)

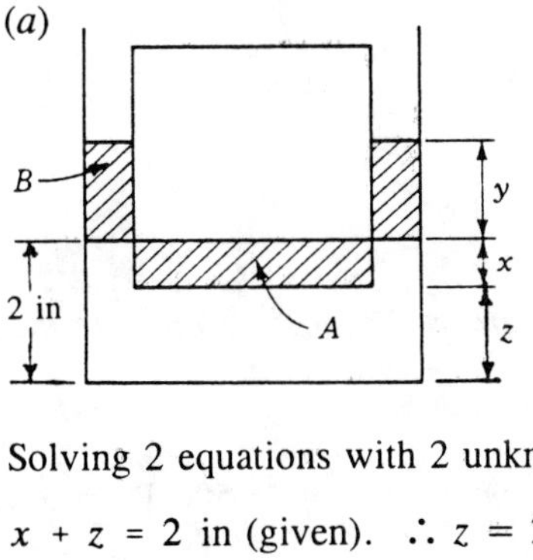

Volume *A* = Volume *B*

$x(4^2) = y(5^2 - 4^2)$

$x = 0.563y$

$F_B = W: \quad \dfrac{(x+y)4^2}{12^3}(55) = 0.9$

$x + y = 1.767$

Solving 2 equations with 2 unknowns: $y = 1.131$ in ; $x = 0.636$ in

$x + z = 2$ in (given). $\therefore z = 2 - x = 2 - 0.636 = 1.364$ in ◄

/cont...

(b)

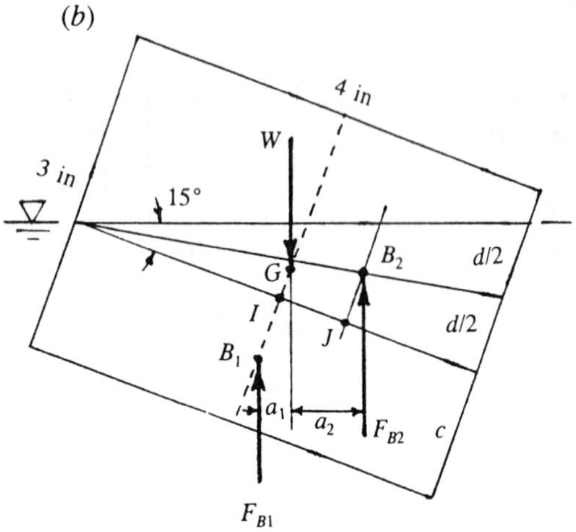

$d = 4 \tan 15° = 1.072$ in

Displaced volume $= (1/2)d(4^2) + c(4^2) = (x + y)4^2$

$c = (x + y) - (d/2) = 1.767 - (1.072/2) = 1.231$ in

Divide buoyancy into two components, F_{B1} and F_{B2} at B_1 and B_2 respectively. Since $W = \Sigma F_B = 0.90$ lb, either can be used to calculate the righting moment.

$F_{B1} = \gamma(Vol) = 55(4^2)c/12^3 = 0.627$ lb

$F_{B2} = 55(4^2)(d/2)/12^3 = 0.273$ lb

$IJ = 4/6 = 0.667$ in, $\quad 1.5 - c = 1.5 - 1.231 = 0.269$ in

$B_2J = (2/3)(d/2) = d/3 = 1.072/3 = 0.357$ in

$B_1G = 1.5 - c/2 = 0.884$ in ; $\quad a_1 = BG_1 \sin 15° = 0.229$ in

$a_2 = IJ \cos 15° + (B_2J - GI)\sin 15° = 0.667$ in

Moments about G: Righting moment $= F_2a_2 - F_1a_1$

$= 0.273(0.667) - 0.627(0.229) = 0.0385$ lb·in counterclockwise ◄

Sec 3.11: Fluid Masses Subjected to Acceleration -- ***Exercises***

3.11.1 *What would be the hydrostatic gage pressure at a depth of 9 in in a bucket of oil (s = 0.88) that is in an elevator being accelerated upward at 12 ft/s²?*

BG

Eq. 3.24: $\partial p/\partial z = -\rho(g + a_z)$; $\quad \Delta p = -\rho(g + a_z)\Delta z = -(\gamma/g)(g + a_z)\Delta z$

$\Delta p = -(62.4 \times 0.88/32.2)(32.2 + 12)(-9/12) = 56.5$ psf $= 0.392$ psi ◄

3.11.2 *What would be the hydrostatic gage pressure at a depth of 300 mm in a bucket of oil (s = 0.86) that is in an elevator being accelerated upward at 2.5 m/s²?*

SI

Eq. 3.24: $\partial p/\partial z = -\rho(g + a_z)$; $\quad \Delta p = -\rho(g + a_z)\Delta z = -(\gamma/g)(g + a_z)\Delta z$

$$\Delta p = -0.86\frac{\text{gm}}{\text{cm}^3}\left(\frac{9807\ \text{N/m}^3}{9.807\ \text{m/s}^2}\right)(9.807 + 2.50)\frac{\text{m}}{\text{sec}^2}(-0.3\ \text{m}) = 3180\ \text{Pa}$$ ◄

3.11.3 *A tank containing water to a depth of 5 ft is accelerated upward at 8 ft/s². Calculate the pressure on the bottom of the tank.*

BG

From Eq. 3.24: $\Delta p = -(\gamma/g)(a_z + g)\Delta z = -(62.4/32.2)(8 + 32.2)(-5) = 390$ psf

$\Delta p = 390/144 = 2.70$ psi ◄

3.11.4 *A tank containing water to a depth of 2.5 m is accelerated upward at 3.6 m/s². Calculate the pressure on the bottom of the tank.*

SI

From Eq. 3.24: $$\Delta p = -\frac{\gamma}{g}(a_z + g)\Delta z = -\left(\frac{9807\ \text{N/m}^3}{9.807\ \text{m/s}^2}\right)(3.60 + 9.81)(-2.5) = 33\,500\ \text{Pa} = 33.5\ \text{kPa}$$ ◄

3.11.5 *Suppose the tank shown in Fig. 3.24 is rectangular and completely open at the top. It is 16 ft long, 6 ft wide, and 5 ft deep. If it is initially filled to the top, how much liquid will be spilled if it is given a horizontal acceleration $a_x = 0.25g$ in the direction of its length?*

BG

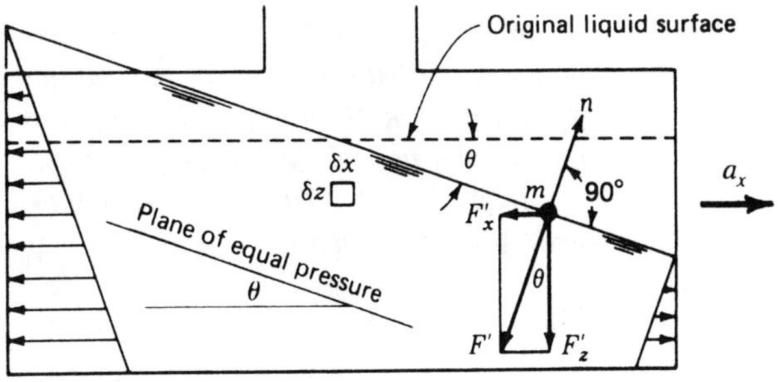

Figure 3.24a

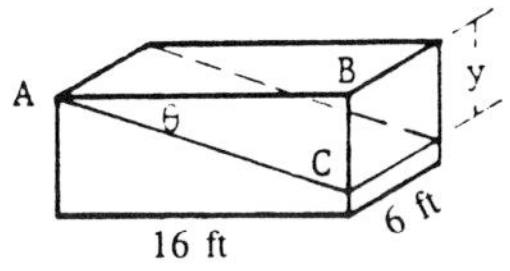

Sec 3.11: $\tan\theta = -a_x/g = -0.25$;

$y = -0.25(16) = -4.0$ ft

If tank is more than 4 ft deep, fluid spilled will be

wedge *ABC* or (1/2)16(4)6 = 192 ft^3 ◄

3.11.6 *Suppose the tank shown in Fig. 3.24 is rectangular and completely open at the top. It is 18 m long, 5 m wide, and 4 m deep. If it is initially filled to the top, how much liquid will be spilled if it is given a horizontal acceleration $a_x = 0.4g$ in the direction of its length?*

See Fig. 3.24 in Solution 3.11.5 above.

SI

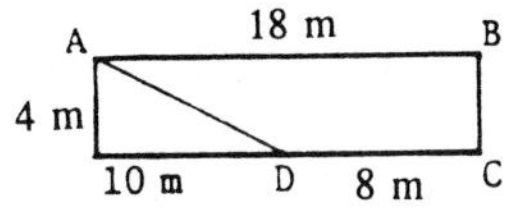

Sec 3.11: $\tan\theta = -a_x/g = -0.4$; $4/0.4 = 10$ m

ABCD is spilled ; $[8(4) + (1/2)10(4)]5 = 260$ m^3 ◄

3.11.7 *If the tank of Exer 3.11.5 is closed at the top and is completely filled, what will be the pressure difference between the left-hand end at the top and the right-hand end at the top if the liquid has a specific weight of 50 lb/ft³ and the horizontal acceleration is $a_x = 0.3g$? Sketch planes of equal pressure, indicating their magnitude; assume zero pressure in the upper right-hand corner.*

Exer. 3.11.5: The rectangular tank (similar to Fig. 3.24 in Solution 3.11.5 above) is 16 ft long, 6 ft wide, and 5 ft deep.

BG

Eq. 3.23: $\partial p/\partial x = -\rho a_x = -(50/32.2)0.3(32.2) = -15$ psf per ft

$\Delta p = p_A - p_B = 15(16) = 240$ psf ◄

Eq. 3.24 with $a_z = 0$: $\partial p/\partial z = -\rho g = -\gamma = -50$ psf per ft

$\Delta p = p_C - p_B = 50(5) = 250$ psf ;

Lines of equal pressure are as follows:

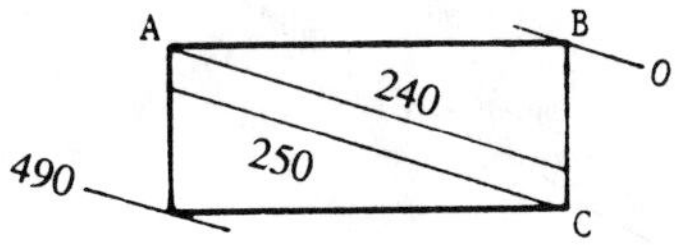

◄

3.11.8 *If the tank of Exer 3.11.6 is closed at the top and is completely filled, what will be the pressure difference between the left-hand end at the top and the right-hand end at the top if the liquid has a specific weight of 8.0 kN/m³ and the horizontal acceleration is $a_x = 0.3g$? Sketch planes of equal pressure, indicating their magnitude; assume zero pressure in the upper right-hand corner.*

Exer. 3.11.6: The rectangular tank (similar to Fig. 3.24 in Solution 3.11.5 above) is 18 m long, 5 m wide, and 4 m deep.

SI

Eq. 3.23: $\partial p/\partial x = -\rho a_x = -(\gamma/g)a_x = -(8.0/9.81)0.3(9.81) = -2.4$ kPa per m

$\Delta p = p_A - p_B = 2.4(18) = 43.2$ kPa ◀

Eq. 3.24 with $a_z = 0$: $\partial p/\partial z = -\rho g = -\gamma = -8.0$ kPa per m

$\Delta p = p_C - p_B = 8.0(4) = 32$ kPa ;

Lines of equal pressure are as follows:

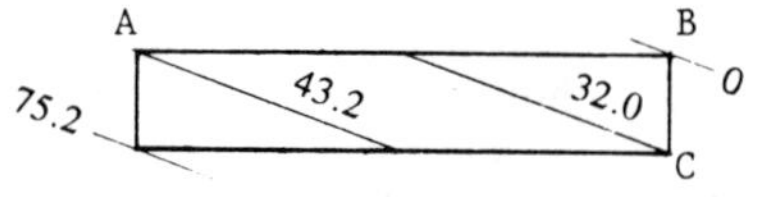

◀

Sec. 3.11: Fluid Masses Subjected to Acceleration -- *Problems*

3.26 *At a particular instant an airplane is traveling upward at a velocity of 180 mph in a direction that makes an angle of 30° with the horizontal. At this instant the airplane is losing speed at the rate of 3.6 mph/sec. Also, it is moving on a concave-upward circular path having a radius of 5000 ft. Determine for the given conditions the slope of the free liquid surface in the fuel tank of this vehicle.*

BG

180 mph = 180(5280 ft/mi)/(3600 sec/hr) = 264 fps ; 3.6 mph = 5.28 fps

$a_n = v^2/r = 264^2/5000 = 13.94$ ft/sec² ; $a_t = -5.28$ ft/sec².

$a_x = -5.28\cos 30° - 13.94\sin 30° = -11.54$ ft/sec²

$a_z = -5.28\sin 30° + 13.94\cos 30° = +9.43$ ft/sec²

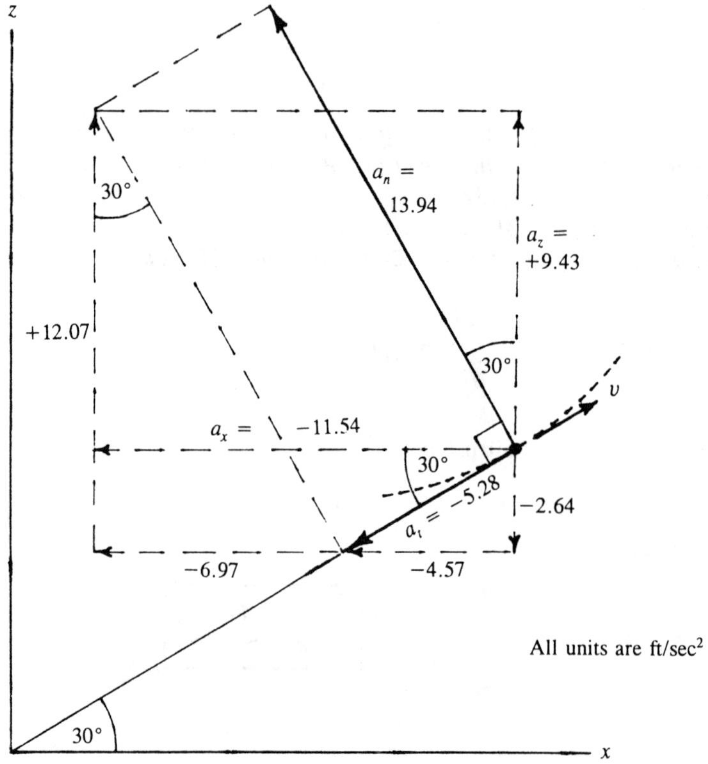

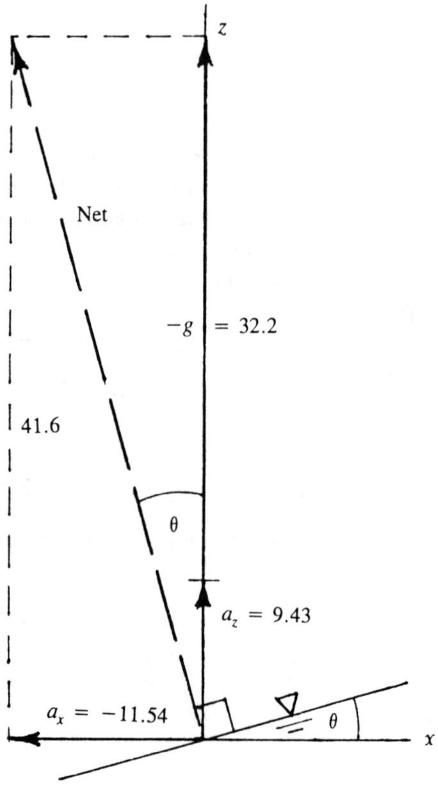

/cont...

Liquid surface is perpendicular to net force, i.e., to net acceleration.

Eq. 3.26: Surface slope $= \tan\theta = \dfrac{dz}{dx} = -\dfrac{(-11.54)}{(9.43 + 32.2)} = 0.277$ ◄

or $\theta = \tan^{-1}0.277 = 15.50°$ from the horizontal. ◄

3.27 *Refer to Sample Prob. 3.10. Suppose the velocity of the airplane is 220 m/s, with all other data unchanged. What then would be the slope of the liquid surface in the tank?*

Sample Prob 3.10: At a particular instant an airplane is travelling upward at 40° to the horizontal, and losing speed at 4 m/s². Also it is moving on a concave-upward circular path of radius 2600 m.

SI

$a_n = v^2/r = 220^2/2600 = 18.62\ \text{m/s}^2$; $a_t = -4\ \text{m/s}^2$ (given)

$a_x = -4\cos 40° - 18.62\sin 40° = -15.03\ \text{m/s}^2$; $a_z = -4\sin 40° + 18.62\cos 40° = +11.69\ \text{m/s}^2$

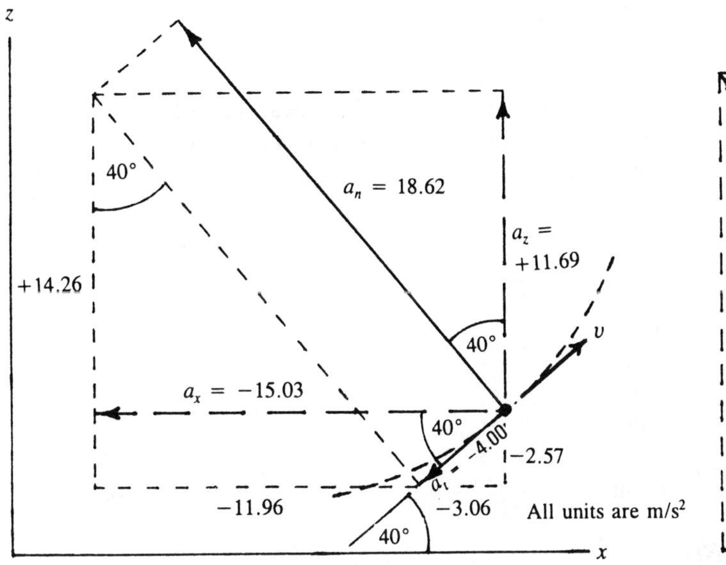

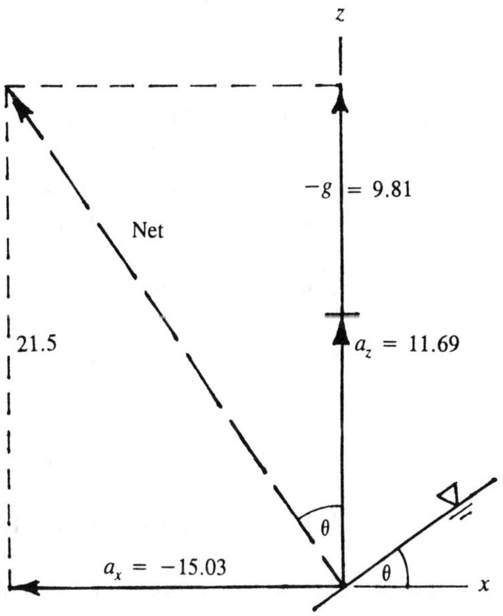

Liquid surface is perpendicular to net force, i.e., to net acceleration.

Eq. 3.26: Surface slope $= \tan\theta = \dfrac{dz}{dx} = -\dfrac{(-15.03)}{(11.69 + 9.81)} = 0.699$ ◄

or $\theta = \tan^{-1}0.699 = 35.0°$ from the horizontal. ◄

Chapter 4

Basics of Fluid Flow

PROBLEM SELECTION GUIDE

Sec.	Exer/Prob	Units	Difficulty	Length	Parts	Similar	Special features
4.3	*Steady Flow and Uniform Flow*						
	X[1]4.3.1	N	Easy	V Short	6		
4.5	*Flow Rate and Mean Velocity*						
	X 4.5.1	N	Easy	Short	1		Integration
	4.5.2	BG	V Easy	V Short	1		
	4.5.3	BG	Easy	Short	1	4.5.4	
	4.5.4	SI	Easy	Short	1	4.5.3	
	4.5.5	BG	Easy	Short	1	4.5.6	
	4.5.6	SI	Easy	Short	1	4.5.5	
	P 4.1	BG	Medium	Long	1	4.2	Graphical solution
	4.2	SI	Medium	Long	1	4.1	Graphical solution
	4.3	BG	Easy	Short	1	4.4	Uses $pv = RT$ (Sec. 2.7)
	4.4	SI	Easy	Short	1	4.3, S4.1	Uses $pv = RT$ (Sec. 2.7)
4.7	*Equation of Continuity*						
	X 4.7.1	SI	Easy	Short	2		
	4.7.2	BG	Easy	Short	1		
	4.7.3	BG	V Easy	V Short	1	4.7.4	
	4.7.4	SI	V Easy	V Short	1	4.7.3	
	P 4.5	BG	Medium	Medium	1		Numerical differentiation
4.10	*Use and Limitations of Flow Net*						
	X 4.10.1	SI	Medium	Short	1		
	4.10.2	SI	Easy	Short	1		† Measure Fig X4.10.2
	4.10.3	BG	Easy	Medium	2		† Measure Fig 4.10
	4.10.4	SI	Easy	Medium	2		† Measure Fig 4.12
	P 4.6	BG	Medium	Medium	1	4.7	† Measure Fig X4.10.2
	4.7	SI	Medium	Medium	1	4.6	† Measure Fig X4.10.2
	4.8	N	Medium	Medium	1		† Measure Fig 4.10; plot
	4.9	BG	Hard	Long	1		Sketch flow net; plot

/cont...

[1] For all Exercises (marked "X"), answers are given in Appendix D of the textbook.
† Answers are sensitive to values which are measured from figures.
X = Exercise, P = (end-of-chapter) Problem, S = Sample Problem.

Sec.	Exer/Prob	Units	Difficulty	Length	Parts	Similar	Special features
4.12	*Velocity and Acceleration in Steady Flow*						
	X 4.12.1	N	V Easy	V Short	1		
	4.12.2	N	Medium	Short	1		Differentiation
	4.12.3	N	Medium	Medium	1	4.12.4	Partial differentiation
	4.12.4	N	Medium	Medium	1	4.12.3	Partial differentiation
	4.12.5	N	Medium	Medium	1		Partial differentiation
	4.12.6	N	Medium	Medium	1		Partial differentiation
	4.12.7	BG	V Easy	V Short	1	4.12.8	
	4.12.8	SI	V Easy	V Short	1	4.12.7	
	4.12.9	BG	Medium	Medium	1		Differentiation
	4.12.10	SI	Medium	Short	1		Differentiation
	4.12.11	BG	Medium	Medium	1		† Meas. Fig 4.10; numerical different'n
	P 4.10	N	Medium	Medium	1		Sketch flow field
	4.11	N	Medium	Medium	1		Sketch flow field
	4.12	N	Medium	Long	1		† Sketch flow field; numerical differ'n
	4.13	SI	Medium	Medium	1		† Measure Fig P4.13; numerical differ'n
4.13	*Velocity and Acceleration in Unsteady Flow*						
	X 4.13.1	N	Medium	Short	1		Differentiation
	4.13.2	N	Medium	Short	1		Partial differentiation
	4.13.3	BG	Easy	Short	1	4.13.4	Differentiation
	4.13.4	SI	Easy	Short	1	4.13.3	Differentiation
	P 4.14	BG	Medium	Medium	1	4.15	Partial differentiation
	4.15	SI	Medium	Medium	1	4.14	Partial differentiation
	4.16	SI	Medium	Medium	1		† Measure Fig P4.13; numerical differ'n
Ch.4	*Miscellaneous*						
	P 4.17	N	Medium	Medium	1		Plot
	4.18	BG	Medium	Medium	1		Plot
	4.19	SI	Medium	Long	3		Plot

Chapter 4
BASICS OF FLUID FLOW

Sec 4.3: Steady Flow and Uniform Flow -- *Exercise*

4.3.1 *Classify the following cases of flow as to whether they are steady or unsteady, uniform of nonuniform: (a) water flowing from a tilted pail; (b) flow from a rotating lawn sprikler; (c) flow through the hose leading to the sprinkler; (d) natural stream during dry-weather flow; (e) natural stream during flood; (f) flow in a city water-distribution main in a straight section of constant diameter and no side connections. (Note: There is room for legitimate argument in some of the above cases, which should stimulate independent thought.)*

N

(*a*) Unsteady, nonuniform; (*b*) Unsteady (although steady with respect to rotating frame of reference), nonuniform; (*c*) Steady, uniform; (*d*) Almost steady, nonuniform; (*e*) Unsteady, nonuniform; (*f*) Unsteady, uniform. ◄◄

Sec 4.5: Flow Rate and Mean Velocity -- *Exercises*

4.5.1 *In the laminar flow of a fluid in a circular pipe the velocity profile is exactly a true parabola. The rate of discharge is then represented by the volume of a paraboloid. Prove that for this case the ratio of the mean velocity to the maximum velocity is 0.5.*

N

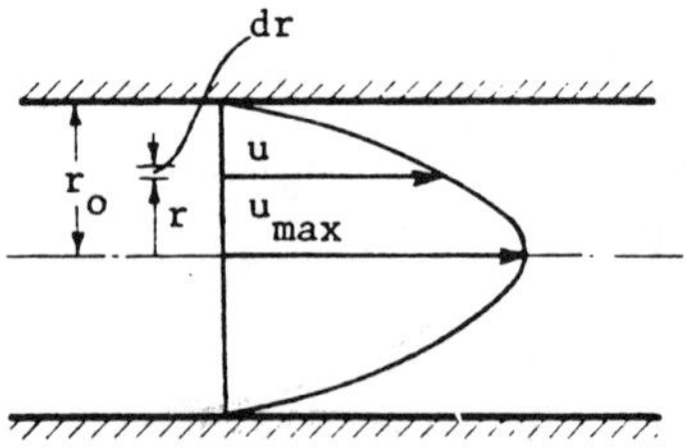

For paraboloid, $u = u_{\max}[1 - (r/r_o)^2]$

Eq. 4.3: $Q = \int u\,dA = 2\pi u_{\max}\int_r^{r_o} [1 - (r/r_o)]^2 r\,dr$

$$= u_{\max}\frac{\pi r_o}{2};\quad V_{\text{mean}} = V = \frac{Q}{A} = \frac{u_{\max}(\pi r_o^2)}{2(\pi r_o^2)} = \frac{u_{\max}}{2}$$

Thus $\dfrac{V_{\text{mean}}}{u_{max}} = 0.5$ Q.E.D. ◄

4.5.2 *A gas (γ = 0.05 lb/ft^3) flows at a rate of 1.0 lb/sec past section A through a long rectangular duct of uniform cross section 1.5 by 2 ft. At section B some distance along the duct the gas weighs 0.085 lb/ft^3. What is the average velocity of flow at sections A and B?*

BG

From Eq. 4.5: $V_A = \dfrac{G}{\gamma A} = \dfrac{1.0}{0.05(1.5 \times 2)} = 6.67$ fps ◄

$$V_B = \frac{1.0}{0.085(1.5 \times 2)} = 3.92 \text{ fps}$$ ◄

4.5.3 *The velocity of a liquid (s = 1.59) in a 3-in pipeline is 1.8 fps. Calculate the rate of flow in cfs, gal/min, slugs/sec, and lb/sec.*

BG

Eq. 4.3: $Q = AV = \pi(1.5/12)^2 \times 1.8 = 0.0884$ cfs ◄

$= 0.0884(448.8 \text{ gpm/cfs}) = 39.7$ gpm ◄

From Eqs. 4.4 and 4.3: $\dot{m} = \rho Q = (1.59 \times 1.940)0.0884 = 0.273$ slug/sec ◄

From Eqs. 4.5 and 4.3: $G = \gamma Q = (1.59 \times 62.4)0.0884 = 8.77$ lb/sec ◄

4.5.4 *The velocity of a liquid (s = 1.59) in a 15-cm pipeline is 0.72 m/s. Calculate the rate of flow in L/s, m³/s, kg/s, and kN/s.*

SI

Eq. 4.3: $Q = AV = \pi(0.075)^2 \times 0.72 = 0.012\,72\ \text{m}^3/\text{s}$ ◄

$= 0.012\,72\ (\text{m}^3/\text{s}) \times 10^3\ (\text{L/m}^3) = 12.72\ \text{L/s}$ ◄

From Eqs. 4.4 and 4.3: $\dot{m} = \rho Q = (1.59 \times 1000)0.012\,72 = 20.2\ \text{kg/s}$ ◄

From Eqs. 4.5 and 4.3: $G = \gamma Q = (1.59 \times 9.81)0.012\,72 = 0.198\,4\ \text{kN/s}$ ◄

4.5.5 *Water flows at 4 gal/min through a small circular hole in the bottom of a large tank. Assuming the water in the tank approaches the hole radially, what is the velocity in the tank at 2, 4, and 8 in from the hole?*

BG

Area through which flow occurs is a hemispherical surface, with $A = 2\pi r^2$ (Table A.8)

$Q = 4\ \text{gal/min} \times (0.1337\ \text{ft}^3/\text{g})/(60\ \text{sec/min}) = 0.00891\ \text{cfs}$

r (in)	A (in²)	A (ft²)	$V = Q/A$
2	25.1	0.174 5	0.051 1
4	100.5	0.698	0.012 77
8	402.1	2.793	0.003 19

▲

4.5.6 *Water flows at 0.25 L/s through a small circular hole in the bottom of a large tank. Assuming the water in the tank approaches the hole radially, what is the velocity in the tank at 50, 100, and 200 mm from the hole?*

SI

Area through which flow occurs is a hemispherical surface, with $A = 2\pi r^2$ (Table A.8)

$Q = 0.25\ \text{L/s} = 0.000\,25\ \text{m}^3/\text{s}$

r (mm)	A (m²)	$V = Q/A$
50	0.015 71	0.015 92
100	0.062 83	0.003 98
200	0.251 33	0.000 995

▲

Sec 4.5: Flow Rate and Mean Velocity -- *Problems*

4.1 *The velocities in a circular conduit 8 in in diameter were measured at radii 0, 1.44, 2.60, and 3.48 in and were found to be 20.3, 19.7, 17.7, and 14.5 fps respectively. Find approximate values (graphically) of the volume flow rate and the mean velocity. Also determine the ratio of the mean velocity to the maximum velocity.*

BG

Note: Answers may vary somewhat due to graphical solution.

Eq. 4.3: $Q = \int u dA = u_1 dA_1 + u_2 dA_2 + \ldots$

Plot the velocity profile and divide the area into a number of equal (annular) parts.

/cont...

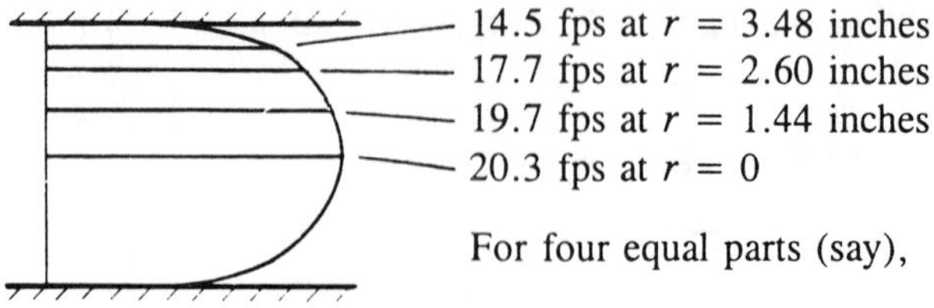

For four equal parts (say),

$$\frac{A}{4} = \frac{\pi r^2}{4} = \pi r_1^2 = \pi(r_2^2 - r_1^2) = \pi(r_3^2 - r_2^2);\ dA = \frac{A}{4} = \frac{\pi 4^2}{4} = 4\pi \text{ in}^2$$

r = 4 inches, so $r_0 = 0$, $r_1 = \sqrt{4}$ = 2.00 inches, $r_2 = \sqrt{8}$ = 2.83 inches, $r_3 = \sqrt{12}$ =3.46 inches

From the sketched profile, $u_1 \approx$ 7.8 fps, $u_2 \approx$ 15.4 fps, $u_3 \approx$ 17.7 fps, $u_4 \approx$ 19.4 fps

$Q = \Sigma(u_i dA_i) \approx (7.8 + 15.4 + 17.7 + 19.4)dA = 60.3(4\pi)$

$= 757$ ft(in²/s) = 5.26 cfs $\qquad Q \approx 5.3$ cfs ◀

$V = Q/A \approx 5.26/(16\pi/144) = 15.075$ fps $\qquad V \approx 15.1$ fps ◀

$V/V_{max} \approx 15.075/20.3 = 0.743$ $\qquad V/V_{max} \approx 0.74$ ◀

If the area had been broken down into more parts, the accuracy would have been improved.

4.2 *The velocities in a circular conduit 200 mm in diameter were measured at radii 0, 36, 65, and 87 mm and were found to be 7.0, 6.8, 6.1, and 5.0 m/s respectively. Find approximate values (graphically) of the volume flow rate and the mean velocity. Also determine the ratio of the mean velocity to the maximum velocity.*

SI

Note: Answers may vary somewhat due to graphical solution.

Eq. 4.3: $Q = \int u dA = u_1 dA_1 + u_2 dA_2 + \ldots$

Plot the velocity profile and divide the area into a number of equal (annular) parts.

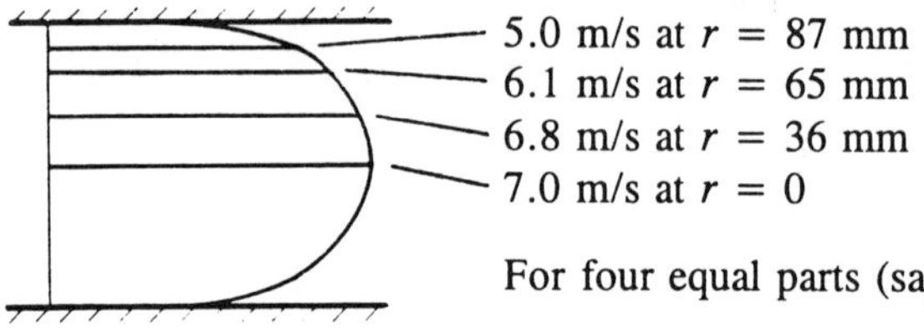

For four equal parts (say).

$$A/4 = \pi r^2/4 = \pi r_1^2 = \pi(r_2^2 - r_1^2) = \pi(r_3^2 - r_2^2)$$

$$dA = A/4 = \pi 100^2/4 = 2500\pi \text{ mm}^2$$

r = 100 mm, so $r_0 = 0$, r_1 = 50.0 mm, $r_2 = \sqrt{5000}$ = 70.7 mm, $r_3 = \sqrt{7500}$ =86.6 mm

From the sketched profile, $u_1 \approx$ 6.7 m/s, $u_2 \approx$ 6.1 m/s, $u_3 \approx$ 5.3 m/s, $u_4 \approx$ 2.7 m/s

$Q = \Sigma(u_i dA_i) \approx (6.7 + 6.1 + 5.3 + 2.7)dA = 20.8(2500\pi)$

$= 163\,400$ m(mm²/s) = 0.1633 m³/s $\qquad Q \approx 0.163$ m³/s ◀

$V = Q/A \approx 163\,400/(4 \times 2500\pi) = 5.20$ m/s $\qquad V \approx 5.2$ m/s ◀

$V/V_{max} \approx 5.20/7.0 = 0.743$ $\qquad V/V_{max} \approx 0.74$ ◀

If the area had been broken down into more parts, the accuracy would have been improved.

4.3 *Carbon dioxide flows in a 2-in by 2-in duct at a pressure of 50 psi and a temperature of 80°F. If the atmospheric pressure is 13.4 psia and the velocity of flow is 15 fps, calculate the weight flow rate.*

BG

Table A.5 for carbon dioxide: $R = 1123$ ft·lb/(slug·R°)

Eq. 2.5: $\gamma = \dfrac{gp}{RT} = \dfrac{32.2(50 + 13.4)144}{1123(460 + 80)} = 0.485$ pcf

Eq. 4.5: $G = \gamma A V = 0.485(2/12)^2 15 = 0.202$ lb/sec ◄

4.4 *Nitrogen at 45°C and under a pressure of 2800 mb abs flows in a 300-mm-diameter conduit at a mean velocity of 10 m/s. Find the mass flow rate.*

SI

Table A.5 for nitrogen: $R = 297\ \text{m}^2/(\text{s}^2\cdot\text{K})$

From Eq. 2.4: $\rho = \dfrac{p}{RT} = \dfrac{2800\ \text{mbars} \times (1\ \text{KN/m}^2)/10\ \text{mbars} \times (1000\ \text{kg·m/s}^2)/\text{kN}}{297\ \text{m}^2\text{s}^{-2}\text{K}^{-1}(273+45)\ \text{K}} = 2.96\ \text{kg/m}^3$

From Eq. 4.4: $\dot{m} = \rho A V = 2.96(\pi \times 0.15^2)10 = 2.10$ kg/s ◄

Sec 4.7: Equation of Continuity -- *Exercises*

4.7.1 *Gas flows at a steady rate in a 150-mm-diameter pipe that enlarges to a 200-mm-diameter pipe. (a) At a certain section of the 150-mm pipe the density of the gas is 175 kg/m³ and the velocity is 18 m/s. At a certain section of the 200-mm pipe the velocity is 12 m/s. What must be the density of the gas at that same section? (b) If these same data were given for the case of unsteady flow at a certain instant, could the problem be solved? Discuss.*

SI

(*a*) Eq. 4.16*a*: $\rho_1 A_1 V_1 = \rho_2 A_2 V_2$; as $A \propto (\text{diam})^2$, we therefore obtain $\rho_1 d_1^2 V_1 = \rho_2 d_2^2 V_2$

$175(150)^2 18 = \rho_2(200)^2 12$ from which $\rho_2 = 147.7\ \text{kg/m}^3$ ◄

(*b*) If the flow were unsteady, the problem could not be solved for no information is given on $\partial\rho/\partial t$. Also, the volume between the two sections is unknown. ◄

4.7.2 *Gas is flowing in a long 8-in-diameter pipe from A to B. At section A the flow is 0.62 lb/sec while at the same instant at section B the flow is 0.68 lb/sec. The distance between A and B is 750 ft. Find the mean value of the time rate of change of the specific weight of the gas between sections A and B at that instant.*

BG

Eq. 4.15 × g: $\gamma_1 A_1 V_1 - \gamma_2 A_2 V_2 = (\partial\gamma/\partial t) \times \forall$;

Eq. 4.16*b*: $\gamma A V = G$

Substituting: $G_1 - G_2 = (\partial\gamma/\partial t) \times \forall$

$(0.62 - 0.68)$ lb/s $= \partial\gamma/\partial t \times 750\pi(4/12)^2$ ft³; $\partial\gamma/\partial t = -0.000\,229$ lb/ft³ per sec ◄

4.7.3 *Water flows in a river. At 8 A.M. the flow past bridge 1 is 2200 cfs. At the same instant the flow past bridge 2 is 1750 cfs. At what rate is water being stored in the river between the two bridges at this instant? Assume zero seepage and negligible evaporation.*

BG

Eq. 4.18: $Q_1 - Q_2 = d\forall/dt$

$(2200 - 1750)$ ft³/s $= d\forall/dt$; $d\forall/dt = 450$ ft³/sec ◄

4.7.4 *Water flows in a river. At 9 A.M. the flow past bridge 1 is 58.4 m^3/s. At the same instant the flow past bridge 2 is 42.7 m^3/s. At what rate is water being stored in the river between the two bridges at this instant? Assume zero seepage and negligible evaporation.*

SI

Eq. 4.18: $Q_1 - Q_2 = d\forall/dt$

$(58.4 - 42.7)\ \text{m}^3/\text{s} = d\forall/\text{dt}$; $d\forall/dt = 15.70\ \text{m}^3/\text{sec}$ ◀

Sec 4.7: Equation of Continuity -- *Problem*

4.5 *A compressible fluid flows in an 18-in-diameter leaky pipe. Measurements are made simultaneously at two points A and B along the pipe that are 36,000 ft apart. Two sets of measurements are taken with an interval of exactly 40 min between them. The data are as follows:*

Time	ρ_1 *(slug/ft³)*	V_1 *(ft/s)*	ρ_2 *(slug/ft³)*	V_2 *(ft/s)*
0	*0.565*	*68*	*0.694*	*52*
40 min	*0.672*	*52*	*0.827*	*40*

Assuming ρ varies linearly with respect to time and distance, compute the approximate average mass rate of leakage between A and B.

BG

Modifying Eq. 4.15 to include for leakage, at any instant

$$\rho_1 A_1 V_1 - \rho_2 A_2 V_2 - \text{leakage rate} = (\partial\rho/\partial t) \times \forall$$

$$\partial\rho/\partial t \text{ at Section 1} = \frac{\Delta\rho_1}{\Delta t} \approx \frac{0.672 - 0.565}{40\times 60} = 4.46 \times 10^{-5}\ \text{slug/(ft}^3\text{-sec)}$$

$$\partial\rho/\partial t \text{ at Section 2} = \frac{\Delta\rho_2}{\Delta t} \approx \frac{0.827 - 0.694}{40\times 60} = 5.54 \times 10^{-5}\ \text{slug/(ft}^3\text{-sec)}$$

During this time interval, and through the 36,000-ft length *AB*,

$$\text{average } \partial\rho/\partial t \approx \tfrac{1}{2}(4.46 + 5.54)\times 10^{-5} = 5.00 \times 10^{-5}\ \text{slug/(ft}^3\cdot\text{sec)}$$

During any time interval, $\bar{\rho}_1 A_1 \bar{V}_1 - \bar{\rho}_2 A_2 \bar{V}_2 - \text{avg leak rate} \approx \overline{(\partial\rho/\partial t)} \times \forall$

$$(0.6185)\frac{\pi}{4}\left(\frac{18}{12}\right)^2 60 - (0.7605)\frac{\pi}{4}\left(\frac{18}{12}\right)^2 46 - \text{avg leak rate} \approx 5.00 \times 10^{-5}\left[36{,}000\times\frac{\pi}{4}\left(\frac{18}{12}\right)^2\right]$$

65.579 − 61.820 − avg leak rate ≈ 3.181 slug/sec; Avg leakage rate ≈ 0.578 slug/sec ◀

Sec 4.10: Use and Limitations of Flow New -- *Exercises*

4.10.1 *An incompressible ideal fluid flows at 10 L/s through a circular 160-mm-diameter pipe into a conically converging nozzle like that of Sample Prob. 4.4 (diameter at B is 80 mm). Determine the average velocity of flow at sections D and B.*

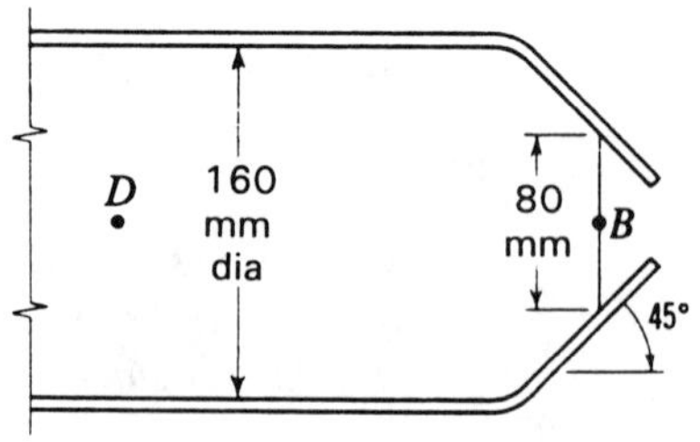

/cont...

SI

Consider the sketched flow net of Sample Prob 4.4. At section D the streamlines are parallel, so the area normal to the velocity vectors is a plane circle.

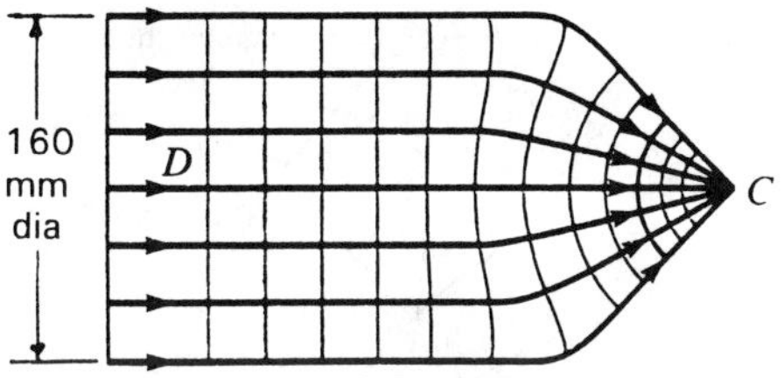

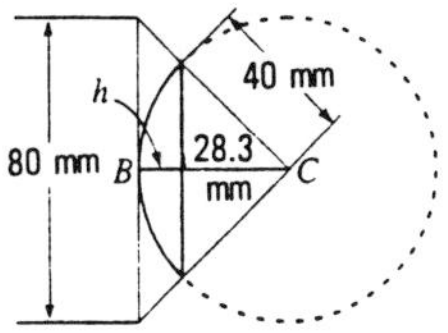

$$\text{Thus, } V_D = \frac{Q}{A_D} = \frac{10 \times 10^{-3}\ \text{m}^3/\text{s}}{(\pi/4)(0.16\ \text{m})^2} = 0.497\ \text{m/s} \quad ◀$$

At section B, however, the area normal to the streamlines is best approximated by the portion of the surface of a sphere of radius 40 mm which is inside the nozzle and passes through B (see right-hand solution figure of Sample Prob. 4.4). By table lookup, by integration, or from Sample Prob 4.4, $A_B = 2\pi rh$

$$h = 40 - 40\cos 45° = 11.72\ \text{mm}$$

$$V_B = \frac{Q}{A_B} = \frac{Q}{2\pi rh} = \frac{10 \times 10^{-3}\ \text{m}^3/\text{s}}{2\pi(0.040)0.011\,72\ \text{m}^2} = 3.40\ \text{m/s} \quad ◀$$

4.10.2 *Figure X4.10.2 shows the flow net for two-dimensional flow from a rounded, long-slotted exit from a tank. If U_0 = 2.5 m/s, what is the approximate flow velocity at A?*

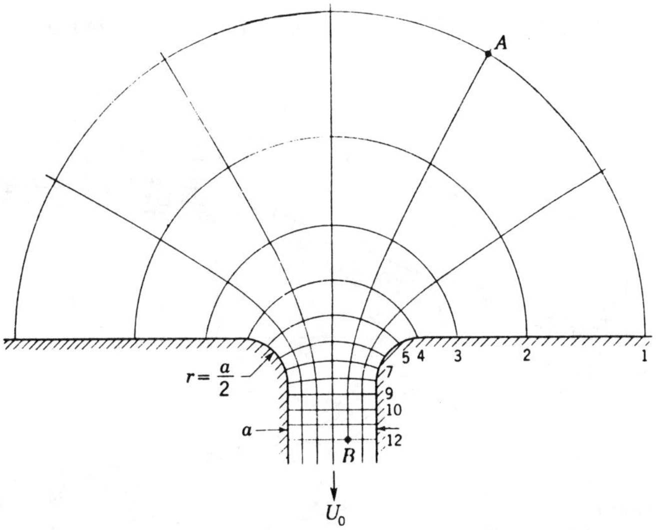

Figure X4.10.2

SI

At B: $\Delta n_o \approx 12.05/6 = 2.01$ mm

At A: $\Delta n = \dfrac{c}{12} = \dfrac{\pi D}{12} \approx \dfrac{\pi(85.3)}{12} = 22.3$ mm

From Eq. 4.19: $V_A = \dfrac{\Delta n_o}{\Delta n} U_o \approx \dfrac{2.01}{22.3}(2.5) = 0.225 \approx 0.23$ m/s ◀

4.10.3 *Given that U_0 in Fig. 4.10 is 5.0 fps, find approximately (a) the maximum velocity in the bend, and (b) the uniform velocity in the downstream section.*

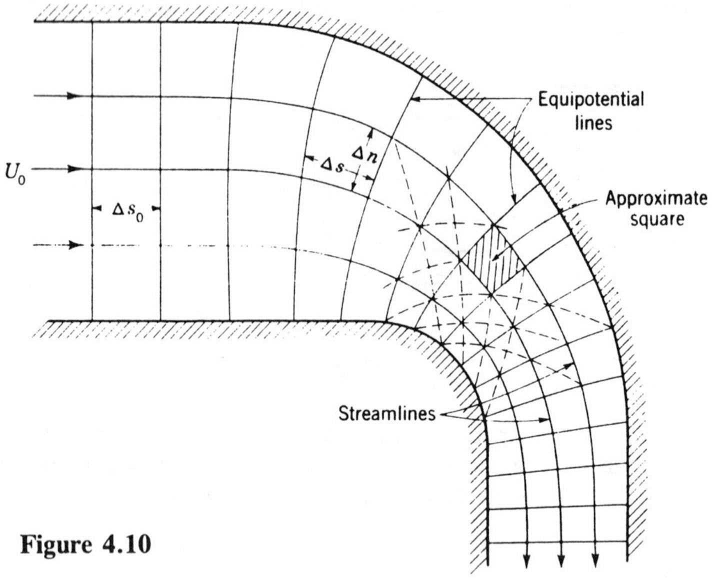

Figure 4.10

BG

(All values approximate)	Δn, mm	$\Delta n_o / \Delta n$	$V = (\Delta n_o / \Delta n) U_o$
In uniform entrance:	38.8/4 = 9.70	1.00	5.00 fps (given)
(*a*) In bend (smallest square)	3.50	2.77	13.9 fps ◄
(*b*) In uniform exit:	18.7/4 = 4.68	2.07	10.4 fps ◄

4.10.4 *Given that U_0 in Fig. 4.12 is 3.5 m/s, find approximately (a) the maximum and (b) the minimum velocity on the body surface.*

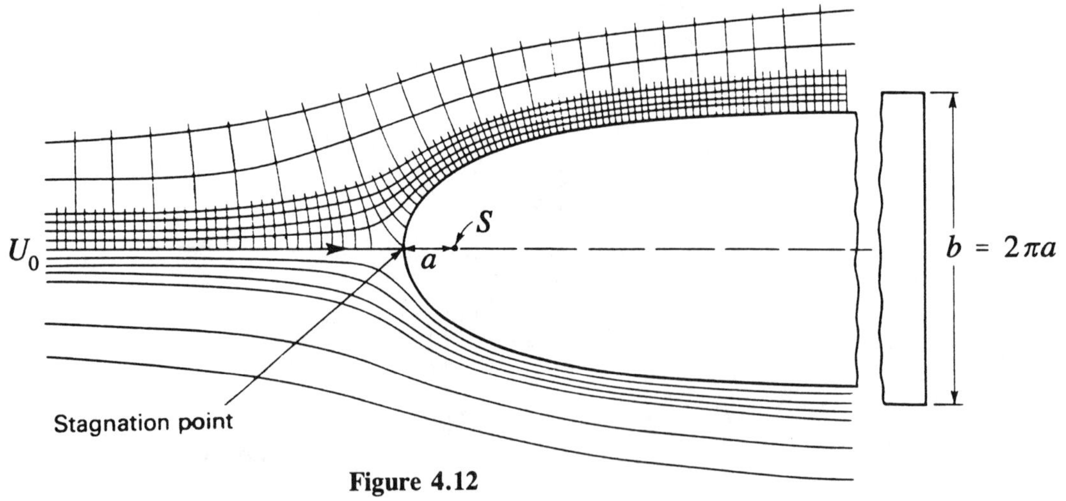

Figure 4.12

SI

(*a*) Measuring four of the small squares, on average:

At uniform entrance: $$\Delta n_o = \frac{1}{2}\left(\frac{L}{4} + \frac{W}{4}\right) \approx \frac{1}{2}\left(\frac{4.0 + 3.7}{4}\right) = 0.962 \text{ mm}$$

At the smallest square, about 0.6*b* or 3.8*a* downstream of the stagnation point:

$$\Delta n = \frac{1}{2}\left(\frac{L}{4} + \frac{W}{4}\right) \approx \frac{1}{2}\left(\frac{2.6 + 3.0}{4}\right) = 0.700 \text{ mm}$$

/cont...

$V_{max} = (\Delta n_0/\Delta n)U_0 \approx (0.962/0.700)3.50 = 4.8$ m/s ◄

Note: Answers may vary somewhat due to scaling

(*b*) At stagnation point, $V_{min} = 0$ ◄

Sec 4.10: Use and Limitations of Flow Net -- ***Problems***

4.6 *Refer to Fig. X4.10.2. If a is 3 in and U_0 is 10 fps, approximately how long will it take a particle to move from point A to point B on the same streamline? (Note: Between each pair of equipotential lines, measure Δs, and then compute the average velocity and time increment.)*

BG

$U_o = 10$ fps and $a = 3$ in. $\therefore\ \Delta s_0 = \dfrac{a}{6} = \dfrac{1}{2}$ in $= \dfrac{1}{24}$ ft

Equipotential space	$\Delta s/\Delta s_0$	$\bar{V} = \dfrac{10}{\Delta s/\Delta s_0}$	$\Delta s = \dfrac{1}{24}\dfrac{\Delta s}{\Delta s_0}$	$\Delta t = \Delta s/\bar{V}$
1-2	7.5	1.33 fps	0.31 ft	0.23 sec
2-3	5.5	1.82 fps	0.23 ft	0.13 sec
3-4	3.4	2.94 fps	0.14 ft	0.048 sec

and similarly on through space 11 - 12, yields $\Sigma\Delta t \approx 0.47$ sec ◄

4.7 *Repeat Prob. 4.6 using the following data: a = 15 cm and U_0 = 0.5 m/s. Find also the approximate velocity where the flow crosses equipotential line 3.*

SI

$U_0 = 0.5$ m/s and $a = 15$ cm $= 0.15$ m. $\therefore\ \Delta s_0 = \dfrac{a}{6} = \dfrac{0.15\text{ m}}{6} = 0.025$ m

Equipotential space	$\Delta s/\Delta s_0$	$\bar{V} = \dfrac{0.5}{\Delta s/\Delta s_0}$	$\Delta s = 0.025\dfrac{\Delta s}{\Delta s_0}$	$\Delta t = \Delta s/\bar{V}$
1-2	7.5	0.067 m/s	0.188 m	2.80 sec
2-3	5.5	0.091 m/s	0.138 m	1.52 sec
3-4	3.4	0.147 m/s	0.085 m	0.58 sec

and similarly on through space 11 - 12, yields $\Sigma\Delta t \approx 5.7$ sec ◄

Area at equipotential line 3, length of arc ≈ 0.60 m. $\therefore\ V_3 \approx \dfrac{0.5}{0.6/0.15} = 0.13$ m/s ◄

4.8 *Make an approximate plot of the frictionless velocity (relative to U_0) along both the inner and the outer boundaries of Fig. 4.10. By what percent is the ideal maximum inner velocity greater than the ideal minimum outer velocity?*

Fig. 4.10 appears with the solution of Exercise 4.10.3

N

In stream tube, $q = \text{const} = AV \propto \Delta sV = \Delta s_0 U_0$, so $\dfrac{V}{U_0} = \dfrac{1}{(\Delta s/\Delta s_0)}$

/cont...

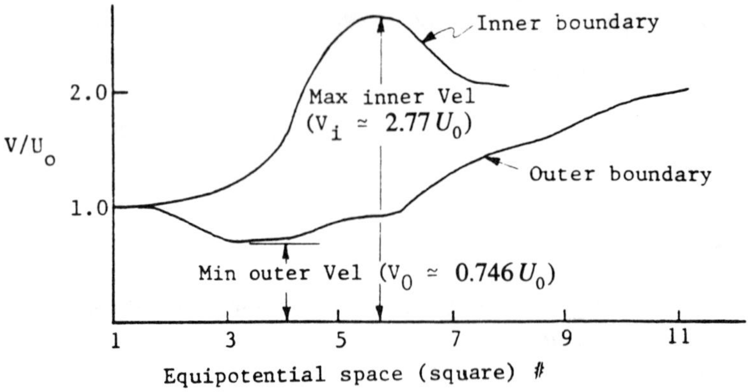

$$\frac{9.77 \text{ mm}}{3.50 \text{ mm}} = 2.77$$

$$\frac{9.7 \text{ mm}}{13.0 \text{ mm}} = 0.746$$

Ratio $\dfrac{\text{max. inner vel.}}{\text{min. outer vel.}} = \dfrac{V_i}{V_o} \approx \dfrac{2.77U_0}{0.764U_0} = 3.74$, or V_i is about 274% greater than V_0 ◄

Note: Answers may vary somewhat due to scaling.

4.9 *Consider the two-dimensional flow about a 2-in-diameter cylinder. Sketch the flow net for the ideal flow around one quarter of the cylinder. Start with a uniform net of ½-in squares, and fill in with ¼-in squares where desirable. (Note: It can be proved by classical hydrodynamics that the velocity tangent to the cylinder at a point 90° from the stagnation point is twice the uniform velocity.) From the flow net, determine the velocities (relative to U_0) along the center streamline from a point upstream where the velocity is uniform to the stagnation point, and then along the boundary of the cylinder from the stagnation point to the 90° point; plot them vs distance. By plotting a second curve on the same graph, compare the result thus obtained with the values given by the equation $V = 2U_0 \sin\theta$, where U_0 is the undisturbed stream velocity and θ is the angle subtended by the arc from the stagnation point to any point on the cylinder where V is desired.*

BG

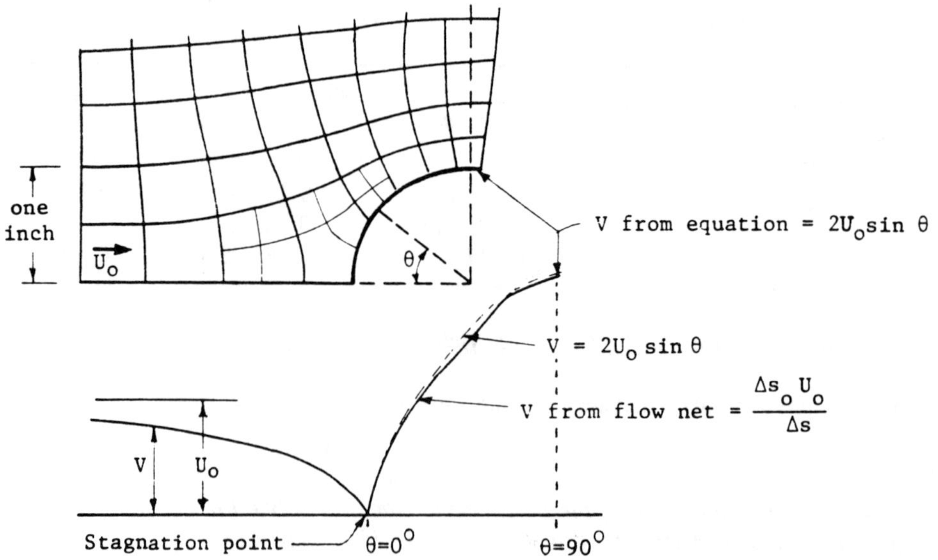

Sec 4.12: Velocity and Acceleration in Steady Flow -- *Exercises*

4.12.1 *A flow field is defined by u = 2, v = 3, w = 4. What is the velocity of flow? Specify units in terms of L and T.*

N

Sec. 4.12: $V = (2^2 + 3^2 + 4^2)^{1/2} = 5.39\ L/T$ ◄

4.12.2 *The velocity along a streamline lying on the x axis is given by $u = 6 + x^{0.4}$. What is the convective acceleration at $x = 5$? Specify units in terms of L and T. Assuming the fluid is incompressible, is the flow converging or diverging?*

N

$u = 6 + x^{0.4}$ ∴ $\partial u/\partial x = 0.4\, x^{-0.6}$

Eq. 4.23*a*: Conv. accel. $= u(\partial u/\partial x) = 2.4x^{-0.6} + 0.4x^{-0.2}$

At $x = 5$: Conv. accel. $= 2.4(5)^{-0.6} + 0.4(5)^{-0.2} = 1.204\ L/T^2$ ◄

If the fluid is incompressible, because the velocity increases the flow must be converging. ◄

4.12.3 *A flow field is defined by $u = 2y$, $v = x$. Derive expressions for the x and y components of acceleration. Find the magnitude of the velocity and acceleration at the point (3,1). Specify units in terms of L and T.*

N

$$\frac{\partial u}{\partial x} = 0;\ \frac{\partial u}{\partial y} = 2;\ a_x = u\frac{\partial u}{\partial x} + v\frac{\partial u}{\partial y} = 0 + 2v = 2x$$ ◄

$$\frac{\partial v}{\partial x} = 1;\ \frac{\partial v}{\partial y} = 0;\ a_y = u\frac{\partial v}{\partial x} + v\frac{\partial v}{\partial y} = u + 0 = 2y$$ ◄

At (3,1): $u = 2y = 2;\ v = x = 3;\ V = \sqrt{2^2 + 3^2} = 3.61\ L/T$ ◄

and $a_x = 2x = 6;\ a_y = 2y = 2;\ a = \sqrt{6^2 + 2^2} = 6.32\ L/T^2$ ◄

4.12.4 *A flow field is defined by $u = 2x$, $v = y$. Derive expressions for the x and y components of acceleration. Find the magnitude of the velocity and acceleration at the point (3,2). Specify units in terms of L and T.*

N

$$\frac{\partial u}{\partial x} = 2;\ \frac{\partial u}{\partial y} = 0;\ a_x = u\frac{\partial u}{\partial x} + v\frac{\partial u}{\partial y} = 2u + 0 = 4x$$ ◄

$$\frac{\partial v}{\partial x} = 0;\ \frac{\partial v}{\partial y} = 1;\ a_y = u\frac{\partial v}{\partial x} + v\frac{\partial v}{\partial y} = 0 + v = y$$ ◄

At (3,2): $u = 2x = 6;\ v = y = 2;\ V = \sqrt{6^2 + 2^2} = 6.32\ L/T$ ◄

and $a_x = 4x = 12;\ a_y = y = 2;\ a = \sqrt{12^2 + 2^2} = 12.17\ L/T^2$ ◄

4.12.5 *A flow field is defined by $u = 2y$, $v = xy$. Derive expressions for the x and y components of acceleration. Find the magnitude of the velocity and acceleration at the point (2,3). Specify units in terms of L and T.*

N

$$\frac{\partial u}{\partial x} = 0;\ \frac{\partial u}{\partial y} = 2;\ a_x = u\frac{\partial u}{\partial x} + v\frac{\partial u}{\partial y} = 0 + 2xy = 2xy$$ ◄

$$\frac{\partial v}{\partial x} = y;\ \frac{\partial v}{\partial y} = x;\ a_y = u\frac{\partial v}{\partial x} + v\frac{\partial v}{\partial y} = 2y^2 + x^2y$$ ◄

At (2,3): $u = 2y = 6;\ v = xy = 6;\ V = \sqrt{6^2 + 6^2} = 8.49\ L/T$ ◄

and $a_x = 2xy = 12;\ a_y = 2y^2 + x^2y = 2(3)^2 + (2)^2(3) = 18 + 12 = 30$

$a = \sqrt{12^2 + 30^2} = 32.3\ L/T^2$ ◄

4.12.6 *A flow field is defined by u = 3y, υ = 2xy, w = 4z. Derive expressions for the x, y, and z components of acceleration. Find the magnitude of the velocity and acceleration at the point (1,2,1). Specify units in terms of L and T.*

N

$$\frac{\partial u}{\partial x} = 0; \quad \frac{\partial u}{\partial y} = 3; \quad \frac{\partial u}{\partial z} = 0; \quad a_x = 0 + 3\upsilon + 0 = 6xy \quad ◄$$

$$\frac{\partial \upsilon}{\partial x} = 2y; \quad \frac{\partial \upsilon}{\partial y} = 2x; \quad \frac{\partial \upsilon}{\partial z} = 0; \quad a_y = 2xu + 2x\,\upsilon + 0 = 6y^2 + 4x^2y \quad ◄$$

$$\frac{\partial w}{\partial x} = 0; \quad \frac{\partial w}{\partial y} = 0; \quad \frac{\partial w}{\partial z} = 4; \quad a_z = 0 + 0 + 4(4z) = 16z \quad ◄$$

At (1,2,1): $u = 3y = 6; \ \upsilon = 2xy = 4; \ w = 4z = 4; \ V = \sqrt{6^2 + 4^2 + 4^2} = 8.25\ L/T$ ◄

and $a_x = 6xy = 12; \ a_y = 6y^2 + 4x^2y = 24 + 8 = 32; \ a_z = 16;$

$$a = \sqrt{12^2 + 32^2 + 16^2} = 37.7\ L/T^2 \quad ◄$$

4.12.7 *The velocity along a circular streamline of radius 4.5 ft is 2.5 fps. Find the normal and tangential components of the acceleration if the flow is steady.*

BG

Eq. 4.19: $a_n = \frac{V^2}{r} = \frac{2.5^2}{4.5} = 1.389\ \text{ft/sec}^2$ ◄

Flow is steady (given), so V_t = const, $a_t = 0$ ◄

4.12.8 *The velocity along a circular streamline of radius 2 m is 0.8 m/s. Find the normal and tangential components of the acceleration if the flow is steady.*

SI

Eq. 4.19: $a_n = \frac{V^2}{r} = \frac{0.8^2}{2} = 0.320\ \text{m/s}^2$ ◄

Flow is steady (given), so V_t = const, $a_t = 0$ ◄

4.12.9 *A large tank contains an ideal liquid which flows out of the bottom of the tank through a 4.5-in-diameter hole. The rate of steady outflow is 6 cfs. Assume that the liquid approaches the center of the hole radially. Find the velocities and convective accelerations at points that are 2 and 3 ft from the center of the hole.*

BG

Area through which (steady) flow occurs is a hemispherical surface, area $2\pi r^2$ (Table A.8)

$$V_r = \frac{Q}{A} = \frac{6}{2\pi r^2}; \quad \frac{\partial V}{\partial r} = -\frac{6}{\pi r^3}$$

At $r_2 = 2$ ft: $V_2 = 0.239$ fps ◄ ; At $r_3 = 3$ ft: $V_3 = 0.1061$ fps ◄

$$a_r = V\left(\frac{\partial V}{\partial s}\right) = V\left(-\frac{\partial V}{\partial r}\right) = \left(\frac{6}{2\pi r^2}\right)\left(\frac{+6}{\pi r^3}\right) = +\frac{18}{\pi^2 r^5}$$

$a_2 = \frac{18}{\pi^2 2^5} = 0.0570\ \text{ft/sec}^2$ ◄; $a_3 = 0.007\ 51\ \text{ft/sec}^2$ ◄

4.12.10 *An ideal liquid flows out the bottom of a large tank though an 8-cm-diameter hole at a steady rate of 0.30 m^3/s. Assume the liquid approaches the center of the hole radially. Find the velocities and convective accelerations at points 0.5 and 1.0 m from the center of the hole.*

SI

Area through which (steady) flow occurs is a hemispherical surface, area $2\pi r^2$ (Table A.8)

$$V = \frac{Q}{A} = \frac{0.30}{2\pi r^2}, \text{ directed only along } r.$$

$$\therefore \text{ From Eq. 4.23: Convective } a = V\left(-\frac{\partial V}{\partial r}\right) = \frac{Q}{2\pi r^2} \times \frac{Q}{\pi r^3} = \frac{0.30^2}{2\pi^2 r^5}$$

At r = 0.5 m: V = 0.1910 m/s ◄; a = 0.1459 m/s^2 ◄

At r = 1.0 m: V = 0.0477 m/s ◄; a = 0.004 56 m/s^2 ◄

4.12.11 *The steady flow rate in each of the four stream tubes of Fig. 4.10 is 40 cfs per foot perpendicular to the plane of the figure. By scaling, the dimensions of the shaded "square" have been found to be 1.65 ft wide on the upstream face, 1.53 ft wide on the downstream face, and 1.67 ft along the flow line through its center; the radius of that flow line measures 11.1 ft. Find the normal, tangential, and resultant accelerations of a fluid particle at the center of the shaded area.*

Fig. 4.10 appears with the solution of Exer. 4.10.3.

BG

At the center of the shaded "square": Width ≈ (1.65 + 1.53)/2 = 1.59 ft

$$V = Q/A \approx 40/1.59 = 25.2 \text{ fps}$$

Eq. 4.25: $a_n = \dfrac{V^2}{r} \approx \dfrac{25.2^2}{11.1} = 57.0 \approx 57$ ft/sec^2 radially inward ◄

In the shaded "square", the tube width changes from 1.65 ft to 1.53 ft over a distance of 1.67 ft, so

$$\frac{\partial V}{\partial s} \approx \frac{(40/1.53) - (40/1.65)}{1.67} = 1.139 \text{ ft/sec}^2$$

Eq. 4.24: $a_t = V\dfrac{\partial V}{\partial s} \approx 25.2(1.139) = 28.6 \approx 29$ ft/sec^2 in the flow direction ◄

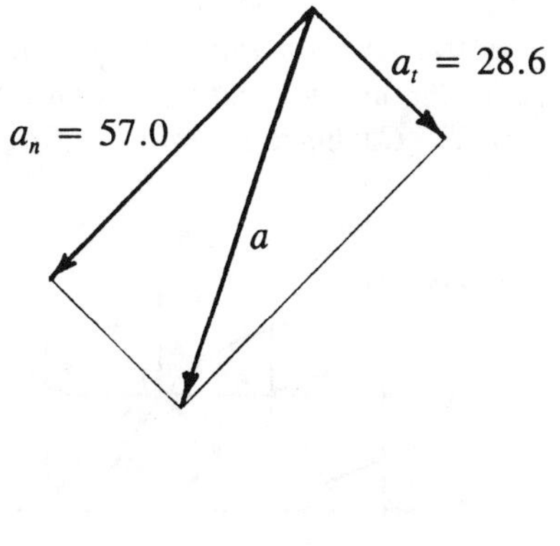

$a \approx \sqrt{57.0^2 + 28.6^2} = 63.8 \approx 64$ ft/sec^2 (downward) ◄

4.10 *Sketch the flow field defined by $u = 0$, $v = 3xy$, and derive expressions for the x and y components of acceleration. Find the acceleration at the point (2,2). Specify units in terms of L and T.*

N

Eq. 4.23*a*: $a_x = 0(0) + 3xy(0) = 0$ ◀

Eq. 4.23*b*: $a_y = 0(3y) + 3xy(3x) = 9\,x^2y$ ◀

At (2,2): $a = 9(2)^2(2) = 72\ L/T^2$ ◀

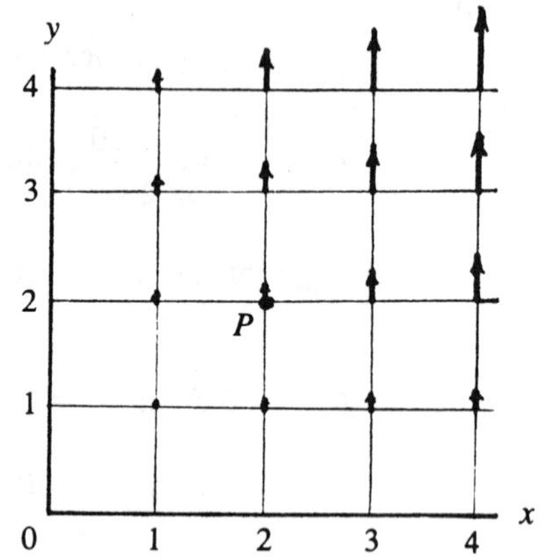

4.11 *Sketch the flow field defined by $u = 3y$, $v = 2$, and derive expressions for the x and y components of acceleration. Find the magnitude of the velocity and acceleration for the point having the coordinates (3,4). Specify units in terms of L and T.*

N

Eq. 4.23*a*: $a_x = 3y(0) + 2(3) = 6\ L/T^2$ ◀

Eq. 4.23*b*: $a_y = 3y(0) + 2(0) = 0$ ◀

At (3,4): $V = \sqrt{(3y)^2 + 2^2}$

$= \sqrt{12^2 + 2^2} = 12.2\ L/T$ ◀

$a = \sqrt{6^2 + 0^2} = 6\ L/T^2$ ◀

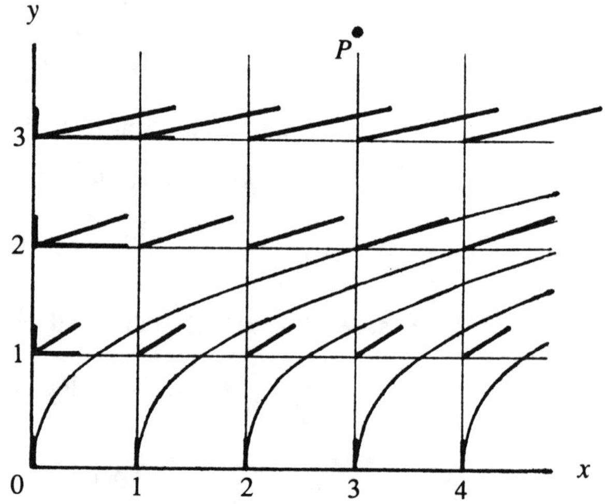

4.12 *Sketch the flow field defined by $u = -2y$, $v = 3x$, and derive expressions for the x and y components of acceleration. As in Sample Prob. 4.3, find approximate values of the normal and tangential accelerations of the particle at the point (2,3). Specify units in terms of L and T. Compare the values of $(a_n^2 + a_t^2)^{1/2}$ with the computed value $(a_x^2 + a_y^2)^{1/2}$.*

N

Eq. 4.23*a*: $a_x = -2y(0) + 3x(-2) = -6x$ ◀

Eq. 4.23*b*: $a_y = -2y(3) + 3x(0) = -6y$ ◀

$V = \sqrt{(2y)^2 + 3x^2} = \sqrt{4y^2 + 9x^2}$

At P(2,3):

$V = \sqrt{4(3)^2 + 9(2)^2} = 8.49\ L/T$ ◀

$a = \sqrt{(-6 \times 2)^2 + (-6 \times 3)^2} = 21.6\ L/T^2$ ◀

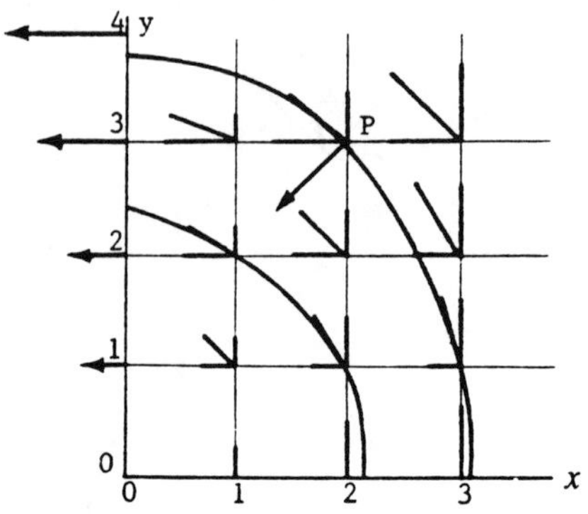

/cont...

$r \approx 3.3L$; Per Sample Prob. 4.3:

$$\frac{\partial V}{\partial s} \approx \frac{V_1 - V_2}{s_1 - s_2} \approx \frac{9.3 - 7.6}{-5.4} = -0.315\ T^{-1}$$

Eq. 4.25: $a_n = \frac{V^2}{r} \approx \frac{8.49^2}{3.3} = 21.8\ L/T^2$ ◀

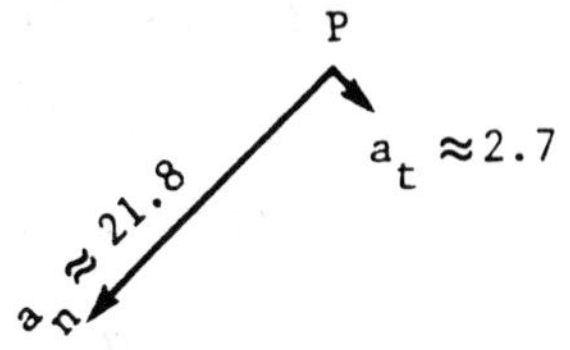

$a_t = V(\partial V/\partial s) \approx 8.49(-0.315) = -2.67 \approx -2.7\ L/T^2$ ◀

$a = \sqrt{a_n^2 + a_t^2} \approx \sqrt{21.8^2 + 2.67^2} = 22.0 \approx 22\ L/T^2$

Cf. $a = 21.6\ L/T^2$, these two accelerations compare well ◀

4.13 *Figure P4.13 shows to scale a two-dimensional stream tube. If the flow rate is 40 m³/s per meter perpendicular to the plane of the sketch, determine approximate values of the normal and tangential accelerations of a fluid particle at A. What is the resultant acceleration of a particle at A?*

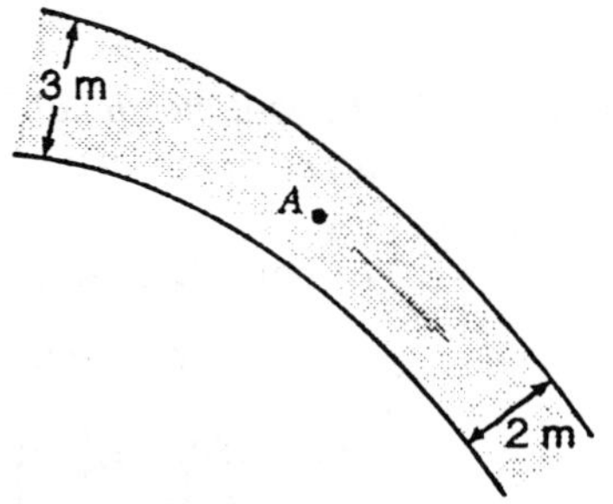

Figure P4.13

SI

Note to instructor: Answers will vary depending on the measured dimensions. The variability of the results can be used to demonstrate the sensitivity to the accuracy of the measurements.

At A, width $\approx$ 2.3 m, $r \approx$ 14.7 m.

$$V_A = \frac{Q}{A} \approx \frac{40}{2.3} = 17.4 \text{ m/s}$$

Eq. 4.25: $a_n = \frac{V_A^2}{r} \approx \frac{17.4^2}{14.7} = 20.6 \approx 21$ m/s² radially inward ◀

In the 2 m surrounding A the width changes from about 2.35 m to 2.27 m, so

$$\frac{\partial V}{\partial s} \approx \frac{(40/2.27) - (40/2.35)}{2} = 0.300 \text{ m/s per m}$$

Eq. 4.24: $a_t = V\frac{\partial V}{\partial s} \approx 17.4(0.300) = 5.22 \approx 5.2$ m/s² tangentially downward to the right ◀

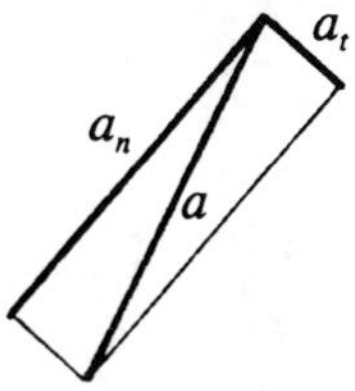

$a \approx \sqrt{20.6^2 + 5.22^2} = 21.2 \approx 21$ m/s² ◀

Sec 4.13: Velocity and Acceleration in Unsteady Flow -- *Exercises*

4.13.1 *A flow is defined by $u = 2(1+t)$, $v = 3(1+t)$, $w = 4(1+t)$. What is the velocity of flow at the point (3,1,4) at $t = 2$? What is the acceleration at that point at $t = 2$? Specify units in terms of L and T.*

N

At $t = 2$ sec: $u = 6\ L/T$, $v = 9\ L/T$, $w = 12\ L/T$

$V_{3,1,4} = \sqrt{6^2 + 9^2 + 12^2} = 16.16\ L/T$ ◄

u, v and w are all not functions of x, y, or z, so all convective accelerations (see Eq. 4.28) are zero

$$\therefore\ a_x = \frac{\partial u}{\partial t} = 2\ L/T;\quad a_y = \frac{\partial v}{\partial t} = 3\ L/T^2;\quad a_z = \frac{\partial w}{\partial t} = 4\ L/T^2$$

and $a = \sqrt{2^2 + 3^2 + 4^2} = 5.39\ L/T^2$ ◄

4.13.2 *A two-dimensional flow field is given by $u = 3 + 2xy + 4t^2$, $v = xy^2 + 3t$. Find the velocity and acceleration of a particle of fluid at point (2,1) at $t = 5$. Specify units in terms of L and T.*

N

At point (2,1) with $t = 5$:

$u = 3 + 4 + 100 = 107\ L/T$, $v = 2 + 15 = 17\ L/T$; $V = \sqrt{u^2 + v^2} = 108.3\ L/T$ ◄

Eq. 4.28*a*: $a_x = (3 + 2xy + 4t^2)2y + (xy^2 + 3t)2x + 8t = 322\ L/T^2$

Eq. 4.28*b*: $a_y = (3 + 2xy + 4t^2)y^2 + (xy^2 + 3t)2xy + 3 = 178\ L/T^2$

Finally, $a = \sqrt{322^2 + 178^2} = 368\ L/T^2$ ◄

4.13.3 *The flow velocity in fps along a circular streamline of radius 4 ft is $0.8 + 1.5t$. Find the normal and tangential components of the acceleration when $t = 1.3$ sec.*

BG

At $t = 1.3$ sec: $V = 0.8 + 1.5(1.3) = 2.75$ fps ; at any t: $\partial V/\partial t = 1.5$ ft/sec^2

Eq. 4.25: $a_n = V^2/r = 2.75^2/4 = 1.891$ ft/sec^2 ◄

Eq. 4.29: $a_t = V\dfrac{\partial V}{\partial s} + \dfrac{\partial V}{\partial t} = V(0) + 1.5 = 1.500$ ft/sec^2 ◄

4.13.4 *The flow velocity in m/s along a circular streamline of radius 1.6 m is $0.25 + 0.5t$. Find the normal and tangential components of the acceleration when $t = 1.4$ s.*

SI

At $t = 1.5$ sec: $V = 0.25 + 0.5(1.5) = 0.950$ m/s

Eq. 4.25: $a_n = V^2/r = 0.950^2/1.6 = 0.564$ m/s^2 ◄

Eq. 4.29: $a_t = V\dfrac{\partial V}{\partial s} + \dfrac{\partial V}{\partial t} = V(0) + 0.5 = 0.500$ m/s^2 ◄

4.14 *A large tank contains an ideal liquid which flows out of the bottom through a 4.5-in-diameter hole. The outflow rate $Q = 6 - 0.3t$, where Q is in cfs and t is in sec. Assume the liquid approaches the center of the hole radially. Find the local acceleration at a point 2 ft from the center of the hole at times $t = 8$ sec and 12 sec. What is the total acceleration at a point 3 ft from the center of the hole at $t = 12$ sec?*

BG

Area through which flow occurs is a hemispherical surface, area $2\pi r^2$ (Table A.8)

$$V = \frac{Q}{A} = \frac{(6 - 0.3t)}{2\pi r^2}; \text{ local acceleration } = \frac{\partial V}{\partial t} = -\frac{0.3}{2\pi r^2}$$

At $r = 2$ ft, local accel $= \left(\frac{\partial V}{\partial t}\right)_2 = -\frac{0.3}{2\pi 2^2} = -0.011\ 94$ ft/sec^2 ◄

Local accel is indep of t, ∴ at $r = 2$, local accel = const.

∴ At $r = 2$ ft, $t = 8$ sec, local accel $= -0.011\ 94$ ft/sec^2 ◄

and at $r = 2$ ft, $t = 12$ sec, local accel $= -0.011\ 94$ ft/sec^2 ◄

$$\frac{\partial V}{\partial s} = -\frac{\partial V}{\partial r} = \frac{-(6 - 0.3t)}{\pi r^3}$$

Eq. 4.29 at $t = 12$ sec, $r = 3$ ft: Total $a = V\frac{\partial V}{\partial s} + \frac{\partial V}{\partial t}$

$$= \frac{6 - 0.3(12)}{2\pi 3^2} \times \frac{6 - 0.3(12)}{\pi 3^3} - \frac{0.3}{2\pi 3^2} = 0.001\ 201 - 0.005\ 31 = -0.004\ 10 \text{ ft/sec}^2$$ ◄

4.15 *An ideal liquid flows out of the bottom of a large tank through an 8-cm-diameter hole. The outflow rate $Q = 0.25 - 0.01t^{0.5}$, where Q is in m^3/s and t is in s. Assume the liquid approaches the center of the hole radially. Find the local and convective accelerations at a point 0.4 m from the center of the hole at time $t = 15$ s; what is the total acceleration?*

SI

Area through which flow occurs is a hemispherical surface, area $2\pi r^2$ (Table A.8)

$$V = \frac{Q}{A} = \frac{0.25 - 0.01t^{0.5}}{2\pi r^2}, \text{ directed only along } r.$$

$$\left(\frac{\partial V}{\partial r}\right) = -\frac{0.25 - 0.01t^{0.5}}{\pi r^3}; \quad \frac{\partial V}{\partial t} = -\frac{0.005t^{-0.5}}{2\pi r^2}$$

∴ From Sec. 4.13, at $r = 0.4$ m, $t = 15$ s: Local accel $= \partial V/\partial t = -0.001\ 284$ m/s^2 ◄

and since $s = \text{const} - r$, convective accel $= V(\partial V/\partial s) = -V(\partial V/\partial r) = 0.221$ m/s^2 ◄

and, from Eq. 4.29: Total accel $= 0.221 - 0.001 = 0.220$ m/s^2 ◄

4.16 *Refer to the two-dimensional stream tube drawn to scale in Fig. P4.13. If the flow rate is $(20 - 3t)$ m^3/s per meter perpendicular to the plane of the sketch, with t in s, find approximate values of the normal, tangential, and total accelerations of a fluid particle at A when $t = 4$ s.*

See Fig. P4.13 with solution to Prob. 4.13

SI

Note to instructor: Answers will vary depending on the measured dimensions. The variability of the results can be used to demonstrate the sensitivity to the accuracy of the measurements.

At A, width ≈ 2.31 m, r ≈ 14.67 m.

/cont...

$$V_A = \frac{Q}{A} \approx \frac{(20 - 3t)}{2.31}; \quad \frac{\partial V_A}{\partial t} \approx \frac{-3}{2.31} = -1.299 \text{ m/s (at any } t\text{)}$$

In the 2 m surrounding A the width changes from about 2.35 m to 2.27 m.

$$\frac{\partial V_A}{\partial s} \approx \frac{(Q/2.27) - (Q/2.35)}{2}$$

When $t = 4$ s: $Q = 8$ m^3/s, $V_A \approx 3.46$ m/s, $\partial V_A/\partial s \approx 0.0600$ m/s per m.

Eq. 4.25: $(a_A)_n = \frac{V_A^2}{r} \approx \frac{3.46^2}{14.67} = 0.818 \approx 0.82$ m/s^2 radially inward ◄

Eq. 4.29: $(a_A)_t = V\left(\frac{\partial V}{\partial s}\right) + \frac{\partial V}{\partial t} \approx 3.46(0.0600) + (-1.299) = -1.091 \approx -1.1$ m/s^2 ◄

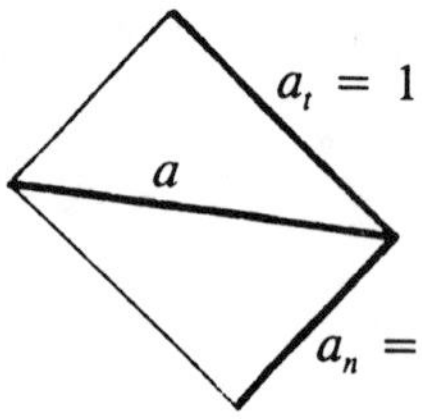

$a \approx \sqrt{1.091^2 + 0.818^2} = 1.363 \approx 1.36$ m/s^2 ◄

Chapter 4: Miscellaneous -- ***Problems***

4.17 *Assume that the streamlines for a two-dimensional flow of a fricitonless incompressible fluid against a flat plate normal to the initial velocity may be represented by the equation xy = constant and that the flow is symmetrical about the plane through x = 0. A different streamline may be plotted for each value of the constant. Using a scale of 1 in = 6 units of distance, plot streamlines for values of the constant of 16, 64, and 128.*

N

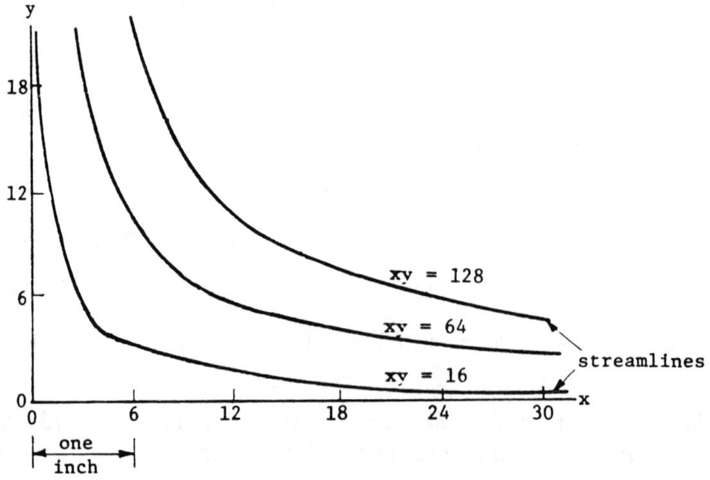

◄

4.18 *For the case in Prob. 4.17, it can be shown that the velocity components at any point are $u = ax$ and $v = -ay$, where a is a constant. Thus the actual velocity is $V = a(x^2+y^2)^{1/2} = ar$, where r is the radius to the origin. Let $a = 1/3$; then if 1 in = 6 ft for the streamlines, 1 in = 2 fps for the velocity scale. Draw curves of equal velocity for values of 2, 4, 6, 8, and 10 fps. How does the velocity vary along the surface of the plate?*

BG

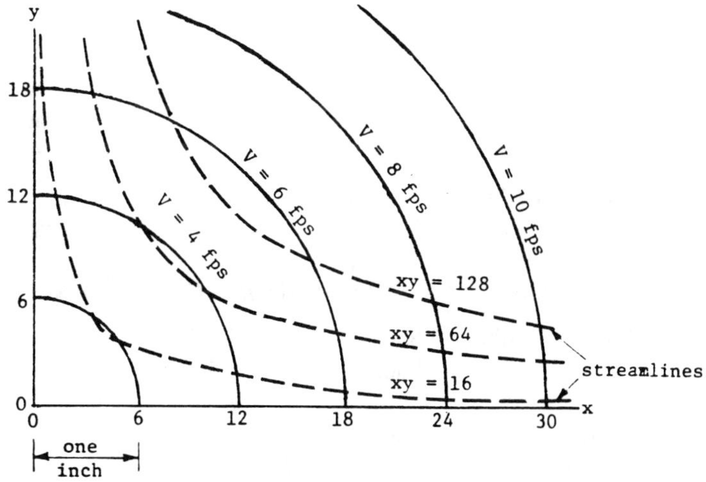

◄

Along surface of plate $V = x/3$ fps, where x ft = distance from origin. ◄

4.19 *For three-dimensional flow with the y axis as the centerline, assume that the equation for the bounding streamline of a jet impinging vertically downward on a flat plate is $x^2y = 64$. (a) Plot the flow showing the centerline and bounding streamlines of the jet. (b) What is the approximate average velocity in the vertical jet at $y = 10$ if the average velocity in the vertical jet is 5.0 m/s at $y = 16$? (c) For the above conditions find the approximate velocity along the plate at $r = 12, 24, 36$.*

SI

$x^2y = 64$

(*a*)

y	x
20	1.79
15	2.07
10	2.53
5	3.58
3	4.62
2	5.66
1	8
0.444	12

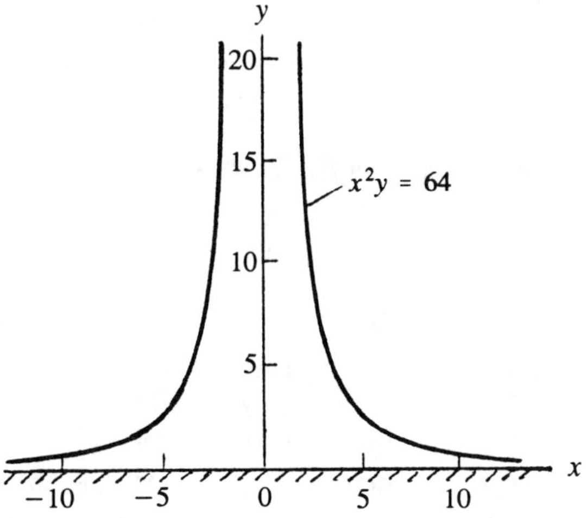

◄

(*b*) At $y = 10$, $x = \sqrt{6.4} = 2.53$

At $y = 16$, $x = 2.00$

$$Q = AV = \pi R_{10}^2 V_{10} = \pi R_{16}^2 V_{16}$$

$$Q = \pi(6.4)V_{10} = \pi(2)^2 5 = 20\pi$$

$V_{10} = 20/6.4 = 3.125$ m/s ◄

(*c*) At $x = 12$, $y = 64/144 = 0.444$; At $x = 24$, $y = 0.111$; At $x = 36$, $y = 0.0494$

$$Q = 20\pi = 2\pi ryV = 2\pi(12)(0.444)V_{12} = 2\pi(24)(0.111)V_{24} = 2\pi(36)(0.0494)V_{36}$$

$V_{12} = 1.875$ m/s ◄ $V_{24} = 3.75$ m/s ◄ $V_{36} = 5.63$ m/s ◄

Chapter 5

Energy Considerations in Steady Flow

PROBLEM SELECTION GUIDE

Sec	Exer/Prob	Units	Difficulty	Length	Parts	Similar	Special features
5.1	*Kinetic Energy of a Flowing Fluid*						
	X[1]5.1.1	N	Medium	Medium	1		Integration
	5.1.2	N	Medium	Medium	1		Integration
	5.1.3	N	Medium	Medium	1		Integration
	P 5.1	N	Medium	Medium	1		Integration
	5.2	N	Hard	Long	1		Integration
5.5	*Energy Equation for Steady Flow of Incompressible Fluids, and Bernoulli's Theorem*						
	X 5.5.1	BG	Medium	Short	1	5.5.2	
	5.5.2	SI	Medium	Short	1	5.5.1	
	5.5.3	BG	Easy	Short	2	5.5.4	
	5.5.4	SI	Easy	Short	2	5.5.3	
	5.5.5	BG	Easy	Short	2		
	5.5.6	BG	Easy	Short	2		
	5.5.7	SI	Easy	Short	1		
	P 5.3	BG	Medium	Medium	2		
	5.4	SI	Medium	Short	1		* Uses Sec. 4.5
	5.5	BG	Easy	Short	1		
	5.6	SI	Easy	Short	1		
5.6	*Energy Equation for Steady Flow of Compressible Fluids*						
	X 5.6.1	BG	Easy	Short	2	5.6.2	Uses $pv = RT$ (Sec. 2.7)
	5.6.2	SI	Easy	Short	2	5.6.1	Uses $pv = RT$ (Sec. 2.7)
	5.6.3	BG	Medium	Medium	2	5.6.4	Uses $pv = RT$ (Sec. 2.7)
	5.6.4	SI	Medium	Medium	2	5.6.3	Uses $pv = RT$ (Sec. 2.7)
5.7	*Euler's Equation for Steady Motion of an Ideal Fluid Along a Streamline*						
	X 5.7.1	BG	Easy	Short	1	5.7.2	
	5.7.2	SI	Easy	Short	1	5.7.1	
5.8	*Pressure in Fluid Flow*						
	X 5.8.1	BG	Easy	Short	1	5.8.2-3	
	5.8.2	SI	Easy	Short	1	5.8.1-3	
	5.8.3	BG	Easy	Short	1	5.8.1-2	
	P 5.7	BG	Hard	Medium	2		Integration
	5.8	BG	Medium	Long	1		
	5.9	SI	Medium	Medium	1		
	5.10	SI	Medium	Short	1		

/cont...

[1] For all Exercises (marked "X"), answers are given in Appendix D of the textbook.
* Note a correction made to early printings of the textbook.
X = Exercise, P = (end-of-chapter) Problem, S = Sample Problem.

Sec	Exer/Prob	Units	Difficulty	Length	Parts	Similar	Special features
5.10	*Power Considerations in Fluid Flow*						
	X 5.10.1	BG	Easy	Medium	1	5.10.2	
	5.10.2	SI	Easy	Medium	1	5.10.1	
	5.10.3	BG	Easy	Medium	1	5.10.4	
	5.10.4	SI	Easy	Medium	1	5.10.3	
	5.10.5	BG	Easy	Short	1		
	P 5.11	BG	Easy	Medium	1	5.12	
	5.12	SI	Easy	Medium	1	5.11	
5.11	*Cavitation*						
	X 5.11.1	BG	Easy	Medium	1	5.11.2	Uses Sec. 3.5
	5.11.2	SI	Easy	Medium	1	5.11.1	Uses Sec. 3.5
	5.11.3	BG	Easy	Medium	1	5.11.4	Uses Sec. 3.5
	5.11.4	SI	Easy	Medium	1	5.11.3	Uses Sec. 3.5
	5.11.5	SI	Medium	Medium	1		Uses Sec. 3.5
5.13	*Loss of Head at Submerged Discharge*						
	P 5.13	BG	Hard	Medium	1	5.14	Confirm assumed temperature; interp.
	5.14	SI	Hard	Medium	1	5.13	Confirm assumed temperature; interp.
	5.15	BG	Medium	Medium	1		
	5.16	BG	Medium	Medium	1	5.17	
	5.17	SI	Medium	Medium	1	5.16	
	5.18	BG	Medium	Medium	1	5.19	
	5.19	SI	Medium	Medium	1	5.18	
5.14	*Applications of Hydraulic Grade Line and Energy Line*						
	X 5.14.1	BG	Easy	Medium	1	5.14.2	Sketch
	5.14.2	SI	Easy	Medium	1	5.14.1	Sketch
	5.14.3	SI	Easy	Short	1	5.14.4	
	5.14.4	BG	Easy	Short	1	5.14.3	
	5.14.5	SI	Easy	Short	1	5.14.6	
	5.14.6	BG	Easy	Short	1	5.14.5	
	5.14.7	BG	Easy	Medium	1	5.14.8	
	5.14.8	SI	Easy	Medium	1	5.14.7	
	P 5.20	SI	Easy	Medium	1		
	5.21	BG	Easy	Medium	2		
	5.22	SI	Easy	Medium	1		
	5.23	BG	Easy	Medium	2	5.24	Cavitation
	5.24	SI	Easy	Medium	2	5.23	Cavitation
	5.25	BG	Easy	Medium	1		Sketch
	5.26	BG	Medium	Long	1	S5.10	Plot
	5.27	BG	Easy	Short	1	5.28	
	5.28	SI	Easy	Short	1	5.27	
	5.29	BG	Medium	Medium	1		Uses $pv = RT$ (Sec. 2.7). Interpolation.
	5.30	BG	Medium	Medium	1		

/cont...

Sec	Exer/Prob	Units	Difficulty	Length	Parts	Similar	Special features
5.15	*Method of Solution of Flow Problems*						
	X 5.15.1	BG	Easy	Short	1	5.15.2	
	5.15.2	SI	Easy	Short	1	5.15.1	
	5.15.3	BG	Medium	Medium	1	5.15.4	T & E (Trial and Error)
	5.15.4	SI	Medium	Medium	1	5.15.3	T & E
5.16	*Jet Trajectory*						
	X 5.16.1	SI	Medium	Short	1		
	5.16.2	SI	Easy	Short	4		
	5.16.3	BG	Medium	Short	1		
	P 5.31	N	Easy	Medium	1		Proof
	5.32	BG	Medium	Medium	1		Differentiation
5.17	*Flow in Curved Path*						
	X 5.17.1	BG	Easy	Medium	1		Integration
	5.17.2	SI	Easy	Short	1		
	P 5.33	BG	Hard	Medium	1	5.34	Integration
	5.34	SI	Hard	Medium	1	5.33	Integration
5.18	*Forced or Rotational Vortex*						
	X 5.18.1	BG	Easy	Short	3		
	5.18.2	SI	Easy	Short	1		
	5.18.3	BG	Easy	Short	1		
	P 5.35	BG	Easy	Short	1	5.36	
	5.36	SI	Easy	Short	1	5.35	
5.19	*Free or Irrotational Vortex*						
	X 5.19.1	SI	Easy	Medium	1	S5.13, 5.19.2	
	5.19.2	BG	Easy	Medium	1	S5.13, 5.19.1	
	P 5.37	BG	Medium	Medium	1	5.38	Integration
	5.38	SI	Medium	Medium	1	5.37	Integration
Ch.5	*Miscellaneous*						
	P 5.39	BG	Medium	Medium	2	5.40	Uses manometry (Sec. 3.6)
	5.40	SI	Medium	Medium	2	5.39	Uses manometry (Sec. 3.6)

Chapter 5
ENERGY CONSIDERATIONS IN STEADY FLOW

Sec. 5.1: Kinetic Energy of a Flowing Fluid -- *Exercises*

5.1.1 *Assume the velocity profile for turbulent flow in a circular pipe to be approximated by a parabola from the axis to a point very close to the wall where the local velocity is $u = 0.7u_m$, where u_m is the maximum velocity at the axis. The equation for this parabola is $u = u_m[1 - 0.3(r/r_0)^2]$. Find α.*

N

$u = u_m[1 - 0.3(r/r_0)^2]$; $dA = 2\pi r\,dr$

$$Q = AV = \int_0^{r_0} u\,dA = \int_0^{r_0} u_m\left[1 - 0.3\frac{r^2}{r_0^2}\right]2\pi r\,dr = 2\pi u_m\int_0^{r_0}\left[r - \frac{0.3r^3}{r_0^2}\right]dr$$

$$= 2\pi u_m\left[\frac{r^2}{2} - \frac{0.3r^4}{4r_0^2}\right]_0^{r_0} = 2\pi u_m\left[\frac{r_0^2}{2} - \frac{3}{40}r_0^2\right] = 2\pi u_m\left[\frac{17}{40}\right]r_0^2$$

$$V = \frac{Q}{A} = \frac{2\pi u_m[17/40]r_0^2}{\pi r_0^2} = 2u_m\left(\frac{17}{40}\right) = 0.85u_m$$

$$\alpha = \frac{\int u^3 dA}{AV^3} = \frac{2\pi u_m^3\int_0^{r_0}[1 - 0.3(r/r_0)^2]^3 r\,dr}{\pi r_0^2(0.85u_m)^3} = \frac{3.26}{r_0^2}\int_0^{r_0}\left[r - 3r(0.3)\frac{r^2}{r_0^2} + 3r(0.3)^2\frac{r^4}{r_0^4} - r(0.3)^3\frac{r^6}{r_0^6}\right]dr$$

$$= \frac{3.26}{r_0^2}\left[\frac{1}{2} - \frac{0.9}{4} + \frac{0.27}{6} - \frac{0.027}{8}\right]r_0^2 = 1.031 \quad ◄$$

5.1.2 *Assume an open rectangular channel with the velocity at the surface twice that at the bottom and with the velocity varying as a straight line from top to bottom. Find α.*

N

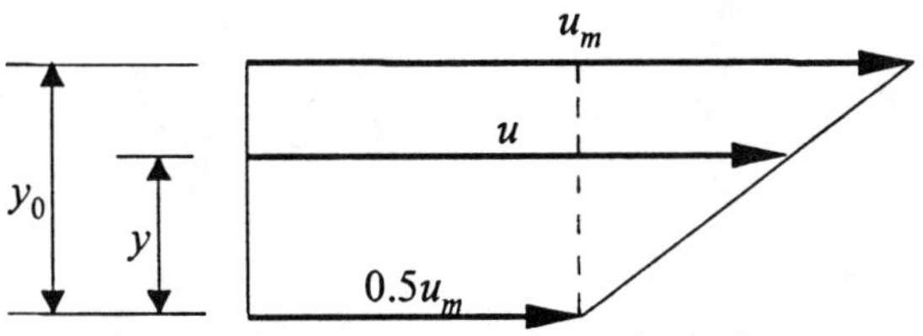

Similar Δ's: $\dfrac{u_m - 0.5u_m}{y_0} = \dfrac{u - 0.5u_m}{y}$

from which $u = \dfrac{u_m}{2}\left(\dfrac{y}{y_0} + 1\right)$

Mean velocity $V = (3/4)u_m$; so $u = \dfrac{2}{3}V\left(\dfrac{y}{y_0} + 1\right)$

Let channel width $= B$; $dA = B\,dy$

$$\int u^3 dA = \int_0^{y_0}\left[\frac{u_m}{2}\left(\frac{y}{y_0} + 1\right)\right]^3 B\,dy = \frac{Bu_m^3}{8}\int_0^{y_0}\left(\frac{y}{y_0} + 1\right)^3 dy = \frac{Bu_m^3}{8}\int_0^{y_0}\left(\frac{y^3}{y_0^3} + 3\frac{y^2}{y_0^2} + 3\frac{y}{y_0} + 1\right)dy$$

$$= \frac{Bu_m^3}{8}\left[\frac{y^4}{4y_0^3} + \frac{3y^3}{3y_0^2} + \frac{3y^2}{2y_0} + y\right]_0^{y_0} = \frac{Bu_m^3}{8}\left[\frac{1}{4} + 1 + \frac{3}{2} + 1\right]y_0 = \frac{Bu_m^3}{8}[3.75]y_0 = \frac{15}{32}By_0u_m^3$$

Eq 5.4: $$\alpha = \frac{\int u^3 dA}{AV^3} = \frac{15By_0u_m^3}{32(By_0)(3u_m/4)^3} = 10/9 \quad ◄$$

5.1.3 *Find α for the case of a two-dimensional laminar flow, as between two flat plates, for which the velocity profile is parabolic.*

N

Let y_0 = half-width of the symmetric flow. Measuring y from the centerline:

$$\left.\begin{array}{l}\text{when } y = 0,\ u = u_m \\ \text{when } y = y_0,\ u = 0\end{array}\right\}\quad u = u_m\left[1 - (y/y_0)^2\right]$$

Letting Q be the flow through the half-width:

$$Q = \int_0^{y_0} u\,dA = \int_0^{y_0} u_m\left[1 - \left(\frac{y}{y_0}\right)^2\right]B\,dy = Bu_m\int_0^{y_0}\left[1 - \frac{y^2}{y_0^2}\right]dy = Bu_m\left[y - \frac{y^3}{3y_0^2}\right]_0^{y_0} = Bu_m\left[1 - \frac{1}{3}\right]y_0$$

$$= \frac{2}{3}By_0u_m\ ;\quad V = \frac{Q}{A} = \frac{(2/3)By_0u_m}{By_0} = \frac{2}{3}u_m$$

$$\int u^3dA = Bu_m^3\int\left[1 - \left(\frac{y}{y_0}\right)^2\right]^3 dy = Bu_m^3\int_0^{y_0}\left[1 - 3\frac{y^2}{y_0^2} + 3\frac{y^4}{y_0^4} - \frac{y^6}{y_0^6}\right]dy = Bu_m^3\left[y - \frac{3y^3}{3y_0^2} + \frac{3y^5}{5y_0^4} - \frac{y^7}{7y_0^6}\right]_0^{y_0}$$

$$= By_0u_m^3\left[1 - 1 + \frac{3}{5} - \frac{1}{7}\right] = \frac{16}{35}By_0u_m^3$$

$$\text{Eq. 5.4:}\quad \alpha = \frac{\int u^3dA}{AV^3} = \frac{(16/35)By_0u_m^3}{(By_0)(2u_m/3)^3} = \frac{54}{35} = 1.543 \quad ◀$$

Sec. 5.1: Kinetic Energy of a Flowing Fluid -- *Problems*

5.1 *Assume the seventh-root law [Eq. (8.38)] for a turbulent-velocity distribution between two smooth flat plates. Find α.*

N

$$\text{From Eq. 8.38:}\quad u = u_m\left(1 - \frac{r}{r_0}\right)^{1/7};\quad dA = B\,dr$$

$$\text{For the half-flow:}\quad Q = AV = \int_0^{r_0} u\,dA = \int_0^{r_0} u_m\left(1 - \frac{r}{r_0}\right)^{1/7} B\,dr = Bu_m\int_0^{r_0}\left(1 - \frac{r}{r_0}\right)^{1/7}dr$$

$$= Bu_m\left(-\frac{7}{8}r_0\right)\left[\left(1 - \frac{r}{r_0}\right)^{8/7}\right]_0^{y_0} = -\frac{7}{8}Bu_mr_0[0 - 1] = \frac{7}{8}Bu_mr_0$$

But the half-area $A = Br_0$, so the mean velocity $V = \dfrac{Q}{A} = \dfrac{(7/8)Bu_mr_0}{Br_0} = \dfrac{7}{8}u_m$

$$\int u^3dA = \int u_m^3\left(1 - \frac{r}{r_0}\right)^{3/7}B\,dr = Bu_m^3\int\left(1 - \frac{r}{r_0}\right)^{3/7}dr$$

$$= Bu_m^3\left(-\frac{7}{10}r_0\right)\left[\left(1 - \frac{r}{r_0}\right)^{10/7}\right]_0^{r_0} = -\frac{7}{10}Bu_m^3r_0[0 - 1] = \frac{7}{10}Bu_m^3r_0$$

$$\text{Eq. 5.4:}\quad \alpha = \frac{\int u^3dA}{AV^3} = \frac{(7/10)Bu_m^3r_0}{(Br_0)(7u_m/8)^3} = \frac{8^3}{10(7)^2} = 1.045 \quad ◀$$

5.2 *Assume the seventh-root law [Eq. (8.38)] for a turbulent-velocity distribution in smooth pipe flow. Find α.*

N

From Eq. 8.38: $u = u_m\left(1 - \frac{r}{r_0}\right)^{1/7}$; $dA = 2\pi r\,dr$

$$Q = AV = \int u\,dA = \int u_m\left(1 - \frac{r}{r_0}\right)^{1/7} 2\pi r\,dr = 2\pi u_m\int r\left(1 - \frac{r}{r_0}\right)^{1/7} dr$$

To integrate by parts, let $u = r$ and $dv = (1 - r/r_0)^{1/7}dr$

then $du = dr$ and $v = -(7/8)r_0(1 - r/r_0)^{8/7}$

so that $$\int_0^{r_0} r\left(1 - \frac{r}{r_0}\right)^{1/7} dr = \left[-\frac{7}{8}rr_0\left(1 - \frac{r}{r_0}\right)^{8/7}\right]_0^{r_0} - \int_0^{r_0}\left[-\frac{7}{8}r_0\left(1 - \frac{r}{r_0}\right)^{8/7}\right]dr$$

$$= -[0 - 0] + \frac{7}{8}r_0\int_0^{r_0}\left(1 - \frac{r}{r_0}\right)^{8/7} dr = \frac{7}{8}r_0\left(-\frac{7}{15}r_0\right)\left[\left(1 - \frac{r}{r_0}\right)^{15/7}\right]_0^{r_0} = -\frac{49}{8(15)}r_0^2[0 - 1] = \frac{49}{120}r_0^2$$

so $AV = 2\pi u_m\frac{49}{120}r_0^2 = \frac{49}{60}\pi r_0^2 u_m$; but $A = \pi r_0^2$; $\therefore\ V = (49/60)u_m$

$$\int u^3 dA = \int u_m^3\left(1 - \frac{r}{r_0}\right)^{3/7} 2\pi r\,dr = 2\pi u_m^3\int r\left(1 - \frac{r}{r_0}\right)^{3/7} dr$$

To integrate by parts, let $u = r$ and $dv = \left(1 - \frac{r}{r_0}\right)^{3/7} dr$; then $du = dr$ and $v = -\frac{7}{10}r_0\left(1 - \frac{r}{r_0}\right)^{10/7}$

so that $$\int_0^{r_0} r\left(1 - \frac{r}{r_0}\right)^{3/7} dr = \left[-\frac{7}{10}rr_0\left(1 - \frac{r}{r_0}\right)^{10/7}\right]_0^{r_0} - \int_0^{r_0}\left[-\frac{7}{10}r_0\left(1 - \frac{r}{r_0}\right)^{10/7}\right]dr$$

$$= [-0 + 0] + \frac{7}{10}r_0\int_0^{r_0}\left(1 - \frac{r}{r_0}\right)^{10/7} dr = \frac{7}{10}r_0\left[-\frac{7}{17}r_0\left(1 - \frac{r}{r_0}\right)^{17/7}\right]_0^{r_0} = -\frac{49}{170}r_0^2(0 - 1) = \frac{49}{170}r_0^2$$

and $\int u^3 dA = 2\pi u_m^3(49/170)r_0^2 = (49/85)\pi r_0^2 u_m^3$

$$\text{Eq. 5.4: } \alpha = \frac{\int u^3 dA}{AV^3} = \frac{(49/85)\pi r_0^2 u_m^3}{(\pi r_0^2)(49u_m/60)^3} = \frac{60^3}{85(49^2)} = 1.058 \quad ◄$$

Sec. 5.5: Energy Equation for Steady Flow of Incompressible Fluids, and Bernoulli's Theorem -- *Exercises*

5.5.1 *A pipeline supplies water to a hydroelectric power plant, the elevation of which is 2200 ft below the level of the water at intake to the pipe. If 8 per cent of this total, or 176 ft, is lost in friction in the pipe, what will be the value of ΔI in Btu/lb; and what will be the rise in temperature if there is no heat transfer?*

BG

Eq. 5.11 with $Q_H = 0$: $\Delta I = h_2 =$ (176 ft·lb/lb)/(778 ft·lb/Btu) = 0.226 Btu/lb ◄

Eq. 5.11: ΔI = 0.226 Btu/lb = $(c/g)\Delta T$

where (from Sec. 5.5, footnote 3) c = 1.0 Btu/(mass of one lb·°R)

$$\therefore\ \Delta T = \frac{\Delta I g}{c} = \frac{0.226\text{ Btu/lb}(32.2\text{ ft/sec}^2)}{1.0\text{ Btu}(32.2\text{ ft/sec}^2)/(\text{lb}\cdot°\text{R})} = 0.226°\text{R} \quad ◄$$

5.5.2 *A pipeline supplies water to a hydroelectric power plant, the elevation of which is 650 m below the level of the water at intake to the pipe. If 8 per cent of this total, or 52 m, is lost in friction in the pipe, what will be the value of ΔI in J/N, and what will be the rise in temperature if there is no heat transfer?*

SI

Eq. 5.11: with $Q_H = 0$: $\Delta I = h_2 = 52\text{ m} = 52\text{ N·m/N} = 52\text{ J/N}$ ◀

Eq. 5.11: $\Delta I = 52\text{ N·m/N} = (c/g)\ \Delta T$

where (from Sec. 5.5, footnote 3) $c = 1\text{ cal/(g·K)} = 4187\text{ m}^2/(\text{s}^2\text{·K})$

$$\therefore\ \Delta T = \frac{\Delta I g}{c} = \frac{(52\text{ N·m/N})(9.81\text{ m/s}^2)}{4187\text{ m}^2/(\text{s}^2\text{·K})} = 0.1218\text{ K}$$ ◀

5.5.3 *A vertical pipe 5 ft in diameter and 64 ft long has a pressure head at the upper end of 20.2 ft of water. When the flow of water through it is such that the mean velocity is 15 fps, the friction loss is $h_L = 3.8$ ft. Find the pressure head at the lower end of the pipe when the flow is (a) downward; (b) upward.*

BG

(*a*) Eq. 5.14: $20.2 + 64 + \dfrac{(15)^2}{2(32.2)} = \dfrac{p_2}{\gamma} + 0 + \dfrac{(15)^2}{2(32.2)} + 3.8$ (velocity heads cancel)

$p_2/\gamma = 80.4$ ft for downflow ◀

(*b*) Eq. 5.14: $\dfrac{p_2}{\gamma} + 0 + \dfrac{(15)^2}{2(32.2)} = 20.2 + 64 = \dfrac{(15)^2}{2(32.2)} + 3.8$ (velocity heads cancel)

$p_2/\gamma = 88.0$ ft for upflow ◀

5.5.4 *A vertical pipe 1.8m in diameter and 22 m long has a pressure head at the upper end of 5.6 m of water. When the flow of water through it is such that the mean velocity is 5 m/s, the friction loss is $h_L = 1.15$ m. Find the pressure head at the lower end of the pipe when the flow is (a) downward (b) upward.*

SI

(*a*) Eq. 5.14: $5.6 + 22 + \dfrac{5^2}{2(9.81)} = \dfrac{p_2}{\gamma} + 0 + \dfrac{5^2}{2(9.81)} + 1.15$ (velocity heads cancel)

$\dfrac{p_2}{\gamma} = 26.5$ m for downflow ◀

(*b*) Eq. 5.14: $\dfrac{p_2}{\gamma} + 0 + \dfrac{5^2}{2(9.81)} = 5.6 + 22 + \dfrac{5^2}{2(9.81)} + 1.15$ (velocity heads cancel)

$\dfrac{p_2}{\gamma} = 28.8$ m for upflow ◀

5.5.5 *A conical pipe has diameters at the two ends of 1.2 and 4.2 ft and is 48 ft long. It is vertical, and the friction loss is $h_L = 7.6$ ft for flow of water in either direction when the velocity at the smaller section is 28 fps. If the smaller section is at the top and the pressure head there is 6.4 ft of water, find the pressure head at the lower end when the flow is (a) downward; (b) upward.*

BG

$Q = AV = (\pi/4)D^2V$; so $V_b = V_t\,(D_t/D_b)^2 = 28(1.2/4.2)^2 = 2.29$ fps

(*a*) Eq. 5.14: $6.4 + 48 + \dfrac{28^2}{2(32.2)} = \dfrac{p_2}{\gamma} + 0 + \dfrac{2.29^2}{2(32.2)} + 7.6$; $p_2/\gamma = 58.9$ ft for downflow ◀

(*b*) Eq. 5.14: $\dfrac{p_2}{\gamma} + 0 + \dfrac{2.29^2}{2(32.2)} = 6.4 + 48 + \dfrac{28^2}{2(32.2)} + 7.6$; $\dfrac{p_2}{\gamma} = 74.1$ ft for upflow ◀

5.5.6 *In Figure X5.5.6, the pipe AB is of uniform diameter. The pressure at A is 25 psi and at B is 36 psi. In which direction is the flow, and what is the friction loss in feet of the fluid if the liquid has a specific weight of (a) 32 lb/ft³; (b) 95 lb/ft³?*

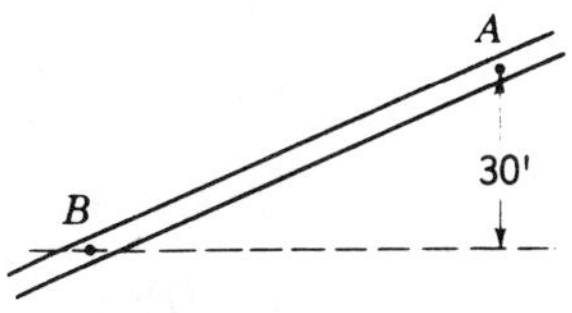

Figure X5.5.6

BG

Assume flow from A to B.

(a) For $\gamma = 32$ lb/ft³, and $V_1 = V_2$

$$\text{Eq. 5.14:} \quad \frac{25(144)}{32} + 30 + \frac{V^2}{2g} = \frac{36(144)}{32} + 0 + \frac{V^2}{2g} + h_L$$

$h_L = -19.50$ ft ◄ ; so flow is from B to A ◄

(b) For $\gamma = 95$ lb/ft³, and $V_1 = V_2$

$$\text{Eq. 5.14:} \quad \frac{25(144)}{95} + 30 + \frac{V^2}{2g} = \frac{36(144)}{95} + 0 + \frac{V^2}{2g} + h_L$$

$h_L = +13.33$ ft ◄ ; so flow is from A to B ◄

5.5.7 *If the difference in elevation between A and B in Fig. X5.5.6 is 10 m and the pressures at A and B are 160 and 260 KN/m² respectively, find the direction of flow and the head loss in meters of liquid. Assume the liquid has a specific gravity of 0.88.*

Fig. X5.5.6: See with Solution 5.5.6 above.

SI

Assume flow is from A to B.

$$\text{Eq. 5.14:} \quad \frac{160 \text{ kN/m}^2}{(0.88)9.81 \text{ N/m}^3} + 10 + \frac{V^2}{2g} = \frac{260 \text{ kN/m}^2}{(0.88)9.81 \text{ N/m}^3} + 0 + \frac{V^2}{2g} + h_L$$

$h_L = 18.53 + 10 - 30.12 = -1.58$ m ◄ so flow is from B to A ◄

Sec. 5.5: Energy Equation for Steady Flow of Incompressible Fluids, and Bernoulli's Theorem -- *Problems*

5.3 *A pipeline supplies water to a hydroelectric plant from a reservoir in which the water temperature is 58.8°F. (a) Suppose that in the length of the pipe there is a total loss of heat to the surrounding air of 0.28 Btu/lb of water and the temperature of the water at the power house is 58.7°F. What is the friction loss per pound of water? (b) With the same flow rate as in (a) what will be the temperature of the water at the power house if there is absorption of heat from hot sunshine at the rate of 3.2 Btu/lb of water?*

BG

From footnote 3 of sec. 5.5: $c = 1.0$ Btu/(mass of one lb·°R) = 25,000 ft²/(s²·°R)

Loss of heat $Q_H = 0.28$ Btu/lb = 0.28(778) = 217.8 ft·lb/lb

(a) Eq. 5.11: $h_L = (c/g)(T_2 - T_1) = Q_H$, where Q_H is negative in this case

$h_L = (25{,}000/32.2)(58.7 - 58.8) + 217.8 = 140.2$ ft·lb/lb ◄

(b) Gain in heat = 3.2(778) = 2490 ft·lb/lb

Eq. 5.11: $(c/g)(T_2 - T_1) = h_L + Q_H = 140.2 + 2490 = 2630$ ft·lb/lb

$(T_2 - T_1) = 2630(32.2)/25{,}000 = 3.39$ °R; $T_2 = 58.8 + 3.4 = 62.2$°F ◄

5.4 *Water is flowing at 15 m^3/s through a long pipe. The temperature of the water *rises 0.23°C *when heat is transferred to the water at the rate of 5600 kN·m/s. Find the head loss in the pipe. [*Note two corrections to the textbook.]*

SI

From Eq. 4.7: $\dot{m}g = G = \gamma Q = (9810\ \text{N/m}^3)(15\ \text{m}^3/\text{s}) = 147\,150\ \text{N/s}$

Q_H = (Heat energy per second)/G = (5600 kN·m/s)/(147 150 N/s) = 38.1 N·m/N

Eq. 5.12: $h_L = (c/g)\Delta T - Q_H = \dfrac{4187\ \text{m}^2/(\text{s}^2\cdot\text{K})}{9.81\ \text{m/s}^2}(0.23\ \text{K}) - 38.1\ ;\quad h_L = 98.2 - 38.1 = 60.1\ \text{m}$ ◄

5.5 *Assume frictionless flow in a long, horizontal, conical pipe, where the diameter is 3.8 ft at one end and 2.2 ft at the other. The pressure head at the smaller end is 12.5 ft of water. If water flows through this cone at the rate of 120 cfs, find the velocities at the two ends and the pressure head at the larger end.*

BG

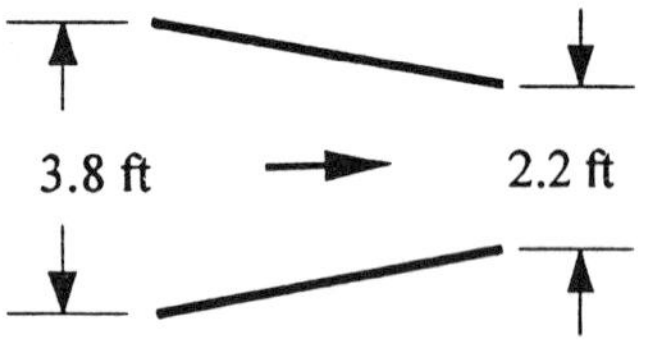

$V_1 = \dfrac{Q}{A_1} = \dfrac{120}{\pi(1.9)^2} = 10.58\ \text{fps}$ ◄

$V_2 = \dfrac{Q}{A_2} = \dfrac{120}{\pi(1.1)^2} = 31.6\ \text{fps}$ ◄

Eq. 5.15 with $z_1 = z_2$: $\dfrac{p_1}{\gamma} + z + \dfrac{10.58^2}{2(32.2)} = 12.5 + z + \dfrac{31.6^2}{2(32.2)}\ ;\quad \dfrac{p_1}{\gamma} = 26.2\ \text{ft}$ ◄

5.6 *Water flows through a long, horizontal, conical diffuser at the rate of 4.5 m^3/s. The diameter of the diffuser changes from 1.2 m to 1.8 m. The pressure at the smaller end is 7.5 kN/m^2. Find the pressure at the downstream end of the diffuser, assuming frictionless flow. Assume also, that the angle of the cone is so small that separation of the flow from the walls of the diffuser does not occur.*

SI

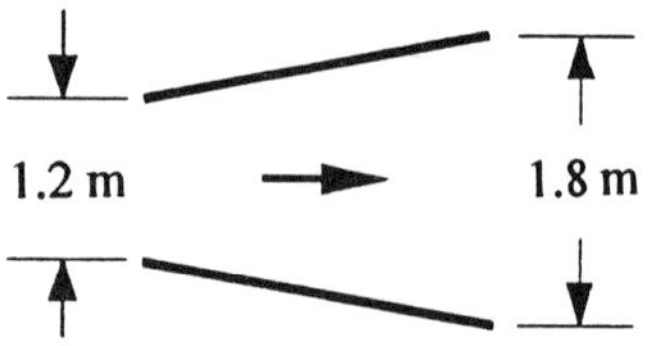

$V_1 = \dfrac{Q}{A_1} = \dfrac{4.5}{\pi(0.6)^2} = 3.98\ \text{m/s}$

$V_2 = \dfrac{Q}{A_2} = \dfrac{4.5}{\pi(0.9)^2} = 1.768\ \text{m/s}$

Eq. 5.16 with $z_1 = z_2$: $p/\gamma + v^2/2g = \text{constant}$

ie: $\dfrac{7.5\ \text{kN/m}^2}{9.81\ \text{kN/m}^3} + \dfrac{3.98^2}{2(9.81)} = \dfrac{p}{\gamma} + \dfrac{1.768^2}{2(9.81)}\ ;\quad 0.765\ \text{m} + 0.807\ \text{m} = p/\gamma + 0.159$

$p/\gamma = 1.412\ \text{m}\ ;\quad p = 1.412(9.81) = 13.85\ \text{kN/m}^2$ ◄

Sec. 5.6: Energy Equation for Steady Flow of Compressible Fluids -- *Exercises*

5.6.1 *Gas, which may be assumed to be perfect, flows at a constant temperature though a uniform, horizontal pipe. At section 1 the pressure is 120 psia and the velocity is 60 fps; at section 2 the pressure is 100 psia and the velocity is 72 fps. Find (a) the change in enthalpy, and (b) the gain or loss of heat per lb; both between sections 1 and 2. Hint: Recall Eq. (2.4).*

BG

(*a*) Sec. 5.3 for a perfect gas: When T = const, $I = f(T)$ = const.

Eq. 2.4 with const T: $p/\rho = RT$ = const. So from Eq. 5.18: h = const and $\Delta h = 0$ ◄

(*b*) From Eq. 5.20: $Q_H = \dfrac{h_2 - h_1}{g} + \dfrac{V_2^2 - V_1^2}{2g} = 0 + \dfrac{72^2 - 60^2}{2(32.2)} = +24.6\ \text{ft·lb/lb}$

$Q_H = 24.6/778 = 0.0316$ Btu/lb gain ◄

5.6.2 *Air flows isothermally (constant temperature) through a horizontal duct of constant cross section. At station 1 the pressure is 800 kPa abs and the velocity is 20 m/s; at station 2 the pressure is 1000 kPa abs and the velocity is 16 m/s. Find (a) the change in enthalpy, and (b) the gain or loss of heat per newton; both between stations 1 and 2. Assume the air to be a perfect gas. Hint: Recall Eq. (2.4).*

SI

(*a*) Sec. 5.3 for perfect gas: When T = const, $I = f(T)$ = const.

Eq. 2.4 with const T: $p/\rho = RT$ = const. So from Eq. 5.18: h = const and $\Delta h = 0$ ◀

(*b*) From Eq. 5.20: $Q_H = \dfrac{h_2 - h_1}{g} + \dfrac{V_2^2 - V_1^2}{2g} = 0 + \dfrac{16^2 - 20^2}{2(9.81)} = -7.34$ N·m/N

$Q_H = 7.34$ J/N loss ◀

5.6.3 *Oxygen flows without a gain or loss of heat through a horizontal pipe of constant cross section. At section 1 the pressure is 170 psia, the velocity is 75 fps, and the temperature is 50°F; at section 2 the pressure is 125 psia, the velocity is 98 fps, and the temperature is 30°F. Find (a) the head (mechanical energy) loss in Btu/lb; (b) the change in enthalpy; both between sections 1 and 2. Assume the oxygen to be a perfect gas. Hint: Recall Eq. (2.4).*

BG

(*a*) From Eq. 5.14: $h_L = \left(\dfrac{p}{\gamma} + z + \dfrac{V^2}{2g}\right)_1 - \left(\dfrac{p}{\gamma} + z + \dfrac{V^2}{2g}\right)_2$

$= \dfrac{p_1}{\rho_1 g} - \dfrac{p_2}{\rho_2 g} + \dfrac{V_1^2 - V_2^2}{2g}$ since $z_1 = z_2$

Eq. 2.4: $p/\rho = RT$; Table A.5 for oxygen: $R = 1554$ ft²/(s²·°R)

So $h_L = \dfrac{R}{g}(T_1 - T_2) + \dfrac{V_1^2 - V_2^2}{2g} = \dfrac{1554}{32.2}(510 - 490) + \dfrac{75^2 - 98^2}{2(32.2)}$

$= 965 - 61.8 = 903$ ft $= 903$ ft·lb/lb ; $h_L = 903/778 = 1.161$ Btu/lb ◀

(*b*) From Eq. 5.20 with $Q_H = 0$ (given): $\dfrac{h_1 - h_2}{g} = \dfrac{V_2^2 - V_1^2}{2g} = \dfrac{98^2 - 75^2}{2(32.2)} = 61.8$ ft $= 61.8$ ft·lb/lb

So $\Delta h = h_1 - h_2 = g(61.8) = 32.2(61.8) = 1990$ ft·lb/slug decrease ◀

5.6.4 *Air, which may be assumed to be a perfect gas, flows without heat gain or loss through a uniform horizontal pipe. At station 1 the pressure is 1135 kPa abs, the velocity is 25 m/s, and the temperature is 10°C; at station 2 the pressure is 830 kPa abs, the velocity is 33 m/s, and the temperature is 0°C. Find (a) the head (mechanical energy) loss in J/N; (b) the change in enthalpy; both between sections 1 and 2. Hint: Recall Eq. (2.4).*

SI

(*a*) From Eq. 5.14: $h_L = \left(\dfrac{p}{\gamma} + z + \dfrac{V^2}{2g}\right)_1 - \left(\dfrac{p}{\gamma} + z + \dfrac{V^2}{2g}\right)_2$

$= \dfrac{p_1}{\rho_1 g} - \dfrac{p_2}{\rho_2 g} + \dfrac{V_1^2 - V_2^2}{2g}$ since $z_1 = z_2$

Eq. 2.4: $p/\rho = RT$; Table A.5 for air: $R = 287$ m²/(s²·K)

So $h_L = \dfrac{R}{g}(T_1 - T_2) + \dfrac{V_1^2 - V_2^2}{2g} = \dfrac{287}{9.81}(283 - 273) + \dfrac{25^2 - 33^2}{2(9.81)} = 293 - 23.6 = 269$ m

/cont...

or $h_L = 269 \text{ N·m/N} = 269 \text{ J/N}$ ◄

(*b*) From Eq. 5.20 with $Q_H = 0$ (given): $\dfrac{h_1 - h_2}{g} = \dfrac{V_2^2 - V_1^2}{2g} = 23.6 \text{ m} = 23.6 \text{ N·m/N}$

So $\Delta h = h_1 - h_2 = g(23.6) = 9.81(23.6) = 232 \text{ N·m/kg}$ decrease ◄

Sec. 5.7: Euler's Equation for Steady Motion of an Ideal Fluid Along a Streamline -- ***Exercises***

5.7.1 *Assume the flow to be frictionless in the siphon shown in Fig. X5.7.1. Find the rate of discharge in cfs and the pressure head at B if the pipe has a uniform diameter of 4 in.*

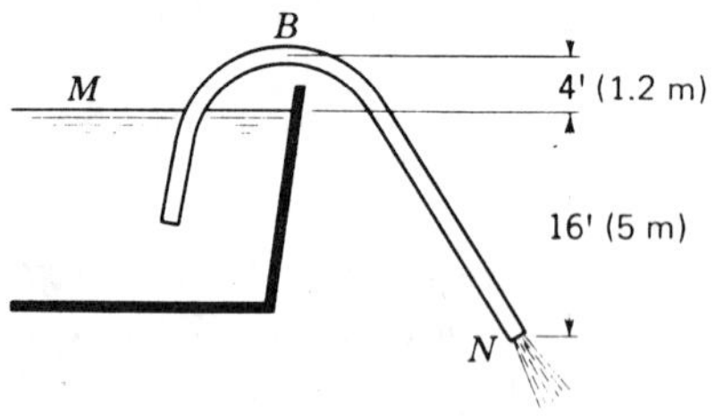

Figure X5.7.1

BG

Eq. 5.24 from M to N (elevation datum at N):

$0 + 0 + 16 = 0 + V_N^2/2g + 0$; $V_N = V_B = 32.1$ fps

$Q = \pi(2/12)^2\, 32.1 = 2.80$ cfs ◄

Eq. 5.24 from M to B:

$0 + 0 + 16 = p_B/\gamma + V_B^2/2g + 20$; $p_B/\gamma = -20.0$ ft ◄

5.7.2 *Referring to Fig. X5.7.1, assume the flow to be frictionless in the siphon. Find the rate of discharge in m³/s and the pressure head at B if the pipe has a uniform diameter of 18 cm.*
Fig. X5.7.1: See with Solution 5.7.1 above.

SI

Eq. 5.24 from M to N (elevation datum at N): $0 + 0 + 5 = 0 + V_N^2/2g + 0$; $V_N = V_B = 9.90$ m/s

$Q = \pi(0.18/2)^2 9.90 = 0.252 \text{ m}^3\text{/s}$ ◄

Eq. 5.24 from M to B: $0 + 0 + 5 = p_B/\gamma + V_B^2/2g + 6.2$; $p_B/\gamma = -6.20$ m ◄

Sec. 5.8: Pressure in Fluid Flow -- ***Exercises***

5.8.1 *Find the stagnation pressure on the nose of a submarine moving at 10.5 knots in seawater ($\gamma = 64$ lb/ft³) when it is 55 ft below the surface.*

BG

From inside cover: 1 knot (1 nautical mile per hour) = 1.688 ft/sec

$V = 10.5$ knots $= 10.5(1.688) = 17.72$ fps

Energy Eq. from the stagnation point to the surface, referring to all quantities to the stagnation point:

$p/\gamma + 0 + 0 = 0 + 55 + V^2/2g$

$p = \gamma(z + V^2/2g) = 64[55 + 17.73^2/(2\times32.2)] = 3830 \text{ psf} = 26.6 \text{ psi}$ ◄

5.8.2 *Find the stagnation pressure on the nose of a submarine moving at 8 m/s in seawater ($\gamma = 10{,}050$ N/m³) when it is 25 m below the surface.*

SI

Energy Eq. from the stagnation point to the surface, referring to all quantities to the stagnation point:

$p/\gamma + 0 + 0 = 0 + 25 + V^2/2g$

$p = \gamma(z + V^2/2g) = 10\,050[25 + 8^2/(2\times9.81)] = 284 \text{ kN/m}^2$ ◄

5.8.3 *Find the stagnation pressure on the nose of a fish swimming at 25 fps in fresh water (γ = 62.4 lb/ft^3) when it is 5 ft below the surface.*

BG

Energy Eq. from the stagnation point to the surface, referring to all quantities to the stagnation point:

$p/\gamma + 0 + 0 = 0 + z + V^2/2g$; $p = \gamma(z + V^2/2g) = 62.4[5 + 25^2/2(32.2)] = 918$ psf

Dividing by 144, $p = 6.37$ psi ◄

Sec. 5.8: Pressure in Fluid Flow -- *Problems*

5.7 *In Fig. P5.7 water is admitted at the center at a rate of 3 cfs and is discharged into the air around the periphery. The upper circular plate in the figure is horizontal and is fixed in position, while the lower annular plate is free to move vertically and is not supported by the pipe in the center. The annular plate weighs 6 lb, and the weight of the water on it should be considered. (a) If the distance between the two plates is to be maintained at 2 in, what is the total weight W that can be supported? (b) What is the pressure head where the radius is 4 in, and what is it at a radius of 8 in?*

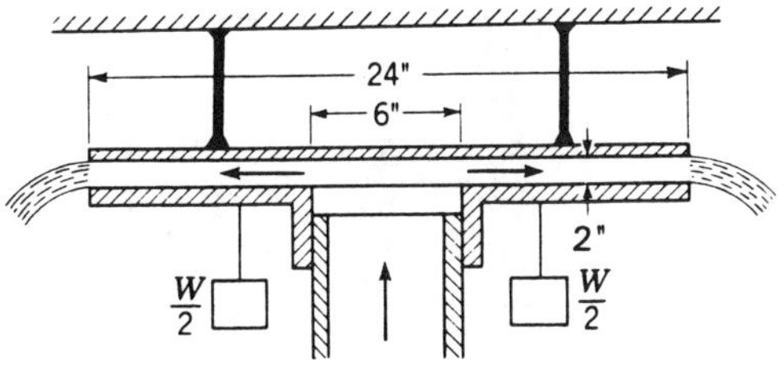

Figure P5.7

BG

(*a*) $V = \dfrac{Q}{2\pi rB} = \dfrac{3}{2\pi r(2/12)} = \dfrac{9}{\pi r}$ fps   At low r, V will be high, so p will be low, providing suction to support the lower plate

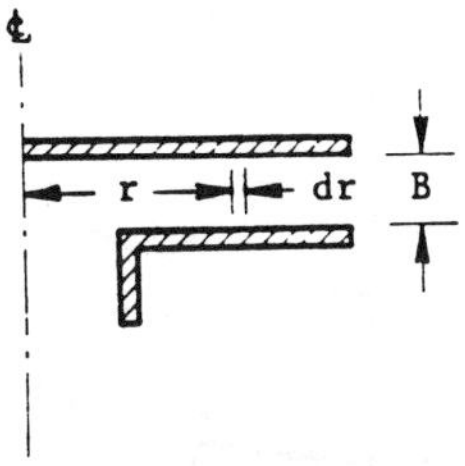

Energy Equation from any point between the plates to the periphery:

$p_r/\gamma + z + V_r^2/2g = 0 + z + V_R^2/2g$; $R = 1$ ft

$$\therefore \quad \frac{p_r}{\gamma} = \frac{V_R^2 - V_r^2}{2g} = \frac{1}{2g}\left[\left(\frac{9}{\pi R}\right)^2 - \left(\frac{9}{\pi r}\right)^2\right] = \frac{81}{2\pi^2 g}\left(1 - \frac{1}{r^2}\right)$$

To find total pressure force on plate,

Eq. 3.14: $$F = \int_A p\,da = \int_{0.25}^{1} p(2\pi r\,dr) = \frac{2\pi(81)62.4}{2\pi^2 g}\int_{0.25}^{1}\left(1 - \frac{1}{r^2}\right) r\,dr$$

$$= 50.0\left[\frac{r^2}{2} - \ln r\right]_{0.25}^{1} = -45.8 \text{ lb}$$

Weight of water = $62.4(2\pi/12)(1 - 1/4^2) = 30.6$ lb. Then $W = 45.8 - 30.6 - 6 = 9.21$ lb ◄

(*b*) At r = 4 in = 1/3 ft, $$\frac{p}{\gamma} = \frac{81}{2\pi^2 g}\left[1 - \frac{1}{(1/3)^2}\right] = -1.020 \text{ ft}$$ ◄

At r = 8 in = 2/3 ft, $$\frac{p}{\gamma} = \frac{81}{2\pi^2 g}\left[1 - \frac{1}{(2/3)^2}\right] = -0.1593 \text{ ft}$$ ◄

5.8 *Plot the stagnation pressure (psia) on an object as it passes through air at sea level (standard atmosphere) as a function of velocity. Repeat for movement through air at 10,000 ft elevation. Let V vary from zero to c using 0, 25, 50, 75, and 100% of c.*

BG

At sea level, from Table A.3: $p = 14.70$ psia, $\rho = 0.002\,38$ slug/ft^3, $c = 1116$ fps

Eq. 5.27: $(p_0)_{abs} = 14.70 + \frac{0.002\,38}{144}\frac{V^2}{2}\left(1 + \frac{V^2}{4c^2}\right) = 14.70 + 8.26 \times 10^{-6}V^2\left(1 + \frac{V^2}{4c^2}\right)$

% of c	V (fps)	V^2	$8.26 \times 10^{-6}\,V^2$	$\left(1 + \frac{V^2}{4c^2}\right)$	$(p_0)_{abs}$ (psia)
0	0	0	0	1.000	14.70
25	279	7.79×10^4	0.643	1.016	15.35
50	558	3.12×10^5	2.57	1.063	17.43
75	837	7.01×10^5	5.79	1.141	21.3
100	1116	12.45×10^5	10.29	1.250	27.6

▲

At 10,000 ft, from Table A.3: $p = 10.11$ psia, $\rho = 0.001\,756$ slug/ft^3, $c = 1077$ fps

Eq. 5.27: $(p_0)_{abs} = 10.11 + 6.10 \times 10^{-6}V^2(1 + V^2/4c^2)$

% of c	V fps	$(p_0)_{abs}$ psia
0	0	10.11
25	269	10.56
50	539	11.99
75	808	14.65
100	1077	18.95

▲

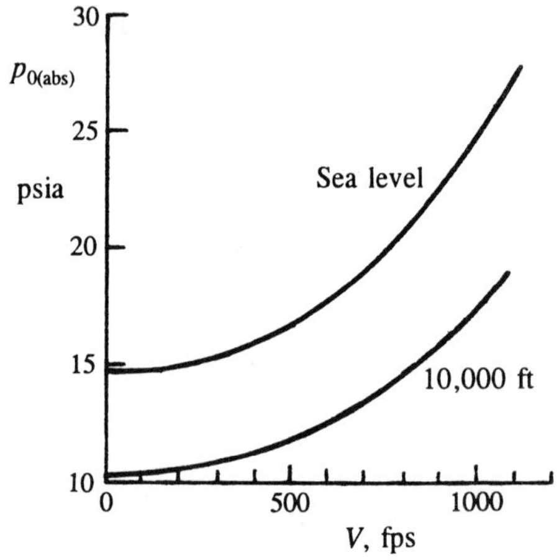

5.9 *Wind blows at a velocity of 20 m/s against the side of a pole at an elevation of 1000 m above sea level. What is the stagnation pressure assuming standard atmospheric conditions? Express answer as a gage pressure and as an absolute pressure in kN/m², Pa, and mmHg.*

SI

Table A.3 at 1000 m: $p = 89.876$ kPa abs, $\rho = 1.112$ kg/m^3, $c = 336$ m/s.

From Eq. 5.27: $(p_0)_{gage} = (p_0)_{abs} - p_{abs} = \frac{1.112(20)^2}{2}\left(1 + \frac{20^2}{4(336)^2}\right) = 223$ N/m^2

$(p_0)_{gage} = 223$ Pa ◄ $= 0.223$ kN/m^2 ◄

$(p_0/\gamma)_{gage} = 0.223\left(\frac{760 \text{ mmHg}}{101.3 \text{ kPa}}\right) = 1.670$ mmHg ◄

$(p_0)_{abs} = (p_0)_{gage} + p_{abs} = 223 + 89\,876 = 90\,100$ Pa ◄ $= 90.1$ kN/m^2 ◄

$(p_0/\gamma)_{abs} = 90.1(760/101.3) = 676$ mmHg ◄

5.10 *Find the stagnation pressure on a tree trunk at an elevation of 2000 m if the wind speed is 15 m/s.*

SI

Table A.3 at 2000 m: $\rho = 1.007\ \text{kg/m}^3,\ c = 333\ \text{m/s}$

From Eq. 5.27: $(p_0)_{\text{gage}} = (p_0)_{\text{abs}} - p_{\text{abs}} = \dfrac{1.007(15)^2}{2}\left(1 + \dfrac{15^2}{4(333)^2}\right) = 113.3\ \text{Pa}$ ◄

Sec. 5.10: Power Considerations in Fluid Flow -- *Exercises*

5.10.1 *Water entering a pump through an 8-in-diameter pipe at 4 psi has a flow rate of 4 cfs. It leaves the pump through a 4-in-diameter pipe at 15 psi. Assuming that the suction and discharge sides of pump are at the same elevation, find the horsepower delivered to the water by the pump.*

BG

$$V_1 = \frac{4}{(\pi/4)(8/12)^2} = 11.46\ \text{fps;}\quad V_2 = \frac{4}{(\pi/4)(4/12)^2} = 45.8\ \text{fps}$$

$$\text{Eq. 5.28:}\quad H_1 = \frac{4(144)}{62.4} + \frac{(11.46)^2}{2(32.2)} + z_1 = 11.27\ \text{ft} + z_1$$

$$\text{Eq. 5.28:}\quad H_2 = \frac{15(144)}{62.4} + \frac{(45.8)^2}{2(32.2)} + z_2 = 67.2\ \text{ft} + z_2\ ;\quad \text{given } z_1 = z_2$$

From Eq. 5.30 with $h_L = 0$: $h_p = +h_m = H_2 - H_1 = 67.2 - 11.27 = 56.0$ ft

Eq. 5.31: $P = Gh_p = (62.4 \times 4)56.0 = 13{,}970$ ft·lb/sec $= 25.4$ hp ◄

5.10.2 *After entering a pump through an 18-cm-diameter pipe at 35 kN/m², oil (s = 0.82) leaves the pump through a 12-cm-diameter pipe at 120 kN/m². The suction and discharge sides of the pump are at the same elevation. Find the rate at which energy is delivered to the oil by the pump if the flow rate is 70 L/s.*

SI

$V_1 = Q/A_1 = 0.070/(\pi 0.09^2) = 2.75$ m/s ; $V_2 = Q/A_2 = 0.070/(\pi 0.06^2) = 6.19$ m/s

Eqs. 5.28 and 5.30 with $z_1 = z_2$ and $h_L = 0$: $p_1/\gamma + V_1^2/2g + h_p = p_2/\gamma + V_2^2/2g$

$$\frac{35}{0.85(9.81)} + \frac{2.75^2}{2(9.81)} + h_p = \frac{120}{0.82(9.81)} + \frac{6.19^2}{2(9.81)}$$

$4.35 + 0.386 + h_p = 14.92 + 1.953$; $h_p = 12.13$ N·m/N of oil

Eq. 5.31: Rate of energy delivery $= \gamma Q h_p = (0.82 \times 9.81)0.070(12.13) = 6.83$ kW ◄

5.10.3 *Water from a reservoir is being supplied to a powerhouse that is located 975 ft below the reservoir surface. Discharging through a nozzle, the water has a jet velocity of 245 fps and jet diameter of 6 in. Find the horsepower lost to friction between the reservoir and the jet, and find the horsepower of the jet.*

BG

Eq. 5.14 from reservoir (1) to jet (2) ,

$$0 + 975 + 0 = 0 + 0 + \frac{(245)^2}{2(32.2)} + h_L\ ;\quad h_L = 975 - 932 = 42.9\ \text{ft}$$

$Q = A_2V_2 = (\pi/4)(6/12)^2 245 = 48.1$ cfs

$$\text{Eq. 5.32: hp in jet} = \frac{\gamma Q V_2^2/2g}{550} = \frac{62.4(48.1)932}{550} = 5090\ \text{hp}$$ ◄

$$\text{Eq. 5.32: hp lost} = \frac{\gamma Q h_L}{550} = \frac{62.4(48.1)42.9}{550} = 234\ \text{hp}$$ ◄

5.10.4 *Water from a reservoir is being supplied to a powerhouse that is located 325 m below the surface of the reservoir. Discharging through a nozzle, the water has a jet velocity of 75 m/s and jet diameter of 25 cm. Find the kW lost to friction between the reservoir and the jet, and find the power of the jet in kW.*

SI

Eq. 5.14: $0 + 325 + 0 = 0 + \dfrac{75^2}{2(9.81)} + h_L\ ;\quad h_L = 325 - 286.7 = 38.3\ \text{m}$

$Q = A_2V_2 = (\pi/4)(0.25)^2 75 = 3.68\ \text{m}^3/\text{s}$

Eq. 5.33: Power in jet $= (3.68\ \text{m}^3/\text{s})(9.81\ \text{kN/m}^3)\dfrac{75^2}{2(9.81)}\ \text{m} = 10\,350\ \text{kN·m/s} = 10\,350\ \text{kW}$ ◄

Eq. 5.33: Power lost $= 9.81(3.68)38.3 = 1383\ \text{kW}$ ◄

5.10.5 *A turbine, located 765 ft below the water surface of the intake, carries a flow of 125 cfs. The pipeline leading to it has a friction loss of 30 ft. Find the horsepower delivered if the turbine efficiency is 90 percent.*

BG

Eq. 5.32: Water hp $= \dfrac{\gamma Q(\Delta z - h_L)}{550} = \dfrac{62.4)125(765 - 30)}{550} = 10{,}420\ \text{hp}$

From Eq. 5.34 Output power $= 10{,}420(0.90) = 9380\ \text{hp}$ ◄

Sec. 5.10: Power Considerations in Fluid Flow -- *Problems*

5.11 *A pump, with an efficiency of 90 per cent, circulates water at the rate of 2500 gpm in a closed circuit that holds 8500 gal. The net head developed by the pump is 360 ft. What is the change in water temperature after one hour, assuming that the bearing friction is negligible and that there is no heat loss from the system?*

BG

Inside cover: 1 cfs = 449 gpm. $\therefore\ Q = 2500/449 = 5.57\ \text{cfs}$

Eq. 5.31: Power input $= \gamma Qh/\eta = 62.4(5.57)360/0.90 = 139{,}000\ \text{ft·lb/s}$

Heat input in 1 hr $= 139{,}000(3600) = 5.00 \times 10^8\ \text{ft·lb}$

Mass heated $= (8500\ \text{gal})(\text{ft}^3/7.48\ \text{gal})(1.940\ \text{slug/ft}^3) = 2200\ \text{slugs}$

Per Eq. 5.12: $c(\Delta T)(\text{mass}) = 5.00 \times 10^8\ \text{ft·lb}$

From Sec. 5.5, footnote 3: For water $c = 25{,}000\ \text{ft·lb/(slug·°R)}$

$\dfrac{25{,}000\ \text{ft·lb}}{(\text{slug·°R})}(\Delta T)2200\ \text{slugs} = 5.00 \times 10^8\ \text{ft·lb}\ ;\quad \Delta T = 9.08\text{°R} = 9.08\text{°F}$ ◄

5.12 *A pump, with an efficiency of 90 per cent, circulates water at the rate of 150 L/s in a closed circuit that holds 50 m³. The net head developed by the pump is 110 m. What is the change in water temperature after one hour, assuming that the bearing friction is negligible and that there is no heat loss from the system?*

SI

Eq. 5.31: Power input $= \gamma Qh/\eta = 9.81(0.150)110/0.90 = 179.9\ \text{kN·m/s}$

Heat input in 1 h $= 179.9(3600) = 647\,000\ \text{kN·m}$; Mass heated $= 50\ \text{m}^3(1000\ \text{kg/m}^3) = 50\,000\ \text{kg}$

Per Eq. 5.12: $c(\Delta T)(\text{mass}) = 647 \times 10^6\ \text{N·m}$

From Sec. 5.5, footnote 3: For water $c = 4187\ \text{N·m/(kg·K)}$

So $\dfrac{4187\ \text{N·m}}{\text{kg·K}}(\Delta T)50\,000\ \text{kg} = 647 \times 10^6\ \text{N·m}\ ;\quad \Delta T = 3.09\ \text{K} = 3.09\text{°C}$ ◄

Sec. 5.11: Cavitation -- *Exercises*

5.11.1 *Water at a temperature of 160°F flows through the horizontal constriction of Fig. S5.8 when the atmospheric pressure is 27.9 in Hg. Neglecting head loss, find the flow rate at which cavitation begins.*

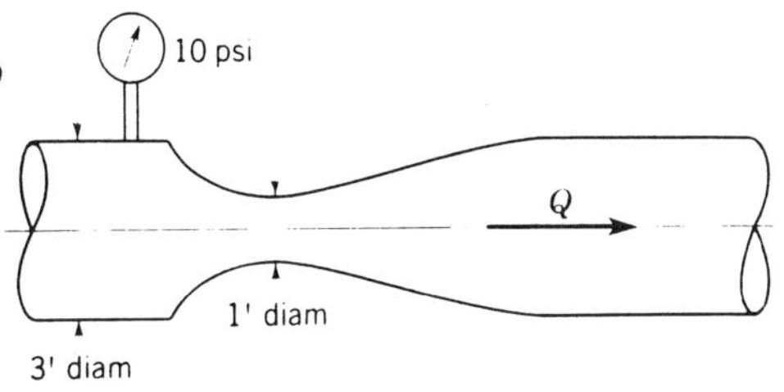

Figure S5.8

BG

Table A.1 at 160°F: $\gamma = 61.00\ \text{lb/ft}^3$, $p_v = 4.74$ psia

From Sec. 3.5: $p_{atm} = \dfrac{27.9\ \text{inHg}}{29.92\ \text{inHg}}(14.70\ \text{psia}) = 13.71$ psia

Cavitation begins when $p_2 = p_c = p_v$;

Eq. 5.36: $\left(\dfrac{p_{crit}}{\gamma}\right)_{gage} = -\left(\dfrac{13.71 - 4.74}{61.0}\right)144 = -21.2$ ft

Eq. 5.15: $\dfrac{10(144)}{61.0} + 0 + \dfrac{(Q/2.25\pi)^2}{2(32.2)}$

$= -21.2 + 0 + \dfrac{(Q/0.25\pi)^2}{2(32.2)}$; $Q = 42.4$ cfs ◄

5.11.2 *Water at a temperature of 50°C flows through a horizontal constriction similar to that in Fig. S5.8 when the atmospheric pressure is 709 mmHg. The gage reading is 30 kN/m², $d_1 = 0.5$ m, and $d_2 = 0.2$ m. Neglecting head loss, find the flow rate at which cavitation begins.*

Fig. S5.8: See with Solution 5.11.1 above.

SI

Table A.1 at 50°C : $\gamma = 9.689\ \text{kN/m}^3$, $p_v = 12.33\ \text{kN/m}^2$ abs

From Sec. 3.5: $p_{atm} = \left(\dfrac{709\ \text{mmHg}}{760\ \text{mmHg}}\right)101.3\ \text{kN/m}^2$ abs $= 94.5\ \text{kN/m}^2$ abs

Cavitation begins when $p_2 = p_c = p_v$. Eq. 5.36: $\left(\dfrac{p_{crit}}{\gamma}\right)_{gage} = -\left(\dfrac{94.5 - 12.33}{9.69}\right) = -8.48$ m

Eq. 5.15: $\dfrac{30}{9.69} + 0 + \dfrac{[Q/(\pi 0.25^2)]^2}{2(9.81)} = -8.48 + 0 + \dfrac{[Q/(\pi 0.1^2)]^2}{2(9.81)}$; $Q = 0.480\ \text{m}^3/\text{s}$ ◄

5.11.3 *Water at a temperature of 80°F flows through the horizontal constriction of Fig. S5.8 at a rate of 65 cfs when the atmospheric pressure is 27.9 inHg. Neglecting head loss, find the largest throat constriction diameter d_2 which will cause cavitation.*

Fig. S5.8: See with Solution 5.11.1 above.

BG

Table A.1 at 80°F : $\gamma = 62.22\ \text{lb/ft}^3$, $p_v = 0.51$ psia

From Sec. 3.5: $p_{atm} = \left(\dfrac{27.9\ \text{inHg}}{29.92\ \text{inHg}}\right)(14.70\ \text{psia}) = 13.71$ psia

Cavitation begins when $p_2 = p_c = p_v$. Eq. 5.36: $\left(\dfrac{p_{crit}}{\gamma}\right)_{gage} = -\left(\dfrac{13.71 - 0.51}{62.2}\right)144 = -30.5$ ft

Eq. 5.15: $\dfrac{10(144)}{62.2} + 0 + \dfrac{[65/(\pi 3^2/4)]^2}{2(32.2)} = -30.5 + 0 + \dfrac{[65/(\pi d_2^2/4)]^2}{2(32.2)}$; $d_2 = 1.179$ ft ◄

5.11.4 *Water at a temperature of 50°C flows at a rate of 2 m³/s through a horizontal constriction similar to that in Fig. S5.8 when the atmospheric pressure is 750 mmHg. The gage reading is 30 kN/m² and d_1 =0.5 m. Neglecting head loss, find the largest throat constriction diameter d_2 which will cause cavitation.*

Fig. S5.8: See with Solution 5.11.1 above.

SI

Table A.1 at 50°C : γ = 9.689 kN/m³, p_v = 12.33 kN/m² abs

From Sec. 3.5: $p_{atm} = \left(\dfrac{750 \text{ mmHg}}{760 \text{ mmHg}}\right)(101.3 \text{ kN/m}^2\text{abs}) = 100.0 \text{ kN/m}^2 \text{ abs}$

Cavitation begins when $p_2 = p_c = p_v$. Eq. 5.36: $\left(\dfrac{p_{crit}}{\gamma}\right)_{gage} = -\left(\dfrac{100.0 - 12.33}{9.69}\right) = -9.05 \text{ m}$

Eq. 5.15: $\dfrac{30}{9.69} + 0 + \dfrac{[2/(\pi 0.25^2)]^2}{2(9.81)} = 0 - 9.05 + \dfrac{[2/(\pi d_2^2/4)]^2}{2(9.81)}$; $d_2 = 0.371$ m ◄

5.11.5 *Referring to Fig. S5.8, find the maximum theoretical flow rate for water at 70°C. Neglect head loss. The diameters are 80 cm and 40 cm respectively, the upstream pressure is 20 kN/m², and the atmospheric pressure is 750 mmHg.*

Fig. S5.8: See with Solution 5.11.1 above.

SI

From Sec. 3.5: p_{atm} = 750(101.3/760) = 100.0 kN/m² abs

Table A.1 at 70°C : γ = 9.589 kN/m³, p_v = 31.16 kN/m² abs

Q is max when p_2 (=p_c) = p_v, i.e., when cavitation begins.

$V_1 = Q/(\pi 0.40^2) = 1.989Q$; $V_2 = Q/(\pi 0.20^2) = 7.96Q$

Eq. 5.15 from 1 to 2 in absolute pressure heads:

$$\frac{100.0 + 20}{9.59} + z_1 + \frac{(1.989Q)^2}{2(9.81)} = \frac{31.16}{9.59} + z_2 + \frac{(7.96Q)^2}{2(9.81)} \; ; \quad z_1 = z_2$$

Maximum theoretical Q = 1.749 m³/s ◄

Sec. 5.13: Loss of Head at Submerged Discharge -- *Problems*

5.13 *In Fig. 5.10 neglect all head losses except at discharge, and assume water is flowing. If h = 15 ft and the water surface in the lower reservoir is 12 ft higher than the constriction, find the highest permissible temperature of the water in order that there be no cavitation. The diameter of the constriction is 80 percent of the diameter of the pipe where it joins the downstream tank. Atmospheric pressure is 13.6 psia.*

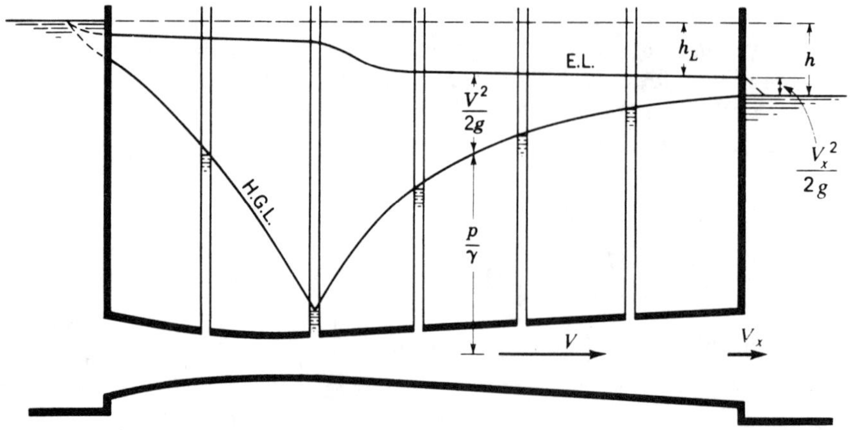

Figure 5.10

/cont...

BG

$V_x^2/2g = 15$, $V_x = 31.1$ fps; $V_c = 31.1(1/0.8)^2 = 48.6$ fps

Energy equation from upper reservoir to constriction:

$15 + 12 = p_c/\gamma + V_c^2/2g = p_c/\gamma + 48.6^2/(2\times 32.2) = p_c/\gamma + 36.6$; $p_c/\gamma = -9.62$ ft

At the maximum temperature, $(p_c/\gamma)_{abs} = (p_v/\gamma)_{abs}$

$$\left(\frac{p_c}{\gamma}\right)_{abs} = \frac{p_{atm}}{\gamma} + \left(\frac{p_c}{\gamma}\right)_{gage} = \frac{13.6(144)}{\gamma} - 9.62 \text{ ft}$$

Try T	γ (from Table A.1)	$\left(\frac{p_c}{\gamma}\right)_{abs}$	T for this p_v/γ (from Table A.1)
60 °F	62.37	21.78 ft	≈ 191 °F
190 °F	60.36	22.82	* 192.5 °F
192.5 °F	*60.30	22.86	* 192.6 °F
192.6 °F	*60.30		

* By interpolation.

Highest permissible temperature = 192.6 °F ◀

5.14 *In Fig. 5.10 neglect all head losses except at discharge, and assume water is flowing. If h = 4 m and the water surface in the lower reservoir is 5.5 m higher than the constriction, find the highest permissible temperature of the water in order that there be no cavitation. The diameter of the constriction is 70 percent of the diameter of the pipe where it joins the downstream tank. Atmospheric pressure is 97 kN/m² abs.*

See Fig. 5.10 with solution to Prob. 5.13 above.

SI

$V_x^2/2g = 4$ m, $V_x = \sqrt{2(9.81)4} = 8.86$ m/s; $V_c = 8.86(1/0.7)^2 = 18.08$ m/s

Energy equation from upper reservoir to constriction:

$4.0 + 5.5 = p_c/\gamma + 18.08^2/(2\times 9.81) = p_c/\gamma + 16.66$; $p_c/\gamma = -7.16$ m

At the maximum temperature, $(p_c/\gamma)_{abs} = (p_v/\gamma)_{abs}$

$$\left(\frac{p_c}{\gamma}\right)_{abs} = \frac{p_{atm}}{\gamma} + \left(\frac{p_c}{\gamma}\right)_{gage} = \frac{97}{\gamma} - 7.16 \text{ m}$$

Try T	γ (from Table A.1)	$\left(\frac{p_c}{\gamma}\right)_{abs}$	T for this p_v/γ (from Table A.1)
15 °C	9.798	2.74 m	≈ 66 °C
70 °C	9.589	2.96	* 67.9 °C
67.9 °C	*9.600	2.94	* 67.8 °C
67.8 °C	*9.601	2.94	

* By interpolation.

Highest permissible temperature = 67.8 °C ◀

5.15 *Redo Prob. 5.13 but this time let the water temperature be 50°F. Find the minimum permissible diameter of the constriction in order to not have cavitation. Express the answer as a fraction of the outlet diameter.*

Prob. 5.13: Water flows in Fig. 5.10 (with Solution 5.13 above). Neglect all losses except discharge. Given: h = 15 ft, the lower water surface is 12 ft higher than the constriction, and p_{atm} = 13.6 psia.

BG

Cavitation begins in constriction (subscript c) when $p_c = p_v$.

Table A.1 at 50°F: $p_v/\gamma = 0.41$ ft, $\gamma = 62.41$ pcf

Express energy equation from upper reservoir to constriction in terms of absolute pressure head:

$$15 + 12 + \frac{13.6(144)}{62.41} = \frac{p_c}{\gamma} + \frac{V_c^2}{2g} \; ; \; 27 + 31.38 = 0.41 + \frac{V_c^2}{2g} \; ; \; \frac{V_c^2}{2g} = 58.0 \text{ ft} \; ; \; V_c = 61.1 \text{ fps}$$

$A_c V_c = A_x V_x$, where subscript x indicates outlet (exit), and $V_x = \sqrt{2g(15)} = 31.1$ fps

So $A_c/A_x = 31.1/61.1 = 0.509 = (D_c/D_x)^2$; $D_c/D_x = 0.713$ ◄

5.16 *Referring to Fig. 5.10, assume water is flowing and neglect all head losses except at discharge. Find the flow rate if h = 8 ft. Assuming that the water surface in the lower reservoir is 10 ft above the constriction, the diameter of the constriction is two-thirds the diameter of the pipe where it joins the downstream tank, and the atmospheric pressure is equal to the standard atmospheric pressure at 10,000 ft elevation, calculate the gage pressure and the absolute pressure in the constriction. The diameter of the constriction is 14 in.*

See Fig. 5.10 with the solution to Prob. 5.13.

BG

Eq. 5.14 from water surface in upper reservoir (=elev datum) to pipe outlet (x) in lower reservoir:

$0 + 0 + 0 = p_x/\gamma - (8 + p_x/\gamma) + V_x^2/2g$; $V_x^2/2g = 8$ ft; $V_x = \sqrt{2(32.2)8} = 22.7$ fps

In constriction: $V_c = (3/2)^2 22.7 = 51.1$ fps ; $Q = A_c V_c = \pi(7/12)^2 51.1 = 54.6$ cfs ◄

Energy equation using absolute pressure head, from constriction (c) to lower reservoir surface:

$(p_c/\gamma)_{abs} + 0 + V_c^2/2g = (p_{atm}/\gamma) + 10 + 0 + V_x^2/2g$ where $V_x^2/2g$ is the discharge loss.

Table A.3 at 10,000 feet altitude: $p_{atm} = 10.1083$ psia

$(p_c/\gamma)_{abs} + 0 + 51.1^2/(2 \times 32.2) = 10.1083(144)/62.4 + 10 + 8$; $(p_c/\gamma)_{abs} = 0.827$ ft

$(p_c)_{abs} = 0.827(62.4) = 51.6$ psfa $= 0.385$ psia ◄ $(p_c)_{gage} = 0.385 - 10.1083 = -9.75$ psi ◄

5.17 *Referring to Fig. 5.10, assume water is flowing and neglect all head losses except at discharge. Find the flow rate if h = 2 m. Assuming that the water surface in the lower reservoir is 4 m above the constriction, the diameter of the constriction is two-thirds the diameter of the pipe where it joins the downstream tank, and the atmospheric pressure is equal to the standard atmospheric pressure at 3000 m elevation, calculate the gage pressure and the absolute pressure in the constriction. The diameter of the constriction is 36 cm.*

See Fig. 5.10 with the solution to Prob. 5.13.

SI

Eq. 5.14 from water surface in upper reservoir (=elev datum) to pipe outlet (x) in lower reservoir:

$0 + 0 + 0 = p_x/\gamma - (2 + p_x/\gamma) + V_x^2/2g$; $V_x^2/2g = 2$ m; $V_x = \sqrt{2(9.81)2} = 6.26$ m/s

In constriction: $V_c = (3/2)^2 6.26 = 14.09$ m/s ; $Q = A_c V_c = \pi(0.18)^2 14.09 = 1.435$ m^3/s ◄

Energy equation using absolute pressure head, from constriction (c) to lower reservoir surface:

$(p_c/\gamma)_{abs} + 0 + V_c^2/2g = (p_{atm}/\gamma) + 4 + 0 + V_x^2/2g$ where $V_x^2/2g$ is the discharge loss.

Table A.3 at 3 km altitude: $p_{atm} = 70.121$ kN/m^2 abs

$(p_c/\gamma)_{abs} + 0 + 14.09^2/(2 \times 9.81) = (70.1/9.81) + 4 + 2$; $(p_c/\gamma)_{abs} = 3.02$ m

$(p_c)_{abs} = 3.02(9.81) = 29.7$ kN/m^2 abs ◄ $(p_c)_{gage} = 29.7 - 70.1 = -40.5$ kN/m^2 ◄

5.18 *Repeat Prob. 5.16, assuming head losses are as follows: 6 inches in the converging section and 30 inches in the diverging section.*

Prob. 5.16: Water flows in Fig. 5.10 (see with Solution 5.13 above). Given h = 8 ft, and the lower water surface is 10 ft above the constriction, find Q. The constriction D = 14 in = (2/3)D at exit. The elevation = 10,000 ft, atmosphere = standard. Find the gage and abs pressures in the constriction.

BG

Eq. 5.14 from water surface in upper reservoir (= elev datum) to pipe outlet in lower reservoir:

$0 + 0 + 0 = p_x/\gamma - (8 + p_x/\gamma) + V_x^2/2g + (6 + 30)/12$; $V_x^2/2g = 5$ ft; $V_x = 17.94$ fps

In constriction: $V_c = (3/2)^2 17.94 = 40.4$ fps; $V_c^2/2g = 25.3$ ft

$Q = A_c V_c = \pi(7/12)^2 40.4 = 43.2$ cfs ◄

Energy equation using absolute pressure head, from constriction (c) to lower reservoir surface:

$(p_c/\gamma)_{abs} + 0 + V_c^2/2g = (p_{atm}/\gamma) + 10 + 0 + V_x^2/2g + 30/12$ where $V_x^2/2g$ is the discharge loss.

Table A.3 at 10,000 feet altitude: $p_{atm} = 10.11$ psia

$(p_c/\gamma)_{abs} + 0 + 25.3 = 10.11(144)/62.4 + 10 + 5 + 2.50$; $(p_c/\gamma)_{abs} = 15.51$ ft

$(p_c)_{abs} = 15.51(62.4) = 968$ psfa $= 6.72$ psia ◄ $(p_c)_{gage} = 6.72 - 10.11 = -3.39$ psi ◄

5.19 *Repeat Prob. 5.17, assuming head losses are as follows: 0.15 m in the converging section and 0.8 m in the diverging section.*

Prob. 5.17: Water flows in Fig. 5.10 (see with Solution 5.13 above). Given h = 2 m, and the lower water surface is 4 m above the constriction, find Q. The constriction D = 36 cm = (2/3)D at exit. The elevation = 3000 m, atmosphere = standard. Find the gage and abs pressures in the constriction.

SI

$0 + 0 + 0 = p_x/\gamma - (2 + p_x/\gamma) + V_x^2/2g + (0.15 + 0.8)$; $V_x^2/2g = 1.05$ m ; $V_x = 4.54$ m/s

In constriction: $V_c = (3/2)^2 4.54 = 10.21$ m/s; $V_c^2/2g = 5.32$ m

$Q = A_c V_c = \pi(0.18)^2 10.21 = 1.039\ m^3/s$ ◄

Energy equation using absolute pressure head, from constriction (c) to lower reservoir surface:

$(p_c/\gamma)_{abs} + 0 + V_c^2/2g = (p_{atm}/\gamma) + 4 + 0 + V_x^2/2g + 0.8$

Table A.3 at 3 km altitude: $p_{atm} = 70.121\ kN/m^2$ abs

$(p_c/\gamma)_{abs} + 0 + 5.32 = (70.1/9.81) + 4 + 1.05 + 0.8$; $(p_c/\gamma)_{abs} = 7.68$ m

$(p_c)_{abs} = 7.68(9.81) = 75.4\ kN/m^2$ abs ◄ $(p_c)_{gage} = 75.4 - 70.1 = 5.24\ kN/m^2$ ◄

/more...

Sec. 5.14: **Applications of Hydraulic Grade Line and Energy Line** -- *Exercises*

5.14.1 *A pump, having an efficiency of 90%, lifts water to a height of 465 ft at the rate of 250 cfs. The friction loss in the pipe is 35 ft. What is the required horsepower? Also provide a sketch of the energy line and the hydraulic grade line of this system.*

BG

Eq. 5.32: Water hp $= \gamma Q(\Delta z + h_L)/550 = (62.4)250(465 + 35)/550 = 14{,}180$ hp

From Eq. 5.34: Input power $= 14{,}180/0.90 = 15{,}760$ hp ◄

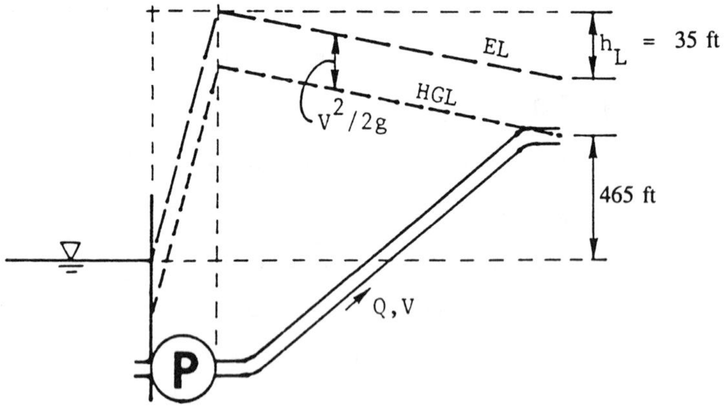

5.14.2 *A pump, having an efficiency of 90%, lifts water to a height of 155 m at the rate of 7.5 m³/s. The friction loss in the pipe is 13 m. What is the required pump power in kW? Also provide a sketch of the energy line and the hydraulic grade line of this system.*

SI

From Eqs. 5.33 and 5.34: Input power $= \gamma Q(\Delta z + h_L)/\eta = (9.81)7.5(155 + 13)/0.90 = 13\,730$ kW ◄

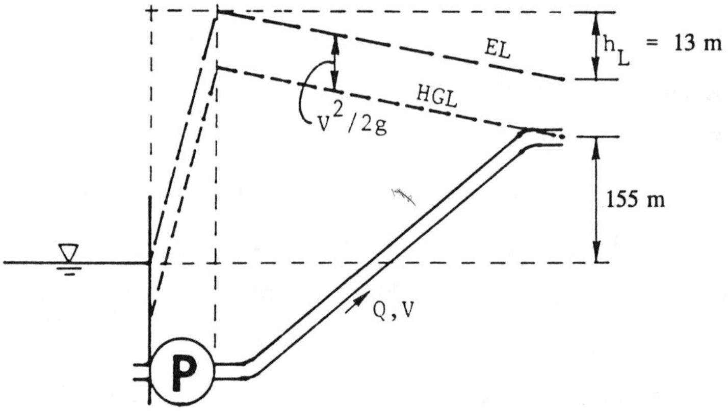

5.14.3 *Assume there is friction loss in the siphon of Fig. X5.7.1. The loss between the intake and B is 0.7 m and between B and N is 1.1 m. What is the rate of discharge and pressure head at B when the diameter is 18 cm?*

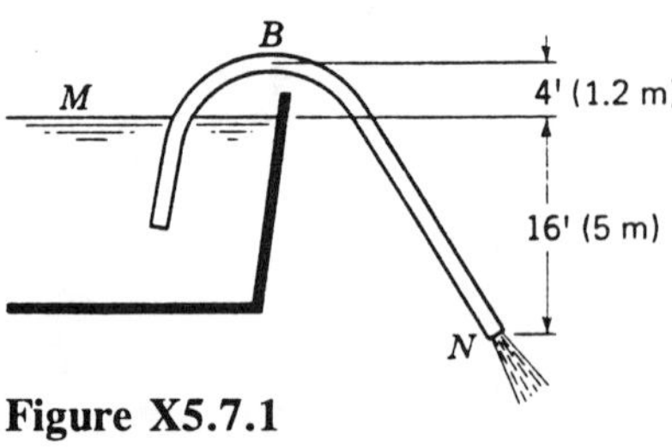

Figure X5.7.1

SI

Energy eq. from M to N: $5 - 0.7 - 1.1$

$= V_N^2/2g = 3.2 \text{ m} = V_B^2/2g$; $V_B = 7.92$ m/s

$Q = \pi(0.18/2)^2 7.92 = 0.202$ m³/s ◄

Energy eq. from M to B: $5 - 0.7 = p_B/\gamma + 6.2 + V_B^2/2g$; $p_B/\gamma = -5.10$ m ◄

5.14.4 *Assume there is friction loss in the siphon of Fig. X5.7.1. The loss between the intake and B is 3 ft and between B and N is 4 ft. What is the rate of discharge and pressure head at B when the diameter is 8 in?*

Fig. X5.7.1: See with Solution 5.14.3 above.

BG

Energy eq. from M to N: $16 - 3 - 4 = V_N^2/2g = V_N^2/2g = 9 \text{ ft} = V_B^2/2g$

$V_B = \sqrt{2(32.2)9} = 24.1$ fps; $Q = \pi(4/12)^2 24.1 = 8.40$ cfs ◀

$16 - 3 = p_B/\gamma + 20 + V_B^2/2g$; $p_B/\gamma = -16.00$ ft ◀

5.14.5 *Refer to Fig. X5.7.1. Find the maximum difference in elevation between* M *and* N *instead of the 5 m shown in the sketch. Assume friction is neglected and the minimum pressure allowable in the siphon is a vacuum of −9.8 m of water.*

Fig. X5.7.1: See with Solution 5.14.3 above.

SI

Let z = the elevation difference between M and N.

Eq. 5.15 for M to N: $0 + z + 0 = 0 + 0 + V_N^2/2g$; $V_N^2/2g = V_B^2/2g = z$

Eq. 5.15 for M to B: $0 + z + 0 = -9.8 + (z + 1.2) + z$; $z = 8.60$ m ◀

5.14.6 *Refer to Fig. X5.7.1. Find the maximum difference in elevation between* M *and* N *instead of the 16 ft shown in the sketch. Assume friction is neglected and the minimum pressure allowable in the siphon is a vacuum of −32.8 ft of water.*

Fig. X5.7.1: See with Solution 5.14.3 above.

BG

Let z = the elevation difference between M and N.

Eq. 5.15 for M to N: $0 + z + 0 = 0 + 0 + V_N^2/2g$; $V_N^2/2g = V_B^2/2g = z$

Eq. 5.15 for M to B: $0 + z + 0 = -32.8 + (z + 4) + z$; $z = 28.8$ ft ◀

5.14.7 *In Fig. X5.14.7 let a = 25 ft, b = 60 ft, c = 40 ft, and d = 2 ft. All the losses of energy are to be ignored when the stream discharging into the air at E has a diameter of 4 in. What are pressure heads at B, C, and D if the diameter of the vertical pipe is 5 in?*

BG

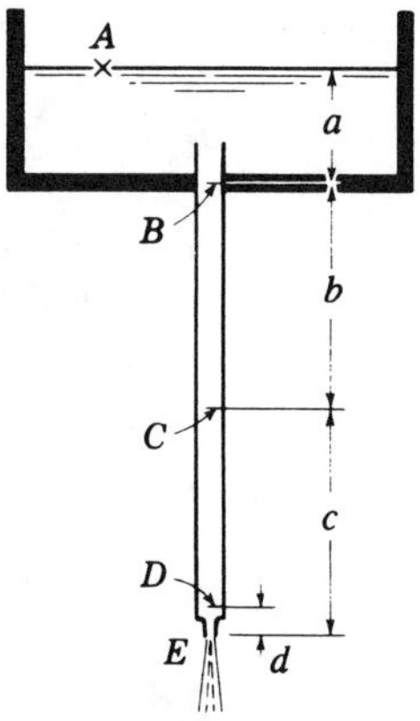

Figure X5.14.7

Eq. 5.15 from A to E: $0 + 125 + 0 = 0 + 0 + V_E^2/2g$; $V_E^2/2g = 125$ ft

$V_B^2/2g = V_C^2/2g = V_D^2/2g = (4/5)^4 V_E^2/2g = 51.2$ ft

From A to B: $0 + 125 + 0 = p_B/\gamma + 100 + 51.2$; $p_B/\gamma = -26.6$ ft ◀

From A to C: $0 + 125 + 0 = p_C/\gamma + 40 + 51.2$; $p_C/\gamma = 33.8$ ft ◀

From D to E: $p_D/\gamma + 2 + 51.2 = 0 + 0 + V_E^2/2g = 125$

$p_D/\gamma = 71.8$ ft ◀

5.14.8 *In Fig. X5.14.7 let a = 7.5 m, b = c = 15 m, and d = 30 cm. All the losses of energy are to be ignored when the stream discharging into the air at E has a diameter of 8 cm. What are pressure heads at B, C, and D if the diameter of the vertical pipe is 12 cm?*

Fig. X5.14.7: See with Solution 5.14.7 above.

SI

Eq. 5.15 from A to E: $0 + 37.5 + 0 = 0 + 0 + V_E^2/2g$; $V_E^2/2g = 37.5$ m

$V_B^2/2g = V_C^2/2g = V_D^2/2g = (8/12)^4 V_E^2/2g = 7.41$ m

From A to B: $0 + 37.5 + 0 = p_B/\gamma + 30 + 7.41$; $p_B/\gamma = 0.0926$ m ◀

From A to C: $0 + 37.5 + 0 = p_C/\gamma + 15 + 7.41$; $p_C/\gamma = 15.09$ m ◀

From D to E: $p_D/\gamma + 0.30 + 7.41 = V_E^2/2g = 37.5$; $p_D/\gamma = 29.8$ m ◀

Sec. 5.14: Applications of Hydraulic Grade Line and Energy Line -- *Problems*

5.20 *In Fig. P5.20 friction loss between A and B is neglected while between B and C it is $0.1(V_B^2/2g)$. Find the pressure heads at A and C if the liquid is flowing from A to C at the rate of 300 L/s.*

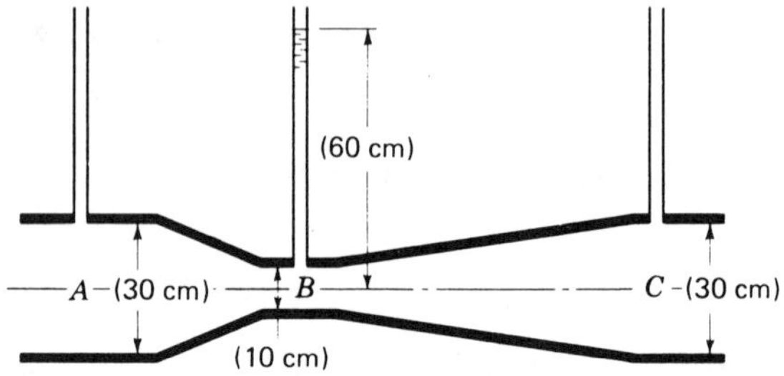

Figure P5.20

SI

$V_A = Q/A_A = (0.30 \text{ m}^3/\text{s})/(\pi 0.15^2) = 4.24 \text{ m/s} = V_c$

$V_B = Q/A_B = 0.30/(\pi 0.05^2) = 38.2$ m/s;

$p_B/\gamma = 0.60$ m (from figure)

$V_A^2/2g = V_C^2/2g = 4.24^2/(2 \times 9.81) = 0.918$ m;

$V_B^2/2g = 38.2^2/(2 \times 9.81) = 74.4$ m

Eq. 5.15 from A to B: $p_A/\gamma + 0 + V_A^2/2g = p_B/\gamma + 0 + V_B^2/2g$

$p_A/\gamma + 0.918 = 0.60 + 74.4$; $P_A/\gamma = 74.0$ m ◀

Eq. 5.15 from B to C: $p_B/\gamma + 0 + V_B^2/2g - 0.1V_B^2/2g = p_C/\gamma + 0 + V_C^2/2g$

$0.60 + (1 - 0.1)74.4 = p_C/\gamma + 0.918$; $p_C/\gamma = 66.6$ m ◀

5.21 *In Fig. P5.21 assume the tube flows full. At B, the diameter of the tube is 3 in and the diameter of the water jet discharging into the air is 4.5 in. (a) If all friction losses are neglected, what are the velocity and the pressure head at B if h = 10 ft. (b) What is the rate of discharge in cfs? And what would it be if the tube were cut off at B?*

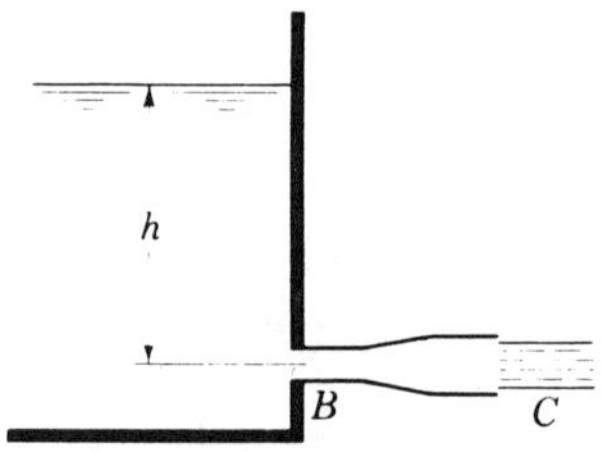

Figure P5.21

BG

(*a*) Eq. 5.15, from water suface to C: $0 + 10 + 0 = 0 + 0 + V_C^2/2g$;

$V_C^2/2g = 10$ ft

Eq. 5.15, from B to C: $p_B/\gamma + 0 + (4.5/3)^4 10 = 0 + 0 + 10$;

$p_B/\gamma = 10(1 - 5.06) = -40.6$ ft ◀

$V_B^2/2g = (4.5/3)^4 10 = 50.6$; $V_B = \sqrt{2(32.2)50.6} = 57.1$ fps ◀

(*b*) $Q = A_B V_B = (\pi/4)(3/12)^2 57.1 = 2.80$ cfs ◀

If tube were cut off at B, $V_B^2/2g = 10$ ft; $V_B = 25.4$ fps ; $Q = (\pi/4)(0.25)^2 25.4 = 1.246$ cfs ◀

5.22 *Referring to Fig. P5.21, assume the tube flows fully and all friction losses are neglected. The diameter at B is 6 cm and the diameter of the jet discharging into the air is 8 cm. If h = 5 m, what is the flow rate? What is the pressure head at B? What would be the flow rate if the tube were cut off at B?*

Fig. P5.21: See with Solution 5.21 above.

SI

Eq. 5.15, from water suface to *C*: $0 + 5 + 0 = 0 + 0 + V_C^2/2g; \; V_C^2/2g = 5$ m

$V_C = \sqrt{2(9.81)5} = 9.90$ m/s ; $Q = A_C V_C = \pi(0.04)^2 9.90 = 0.0498$ m³/s = 49.8 L/s ◄

$V_B = 9.90(8/6)^2 = 17.61$ m/s; $V_B^2/2g = 15.80$ m

Eq. 5.15 from *B* to *C*: $p_B/\gamma + 0 + V_B^2/2g = p_C/\gamma + 0 + V_C^2/2g$

$p_B/\gamma + 15.80 = 0 + 5; \; p_B/\gamma = -10.80$ m ◄

If the tube were cut off at *B*: $V_B^2/2g = 5$ m; $V_B = 9.90$ m/s

$Q = A_B V_B = \pi(0.03)^2 9.90 = 0.0280$ m³/s = 28.0 L/s ◄

5.23 *In Fig. P5.23 friction losses in the pipe are $V^2/2g$ with the barometer pressure at 12.50 psia. The liquid in the suction pipe has a velocity of 7 fps. What would be the maximum allowable value of z if the liquid were (a) water at 70°F; (b) gasoline at a vapor pressure of 9 psia with a specific weight of 47 lb/ft³ ?*

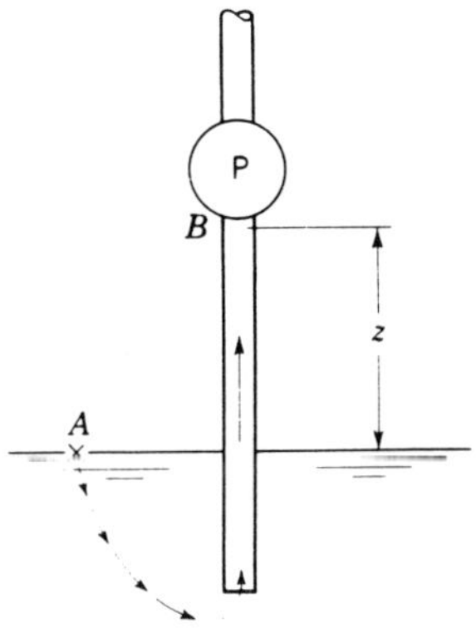

Figure P5.23

BG

Eq. 5.14 (using absolute pressure heads) from *A* to *B*, where, to prevent cavitation, $p_B = p_{min} = p_v$:

$$\frac{12.50(144)}{\gamma} + 0 + 0 = 144\frac{p_v}{\gamma} + z_{max} + \frac{7^2}{2(32.2)} + \frac{7^2}{2(32.2)}$$

$$= 144\frac{p_v}{\gamma} + z_{max} + 1.522$$

from which $z_{max} = (12.50 - p_v)144/\gamma - 1.522$

(*a*) Table A.1 for water at 70°F: $\gamma = 62.30$ pcf, $p_v = 0.36$ psia from which $z_{max} = 26.5$ ft ◄

(*b*) For gasoline ($\gamma = 47$ pcf, $p_v = 9$ psia, given) : $z_{max} = 9.20$ ft ◄

5.24 *In Fig. P5.23 friction losses in the pipe are $V^2/2g$ with the barometer pressure at 90 kN/m². The liquid in the suction pipe has a velocity of 1.8 m/s. What would be the maximum allowable value of z if the liquid were (a) water at 20°C; (b) gasoline at a vapor pressure of 49 kN/m² abs, with a specific weight of 8 kN/m³?*

Fig. P5.23: See with Solution 5.23 above.

SI

Eq. 5.14 from *A* to *B*, where, to prevent cavitation, $p_B = p_{min} = p_v$:

$90/\gamma + 0 + 0 = p_v/\gamma + z_{max} + 1.8^2/(2 \times 9.81)(1 + 1) = p_v/\gamma + z_{max} + 0.330$

from which $z_{max} = (90 - p_v)/\gamma - 0.330$

(*a*) Table A.1 for water at 20°C: $\gamma = 9.789$ kN/m³, $p_v = 2.34$ kN/m² abs

from which $z_{max} = 8.62$ m ◄

(*b*) For gasoline ($\gamma = 8$ kN/m³, $p_v = 49$ kN/m² abs, given) : $z_{max} = 4.79$ m ◄

5.25 *A discharge pressure gage reading, taken at a point 6.5 ft above the center line of a pump, is 25 psi. A suction pressure gage reading, taken 2.5 ft below the center line, indicates a vacuum of 12 inHg when gasoline (s = 0.75) is pumped at the rate of 1.5 cfs. The diameter of the suction and discharge pipes of the pump are 8 and 6 in, respectively. What is the power delivered to the fluid? Sketch the energy line and the hydraulic grade line.*

BG

$$V_1 = \frac{Q}{A_1} = \frac{1.5}{(\pi/4)(8/12)^2} = 4.30 \text{ fps} ; \quad V_2 = \frac{Q}{A_2} = \frac{1.5}{(\pi/4)(6/12)^2} = 7.64 \text{ fps}$$

Eq. 5.30 (with $h_M = +h_p$, $h_L = 0$) and Eq. 5.28, from suction to discharge in feet of gasoline:

$$-\left(\frac{12}{12}\right)\frac{13.56}{0.75} - 2.5 + \frac{(4.30)^2}{2(32.2)} + h_p = \frac{25(144)}{0.75(62.4)} + 6.5 + \frac{(7.64)^2}{2(32.2)}$$

$$h_p = 18.08 + 2.5 - 0.287 + 76.9 + 6.5 + 0.906 = 104.6 \text{ ft}$$

Eq. 5.31: Power $= \gamma Q h_p = (0.75 \times 62.4)1.5(104.6) = 7345$ ft·lb/sec $= 13.35$ hp ◀

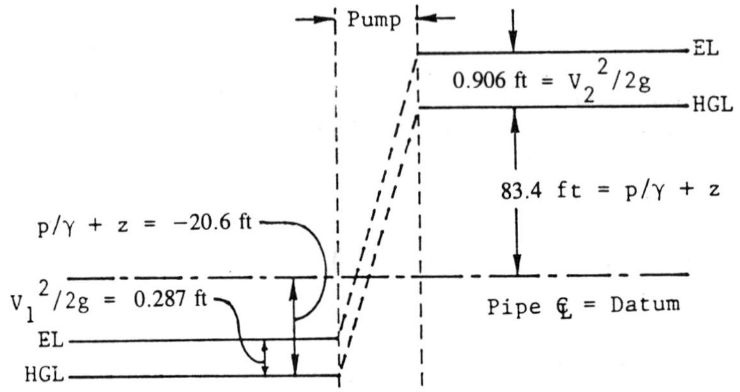

5.26 *For this problem, use the same data as in Sample Prob. 5.10, except that instead of the pump developing 80 ft of head, it delivers 110 hp to the water. Find the new flow rate. Plot the energy line and hydraulic grade line. Calculate the pressure on the suction side of the pump.*

Sample Prob. 5.10: In Fig. S5.10, $h_L = 5V_6^2/2g$ in the 6-in-dia suction line, $h_L = 12V_4^2/2g$ in the 4-in-dia discharge line.

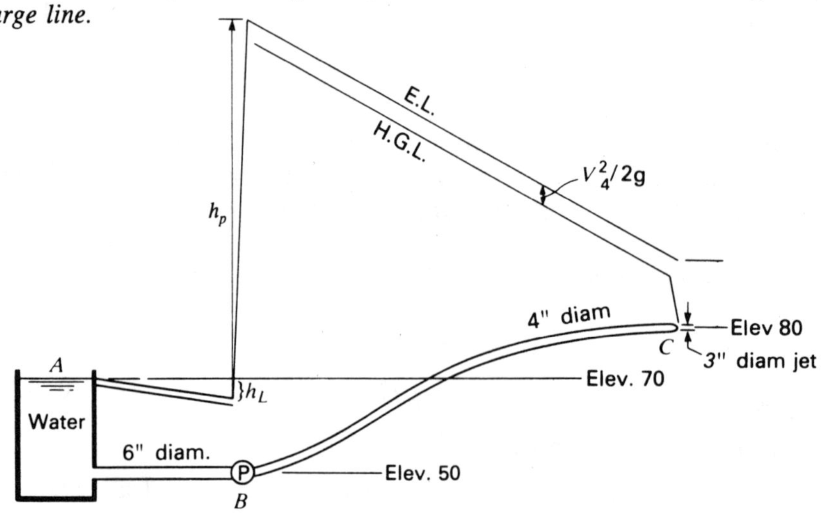

Figure S5.10

/cont...

BG

$V_6 = Q/A_6 = 5.09Q$; $V_4 = Q/A_4 = 11.46Q$; $V_3 = Q/A_3 = 20.4Q$

Energy equation from reservoir surface to jet:

$(0 + 70 + 0) - 5V_6^2/2g + h_p - 12V_4^2/2g = 0 + 80 + V_3^2/2g$

Expressing all velocities in terms of Q:

$70 - 2.01Q^2 + h_p - 24.5Q^2 = 80 + 6.44Q^2$, i.e., $h_p = 10 + 32.9Q^2$

where, from Eq. 5.32, HP = 110 = $62.4Qh_p/550$, i.e., $h_p = 970/Q$

Thus $970/Q = 10 + 32.9Q^2$; by trial $Q = 3.06$ cfs ◀

$(0 + 70 + 0) - 2.01Q_2 = p_s/\gamma + 50 + V_6^2/2g$ where $V_6 = 5.09Q = 15.56$ fps, $V_6^2/2g = 3.76$ ft

∴ on the suction side of the pump, $p_s/\gamma = 70 - 18.80 - 50 - 3.76 = -2.56$ ft

$p_s = -2.56(62.4)/144 = 1.108$ psi suction ◀

Alternatively (from Sec. 3.5): $p_s/\gamma = -2.56(29.92 \text{ inHg}/33.91 \text{ ft}) = 2.26$ inHg vacuum ◀

The energy line and hydraulic grade line are similar to those in the figure of Sample Prob. 5.10, but with the following elevations:

	EL	HGL
Reservoir surface	70.00 ft	70.00 ft
6 inch pipe entrance	70.00	66.24
Pump entrance	51.20	47.44
Pump exit	368.58	349.55
Before nozzle	140.22	121.19
After nozzle	140.22	80.00
	▲	▲

5.27 *In Fig. 4.12 the velocity of the undisturbed field is 22 fps and the velocities very near the surface at radii from the "source" making angles with the axis of 0, 60, 120, 150° are 0, 17.5, 23.7 and 21.9 fps, respectively. What will be the elevation of the surface of a liquid relative to that of the free surface of the undisturbed field? (This problem illustrates the way in which the water surface drops alongside a bridge pier or past the side of a moving ship.)*

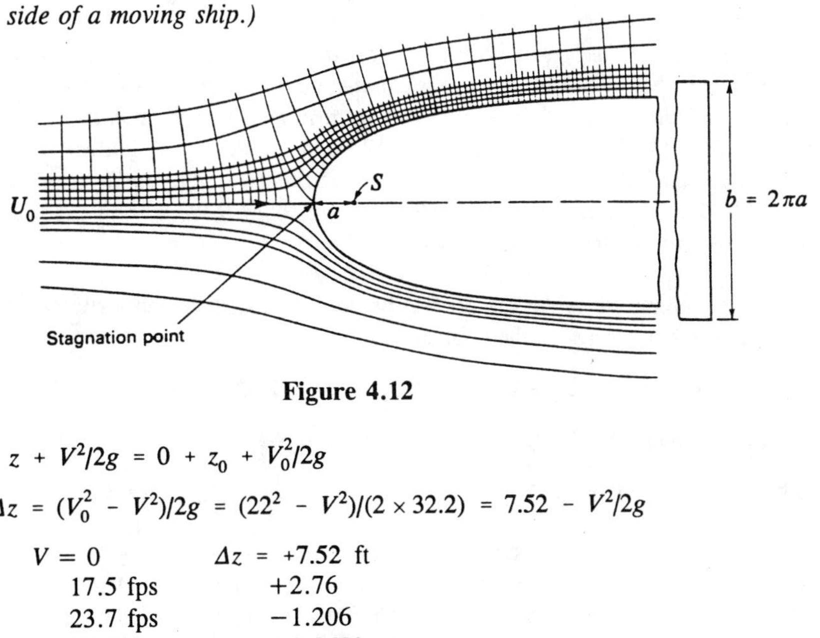

Figure 4.12

BG

Eq. 5.15: $0 + z + V^2/2g = 0 + z_0 + V_0^2/2g$

So $z - z_0 = \Delta z = (V_0^2 - V^2)/2g = (22^2 - V^2)/(2 \times 32.2) = 7.52 - V^2/2g$

At			
	0°	$V = 0$	$\Delta z = +7.52$ ft
	60°	17.5 fps	+2.76
	120°	23.7 fps	−1.206
	150°	21.9 fps	+0.0682
			▲

5.28 *In Fig. 4.12 the velocity of the undisturbed field is 6 m/s and the velocities very near the surface at radii from the "source" making angles with the axis of 0, 60, 120, 150° are 0, 4.8, 6.5, 6.0 m/s respectively. What will be the elevation of the surface of a liquid relative to that of the free surface of the undisturbed filed? (This problem illustrates that way in which the water surface drops alongside a bridge pier or past the side of a moving ship.)*

Fig. 4.12: See with Solution 5.27 above.

SI

Eq. 5.15: $0 + z + V^2/2g = 0 + z_0 + V_0^2/2g$

So $z - z_0 = \Delta z = (V_0^2 - V^2)/2g = (6^2 - V^2)/(2 \times 9.81) = 1.835 - V^2/2g$

At			
	0°	$V = 0$	Δz = +1.835 m
	60°	4.80 m/s	+0.661
	120°	6.50 m/s	−0.319
	150°	6.00 m/s	0.0

▲

5.29 *If the body shown in Fig. 4.12 is not two-dimensional but is a solid of revolution about a horizontal axis, the flow will be three-dimensional and the streamlines will be differently spaced. Also, the distance between the stagnation point and the "source" will be d/4, where d is the diameter at a great distance from the stagnation point. At points very near the surface at radii from the source making angles with the axis of 0, 60, 120, and 150°, the velocities are 0, 14.0, 21.3, and 19.8 fps respectively when the velocity of the undisturbed field is 19 fps. If the body is an airship and the atmospheric pressure in the undisturbed field is 14 psia, what will be the pressure at these points, for air temperature of 53.9°F?*

Fig. 4.12: See with Solution 5.27 above.

BG

Table A.3 for standard atmosphere at 14 psia: T = 54.0°F (by linear interpolation)

Table A.5 for air: R = 1715 ft²/(s²·°R) ; Sec. 2.7: $\gamma = \dfrac{gp}{RT} = \dfrac{32.2(14 \times 144)}{1715(460 + 54)} = 0.0736$ lb/ft³

From Eq. 5.15: $\Delta p/\gamma = (V_0^2 - V^2)/2g = (19^2 - V^2)/(2 \times 32.2) = 5.61 - V^2/2g$

At				
	0° F	$V = 0$ fps	$\Delta p/\gamma$ = +5.61 ft	= +0.413 psf
	60°	14.0	+2.56	+0.1887
	120°	21.3	−1.439	−0.1060
	150°	19.8	−0.482	−0.0355

▲

5.30 *In Prob. 5.29 assume the body is a submarine with diameters at the four points of 0, 8.24, 14.28, and 15.90 ft respectively. If the submarine is submerged in the ocean (γ = 64.1 lb/ft³) with its axis 50 ft below the surface, find the pressures in pounds per square inch at these points along the top and along the bottom.*

Prob. 5.29: The body shown in Fig. 4.12 (with Solution 5.27 above) is a solid of revolution about a horizontal axis, and a = b/4. At points very near the surface at radii from the source making angles with the axis of 0, 60, 120, and 150°, the velocities are 0, 14.0, 21.3, and 19.8 fps respectively when the velocity of the undisturbed field is 19 fps.

BG

Take datum at submarine centerline 50 feet below water surface and write energy equation 5.15 along the upper and lower boundaries of the submarine.

$p/\gamma + z + V^2/2g = p_0/\gamma + z_0 + V_0^2/2g$; $p/\gamma \pm \Delta z + V^2/2g = 50 + 0 + 19^2/(2 \times 32.2)$; Δz = diam/2

$p/\gamma = 55.61 \mp$ diam/2 $- V^2/2g$; $p = (p/\gamma$ ft$)64.1/144$ psi

θ	V fps (given)	diam ft (given)	Top p/γ ft	Bot p/γ ft	Top p psi	Bot p psi
0°	0	0	55.61	55.61	24.8	24.8
60°	14.0	8.24	48.44	56.68	21.6	25.2
120°	21.3	14.28	41.42	55.70	18.4	24.8
150°	19.8	15.90	41.57	57.47	18.5	25.6

▲ ▲

Sec. 5.15: Method of Solution of Flow Problems -- *Exercises*

5.15.1 *Refer to Sample Problem 5.11. If the depths upstream and downstream of the gate were 8.0 ft and 4.0 ft respectively, find the flow rate per foot of channel width.*

BG

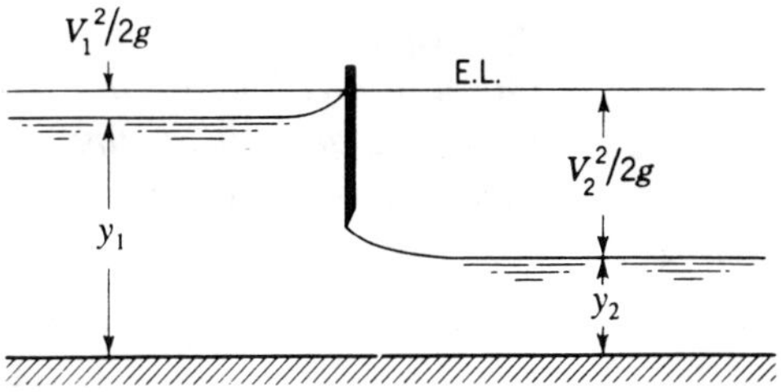

Figure S5.11

Energy Eq: $8 + 0 + V_1^2/2g = 4 + 0 + V_2^2/2g$

Continuity: $A_1V_1 = A_2V_2$ yields $V_2 = (8/4)V_1$

Combining these two equations gives $V_1 = 9.27$ fps

Hence $Q = A_1V_1 = (8 \times 1)9.27 = 74.1$ cfs per ft of channel width ◄

5.15.2 *Refer to Sample Prob. 5.11. If the depths upstream and downstream of the gate were 2.0 m and 0.6 m respectively, find the flow rate per meter of channel width.*
See Fig. S5.11 with Solution 5.15.1.

SI

Energy Eq. 5.15: $2 + 0 + V_1^2/2g = 0.6 + 0 + V_2^2/2g$

Continuity: $A_1V_1 = A_2V_2$ yields $V_2 = (2/0.6)V_1$

Combining these two equations gives $V_1 = 1.648$ m/s

Hence $Q = A_1V_1 = (2 \times 1)1.648 = 3.30$ m^3/s per m ◄

5.15.3 *Refer to Sample Prob. 5.11. Suppose the gate opening is reduced so the depth downstream is 2.0 ft. Find the upstream depth under these conditions if the flow rate remains constant at 45 ft^3/sec per ft of width.*
See Fig. S5.11 with Solution 5.15.1.

BG

Energy eq. 5.15: $y_1 + V_1^2/2g = y_2 + V_2^2/2g$, where z's cancel and $y = p/\gamma$

where $V_1 = Q/A_1 = 45/(y_1 \times 1) = 45/y_1$ and $V_2 = Q/A_2 = 45/(2.0 \times 1) = 22.5$ ft/sec

so: $y_1 + \dfrac{(45/y_1)^2}{2(32.2)} = 2.0 + \dfrac{22.5^2}{2(32.2)}$ or $y_1 + \dfrac{31.4}{y_1^2} = 9.86$ ft ; By trial $y_1 = 9.51$ or 2.00 ft

2.0 ft is the given downstream depth, so upstream depth = 9.51 ft ◄

5.15.4 *Refer to Sample Prob. 5.11. Suppose the gate opening is reduced so the depth downstream is 0.7 m. Find the upstream depth under these conditions if the flow rate remains constant at 4.24 m^3/s per m of width.*
See Fig. S5.11 with Solution 5.15.1.

SI

$Q = 4.24$ m^3/s per meter of width (Sample Prob. 5.11)

Energy Eq. 5.15: $y_1 + V_1^2/2g = y_2 + V_2^2/2g$, where z's cancel and $y = p/\gamma$,

and where $V_1 = Q/A_1 = 4.24/(y_1 \times 1) = 4.24/y_1$ and $V_2 = Q/A_2 = 4.24/(0.7 \times 1) = 6.06$ m/s

So: $y_1 + \dfrac{(4.24/y_1)^2}{2(9.81)} = 0.7 + \dfrac{6.06^2}{2(9.81)}$ or $y_1 + \dfrac{0.916}{y_1^2} = 2.57$ m. By trial, $y_1 = 0.700$ or 2.41 m

0.7 m is the given downstream depth. So upstream depth = 2.41 m ◄

Sec. 5.16: Jet Trajectory -- *Exercises*

5.16.1 *It is required to throw a fire steam so as to reach the window in the wall shown in Fig. X5.16.1 (h = 14 m). Assuming a jet velocity of 25 m/s and neglecting air friction, find the angle (or angles) of inclination which will achieve this result, given d = 23 m and a = 2 m.*

SI

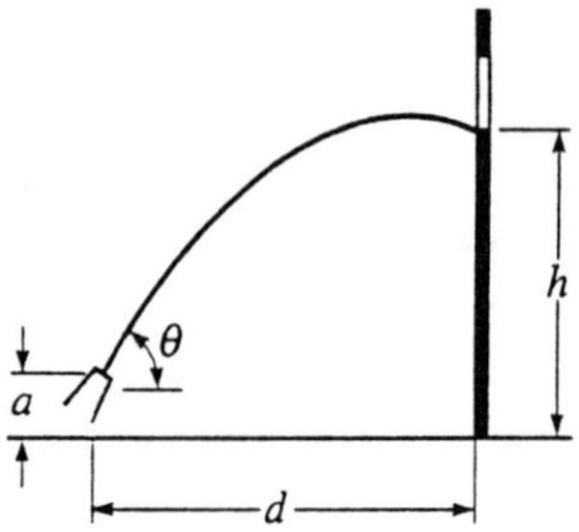

Figure X5.16.1

Eq. 5.37: $(14 - 2) = \frac{25 \sin\theta}{25 \cos\theta}(23) - \frac{9.81}{2(25)^2 \cos^2\theta}(23)^2$

$12 = 23 \tan\theta - 4.15(1 + \tan^2\theta)$

$\tan^2\theta - 5.54 \tan\theta + 3.89 = 0$, quadratic; $\tan\theta = 0.825$ or 4.71

$\theta = 39.5°$ or $78.0°$ (either angle will work) ◄

5.16.2 *A jet issues horizontally from an orifice in the vertical wall of a large tank (Fig. X5.16.2). Neglecting air resistance, determine the velocity of the jet for the following variety of trajectories: (a) x = 1.0 m, y = 1.0 m; (b) x = 2.0 m, y = 2.0 m; (c) x = 3.0 m, y = 3.0 m; (d) x = 4.0 m, y = 4.0 m. Express the answers in m/s.*

SI

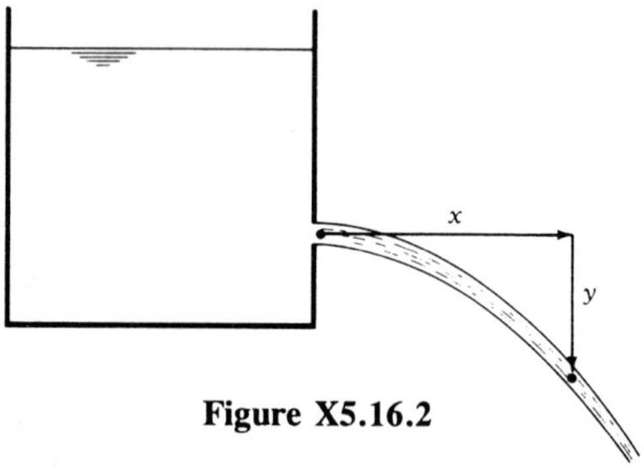

Figure X5.16.2

$x = V_0 t$; $y = (1/2)gt^2 = (1/2)g(x/V_0)^2$

Eq. 5.38: $V_0 = x(g/2y)^{1/2}$

(*a*) $x = 1$ m, $y = 1$ m: $V_0 = 2.21$ m/s ◄
(*b*) $x = 2$ m, $y = 2$ m: $V_0 = 3.13$ m/s ◄
(*c*) $x = 3$ m, $y = 3$ m: $V_0 = 3.84$ m/s ◄
(*d*) $x = 4$ m, $y = 4$ m: $V_0 = 4.43$ m/s ◄

5.16.3 *Freshwater sewage effluent is discharged from a horizontal outfall pipe on the floor of the ocean at a point where the depth is 120 ft. When the ocean is still, the jet is observed to rise to the surface at a point 95 ft horizontally from the end of the pipe. Assuming the ocean water to have a specific gravity of 1.03 and neglecting fluid friction and mixing of the jet with the ocean water, find the velocity at the end of the outfall. [Note: In this case the jet is submerged, and it is no longer possible to neglect the density of the surrounding medium; hence the value of g is Eqs. (5.37) and (5.38) must be adjusted accordingly.]*

BG

Replace g in trajectory equations with the force per unit mass ($F/m = a = g'$) of the jet fluid:

Eq. 5.39: $g' = 32.2\left(1 - \frac{1.03}{1.00}\right) = -0.966$ lb/slug $= -0.966$ ft/sec^2

Eq. 5.38, modified: $V_0 = x\sqrt{\frac{g'}{2z}} = 95\sqrt{\frac{-0.966}{2(-120)}} = 6.03$ fps ◄

Sec. 5.16: Jet Trajectory -- *Problems*

5.31 *By manipulation of Eq. (5.37), demonstrate that it represents a standard parabola of the form $z - z_0 = a(x - x_0)^2$, where a is a constant and x_0 and z_0 are the coordinates of the vertex.*

N

$$\text{Eq. 5.37:}\quad z = \frac{V_{z0}}{V_{x0}}x - \frac{g}{2V_{x0}^2}x^2 = -\frac{g}{2V_{x0}^2}\left(x^2 - \frac{2V_{x0}V_{z0}}{g}x\right)$$

$$\text{Completing the square,}\quad z = -\frac{g}{2V_{x0}^2}\left[x^2 - \frac{2V_{x0}V_{z0}}{g}x + \frac{V_{x0}^2V_{z0}^2}{g^2}\right] + \frac{V_{z0}^2}{2g}$$

$$\therefore\ z - \frac{V_{z0}^2}{2g} = -\frac{g}{2V_{x0}^2}\left[x - \left(\frac{V_{x0}V_{z0}}{g}\right)\right]^2 \quad \text{or} \quad (z - z_0) = +a(x - x_0)^2 \quad ◀$$

where $z_0 = V_{z0}^2/2g$, $x_0 = V_{x0}V_{z0}/g$ and $a = -g/(2V_{x0}^2)$

5.32 *Find the maximum ideal horizontal range of a jet having an initial velocity of 85 fps. At what angle of inclination is this obtained?*

BG

$$\text{Eq. 5.37 for horizontal range } (z = 0):\quad \frac{V_{z0}}{V_{x0}}x = \frac{g}{2V_{x0}^2}x^2 \qquad (1)$$

From Fig. 5.11: $V_{z0}/V_{x0} = \tan\theta$, $V_{x0} = V_0\cos\theta$

Substituting these into (1): $x\tan\theta = \dfrac{gx^2}{2V_0^2\cos^2\theta}$

Thus for $z = 0$, $x = (2V_0^2/g)\cos^2\theta\tan\theta = (V_0^2/g)\sin 2\theta$

For x_{max}, $dx/d\theta = (2V_0^2/g)\cos 2\theta = 0$; $2\theta = 90°$

$\theta = 45°$ ◀ and $x_{max} = V_0^2/g = (85)^2/32.2 = 224$ ft ◀

Sec. 5.17: Flow in Curved Path -- *Exercises*

5.17.1 *Figure X5.17.1 shows a two-dimensional ideal flow in a vertical plane. Data are as follows: r = 12 ft, b = 5 ft, γ = 62.4 lb/ft³, V = 24 fps. If the pressure at A is 6 psi, find the pressure at B.*

BG

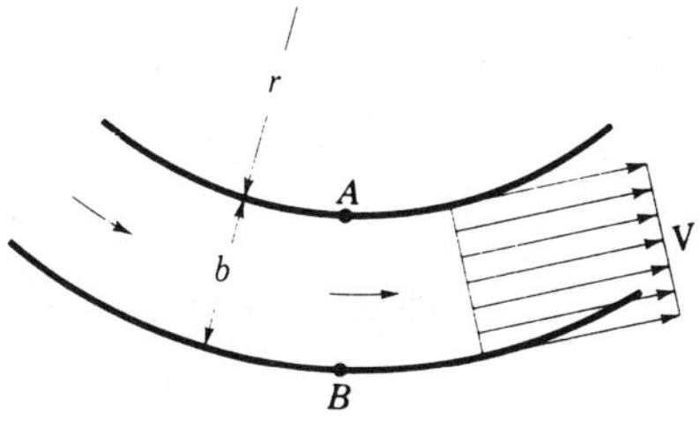

Figure X5.17.1

Pressure increase:

(*a*) Elevation effect: $\Delta p = \gamma(\Delta h) = 5\gamma$

(*b*) Curved path effect (Eq. 5.40): $dp = \rho\dfrac{V^2}{r}dr$

$$\text{Integrating:}\quad \int_{12}^{17} dp = \rho V^2\int_{12}^{17}\frac{dr}{r} = \frac{\gamma}{g}V^2\int_{12}^{17}\frac{dr}{r}$$

$$\text{So}\quad \Delta p = p_{17} - p_{12} = \gamma\left(\frac{V^2}{g}\right)[\ln r]_{12}^{17} = \gamma\left(\frac{24^2}{32.2}\right)\ln(17/12) = 6.23\gamma$$

$$\left[\text{Approx. method:}\quad \Delta p \approx \frac{\gamma V^2}{gr}\Delta r = \frac{24^2}{32.2(12 + 2.5)}\gamma(5) = 6.17\gamma\right]$$

Thus p_B = 6 psi + (5 + 6.23)62.4/144 = 10.87 psi ◀

5.17.2 *Refer to Fig. X5.17.1. Flow occurs in a vertical plane. Data are as follows: $r = 7$ m, $b = 3$ m, $\gamma = 9.81$ kN/m^3, $V = 5$ m/s. Find the pressure at A if the pressure at B is 150 kN/m^2.*

Fig. X5.17.1: See with Solution 5.17.1 above.

SI

Elevation effect: $\Delta p = \gamma(\Delta h) = 9.81(3) = 29.43$ kN/m^2

Curved path effect (Eq 5.40): $\Delta p \approx \rho\frac{V^2}{r}\Delta r = \frac{\gamma V^2}{gr}\Delta r = \frac{9.81}{9.81}\frac{5^2}{(7 + 1.5)}3 = 8.82$ kN/m^2

Thus $p_A = 150 - 29.43 - 8.82 = 111.7$ kN/m^2 ◄

Sec. 5.17: Flow in Curved Path -- *Problems*

5.33 *Repeat Exer. 5.17.1. Let $V = Q/A = 24$ fps, but assume a parabolic velocity profile.*

Exer. 5.17.1: The two-dimensional ideal flow in Fig. X5.17.1 (see with Solution 5.17.1) occurs in a vertical plane. Given $r = 12$ ft, $b = 5$ ft $\gamma = 62.4$ pcf, $p_A = 6$ psi. Find p_B.

BG

Table A.7 for parabola: Area = 2/3 of enclosing rectangle.

So $V = (2/3)u_{max}$ and $u_{max} = 1.5V = 1.5(24) = 36$ fps.

Equation of parabola: $(r - 14.5)^2 = k(36 - u)$

Boundary conditions: when $r = 12$ or 17, $u = 0$; when $r = 14.5$, $u = 36$

Substituting into parabola, $k = (2.5)^2/36 = 1/5.76$. Hence $u = 36 - 5.76(r - 14.5)^2$

Elevation effect: $\Delta p = \gamma(\Delta h) = 5\gamma$

Curved path effect:

From Eq. 5.40: $\Delta p = \int_{12}^{17}\rho\frac{u^2}{r}dr = \frac{\gamma}{g}\int_{12}^{17}[36 - 5.76(r - 14.5)^2]^2\frac{dr}{r}$

$$= \frac{\gamma}{g}\int_{12}^{17}\left(33.2r^3 - 1924r^2 + 41{,}400r - 393{,}000 + \frac{1{,}381{,}000}{r}\right)dr$$

$$= (\gamma/g)[33.2r^4/4 - 1924r^3/3 + 41{,}400r^2/2 - 393{,}000r + 1{,}381{,}000 \ln r]_{12}^{17}$$

$$\Delta p = 239\gamma/g = (239/32.2)\gamma = 7.43\gamma$$

Thus $p_B = 6\text{psi} + (5 + 7.43)62.4/144 = 11.39$ psi ◄

/more...

5.34 *Using Fig. X5.17.1, which depicts a two-dimensional ideal flow in a vertical plane, find the pressure at B if the pressure at A is 36 kN/m². Data are as follows: r = 3 m, b = 1.2 m, γ = 9.81 kN/m³, V = Q/A= 6 m/s. Assume a parabolic velocity profile.*

Fig. X5.17.1: See with Solution 5.17.1 above.

SI

Table A.7 for parabola: Area = 2/3 of enclosing rectangle.

So $V = (2/3)u_{max}$ and $u_{max} = 1.5V = 1.5(6) = 9$ m/s

Equation of parabola: $(r - 3.6)^2 = k(9 - u)$

Boundary conditions: when $r = 3$ or 4.2, $u = 0$; when $r = 3.6$, $u = 9$.

Substituting into parabola, $k = 0.04 = 1/25$; hence $u = 9 - 25(r - 3.6)^2$

Elevation effect: $\Delta p = \gamma(\Delta h) = 1.2\gamma$

Curved path effect:

$$\text{From Eq. 5.40:}\quad \Delta p = \int_{3.0}^{4.2} \rho\frac{u^2}{r}dr = \frac{\gamma}{g}\int_{3.0}^{4.2}[9 - 25(r - 3.6)^2]^2\frac{dr}{r}$$

$$= \frac{\gamma}{g}\int_{3.0}^{4.2} 25\left(25r^3 - 360r^2 + 1926r - 4536 + \frac{3969}{r}\right)dr$$

$$= (\gamma/g)25\left[25r^4/4 - 360r^3/3 + 1926r^2/2 - 4536r + 3969\ln r\right]_{3.0}^{4.2}$$

$$= \Delta p = 14.45\gamma/g = (14.45/9.81)\gamma = 1.473\gamma$$

Thus, $p_B = 36 + (1.2 + 1.473)9.81 = 62.2$ kN/m² ◄

Sec. 5.18: Forced or Rotational Vortex -- *Exercises*

5.18.1 *A closed vessel 16 in in diameter completely filled with fluid is rotated at 1800 rpm. What will be the pressure difference between the circumference and the axis of rotation in feet of the fluid and in pounds per square inch if the fluid is (a) air with a specific weight of 0.076 lb/ft³; (b) water at 70°F; (c) oil with a specific weight of 46 lb/ft³?*

BG

$$\text{Eq. 5.41:}\quad \frac{p_2}{\gamma} - \frac{p_1}{\gamma} = \left(\frac{1800\times 2\pi}{60}\right)^2\frac{1}{2(32.2)}\left[\left(\frac{16}{2\times 12}\right)^2 - 0\right] = 245 \text{ ft of any fluid} \quad ◄$$

(*a*) With air at $\gamma = 0.076$ lb/ft³, $\Delta p = (245\times 0.076)/144 = 0.1294$ psi ◄

(*b*) Table A.1 for water at 70°F: $\gamma = 62.30$ lb/ft³

∴ With water at 70°F, $\Delta p = (245\times 62.3)/144 = 106.1$ psi ◄

(*c*) With oil at $\gamma = 46$ lb/ft³, $\Delta p = (245\times 46)/144 = 78.3$ psi ◄

5.18.2 *A closed vessel 120 cm in diameter is completely filled with oil (γ = 8.3 kN/m³) and is rotated at 400 rpm. What will be the pressure difference between the circumference and axis of rotation? Express answer in Pa.*

SI

$r_1 = 0$; $r_2 = 1.20/2 = 0.60$ m

$$\text{Eq. 5.41:}\quad \frac{p_2}{\gamma} - \frac{p_1}{\gamma} = \left(\frac{400\times 2\pi}{60}\right)^2\frac{1}{2(9.81)}(0.60^2 - 0) = 32.2 \text{ m of fluid}$$

$\Delta p = \gamma(32.2) = 8.3(32.2) = 267$ kN/m² = 267 kPa ◄

5.18.3 *An open cylindrical vessel partially filled with water is 2 ft in diameter and rotates about its axis, which is vertical. How many revolutions per minute would cause the water surface at the periphery to be 5 ft higher than the water surface at the axis? What would be the necessary speed for the same conditions if the fluid were mercury?*

BG

$r_1 = 0$; $r_2 = 2 / 2 = 1.0$ ft

Eq. 5.43: $z_2 - z_1 = 5 = \dfrac{\omega^2}{2g}(1.0^2 - 0)$; $\omega^2 = 5(2g) = 10(32.2) = 322$

So $\omega = 17.94$ radians/sec or $(17.94 \times 60)/(2\pi) = 171.4$ rpm ◄

For mercury or any other fluid, the speed would be the same. ◄

Sec. 5.18: Forced or Rotational Vortex -- *Problems*

5.35 *In Fig. 5.14 suppose that the vanes are all straight and radial, that $r_1 = 0.3$ ft, $r_2 = 9$ ft, and that the of the figure is constant at $b = 25$ ft. Then $a = 2\pi rb$. If the speed is 1000 rpm and the flow of liquid is 9.6 cfs, find the difference in the pressure head between the outer and the inner circumferences, neglecting friction losses. Does it make any difference whether the flow is outward or inward?*

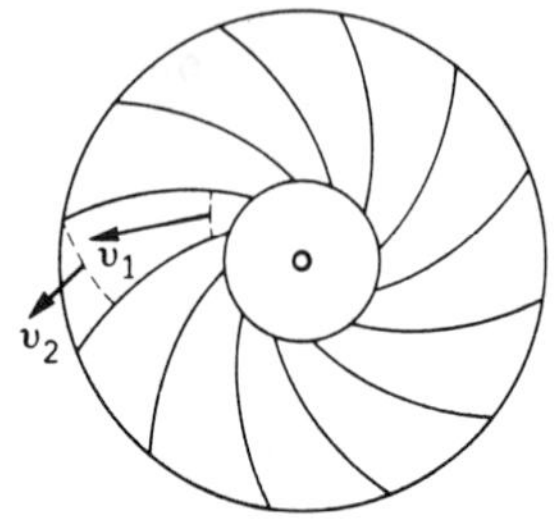

Figure 5.14

BG

$$V_1 = \frac{Q}{a_1} = \frac{9.6}{2\pi(0.30)0.25} = 20.4 \text{ fps}; \quad V_2 = \frac{Q}{a_2} = \frac{9.6}{2\pi(0.90)0.25} = 6.79 \text{ fps}$$

Eq. 5.46: $$\frac{p_2}{\gamma} - \frac{p_1}{\gamma} = \left(\frac{1000 \times 2\pi}{60}\right)^2 \frac{1}{2\,(32.2)}(0.90^2 - 0.30^2) + \frac{20.4^2 - 6.79^2}{2(32.2)}$$

$$= 128.3 \text{ ft}$$ ◄

Neglecting friction, the answer is the same for flow in either direction. ◄

5.36 *In Fig. 5.14 suppose that the vanes are all straight and radial, that $r_1 = 10$ cm, $r_2 = 20$ cm, and that the height perpendicular to the plane of the figure is constant at $b = 8$ cm. Then $a = 2\pi rb$. If the speed is 1000 rpm and the flow of liquid is 0.3 m³/s, find the difference in the pressure head between the outer and the inner circumferences, neglecting friction losses. Does it make any difference whether the flow is outward or inward?*

Fig. 5.14: See with Solution 5.35 above.

SI

$$V_1 = \frac{Q}{a_1} = \frac{Q}{2\pi r_1 b} = \frac{0.3}{2\pi(0.10)0.08} = 5.97 \text{ m/s}; \quad V_2 = \frac{Q}{a_2} = \frac{Q}{2\pi r_2 b} = \frac{0.3}{2\pi(0.20)0.08} = 2.98 \text{ m/s}$$

Eq. 5.46: $$\frac{p_2}{\gamma} - \frac{p_1}{\gamma} = \left(\frac{1000 \times 2\pi}{60}\right)^2 \frac{1}{2(9.81)}(0.20^2 - 0.10^2) + \frac{5.97^2 - 2.98^2}{2(9.81)} = 18.13 \text{ m}$$ ◄

Neglecting friction, the answer is the same for flow in either direction. ◄

/more...

Sec. 5.19: Free or Irrotational Vortex -- *Exercises*

5.19.1 *Refer to Sample Prob. 5.13. If the impeller diameter is 22 cm, the casing height is 4 cm between a and b, and water leaves the impeller with a velocity of 18 m/s at an angle of 16° with the tangent, find the flow rate, the magnitude and direction of the velocity at b (where r = 16 cm), and the pressure increase from a to b. Neglect friction.*

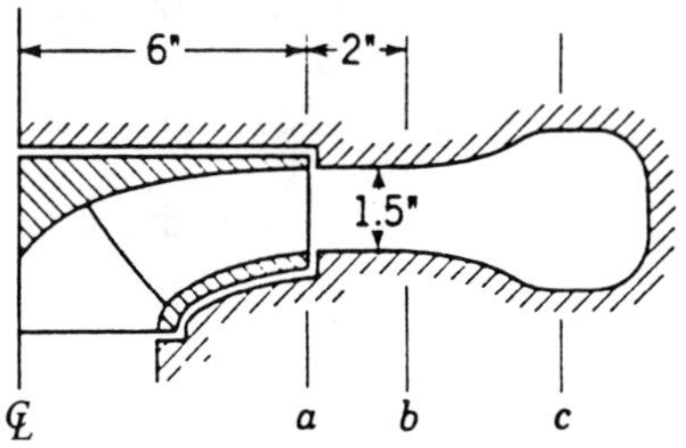

Figure S5.13

SI

Sec. 5.19: $Q = 2\pi r_a B(V_r)_a$ where $(V_r)_a = 18\sin 16° = 4.96$ m/s

So $Q = 2\pi(0.22/2)0.04(4.96) = 0.1372\ \text{m}^3/\text{s}$ ◄

From continuity: $(V_r)_b/(V_r)_a = r_a/r_b = 110/160$

∴, as in Sample Prob. 5.13, the angle remains at 16° ◄ $V_b = (110/160)18 = 12.38$ m/s ◄

From Eq. 5.15: $$\frac{p_b}{\gamma} - \frac{p_a}{\gamma} = \frac{V_a^2}{2g} - \frac{V_b^2}{2g} = \frac{18^2 - 12.38^2}{2(9.81)} = 8.71\ \text{m}$$

$\Delta p = 8.71(9810\ \text{N/m}^3) = 85\,400\ \text{N/m}^2 = 85.4$ kPa ◄

5.19.2 *Refer to Sample Prob. 5.13. If the impeller diameter is 10 in, the casing height is 1.8 in between a and b, and water leaves the impeller with a velocity of 50 fps at an angle of 16° with the tangent, find the flow rate, the magnitude and direction of the velocity at b (where r = 7 in), and the pressure increase from a to b. Neglect friction.*

Sample Prob. 5.13: See Fig. S5.13 with Solution 5.19.1 above.

BG

Sec 5.19: $Q = 2\pi r_a B(V_r)_a$ where $(V_r)_a = 50\sin 16° = 13.78$ fps

So $Q = 2\pi(5/12)(1.8/12)13.78 = 5.41$ cfs ◄ From continuity: $(V_r)_b/(V_r)_a = r_a/r_b = 5/7$

∴, as in Sample Prob. 5.13, the angle remains at 16° ◄ $V_b = (5/7)50 = 35.7$ fps ◄

From Eq. 5.15: $$\frac{p_b}{\gamma} - \frac{p_a}{\gamma} = \frac{V_a^2}{2g} - \frac{V_b^2}{2g} = \frac{50^2 - 35.7^2}{2(32.2)} = 19.01\ \text{ft}$$

$\Delta p = 19.01(62.4\ \text{lb/ft}^3) = 1186$ psf $= 8.24$ psi ◄

Sec. 5.19: Free or Irrotational Vortex -- *Problems*

5.37 *An air duct of 2.5 by 2.5 ft square cross section turns a bend of radius 5 ft as measured to the center line of the duct. If the measured pressure difference between the inside and the outside walls of the bend is 1.5 in of water, estimate the rate of air flow in the duct. Assume standard sea-level conditions in the duct and assume ideal flow around the bend.*

BG

Flow = cylindrical free vortex. Table A.3 for air at sea level: $\gamma = 0.0765$ pcf

Given $\dfrac{p_2}{\gamma} - \dfrac{p_1}{\gamma} = (1.5/12)62.4/0.0765 = 102.0$ ft of air ; Eq. 5.47: $\dfrac{p_2}{\gamma} - \dfrac{p_1}{\gamma} = \dfrac{C^2}{2g}\left(\dfrac{1}{r_1^2} - \dfrac{1}{r_2^2}\right)$

Thus, with $r_1 = 3.75$ ft and $r_2 = 6.25$ ft: $102.0 = \dfrac{C^2}{2(32.2)}\left(\dfrac{1}{3.75^2} - \dfrac{1}{6.25^2}\right)$; $C = 380\ \text{ft}^2/\text{sec}$

Thus, with $Q = \int V\,dA$ and $V = C/r$, while $dA = B\,dr$, where B is the width of the duct

$Q = BC\int_{r_1}^{r_2} dr/r = BC\ln(r_2/r_1) = 2.5(380)\ln(6.25/3.75) = 485\ \text{cfs} = 29{,}100\ \text{ft}^3/\text{min}$ ◄

5.38 *An air duct of 1.2 m by 1.2 m square cross section turns a bend of radius 2.4 m as measured to the center line of the duct. If the measured pressure difference between the inside and the outside walls of the bend is 5 cm of water, estimate the rate of air flow in the duct. Assume standard sea-level conditions in the duct and assume ideal flow around the bend.*

SI

Flow = cylindrical free vortex. Table A.3 for air at sea level: γ = 12.01 N/m^3

Given $\dfrac{p_2}{\gamma} - \dfrac{p_1}{\gamma}$ = 0.05(9810/12.01) = 40.8 m of air ; Eq. 5.47: $\dfrac{p_2}{\gamma} - \dfrac{p_1}{\gamma} = \dfrac{C^2}{2g}\left(\dfrac{1}{r_1^2} - \dfrac{1}{r_2^2}\right)$

Thus, with r_1 = 1.8 m and r_2 = 3.0 m: $40.8 = \dfrac{C^2}{2(9.81)}\left(\dfrac{1}{1.8^2} - \dfrac{1}{3.0^2}\right)$; C = 63.7 m^2/s

Thus noting $V = C/r$ and $dA = B\,dr$, where B is the width of the duct,

$Q = \int V\,dA = BC\int_{r_1}^{r_2} dr/r = BC\ln(r_2/r_1)$ = 1.2(63.7) ln (3.0/1.8) = 39.0 m^3/s = 2340 m^3/min ◄

Chapter 5: Miscellaneous -- *Problems*

5.39 *Assume ideal fluid. The pressure at section 1 in Fig. P5.39 is 10 psi, V_1 = 15 fps, V_2 = 50 fps, and γ = 60 lb/ft^3. (a) Determine the reading on the manometer. (b) If the downstream piezometer were replaced with a pitot tube, what would be the manometer reading? Comment on the practicality of these arrangements.*

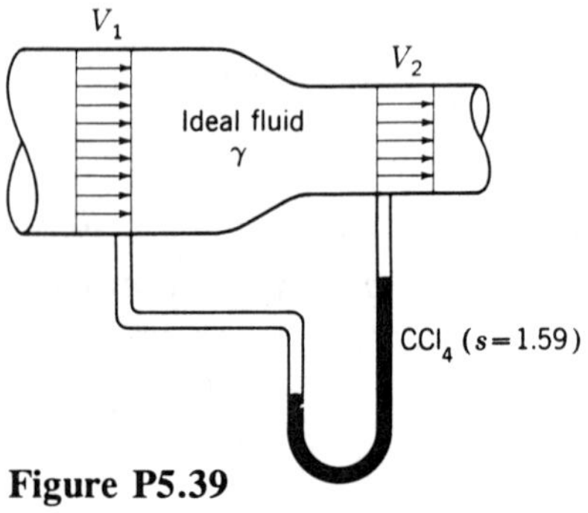

Figure P5.39

BG

(*a*) Flowing fluid: s_F = 60/62.4 = 0.9615

Energy Eq. 5.15 with $z_1 = z_2$: $\dfrac{15^2}{2g} + \dfrac{p_1}{\gamma} = \dfrac{50^2}{2g} + \dfrac{p_2}{\gamma}$

$\dfrac{p_1}{\gamma} - \dfrac{p_2}{\gamma} = \dfrac{50^2 - 15^2}{2(32.2)}$ = 35.3 ft

Using Eq. 3.12 with z_{AB} = 0: $\dfrac{p_1}{\gamma} - \dfrac{p_2}{\gamma}$ = 35.3 ft = $\left(\dfrac{s_M}{s_F} - 1\right)MR = \left(\dfrac{1.59}{0.9615} - 1\right)MR$

from which manometer reads MR = 54.1 ft = 649 in ◄

(*b*) $\left[\left(\dfrac{50^2}{2g} + \dfrac{p_2}{\gamma}\right) - \dfrac{p_1}{\gamma}\right] = \dfrac{15^2}{2(32.2)} = 3.49 = \left(\dfrac{1.59}{0.9615} - 1\right)MR$, so MR = 5.35 ft = 64.1 in ◄

The first reading is too large for practical use. ◄

5.40 *Refer to Fig. P5.39. Assume an ideal fluid with ρ = 900 kg/m^3. The pressure at section 1 is 100 kN/m^2; V_1 = 10 m/s, V_2 = 20 m/s. (a) Determine the reading on the manometer. (b) If the downstream piezometer were replaced with a pitot tube, what would be the manometer reading? Comment on the practicality of these arrangements?*

Fig. P5.39: See with Solution 5.39 above.

SI

(*a*) Flowing fluid: s_F = 900/1000 = 0.90

Energy Eq. 5.15 with $z_1 = z_2$: $\dfrac{10^2}{2g} + \dfrac{p_1}{\gamma} = \dfrac{20^2}{2g} + \dfrac{p_2}{\gamma}$; $\dfrac{p_1}{\gamma} - \dfrac{p_2}{\gamma} = \dfrac{20^2 - 10^2}{2(9.81)}$ = 15.29 m

Using Eq. 3.12 with z_{AB} = 0: $\dfrac{p_1}{\gamma} - \dfrac{p_2}{\gamma}$ = 15.29 m = $\left(\dfrac{s_M}{s_F} - 1\right)MR = \left(\dfrac{1.59}{0.90} - 1\right)MR$

from which the manometer reads MR = 19.94 m ◄

(*b*) $\left[\left(\dfrac{20^2}{2g} + \dfrac{p_2}{\gamma}\right) - \dfrac{p_1}{\gamma}\right] = \dfrac{10^2}{2(9.81)} = 5.097 = \left(\dfrac{1.59}{0.90} - 1\right)MR$, so MR = 6.65 m ◄

Both readings are too large for practical use. ◄

Chapter 6

Momentum and Forces in Fluid Flow

PROBLEM SELECTION GUIDE

Sec	Exer/Prob	Units	Difficulty	Length	Parts	Similar	Special features
6.3	*Momentum Correction Factor*						
	X[1]6.3.1	N	Medium	Medium	1		Uses Sec. 4.5; integration
	6.3.2	N	Medium	Medium	1		Uses Sec. 4.5; integration
	P 6.1	N	Medium	Long	1		Uses Sec. 4.5; integration
6.4	*Applications of the Momentum Principle*						
	X 6.4.1	BG	Medium	Short	1		
	6.4.2	SI	Medium	Medium	1	6.4.3	
	6.4.3	BG	Medium	Medium	1	6.4.2	
	6.4.4	BG	Medium	Medium	1	6.4.5	
	6.4.5	SI	Medium	Medium	1	6.4.4	
	P 6.2	BG	Medium	Medium	1	6.3	
	6.3	SI	Medium	Medium	1	6.2	
	6.4	BG	Hard	Long	2		Uses power (Sec. 5.10)
6.5	*Force Exerted on Pressure Conduits*						
	X 6.5.1	BG	Easy	Short	1	6.5.2	Uses Sec. 4.5
	6.5.2	SI	Easy	Short	1	6.5.1	Uses Sec. 4.5
	6.5.3	BG	Easy	Medium	1		
	6.5.4	BG	Easy	Short	1	6.5.5	
	6.5.5	SI	Easy	Short	1	6.5.4	
	6.5.6	BG	Easy	Short	1		
	P 6.5	BG	Easy	Medium	3	6.6	
	6.6	SI	Easy	Medium	3	6.5	
	6.7	BG	Easy	Medium	1	6.8	Uses Sec. 4.5
	6.8	SI	Easy	Medium	1	6.7	
	6.9	BG	Medium	Medium	1		
	6.10	BG	Medium	Medium	1	6.11	Includes manometer
	6.11	SI	Medium	Medium	1	6.10	Includes manometer

/cont...

[1] For all Exercises (marked "X"), answers are given in Appendix D of the textbook.

* Note a correction made to early printings of the textbook.

† Answers are sensitive to values which are measured from figures.

X = Exercise, P = (end-of-chapter) Problem, S = Sample Problem.

Sec	Exer/Prob	Units	Difficulty	Length	Parts	Similar	Special features
6.6	*Force Exerted on a Stationary Vane or Blade*						
	X 6.6.1	N	Easy	Short	1		Derivation
	6.6.2	BG	Easy	Short	3	6.6.3	
	6.6.3	SI	Easy	Short	3	6.6.2	
	6.6.4	BG	Easy	Short	3	6.6.5	
	6.6.5	SI	Easy	Short	3	6.6.4	
	6.6.6	BG	V Easy	V Short	1	6.6.7	
	6.6.7	SI	V Easy	V Short	1	6.6.6	
	6.6.8	BG	Easy	Short	1	6.6.9	
	6.6.9	SI	Easy	Short	1	6.6.8	
	P 6.12	N	Easy	Short	1		Derivation
	6.13	SI	Hard	Medium	1	6.14	† Scaling from figure: Pappus' theorem
	6.14	BG	Medium	Medium	1	6.13	† Scaling from figure; Pappus' theorem
	6.15	BG	Medium	Medium	3	6.16	
	6.16	SI	Medium	Medium	3	6.15	
	6.17	BG	Medium	Medium	1	6.18	
	6.18	SI	Medium	Medium	1	6.17	
	6.19	BG	Medium	Medium	1		* Integration
6.8	*Force Exerted by a Jet on a Moving Vane or Blade*						
	X 6.8.1	N	V Easy	V Short	1		Proof
	6.8.2	BG	Easy	Short	2	6.8.3	
	6.8.3	SI	Easy	Short	2	6.8.2	
	6.8.4	BG	Medium	Short	1	S6.4	
	6.8.5	BG	Easy	Medium	1		
	P 6.20	BG	Medium	Medium	1	6.21	
	6.21	SI	Medium	Medium	1	6.20	
	6.22	BG	Medium	Short	1	6.23	
	6.23	SI	Medium	Short	1	6.22	
	6.24	BG	Easy	Long	1	6.25-27	
	6.25	BG	Easy	Long	1	6.24-27	
	6.26	BG	Easy	Long	1	6.24-27	
	6.27	BG	Easy	Long	1	6.24-26	
	6.28	BG	Medium	Medium	1	6.29	
	6.29	SI	Medium	Medium	1	6.28	
	6.30	BG	Hard	Medium	2	6.31	Uses power (Sec. 5.10)
	6.31	SI	Hard	Medium	2	6.30	Uses power (Sec. 5.10)
	6.32	BG	Medium	Medium	3	6.33	Uses power (Sec. 5.10)
	6.33	SI	Medium	Medium	3	6.32	Uses power (Sec. 5.10)
6.9	*Reaction of a Jet*						
	X 6.9.1	BG	Easy	Short	1	6.9.2	
	6.9.2	SI	Easy	Short	1	6.9.1	
	P 6.34	BG	Easy	Short	1		
	6.35	BG	Easy	Medium	1	6.36	
	6.36	SI	Easy	Medium	1	6.35	

/cont...

Sec	Exer/Prob	Units	Difficulty	Length	Parts	Similar	Special features
6.10	***Jet Propulsion***						
	X 6.10.1	BG	V Easy	V Short	1	6.10.2	
	6.10.2	SI	Easy	V Short	1	6.10.1	
	P 6.37	BG	Easy	Short	3		
6.11	***Rotating Machines: Continuity, Velocities, Torque***						
	X 6.11.1	BG	Easy	Short	1	6.11.2	Uses power (Sec. 5.10)
	6.11.2	SI	Easy	Short	1	6.11.1	Uses power (Sec. 5.10)
	P 6.38	SI	Medium	Long	3	6.39	Uses power (Sec. 5.10)
	6.39	BG	Medium	Long	3	6.38	Uses power (Sec. 5.10)
6.13	***Flow Through Rotating Channel***						
	X 6.13.1	N	Medium	Medium	1		Derivation
	P 6.40	BG	Medium	Long	5	6.41	Uses power (Sec. 5.10)
	6.41	SI	Medium	Long	5	6.40	Uses power (Sec. 5.10)
6.14	***Reaction with Rotation***						
	X 6.14.1	SI	Easy	Short	1		
	6.14.2	BG	Easy	Short	1	6.14.3	Uses power (Sec. 5.10)
	6.14.3	SI	Easy	Short	1	6.14.2	Uses power (Sec. 5.10)
	6.14.4	BG	Medium	Short	1	6.14.5	
	6.14.5	SI	Medium	Short	1	6.14.4	
	P 6.42	BG	Medium	Medium	1	6.43	Uses power (Sec. 5.10)
	6.43	SI	Medium	Medium	1	6.42	Uses power (Sec. 5.10)
	6.44	N	Medium	Medium	1		Derivation
	6.45	BG	Medium	Long	1		Option to use differentiation
6.15	***Momentum Principle Applied to Propellers and Windmills***						
	X 6.15.1	BG	Easy	Medium	3	6.15.2	Uses power (Sec. 5.10)
	6.15.2	SI	Easy	Medium	3	6.15.1	Uses power (Sec. 5.10)
	6.15.3	B	Medium	Medium	2	6.15.4	Uses power (Sec. 5.10)
	6.15.4	SI	Medium	Medium	2	6.15.3	Uses power (Sec. 5.10)
	6.15.5	BG	Medium	Short	2		
	P 6.46	N	Hard	Long	1		Proof; differentiation

Chapter 6
MOMENTUM AND FORCES IN FLUID FLOW

Sec. 6.3: Momentum Correction Factor -- *Exercises*

6.3.1 *For laminar flow as in Sample Prob 5.1, find β.*

Sample Prob. 5.1: For laminar flow through a circular pipe the velocity profile is parabolic with equation $u = u_m[1 - (r/r_0)^2]$, where u is the velocity at any radius r, u_m is the maximum (centerline) velocity where $r = 0$, and r_0 is the radius of the pipe wall.

N

From Eq. 4.3: $V = \frac{1}{A}\int u\,dA = \frac{1}{\pi r_0^2}\int_0^{r_0} u_m[1 - (r/r_0)^2]2\pi r\,dr = u_m/2$

Eq. 6.9: $\beta = \frac{1}{AV^2}\int u^2 dA = \frac{4}{\pi r_0^2 u_m^2}\int_0^{r_0} u_m^2[1 - (r/r_0)^2]^2 2\pi r\,dr = 4/3$ ◄

6.3.2 *For the turbulent-flow case as approximated in Exer. 5.1.1, find β.*

Exer. 5.1.1: $u = u_m[1 - 0.3(r/r_o)^2]$.

N

Using Eq. 4.3: $V = 0.85u_m$.

By Eq. 6.9: $\beta = \frac{1}{\pi r_0^2(0.85u_m)^2}\int_0^{r_0} u_m^2[1 - 0.3(r/r_0)^2]^2 2\pi r\,dr = 1.014$ ◄

Sec. 6.3: Momentum Correction Factor -- *Problem*

6.1 *For laminar flow between two stationary parallel plates such as to give two-dimensional flow, find (a) the ratio of mean velocity to maximum velocity; (b) α; (c) β. The velocity profile is parabolic as in Exer. 6.3.1.*

Exer. 6.3.1 (replacing r by y): The parabolic velocity profile has equation $u = u_m[1 - (y/y_0)^2]$, where u is the velocity at any distance y, u_m is the maximum (centerline) velocity where $y = 0$, and $2y_0$ is the distance between the plates.

N

$u = u_m[1 - (y/y_0)^2]$; $dA = 2dy$

Eq. 4.3: $V = \frac{1}{A}\int u\,dA = \frac{\int_0^{y_0} u_m[1 - (y/y_0)^2]2dy}{\int_0^{y_0} 2dy} = \frac{2}{3}u_m$; $\therefore V/u_m = 2/3$ ◄

By Eq. 5.4: $\alpha = \frac{1}{2y_0[(2/3)u_m]^3}\int_0^{y_0} u_m^3[1 - (y/y_0)^2]^3 2dy = 54/35 = 1.543$ ◄

By Eq. 6.9: $\beta = \frac{1}{2y_0[(2/3)u_m]^2}\int_0^{y_0} u_m^2[1 - (y/y_0)^2]^2 2dy = 6/5 = 1.200$ ◄

Sec. 6.4: Applications of the Momentum Principle -- *Exercises*

6.4.1 *A cylindrical drum of radius 2 ft is securely held in position in an open channel of rectangular section. The channel is 10 ft wide, and the flow rate is 240 cfs. Water flows beneath the drum as shown in Fig. X6.4.1. Determine the horizontal thrust on the cylinder using impulse-momentum. Neglect fluid friction.*

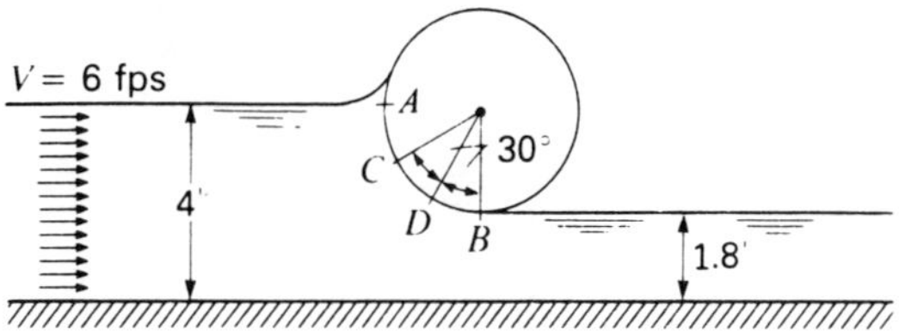

Figure X6.4.1

BG

$V_1 = 6$ fps ;

from continuity $V_2 = 6(4/1.8) = 13.33$ fps

Eq. 6.7*a*: $\Sigma F_x = F_1 - F_2 - F_{C/W} = \rho Q(V_{2x} - V_{1x})$

$[62.4(4^2)10]/2 - [62.4(1.8^2)10]/2 - F_{C/W}$

$= 1.940(240)(13.33 - 6.0)$

$4992 - 1011 - F_{C/W} = 3414$;

$F_{C/W} = +567$ lb to left ;

$F_{W/C} = 567$ lb to right ◄

6.4.2 *Find the horizontal thrust of the water on each meter of width of the sluice gate shown in Fig. X6.4.2, given $y_1 = 2.2$ m, $y_2 = 0.4$ m, and $y_3 = 0.5$ m. Neglect friction.*

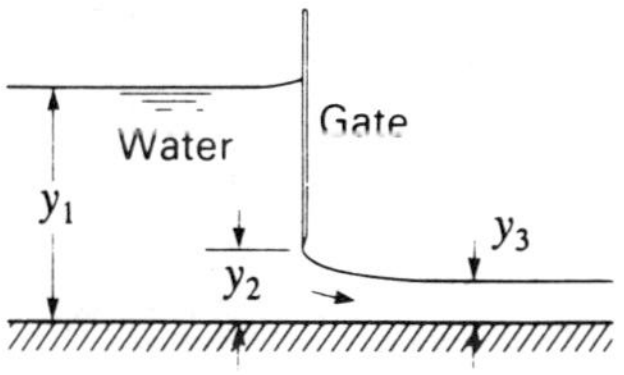

Figure X6.4.2

SI

Let q be the flow rate per meter of width, m^3/s per m.

Continuity, per meter of width: $V_1 = q/2.2, \quad V_2 = q/0.4$

Energy, neglecting friction: $2.2 + (q/2.2)^2/2g = 0.4 + (q/0.4)^2/2g$

from which $q = 2.42\ m^3/s$ per m width ; $V_1 = 1.099$ m/s, $V_2 = 6.04$ m/s

Eq. 6.7*a*: $\Sigma F_x = F_1 - F_2 - F_{G/W} = \rho Q(V_2 - V_1)$

$9810(2.2^2)/2 - 9810(0.4^2)/2 - F_{G/W} = 1000(2.42)(6.04 - 1.099)$

$23\,700 - 785 - F_{G/W} = 11\,950$; $F_{G/W} = +11\,000$ N/m to the left ; $F_{W/G} = 11.00$ kN/m to the right ◄

6.4.3 *Refer to Fig. X6.4.2. Find the horizontal thrust of the water on each foot of width of the sluice gate, given $y_1 = 7$ ft, $y_2 = 1.2$ ft, and $y_3 = 1.4$ ft. Neglect friction.*
For Fig. X6.4.2 refer to Solution 6.4.2.

BG

Let q be the flow rate, per foot of width, cfs per ft.

Continuity, per foot of width: $V_1 = q/7.0, \quad V_2 = q/1.2$

Energy, neglecting friction: $7.0 + (q/7.0)^2/2g = 1.2 + (q/1.2)^2/2g$

from which $q = 23.5$ cfs per foot width ; $V_1 = 3.36$ fps, $V_2 = 19.62$ fps

Eq. 6.7*a*: $\Sigma F_x = F_1 - F_2 - F_{G/W} = \rho Q(V_{2x} - V_{1x})$

$62.4(7.0^2)/2 - 62.4(1.2^2)/2 - F_{G/W} = 1.940(23.5)(19.62 - 3.36)$

$1529 - 44.9 - F_{G/W} = 742$; $F_{G/W} = +742$ lb/ft to the left ; $F_{W/G} = 742$ lb/ft to the right ◄

6.4.4 *Flow occurs over a spillway of constant section as shown in Fig. X6.4.4. Assuming ideal flow, determine the resultant horizontal force on the spillway per foot of spillway width (perpendicular to the spillway section), given that y_1 = 4.2 ft and y_2 = 0.7 ft.*

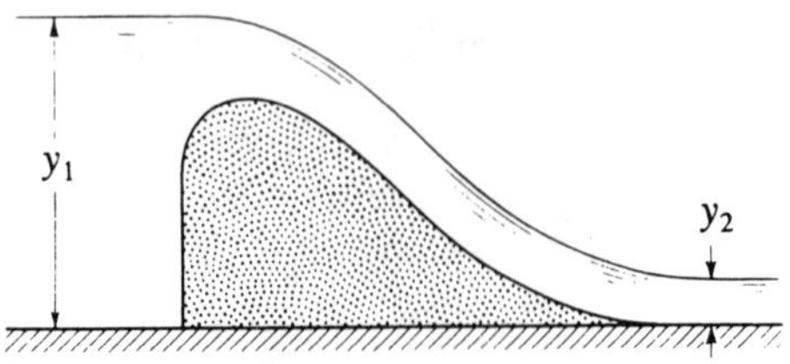

Figure X6.4.4

BG

Energy: $4.2 + V_1^2/(2\times 32.2) = 0.7 + V_2^2/(2\times 32.2)$ (1)

Continuity per foot: $4.2V_1 = 0.7V_2$ (2)

Substituting (2) into (1) yields:

$V_1 = 2.54$ fps, $V_2 = 15.23$ fps, $Q = A_1V_1 = 10.66$ cfs/ft

Eq. 6.7*a*: $\Sigma F_x = F_1 - F_2 - (F_{S/W})_x = \rho Q(V_{2x} - V_{1x})$

$62.4(2.1)4.2 - 62.4(0.35)0.7 - (F_{S/W})_x = 1.940(10.66)(15.23 - 2.54)$

$(F_{S/W})_x = +273$ lb/ft to left. Water $(F_{W/S})_x$ acts on spillway to the right with 273 lb/ft. ◄

6.4.5 *Flow occurs over the spillway of constant section as shown in Fig. X6.4.4. Given that y_1 = 4.2 m and y_2 = 0.7 m, determine the horizontal force on the spillway per meter of spillway width (perpendicular to the spillway section). Assume ideal flow.*

For Fig. X6.4.4 refer to Solution 6.4.4.

SI

Energy: $4.2 + V_1^2/(2\times 9.81) = 0.7 + V_2^2/(2\times 9.81)$ (1)

Continuity per m: $4.2V_1 = 0.7V_2$ (2)

Substituting (2) into (1) yields:

$V_1 = 1.401$ m/s, $V_2 = 8.40$ m/s ; $Q = A_1V_1 = 5.88$ m^3/s per meter

Eq. 6.7*a*: $\Sigma F_x = F_1 - F_2 - (F_{S/W})_x = \rho Q(V_{2x} - V_{1x})$

$9810(2.1)4.2 - 9810(0.35)0.7 - (F_{S/W})_x = 1000(5.88)(8.40 - 1.401)$

$(F_{S/W})_x = +42\,900$ N/m to the left.

Water $(F_{W/S})_x$ acts on spillway to the right with 42.9 kN/m. ◄

Sec. 6.4: Applications of the Momentum Principle -- *Problems*

6.2 *In Sample Prob. 6.1 suppose the passage narrowed down to a width of 9 ft at the second section. With the same depths find the flow rate and the horizontal force on the structure.*

Sample Prob. 6.1: The passage at section (1) of Fig. S6.1 is 10 ft wide. Assume ideal flow.

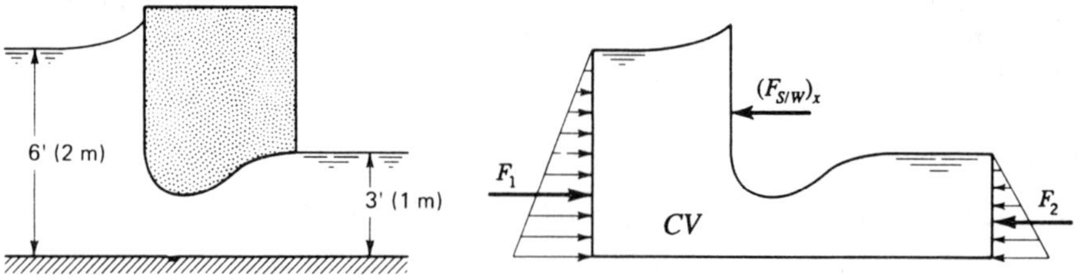

Figure S6.1

BG

Energy: $6 + V_1^2/(2\times 32.2) = 3 + V_2^2/(2\times 32.2)$ (1)

Continuity: $6(10)V_1 = 3(9)V_2$ (2)

/cont...

Substituting (2) into (1) yields: $V_1 = 7.00$ fps, $V_2 = 15.56$ fps, $Q = A_1V_1 = 60(7.00) = 420$ cfs ◄

Eq. 6.7*a*: $\Sigma F_x = F_1 - F_2 - (F_{S/W})_x = \rho Q(V_{2x} - V_{1x})$

$62.4(3)(10 \times 6) - 62.4(1.5)(9 \times 3) - (F_{S/W})_x = 1.940(420)(15.56 - 7.00)$

$(F_{S/W})_x = +1730$ lb to the left. $(F_{W/S})_x = 1730$ lb to the <u>right</u> ◄

6.3 *In Sample Prob. 6.1 suppose the passage narrowed down to a width of 2.0 m at the second section. With the same depths find the flow rate and the horizontal force on the structure.*

Sample Prob 6.1: For Fig. S6.1 refer to Solution 6.2. The passage at section (1) is 3 m wide. Assume ideal flow.

SI

Energy: $2 + V_1^2/(2\times 9.81) = 1 + V_2^2/(2\times 9.81)$ (1)

Continuity: $2(3)V_1 = 1(2.0)V_2$ (2)

Substituting (2) into (1) yields: $V_1 = 1.57$ m/s, $V_2 = 4.70$ m/s ; $Q = A_1V_1 = 6(2.03) = 9.40\ \text{m}^3/\text{s}$ ◄

Eq. 6.7*a*: $\Sigma F_x = F_1 - F_2 - (F_{S/W})_x = \rho Q(V_{2x} - V_{1x})$

$9810(1)(2 \times 3) - 9810(0.5)(1 \times 2) - (F_{S/W})_x = 1000(9.40)(4.70 - 1.57)$

$(F_{S/W})_x = +19\,620$ N to the left. $(F_{W/S})_x = 19.62$ kN to the <u>right</u>. ◄

6.4 *A hydraulic jump (Sec. 10.18) occurs in a "diamond shaped" transparent closed conduit as shown in the figure. The conduit is horizontal, and the water depth just upstream of the jump is 2.0 ft. The conduit is completely full of water downstream of the jump. Pressure-gage readings are as shown in the figure. (a) Compute the flow rate. Note that, because of turbulence in the jump, there is a substantial energy loss. Hence ideal flow cannot be assumed. Shear forces along the boundary may be neglected however. (b) Determine the horsepower loss in the jump.*

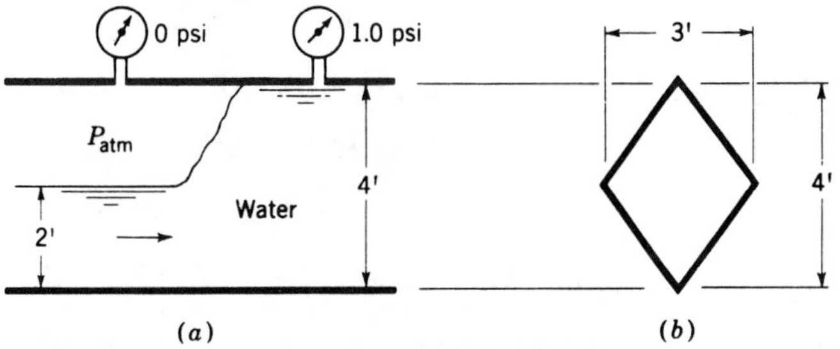

Figure P6.4

BG

(*a*) Since energy is lost, the momentum principle must be used.

Eq. 6.7*a*: $F_1 - F_2 - F_x = \rho Q(V_{2x} - V_{1x})$; $F_1 = 62.4(2/3)3 = 124.8$ lb

$-F_2 = -62.4(2)6 - 144(6) = -1613$ lb ; $F_x = 0$ neglecting shear forces along the boundary

Thus $124.8 - 1613 = 1.940(3V_1)(V_1/2 - V_1)$; $V_1 = 22.6$ fps, $V_2 = 11.31$ fps

$Q = A_1V_1 = 3(22.6) = 67.8$ cfs ◄

(*b*) Taking the channel bed as datum:

Eq. 5.28: $H_1 = 2 + 0 + 22.6^2/(2\times 32.2) = 2 + 0 + 7.94 = 9.94$ ft

$H_2 = 4 + 1(144)/62.4 + 0 + 11.3^2/(2\times 32.2) = 4 + 2.31 + 0 + 1.985 = 8.29$ ft

From Eq. 5.14: $h_L = H_1 - H_2 = 1.648$ ft

Eq. 5.32: Power loss $= \gamma Q h_L/550 = 62.4(67.8)1.648/550 = 12.69$ hp ◄

Sec. 6.5: Force Exerted on Pressure Conduits -- *Exercises*

6.5.1 *A nozzle which discharges a 4-in-diameter water jet into the air is on the right end of a horizontal 10-in-diameter pipe. In the pipe the water has a velocity of 15 fps and a gage pressure of 65 psi. Find the magnitude and direction of the resultant axial force exerted upon the nozzle, and the head loss in the nozzle.*

BG

Eq. 4.5: $G = \gamma A_1 V_1 = 62.4(0.5454)15 = 510.5$ lb/sec

From Eq. 4.6: $V_2 = V_1(A_1/A_2) = 15(10/4)^2 = 93.75$ fps

By Eq. 6.11: $F_x = 65(144)0.5454 - 0 - (510.5/32.2)(93.75 - 15) = 3860$ lb to the right ◀

From Eq. 5.14: $h_L = p_1/\gamma + V_1^2/2g - V_2^2/2g = 150.5 + 3.50 - 136.5 = 17.02$ ft ◀

6.5.2 *A nozzle which discharges a 6-cm-diameter water jet into the air is on the right end of a horizontal 12-cm-diameter pipe. In the pipe the water has a velocity of 4 m/s and a gage pressure of 400 kN/m². Find the magnitude and direction of the resultant axial force exerted upon the nozzle, and the head loss in the nozzle.*

SI

Eq. 4.5: $G = 9.81(\pi 0.06^2)4 = 0.444$ kN/s

From Eq. 4.6: $V_2 = 4(12/6)^2 = 16$ m/s ; $\rho Q(\Delta V) = (G/g)(\Delta V)$

By Eq. 6.11: $F_x = 400(\pi 0.06^2) - (0.444/9.81)(16 - 4) = 3.98$ kN to the right ◀

From Eq. 5.14: $h_L = p_1/\gamma + V_1^2/2g - V_2^2/2g = 40.77 + 0.815 - 13.05 = 28.5$ m ◀

6.5.3 *A diverging nozzle which discharges an 8-in-diameter water jet into the air is on the right end of a horizontal 6-in-diameter pipe. If the velocity in the pipe is 12 fps, find the magnitude and direction of the resultant axial force exerted upon the nozzle. Neglect fluid friction.*

BG

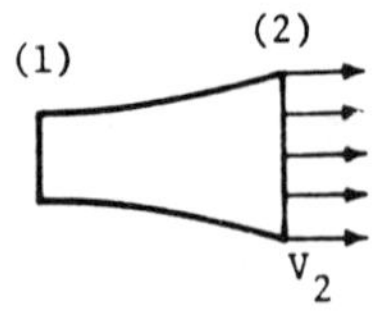

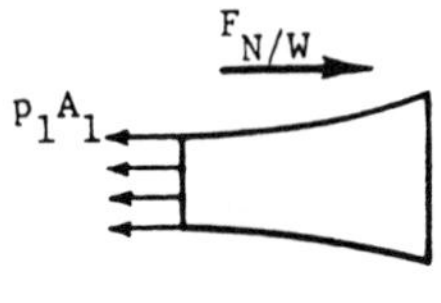

Eq. 5.15: $p_1/\gamma + V_1^2/2g = V_2^2/2g$

$V_1 = 12$ fps, $V_2 = 12(6/8)^2 = 6.75$ fps

$p_1/\gamma = 6.75^2/(2\times 32.2) - 12^2/(2\times 32.2) = -1.53$ ft

$p_1 = -(1.53\times 62.4)/144 = -0.662$ psi

From Eq. 6.11: $F_{N/W} - 0.662(\pi 3^2) = 1.940(9\pi/144)12(6.75 - 12)$

$F_{N/W} = 18.73 - 24.00 = -5.27$ lb to right $= +5.27$ lb to the left

$F_{W/N} = 5.27$ lb to the right ◀

6.5.4 *Water under a gage pressure of 60 psi flows with a velocity of 12 fps through a right-angle bend that has a uniform diameter of 10 inches. The bend lies in a horizontal plane and water enters from the west and leaves towards the north. Assuming no drop in pressure, what is the magnitude and direction of the resultant force acting on the bend?*

BG

$A = 0.545$ ft² $= 78.5$ in² ; Eq. 4.5: $G = 62.4(0.545)12 = 408$ lb/sec

By Eq. 6.12: $N_x = 60(78.5) - 0 - (408/32.2)(0 - 12) = 4860$ lb

By Eq. 6.15: $N_y = 60(78.5) + (408/32.2)12 = 4860$ lb

$N = 4860\sqrt{2} = 6880$ lb at 45° towards the SE ◀

6.5.5 *Water under a gage pressure of 400 kN/m² flows with a velocity of 4 m/s through a right-angle bend that has a uniform diameter of 26-cm. The bend lies in a horizontal plane and water enters from the west and leaves towards the north. Assuming no drop in pressure, what is the magnitude and direction of the resultant force acting on the bend?*

SI

By Eq. 6.12: $N_x = 400(\pi 0.13^2) - 0 - (9.81/9.81)(4\pi 0.13^2)(0 - 4) = 22.1$ kN

$N_y = 400(\pi 0.13^2) + (4\pi 0.13^2)4 = 22.1$ kN

$N = 22.1\sqrt{2} = 31.2$ kN at 45° towards the SE. ◄

6.5.6 *Refer to the double nozzle of Prob. 6.9, in which both water jets have a velocity of 35 fps. What angle should the 6-in jet make with the axis of the 8-in-dia pipe so that the resultant force is along the axis of the 8-in-dia pipe?*

For Fig. P6.9 refer to Solution 6.9 below.

BG

$Q_2 = A_2V_2 = [\pi(4/12)^2/4]35 = 3.05$ cfs ; $Q_3 = A_3V_3 = [\pi(6/12)^2/4]35 = 6.87$ cfs

$\Sigma F_z = 0 = \rho Q_2V_2\sin 25° - \rho Q_3V_3\sin\theta = 1.940(3.05)35\sin 25° - 1.940(6.87)35\sin\theta$

from which $\sin\theta = 0.1878$; $\theta = 10.83°$ ◄

Sec. 6.5: Force Exerted on Pressure Conduits -- *Problems*

6.5 *The diameters in Fig. 6.2 are 42 in and 30 in. At the larger end the pressure is 95 psi and the velocity is 10 fps. Neglecting friction, find the resultant force on the conical reducer if water flows (a) to the right; and (b) to the left. (c) What would happen to the two previous answers if friction were not neglected?*

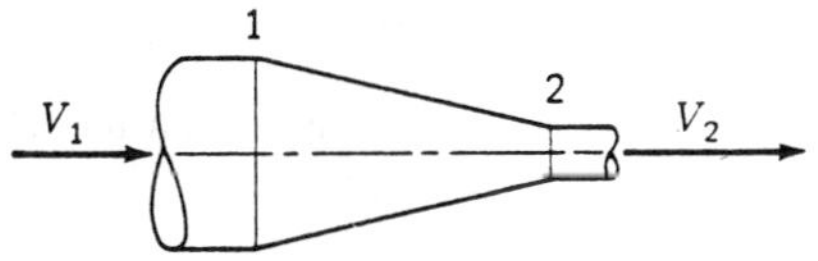

Figure 6.2(a)

BG

(*a*) $V_2 = 10(42/30)^2 = 19.60$ fps ;

By Eq. 4.10: $$\frac{(95\times144)}{62.4} + \frac{10^2}{2(32.2)} = \frac{p_2}{\gamma} + \frac{19.60^2}{2(32.2)} \; ; \quad \frac{p_2}{\gamma} = 215 \text{ ft}, \quad p_2 = 93.1 \text{ psi}$$

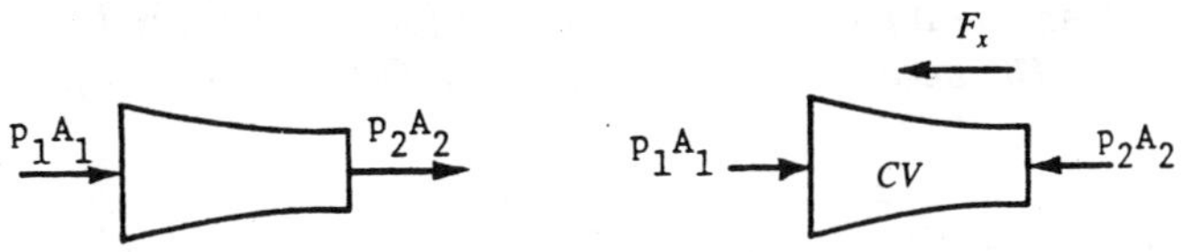

$A_1 = (\pi/4)42^2 = 1385$ in² ; $A_2 = (\pi/4)30^2 = 707$ in²

By Eq. 6.10: $F_x = 95(1385) - 93.1(707) - 186.4(19.60 - 10) = 64{,}000$ lb to the right on reducer. ◄

(*b*) Reversing the entire diagram gives the identical result, in the same direction as before if no friction is considered. ◄

(*c*) Friction will increase F_x in the first case and decrease it in the second case, (assuming the same pressure at the larger section in both cases). ◄

6.6 *The diameters in Fig. 6.2 are 80 cm and 50 cm. At the larger end the pressure is 650 kN/m² and the velocity is 3 m/s. Neglecting friction, find the resultant force on the conical reducer if water flows (a) to the right; and (b) to the left. (c) What would happen to the two previous answers if friction were not neglected?*

See Fig. 6.2 and solution figures with Solution 6.5.

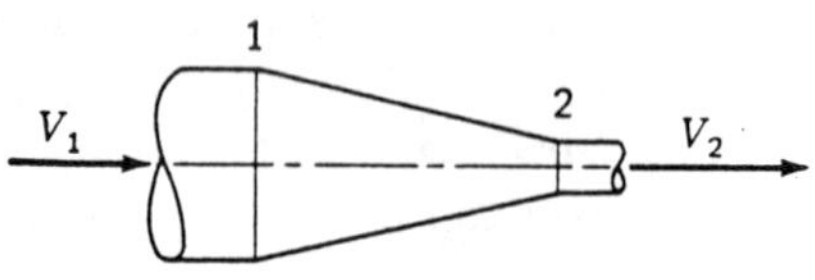

Figure 6.2(*a*)

SI

(*a*) $V_2 = 3(80/50)^2 = 7.68$ m/s ; $\dfrac{650}{9.81} + \dfrac{3^2}{2(9.81)} = \dfrac{p_2}{\gamma} + \dfrac{7.68^2}{2(9.81)}$

$p_2/\gamma = 66.26 + 0.459 - 3.01 = 63.7$ m ; $p_2 = 625$ kN/m²

Eq. 6.10: $650(\pi\, 0.40^2) - F_{N/W} - 625(\pi\, 0.25^2) = \rho Q(\Delta V)$

$\rho Q(\Delta V) = (\gamma/g)A_1V_1(V_2 - V_1) = (9.81/9.81)\pi\, 0.40^2(3)(7.68 - 3) = 7.06$ kN

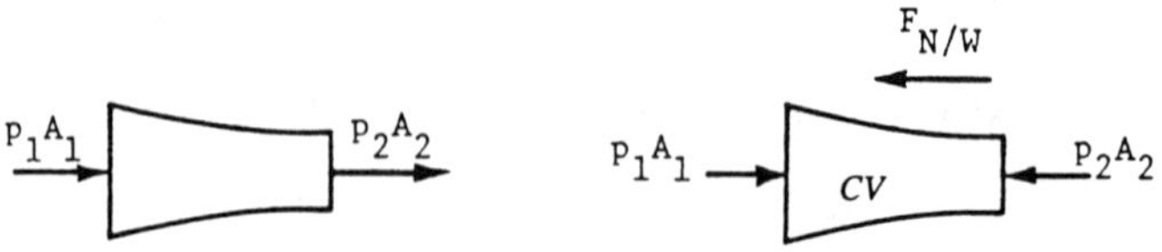

$F_{N/W} = 326.7\text{ kN} - 122.7\text{ kN} - 7.06\text{ kN} = +196.9$ kN to the left

Water ($F_{W/N}$) acts on the reducer to the right with a force of 196.9 kN. ◀

(*b*) Reversing the entire diagram gives the identical result, in the same direction as before if no friction is considered. ◀

(*c*) Friction will increase F_x in the first case and decrease it in the second case (assuming the same pressure at the larger section in both cases). ◀

6.7 *A reducing right-angle bend has the same orientation as in Exer. 6.5.4. Water enters with a velocity of 6 fps and a pressure of 4 psi. The diameter at entrance is 22 in and at exit is 20 in. Neglecting any friction loss, find the magnitude and direction of the resultant force on the bend.*

Exer. 6.5.4: The bend lies in a horizontal plane, water enters from the west and leaves towards the north.

BG

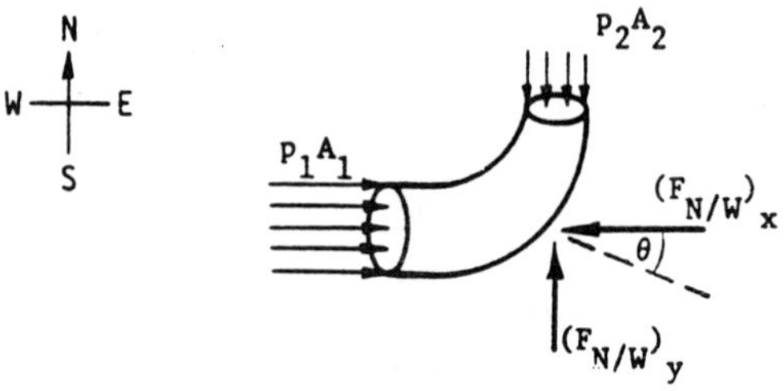

$V_2 = (22/20)^2 6 = 7.26$ fps.

If $p_1 = 4$ psi, by Eq. 5.15 with $z_1 = z_2$:

$4(144)/62.4 + 6^2/(2\times 32.2) = p_2/\gamma + 7.26^2/(2\times 32.2)$

$p_2/\gamma = 9.23 + 0.559 - 0.818 = 8.97$ ft ;

$p_2 = 560$ psf $= 3.89$ psi

Eq. 4.5: $G = 62.4\pi(11/12)^2 6 = 998$ lb/sec ;

$\rho Q = \dot{m} = G/g = 988/32.2 = 30.7$ slugs/sec

By Eq. 6.12: $4(\pi 11^2) - 0 + F_x = 30.7(0 - 6)$; $F_x = -1705$ lb

By Eq. 6.15: $0 - 3.89(\pi 10^2) + F_y = 30.7(7.26 - 0)$; $F_y = +1444$ lb

$N = \sqrt{1705^2 + 1444^2} = 2230$ lb at 40.3° S of E ◀

6.8 *A reducing right-angled bend has the same orientation as in Exer. 6.5.4. Water enters with a velocity of 3 m/s and a pressure of 30 kN/m². The diameter at entrance is 50 cm and at exit is 40 cm. Neglecting any friction loss, find the magnitude and direction of the resultant force on the bend.*

Exer. 6.5.4: The bend lies in a horizontal plane, water enters from the west and leaves towards the north. See solution figures with Solution 6.7.

SI

$V_2 = 3(50/40)^2 = 4.69$ m/s

Eq. 5.15: $\dfrac{30}{9.81} + \dfrac{3^2}{2(9.81)} = \dfrac{p_2}{\gamma} + \dfrac{4.69^2}{2(9.81)}$; $\dfrac{p_2}{\gamma} = 3.06 + 0.459 - 1.12 = 2.40$ m ;

$p_2 = 9.81(2.40) = 23.5$ kN/m²

Eq. 6.12: $30(\pi 0.25^2) - 0 - (F_{N/W})_x = \rho Q(\Delta V) = 10^3(\pi 0.25^2)3(0-3)/1000$

$(F_{N/W})_x = 5.89 + 1.767 = 7.66$ kN to the west.

Eq. 6.15: $0 - 23.5(\pi 0.20^2) + (F_{N/W})_y = 10^3(\pi 0.25^2)3(4.69-0)/1000$

$(F_{N/W})_y = 2.76 + 2.95 = 5.72$ to the north.

$F_{W/N} = \sqrt{7.66^2 + 5.72^2} = 9.56$ kN at 36.8° S of E ◀

6.9 *Referring to the figure, both nozzle jets are in the atmosphere and have a velocity of 35 fps. The liquid has a specific weight of 62.4 lb/ft³. The axes of the pipe and both nozzles all lie in a horizontal plane. Find the magnitude and direction of the resultant force on this double nozzle while neglecting friction.*

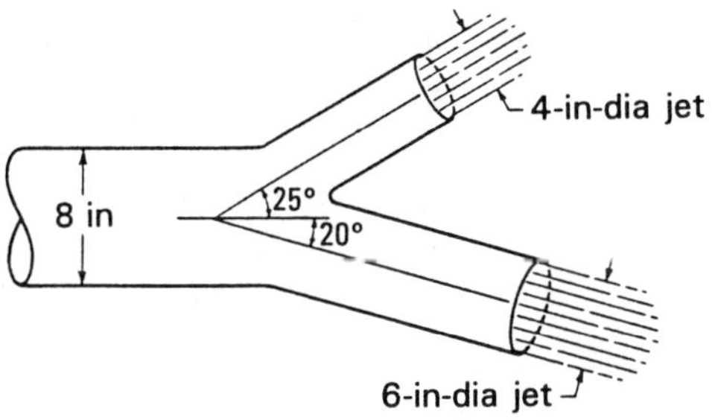

Figure P6.9

BG

Continuity:

$A_1V_1 = A_2V_2 + A_3V_3$;

$64V_1 = 16(35) + 36(35)$; $V_1 = 28.4$ fps

Eq. 5.14: $p_1/\gamma + V_1^2/2g = 0 + 35^2/2g$;

$p_1/\gamma = 19.0 - 12.56 = 6.46$ ft

$p_1 = 2.80$ psi, $p_1A_1 = 140.8$ lb

$\rho Q_1 = 1.940\pi(4/12)^2 28.4 = 19.26$ slugs/sec,

$\rho Q_2 = 5.92$ slugs/sec, $\rho Q_3 = 13.33$ slugs/sec

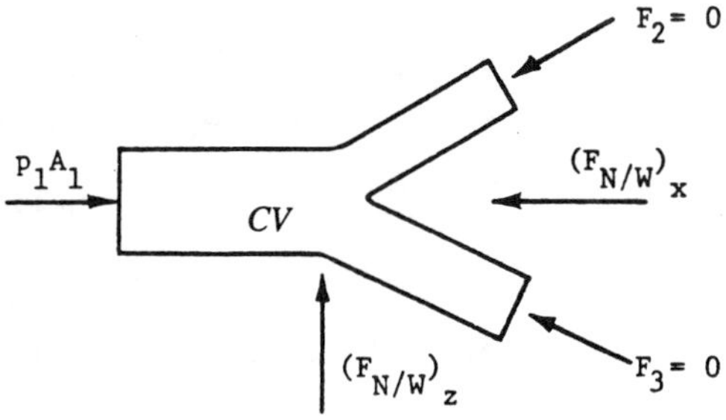

From Eq. 6.12:

$p_1A_1 - (F_{N/W})_x = \rho Q_2 V_{2x} + \rho Q_3 V_{3x} - \rho Q_1 V_{1x}$

$140.8 - (F_{N/W})_x = 5.92(35\cos 25°) + 13.33(35\cos 20°) - 19.26(28.4)$

$(F_{N/W})_x = 140.8 - 187.9 - 438.5 + 547.6 = 62.0$ lb

$(F_{N/W})_z = \rho Q_2 V_{2z} + \rho Q_3 V_{3z} - \rho Q_1 V_{1z} = 5.92(35\sin 25°) + 13.33(-35\sin 20°) - 0 = 87.6 - 159.6 = -71.9$ lb

Force of water on nozzle: 62.0 lb to the right and 71.9 lb upwards ◀

6.10 *Assuming ideal flow, determine the total pull on the bolts in the figure. Given y = 6 ft, d_1 = 2 in, d_2 = 4 in, and d_3 = 1 in, and γ_M of the manometer fluid (oil) is 52 pcf.*

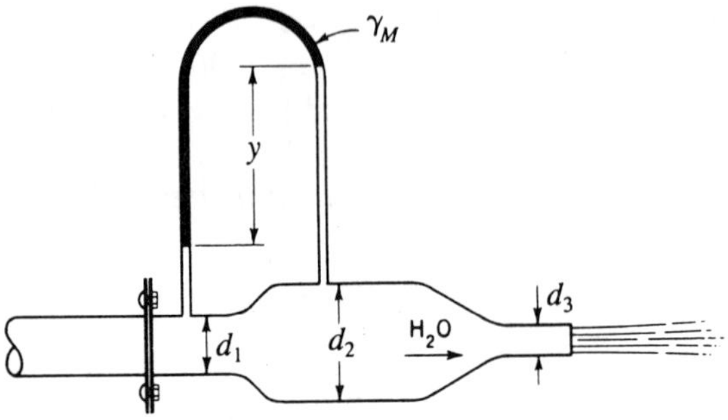

Figure P6.10

BG

Manometer: $6\left(\dfrac{52}{62.4}-1\right) = \dfrac{V_{4''}^2}{2g} - \dfrac{V_{2''}^2}{2g}$

(Note: $V_{1''} = 16V_{4''}$, $V_{2''} = 4V_{4''}$)

So: $6(-0.1667) = \dfrac{V_{4''}^2}{2g} - \dfrac{(4V_{4''})^2}{2g} = -\dfrac{15V_{4''}^2}{2g}$

$V_{4''}$ = 2.07 fps ; $V_{2''}$ = 8.29 fps, $V_{1''}$ = 33.2 fps

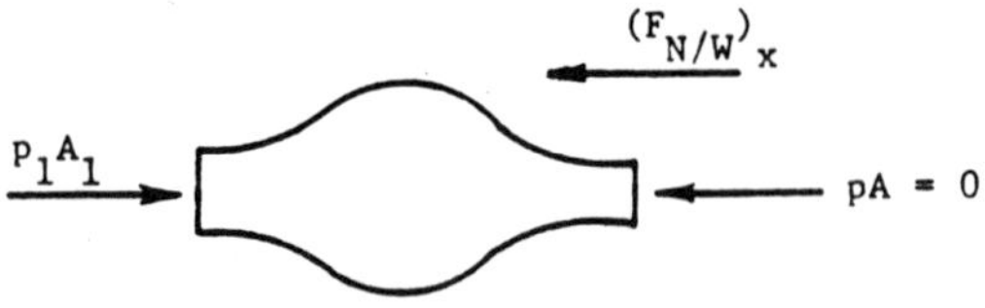

$\dfrac{p_1}{62.4} + \dfrac{8.29^2}{2(32.2)} = 0 + \dfrac{33.2^2}{2(32.2)}$;

p_1 = 998 psf = 6.93 psi

Eq. 6.10: $p_1A_1 - (F_{N/W})_x = \rho Q(V_{2x} - V_{1x})$

$6.93\pi(1)^2 - (F_{N/W})_x = 1.940\pi(1/12)^2 8.29(33.2 - 8.29)$; $(F_{N/W})_x$ = 13.06 lb

Pull on bolts is 13.06 lb. ◄

6.11 *Repeat Prob. 6.10 for the case where y = 180 cm, d_1 = 5 cm, d_2 = 10 cm, d_3 = 2.5 cm, and the manometer liquid has a specific gravity of 0.80.*

Refer to Solution 6.10 for figures.

SI

Manometer: $1.80(0.8 - 1.0) = V_{10}^2/2g - V_5^2/2g$;

Continuity: $V_5 = 4V_{10} = (1/4)V_{2.5}$; So: $-0.36 = V_{10}^2/2g - (4V_{10})^2/2g = -15V_{10}^2/2g$

$V_{10}^2 = 2(9.81)(0.36/15)$; V_{10} = 0.686 m/s, V_5 = 2.74 m/s, $V_{2.5}$ = 10.98 m/s

$\dfrac{p_5}{9.81} + \dfrac{2.74^2}{2(9.81)} = \dfrac{10.98^2}{2(9.81)}$; $p_5 = 56.5\ \text{kN/m}^2$

Eq. 6.10: $p_1A_1 - (F_{N/W})_x = \rho Q(V_{2x} - V_{1x})$; $A_1 = \pi 0.025^2 = 0.001963\ \text{m}^2$

$56.5(0.001963) - (F_{N/W})_x = (9.81/9.81)(0.001963 \times 2.74)(10.98 - 2.74)$

$(F_{N/W})_x$ = 0.1109 kN − 0.0444 kN = 0.0666 kN = 66.6 N to the left

Equal and opposite is the pull on the bolts = 66.6 N. ◄

Sec. 6.6: Force Exerted on a Stationary Vane or Blade -- *Exercises*

6.6.1 *A jet containing any type of fluid with a velocity V and an area A is deflected through an angle θ without changing the velocity magnitude. Derive an equation for the dynamic force exerted.*

N

Choose axes so that the y axis is the axis of symmetry. Then

$\Delta V_x = 0$; $\Delta V_y = V\sin\dfrac{\theta}{2} - \left(-V\sin\dfrac{\theta}{2}\right) = 2V\sin\dfrac{\theta}{2}$

From Eq. 6.6: $F = \dot{m}\,\Delta V = \left(\dfrac{\gamma AV}{g}\right)2V\sin\dfrac{\theta}{2} = \dfrac{2\gamma A}{g}V^2\sin\dfrac{\theta}{2}$ ◄

6.6.2 *Refer to Fig. S6.3. Assume that friction is negligible, that $\theta = 115°$, and that the water jet has a velocity of 95 fps and a diameter of 1 in. Find (a) the component of the force in the direction of the jet; (b) the force component normal to the jet; and (c) the magnitude and direction of the resultant force exerted on the body.*

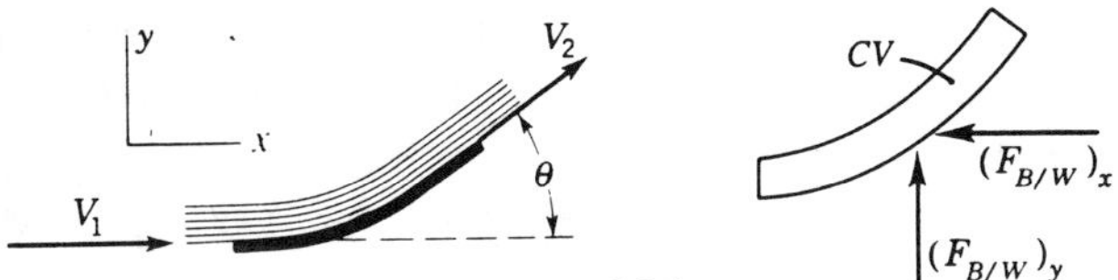

Figure S6.3

BG

(*a*) From Eq. 6.7*a*: $F_x = \dot{m}\Delta V = \rho A V_1(V_1 - V_2\cos\theta)$; $A = \pi(1/12)^2/4 = 0.00545 \text{ ft}^2$

$F_x = 1.940(0.00545)95(95 - 95\cos 115°) = 1.005(135.1) = 135.9 \text{ lb}$ ◄

(*b*) From Eq. 6.7*b*: $F_y = \dot{m}V_2\sin\theta = 1.004(95\sin 115°) = 86.5 \text{ lb}$ ◄

(*c*) $F = \sqrt{135.9^2 + 86.5^2} = 161.1 \text{ lb}$ ◄ ; $\alpha = \tan^{-1}(86.5/135.9) = 32.5°$ (below horizontal) ◄

Note: The direction of the resultant can also be determined by inspection to be the extension of the line bisecting the angle between entering and leaving jets.

6.6.3 *Refer to Fig. S6.3. Assume that friction is negligible, that $\theta = 115°$, and that the water jet has a velocity of 25 m/s and a jet diameter of 4 cm. Find (a) the component of the force in the direction of the jet; (b) the force component normal to the jet; and (c) the magnitude and direction of the resultant force exerted on the vane.*

See Fig S6.3 accompanying Solution 6.6.2.

SI

(*a*) From Eq. 6.7*a*: $F_x = \dot{m}\Delta V = \rho A V_1(V_1 - V_2\cos\theta) = 1000(\pi 0.04^2/4)25(25 - 25\cos 115°)$

$= 31.4(35.6) = 1117 \text{ N}$ ◄ (*b*) From Eq. 6.7*b*: $F_y = \dot{m}V_2\sin\theta = 31.4(25\sin 115°) = 712 \text{ N}$ ◄

(*c*) $F = \sqrt{1117^2 + 712^2} = 1325 \text{ N}$ ◄ $\alpha = \tan^{-1}(712/1117) = 32.5°$ (below horizontal) ◄

Note: The direction of the resultant can also be determined by inspection to be the extension of the line bisecting the angle between entering and leaving jets.

6.6.4 *Solve Exer. 6.6.2 assuming that friction reduces V_2 to 70 fps.*

Exer. 6.6.2: Given $\theta = 115°$, $V_1 = 95$ fps, and $D_1 = 1$ in. Find the force components (a) parallel to and (b) perpendicular to the jet, and (c) find the magnitude and direction of the resultant force acting on the body. See Fig. S6.3 accompanying Solution 6.6.2.

BG

(*a*) From Eq. 6.7*a*: $F_x = \dot{m}\Delta V = \rho A V_1(V_1 - V_2\cos\theta)$; $A = \pi(1/12)^2/4 = 0.00545 \text{ ft}^2$

$F_x = 1.940(0.00545)95(95 - 70\cos 115°) = 1.005(124.6) = 125.2 \text{ lb}$ ◄

(*b*) From Eq. 6.7*b*: $F_y = \dot{m}V_2\sin\theta = 1.005(70\sin 115°) = 63.8 \text{ lb}$ ◄

(*c*) $F = \sqrt{125.2^2 + 63.8^2} = 140.5 \text{ lb}$ ◄ $\alpha = \tan^{-1}(63.7/125.1) = 27.0°$ (below horizontal) ◄

6.6.5 *Solve Exer. 6.6.3 assuming that friction reduces V_2 to 20 m/s.*

Exer. 6.6.3: Given $\theta = 115°$, $V_1 = 25$ m/s, and $D_1 = 4$ cm. Find the force components (a) parallel to and (b) perpendicular to the jet, and (c) find the magnitude and direction of the resultant force acting on the body. See Fig. S6.3 accompanying Solution 6.6.2.

SI

(*a*) From Eq. 6.7*a*: $F_x = \dot{m}\Delta V = \rho A V_1(V_1 - V_2\cos\theta) = 1000(\pi 0.04^2/4)25(25 - 20\cos 115°)$

$= 31.4(33.5) = 1051 \text{ N}$ ◄ (*b*) From Eq. 6.7*b*: $F_y = \dot{m}V_2\sin\theta = 31.4(20\sin 115°) = 569 \text{ N}$ ◄

(*c*) $F = \sqrt{1051^2 + 569^2} = 1195 \text{ N}$ ◄ $\alpha = \tan^{-1}(569/1051) = 28.5°$ ◄

6.6.6 *Calculate the resultant force on a large flat plate if the jet in Exer. 6.6.2 were to strike it normally.*
Exer. 6.6.2: Given V_1 = 95 fps and D_1 = 1 in. See Fig. S6.3 accompanying Solution 6.6.2.

BG

Eq. 6.7*a*: $F = F_x = \dot{m}\Delta V = \rho A V_1(V_1 - V_2\cos\theta)$; $A = \pi(1/12)^2/4 = 0.00545 \text{ ft}^2$; $\theta = 90°$ ◀

$F = 1.940(0.00545)95(95 - 0) = 1.005(95) = 95.5 \text{ lb}$

6.6.7 *Suppose the jet in Exer. 6.6.3 were to strike a large flat plate normally. What would be the resultant force on the plate?*
Exer. 6.6.3: Given V_1 = 25 m/s and D_1 = 4 cm. See Fig. S6.3 accompanying Solution 6.6.2.

SI

Eq. 6.7*a*: $F = F_x = \dot{m}\Delta V = \rho A V_1(V_1 - V_2\cos\theta)$; $\theta = 90°$

$F = 1000(\pi 0.04^2/4)25(25 - 0) = 31.4(25) = 785 \text{ N}$ ◀

6.6.8 *In Exer. 6.6.6 assume the center of the jet is coincident with the center of the circular plate. Find (a) the stagnation pressure, and (b) the average pressure on the plate if the area of the plate is 22 times the area of the jet.*
Exer. 6.6.6: The jet, with V_1 = 95 fps and D_1 = 1 in, strikes the plate normally. See Fig S6.3 accompanying Solution 6.6.2.

BG

(*a*) From Eq. 5.15: $p_O = \gamma V^2/2g = 62.4\times 95^2/(2\times 32.2) = 8744 \text{ psf} = 60.7 \text{ psi}$ ◀

(*b*) Eq. 6.7*a*: $F = F_x = \dot{m}\Delta V = \rho A V_1(V_1 - V_2\cos\theta)$; $A = \pi(1/12)^2/4 = 0.00545 \text{ ft}^2$; $\theta = 90°$

$F = 1.940(0.00545)95(95 - 0) = 1.005(95) = 95.5 \text{ lb}$

$$\text{Average pressure} = F/A = \frac{95.5}{22(0.00545)} = 796 \text{ psf} = 5.53 \text{ psi}$$ ◀

6.6.9 *In Exer. 6.6.7 assume the center of the jet is coincident with the center of the circular plate. Find (a) the stagnation pressure, and (b) the average pressure on the plate if the area of the plate is 25 times the area of the jet.*
Exer. 6.6.7: The jet, with V_1 = 25 m/s and D_1 = 4 cm, strikes the plate normally. See Fig. S6.3 accompanying Solution 6.6.2.

SI

(*a*) From Eq. 5.15: $p_O = \gamma V^2/2g = 9.81(25^2)/(2\times 9.81) = 313 \text{ kN/m}^2$ ◀

(*b*) Eq. 6.7*a*: $F = F_x = \dot{m}\Delta V = \rho A V_1(V_1 - V_2\cos\theta)$; $\theta = 90°$

$F = 1000(\pi 0.04^2/4)25(25 - 0) = 31.4(25) = 785 \text{ N}$

$$\text{Average pressure} = F/A = \frac{785}{25(\pi/4)(4/100)^2} = 25\,000 \text{ N/m}^2 = 25.0 \text{ kN/m}^2$$ ◀

Sec. 6.6: Force Exerted on a Stationary Vane or Blade -- *Problems*

6.12 *If a jet of any fluid has its velocity V_2 reduced to $0.7V_1$ by friction while it is deflected through an angle θ, derive an equation for the dynamic force exerted in terms of $\dot{m}$, V_1, and θ.*

N

From Eq. 6.7: $F_x = \dot{m}(V_1 - V_2\cos\theta) = \dot{m}V_1(1 - 0.7\cos\theta)$; $F_y = \dot{m}V_2\sin\theta = \dot{m}\,0.7\,V_1\sin\theta$

$$F = \sqrt{F_x^2 + F_y^2} = \dot{m}\sqrt{V_1^2(1 - 0.7\cos\theta)^2 + 0.49\,V_1^2\sin^2\theta} = \dot{m}V_1\sqrt{1.49 - 1.4\cos\theta}$$ ◀

6.13 *Plotted to scale in Fig. P6.13 are streamlines in the plane of the center of a free jet impinging vertically on a horizontal circular plate. By scaling off the pertinent dimensions determine the velocity of the water as it leaves the plate and the total resultant force exerted by the water on the plate. Given the jet diameter is 28 cm and stagnation pressure at point S is 5.5 kN/m².*

SI

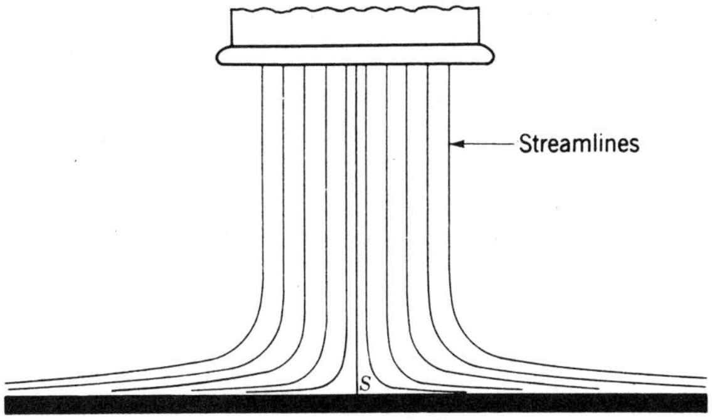

Figure P6.13

Energy equation from where streamlines diverge (z = 14 cm) to stagnation point:

$$0 + 0.14 + \frac{V^2}{2g} = \frac{p_S}{\gamma} + 0 + 0 = \frac{5.5}{9.81} = 0.561 \text{ m};$$

$$\frac{V^2}{2g} = 0.421 \text{ m} \; ; \quad V = \sqrt{2(9.81)0.421} = 2.87 \text{ m/s}$$

$$Q = AV = (\pi 0.14^2)2.87 = 0.1769 \text{ m}^3\text{/s}$$

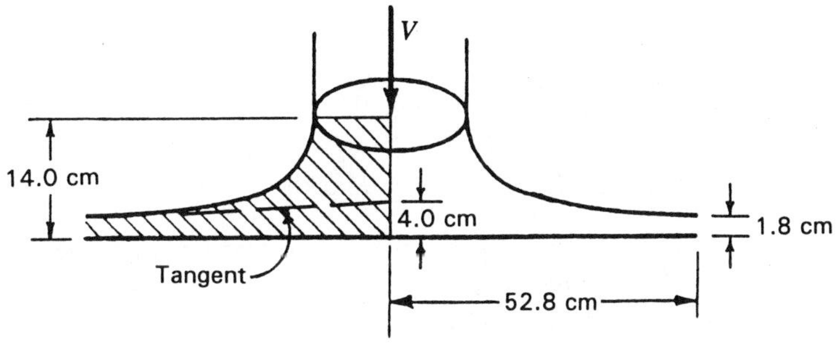

By scaling from Fig. P6.13: Water leaves the plate (r = 52.8 cm) at an average angle of about 1.2° and at a depth of about 1.8 cm, for example. (Scaling at different points will give rise to somewhat different answers.)

$V_2 = Q/A_2 = 0.1769/[(2\pi(0.528)0.018)] = 2.96$ m/s ◄

$\rho Q(\Delta V) = (9.81/9.81)0.1769(2.87 - 2.96 \sin 1.2°) = 0.497$ kN

By scaling from Fig. P6.13, and by calculation, for water to one side of the vertical centerline and below the point where the jet first begins to widen: Area = A = 333 cm² = 0.0333 m², distance from centerline to centroid = r_c = 16.57 cm = 0.1657 m.

Using Pappus' theorem: $W = \gamma \maltese = 9.81(2\pi r_c A) = 19.62\pi(0.1657)0.0333 = 0.340$ kN

Total Force = $W + \rho Q(\Delta V) = 0.340 + 0.497 = 0.837$ kN ◄

6.14 *Repeat Prob. 6.13 with a jet diameter of 4.2 in and a jet velocity of 23 fps.*

Plotted to scale in Fig. P6.13 (see with Solution 6.13) are streamlines in the plane of the center of a free jet impinging vertically on a horizontal circular plate. By scaling off the pertinent dimensions determine the velocity of the water as it leaves the plate and the total resultant force exerted by the water on the plate.

BG

$Q = \pi(2.1/12)^2 23 = 2.21$ cfs

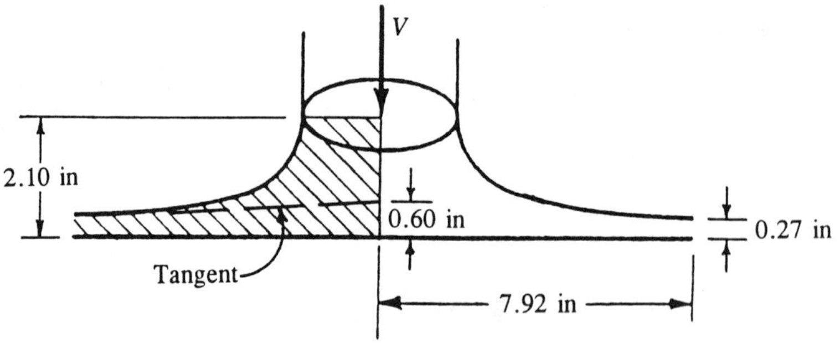

By scaling from Fig. P6.13: Water leaves the plate (r = 7.92 in) at an average angle of 1.2° and at a depth of about 0.27 in, for example. (Scaling at different points will give rise to somewhat different answers.)

$$V_2 = Q/A_2 = 2.21/\left(2\pi\frac{7.92}{12}\frac{0.27}{12}\right) = 23.7 \text{ fps} \quad \blacktriangleleft$$

$$\rho Q(\Delta V) = (1.940)2.21(23 - 23.7\sin 1.2°) = 96.6 \text{ lb}$$

By scaling from Fig. P6.13, and by calculation, for water to one side of the vertical centerline and below the point where the jet first begins to widen: Area $= A = 7.49$ in^2 $= 0.0520$ ft^2, distance from centerline to centroid $= r_c = 2.49$ in $= 0.207$ ft.

Using Pappus' theorem: $W = \gamma V = 62.4(2\pi r_c A) = 124.8\pi(0.207)0.0520 = 4.22$ lb

Total Force $= W + \rho Q(\Delta V) = 4.22 + 96.6 = 100.8$ lb ◄

6.15 *A horizontal jet of water issues from an orifice in the side of a tank under a head h_1 and strikes a large plate a short distance away which covers the end of a horizontal tube in the side of a second tank. The second tank contains oil of specific weight 50 lb/ft^3 at rest. The height of the oil above the tube is h_2. The jet diameter is three-fourths of the inside diameter of the tube. The jet and the tube are the same elevation. (a) If the impact of the water is just sufficient to hold the plate in place, find the relation between h_1 and h_2. Neglect the weight of the plate and assume ideal flow. (b) Consider the effect of the weight of the plate. Find h_2 if $h_1 = 9$ ft, the weight of the plate = 45 lb, the jet diameter = 1.4 in, and the coefficient of friction between the plate and tube = 0.5. (c) Repeat part (b) with the plate weighing only 5 lb.*

BG

(a) $F_j = \rho_1 Q_1 \Delta V_1 = \rho_1(A_1 V_j)V_j = \rho_1(\pi/4)D_j^2 V_j^2$ where $V_j = \sqrt{2gh_1}$

$\therefore\ F_j = (\gamma_1/g)(\pi/4)D_j^2(2gh_1) = (\pi/2)\gamma_1 h_1 D_j^2$

$F_{oil} = A_2 p_2 = (\pi/4)D_t^2(\gamma_2 h_2)$

To hold (weightless) plate in place (with $D_j = 0.75D_t$), $F_j = F_{oil}$

$\therefore\ (\pi/2)\gamma_1 h_1 (0.75D_t)^2 = (\pi/4)D_t^2\gamma_2 h_2$

$h_2 = 2(\gamma_1/\gamma_2)0.75^2 h_1 = 2(62.4/50)0.75^2 h_1 = 1.404 h_1$ ◄

(b) $F_j = (\pi/2)\gamma_1 h_1 D_j^2 = (\pi/2)62.4(9)(1.4/12)^2 = 12.01$ lb

For equilibrium we need friction force $F = \mu N = \mu(F_j - F_{oil}) = W$ /cont...

i.e., $0.5(12.01 - F_{oil}) = 45$ or $12.01 - F_{oil} = 90$

F_{oil} cannot be negative, so this is impossible, the 45-lb plate cannot be held in place. ◀

(*c*) If $W = 5$ lb we need: $0.5(12.01 - F_{oil}) = 5$

$$\therefore\ F_{oil} = 12.01 - 10 = 2.01\ \text{lb} = \frac{\pi}{4}\left(\frac{D_j}{0.75}\right)^2 \gamma_2 h_2 = \frac{\pi}{4}\left(\frac{1.4}{12 \times 0.75}\right)^2 50 h_2$$

from which $h_2 = 2.11$ ft ◀

6.16 *Repeat Prob. 6.15 with the following data: (a) oil specific weight = 7500 N/m³ ; (b) h_1 = 2.8 m, W = 180 N, jet diameter = 3.5 cm, μ = 0.5; (c) W = 20 N.*

SI

(*a*) $F_j = \rho_1 Q_1 \Delta V_1 = \rho_1 (A_1 V_j) V_j = \rho_1 (\pi/4) D_j^2 V_j^2$ where $V_j = \sqrt{2gh_1}$

$\therefore\ F_j = (\gamma_1/g)(\pi/4)D_j^2(2gh_1) = (\pi/2)\gamma_1 h_1 D_j^2$; $F_{oil} = A_2 p_2 = (\pi/4)D_t^2(\gamma_2 h_2)$

To hold (weightless) plate in place (with $D_j = 0.75 D_t$), $F_j = F_{oil}$

$\therefore\ (\pi/2)\gamma_1 h_1 (0.75 D_t)^2 = (\pi/4) D_t^2 \gamma_2 h_2$

$h_2 = 2(\gamma_1/\gamma_2)0.75^2 h_1 = 2(9810/7500)0.75^2 h_1 = 1.472 h_1$ ◀

(*b*) $F_j = (\pi/2)\gamma_1 h_1 D_j^2 = (\pi/2)9810(2.8)(0.035)^2 = 52.9$ N

For equilibrium we need friction force $F = \mu N = \mu(F_j - F_{oil}) = W$

i.e., $0.5(52.9 - F_{oil}) = 180$ or $52.9 - F_{oil} = 360$

F_{oil} cannot be negative, so this is impossible: the 180-N plate cannot be held in place ◀

(*c*) If $W = 20$ N we need: $0.5(52.9 - F_{oil}) = 20$

$$\therefore\ F_{oil} = 52.9 - 40 = 12.85\ \text{N} = \frac{\pi}{4}\left(\frac{D_j}{0.75}\right)^2 \gamma_2 h_2 = \frac{\pi}{4}\left(\frac{0.035}{0.75}\right)^2 7500 h_2$$

from which $h_2 = 1.002$ m ◀

6.17 *At section P a 5-in-diameter water jet with a velocity of 28 fps is directed vertically upward against the cone shown in Fig. P6.17. Neglecting friction and assuming the streamlines at Q are parallel, find the weight of the cone if a = 1.5 ft, b = 0.6 ft, and c = 4 ft.*

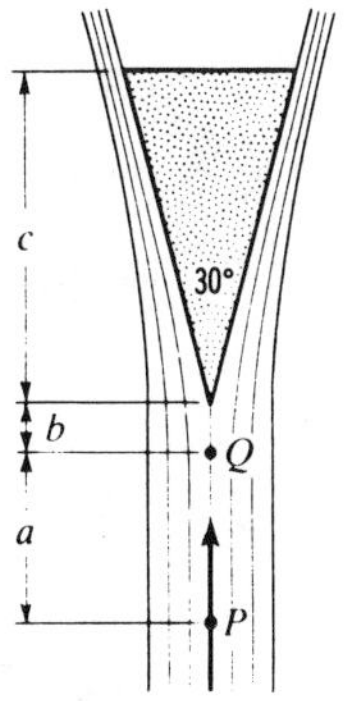

Figure P6.17

BG

Find V at section Q: Eq. 5.15: $V_P^2/2g = 1.5 + V_Q^2/2g$

$V_Q^2/2g = 28^2/2g - 1.5 = 10.67$ ft ; $V_Q = 26.2$ ft/sec

Find V_d at discharge: $V_Q^2/2g = 4.6 + V_d^2/2g$

$V_d^2/2g = 26.2^2/2g - 4.6 = 6.07$; $V_d = 19.78$ fps

From Eq. 6.10: $-(F_{C/W})_z - W_w = \rho Q(V_{2z} - V_{1z})$

where W_w = weight of water = $\gamma Q \times$ time to flow about 4.6 ft

$$\therefore\ \text{flow time} = \frac{4.6}{(26.2 + 19.78)/2} = 0.200\ \text{sec} ;$$

$W_w = \gamma Q t = 62.4(\pi/4)(5/12)^2 28(0.200) = 47.7$ lb

$W_{cone} = (F_{C/W})_z = -W_w - \rho Q(V_{2z} - V_{1z})$

$W_{cone} = -47.7 - 1.940(\pi/4)(5/12)^2 28(19.78 \cos 15° - 26.2) = -47.7 + 52.7 = 5.04$ lb ◀

6.18 *Repeat Prob. 6.17 if the jet diameter is 0.6 m, and the jet velocity is 22 m/s at P, and a = 1.6 m, b = 0.6 m, and c = 3.5 m.*

Refer to the problem statment and figure with Solution 6.17.

SI

Find V at section Q: Eq. 5.15: $22^2/(2\times9.81) = 1.6 + V_Q^2/2g$; $V_Q = 21.27$ m/s

Find V_d at discharge: $21.27^2/(2\times9.81) = 4.1 + V_d^2/2g$; $V_d = 19.29$ m/s

From Eq. 6.10: $-(F_{C/W})_z - W_w = \rho Q(V_{2z} - V_{1z})$

where W_w = weight of water = $\gamma Q \times$ time to flow about 4.1 m

$$\therefore \text{ flow time} = \frac{4.1}{(19.29+21.27)/2} = 0.202 \text{ s} ; \quad W_w = \gamma Qt = 9.81(\pi 0.30^2)22(0.202) = 12.33 \text{ kN}$$

$$W_{\text{cone}} = (F_{C/W})_z = -W_w - \rho Q(V_{2z} - V_{1z})$$

$$W_{\text{cone}} = -12.33 - (9.81/9.81)(\pi 0.30^2)22(19.29\cos 15° - 21.27) = -12.33 + 16.42 = 4.09 \text{ kN}$$ ◄

6.19 *Repeat Prob. 6.17 with the jet velocity at *19 fps and a = 0 , b = 2 ft, and c = 4 ft.*

*Refer to the problem statement and figure with Solution 6.17. [*Note this correction to the textbook.]*

BG

$$\frac{19^2}{2g} = 6 + \frac{V_d^2}{2g} ; \quad \frac{V_d^2}{2g} = 5.61 - 6 = -0.394 \text{ ft}$$

Thus jet does not have enough energy to rise to top of cone.

$$\frac{19^2}{2g} = z + 0 ; \quad z = 5.61 \text{ ft} = \text{maximum rise above point Q}$$

Thus the jet rises up 5.61 − 2 = 3.61 ft in contact with cone, when $\frac{19^2}{2g} = z + \frac{V^2}{2g}$ or

$$V = \sqrt{2g(5.61 - z)} ; \quad V = \frac{dz}{dt}, \quad \therefore t = \int_0^t dt = \int_0^{5.61} \frac{dz}{V} = \int_0^{5.61} \frac{dz}{\sqrt{2g(5.61 - z)}}$$

$$= -\frac{2\sqrt{5.61 - z}}{\sqrt{2g}}\Bigg|_0^{5.61} = 0.590 \text{ sec}$$

$$W_w = \gamma Qt = \gamma AVt = 62.4(\pi/4)(5/12)^2 19(0.590) = 95.4 \text{ lb}$$

$$\rho Q(V_{2z} - V_{1z}) = 1.940(\pi/4)(5/12)^2 19(0 - 19) = -95.5 \text{ lb}$$

From Eq. 6.10: $-(F_{C/W})_z - W_w = \rho Q \Delta V$

$$\therefore W_{\text{cone}} = (F_{C/W})_z = -W_w - \rho Q\Delta V = -95.4 - (-95.5) = 0.1040 \text{ lb}$$ ◄

Sec. 6.8: Force Exerted by a Jet on a Moving Vane or Blade -- *Exercises*

6.8.1 *If a jet of fluid strikes a single body moving in the same direction with a velocity u, flows over it without friction loss, and leaves with a relative velocity in the same direction of β_2, prove that $F_u = (\gamma A_1/g)(1 - \cos\beta_2)(V_1 - u)^2$.*

N

By Eq 6.23: $F_u = \frac{\gamma A_1}{g} v_1 \Delta v_u = \frac{\gamma A_1}{g} v_1(v_1 - v_2\cos\beta_2)$

But, $v_1 = v_2 = V_1 - u$, and $\therefore$ $F_u = (\gamma A_1/g)(1 - \cos\beta_2)(V_1 - u)^2$ QED ◄

6.8.2 *A jet of water strikes a single vane which reverses it through 180° without friction loss. If the jet has an area of 3.5 in² and a velocity of 175 fps, find the force exerted if the vane moves (a) in the same direction as the jet with a velocity of 75 fps, or (b) in a direction opposite to that of the jet with a velocity of 75 fps.*

BG

(*a*) $v_1 = 175 - 75 = 100$ fps ; Eq 6.20: $\rho Q' = (62.4/32.2)(3.5/144)100 = 4.71$ slugs/sec

By Eq. 6.22: $F = 4.71[100 - (-100)] = 942$ lb ◀

(*b*) $v_1 = 175 - (-75) = 250$ fps ; Eq. 6.22: $F = 4.71(250/100)[250 - (-250)] = 5890$ lb ◀

6.8.3 *A jet of water strikes a single vane which reverses it through 180° without friction loss. If the jet has an area of 25 cm² and a velocity of 55 m/s, find the force exerted if the vane moves (a) in the same direction as the jet with a velocity of 20 m/s; (b) in a direction opposite to that of the jet with a velocity of 20 m/s.*

SI

(*a*) $v_1 = 55 - 20 = 35$ m/s ; Eq. 6.20: $\rho Q' = (9810/9.81)(25/10\,000)35 = 87.5$ N·sec/m

By Eq. 6.22: $F = 87.5[35 - (-35)] = 6125$ N $= 6.13$ kN ◀

(*b*) $v_1 = 55 - (-20) = 75$ m/s ; Eq. 6.22: $F = 87.5(75/35)[75 - (-75)] = 28\,100$ N $= 28.1$ kN ◀

6.8.4 *What would be the resultant force components on the single vane of Sample Prob. 6.4 if it were traveling to the left toward the nozzle at 15 fps?*
Sample Prob. 6.4: A 2-in-diameter water jet, $V = 100$ fps, strikes a vane with $\beta_2 = 150°$. Friction causes $v_2 = 0.9\,v_1$.

BG

At entrance $V_1 = 100$, $u = -15$, $\alpha_1 = \beta_1 = 180°$, $\therefore\ v_1 = V_1 - u = 115$;

Hence $v_2 = 0.9(115) = 103.5$ fps. $A_j = \pi(2/12)^2/4 = 0.0218$ ft²

At exit, Eq. 6.17: $V_2\cos\alpha_2 = v_2\cos\beta_2 + u_2 = 103.5(\cos 30°) + 15 = 104.6$ fps to the left

$\Delta V_x = 100 - (-104.6) = 204.6$ fps

Eqs. 6.19 and 6.21: $F_x = \rho Q'(\Delta V_x) = 1.940(0.0218 \times 115)204.6 = 996$ lb ◀

Eq. 6.18: $V_{2y} = V_2\sin\alpha_2 = v_2\sin\beta_2 = 103.5\sin 30° = 51.8$ fps ;

$\Delta V_y = V_{2y} - V_{1y} = 51.8 - 0 = 51.8$ fps

From Eqs 6.19 and 6.21: $F_y = \rho Q'(\Delta V_y) = 1.940(0.0218 \times 115)51.8 = 252$ lb ◀

6.8.5 *A series of vanes is acted on by a 4-in water jet having a velocity of 105 fps, with $\alpha_1 = \beta_1 = 0$. Neglect friction and find the required blade angle β_2 in order that the resultant force acting on the vane in the direction of the jet is 200 lb. Solve using vane velocities of 0, 15, 45, and 75 fps. Also find the maximum possible vane velocity.*

BG

$A = \pi(4/12)^2/4 = 0.0873$ ft² ; $Q = AV = (0.0873)105 = 9.16$ cfs ; $v_2 = v_1 = v = 105 - u$

From Eqs. 6.24 and 6.17: $F = -200 = \rho Q(V_{2x} - V_{1x}) = 1.940(9.16)(u + v\cos\beta_2 - 105)$

i.e. $93.7 = u + (105 - u)\cos\beta_2$; $\cos\beta_2 = (93.7 - u)/(105 - u)$

For $u = 0$ fps ; $\beta_2 = 26.8°$ ◀
$u = 15$ fps ; $\beta_2 = 29.0°$ ◀
$u = 45$ fps ; $\beta_2 = 35.7°$ ◀
$u = 75$ fps ; $\beta_2 = 51.3°$ ◀

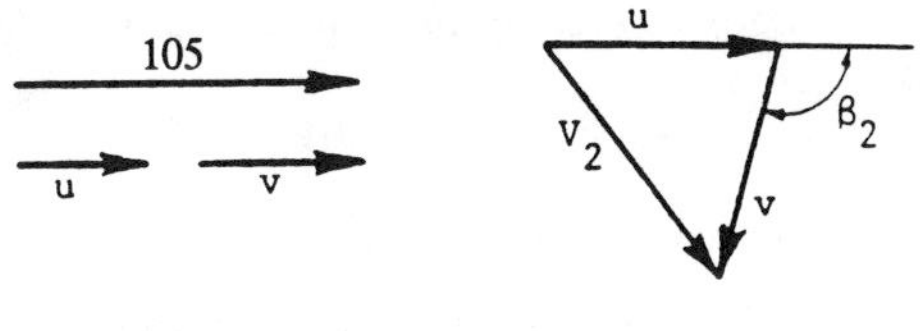

For $u = u_{max}$, $\beta_2 = 180°$,

yielding $u_{max} = 99.4$ fps ◀

6.20 *Assuming ideal flow in a horizontal plane, calculate the magnitude and direction of the resultant force on the stationary blade in Fig. P6.20, knowing that the velocity of the water jet is 40 fps. Note that the jet is divided by the splitter so that one-third of the water is diverted toward A.*

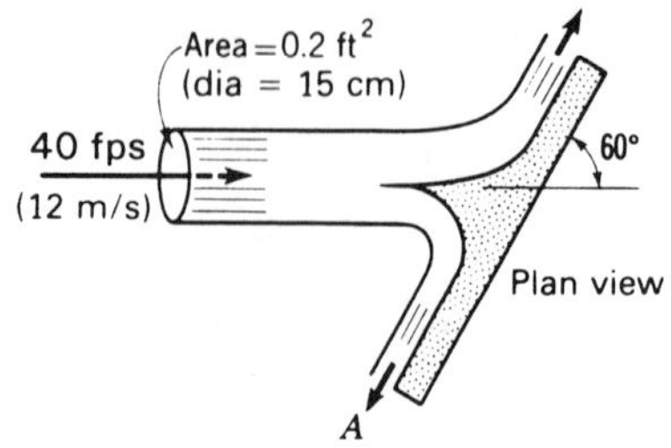

Figure P6.20

BG

$Q = 0.2(40) = 8$ cfs ; $Q_A = Q/3 = 2.67$ cfs, $Q_B = 5.33$ cfs

Eq. 6.7*a* for CV: $(F_{B/W})_x = \rho Q_A(V_{2x} - V_{1x}) + \rho Q_B(V_{2x} - V_{1x})$

$= 1.940(2.67)(-40\cos60° - 40) + 1.940(5.33)(40\cos60° - 40)$

$= -310 - 206 = -517$ lb

Eq. 6.7*b* for CV:

$(F_{B/W})_y = \rho Q_A(V_{2y} - V_{1y}) + \rho Q_B(V_{2y} - V_{1y}) = 1.940(2.67)(-40\cos30° - 0) + 1.940(5.33)(40\cos30° - 0)$

$= -179 + 358 = 179.2$ lb

Water acts on the blade with a force $(F_{W/B})_x$ of 517 lb along the centerline to the right; and perpendicularly with a force $(F_{W/B})_y$ of 179.2 lb (towards *A*).

Resultant force on blade = 547 lb at 19.11° from the centerline (towards *A*) ◄

6.21 *Refer to Fig. P6.20. Assuming ideal flow in a horizontal plane, calculate the magnitude and direction of the resultant force on the stationary blade knowing that the velocity of the water jet is 12 m/s. Note that the jet is divided by the splitter so that one-third of the water is diverted towards A.*
Fig. P6.20 is provided with Solution 6.20.

SI

$Q = \pi(0.075)^2 12 = 0.212$ m³/s ; $Q_A = Q/3 = 0.0707$ m³/s, $Q_B = 0.1414$ m³/s

Eq. 6.7*a* for CV: $(F_{B/W})_x = (9.81/9.81)0.0707(12\cos60° - 12) + (9.81/9.81)0.1414(12\cos60° - 12)$

$= -1.272 - 0.848 = -2.12$ kN

Eq 6.7*b* for CV: $(F_{B/W})_y = 1(0.0707)(-12\cos30° - 0) + 1(0.1414)(12\cos30° - 0) = 0.735$ kN

Water acts on the blade with a force $(F_{W/B})_x$ of 2.12 kN along the centerline to the right, and perpendicularly with a force $(F_{W/B})_y$ of 0.735 kN (towards *A*).

Resultant force on blade = 2.24 kN at 19.11° from the centerline (toward *A*). ◄

6.22 *A locomotive tender running at 25 mph scoops up water from a trough between the rails, as shown in the figure. The scoop delivers water at a height h = 7.5 ft above its origional level and in the direction of motion. The area of the stream of water at entrance is 40 in². The water is everywhere under atmospheric pressure. Neglecting all losses, what is the absolute velocity of the water as it leaves the scoop? What is the force acting on the tender caused by this? What is the minimum speed at which water will be delivered to the height h above the origional level?*

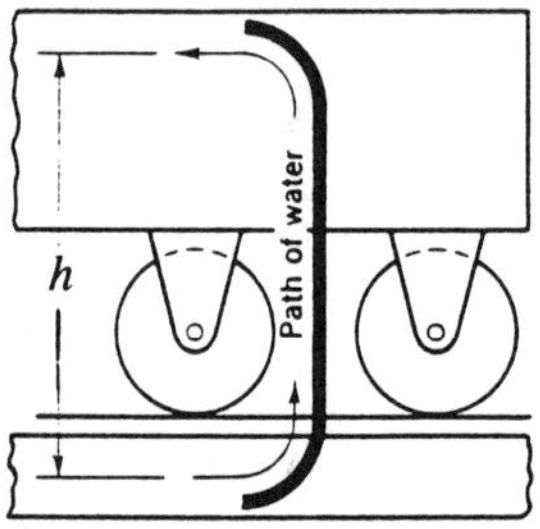

Figure 6.22

BG

$u_1 = u_2 = -25(88/60) = -36.7$ fps ; $\upsilon_1 = 36.7$ fps

Relative to moving car,

Eq. 5.15: $0 + 0 + 36.7^2/(2\times32.2) = 0 + 7.5 + \upsilon_2^2/(2\times32.2)$

$\upsilon_2 = -29.4$ fps, $V_2 = u_2 + \upsilon_2 = -36.7 - 29.4 = -66.0$ fps ◄ /cont...

Eq. 6.23: $F = -\frac{62.4}{32.2}\frac{40}{144}36.7(-66.0 - 0) = 1303$ lb ◄

Minimum tender speed $= \sqrt{2gh} = \sqrt{2(32.2)7.5} = 22.0$ fps $= 14.98$ mph ◄

6.23 *Repeat Prob. 6.22 for the following data: locomotive speed = 12 m/s, h = 2.3 m, stream area = 300 cm²*.
Refer to the problem statement and figure with Solution 6.22

SI

$u_1 = u_2 = -12$ m/s ; $\upsilon_1 = 12$ m/s

Relative to moving car, Eq. 5.15: $0 + 0 + 12^2/(2\times 9.81) = 0 + 2.3 + \upsilon_2^2/(2\times 9.81)$; $\upsilon_2 = -9.94$ m/s

$V_2 = u_2 + \upsilon_2 = -12 - 9.94 = -21.94$ m/s ◄

Eq. 6.23: $F = -(9.81/9.81)(0.03)12(-21.94 - 0) = 7.90$ kN ◄

Minimum tender speed $= \sqrt{2gh} = \sqrt{2(9.81)2.3} = 6.72$ m/s ◄

6.24 *A 4-in-dia water jet has a velocity of 130 fps. It strikes a single vane, which has an angle $\beta_2 = 90°$ and which is moving in the same direction as the jet with a velocity u. When u has values of 0, 45, 65, 85, 105, and 130 fps, find the values of (a) $\dot{m}'g$; (b) $V_2\cos\alpha_2$; (c) ΔV_u; (d) $\Delta\upsilon_u$; (e) F_u. Assume $\upsilon_2 = 0.9\upsilon_1$. Present answers in a neat tabular form or on a spreadsheet.*

BG

Eq. 6.17: $V_u = u + \upsilon_u$ $\quad\therefore\quad V_1 = u + \upsilon_1$

$\therefore\quad \upsilon_1 = V_1 - u = 130 - u = \upsilon_{1u}$ (since $\beta_1 = 0$) (for Col. 2)

From Eq. 6.20: $\dot{m}'g = (\rho A_1\upsilon_1)g = \gamma A_1\upsilon_1 = 62.4[\pi(4/12)^2/4]\upsilon_1 = 5.45\,\upsilon_1$ (for Col. *a*)

Eq. 6.17: $V_2\cos\alpha_2 = u + \upsilon_2\cos\beta_2 = u + \upsilon_2\cos 90° = u$ (for Col. *b*)

Given $\upsilon_2 = 0.9\,\upsilon_1$ at $\beta_2 = 90°$; $\upsilon_{2u} = \upsilon_2\cos\beta_2 = \upsilon_2\cos 90° = 0$

$\therefore\ \Delta\upsilon_u = \upsilon_{2u} - \upsilon_{1u} = 0 - \upsilon_1 = -\upsilon_1$ (for Col. *d*)

Sec. 6.8: $\Delta V = \Delta\upsilon, \quad\therefore\quad \Delta V_u = \Delta\upsilon_u$ (for Col. *c*)

From Eqs. 6.23 and 6.20: $F_u = \pm(\gamma/g)A_1\upsilon_1\Delta\upsilon_u = (\rho A_1\upsilon_1)\Delta\upsilon_u = \dot{m}'\Delta\upsilon_u$

$=$ Col. (*a*) $\times$ Col. (*d*)$/g$ (for Col. *e*)

u fps	υ_1 fps	(a) $\dot{m}'g$ lb/sec	(b) $V_2\cos\alpha_2$ fps	(c) ΔV_u fps	(d) $\Delta\upsilon_u$ fps	(e) $F_u = \pm\dot{m}'\Delta\upsilon_u$ lb
0	130	708	0	−130	−130	2860
45	85	463	45	−85	−85	1222
65	65	354	65	−65	−65	715
85	45	245	85	−45	−45	342
105	25	136.1	105	−25	−25	105.7
130	0	0	130	0	0	0
		▲	▲	▲	▲	▲

6.25 *Assume all the data in Prob. 6.24 are the same except that $\beta_2 = 180°$. Find the values of (a) $\dot{m}'g$; (b) υ_2; (c) V_2; (d) ΔV; (e) $\Delta \upsilon$; (f) F_u. Assume $\upsilon_2 = 0.8\upsilon_1$. Present answers in a neat tabular form or on a spreadsheet.*

Prob. 6.24: A jet with V = 130 fps, diameter = 4 in, strikes a single vane moving in the same direction at velocity u. Solve for u = 0, 45, 65, 85, 105, and 130 fps.

BG

Eq. 6.17: $V_u = u + \upsilon_u \quad \therefore \quad V_1 = u + \upsilon_1$

$\therefore \quad \upsilon_1 = V_1 - u = 130 - u = \upsilon_{1u}$ (since $\beta_1 = 0$) (for Col. 2)

From Eq. 6.20: $\dot{m}'g = (\rho A_1 \upsilon_1)g = \gamma A_1 \upsilon_1 = 62.4[\pi(4/12)^2/4]\upsilon_1 = 5.45\,\upsilon_1$ (for Col. *a*)

Given: $\beta_2 = 180°$ and $\upsilon_2 = 0.8\,\upsilon_1$ (for Col. *b*)

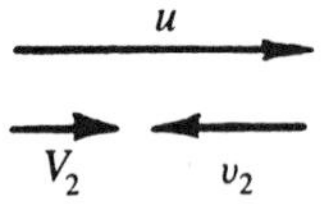

Eq. 6.17: $V_2 = V_{2u} = u + \upsilon_{2u} = u + \upsilon_2 \cos 180° = u - \upsilon_2$ (for Col. *c*)

$\Delta V = V_2 - V_1 = V_2 - 130$ (for Col. *d*). Sec. 6.8: $\Delta V = \Delta \upsilon$ (for Col. *e*)

From Eqs. 6.23 and 6.20: $F_u = \pm(\gamma/g)A_1\upsilon_1\Delta\upsilon_u = (\rho A_1 \upsilon_1)\Delta\upsilon_u = \dot{m}'\Delta\upsilon_u$

$=$ Col. (*a*) $\times$ Col. (*e*)$/g$ (for Col. *f*)

u fps	υ_1 fps	(*a*) $\dot{m}'g$ lb/sec	(*b*) υ_2 fps	(*c*) V_2 fps	(*d, e*) $\Delta V = \Delta\upsilon$ fps	(*f*) $F_u = \pm\dot{m}'\Delta\upsilon_u$ lb
0	130	708	104	−104	−234	5140
45	85	463	68	−23	−153	2200
65	65	354	52	13	−117	1286
85	45	245	36	49	−81	616
105	25	136.1	20	85	−45	190.3
130	0	0	0	130	0	0
		▲	▲	▲	▲	▲

6.26 *Repeat Prob. 6.25, except that $\upsilon_2 = 0.7\upsilon_1$*

Prob. 6.25: A jet with V = 130 fps, diameter = 4 in, strikes a single vane ($\beta_2 = 180°$) moving in the same direction at velocity u. Find the values of (a) $\dot{m}'g$; (b) υ_2; (c) V_2; (d) ΔV; (e) $\Delta\upsilon$; (f) F_u. Solve for u = 0, 45, 65, 85, 105, and 130 fps. Present answers in a neat tabular form or on a spreadsheet.

BG

Eq. 6.17: $V_u = u + \upsilon_u \quad \therefore \quad V_1 = u + \upsilon_1$

$\therefore \quad \upsilon_1 = V_1 - u = 130 - u = \upsilon_{1u}$ (since $\beta_1 = 0$) (for Col. 2)

From Eq. 6.20: $\dot{m}'g = (\rho A_1 \upsilon_1)g = \gamma A_1 \upsilon_1 = 62.4[\pi(4/12)^2/4]\upsilon_1 = 5.45\,\upsilon_1$ (for Col. *a*)

Given: $\beta_2 = 180°$ and $\upsilon_2 = 0.7\,\upsilon_1$ (for Col. *b*)

Eq. 6.17: $V_2 = V_{2u} = u + \upsilon_{2u} = u + \upsilon_2 \cos 180° = u - \upsilon_2$ (for Col. *c*)

$\Delta V = V_2 - V_1 = V_2 - 130$ (for Col. *d*). Sec. 6.8: $\Delta V = \Delta \upsilon$ (for Col. *e*)

From Eqs. 6.23 and 6.20: $F_u = \pm(\gamma/g)A_1\upsilon_1\Delta\upsilon_u = (\rho A_1 \upsilon_1)\Delta\upsilon_u = \dot{m}'\Delta\upsilon_u$

$=$ Col. (*a*) $\times$ Col. (*e*)$/g$ (for Col. *f*)

u fps	v_1 fps	(a) $\dot{m}'g$ lb/sec	(b) v_2 fps	(c) V_2 fps	(d, e) $\Delta V = \Delta v$ fps	(f) $F_u = \pm\dot{m}'\Delta v_u$ lb
0	130	708	91.0	−91.0	−221	4870
45	85	463	59.5	−14.5	−144.5	2080
65	65	354	45.5	19.5	−110.5	1215
85	45	245	31.5	53.5	−76.5	582
105	25	136.1	17.5	87.5	−42.5	179.7
130	0	0	0	130	0	0

▲ ▲ ▲ ▲ ▲

6.27 *Assume that all the data are the same as in Prob. 6.24, except that $\beta_2 = 145°$ and $v_2 = 0.7v_1$. Find the values of (a) $v_2\cos\beta_2$; (b) $V_2\cos\alpha_2$; (c) ΔV_u; (d) Δv_u; (e) F_u. Present answers in a neat tabular form or on a spreadsheet.*

Prob 6.24: A jet with V = 130 fps, diameter = 4 in, strikes a single vane moving in the same direction at velocity u. Solve for u = 0, 45, 65, 85, 105, and 130 fps.

BG

Eq. 6.17: $V_u = u + v_u \quad \therefore \quad V_1 = u + v_1$

$\therefore \quad v_1 = V_1 - u = 130 - u = v_{1u}$ (since $\beta_1 = 0$) (for Col. 2)

From Eq. 6.20: $\dot{m}'g = (\rho A_1 v_1)g = \gamma A_1 v_1 = 62.4[\pi(4/12)^2/4]v_1 = 5.45\,v_1$

$v_2\cos\beta_2 = 0.7\,v_1\cos 145° = -0.573\,v_1$ (for Col. *a*)

Eq. 6.17: $V_2\cos\alpha_2 = u + v_2\cos\beta_2$ (for Col. *b*)

Using Eq. 6.17: $\Delta V_u = V_{2u} - V_{1u} = V_2\cos\alpha_2 - V_1 = V_2\cos\alpha_2 - 130$ (for Col. *c*)

Sec. 6.8: $\Delta V = \Delta v, \quad \therefore \quad \Delta V_u = \Delta v_u$ (for Col. *d*)

From Eqs. 6.23 and 6.20: $F_u = \pm(\gamma/g)A_1 v_1 \Delta v_u = (\rho A_1 v_1)\Delta v_u = \dot{m}'\Delta v_u$

$= $ Col. (3) × Col. $(d)/g$ (for Col. *e*)

u fps	v_1 fps	$\dot{m}'g$ lb/sec	(a) $v_2\cos\beta_2$ fps	(b) $V_2\cos\alpha_2$ fps	(c, d) $\Delta V_u = \Delta v_u$ fps	(e) $F_u = \pm\dot{m}'\Delta v_u$ lb
0	130	708	−74.5	−74.5	−205	4500
45	85	463	−48.7	−3.74	−133.7	1922
65	65	354	−37.3	27.7	−102.3	1124
85	45	245	−25.8	59.2	−70.8	539
105	25	136.1	−14.34	90.7	−39.3	166.3
130	0	0	0	130	0	0

▲ ▲ ▲ ▲

6.28 *For the conditions of Prob. 6.20, compute the magnitude and direction of the resultant force on the single blade if it is moving to the right at a velocity of 15 fps.*

Prob. 6.20: The water jet shown in Fig. P6.20 (given with Solution 6.20) strikes the blade and is divided so that one third of the water is diverted toward A. Assume ideal flow in a horizontal plane.

BG

$v_1 = V_1 - u = 40 - 15 = 25$ fps

Eq. 6.19: $Q' = A_1 v_1 = 0.2(25) = 5$ cfs ; $Q'_A = 1.667$ cfs, $Q'_B = 3.33$ cfs

Eq. 6.23: $-F_x = 1.940(1.667)(-25\cos 60° - 25) + 1.940(3.33)(25\cos 60° - 25)$

$-F_x = -121.3 - 80.8 = -202$ lb ; $F_x = 202$ lb

Similarly: $F_y = 1.940(1.667)(-25\sin 60° - 0) + 1.940(3.33)(25\sin 60° - 0)$

$F_y = -70.0 + 140 = 70.0$ lb

Water acts on the blade with force of 202 lb along the centerline to the right, and perpendicularly with force of 70.0 lb (towards *A*).

Resultant force = 214 lb at 19.11° from the centerline (toward *A*). ◀

6.29 *For the conditions of Prob. 6.21, compute the magnitude and direction of the resultant force on the single blade if it is moving to the right at a velocity of 5 m/s.*

Prob. 6.21: Fig. P6.20 (included with Solution 6.20) represents ideal flow in a horizontal plane. The jet (V = 12 m/s, D = 15 cm) is divided by the splitter so that one third of the water is diverted toward A.

SI

$A_1 = \pi(0.15)^2/4 = 0.01767$ m^2; $v_1 = V_1 - u = 12 - 5 = 7$ m/s

Eq. 6.19: $Q' = A_1 v_1 = 0.01767(7) = 0.1237$ m^3/s; $Q'_A = 0.0412$ m^3/s, $Q'_B = 0.0825$ m^3/s

Eq. 6.23: $-F_x = (9.81/9.81)(0.0412)(-7\cos 60° - 7) + (9.81/9.81)(0.0825)(7\cos 60° - 7)$

$-F_x = -0.433 - 0.289 = -0.722$ kN

Similarly: $F_y = 1(0.0412)(-7\sin 60° - 0) + 1(0.0825)(7\sin 60° - 0) = -0.250 + 0.500 = 0.250$ kN

Water acts on the blade with a force of 0.722 kN along the centerline to the right, and perpendicularly with a force of 0.250 kN (toward *A*).

Resultant force = 0.764 kN at 19.11° from the centerline (toward *A*). ◀

6.30 *A 3-in-diameter air jet impinges on a series of blades, entering smoothly, and having the absolute velocities (fps) shown in Fig. P6.30. Assume $\gamma = 0.075$ lb/ft^3, that the pressure is the same on both sides, and neglect friction. (a) What is the velocity of the blades and the power being transmitted to them? (b) Determine the necessary blade angles at entrance and exit.*

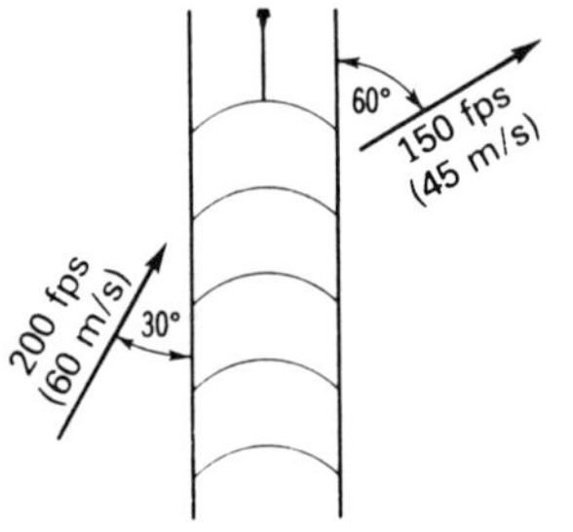

Figure P6.30

BG

(a) $H = \dfrac{200^2}{2g} - \dfrac{150^2}{2g} = 272$ ft·lb/lb

$\gamma Q = \gamma A V = 0.075(\pi/4)(3/12)^2 200 = 0.736$ lb/sec

Eq. 5.32: $\text{hp} = \dfrac{\gamma Q H}{550} = \dfrac{0.736(272)}{550} = 0.364$ ◀

Eq. 6.23: $F_x = \rho Q(V_{2x} - V_{1x}) = (0.736/32.2)(75 - 200\sin 60°) = -2.25$ lb

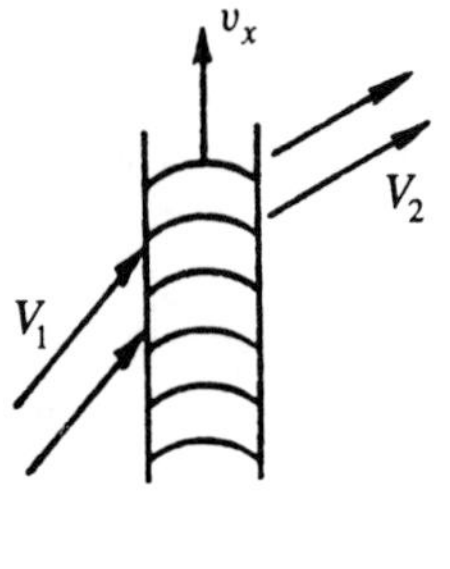

$$hp = 0.364 = \frac{F_x v_x}{550} \; ; \quad v_x = \frac{0.364(550)}{2.25} = 89.1 \text{ fps}$$ ◄

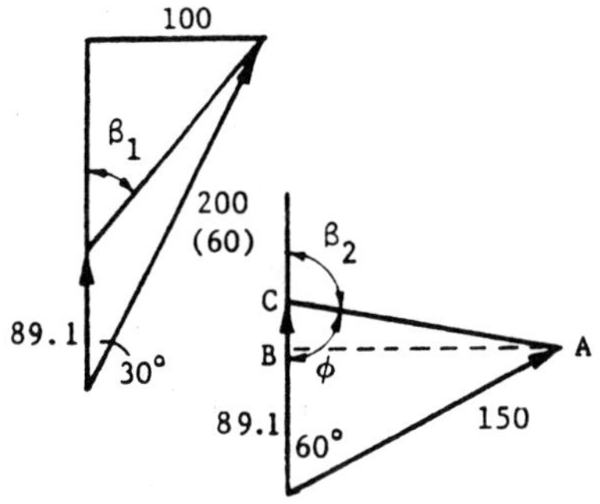

(b) $$\tan \beta_1 = \frac{100}{173.2 - 89.1} = 1.189 \; ; \quad \beta_1 = 49.9°$$

$$\tan\phi = \frac{AB}{BC} = \frac{150(\sin 60°)}{89.1 - 75} = 9.21 \; ; \quad \phi = 83.8°$$

$$\beta_2 = 180 - 83.8 = 96.2°$$ ◄

6.31 *A 6-cm-diameter air jet impinges on a series of blades, entering smoothly, and having the absolute velocities (m/s) shown in Fig. P6.30. Assume $\gamma = 11$ N/m³, that the pressure is the same on both sides, and neglect friction. (a) What is the velocity of the blades and the power being transmitted to them? (b) Determine the necessary blade angles at entrance and exit.*

SI

(a) $$H = \frac{60^2}{2(9.81)} - \frac{45^2}{2(9.81)} = 80.3 \text{ m}$$

$\gamma Q = \gamma AV = 11(\pi/4)(6/100)^2 60 = 1.866$ N/s

Eq. 5.31: Power = $\gamma QH = 1.866(80.3) = 149.8$ N·m/s = 149.8 W ◄

Eq. 6.23: $F_x = \rho Q(V_{2x} - V_{1x}) = (1.866/9.81)(45\cos 60° - 60\cos 30°) = -5.60$ N

Power = 149.8 N·m/s = $F_x v_x$; $\therefore v_x = 149.8/5.60 = 26.7$ m/s ◄

(b) $\tan\beta_1 = 30/(60\cos 30° - 26.7) = 1.189$; $\beta_1 = 49.9°$

$\tan\phi = AB/BC = 45\sin 60°/(26.7 - 45\cos 60°) = 9.21$; $\phi = 83.8°$

$\beta_2 = 180 - 83.8 = 96.2°$ ◄

6.32 *Suppose the blade of Prob. 6.20 is one of a series of blades which are moving to the right at 10 fps. (a) Determine the resultant horizontal force on the blade system, and (b) compute the power transferred to the blades. (c) Compute the power of the jet and of the water leaving the blade system to verify an energy balance (Sample Prob. 6.4).*

Refer to Fig. P6.20 included with Solution 6.20. Assume ideal flow in a horizontal plane. Given V_{1x} = 40 fps, and that one-third of the water is directed by the splitter towards A.

BG

$Q = AV = 0.2(40) = 8$ cfs; $Q_A = 8/3 = 2.67$ cfs, $Q_B = 5.33$ cfs

(a) From Eq. 6.17: $v_1 = v_{1x} = V_{1x} - u = 40 - 10 = 30$ fps. Ideal flow, $\therefore v_2 = v_1 = 30$ fps

Eq. 6.24: $F_u = 1.940(2.67)(-30\cos 60° - 30) + 1.940(5.33)(30\cos 60° - 30)$

$F_u = -233 - 155.2 = -388$ lb. Horizontal force on blades is 388 lb to the right ◄

(b) hp = $Fu/550 = 388(10)/550 = 7.05$ hp transferred to blades ◄

(c) Energy balance check: Velocities at exit:

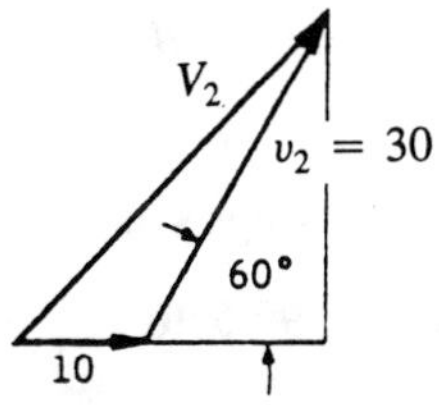

$V_{2x} = 10 + 30\cos 60° = 25$ fps

$V_{2y} = 30\sin 60° = 26.0$ fps

$V_2 = 36.1$ fps

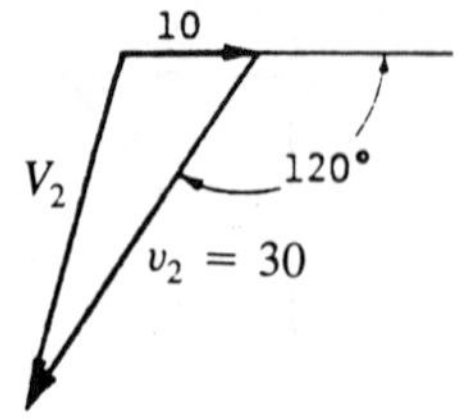

$V_{2x} = 10 - 30\cos 60° = -5$ fps

$V_{2y} = -26.0$ fps

$V_2 = 26.5$ fps

Energy entering: $H_1 = 40^2/2g = 24.8$ ft ;

Energy leaving: $H_2 = (2/3)36.1^2/2g + (1/3)26.5^2/2g = 17.08$ ft

Power transferred to blades: hp $= \gamma QH/550 = 62.4(8)(24.8 - 17.08)/550 = 7.05$ ◄

6.33 *Suppose the blade of Prob. 6.21 is one of a series of blades which are moving to the right at 4 m/s. (a) Determine the resultant horizontal force on the blade system, and (b) compute the power transferred to the blades. (c) Compute the power of the jet and of the water leaving the blade system to verify an energy balance (Sample Prob. 6.4).*

Refer to Fig. P6.20 included with Solution 6.20. Assume ideal flow in a horizontal plane. Given V_{1x} = 12 m/s, and that one-third of the water is directed by the splitter towards A.

SI

$Q = AV = \pi(0.075)^2 12 = 0.212$ m³/s; $Q_A = Q/3 = 0.0707$ m³/s; $Q_B = 0.1414$ m³/s

(*a*) From Eq. 6.17: $v_1 = v_{1x} = V_{1x} - u = 12 - 4 = 8$ m/s. Ideal flow, ∴ $v_2 = v_1 = 8$ m/s

Eq.6.24: $F_u = 1000(0.0707)(-8\cos 60° - 8) + 1000(0.1414)(8\cos 60° - 8)$

$F_u = -848 - 565 = -1414$ N. Horizontal force on blades is 1414 N to the right ◄

(*b*) Power transferd to blades $= F_u/1000 = 1414(4)/1000 = 5.65$ kW ◄

(*c*) Energy balance check: Velocities at exit:

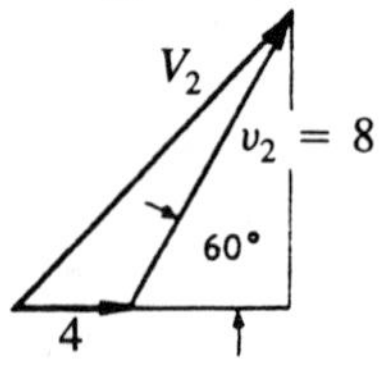

$V_{2x} = 4 + 8\cos 60° = 8$ m/s

$V_{2y} = 8\sin 60° = 6.93$ m/s

$V_2 = 10.58$ m/s

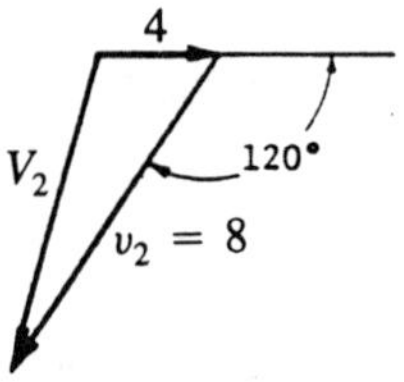

$V_{2x} = 4 - 8\cos 60° = 0$

$V_{2y} = -6.93$ m/s

$V = 6.93$ m/s

Energy entering: $H_1 = 12^2/2g = 7.34$ m

Energy leaving: $H_2 = (2/3)10.58^2/2g + (1/3)6.93^2/2g = 4.62$ m

Power transferred to blades: $\gamma QH/1000 = 9810(0.212)(7.34-4.62)/1000 = 5.65$ kW ◄

Sec. 6.9: Reaction of a Jet -- ***Exercises***

6.9.1 *Find the thrust developed when water is pumped through an 8-in-diameter pipe in the bow of a boat at v = 7 fps and emitted through a 4-in-diameter pipe in the stern of the boat.*

BG

Eq. 6.7*a*: $F = \rho Q(V_{2x} - V_{1x})$, where $\rho Q = 1.940(\pi/4)(8/12)^2 7 = 4.74$ slugs/sec

$V_{2x} = 7(8/4)^2 = 28.0$ fps ; ∴ $F = 4.74(28.0 - 7) = 99.5$ slug·ft/sec² $= 99.5$ lb ◄

6.9.2 *Find the thrust developed when water is pumped through a 20-cm-diameter pipe in the bow of a boat at $v = 2$ m/s and emitted through a 10-cm-diameter pipe in the stern of the boat.*

SI

$\rho Q = 1000(\pi\, 0.1^2)2 = 62.8$ kg/s

$V_{2x} = 2(20/10)^2 = 8$ m/s ; Eq. 6.7*a*: $F = 62.8(8-2) = 377\ \mathrm{kg \cdot m/s^2} = 377$ N ◄

Sec. 6.9: Reaction of a Jet -- *Problems*

6.34 *An ideal liquid ($\gamma = 62.4\ lb/ft^3$) flows from a 2-ft-diameter tank as shown in Fig. P6.34. The jet diameter is 3 in. If the static coefficient of friction between the tank and floor is 0.56, determine the minimum value of h at which the tank will commence to move to the left. The tank itself weighs 100 lb.*

BG

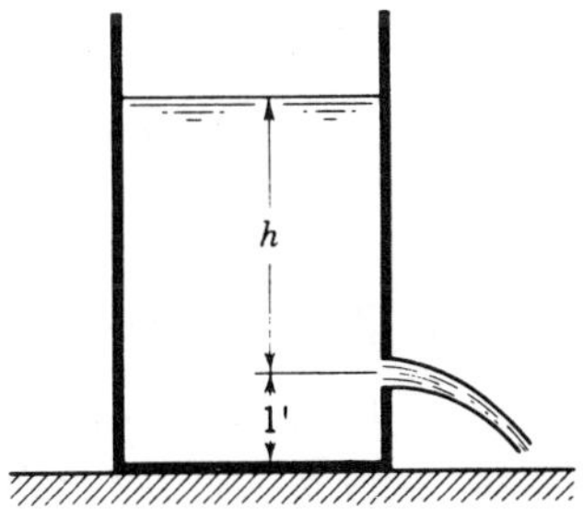

Figure P6.34

$W = 100 + \pi(2^2/4)(1+h)62.4$; $\mu W = 0.56W = 165.8 + 109.8h$

Eq. 6.25: $F_x = \rho Q V = \rho A V^2 = 1.940\,\pi(1.5/12)^2(2gh) = 6.13h$

$F_x = \mu W$; Thus $166 + 110h = 6.13h$; $h = -1.599$ ft

h is negative, therefore the tank will not move regardless of the value of h. ◄

6.35 *Find the magnitude and direction of the resultant force of the fluid on the compressor shown in the figure. Air ($\gamma = 0.075\ lb/ft^3$) enters at A through a 4-ft² area at a velocity of 15 fps. Air is discharged at B through a 3-ft² area at a velocity of 17 fps.*

BG

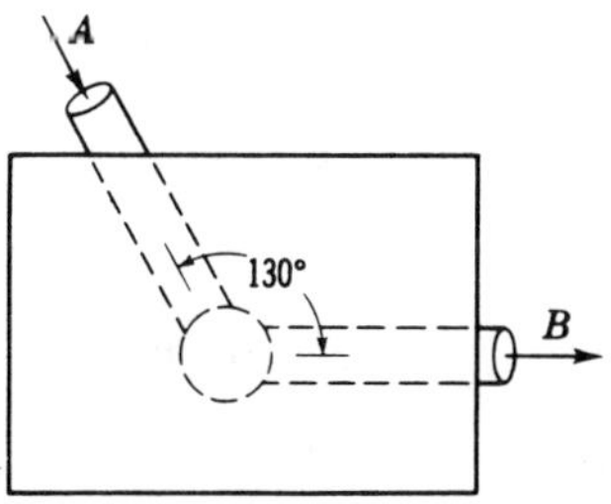

Figure P6.35

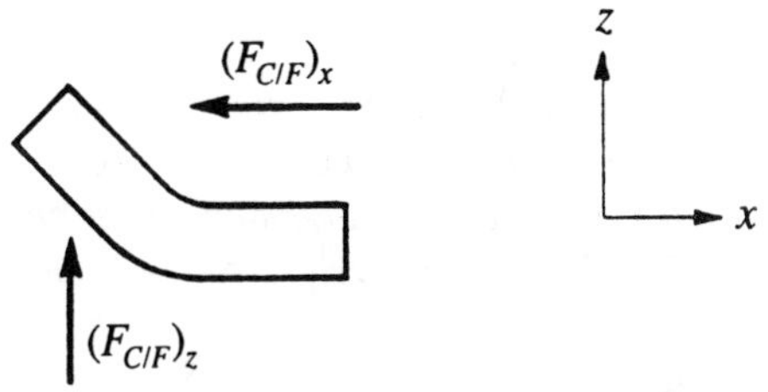

$\rho Q = (\gamma/g)Q = (0.075/32.2)(4 \times 15) = 0.1398$ slugs/sec

Eq. 6.23: $-(F_{C/F})_x = \rho Q(V_{2x} - V_{1x}) = 0.1398(17 - 15\cos 50°) = 1.028$ slugs/sec

$\therefore\ (F_{C/F})_x = -1.028$ lb. Minus sign indicates that compressor pushes on fluid to the right.

Eq. 6.23: $(F_{C/F})_z = \rho Q(V_{2z} - V_{1z}) = 0.1398(0 + 15\cos 40°) = 1.606$ lb

Positive sign indicates that assumed direction of compressor force on fluid is correct.

Force of fluid on compressor: 1.028 lb to the left, 1.606 lb downward

Resultant force = 1.907 lb at 57.4° to the horizontal. ◄

6.36 *Repeat Prob. 6.35 for the case where a gas (γ = 12.1 N/m³) enters at A through a 60-cm-diameter pipe at 5 m/s and leaves at B through a 50-cm-diameter pipe at 7 m/s.*

Prob. 6.35: Refer to figures with Solution 6.35.

SI

$\rho = \gamma/g = (12.1 \text{ N/m}^3)/(9.81 \text{ m/s}^2) = 1.233 \text{ kg/m}^3$; $Q = \pi(0.30^2)5 = 1.414 \text{ m}^3/\text{s}$

$\dot{m} = \rho Q = 1.233(1.414) = 1.744 \text{ kg/s}$

$\Delta V_x = V_{2x} - V_{1x} = 7 - 5\cos 50° = 3.79 \text{ m/s}$; $\Delta V_z = V_{2z} - V_{1z} = 0 - 5\cos 40° = -3.83 \text{ m/s}$

$-(F_{C/F})_x = \rho Q(\Delta V_x) = 1.744(3.79) = 6.60 \text{ N}$; $(F_{C/F})_z = \rho Q(\Delta V_z) = 1.744(-3.83) = -6.68 \text{ N}$

Force on fluid on compressor: 6.60 N to the left, 6.68 N downward.

Resultant force = 9.39 N at 45.3° to the horizontal. ◀

Sec. 6.10: Jet Propulsion -- *Exercises*

6.10.1 *Find the thrust of a turbojet whose speed is 700 fps and whose air intake rate is 40 lb/sec. The air/fuel ratio is 30:1 and the exhaust velocity is 1800 fps.*

BG

$(\dot{m}g)_{\text{fuel}} = 40/30 = 1.333 \text{ lb/sec}$; Eq. 6.27: $F = [(40+1.333)1800 - 40(700)]/32.2 = 1441 \text{ lb}$ ◀

6.10.2 *Find the thrust of a turbojet whose speed is 300 m/s and whose air intake rate is 10 kg/s. The air/fuel ratio is 22:1 and the exhaust velocity is 500 m/s.*

SI

$\dot{m}_{\text{fuel}} = 10/22 = 0.455 \text{ kg/s}$

Eq. 6.27 can be expressed as: $F = (\dot{m}_{\text{air}} + \dot{m}_{\text{fuel}})v_2 - \dot{m}_{\text{air}}u = 10.455(500) - 10(300) = 2230 \text{ N}$ ◀

Sec. 6.10: Jet Propulsion -- *Problem*

6.37 *A rocket has a propellant flow rate of 21.6 lb/sec through a nozzle with a throat area of 9.3 in². The nozzle is designed to expand the gases down to 14.7 psia at exit. The exit area of the nozzle is 48.5 in², the exhaust velocity is 6370 fps. Find the rocket's thrust (a) at sea level and (b) at an elevation of 20,000 ft where the barometer pressure is 6.75 psia; (c) find the specific weight of the exhaust gas.*

BG

(*a*) Eq. 6.26 with $p_0 = p_a = 14.7$ psia at sea level:

$F = \rho A v_2^2 = \dot{m} v_2 = (21.6/32.2)6370 = 4270 \text{ lb}$ at sea level ◀

(*b*) $p_0 = 14.7$ psia, $p_a = 6.75$ psia (given) at 20,000 ft elev.

From Eq. 6.26: $F = \dot{m} v_2 + (p_0 - p_a)A_2$

$F = 4270 + (14.7 - 6.75)\,48.5 = 4660 \text{ lb}$ at 20,000 ft ◀

(*c*) $\dot{m}g = 21.6 \text{ lb/sec} = \gamma A v = \gamma(48.5/144)6370$; $\gamma = 0.0101 \text{ pcf}$ ◀

Sec. 6.11: Rotating Machines: Continuity, Velocities, Torque -- *Exercises*

6.11.1 *The absolute velocity of a jet of steam impinging upon the blades of a steam turbine is 3800 fps, and that leaving is 2600 fps. α_1 = 20°, α_2 = 150°, $u_1 = u_2$ = 500 fps, and $r_1 = r_2$ = 0.5 ft. Find the torque exerted on the rotor and the power delivered to it if the steam flows at 0.4 lb/sec.*

BG

Noting that $\rho Q = \dot{m} = G/g = 0.4/32.2$ slug/sec

By Eq. 6.28: $T = (0.4/32.2)0.5(3800\cos 20° - 2600\cos 150°) = 36.2$ ft·lb ◄

Eqs. 5.32 and 5.35: Power $= T\omega/550 = T(u/r)/550 = 36.2(500/0.5)/550 = 65.8$ hp ◄

6.11.2 *The absolute velocity of a jet of steam impinging upon the blades of a steam turbine is 1200 m/s, and that leaving is 950 m/s. α_1 = 20°, α_2 = 150°, $u_1 = u_2$ = 180 m/s, and $r_1 = r_2$ = 12 cm. Find the torque exerted on the rotor and the power delivered to it if the steam flows at 2 N/s.*

SI

Noting that $\rho Q = \dot{m} = G/g = 2/9.81$ kg/s

By Eq. 6.28: $T = (2/9.81)0.12(1200\cos 20° - 950\cos 150°) = 47.7$ N·m ◄

Eq. 5.35: Power $= T\omega = Tu/r = 47.7(180/0.12) = 71\,600$ N·m/s $= 71.6$ kW ◄

Sec. 6.12: Head Equivalent of Mechanical Work -- *Problems*

6.38 *A radial-flow turbine has the following dimensions: r_1 = 0.5 m, r_2 = 0.3 m, β_1 = 74°, and β_2 = 126°. The width of the flow passage between the two sides of the turbine is 0.2 m. When operating at 160 rpm the flow rate through the turbine is 1.5 m³/s. Find (a) the torque exerted by the water, (b) the power delivered to the shaft, and (c) the head converted into mechanical work.*

SI

At the outer periphery (r_1 = 0.5 m, β_1 = 74°):

$u_1 = \omega r_1 = (2\pi/60)160(0.5) = 16.76(0.5) = 8.38$ m/s

From continuity: $Q = 1.5 = 2\pi(0.5)0.2V_{r1}$; $V_{r1} = 2.39$ m/s

$V_1\cos\alpha_1 = u_1 + V_{r1}/\tan\beta_1 = 8.38 + 2.39/\tan 74° = 9.06$ m/s

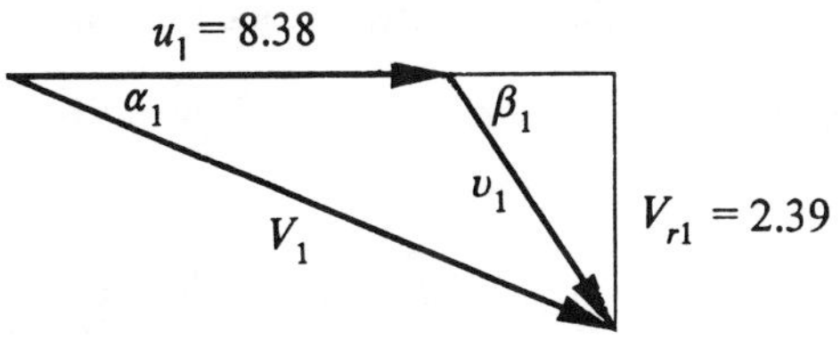

At the inner periphery (r_2 = 0.3 m, β_2 = 126°):

$u_2 = \omega r_2 = u_1(r_2/r_1) = 8.38(0.3/0.5) = 5.03$ m/s

$V_{r2} = V_{r1}(r_1/r_2) = 2.39(0.5/0.3) = 3.98$ m/s

$v_2\sin 126° = v_{r2} = 3.98$, $v_2 = 4.92$ m/s

$v_2\cos\alpha_2 = u_2 + v_2\cos\beta_2 = 5.03 + 4.92\cos 126° = 2.14$ m/s

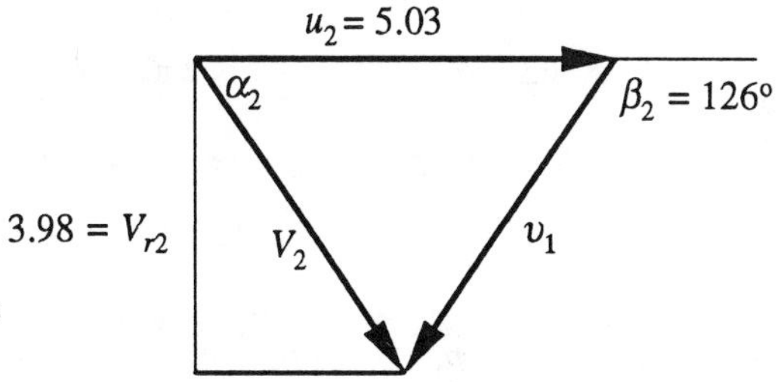

(a) Eq. 6.28: $T = 1000(1.5)(0.5\times 9.06 - 0.3\times 2.14) = 5840$ N·m ◄

(b) Eq. 5.35: Power $= T\omega/1000 = 5840(2\pi/60)160/1000 = 97.8$ kW ◄

(c) Eq. 6.29: $h_m = (8.38\times 9.06 - 5.03\times 2.14)/9.81 = 6.64$ m ◄

Check: $\gamma Q h_m = 9810(1.5)6.64/1000 = 97.8$ kW. Good!

6.39 *A radial-flow turbine has the following dimensions: r_1 = 3.2 ft, r_2 = 1.6 ft, β_1 = 76°, and β_2 = 135°. The width of the flow passage between the two sides of the turbine is 0.65 ft. When operating at 150 rpm the flow rate through the turbine is 60 cfs. Find (a) the torque exerted by the water, (b) the horsepower delivered to the shaft, and (c) the head converted into mechanical work.*

BG

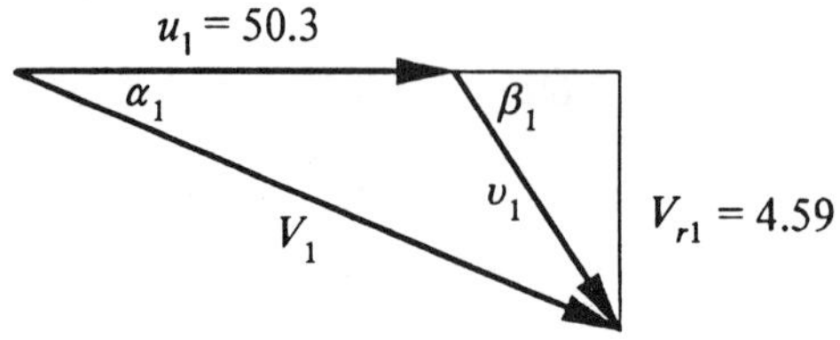

At the outer periphery (r_1 = 3.2 ft, β_1 = 76°):

$u_1 = \omega r_1 = (2\pi/60)150(3.2) = 15.71(3.2) = 50.3$ fps

From continuity: $Q = 60 = 2\pi(3.2)0.65V_{r1}$, $V_{r1} = 4.59$ fps

$V_1\cos\alpha_1 = u_1 + V_{r1}/\tan\beta_1 = 50.3 + 4.59/\tan 76° = 51.4$ fps

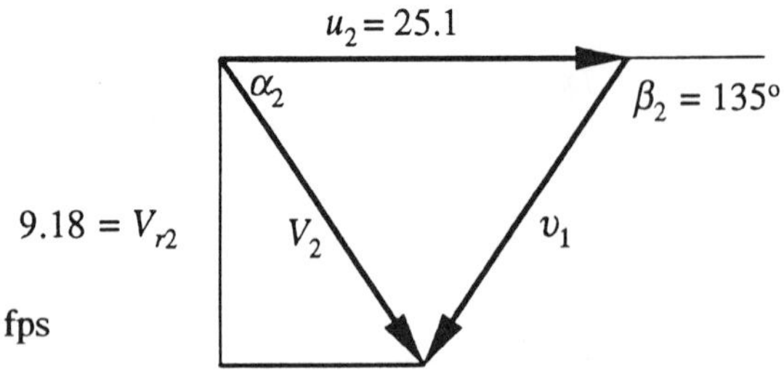

At the inner periphery (r_2 = 1.6 ft, β_2 = 135°):

$u_2 = \omega r_2 = u_1(r_2/r_1) = 50.3(1.6/3.2) = 25.1$ fps

$V_{r2} = V_{r1}(r_1/r_2) = 4.59(2.2/1.6) = 9.18$ fps

$v_2 \sin 135° = v_{r2} = 9.18$, $v_2 = 12.98$ fps

$v_2\cos\alpha_2 = u_2 + v_2\cos\beta_2 = 25.1 + 12.98\cos 135° = 15.95$ fps

(*a*) Eq. 6.28: $T = 1.940(60)(3.2\times 51.4 - 1.6\times 15.95)$

$= 16{,}180$ ft·lb ◄

(*b*) Eq. 5.35: HP $= T\omega/550 = 16{,}180(2\pi/60)150/550 = 462$ hp ◄

(*c*) Eq. 6.29: $h_m = (50.3\times 51.4 - 25.1\times 15.95)/32.2 = 67.8$ ft ◄

Check: $\gamma Q h_m/550 = 62.4(60)67.8/550 = 462$ hp. Good!

Sec. 6.13: Flow Through Rotating Channel -- *Exercise*

6.13.1 *Develop Eq. (6.31) by making the substitutions indicated in the text.*

N

Given Eq. 6.30: $$\left(\frac{p_1}{\gamma} + z_1 + \frac{V_1^2}{2g}\right) - \left(\frac{p_2}{\gamma} + z_2 + \frac{V_2^2}{2g}\right) = h_L + \frac{u_1V_1\cos\alpha_1 - u_2V_2\cos\alpha_2}{g}$$

But $V^2 = v^2 + u^2 + 2vu\cos\beta$ and $V\cos\alpha = u + v\cos\beta$

Substituting: $$p_1/\gamma + z_1 + (v_1^2 + u_1^2 + 2v_1u_1\cos\beta)/2g - p_2/\gamma - z_2 - (v_2^2 + u_2^2 + 2v_2u_2\cos\beta)/2g$$
$$= h_L + (2u_1^2 + 2u_1v_1\cos\beta - 2u_2^2 - 2u_2v_2\cos\beta)/2g$$

i.e., $$\frac{p_1}{\gamma} + z_1 - \left(\frac{p_2}{\gamma} + z_2\right) + \frac{v_1^2 + u_1^2 - v_2^2 - u_2^2 - 2u_1^2 + 2u_2^2}{2g} = h_L$$

i.e., $$\left(\frac{p_1}{\gamma} + z_1 + \frac{v_1^2 - u_1^2}{2g}\right) - \left(\frac{p_2}{\gamma} + z_2 + \frac{v_2^2 - u_2^2}{2g}\right) = h_L$$ which is Eq. 6.31. QED ◄

Sec. 6.13: Flow Through Rotating Channel -- *Problems*

6.40 *A paddle wheel with vanes that are all straight and radial is to be used as a crude centrifugal pump for water (Fig. P6.40): $r_1 = 3$ in, $r_2 = 9$ in, and the height z perpendicular to the plane of the figure is 0.2 ft. If the speed is 1200 rpm and the flow is 3380 gpm, (a) at the centerline elevation find the difference in pressure (psi) between the inner and outer circumferences, neglecting friction losses, (b) find which of these two points has the higher pressure, (c) compute the torque required to drive the pump, (d) calculate the horsepower requirement, and (e) verify that the horsepower requirement is equal to the difference between the horsepower of the outflow minus the horsepower of the inflow.*

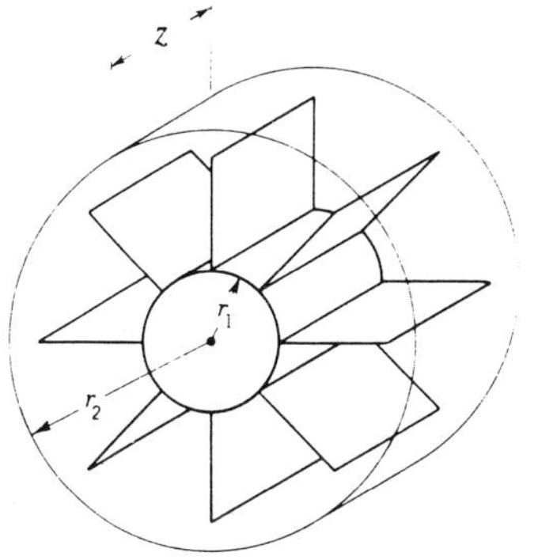

Figure P6.40

BG

$\omega = 1200(2\pi/60) = 40\pi$ rad/sec

$u_1 = \omega r_1 = 40\pi(3/12) = 10\pi$ fps ; $u_2 = \omega r_2 = 30\pi$ fps

$a_1 = 2\pi r_1 h = 2\pi(3/12)(0.2) = \pi/10$ ft^2

$a_2 = 2\pi(9/12)(0.2) = 0.3\pi$ ft^2

$Q = 3380$ gpm $= 7.53$ cfs

$v_1 = 7.53/(0.10\pi) = 24.0$ fps ; $v_2 = 7.53/(0.30\pi) = 7.99$ fps

(*a*) Eq. 6.31: $$\left[\frac{p_1}{\gamma} + \frac{24.0^2 - (10\pi)^2}{2g}\right] - \left[\frac{p_2}{\gamma} + \frac{7.99^2 - (30\pi)^2}{2g}\right] = 0$$

$$p_1/\gamma - p_2/\gamma = (63.8 - 8880 - 575 + 987)/(2g) = -130.5 \text{ ft}$$

$$p_1 - p_2 = 62.4(-130.5)/144 = -56.6 \text{ psi} \quad ◀$$

(*b*) p_2 is 56.6 psi greater than p_1. ◀

(*c*) Eq. 6.28: $T = \rho Q(r_1 u_1 - r_2 u_2) = 1.940(7.53)[(3/12)10\pi - (9/12)30\pi] = -918$ ft·lb (on water)

Thus torque required is 918 ft·lb

(*d*) Eq. 5.35: Power $= T\omega/550 = 918(40\pi)/550 = 210$ hp ◀

(*e*) $V_1 = \sqrt{u_1^2 + v_1^2} = 39.5$ fps ; $V_2 = \sqrt{u_2^2 + v_2^2} = 94.6$ fps

Head $= (p_2/\gamma + V_2^2/2g) - (p_1/\gamma + V_1^2/2g) = 130.5 + 138.9 - 24.3 = 245$ ft

Eq. 5.32: Power $= \gamma QH/550 = 62.4(7.53)245/550 = 210$ hp, checks out ◀

6.41 *Repeat Prob. 6.40 where the data are given in SI units as follows: $r_1 = 6.5$ cm, $r_2 = 21.5$ cm, the height z perpendicular to the plane of figure = 4.8 cm. The speed is 1200 rpm and the flow is 150 L/s. Express pressure difference in kN/m^2 and the power in kW.*

Refer to the figure and requirements with Solution 6.40.

SI

$\omega = 1200(2\pi/60) = 40\pi$ rad/s ; $u_1 = \omega r_1 = 2.6\pi$ m/s ; $u_2 = \omega r_2 = 8.6\pi$ m/s

$a_1 = 2\pi r_1 h = 2\pi(0.065)0.048 = 0.01960$ m^2, $a_2 = 0.0648$ m^2

$Q = 0.15$ m^3/s ; $v_1 = Q/a_1 = 7.65$ m/s ; $v_2 = 2.31$ m/s

/cont...

(*a*) Eq. 6.23: $\left[\dfrac{p_1}{\gamma}+\dfrac{7.65^2-(2.6\pi)^2}{2(9.81)}\right]-\left[\dfrac{p_2}{\gamma}+\dfrac{2.31^2-(8.6\pi)^2}{2(9.81)}\right] = 0$

$$\frac{p_1}{\gamma}-\frac{p_2}{\gamma} = \frac{-58.5 + 66.7 + 5.35 - 730}{2(9.81)} = -36.5 \text{ m}$$

$\Delta p = p_2 - p_1 = -36.5(9.81) = -358 \text{ kN/m}^2 = -358 \text{ kPa}$ ◄

(*b*) p_2 is 358 kPa greater than p_1 ◄

(*c*) Eq. 6.28: $T = \rho Q(r_1u_1 - r_2u_2) = (1)0.15(0.065\times 2.6\pi - 0.215\times 8.6\pi) = -0.792 \text{ kN·m}$

Thus torque required is 0.792 kN·m ◄

(*d*) Eq. 5.35: Power $= T\omega = 0.792(40\pi) = = 99.5$ kW ◄

(*e*) $V_1 = \sqrt{u_1^2+v_1^2} = 11.19$ m/s ; $V_2 = \sqrt{u_2^2 + v_2^2} = 27.1$ m/s

Head $= (p_2/\gamma + V_2^2/2g) - (p_1/\gamma + V_1^2/2g) = 36.5 + 37.5 - 6.38 = 67.6$ m

Power $= \gamma QH = 9.81(0.15)67.6 = 99.5$ kW, checks out ◄

Sec. 6.14: Reaction with Rotation -- ***Exercises***

6.14.1 *The flow from a lawn sprinkler such as Fig. 6.13 is 140 L/min, $\beta_2 = 180°$, and the total area of the jets is 120 mm^2. The jets are located 20-cm from the center of rotation. Determine the speed of rotation if there is no friction.*

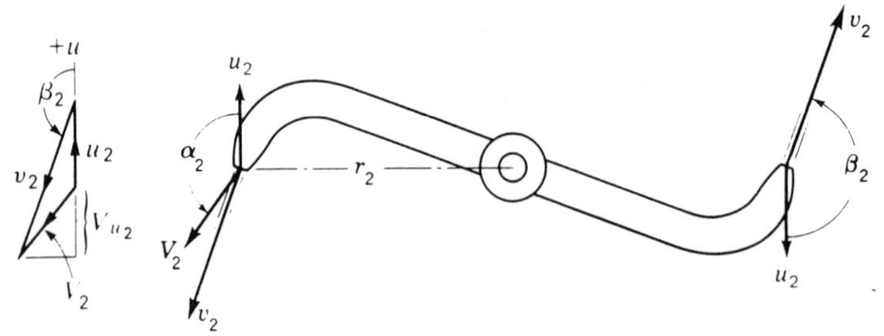

Figure. 6.13

SI

$Q = 0.14/60 = 0.00233 \text{ m}^3/\text{s} = a_2v_2 = (120\times10^{-6})v_2$; $v_2 = 19.44$ m/s

Eq.6.33: $T = F_u r_2 = -\rho a_2 v_2 r_2(u_2 + v_2\cos\beta_2) = 0$ for no friction

Thus $u_2 + v_2\cos 180° = 0$; $u_2 - 19.44 = 0$, $u_2 = 19.44$ m/s

$u_2 = 19.44 = \omega r = \omega(0.20)$; thus $\omega = 97.2$ rad/s; $n = 97.2(60/2\pi) = 928$ rpm ◄

6.14.2 *A lawn sprinkler like that of Fig. 6.13 with $\beta_2 = 156°$ has a total jet area of 0.00086 ft^2 and a radius of 16 inches. Compute the discharge rate, the torque exerted by the water, and the power developed, when h = 120 ft and the sprinkler is prevented from rotating.*

Refer to Fig. 6.13 with Solution 6.14.1.

BG

Eq. 6.32 with $u_2 = 0$: $v_2 = \sqrt{2(32.2)120} = 87.9$ fps; $Q = a_2v_2 = 0.0756$ cfs ◄

Eq. 6.33: $T = -1.940(0.0756)(16/12)(0 + 87.9\cos 156°) = 15.70$ ft·lb ◄

Since no motion occurs, hp = zero. ◄

6.14.3 *Repeat Exer. 6.14.2 with the following changes: total jet area = 0.75 cm², radius = 36 cm, h = 45 m.*
Refer to Fig. 6.13 with Solution 6.14.1. Exer. 6.14.2: The sprinkler has β_2 = 156° *and is prevented from rotating. Compute Q, T, and the power developed.*

SI

Eq. 6.32 with $u_2 = 0$: $v_2 = \sqrt{2(9.81)45} = 29.71$ m/s

$Q = a_2 v_2 = 0.000\,075(29.71) = 0.002\,23$ m³/s ◀

Eq. 6.33: $T = -1000(0.00223)0.36(0 + 29.71\cos 156°) = 21.8$ N·m ◀

As $\omega = 0$, Power = 0 ◀

6.14.4 *How fast would the sprinkler of Exer. 6.14.2 rotate if there were no mechanical friction or air resistance (i.e., consider the case where T = 0)? This is known as runaway speed.*
Refer to Fig. 6.13 with Solution 6.14.1. Exer. 6.14.2: The sprinkler has β_2 = 156° , *a = 0.000 86 ft², r = 16 in, h = 120 ft.*

BG

$u_2 = \omega r_2 = (16/12)\omega = 1.333\omega$; Eq. 6.32: $v_2 = \sqrt{2(32.2)120 + (16\omega/12)^2}$; $Q = a_2 v_2$

Eq. 6.33: $T = -\rho Q r_2(u_2 + v_2\cos\beta_2) = 0$ for runaway speed

$\therefore \quad u_2 = -v_2\cos\beta_2 = -v_2\cos 156° = 0.914\, v_2$

$(16/12)\omega = 0.914\sqrt{2(32.2)120 + (16\omega/12)^2}$; $\omega = 148$ rad/s = 1414 rpm ◀

6.14.5 *How fast would the sprinkler of Ex. 6.14.3 rotate if there were no mechanical friction or air resistance (i.e. consider the case where T = 0)? This is known as runaway speed.*
Refer to Fig. 6.13 with Solution 6.14.1. Exer. 6.14.3: The sprinkler has β_2 = 156°, *a = 0.75 cm², r = 36 cm, h = 45 m.*

SI

$u_2 = \omega r_2 = 0.36\omega$; Eq. 6.32: $v_2 = \sqrt{2g(45) + (0.36\omega)^2}$; $Q = a_2 v_2$

Eq. 6.33: $T = -(\gamma/g)Q r_2(u_2 + v_2\cos\beta_2) = 0$ for runaway speed

$\therefore\ u_2 = -v_2\cos\beta_2 = -v_2\cos 156° = 0.914 v_2$

$\therefore\ 0.36\omega = 0.914\sqrt{2g(45) + (0.36\omega)^2}$; $\omega = (185.4)60/(2\pi) = 1770$ rpm ◀

Sec. 6.14: Reaction with Rotation -- *Problems*

6.42 *Given a lawn sprinkler such as that in Fig. 6.13 with* β_2 = 160°, *and the total area of the jets at a radius of 15 in being 0.0008 ft². When h = 144 ft, compute the rate of discharge, the torque exerted by the water, and the power developed if the rotative speed of the sprinkler is 400 rpm. Neglect fluid friction, but note that the calculated torque is that required to overcome mechanical friction and air resistance.*
Refer to Fig. 6.13 with Solution 6.14.1.

BG

$\omega = 400(2\pi/60) = 41.9$ rad/sec ; $u_2 = \omega r_2 = 41.9(15/12) = 52.4$ fps

Eq. 6.32: $v_2 = \sqrt{2(32.2)144 + (52.4)^2} = 109.6$ fps

$Q = a_2 v_2 = (0.0008)109.6 = 0.0877$ cfs ◀

Eq. 6.33: $T = -1.940(0.0877)(15/12)(52.4 + 109.6\cos 160°) = 10.77$ ft·lb ◀

Eq. 5.35: Power = $T\omega/550 = 10.76(41.9)/550 = 0.820$ hp ◀ /cont...

Check on hp: Eq. 5.32: hp $= \gamma QH/550$ where $H = 144 - V_2^2/2g$

From Fig. 6.13: $V^2 = (v_2\cos 20° - u_2)^2 + (v_2\sin 20°)^2 = (109.6\cos 20° - 52.4)^2 + (109.6\sin 20°)^2$

$= 50.6^2 + 37.5^2 = 3970$; $V = 63.0$ fps

$H = 144 - 63.0^2/(2\times 32.2) = 82.4$ ft

hp $= 62.4(0.088)82.4/550 = 0.819$, checks closely

6.43 *Repeat Prob 6.42 with the following changes: radius = 40 cm, total jet area = 0.7 cm², h = 42 m.*

Refer to Fig. 6.13 with Solution 6.14.1. Prob. 6.42: The sprinkler has $\beta_2 = 160°$ and a rotative speed of 400 rpm. Compute Q, T, and the paver developed. Neglect fluid friction, but note that the calculated torque is that required to overcome mechanical friction and air resistance.

SI

$\omega = 400(2\pi/60) = 41.9$ rad/s ; $u_2 = \omega r_2 = 41.9(0.4) = 16.76$ m/s

Eq. 6.32: $v_2 = \sqrt{2(9.81)42 + 16.76^2} = 33.2$ m/s ; $Q = a_2 v_2 = 0.000\,07(33.2) = 0.002\,33$ m³/s ◄

Eq. 6.33: $T = -(1000)0.00233(0.4)(16.76 + 33.2\cos 160°) = 13.47$ N·m ◄

Eq. 5.35: Power $= T\omega = 13.47(41.9) = 564$ N·m/s $= 564$ W ◄

6.44 *For a lawn sprinkler like that of Fig. 6.13, develop an expression for the runaway speed ω in terms of h, r, and β_2. This would occur if there were no mechanical friction or air resistance, i.e., zero torque.*

Refer to Fig. 6.13 with Solution 6.14.1.

N

From Eq. 6.33 with $T = 0$: $u_2 = -v_2\cos\beta_2$. ∴ using Eq. 6.32: $u_2 = -\cos\beta_2\sqrt{2gh + u_2^2}$

i.e. $\dfrac{u_2^2}{\cos^2\beta_2} = 2gh + u_2^2$ or $u_2^2\left(\dfrac{1}{\cos^2\beta_2} - 1\right) = 2gh$ or $u_2^2\left(\dfrac{1 - \cos^2\beta_2}{\cos^2\beta_2}\right) = u_2^2\tan^2\beta_2 = 2gh$

∴ $\omega r_2|\tan\beta| = \sqrt{2gh}$; $\omega = \dfrac{\sqrt{2gh}}{r_2|\tan\beta|}$ ◄

6.45 *At what approximate speed will the sprinkler of Prob. 6.42 develop maximum horsepower?*

Refer to Fig. 6.13 with Solution 6.14.1. Prob. 6.42: The sprinkler has $\beta_2 = 160°$, $a = 0.0008$ ft², $r = 15$ in, $h = 144$ ft. Neglect fluid friction, but note that the calculated torque is that required to overcome mechanical friction and air resistance.

BG

Two methods of solution may be used:

(*a*) Express power as a function of ω. Take ∂ Power$/\partial\omega = 0$ and solve for ω.

(*b*) Assume various ω's and compute the corresponding Powers.

Method (*b*) will be used here.

$\omega = n(2\pi/60)$ rad/sec; $u_2 = \omega r_2 = \omega(15/12)$ fps

Eq. 6.32: $v_2 = \sqrt{2gh + u_2^2} = \sqrt{2(32.2)144 + u_2^2}$ fps ; $Q = a_2 v_2 = 0.0008\,v_2$ cfs

$\rho Q r_2 = 1.940Q(15/2) = 2.425Q$ lb·sec

Sec. 6.14: $\Delta V_u = u_2 + v_2\cos\beta_2 = u_2 + v_2\cos 160° = u_2 - 0.940v_2$ fps

From Eq. 6.33: $T = -\rho Q r_2(\Delta V_u) = -2.425Q(u_2 - 0.940v_2)$ ft·lb

Eqs. 5.32 and 5.35: Power $= T\omega/550$ hp /cont...

n rpm	ω rad/sec	u_2 fps	v_2 fps	Q cfs	$\rho Q r_2$ lb·sec	ΔV_u fps	T ft·lb	Power hp	
200	20.9	26.2	99.8	0.0798	0.1936	-67.6	13.09	0.498	
600	62.8	78.5	124.3	0.0994	0.241	-38.2	9.22	1.053	
1000	104.7	130.9	162.5	0.1300	0.315	-21.8	6.87	1.309	
1200	125.7	157.1	184.2	0.1474	0.357	-16.06	5.74	1.311	(max)
1400	146.6	183.3	207	0.1656	0.402	-11.28	4.53	1.207	
1500	157.1	196.3	219	0.1750	0.424	-9.15	3.88	1.109	
1980	207	259	276	0.221	0.536	-0.637	0.342	0.1289	

Maximum power occurs at a rotative speed of approximately 1200 rpm ◀

Sec. 6.15: Momentum Principle Applied to Propellers and Windmills -- ***Exercises***

6.15.1 *A 18-in-diameter household fan drives air (γ = 0.076 lb/ft^3) at a rate of 1.80 lb/sec. (a) Find the thrust exerted by the fan. (b) What is the pressure difference on the two sides of the fan? (c) Find the required horsepower to drive the fan. Neglect losses.*

BG

Sec. 6.15 for fan: $V_1 = 0$

Eq. 6.37: $\gamma Q = \gamma A(0 + \Delta V/2) = 1.8$ lb/sec ; $\quad \Delta V = \dfrac{1.8(2)}{(0.076)\dfrac{\pi \times 18^2}{4 \times 144}} = 26.8$ fps

(*a*) Eq. 6.35: $F_T = \rho Q(\Delta V) = (1.8/32.2)26.8 = 1.498$ lb ◀

(*b*) Eq. 6.36 with $V_1 = 0$: $\dfrac{\Delta p}{\gamma} = \dfrac{(\Delta V)^2}{2g} = \dfrac{26.8^2}{2(32.2)} = 11.16$ ft ; $\quad \Delta p = 0.076(11.16) = 0.848$ psf ◀

(*c*) Eqs. 5.35 and 5.32: hp required $= \dfrac{F_T V}{550} = \dfrac{F_T(\Delta V/2)}{550} = 0.0365$ hp ◀

or (Eq. 5.32) hp required $= \dfrac{\gamma QH}{550} = \dfrac{1.8(11.16)}{550} = 0.0365$ hp ◀

6.15.2 *A 1.8-m-diameter household fan drives air (γ = 12 N/m^3) at a rate of 50 N/s. (a) Find the thrust exerted by the fan. (b) What is the pressure difference on the two sides of the fan? (c) Find the required power to drive the fan. Neglect losses.*

SI

Sec. 6.15 for fan: $V_1 = 0$.

Eq. 6.37: $\gamma Q = \gamma A(0 + \Delta V/2) = 50$ N/s $= 12(\pi/4)1.8^2(\Delta V/2)$; $\quad \Delta V = 3.27$ m/s

(*a*) Eq. 6.35: $F_T = \rho Q(\Delta V) = (50/9.81)3.27 = 16.69$ N ◀

(*b*) Eq. 6.36 with $V_1 = 0$: $\dfrac{\Delta p}{\gamma} = \dfrac{(\Delta V)^2}{2g} = \dfrac{3.27^2}{2(9.81)} = 0.547$m ; ◀

$\Delta p = 12(0.547) = 6.56$ N/m^2

(*c*) Eq. 5.35: Power $= F_T V = F_T(\Delta V/2) = 16.69(3.27/2) = 27.3$ watts ◀

or (Eq. 5.31) Power $= \gamma QH = 50(0.547) = 27.3$ watts ◀

6.15.3 *A 12-in electric fan is placed on a frictionless mount and is observed to exert a thrust of 0.8 lb. (a) Find the approximate velocity of the slipstream of standard air (sea level) which it produces. (b) If 45 percent of the power supplied to the blades is lost in eddies and friction and if the driving motor has an efficiency of 60 percent, find the required electrial input in watts.*

B

(*a*) Eq. 6.34: $F_T = 0.8 \text{ lb} = (\Delta p)\pi 1^2/4 = 0.785\Delta p$; $\Delta p = 1.019$ psf

Table A.3 for standard air at sea level: $\gamma = 0.0765 \text{ lb/ft}^3$

Sec 6.15 for fan: $V_1 = 0$. ∴ Eq. 6.36 becomes: $\Delta p/\gamma = (\Delta V)^2/2g$

from which $\Delta V = \sqrt{2g(\Delta p/\gamma)} = \sqrt{2(32.2)1.019/0.0765} = 29.3$ fps

Slipstream velocity $= V_1 + \Delta V = 0 + \Delta V = 29.3$ fps ◄

(*b*) Eq. 6.37 with $V_1 = 0$: $Q = A(\Delta V/2) = 0.785(29.3/2) = 11.50$ cfs

Eq. 5.31: $\text{Power}_{\text{blades}} = \gamma Q h = \gamma Q(\Delta p/\gamma) = Q\Delta p = 11.50(1.019) = 11.71$ ft·lb/sec

Eqs. 5.31 and 5.34: Required power

$$= \frac{11.71}{(1-0.45)0.60} \text{ ft·lb/sec} = \frac{11.71}{(1-0.45)0.60} \times \frac{745.7}{550} = 48.1 \text{ watts}$$ ◄

6.15.4 *A 30-cm electric fan is placed on a frictionless mount and is observed to exert a thrust of 2.5 N. (a) Find the approximate velocity of the slipstream of standard air (sea level) which it produces. (b) If 45 percent of the power supplied to the blades is lost in eddies and friction and if the driving motor has an efficiency of 60 percent, find the required electrial input in watts.*

SI

(*a*) Eq. 6.34: $F_T = 2.5 \text{ N} = \Delta p(\pi 0.15^2) = 0.0707\Delta p$; $\Delta p = 35.4 \text{ N/m}^2$

Table A.3 for standard air at sea level: $\gamma = 12.01 \text{ N/m}^3$

Sec 6.15 for fan: $V_1 = 0$. ∴ Eq. 6.36 becomes: $\Delta p/\gamma = (\Delta V)^2/2g$

from which $\Delta V = \sqrt{2g(\Delta p/\gamma)} = \sqrt{2(9.81)35.4/12.01} = 7.60$ m/s

Slipstream velocity $= V_1 + \Delta V = 0 + \Delta V = 7.60$ m/s ◄

(*b*) Eq. 6.37 with $V_1 = 0$: $Q = A(\Delta V/2) = 0.707(7.60/2) = 0.269$ m/s

Eq. 5.31: $\text{Power}_{\text{blades}} = \gamma Q h = \gamma Q(\Delta p/\gamma) = Q\Delta p = 0.269(35.4) = 9.50$ watts

From Eq. 5.34: Required power $= \dfrac{9.50}{(1-0.45)0.60} = 28.8$ watts ◄

6.15.5 *A fan sucks air from outside to inside a building through a 20-in-diameter duct. The density of air is 0.0022 slug/ft³. (a) If the pressure difference across the two sides of the fan is 4.0 in of water, determine the flow rate of the air in cubic feet per second. (b) What thrust must the fan support be designed to withstand?*

BG

(*b*) $\Delta p = 4.0$ inches of water $= 0.333$ ft of water ; $\Delta p = 0.333(62.4/144) = 0.1444$ psi

Eq. 6.34: $F_T = A\Delta p = (\pi 20^2/4)0.1444 = 45.4$ lb ◄

(*a*) Sec. 6.15 for fan: $V_1 = 0$. Eq. 6.37 with $V_1 = 0$: $Q = A(\Delta V/2)$

Eq. 6.35: $F_T = \rho Q(\Delta V) = \rho[A(\Delta V/2)](\Delta V)$; $45.4 = 0.0022(\pi/4)(20/12)^2(\Delta V^2/2)$;

$\Delta V = 137.5$ fps ; $Q = A(\Delta V/2) = 2.18(137.5/2) = 150.0$ cfs ◄

6.46 *Apply the momentum and energy principles to the case of a windmill (essentially the opposite of a propeller), to determine the maximum theoretical effciency based on an input energy available from the wind velocity in a stream tube having a cross section equivalent to that of the windmill blade circle.*

N

Wind velocity V_1 ; Velocity through windmill = $V_1 - \Delta V/2$ where $\Delta V = V_1 - V_4$

Eq. 5.31: $P_{out} = \gamma Q\left[\frac{V_1^2}{2g} - \frac{(V_1 - \Delta V)^2}{2g}\right]$ where $Q = A\left(V_1 - \frac{\Delta V_1}{2}\right)$

Eq. 5.31: $P_{in} = \gamma Q V_1^2/2g$ where $Q = AV_1$

$$\frac{P_{out}}{P_{in}} = \text{efficiency} = \frac{[V_1 - (\Delta V/2)][V_1^2 - (V_1 - \Delta V)^2]}{V_1(V_1^2)} = \frac{[V_1 + (V_1 - \Delta V)][V_1^2 - (V_1 - \Delta V)^2]}{2V_1^3}$$

$$= \text{efficiency} = \frac{[1 + (V_1 - \Delta V)/V_1]^2[1 - (V_1 - \Delta V)/V_1]}{2}$$

$$\frac{\partial(\text{eff})}{\partial\left(\frac{V_1 - \Delta V}{V_1}\right)} = \frac{\left[1 + \left(\frac{V_1 - \Delta V}{V_1}\right)\right]^2(-1) + \left[1 - \left(\frac{V_1 - \Delta V}{V_1}\right)\right]2\left[1 + \left(\frac{V_1 - \Delta V}{V_1}\right)\right]}{2} = 0$$

This simplifies to: $3\left(\frac{1 - \Delta V}{V_1}\right)^2 + 2\left(\frac{1 - \Delta V}{V_1}\right) - 1 = 0$; $\frac{V_1 - \Delta V}{V_1} = \frac{-2 \pm \sqrt{4 + 12}}{6} = 1/3$ or -1

Hence efficiency is maximum when $\frac{V_1 - \Delta V}{V_1} = 1/3$ or when $\Delta V = (2/3)V_1$

$$(\text{eff})_{max} = \frac{(1 + 1/3)^2(1 - 1/3)}{2} = \frac{16}{27} = 0.593$$ ◄

Chapter 7

Similitude and Dimensional Analysis

PROBLEM SELECTION GUIDE

Sec	Exer/Prob	Units	Difficulty	Length	Parts	Similar	Special features
7.4	*Dynamic Similarity*						
	X[1]7.4.1	BG	Easy	V Short	1		Interpolation
	7.4.2	SI	V Easy	V Short	1		
	7.4.3	N	Easy	Medium	1		Uses $\nu = \mu/\rho$ (Sec. 2.11)
	7.4.4	BG	V Easy	Medium	1		Uses $\nu = \mu/\rho$ (Sec. 2.11)
	7.4.5	BG	Easy	Short	1		
	7.4.6	SI	Medium	Medium	1		
	7.4.7	BG	Easy	Short	1	P7.12	
	7.4.8	BG	Medium	Medium	1		
	P 7.1	BG	Easy	Short	1	7.2	Interpolation; uses $\nu = \mu/\rho$ (Sec. 2.11)
	7.2	SI	Easy	Short	1	7.1	Interpolation; uses $\nu = \mu/\rho$ (Sec. 2.11)
	7.3	SI	Easy	Short	1		Uses $pv = RT$ (Sec. 2.7)
	7.4	N	Medium	Short	8		
	7.5	BG	Easy	Short	1		
	7.6	BG	Medium	Medium	1	7.8	Uses Secs. 2.7 and 4.5
	7.7	SI	Medium	Medium	1		Uses $pv = RT$ (Sec. 2.7)
	7.8	SI	Medium	Medium	1	7.6	Uses Secs. 2.7 and 4.5
	7.9	BG	Medium	Medium	1		
	7.10	B	Hard	Medium	1		Unit conversions; uses $\nu = \mu/\rho$ (§ 2.11)
	7.11	N	Medium	Medium	1		
	7.12	BG	Medium	Short	1	X7.4.7	
	7.13	BG	Medium	Short	1		
	7.14	SI	Medium	Short	1		
	7.15	BG	Medium	Medium	2	7.16	Uses Secs. 2.8 and 13.3
	7.16	SI	Medium	Medium	1	7.15	
	7.17	BG	Medium	Medium	1	7.18	Interpolation: uses Secs. 2.7-8
	7.18	SI	Medium	Medium	1	7.17	Interpolation: uses Secs. 2.7-8
	7.19	BG	Medium	Short	1		
7.5	*Scale Ratios*						
	P 7.20	N	Medium	Medium	1		
	7.21	N	Medium	Medium	1		
	7.22	N	Medium	Medium	1		
	7.23	BG	Hard	Medium	1	7.24	Uses Sec. 4.5
	7.24	SI	Hard	Medium	1	7.23	Uses Sec. 4.5
	7.25	N	Medium	Short	1		
	7.26	N	Medium	Short	1		

/cont...

[1] For all Exercises (marked "X"), answers are given in Appendix D of the textbook.
X = Exercise, P = (end-of-chapter) Problem, S = Sample Problem.

Sec	Exer/Prob	Units	Difficulty	Length	Parts	Similar	Special features
7.7	***Dimensional Analysis***						
	X 7.7.1	N	Medium	Medium	2		
	7.7.2	N	Medium	Medium	2		
	7.7.3	N	Medium	Long	1		
	7.7.4	N	Medium	Long	1		
	7.7.5	N	Medium	Long			
	P 7.27	N	Hard	Long	1		
	7.28	N	Medium	Long	1		
	7.29	N	Medium	Long	1		
	7.30	N	Medium	Long	1		Uses Sec. 2.12

Chapter 7
SIMILTUDE AND DIMENSIONAL ANALYSIS

Sec. 7.4: Dynamic Similarity -- *Exercises*

7.4.1 *Water at 65°F in a 6-in-diameter pipe flows with a velocity of 6 fps. What is the Reynolds number? Note that L = D and that the physical properties of water are given in Appendix A.*

BG

Table A.1 for water at 65°F: $\nu = 11.38 \times 10^{-6}$ ft²/sec (by interpolation)

Eq 7.6: $\mathbf{R} = LV/\nu = DV/\nu = (6/12)6/(11.38 \times 10^{-6}) = 264{,}000$ ◀

7.4.2 *Oil (s = 0.85 and μ = 0.24 N·s/m²) in a 12-cm-diameter pipe flows with a velocity of 4.0 m/s. What is the Reynolds number?*

SI

Eq 7.6: $\mathbf{R} = LV/\nu = DV\rho/\mu = \dfrac{(0.12\text{ m})(4.0\text{ m/s})(0.85\times10^3\text{ kg/m}^3)}{0.24\text{ N}\cdot\text{s/m}^2}\left(\dfrac{\text{N}}{\text{kg}\cdot\text{m/s}^2}\right) = 1700$ ◀

7.4.3 *The dimensions of a model airplane are 1/25 those of its prototype. The model is to be tested in a pressure wind tunnel at the same speed and air temperature as the prototype. What must the pressure in the wind tunnel be relative to the atmospheric pressure if the Reynolds number remains the same?*

N

Given $L_r = L_m/L_p = 1/25$ and $V_m = V_p$. Substituting these into

$L_mV_m/\nu_m = L_pV_p/\nu_p$ (since $\mathbf{R}_m = \mathbf{R}_p$, given) we obtain $\nu_m/\nu_p = (L_m/L_p)(V_m/V_p) = (1/25)1 = 1/25$

Using Eq. 2.11: $\dfrac{\nu_m}{\nu_p} = \dfrac{\mu_m/\rho_m}{\mu_p/\rho_p} = \dfrac{\mu_m\,\rho_p}{\mu_p\,\rho_m} = \dfrac{1}{25}$; for air temp. $T_m = T_p$, assume $\mu_m = \mu_p$, $\therefore \dfrac{\rho_p}{\rho_m} = \dfrac{1}{25}$

Using Eqs. 2.1 and 2.5: $\rho_p/\rho_m = \gamma_p/\gamma_m = p_p/p_m = 1/25$ or $p_m = 25p_p = 25$ atmospheres ◀

7.4.4 *A wind tunnel test on a 1:40 scale model of a submarine is to be performed to find the drag on the submarine when it is moving at 10 knots through 50°F ocean water. At what velocity should the test be conducted in the wind tunnel if it contains 60°F air at atmospheric pressure? If the drag on the model is 75 lb, what will be the drag on the prototype?*

BG

R governs because the significant forces are inertia and fluid friction due to viscosity.

∴ (Eq. 7.6) $\mathbf{R} = DV/\nu =$ constant. Table A.1 for fresh water at 50°F: $\nu = 14.10 \times 10^{-6}$ ft²/sec.

Using Table A.4 and Eq. 2.11: $\dfrac{\nu_{\text{ocean}}}{\nu_{\text{fresh}}} = \dfrac{\mu_o\,\rho_f}{\rho_o\,\mu_f} = \dfrac{(2.25\times10^{-6})1.936}{1.985(2.10\times10^{-6})} = 1.045$

∴ at 50°F, $\nu_{\text{ocean}} = \nu_p = 1.045(14.10\times10^{-6}) = 14.73\times10^{-6}$ ft²/sec

Table A.2 for air at 60°F: $\nu_m = 0.158\times10^{-3}$ ft²/sec ; From inside cover: 10 knots = 16.88 fps = V_p

$\mathbf{R}_m = \mathbf{R}_p$: $\dfrac{(1/40)V_m}{0.158\times10^{-3}} = \dfrac{(1)16.88}{14.73\times10^{-6}}$; $V_m = 7240$ fps $= 4940$ mph ◀

Since V_m greatly exceeds sonic velocity (Table A.3), the model will not properly indicate prototype behavior, including drag, due to compressibility effects (Sec. 2.4). ◀

Subscript *m* denotes model, subscript *p* denotes prototype, and subscript *r* denotes relative or ratio in this chapter.

7.4.5 *A 500-ft-long ship is to operate at a speed of 20 mph. If a model of this ship is 10 ft long, what should be its speed in fps to give the same Froude number? What is the value of this Froude number?*

BG

Given $\mathbf{F}_p = \mathbf{F}_m$; Eq 7.8: $\mathbf{F} = V/(gL)^{1/2}$

$[V/(gL)^{1/2}]_p = [V/(gL)^{1/2}]_m$; $V_m = V_p(g_m/g_p)^{1/2}(L_m/L_p)^{1/2} = V_p(10/500)^{1/2}$

$V_m = 20\text{ mph}/7.07 = 2.83\text{ mph} = 4.15\text{ fps}$ ◄ $\mathbf{F} = 4.15/(32.2 \times 10)^{1/2} = 0.231$ ◄

7.4.6 *Water flows over the crest of a 1:30 model spillway, and the velocity measured at a particular point is 0.4 m/s. What velocity does this represent in the prototype? The force exerted on a certain area of the model is measured to be 0.15 N. What would be the force on the corresponding area in the prototype? Develop your own dimensionless ratios.*

SI

Gravity and inertia govern this flow, so the Froude criterion must be satisfied.

Equating Froude numbers, Eq. 7.9: $[V/(gL)^{1/2}]_m = [V/(g/L)^{1/2}]_p$

$\therefore\ V_r = V_p/V_m = (L_p/L_m)^{1/2} = 30^{1/2} = 5.48;\ V_p = V_rV_m = 5.48(0.4) = 2.19\text{ m/s}$ ◄

$$F_r = \frac{(F_I)_p}{(F_I)_m} = \frac{(\rho V^2L^2)_p}{(\rho V^2L^2)_m} = \rho_r V_r^2 L_r^2 = (1)L_r(L_r^2) = L_r^3$$

$$\text{or: } F_r = \frac{(F_G)_p}{(F_G)_m} = \frac{(\rho L^3 g)_p}{(\rho L^3 g)_m} = \rho_r L_r^3 g_r = (1)L_r^3(1) = L_r^3.\quad \therefore\ F_p = L_r^3F_m = 30^3(0.15) = 4050\text{ N}$$ ◄

7.4.7 *A vertical jet of water issuing upward from a nozzle at a velocity of 76 fps will rise to a height of approximately 90 ft on the earth. To get a water jet to rise to a height of 120 ft on the moon, where gravity is one-sixth of that on earth, what must be the jet velocity? Neglect atmospheric resistance.*

BG

Gravity and inertia dominate, so $\mathbf{F}$ governs, and $\mathbf{F}_e = \mathbf{F}_m$.

Eq. 7.1: $L_r = L_e/L_m = 90/120 = 0.75$; $g_r = g_e/g_m = 6$

Eq. 7.9: $[V/(gL)^{1/2}]_e = [V/(gL)^{1/2}]_m$; $\therefore\ V_e/V_m = \sqrt{(g_e/g_m)(L_e/L_m)} = \sqrt{6(0.75)} = 2.12 = V_r$ (Eq. 7.2)

$V_m = V_e/V_r = 76/2.12 = 35.8\text{ fps}$ ◄

7.4.8 *The flow about a missile which travels at 1500 mph through the atmosphere at an elevation of 15,000 ft is to be modeled in a wind tunnel at standard atmospheric conditions with 70°F air. What is the required air speed in the wind tunnel to achieve dynamic similarity?*

BG

Compressibility is important at supersonic speeds, so $\mathbf{M}$ governs, and $\mathbf{M}_p = \mathbf{M}_m$.

Table A.3 at 15,000 ft: $p = 8.30$ psia, $\rho = 0.001\,496\text{ slug/ft}^3$

Table A.2 at 70°F: $\rho = 0.002\,33\text{ slug/ft}^3$. Eq. 7.10: $\mathbf{M} = V/(E_v/\rho)^{1/2}$ where (Eq. 2.8) $E_v = np$

Thus $[V/(p/\rho)^{1/2}]_p = [V/(p/\rho)^{1/2}]_m$ as n is the same for both.

$$\frac{1500\text{ mph}}{(8.30/0.001\,496)^{1/2}} = \frac{V_m}{(14.70/0.002\,33)^{1/2}};\quad V_m = 1600\text{ mph} = 2350\text{ fps}$$ ◄

Sec. 7.4: Dynamic Similarity -- *Problems*

7.1 *What is the Reynolds number for 130°F air at a pressure of 120 psia when it flows at a velocity of 100 fps through a 6-in-diameter pipe? Note that physical properties of air are given in Appendix A.*

BG

Table A.2 for air at 130°F and 14.7 psia: $\nu = 0.195 \times 10^{-3}$ ft²/sec (by interpolation)

From Sec. 2.11 and Eq. 2.11: $\nu \propto 1/\rho$; from Eq. 2.4 at const. $T°$: $\rho \propto p$

$\therefore\ \nu_2 = (p_1/p_2)\nu_1 = (14.7/120)0.195 \times 10^{-3} = 23.9 \times 10^{-6}$ ft²/sec

Eq 7.6: $\mathbf{R} = LV/\nu = DV/\nu = (6/12)100/(23.9 \times 10^{-6}) = 2{,}090{,}000$ ◄

7.2 *What is the Reynolds number for 55°C air at a pressure of 800 kN/m² abs when it flows at a velocity of 30 m/s through a 15-cm-diameter pipe? Note that physical properties of air are given in Appendix A.*

SI

Table A.2 for air at 55°C and 101.3 kN/m²: $\nu = 18.2 \times 10^{-6}$ m²/s (by interpolation)

From Sec. 2.11 and Eq. 2.11: $\nu \propto 1/\rho$; from Eq. 2.4 at const. $T°$: $\rho \propto p$

$\therefore\ \nu_2 = (p_1/p_2)\nu_1 = (101.3/800)18.2 \times 10^{-6} = 2.31 \times 10^{-6}$ m²/s

Eq 7.6: $\mathbf{R} = LV/\nu = DV/\nu = (0.15)30/(2.31 \times 10^{-6}) = 1\ 950\ 000$ ◄

7.3 *Air at 150°C and a pressure of 240 kN/m² abs flows at a velocity of 16 m/s through an 18-cm-diameter pipe. What is the Reynolds number?*

SI

Table A.5 for air: $R = 287$ N·m/(kg·K). From Eq. 2.4: $\rho = \dfrac{p}{RT} = \dfrac{240 \times 10^3}{287(150 + 273)} = 1.977$ kg/m³

From Table A.2 by interpolation (or Fig. 2.3): $\mu = 23.8 \times 10^{-6}$ N·s/m²

Eq 7.6: $\mathbf{R} = LV\rho/\mu = DV\rho/\mu = 0.18(16)1.977/(23.8 \times 10^{-6}) = 239\ 000$ ◄

7.4 *Models are to be built of the following prototypes: (a) tides; (b) oil flowing through a full pipeline; (c) a water jet; (d) flow over the spillway of a dam; (e) a deep submersible vehicle; (f) an airplane flying at low speed; (g) a supersonic aircraft; (h) a supersonic missile. For dynamic similarity, indicate which single dimensionless ratio will govern, and give reasons why.*

N

Governing dimensionless ratios are:

R for parts (*b*), (*e*), and (*f*), because for these the significant forces are inertia and fluid friction due to viscosity (air compressibility is not appreciable at low airplane speeds). ◄

F for parts (*a*), (*c*), and (*d*), because for these the significant forces are inertia and gravity. ◄

M for parts (*g*) and (*h*), because for these compressibility is important. ◄

7.5 *The linear dimensions of a model airplane are 1/18 those of its prototype. If the protoype is to fly at 500 mph, what must be the air velocity in a wind tunnel for the same Reynolds number if the air temperature and pressure are the same?*

BG

$L_r = L_m/L_p = 1/18$; $\nu_m = \nu_p$ due to same air temperature and pressure.

$\mathbf{R}_m = \mathbf{R}_p$, so $L_mV_m/\nu_m = L_pV_p/\nu_p$; $V_m = V_p(L_p/L_m) = 500(18/1) = 9000$ mph ◄

This answer, which indicates a velocity about 12 times the speed of sound, illustrates why it is not generally possible to test model aircraft by the Reynolds model law in atmospheric wind tunnels.

7.6 *Air at 68°F and 60 psia flows in a 1.5-in-diameter pipe. What weight flow rate of this air will give dynamic similarity to 60°F water flowing at 200 gpm through a 3-in-diameter pipe?*

BG

Table A.1 for water at 60°F: $\nu = 12.17 \times 10^{-6}$ ft²/sec

Velocity of water, $V = Q/A = (200/449)/[\pi(3/12)^2/4] = 9.08$ fps

Table A.5 for air: $R = 1715$ ft²/(sec²·°R)

Eq. 2.5 for air at 68°F and 60 psia: $\gamma = \dfrac{gp}{RT} = \dfrac{32.2(60)144}{1715(68+460)} = 0.307\ \text{lb/ft}^3$

Table A.2 for air at 68°F and any pressure (Sec 2.11), by interpolation: $\mu = 0.380 \times 10^{-6}$ lb·sec/ft²

$\therefore\ \nu_{air} = \mu/\rho = 0.380 \times 10^{-6}/(0.307/32.2) = 39.9 \times 10^{-6}$ ft²/sec

For dynamic similarity: $\mathbf{R}_{water} = \mathbf{R}_{air}$.

Using Eq. 7.6: $\dfrac{(3/12)9.08}{12.17 \times 10^{-6}} = \dfrac{(1.5/12)Q/[(\pi/4)(1.5/12)^2]}{39.9 \times 10^{-6}}$; $Q = 0.730$ cfs ◄

Eq 4.5: Weight flowrate of air, $G = \gamma Q = 0.307(0.730) = 0.224$ lb/sec ◄

7.7 *When a submerged sphere moves through 20°C water at 2.0 m/s, the drag force exerted on it is 15 N. Another sphere of three times the diameter is tested in a wind tunnel where the air pressure and temperature are 2.0 MN/m² abs and 300 K respecitively. What air velocity is required for dynamic similarity, and what then will be the drag force on the larger sphere?*

SI

Table A.2 for air at 27°C (=300K) and any pressure (Sec 2.11), by interpol'n: $\mu = 18.45 \times 10^{-6}$ N·s/m²

Table A.5 for air: $R = 287$ N·m/(kg·K)

For air, from Eq. 2.4: $\rho = \dfrac{p}{RT} = \dfrac{2.0 \times 10^6\ \text{N/m}^2}{(287\ \text{N·m/kg·K})300\ \text{K}} = 23.2\ \text{kg/m}^3$

Table A.1 for water at 20°C: $\rho = 998$ kg/m³, $\mu = 0.001\,002$ N·s/m²

For dynamic similarity: $\mathbf{R}_{water} = \mathbf{R}_{air}$.

Using Eq. 7.6: $\therefore\ \dfrac{D(2.0)998}{0.001\,002} = \dfrac{3D(V_a)23.2}{18.45 \times 10^{-6}}$; $V_a = 0.528$ m/s ◄

$$F_r = \frac{(F_l)_a}{(F_l)_w} = \frac{(\rho V^2 L^2)_a}{(\rho V^2 L^2)_w} = \left(\frac{23.2}{998}\right)\left(\frac{0.528}{2.0}\right)^2\left(\frac{3D}{D}\right)^2 = 0.01457$$

$\therefore\ F_a = F_r F_w = 0.01457(15) = 0.219$ N ◄

7.8 *Air at 80°C and 500 kN/m² abs pressure flows in a 6-cm-diameter pipe. What air flow rate (kg/s) will give dynamic similarity to 60 L/s of 60°C water flowing in a 50-cm-diameter pipe?*

SI

Table A.5 for air: $\mathbf{R} = 287$ N·m/(kg·K)

Air, Eq. 2.4: $\rho_a = \left(\dfrac{p}{RT}\right)_a = \dfrac{500 \times 10^3\ \text{N/m}^2}{(287\ \text{N·m/kg·K})(80+273)\ \text{K}} = 4.94\ \text{kg/m}^3$

Table A.2 for air at 80°C and any pressure (Sec. 2.11): $\mu_a = 20.9 \times 10^{-6}$ N·s/m²

Water, Eq. 4.7: $V_w = Q_w/A_w = 0.060/(\pi 0.25^2) = 0.306$ m/s

Table A.1 for water at 60°C: $\nu_w = 0.474 \times 10^{-6}$ m²/s /cont...

For dynamic similarity: $\mathbf{R}_{air} = \mathbf{R}_{water}$.

Using Eq. 7.6: $\therefore \quad \dfrac{0.06(V_a)4.94}{20.9 \times 10^{-6}} = \dfrac{0.50(0.306)}{0.474 \times 10^{-6}}$; $V_a = 22.8$ m/s

Eq. 4.4: $\dot{m}_a = (\rho A V)_a = 4.94(\pi 0.03^2)22.8 = 0.317$ kg/s ◀

7.9 *Water flows over a spillway at 5000 cfs. For dynamic similarity, what should be the model scale if the flow rate over the model is to be 45 cfs? The force exerted on a certain area of the model is measured to be 1.0 lb. What would be the force on the corresponding area of the prototype?*

BG

Gravity and inertia govern, so satisfies Froude's number. Equating these, we get

Eq 7.2: $V_r = V_p/V_m = (L_p/L_m)^{1/2} = L_r^{1/2}$ assuming $g_m = g_p$

$Q_r = (A \times V)_r = (L^2 \times L^{1/2})_r = L_r^{5/2}$; $(5000/45) = L_r^{5/2}$;

$L_r = L_p/L_m = 6.58$; $\lambda = L_m/L_p = 1:6.58$ ◀

$F_r = (\rho V^2 L^2)_r = \rho_r V_r^2 L_r^2 = 1(L_r)L_r^2 = L_r^3$ assuming $\rho_p = \rho_m$. [Alternatively, $F_r = (\rho Q V)_r$.]

$(F_p/1.0) = (6.58)^3 = 285$; $F_p = 1.0(285) = 285$ lb ◀

7.10 *A 500 ft long ship is to operate at a speed of 20 mph in ocean water whose viscosity is 1.2 cP and specific weight is 64 lb/ft³. What should be the kinematic viscosity of the liquid used with a 10-ft-long model of the ship so that both the Reynolds number and the Froude number would be the same? Does such a liquid exist?*

B

$\mathbf{R}_p = \mathbf{R}_m$; $\mathbf{F}_p = \mathbf{F}_m$; $L_r V_r / \nu_r = 1$; $V_r/(g_r L_r)^{1/2} = 1$

To satisfy both **R** and **F**, $[LV/\nu]_r = [V/(gL)^{1/2}]_r$, i.e. $\nu_r = [(gL)^{1/2} L]_r = L_r^{3/2}$ assuming $g_r = 1.0$

$L_r = L_p/L_m = 500/10 = 50$; $\nu_r = \nu_p/\nu_m = (50)^{3/2} = 354$

$\mu_p = 1.2\text{ cP} = 1.2 \times 10^{-3}\text{ N·s/m}^2 = 0.0209(1.2 \times 10^{-3})\text{ lb·sec/ft}^2 = 2.51 \times 10^{-5}\text{ lb·sec/ft}^2$

Eq 2.11: $\nu_p = \mu_p/\rho = 2.51 \times 10^{-5}/(64/32.2) = 1.261 \times 10^{-5}\text{ ft}^2\text{/sec}$

$\nu_m = \nu_p/\nu_r = 1.261 \times 10^{-5}/354 = 3.57 \times 10^{-8}\text{ ft}^2\text{/sec}$ ◀

Fig 2.4: There is no such liquid available. ◀

7.11 *A 1:600 scale model is built to study tides. What length of time in the model corresponds to one day in the prototype? Suppose this model could be tested on the moon where g is one-sixth of that on earth. What then would be the time relationship between the model and prototype?*

N

Gravity and inertia dominate, so **F** governs, and $\mathbf{F}_p = \mathbf{F}_m$ (Eq. 7.9).

$[V/(gL)^{1/2}]_p = [V/(gL)^{1/2}]_m$; $\therefore V_r = (gL)_r^{1/2}$. Eq. 7.3: $T_r = L_r/V_r = L_r/(gL)_r^{1/2} = \sqrt{L_r/g_r}$

$T_m = T_p/T_r = T_p/\sqrt{L_r/g_r} = T_p\sqrt{g_r/L_r} = T_p\sqrt{(g_p/g_m)/(L_p/L_m)}$

On earth: $T_m = (24\text{ hr})\sqrt{(1/1)/(600/1)} = 24\text{ hr}/24.5 = 0.980$ hr ◀

On the moon: $T_m = (24\text{ hr})\sqrt{(6/1)/(600/1)} = 24\text{ hr}/10 = 2.40$ hr ◀

7.12 *A vertical jet of water issuing upward from a nozzle at a velocity of 44 fps will rise to a height of approximately 30 ft on the earth. To get a water jet to rise to a height of 120 ft on the moon, where the gravity in one-sixth of that on earth, what must be the jet velocity? Neglect atmospheric resistance.*

BG

Gravity and inertia dominate, so **F** governs, and $\mathbf{F}_e = \mathbf{F}_m$.

Eq. 7.1: $L_r = L_e/L_m = 30/120 = 0.25; \; g_r = g_e/g_m = 6$

Eq. 7.9: $[V/(gL)^{1/2}]_e = [V/(gL)^{1/2}]_m \; ; \; \therefore \; V_e/V_m = \sqrt{(g_e/g_m)(L_e/L_m)} = \sqrt{6(0.25)} = 1.225 = V_r$ (Eq. 7.2)

$V_m = V_e/V_r = 44/1.225 = 35.9$ fps ◄

7.13 *A 3-ft-high sectional model of a spillway is built in a 1-ft-wide laboratory flume. The flow is 0.86 cfs under a head of 0.380 ft. If the model scale is 1:20 and the prototype spillway is 600 ft long, what flow does this represent in the prototype?*

BG

$L_r = L_p/L_m = 20 \; ; \; g_r = 1.0.$ Gravity and inertia dominate, so **F** governs, and $\mathbf{F}_m = \mathbf{F}_p$ (Eq. 7.9).

$[V/(gL)^{1/2}]_p = [V/(gL)^{1/2}]_m$ so $V_r = (gL)_r^{1/2} = \sqrt{g_r L_r} = \sqrt{1(20)} = 4.47 \; ; \quad A_r = L_r^2 = 20^2 = 400$

$Q_r = A_r V_r = 400(4.47) = 1789 \; ; \quad Q_m = (0.86 \text{ cfs/ft})(600/20\text{ft}) = 25.8$ cfs

$Q_p = Q_m Q_r = (25.8 \text{ cfs})1789 = 46{,}200$ cfs ◄

7.14 *A model spillway has a flow of 0.10 m³/s per m of width. What is the actual flow for the prototype spillway if the model scale is 1:20?*

SI

$L_r = L_p/L_m = 20 \; ; \; g_r = 1.0.$ Gravity and inertia dominate, so F governs, and $\mathbf{F}_m = \mathbf{F}_p$ (Eq. 7.9).

$[V/(gL)^{1/2}]_p = [V/(gL)^{1/2}]_m$ so $V_r = (gL)_r^{1/2} = \sqrt{g_r L_r} = \sqrt{1 \times 20} = 4.47$

$q = Q/b = AV/b = byV/b = yV \; ; \; q_r = y_r V_r = L_r V_r = 20(4.47) = 89.4$

$q_p = q_r q_m = 89.4(0.1 \text{ m}^3\text{/s per m}) = 8.94 \text{ m}^3\text{/s per m}$ ◄

7.15 *An aircraft company is investigating the flow about a model of a supersonic plane in a variable-density wind tunnel at 1500 fps. The air, at 80°F, has a pressure of 20 psia. At what velocity should the model be tested to maintain dynamic similarity if the air temperature is raised to 90°F and the pressure is increased to 30 psia? Solve this two ways: (a) using and (b) not using specific weights and densities.*

BG

Compressibility is important at supersonic speeds, so **M** governs, and $\mathbf{M}_1 = \mathbf{M}_2$.

(*a*) Table A.5 for air: $R = 1715 \text{ ft}^2/(\text{sec}^2 \cdot {}^\circ\text{R})$. Eq. 2.5: $\gamma = pg/RT$

$\gamma_1 = 20(144)32.2/[1715(460 + 80)] = 0.1001 \text{ lb/ft}^3$

$\gamma_2 = 30(144)32.2/[1715(460 + 90)] = 0.1475 \text{ lb/ft}^3$

Eq. 7.10: $\mathbf{M} = V/(E_\upsilon/\rho)^{1/2} = V\sqrt{\rho/E_\upsilon}$ where $\rho = \gamma/g$ and $E_\upsilon = np$ (Eq. 2.8)

Thus $\left[V\sqrt{\gamma/p}\right]_1 = \left[V\sqrt{\gamma/p}\right]_2$ as g and n are the same for both.

$1500\sqrt{0.1001/20} = V_2\sqrt{0.1475/30} \; ; \; V_2 = 1514$ fps ◄

(*b*) Eq. 7.10: $\mathbf{M} = V/c$ where (Eq. 13.15) $c = \sqrt{kRT}$

Thus $\left[V/\sqrt{T}\right]_1 = \left[V/\sqrt{T}\right]_2$ as k and R are the same for both.

$1500/\sqrt{460 + 80} = V_2/\sqrt{460 + 90} \; ; \; V_2 = 1514$ fps ◄

7.16 *An aircraft company is investigating the flow about a model of a supersonic plane in a variable density wind tunnel at 400 m/s. The air, at 40°C, has a pressure of 150 kN/m² abs. At what velocity should the model be tested to maintain dynamic similarity if the air temperature is raised to 75°C and the pressure is increased to 200 kN/m² abs?*

SI

Compressibility is important at supersonic speeds, so **M** governs, and $\mathbf{M}_1 = \mathbf{M}_2$.

Table A.5 for air: $R = 287\ \text{m}^2/(\text{s}^2\cdot\text{K})$. Eq. 2.5: $\gamma = pg/RT$

$\gamma_1 = 150\,000(9.81)/[287(273+40)] = 16.38\ \text{N/m}^3$; $\gamma_2 = 200\,000(9.81)/[287(273+75)] = 19.64\ \text{N/m}^3$

Eq. 7.10: $\mathbf{M} = V/(E_v/\rho)^{1/2} = V\sqrt{\rho/E_v}$ where $\rho = \gamma/g$ and $E_v = np$ (Eq. 2.8)

Thus $[V\sqrt{\gamma/p}]_1 = [V\sqrt{\gamma/p}]_2$ as g and n are the same for both.

$400\sqrt{16.38/150\,000} = V_2\sqrt{19.64/200\,000}$; $V_2 = 422\ \text{m/s}$ ◄

7.17 *A 1:13 model of a ballistic missile is to be tested in a high-speed wind tunnel. The prototype missile will travel at 1250 fps through air at 68°F and 14.5 psia. If the air in the wind tunnel test section has a temperature of 32°F and a pressure of 12.1 psia, what velocity is needed there? Estimate the drag force on the prototype if the drag force on the model is 95 lb?*

BG

Compressibility is important at supersonic speeds, so **M** governs, and $\mathbf{M}_p = \mathbf{M}_m$.

Table A.5 for air: $R = 1715\ \text{ft}^2/(\text{sec}^2\cdot{}^\circ\text{R})$. Eq. 2.5: $\gamma = pg/RT$

$\gamma_p = 14.5(144)32.2/[1715(460+68)] = 0.0742\ \text{lb/ft}^3$

$\gamma_m = 12.1(144)32.2/[1715(460+32)] = 0.0665\ \text{lb/ft}^3$

Eq. 7.10: $\mathbf{M} = V/(E_v/\rho)^{1/2} = V\sqrt{\rho/E_v}$ where $\rho = \gamma/g$ and $E_v = np$ (Eq. 2.8)

Thus $[V\sqrt{\gamma/p}]_p = [V\sqrt{\gamma/p}]_m$ as g and n are the same for both.

$1250\sqrt{0.0742/14.5} = V_m\sqrt{0.0665/12.1}$; $V_m = 1207\ \text{fps}$ ◄

$F_r = (\rho V^2 L^2)_r = \rho_r V_r^2 L_r^2 = (\gamma/g)_r V_r^2 L_r^2 = (0.0742/0.0665)(1250/1207)^2(13/1)^2 = 203$

$\therefore\ F_p = F_r F_m = 203(95\ \text{lb}) = 19{,}240\ \text{lb}$ ◄

7.18 *A 1:13 model of a ballistic missile is to be tested in a high-speed wind tunnel. The prototype missile will travel at 380 m/s through air at 23°C and 95.0 kN/m² abs. If the air in the wind tunnel has a temperature of -20°C at a pressure of 89 kN/m² abs, what velocity is needed there? Estimate the drag force on the prototype if the drag force on the model is 400 N?*

SI

Compressibility is important at supersonic speeds, so **M** governs, and $\mathbf{M}_p = \mathbf{M}_m$.

Table A.5 for air: $R = 287\ \text{m}^2/(\text{s}^2\cdot\text{K})$. Eq. 2.5: $\gamma = pg/RT$

$\gamma_p = 95\,000(9.81)/[287(273+23)] = 10.97\ \text{N/m}^3$; $\gamma_m = 89\,000(9.81)/[287(273-20)] = 12.02\ \text{N/m}^3$

Eq. 7.10: $\mathbf{M} = V/(E_v/\rho)^{1/2} = V\sqrt{\rho/E_v}$ where $\rho = \gamma/g$ and $E_v = np$ (Eq. 2.8)

Thus $[V\sqrt{\gamma/p}]_p = [V\sqrt{\gamma/p}]_m$ as g and n are the same for both.

$380\sqrt{10.97/95} = V_m\sqrt{12.02/89}$; $V_m = 351\ \text{m/s}$ ◄

$F_r = (\rho V^2 L^2)_r = \rho_r V_r^2 L_r^2 = (\gamma/g)_r V_r^2 L_r^2 = (10.97/12.02)(380/351)^2(13/1)^2 = 180.4$

$\therefore\ F_p = F_r F_m = 180.4(400\ \text{N}) = 72\,200\ \text{N}$ ◄

7.19 *When traveling at a velocity of 2.7 fps a 1:64 scale model of a ship has a wave resistance of 0.08 lb. This is kinematically similar to the design velocity of the prototype ship. What is the design velocity of the prototype and what is its wave resistance at that velocity?*

BG

Gravity and inertia dominate, so **F** governs, and $\mathbf{F}_p = \mathbf{F}_m$. Eq. 7.9: $[V/(gL)^{1/2}]_p = [V/(gL)^{1/2}]_m$

$\therefore\ V_p/V_m = \sqrt{(g_p/g_m)(L_p/L_m)} = \sqrt{(1/1)(64/1)} = 8.00 = V_r$ (Eq. 7.2)

$V_p = V_r V_m = 8.00(2.7\text{ fps}) = 21.6\text{ fps}$ ◄

$F_r = (\rho V^2 L^2)_r = \rho_r V_r^2 L_r^2 = 1(8)^2(64/1)^2 = 262{,}000$ assuming $\rho_p = \rho_m$.

$\therefore\ F_p = F_r F_m = 262{,}000(0.08\text{ lb}) = 21{,}000\text{ lb}$ ◄

Sec. 7.5: Scale Ratios -- ***Problems***

7.20 *Develop the scale ratios given in Table 7.1 for the case where prototype and model Reynolds numbers are the same.*

N

Using Eq 7.6: $V_r = V_p/V_m = [\mathbf{R}\nu/D]_p \times [D/(\mathbf{R}\nu)]_m = (\nu/L)_r = (\mu/L\rho)_r$ ◄

$T_r = (L/V)_r = (L^2/\nu)_r = (L^2\rho/\mu)_r$ ◄ $Q_r = A_r V_r = L_r^2(\nu/L)_r = (L\nu)_r = (L\mu/\rho)_r$ ◄

$F_r = (\rho V^2 L^2)_r = (\rho L^2)_r(\mu/L\rho)_r^2 = (\mu^2/\rho)_r$ ◄ $P_r = F_r V_r = (\mu^2/\rho)_r(\mu/L\rho)_r = (\mu^3/L\rho^2)_r$ ◄

The other relationships (for acceleration, mass, pressure, impulse and momentum, energy and work) can be determined using their dimensions, the above, and algebra.

7.21 *Develop the scale ratios given in Table 7.1 for the case where prototype and model Froude numbers are identical.*

N

Using Eq 7.8: $V_r = V_p/V_m = [\mathbf{F}(gL)^{1/2}]_p/[\mathbf{F}(gL)^{1/2}]_m = (L^{1/2}g^{1/2})_r$ ◄

$T_r = (L/V)_r = [L/(L^{1/2}g^{1/2})]_r = (L^{1/2}g^{-1/2})_r$ ◄ $Q_r = A_r V_r = L_r^2(L^{1/2}g^{1/2})_r = (L^{5/2}g^{1/2})_r$ ◄

$F_r = (\rho V^2 L^2)_r = (\rho L^2)_r(Lg)_r = (L^3\rho g)_r$ ◄ $P_r = F_r V_r = (L^3\rho g)_r(L^{1/2}g^{1/2})_r = (L^{7/2}\rho g^{3/2})_r$ ◄

The other relationships (for acceleration, mass, pressure, impulse and momentum, energy and work) can be determined using their dimensions, the above, and algebra.

7.22 *Develop the scale ratios given in Table 7.1 for the case where prototype and model Mach numbers are the same.*

N

Using Eq. 7.10: $V_r = V_p/V_m = (\mathbf{M}E_\upsilon^{1/2}/\rho^{1/2})_p(\rho^{1/2}/\mathbf{M}E_\upsilon^{1/2})_m = (E_\upsilon/\rho)_r^{1/2}$ ◄

$T_r = (L/V)_r = L_r(E_\upsilon/\rho)_r^{-1/2} = (L\rho^{1/2}/E_\upsilon^{1/2})_r$ ◄ $Q_r = A_r V_r = L_r^2(E_\upsilon/\rho)_r^{1/2} = (L^2E_\upsilon^{1/2}/\rho^{1/2})_r$ ◄

$F_r = (\rho V^2 L^2)_r = (\rho L^2)_r(E_\upsilon/\rho)_r = (L^2E_\upsilon)_r$ ◄ $P_r = F_r V_r = (L^2E_\upsilon)_r(E_\upsilon/\rho)_r^{1/2} = (L^2E_\upsilon^{3/2}/\rho^{1/2})_r$ ◄

The other relationships (for acceleration, mass, pressure, impulse and momentum, energy and work) can be determined using their dimensions, the above, and algebra.

7.23 *Gas (γ = 0.32 lb/ft^3, μ = 2.0 × 10^{-6} lb·sec/ft^2) is flowing in a 0.75-in-diameter pipe. When a gas flowmeter measures the flow as being 0.13 lb/sec, it registers a pressure drop of 1.17 psi. An enlarged model that is geometrically similar is to be tested in a 6-in-diameter pipe. What flow rate of 80°F water will achieve dynamic similarity? What would be the pressure drop across the water meter?*

BG

Inertia and fluid friction due to viscosity dominate for pipe flow, so it must satisfy Reynolds number.

From Table 7.1 for Reynolds number: $Q_r = (L\mu/\rho)_r = L_r v_r$; $L_r = L_p/L_m = 0.75/6 = 0.125$

$\nu_p = \mu_p/\rho_p = \mu_p/(\gamma/g)_p = 2.0 \times 10^{-6}/(0.32/32.2) = 2.01 \times 10^{-4}$ ft^2/sec

Table A.1 for water at 80°F: $\nu_m = 9.30 \times 10^{-6}$ ft^2/sec, $\rho_m = 1.934$ slugs/ft^3

$\nu_r = \nu_p/\nu_m = (2.01 \times 10^{-4})/(9.30 \times 10^{-6}) = 21.6$; $Q_r = Q_p/Q_m = L_r v_r = 0.125(21.6) = 2.70$

From Eq. 4.5: $Q_p = G/\gamma = 0.13/0.32 = 0.406$ cfs ; $Q_m = Q_p/Q_r = 0.406/2.70 = 0.1502$ cfs ◄

From Table 7.1 , $p_r = (\mu^2/L^2\rho)_r = (\rho v^2/L^2)_r$; $p_r = [(0.32/32.2)/1.934](21.6/0.125)^2 = 154.0$

$\Delta p_r = \Delta p_p/\Delta p_m = 154.0$; $\Delta p_m = \Delta p_p/\Delta p_r = 1.17/154.0 = 0.007\,60$ psi ◄

7.24 *Gas (ρ = 5.25 kg/m^3, ν = 2.0 × 10^{-5} m^2/s) is flowing in a 2-cm-diameter pipe. When a gas flowmeter measures the flow as being 0.064 kg/s, it registers a pressure drop of 8.5 kPa. An enlarged model that is geometrically similar is to be tested in a 18-cm-diameter pipe. What flow rate of 25°C water will achieve dynamic similarity? What would be the pressure drop across the water meter?*

SI

Inertia and fluid friction due to viscosity dominate for pipe flow, so it must satisfy Reynolds number.

From Table 7.1 for Reynolds number: $Q_r = (L\mu/\rho)_r = L_r v_r$; $L_r = L_p/L_m = 2/18 = 0.1111$

Table A.1 for water at 25°C: $\nu_m = 0.893 \times 10^{-6}$ m^2/s, $\rho_m = 997.0$ kg/m^3

$\nu_r = \nu_p/\nu_m = (2.0 \times 10^{-5})/(0.893 \times 10^{-6}) = 22.4$; $Q_r = Q_p/Q_m = L_r v_r = 0.1111(22.4) = 2.49$

From Eq. 4.4: $Q_p = \dot{m}/\rho = 0.064/5.25 = 0.012\,19$ m^3/s ;

$Q_m = Q_p/Q_r = 0.001\,219/2.49 = 0.004\,90$ m^3/s ◄

From Table 7.1: $p_r = (\mu^2/L^2\rho)_r = (\rho v^2/L^2)_r$; $p_r = (5.25/997.0)(22.4/0.1111)^2 = 214$

$\Delta p_r = \Delta p_p/\Delta p_m = 214$; $\Delta p_m = \Delta p_p/\Delta p_r = 8.5/214 = 0.0397$ kPa = 39.7 Pa ◄

7.25 *Find the dimensions of force, energy, and momentum in the FLT system. Repeat for the MLT system.*

N

	FLT	MLT	Eq. No.	Dimens. Equiv.
Force	F	MLT^{-2}	3.14	pA
Energy	FL	ML^2T^{-2}	5.7	$\frac{1}{2}mV^2$
Momentum	FT	MLT^{-1}	6.5	ρQV
	▲	▲		

7.26 *Find the dimensions of torque, pressure, and power, in the FLT system. Repeat for the MLT system.*

N

	FLT	MLT	Eq. No.	Dimens. Equiv.
Torque	FL	ML^2T^{-2}	6.28	$\rho Q_r V$
Pressure	FL^{-2}	$ML^{-1}T^{-2}$	3.4	γh
Power	FLT^{-1}	ML^2T^{-3}	5.31	γQh
	▲	▲		

7.7.1 *Use dimensional analysis to arrange the following groups into dimensionless parameters: (a) τ, V, ρ; (b) V, L, ρ, σ. Use the MLT system.*

N

(a) $\Pi = \tau^a V^b \rho^c = (ML^{-1}T^{-2})^a (LT^{-1})^b (ML^{-3})^1$

M: $0 = a + 1$; L: $0 = -a + b - 3$; T: $0 = -2a - b$. Solving: $a = -1,\ b = 2$

$\Pi = \tau^{-1}V^2\rho$; $(1/\Pi) = \tau/\rho V^2$ ◄

(b) $\Pi = V^a L^b \rho^c \sigma = (LT^{-1})^a L^b (ML^{-3})^c (MT^{-2})^1$

M: $0 = c + 1$; L: $0 = a + b - 3c$; T: $0 = -a - 2$. Solving: $a = -½,\ b = ½,\ c = 1$

$\Pi = V^{-2}L^{-1}\rho^{-1}\sigma = \sigma/(\rho L V^2)$ ◄

7.7.2 *Use dimensional analysis to arrange the following groups into dimensionless parameters: (a) Δp, V, γ, g; (b) F, ρ, L, V. Use the MLT system.*

N

(a) $\Pi = \Delta p^a V^b \gamma^c g = (ML^{-1}T^{-2})^a (LT^{-1})^b (ML^{-2}T^{-2})^c (LT^{-2})^1$

M: $0 = a + c$; L: $0 = -a + b - 2c + 1$; T: $0 = -2a - b - 2c - 2$

Solving: $a = 1,\ b = ½,\ c = 1$. $\Pi = \Delta p V^{-2}\gamma^{-1}g = \Delta p g/(\gamma V^2)$ ◄

(b) $\Pi = F^a \rho^b L^c V = (MLT^{-2})^a (ML^{-3})^b L^c (LT^{-1})^1$

M: $0 = a + b$; L: $0 = a - 3b + c + 1$; T: $0 = -2a - 1$

Solving: $a = -½,\ b = ½,\ c = 1$

$\Pi = F^{-1/2}\rho^{1/2}LV$; $(1/\Pi^2) = F/(\rho L^2 V^2)$ ◄

7.7.3 *Use dimensional analysis to derive an expression for the power developed by an engine in terms of the torque and rotative speed.*

N

Step 1: No. of variables (P, T, ω), $n = 3$

Step 2: Their dimensions are: $P = [FL/T]$, $T = [FL]$, $\omega = [1/T]$

So no. of fundamental dimensions (M,L,T) involved, $m = 3$.

Step 3: Our 3 variables can be formed into a dimensionless group,

$\therefore\ k \neq m = 3$. Try $k = 2$. Any 2 variables can not be so formed, so $k = 2$.

Step 4: No. of dimensionless Π groups needed $= n - k = 3 - 2 = 1$; name it Π. $\therefore\ \phi(\Pi) = 0$

Steps 5 and 6: $\Pi = P^a T^b \omega$; $M^0L^0T^0 = \left(\dfrac{FL}{T}\right)^a (FL)^b \left(\dfrac{1}{T}\right)^1$

F: $0 = a + b$; L: $0 = a + b$; T: $0 = -a - 1$

Solving: $a = -1,\ b = 1$. So $\Pi = P^{-1}T\omega = T\omega/P$

Step 7: $\phi(\Pi) = 0$, i.e., $1/\Pi$ = const or $P/T\omega = C$, $P = CT\omega$ ◄

7.7.4 *Derive an expression for the shear stress at the pipe wall when an incompressible fluid flows through a pipe under pressure. Use dimensional analysis with the following significant parameters: pipe diameter D, flow velocity V, and viscosity μ and density ρ of the fluid.*

N

Step 1: No. of variables (D, V, μ, ρ, τ), $n = 5$

Step 2: Their dimensions are: $D = [L]$, $V = \left[\frac{L}{T}\right]$, $\mu = \left[\frac{M}{LT}\right]$, $\rho = \left[\frac{M}{L^3}\right]$, $\tau = \left[\frac{F}{L^2} = \frac{M}{LT^2}\right]$

So no. of fundamental dimensions (M,L,T) involved, $m = 3$.

Step 3: The 3 variables $[L]$, $[L/T]$, and $[M/L^3]$ can not be formed into a dimensionless group, so the reduction number $k = 3$.

Step 4: No. of dimensionless Π groups needed $= n - k = 5 - 3 = 2$; name them Π_1 and Π_2.

$\therefore f(\Pi_1, \Pi_2) = 0$

Steps 5 and 6: Select the following 3 (= k) variables to be primary (repeating): V, D, ρ.

$$\text{For } \Pi_1 = V^{a_1}D^{b_1}\rho^{c_1}\mu: \quad M^0L^0T^0 = \left(\frac{L}{T}\right)^{a_1}(L)^{b_1}\left(\frac{M}{L^3}\right)^{c_1}\left(\frac{M}{LT}\right)^1$$

L: $0 = a_1 + b_1 - 3c_1 - 1$; M: $0 = c_1 + 1$; T: $0 = -a_1 - 1$

Solving: $a_1 = -1$, $b_1 = -1$, $c_1 = -1$. So $\Pi_1 = V^{-1}D^{-1}\rho^{-1}\mu = \mu/VD\rho = 1/\mathbf{R}$

$$\text{For } \Pi_2 = V^{a_2}D^{b_2}\rho^{c_2}\tau: \quad M^0L^0T^0 = \left(\frac{L}{T}\right)^{a_2}(L)^{b_2}\left(\frac{M}{L^3}\right)^{c_2}\left(\frac{M}{LT^2}\right)^1$$

L: $0 = a_2 + b_2 - 3c_2 - 1$; M: $0 = c_2 + 1$; T: $0 = -a_2 - 2$

Solving: $a_2 = -2$, $b_2 = 0$, $c_2 = -1$. So $\Pi_2 = V^{-2}D^0\rho^{-1}\tau = \tau/\rho V^2$

Step 7: $f(\Pi_1, \Pi_2) = 0$, i.e. $f(1/\mathbf{R}, \tau/\rho V^2) = 0$

so also $\Pi_2 = \phi(\Pi_1^{-1})$ or $\tau/\rho V^2 = \phi(\mathbf{R})$, $\tau = \rho V^2\phi(\mathbf{R})$ ◄

7.7.5 *Refer to the example of the drag on a submerged sphere, used to illustrate the seven-step procedure of Sec. 7.7. Using the same parameters, but also including the acceleration due to gravity g to account for the effect of wave action, derive an expression for the drag on a surface vessel.*

N

Step 1: $f'(F_D, D, V, \rho, \mu, g) = 0$, so no. of variables, $n = 6$.

Step 2: $F_D = \left[\frac{ML}{T^2}\right]$, $D = [L]$, $V = \left[\frac{L}{T}\right]$, $\rho = \left[\frac{M}{L^3}\right]$, $\mu = \left[\frac{M}{LT}\right]$, $g = \left[\frac{L}{T^2}\right]$

No. of fundamental dimensions (M,L,T) involved, $m = 3$.

Step 3: The 3 variables D, V, and ρ can not be formed into a dimensionless group, so the reduction number $k = 3$.

Step 4: No. of dimensionless Π groups needed $= n - k = 6 - 3 = 3$.. $\therefore f(\Pi_1, \Pi_2, \Pi_3) = 0$

Steps 5 and 6: Select the following 3 (= k) variables to be primary (repeating): ρ, D, V

$$\text{For } \Pi_1 = \rho^{a_1}D^{b_1}V^{c_1}\mu: \quad M^0L^0T^0 = \left(\frac{M}{L^3}\right)^{a_1}(L)^{b_1}\left(\frac{L}{T}\right)^{c_1}\left(\frac{M}{LT}\right)^1$$

M: $0 = a_1 + 1$; L: $0 = -3a_1 + b_1 + c_1 - 1$; T: $0 = -c_1 - 1$

/cont...

Solving: $a_1 = -1,\ b_1 = -1,\ c_1 = -1.$ So $\Pi_1 = \rho^{-1}D^{-1}V^{-1}\mu = \mu/(DV\rho) = \mathbf{R}^{-1}$

For $\Pi_2 = \rho^{a_2}D^{b_2}V^{c_2}F_D$: $M^0L^0T^0 = \left(\dfrac{M}{L^3}\right)^{a_2}(L)^{b_2}\left(\dfrac{L}{T}\right)^{c_2}\left(\dfrac{ML}{T^2}\right)^1$

M: $0 = a_2 + 1$; L: $0 = -3a_2 + b_2 + c_2 + 1$; T: $0 = -c_2 - 2$

Solving: $a_2 = -1,\ b_2 = -2,\ c_2 = -2.$ So $\Pi_2 = \rho^{-1}D^{-2}V^{-2}F_D = F_D/(\rho D^2V^2)$

For $\Pi_3 = \rho^{a_3}D^{b_3}V^{c_3}g$: $M^0L^0T^0 = \left(\dfrac{M}{L^3}\right)^{a_3}(L)^{b_3}\left(\dfrac{L}{T}\right)^{c_3}\left(\dfrac{L}{T^2}\right)^1$

M: $0 = a_3$; L: $0 = -3a_3 + b_3 + c_3 + 1$; T: $0 = -c_3 - 2$

Solving: $a_3 = 0,\ b_3 = 1,\ c_3 = -2.$ So $\Pi_3 = \rho^0DV^{-2}g = gD/V^2 = \mathbf{F}^{-2}$

Step 7: $f(\Pi_1, \Pi_2, \Pi_3) = 0$, i.e. $f(\mathbf{R}^{-1}, F_D/(\rho D^2V^2), \mathbf{F}^{-2}) = 0$

so also $\Pi_2 = \phi(\Pi_1^{-1}, \Pi_3^{-1/2})$, $F_D/(\rho D^2V^2) = \phi(\mathbf{R}, \mathbf{F})$, $F_D = \rho D^2V^2\phi(\mathbf{R}, \mathbf{F})$ ◄

Sec. 7.7: Dimensional Analysis -- *Problems*

7.27 *Derive an expression for the velocity of rise of an air bubble in a stationary liquid. Consider the effect of surface tension as well as other variables.*

N

Bubble velocity will be governed by its depth h (pressure and volume) and the properties including surface tension (given).

Step 1: $f'(V, h, \mu, \rho, \sigma) = 0$, so no. of variables, $n = 5$

Step 2: $V = \left[\dfrac{L}{T}\right]$, $h = [L]$, $\mu = \left[\dfrac{M}{LT}\right]$, $\rho = \left[\dfrac{M}{L^3}\right]$, $\sigma = \left[\dfrac{M}{T^2}\right]$; $m = 3$

Step 3: The 3 variables V, h, and ρ can <u>not</u> be formed into a dimensionless group,

so the reduction number $k = 3$.

Step 4: No. of Π groups needed $= n - k = 2$, $\therefore f(\Pi_1, \Pi_2) = 0$.

Steps 5 and 6: Select for the 3 (= k) primary (repeating) variables: V, h, ρ.

For $\Pi_1 = V^ah^b\rho^c\mu$: $M^0L^0T^0 = \left(\dfrac{L}{T}\right)^a(L)^b\left(\dfrac{M}{L^3}\right)^c\left(\dfrac{M}{LT}\right)^1$

M: $0 = c + 1$; L: $0 = a + b - 3c - 1$; T: $0 = -a - 1$

Solving: $a = -1,\ b = -1,\ c = -1.$ So $\Pi_1 = V^{-1}h^{-1}\rho^{-1}\mu = \mu/(hV\rho) = \mathbf{R}^{-1}$

For $\Pi_2 = V^ah^b\rho^c\sigma$: $M^0L^0T^0 = \left(\dfrac{L}{T}\right)^a(L)^b\left(\dfrac{M}{L^3}\right)^c\left(\dfrac{M}{T^2}\right)^1$

M: $0 = c + 1$; L: $0 = a + b - 3c$; T: $0 = -a - 2$

Solving: $a = -2,\ b = -1,\ c = -1.$ So $\Pi_2 = V^{-2}h^{-1}\rho^{-1}\sigma = \sigma/(V^2h\rho) = \mathbf{W}^{-2}$

Step 7: Thus we can write $\Pi_2^{-1} = \phi(\Pi_1^{-1})$, i.e. $V^2h\rho/\sigma = \phi(\mathbf{R})$ or $V = \sqrt{\sigma/h\rho}\,\phi(\mathbf{R})$ ◄

Alternatively $\Pi_1^{-1} = \phi(\Pi_2^{-1/2})$, i.e. $hV\rho/\mu = \phi(\mathbf{W})$ or $V = (\mu/h\rho)\phi(\mathbf{W})$ ◄

7.28 *Derive an expression for the drag on an aircraft flying at supersonic speed.*
N

At supersonic speed the volume modulus of elasticity E_v is also a factor (see Eq 7.10). $\therefore$ we expect both **M** and **R** to govern.

Step 1: $f'(F_D, L, V, \rho, \mu, E_v) = 0$, so no. of variables, $n = 6$. Using FLT:

Step 2: $F_D = [F]$, $L = [L]$, $V = \left[\dfrac{L}{T}\right]$, $\rho = \left[\dfrac{FT^2}{L^4}\right]$, $\mu = \left[\dfrac{FT}{L^2}\right]$, $E_v = \left[\dfrac{F}{L^2}\right]$; $m = 3$

Step 3: The 3 variables V, L, and ρ can <u>not</u> be formed into a dimensionless group,

so the reduction number $k = 3$.

Step 4: No. of Π groups $= n - k = 3$, $\therefore f(\Pi_1, \Pi_2, \Pi_3) = 0$

Steps 5 and 6: Select for the 3 ($=k$) primary (repeating) variables: ρ, L, V

For $\Pi_1 = \rho^a L^b V^c \mu$: $F^0 L^0 T^0 = \left(\dfrac{FT^2}{L^4}\right)^a (L)^b \left(\dfrac{L}{T}\right)^c \left(\dfrac{FT}{L^2}\right)^1$

F: $0 = a + 1$; L: $0 = -4a + b + c - 2$; T: $0 = 2a - c + 1$

Solving: $a = -1$, $b = -1$, $c = -1$. So $\Pi_1 = \rho^{-1}L^{-1}V^{-1}\mu = \mu/(LV\rho) = \mathbf{R}^{-1}$

For $\Pi_2 = \rho^a L^b V^c F_D$: $F^0 L^0 T^0 = \left(\dfrac{FT^2}{L^4}\right)^a (L)^b \left(\dfrac{L}{T}\right)^c (F)^1$

F: $0 = a + 1$; L: $0 = -4a + b + c$; T: $0 = 2a - c$

Solving: $a = -1$, $b = -2$, $c = -2$. So $\Pi_2 = \rho^{-1}L^{-2}V^{-2}F_D = F_D/(\rho L^2 V^2)$

For $\Pi_3 = \rho^a L^b V^c E_v$: $F^0 L^0 T^0 = \left(\dfrac{FT^2}{L^4}\right)^a (L)^b \left(\dfrac{L}{T}\right)^c \left(\dfrac{F}{L^2}\right)^1$

F: $0 = a + 1$; L: $0 = -4a + b + c - 2$; T: $0 = 2a - c$

Solving: $a = -1$, $b = 0$, $c = -2$. So $\Pi_3 = \rho^{-1}L^0V^{-2}E_v = E_v/(\rho V^2) = \mathbf{M}^{-2}$

Step 7: We can write $\Pi_2 = \phi(\Pi_1^{-1}, \Pi_3^{-1/2})$, i.e. $F_D/(\rho L^2 V^2) = \phi(\mathbf{R},\mathbf{M})$, or $F_D = \rho L^2 V^2 \phi(\mathbf{R},\mathbf{M})$ ◄

7.29 *Derive an expression for small flow rates over a spillway, in the form of a function including dimensionless quantities. Use dimensional analysis with the following parameters: height of spillway P, head on the spillway H, viscosity of liquid μ, density of liquid ρ, surface tension σ, and acceleration due to gravity g.*
N

Step 1: $f'(q, P, H, g, \mu, \rho, \sigma) = 0$, so no. of variables, $n = 7$. Using MLT:

Step 2: $q = \left[\dfrac{L^2}{T}\right]$, $\rho = [L]$, $H = [L]$, $g = \left[\dfrac{L}{T^2}\right]$, $\mu = \left[\dfrac{M}{LT}\right]$, $\rho = \left[\dfrac{M}{L^3}\right]$, $\sigma = \left[\dfrac{M}{T^2}\right]$; $m = 3$

Step 3: The 3 variables q, H, and ρ can <u>not</u> be formed into a dimensionless group,

so the reduction number $k = 3$.

Step 4: No. of Π groups needed $= n - k = 4$, $\therefore f(\Pi_1, \Pi_2, \Pi_3, \Pi_4) = 0$.

/cont...

Steps 5 and 6: Select for the 3 (=k) primary (repeating) variables: $q,\ H,\ \rho$.

For $\Pi_1 = q^a H^b \rho^c P$: $M^0L^0T^0 = \left(\frac{L^2}{T}\right)^a (L)^b \left(\frac{M}{L^3}\right)^c (L)^1$

M: $0 = c$; L: $0 = 2a + b - 3c + 1$; T: $0 = -a$

Solving: $a = 0,\ b = -1,\ c = 0$. So $\Pi_1 = q^0 H^{-1} \rho^0 P = P/H$

For $\Pi_2 = q^a H^b \rho^c g$: $M^0L^0T^0 = \left(\frac{L^2}{T}\right)^a (L)^b \left(\frac{M}{L^3}\right)^c \left(\frac{L}{T^2}\right)^1$

M: $0 = c$; L: $0 = 2a + b - 3c + 1$; T: $0 = -a - 2$

Solving: $a = -2,\ b = 3,\ c = 0$. So $\Pi_2 = q^{-2} H^3 g = gH^3/q^2$

For $\Pi_3 = q^a H^b \rho^c \mu$: $M^0L^0T^0 = \left(\frac{L^2}{T}\right)^a (L)^b \left(\frac{M}{L^3}\right)^c \left(\frac{M}{LT}\right)^1$

M: $0 = c + 1$; L: $0 = 2a + b - 3c - 1$; T: $0 = -a - 1$

Solving: $a = -1,\ b = 0,\ c = -1$. So $\Pi_3 = q^{-1} H^0 \rho^{-1} \mu = \mu/q\rho = \mathbf{R}^{-1}$ (as $q = HV$)

For $\Pi_4 = q^a H^b \rho^c \sigma$: $M^0L^0T^0 = \left(\frac{L^2}{T}\right)^a (L)^b \left(\frac{M}{L^3}\right)^c \left(\frac{M}{T^2}\right)^1$

M: $0 = c + 1$; L: $0 = 2a + b - 3c$; T: $0 = -a - 2$

Solving: $a = -2,\ b = 1,\ c = -1$. So $\Pi_4 = q^{-2} H \rho^{-1} \sigma = \sigma H/(\rho q^2) = \mathbf{W}^{-2}$ (*as* $q = HV$)

Step 7: We can write $\Pi_2^{-1/2} = \phi(\Pi_1,\ \Pi_3^{-1},\ \Pi_4^{-1/2})$

i.e., $q/(g^{1/2}H^{3/2}) = \phi(H/P,\ \mathbf{R},\ \mathbf{W})$ or $q = g^{1/2}H^{3/2}\phi(H/P,\ \mathbf{R},\ \mathbf{W})$ ◄

7.30 *Use dimensional analysis to derive an expression for the height of capillary rise in a glass tube.*

N

Step 1: $f'(h, r, \sigma, \gamma) = 0$, so no. of variables, $n = 4$. Using FLT:

Step 2: $h = [L],\ r = [L],\ \sigma = \left[\frac{F}{L}\right],\ \gamma = \left[\frac{F}{L^3}\right]$; $m = 2$

Step 3: The 3 variables h, r, and σ can <u>not</u> be formed into a dimensionless group, so the reduction number $k = 3$.

Step 4: No. of Π groups $= n - k = 1$, $\therefore f(\Pi) = 0$.

Steps 5 and 6: For $\Pi = r^a \sigma^b \gamma^c h$: $F^0L^0T^0 = (L)^a \left(\frac{F}{L}\right)^b \left(\frac{F}{L^3}\right)^c (L)^1$

F: $0 = b + c$; L: $0 = a - b - 3c + 1$; T: $0 = 0$

Solving: $a = 2c - 1,\ b = -c$. So $\Pi = r^{2c-1}\sigma^{-c}\gamma^c h = (h/r)(\gamma r^2/\sigma)^c$

Step 7: We can write $f(\Pi) = 0$, or $\Pi = C =$ const. Then $h = Cr\left(\frac{\sigma}{\gamma r^2}\right)^c$ ◄

Experimental observation suggests that $c = 1$ (see Sec 2.12), in which case $h = \dfrac{C\sigma}{\gamma r}$

Chapter 8

Steady Incompressible Flow in Pressure Conduits

PROBLEM SELECTION GUIDE

Sec	Exer/Prob	Units	Difficulty	Length	Parts	Similar	Special features
8.2	***Critical Reynolds Number***						
	X[1]8.2.1	BG	V Easy	V Short	1	8.2.2	
	8.2.2	SI	V Easy	V Short	1	8.2.1	
	8.2.3	BG	V Easy	V Short	1		
	P 8.1	BG	Medium	Short	1		Uses $pv = RT$ (Sec 2.7)
	8.2	SI	Medium	Medium	1		Uses $pv = RT$ (Sec 2.7)
8.3	***Hydraulic Radius, Hydraulic Diameter***						
	X 8.3.1	BG	V Easy	V Short	1		
	8.3.2	SI	V Easy	V Short	1		
	P 8.3	N	Easy	Short	1		
8.5	***Friction in Circular Conduits***						
	X 8.5.1	BG	Easy	Short	1	8.5.2	Laminar flow
	8.5.2	SI	Easy	Short	1	8.5.1	Laminar flow
	8.5.3	BG	V Easy	V Short	1	8.5.4	
	8.5.4	SI	V Easy	V Short	1	8.5.3	
	P 8.4	N	Easy	Short	1		
	8.5	N	Easy	Short	1		
	8.6	BG	Easy	Short	1		
8.6	***Laminar Flow in Circular Pipes***						
	X 8.6.1	BG	Easy	Short	1		f varies
	8.6.2	N	Easy	Short	1		
	8.6.3	N	Medium	Medium	1		Integration
	P 8.7	N	Easy	Short	1		Plot
	8.8	N	Medium	Medium	1		Integration
	8.9	N	Medium	Medium	1		Integration
	8.10	SI	Easy	Short	1		f varies. Power (Sec 5.10)
8.7	***Entrance Conditions in Laminar Flow***						
	X 8.7.1	BG	Easy	Short	1	8.7.2	
	8.7.2	SI	Easy	Short	1	8.7.1	

/cont...

[1] For all Exercises (marked "X"), answers are given in Appendix D of the textbook.
* Note a correction made to early printings of the textbook.
† Answers are sensitive to values (f, μ, ν, k') which are or may be read from a graph.
X = Exercise, P = (end-of-chapter) Problem, S = Sample Problem. ☐ = could use computing aids.

Sec	Exer/Prob	Units	Difficulty	Length	Parts	Similar	Special features
8.8	***Turbulent Flow***						
	X 8.8.1	BG	Medium	Medium	2		
8.9	***Viscous Sublayer in Turbulent Flow***						
	X 8.9.1	BG	V Easy	V Short	1		
	8.9.2	BG	Easy	Short	2	8.9.3	Given const *f*
	8.9.3	SI	Easy	Short	2	8.9.2	Given const *f*
	8.9.4	BG	Easy	Short	1	8.9.5	Given const *f*
	8.9.5	SI	Easy	Short	1	8.9.4	Given const *f*
	8.9.6	SI	Easy	Short	2		Given const *f*
	P 8.11	BG	Easy	Short	2		
8.10	***Velocity Profile in Turbulent Flow***						
	X 8.10.1	SI	Easy	Short	1		
	8.10.2	BG	Medium	Medium	1		† *f* varies
	8.10.3	N	Easy	Short	1		*f* cancels out
	P 8.12	BG	Medium	Medium	1	8.13	† *f* varies
	8.13	SI	Medium	Medium	1	8.12	† *f* varies
	8.14	BG	Medium	Medium	1	8.15	† *f* varies. Laminar, turbulent
	8.15	SI	Medium	Medium	1	8.14	† *f* varies. Laminar, turbulent
	8.16	SI	Medium	Medium	1		† *f* varies
	8.17	BG	Medium	Short	1		*f* varies. Two simultaneous equations.
	8.18	BG	Medium	Long	1	8.19	* *f* varies. Plot
	8.19	SI	Medium	Long	1	8.18	*f* varies. Plot
8.11	***Pipe Roughness***						
	X 8.11.1	N	Medium	Medium	4		□ *f* varies
	8.11.2	SI	Medium	Medium	1	8.11.3	Given const *f*
	8.11.3	SI	Medium	Medium	1	8.11.2	Given const *f*
	P 8.20	N	Medium	Medium	2		*f* varies. Laminar, turbulent
	8.22	SI	Hard	Medium	1		□ *f* varies. T & E (Trial & Error)
	8.23	BG	Medium	Medium	2	8.24	* † *f* varies, Blasius' Eq.
	8.24	SI	Medium	Medium	2	8.23	† *f* varies, Blasius' Eq.
	8.25	SI	Medium	Medium	1		Given const *f*
8.12	***Chart for Friction Factor***						
	X 8.12.1	SI	Medium	Medium	2		† *f* varies
	8.12.2	BG	Medium	Short	2		† *f* varies. Interpolation
	P 8.26	BG	Medium	Long	1		† □ *f* varies. Laminar, turbulent
	8.27	N	Easy	Long	1		† □ *f* varies. L, T; uses Secs 5.1, 6.3
	8.28	BG	Medium	Medium	1		† *f* varies. Laminar
	8.29	BG	Easy	Short	1	8.30	† *f* varies
	8.30	SI	Easy	Short	1	8.29	† *f* varies
	8.31	SI	Medium	Medium	1		† *f* varies. Power (Sec 5.10)
	8.32	BG	Medium	Medium	1		† *f* varies. Uses Secs 2.7, 2.11, 3.2, 4.5
	8.33	SI	Medium	Medium	1	8.34	† *f* varies
	8.34	SI	Medium	Medium	1	8.33	† *f* varies
	8.35	BG	Easy	Short	1	8.36	† *f* varies
	8.36	SI	Easy	Short	1	8.35	† *f* varies

/cont...

Sec	Exer/Prob	Units	Difficulty	Length	Parts	Similar	Special features
8.13	***Solution of Pipe Flow Problems by Trials***						
	X 8.13.1	BG	Medium	Medium	6		† f varies. Find h_L
	8.13.2	BG	Medium	Medium	3		† f varies. Find h_L
	8.13.3	BG	Hard	Long	1	8.37	† f varies, find Q. T&E (Trial & Error)
	8.13.4	SI	Hard	Long	1	8.38	† f varies, find Q. T & E.
	P 8.37	SI	Hard	Long	1	8.13.3	† f varies, find Q. T & E.
	8.38	BG	Hard	Long	1	8.13.4	† f varies, find Q. T & E.
	8.39	BG	Hard	Long	1	8.40	† f varies, assume regime, find D. T&E
	8.40	SI	Hard	Long	1	8.39	† f varies, assume regime, find D. T&E
	8.41	BG	Hard	Long	1	8.42	† f varies, find Q. T & E.
	8.42	SI	Hard	Long	1	8.41	† f varies, find Q. T & E.
	8.43	BG	Medium	Long	1	8.44	† f varies, find D. T & E.
	8.44	SI	Medium	Long	1	8.43	† f varies, find D. T & E.
8.14	***Rigorous Solution of Pipe Flow Problems***						
	X 8.14.1	SI	Medium	Long	2		□ f varies, find Q
	8.14.2	BG	Medium	Long	3	8.14.3	□ f varies, find Q
	8.14.3	SI	Medium	Long	3	8.14.2	□ f varies, find Q
	P 8.45	SI	Medium	Medium	1	8.37	f varies. Find Q, check **R**.
	8.46	SI	Medium	Long	1	8.40	f varies. Find D, check 2 flow regimes.
	8.47	SI	Medium	Medium	1	8.42	f varies. Find Q, check **R**.
	8.48	SI	Medium	Long	1	8.44	f varies. Find D, check 2 flow regimes.
	8.49	B	Medium	Long	4	8.50	□ f varies. Find Q, check **R**.
	8.50	B	Medium	Long	4	8.49	* □ f varies. Find D, check **R**.
8.15	***Empirical Equations for Pipe Flow***						
	X 8.15.1	BG	Easy	V Short	1	8.15.2	C_{HW}
	8.15.2	SI	Easy	V Short	1	8.15.1	C_{HW}
	P 8.51	BG	Medium	Medium	1	8.52	† Manning's n.
	8.52	SI	Medium	Medium	1	8.51	† Manning's n.
8.16	***Fluid Friction in Noncircular Conduits***						
	X 8.16.1	BG	Easy	Short	1	8.16.2	Power (Sec 5.10)
	8.16.2	SI	Easy	Short	1	8.16.1	Power (Sec 5.10)
	P 8.53	BG	Medium	Medium	1		† f varies. Lam'r; uses $pv = RT$ (§ 2.7).
	8.54	BG	Hard	Medium	1	8.55	† f varies. T & E
	8.55	SI	Hard	Medium	1	8.54	† f varies. T & E
8.19	***Loss of Head at Submerged Discharge***						
	X 8.19.1	BG	Easy	Short	3	8.19.2	Given const f. h' (minor losses).
	8.19.2	SI	Easy	Short	3	8.19.1	Given const f. h'
	8.19.3	BG	Medium	Medium	1	8.19.4	† f varies. h'
	8.19.4	SI	Mediium	Medium	1	8.19.3	† f varies. h'
	8.19.5	BG	Mediium	Medium	1	8.19.6	† f varies. h'
	8.19.6	SI	Medium	Medium	1	8.19.5	† f varies. h'

/cont...

Sec	Exer/Prob	Units	Difficulty	Length	Parts	Similar	Special features
8.21	***Loss Due to Expansion***						
	X 8.21.1	SI	Easy	Medium	3		† h' (minor losses)
	8.21.2	BG	Medium	Medium	2		† Given const f. h'
	P 8.56	BG	Medium	Medium	1	8.57	† ☐ f varies. h'
	8.57	SI	Medium	Medium	1	8.56	† f varies. h'
8.23	***Loss in Bends and Elbows***						
	X 8.23.1	BG	Medium	Medium	1	8.23.2	† f varies. h'. Interpolation
	8.23.2	SI	Medium	Medium	1	8.23.1	† f varies. h'. Interpolation
8.24	***Solution of Single-Pipe Flow Problems with Minor Losses***						
	X 8.24.1	BG	Medium	Medium	1	8.24.2	Given const f. h'. Jet, find Q.
	8.24.2	SI	Medium	Medium	1	8.24.1	Given const f. h'. Jet, find Q.
	8.24.3	BG	Medium	Medium	2	8.24.4	Given const f. h'. Find p.
	8.24.4	SI	Medium	Medium	2	8.24.3	Given const f. h'. Find p.
	8.24.5	BG	Medium	Medium	2		Given const f. h'. Find Q.
	8.24.6	BG	Medium	Long	1		† ☐ f varies. h'. Find Q. T & E.
	P 8.58	BG	Medium	Medium	2	8.59	Given const f. h'. Find p.
	8.59	SI	Medium	Medium	2	8.58	Given const f. h'. Find p.
	8.60	BG	Medium	Medium	2	8.61	Given const f. h'. Find Q, p.
	8.61	SI	Medium	Medium	2	8.60	Given const f. h'. Find Q, p.
	8.62	BG	Medium	Medium	1	8.63	Given const f. Jet, find p.
	8.63	SI	Medium	Medium	1	8.62	Given const f. Jet, find p.
	8.64	BG	Medium	Medium	2		Given const f. h'. Jet, find V, p.
	8.65	BG	Medium	Long	1	8.66	☐ Given const f. h', jet, find D, power
	8.66	SI	Medium	Long	1	8.65	☐ Given const f. h', jet, find D, power
	8.67	BG	Hard	Long	2	8.68	† f varies. h'. Find Q. T & E.
	8.68	BG	Hard	Long	2	8.67	☐ f varies. h'. Find Q.
	8.69	SI	Hard	Long	1		† f varies. h'. Find D. T & E.
	8.70	SI	Hard	Long	1	8.69	† ☐ f varies. h'. Find D.
8.25	***Pipeline with Pump or Turbine***						
	X 8.25.1	BG	Easy	Short	1		Given const f. Power (Sec 5.10).
	8.25.2	BG	Easy	Short	1	8.25.3	Given const f. Power (Sec 5.10), pump.
	8.25.3	SI	Easy	Short	1	8.25.2	Given const f. Power (Sec 5.10), pump.
	8.25.4	BG	Medium	Medium	1	8.25.5	Given const f. Power (§5.10), pump, jet
	8.25.5	SI	Medium	Medium	1	8.25.4	Given const f. Power (§5.10), pump, jet
	8.25.6	BG	Medium	Medium	1	8.25.7	Power (Sec 5.10), pump.
	8.25.7	SI	Medium	Medium	1	8.25.6	Power (Sec 5.10), pump.
	8.25.8	BG	Medium	Short	1		Const f. h', power, cavit (§5.11), pump
	8.25.9	BG	Medium	Short	2	8.25.10	Const f. h', power (§ 5.10), turbine.
	8.25.10	SI	Medium	Short	2	8.25.9	Const f. h', power (§ 5.10), turbine.
	P 8.71	BG	Hard	Long	3		* Const f. h', power, pump; sketch.
	8.72	SI	Medium	Medium	1		Given const f. h', cavit (§5.11), pump.
	8.73	BG	Medium	Medium	2		Given const f. h', power (§5.10), pump
	8.74	BG	Hard	Long	1		☐ Given const f. h', power, cavitation,
	″						pump, jet. Plot graph.
	8.75	BG	Hard	Medium	1	8.76	Given const f. h', power, turbine.
	8.76	SI	Hard	Medium	1	8.75	Given const f. h', power, turbine.
	8.77	BG	Hard	Medium	1		Given const f. h', turbine.
	8.78	BG	Hard	Medium	1		Given const f. Power (Sec 5.10), pump.

/cont...

Sec	Exer/Prob	Units	Difficulty	Length	Parts	Similar	Special features
8.26	***Branching Pipes***						
	P 8.79	BG	Hard	Long	1	8.80	† *f* varies. Case 1
	8.80	SI	Hard	Long	1	8.79	† *f* varies. Case 1
	8.81	SI	Hard	Long	1	8.80	□ *f* varies. Case 1
	8.82	BG	Hard	Long	1	8.85	*f* varies. T&E, interp, check **R**. Case 2
	8.83	BG	Hard	Long	1	8.82	□ *f* varies Case 2
	8.84	SI	Hard	Long	1	8.86	*f* varies. T&E, interp, check **R**. Case 2
	8.85	BG	Hard	Long	1	8.82	*f* varies. T&E, interp, check **R**. Case 2
	8.86	SI	Hard	Long	1	8.84	*f* varies. T&E, interp, check **R**. Case 2
	8.87	SI	Hard	Long	1		*f* varies. T&E, interp, check **R**. Case 3
	8.88	SI	Hard	Long	1	8.87	□ *f* varies. Case 3
	8.89	BG	Hard	Long	1		*f* varies. T&E, interp, check **R**. Case 3
	8.90	BG	Hard	Long	1	8.89	□ *f* varies. Case 3
	8.91	BG	Hard	Long	1		□ *f* varies. Check **R**. Special case
8.27	***Pipes in Series***						
	P 8.92	BG	Hard	Long	1	8.93	† □ *f* varies. T & E
	8.93	SI	Hard	Long	1	8.92	† □ *f* varies. T & E
	8.94	BG	Hard	Long	1		† □ *f* varies. T & E
	8.95	BG	Hard	Long	1	8.96	* *f* varies. Laminar flow, check **R**.
	8.96	SI	Hard	Long	1	8.95	*f* varies. Laminar flow, check **R**.
	8.97	BG	Medium	Medium	1	8.98, 99	Given const *f*. h', jet.
	8.98	SI	Medium	Medium	1	8.97, 100	Given const *f*. h', jet, interpolation.
	8.99	BG	Medium	Medium	1	8.97, 100	Given const *f*. Jet.
	8.100	SI	Medium	Medium	1	8.98, 99	Given const *f*. Jet.
	8.101	BG	Hard	Long	1		† □ *f* varies. h', jet. T & E.
8.28	***Pipes in Parallel***						
	P 8.102	BG	Hard	Medium	1	8.105	* † *f* varies. Assume/verify Blasius' Eq
	8.103	BG	Medium	Medium	1	8.106	† *f* varies. Laminar flow, check **R**.
	8.104	BG	Hard	Long	1	8.107	† □ *f* varies. T & E.
	8.105	SI	Hard	Medium	1	8.102	† *f* varies. Assume/verify Blasius' Eq.
	8.106	SI	Medium	Medium	1	8.103	* † *f* varies. Make/verify assump; Lam.
	8.107	SI	Hard	Long	1	8.104	† □ *f* varies. T & E.
8.29	***Pipe Networks***						
	P 8.108	BG	Hard	Long	2		† □ *f* varies. T & E.
	8.109	BG	Medium	Long	2		□ Given const *f*
	8.110	BG	Medium	Long	2	8.111	□ Given const *f*
	8.111	SI	Medium	Long	2	8.110	□ Given const *f*. See ending comment.
	8.112	BG	Medium	Medium	2	8.113,114	□ Given const *f*. With pump, jet.
	8.113	BG	Medium	Medium	2	8.112,114	□ Given const *f*. With pump, jet.
	8.114	BG	Medium	Medium	2	8.112,113	□ Given const *f*. With pump, jet.
	8.115	BG	Medium	Long	1	8.116	† *f* varies.
	8.116	SI	Medium	Long	1	8.115	† *f* varies.
	8.117	BG	Hard	Long	1	8.118	□ Given *K* values. Repet Cross apprxs.
	8.118	SI	Hard	Long	1	8.117	□ Given *K* values. Repet Cross apprxs.
	8.119	SI	Hard	V Long	1		† □ Const *f*. Repet'v H. Cross apprxns.
	8.120	BG	Hard	V Long	1		† □ Const *f*. Repet'v H. Cross apprxns.

Chapter 8
STEADY INCOMPRESSIBLE FLOW IN PRESSURE CONDUITS

Sec. 8.2: Critical Reynolds Number -- *Exercises*

8.2.1 *Oil with a kinematic viscosity of 0.00018 ft²/s is flowing through a 4-in-diameter pipe. Below what velocity will the flow be laminar?*

BG

Eq. 8.1: $\mathbf{R} = (4/12)V/0.000\,18 < 2000 = \mathbf{R}_{crit}$; so V < 1.080 fps ◄

8.2.2 *Oil with a kinematic viscosity of 0.142 St is flowing through a 10-cm-diameter pipe. Below what velocity will the flow be laminar?*

SI

Inside cover: $\nu = 0.142$ St $= 0.142 \times 10^{-4}$ m²/s

Eq. 8.1: $\mathbf{R} = 0.10V/(0.142 \times 10^{-4}) < 2000 = \mathbf{R}_{crit}$; so $V < 0.284$ m/s ◄

8.2.3 *Oil with a kinematic viscosity of 0.004 ft²/s flows through a 4-in-diameter pipe with a velocity of 13.5 fps. Is the flow laminar or turbulent?*

BG

Eq. 8.1: $\mathbf{R} = (4/12)13.5/0.004 = 1125 < \mathbf{R}_{crit} = 2000$, so flow is laminar ◄

Sec. 8.2: Critical Reynolds Number -- *Problems*

8.1 *Hydrogen at 60°F and atmospheric pressure has a kinematic viscosity of 0.0012 ft²/s. Determine the minimum turbulent flow rate in pounds per second through a 3-in-diameter pipe. At this flow rate what is the average velocity?*

BG

Eq. 8.1: $\mathbf{R} = (3/12)V/0.0012 < 2000 = \mathbf{R}_{crit}$, so max. avg. $V = 9.6$ fps ◄

Table A.5 for hydrogen: γ at 68°F = 0.005 22 lb/ft³

At constant (atmos) p, from Eq. 2.5: $\gamma = \text{const}/T$

$\therefore$ at 60°F: $\gamma = 0.005\,22(460 + 68)/(460+60) = 0.00530$ lb/ft³

Eq. 4.5: $G = \gamma AV = (0.00530)(\pi/4)(3/12)^2 9.6 = 0.00250$ lb/sec ◄

8.2 *Air at 80°C and a pressure of approximately 1350 kN/m² abs flows in a 2-cm-diameter tube. What is the minimum turbulent flow rate? Express the answer in liters per second, newtons per second, and kilograms per second. At this flow rate what is the average velocity?*

SI

From Sec. 2.7: $p/\rho = RT$, $\rho = p/RT$; Eq. 8.1: $\mathbf{R} = DV\rho/\mu = DVp/\mu RT$

Table A.2: $\mu = 2.09 \times 10^{-5}$ N·s/m²; Table A.5: $R = 287$ m²/(s²·K)

For laminar flow, max $\mathbf{R} = \dfrac{(0.02 \text{ m})(V)(1350 \times 10^3 \text{ N/m}^2)}{(2.09 \times 10^{-5} \text{ N·s/m}^2)[287 \text{ m}^2/(\text{s}^2\text{·K})](273 + 80 \text{ K})} = 2000$

i.e. 12 750 V m/s = 2000; max $V = 0.1569$ m/s ◄

Eq. 4.3: $Q = AV = \pi(0.01)^2 0.1569 = 0.000\,0493$ m³/s = 0.0493 L/s ◄

From Sec. 2.7: $\rho = p/RT = 1350 \times 10^3/[287(273 + 80)] = 13.32$ kg/m³

Eq. 4.4: $\dot{m} = \rho Q = (13.32)0.0493 \times 10^{-3} = 0.657 \times 10^{-3}$ kg/s ◄

Eq. 3.9b: $G = \gamma Q = \rho g Q = 13.32(9.81)0.0493 \times 10^{-3} = 6.44 \times 10^{-3}$ N/s ◄

Sec. 8.3: Hydraulic Radius, Hydraulic Diameter -- *Exercises*

8.3.1 *What is the hydraulic radius of a 9 in by 16 in rectangular air duct?*

BG

Eq. 8.2: $R_h = A/P = (9 \times 16)/[2(9 + 16)] = 2.88$ inch $= 0.240$ feet ◄

8.3.2 *What is the percentage difference between the hydraulic radii of a 30-cm-diameter duct and a 30-cm-square duct?*

SI

Eq. 8.2: Circular duct: $R_h = A/P = (\pi/4)(30^2)/(30\pi) = 7.5$ cm

Square duct: $R_h = A/P = 30^2/(4 \times 30) = 7.5$ cm; Therefore % difference = zero ◄

Sec. 8.3: Hydraulic Radius, Hydraulic Diameter -- *Problem*

8.3 *Two pipes, one circular and one square, have the same cross-sectional area. Which has the larger hydraulic radius, and by what percentage?*

N

Circular pipe, diameter D, and square pipe, side a, have same cross-sectional area so $a^2 = (\pi/4)D^2$, i.e., $a = \pi^{1/2}D/2$

Eq. 8.2, circular: $R_h = A/P = (\pi/4)D^2/\pi D = D/4 = 0.25D$

Square: $R_h = A/P = a^2/4a = a/4 = \pi^{1/2}D/8 = 0.222D$

R_h ratio $= 0.25D/0.222D = 1.128$; so circular pipe has larger R_h, by 12.8% ◄

Sec. 8.5: Friction in Circular Conduits -- *Exercises*

8.5.1 *Oil (s = 0.93) of viscosity of 0.004 ft²/s flows in a 4-in-diameter pipe at a rate of 6.5 gpm. Find the head loss per unit length.*

BG

$Q = 6.5$ gpm $= 0.014\,48$ cfs; Eq. 4.7: $V = Q/A = 0.014\,48/[\pi(4/12)^2/4] = 0.1660$ fps

Eq. 8.1: $\mathbf{R} = [(4/12)(0.1660)]/(0.004) = 13.83$, flow is laminar, $f = 64/13.83 = 4.63$

Eq. 8.10: $h_L = 4.63(12/4)(0.1660)^2/[2(32.2)] = 0.005\,94$ feet per feet ◄

8.5.2 *Oil (s = 0.93) of viscosity of 0.00034 m²/s flows in a 10-cm-diameter pipe at a rate of 0.45 L/s. Find the head loss per unit length.*

SI

Eq. 4.7: $V = Q/A = 0.000\,45/(\pi 0.05^2) = 0.0573$ m/s

Eq. 8.1: $\mathbf{R} = [(0.1)0.0573]/(0.000\,34) = 16.85$, flow is laminar, $f = 64/16.85 = 3.80$

Eq. 8.10: $h_L = 3.80(1/0.05)(0.0573)^2/[2(9.81)] = 0.012\,71$ m per meter ◄

8.5.3 *Steam with a specific weight of 0.32 lb/ft³ is flowing with a velocity of 94 fps through a circular pipe with f = 0.0171. What is the shear stress at the pipe wall?*

BG

Eq. 8.14: $\tau_0 = (0.0171/4)0.32(94)^2/[2(32.2)] = 0.1877$ psf ◄

8.5.4 *Steam with a specific weight of 38 N/m³ is flowing with a velocity of 35 m/s through a circular pipe with f = 0.0154. What is the shear stress at the pipe wall?*

SI

Eq. 8.14: $\tau_0 = (0.0154/4)38(35)^2/[2(9.81)] = 9.13$ N/m² ◄

Sec. 8.5: Friction in Circular Conduits -- *Problems*

8.4 *Prove that for a constant rate of discharge and a constant value of f the friction head loss in a pipe varies inversely as the fifth power of the diameter.*

N

Eq. 8.10: $h_L = f(L/D)V^2/(2g)$ where $V = Q/A = Q/(\pi D^2/4) = 4Q/(\pi D^2)$

$h_L = f(L/D)[4Q/(\pi D^2)]^2/(2g) = f(L/D^5)16Q^2/(2g\pi^2)$

Thus for f = constant and Q = constant, $h_L = 1/D^5$ Q.E.D. ◀

8.5 *Two long pipes convey water between two reservoirs whose water surfaces are at different elevations. One pipe has a diameter twice that of the other; both pipes have the same length and the same value of f. If minor losses are neglected, what is the ratio of the flow rates through the two pipes?*

N

Eq. 8.10: $\Delta\text{elev} = h_L = f(L/D)V^2/(2g)$ where $V = Q/A = Q/(\pi D^2/4)$

$\therefore \quad h_L = f(L/D)[Q/(\pi D^2/4)]^2/2g = fL4^2Q^2/(2gD\pi^2 D^4)$

Thus $h_L \propto Q^2/D^5$; $(h_L)_1 = (h_L)_2$; $\therefore$ $Q_1^2/D_1^5 = Q_2^2/D_2^5$ and $Q_2/Q_1 = (D_2/D_1)^{5/2} = 2^{5/2} = 5.66$

The flow in the larger pipe will be 5.66 times that in the smaller pipe. ◀

8.6 *Tests were made with 70°F water flowing in a 10-in-diameter pipe. They showed that, when V = 13 fps, f = 0.0162. Find the unit shear at the pipe wall and at radii of 0, 0.25, 0.4, 0.6, 0.85 times the pipe radius.*

BG

Table A.1 for water at 70°F: ρ = 1.936 slugs/ft^3

(*a*) Eq. 8.14: $\tau_0 = (0.0162/4)1.936(13^2/2) = 0.663$ psf, at wall ◀

(*b*) Stress distribution is linear (Eq. 8.13):

r/r_0	τ (psf)
0	0
0.25	0.1656
0.4	0.256
0.6	0.398
0.85	0.563

▲

Sec. 8.6: Laminar Flow in Circular Pipes -- *Exercises*

8.6.1 *An oil with kinematic viscosity 0.004 ft²/s weighs 62 lb/ft³. What will be its flow rate and head loss in a 2750-ft length of a 3-in-diameter pipe when the Reynolds number is 950?*

BG

Eq. 8.1: $\mathbf{R} = DV/\nu = (3/12)V/0.004 = 950$, $V = 15.20$ fps; $\mathbf{R} < 2000$, so flow is laminar

Eq. 4.3: $Q = AV = (\pi/4)(3/12)^2 15.20 = 0.746$ cfs ◀

Eq. 8.20: $f = 64/\mathbf{R} = 64/950 = 0.0673$; Eq. 8.10: $h_L = 0.0673(2750 \times 12/3)15.20^2/(2g) = 2660$ ft ◀

8.6.2 *With laminar flow in a circular pipe, at what distance from the centerline (in terms of the pipe radius) does the average velocity occur?*

N

Sec. 8.6: Velocity profile is a paraboloid with $V_{\text{mean}} = 0.5V_c$, so

Eq. 8.16: $u = V_c(1 - r^2/r_0^2)$; $(1/2)V_c = V_c(1 - r^2/r_0^2)$; $(r/r_0)^2 = 0.5$; $r/r_0 = 0.707$; $r = 0.707r_0$ ◀

8.6.3 *For laminar flow in a two-dimensional passage, find the relation between the average and maximum velocities.*

N

Let passage width be $2y_0$, so that $u = u_{max} = V_c$ at $y = y_o$.

Eq. 4.3: $Q = AV = (2y_0)V$. But from Eq. 8.16: $Q = 2\int_0^{y_0} u\,dy = 2\int_0^{y_0} V_c(1 - y^2/y_0^2)dy$

By equating the two: $V = 2V_c[y - (y^3/y_0^2)]_0^{y_0}/(2y_0) = (2/3)V_c$ ◄

Sec. 8.6: Laminar Flow in Circular Pipes -- *Problems*

8.7 *For laminar flow in a circular pipe, find the velocities at 0.1r, 0.3r, 0.5r, 0.7r, and 0.9r. Plot the velocity profile.*

N

From Eq. 8.16:

r/r_0	u/V_c
0.1	0.99
0.3	0.91
0.5	0.75
0.7	0.51
0.9	0.19

▲

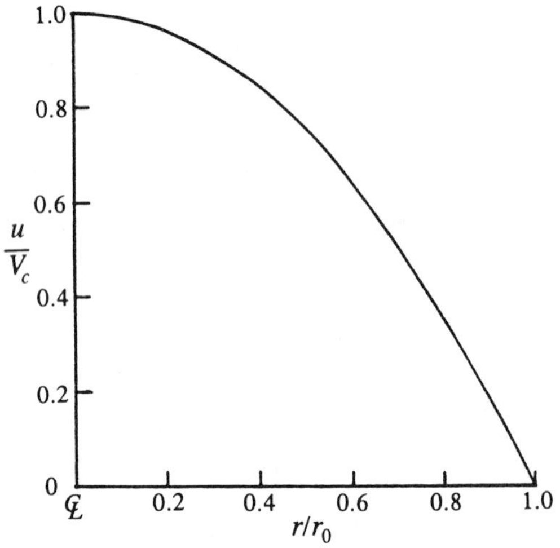

8.8 *Prove that the centerline velocity is twice the average velocity when laminar flow occurs in a circular pipe.*

N

Eq. 8.16: $u = V_c(1 - r^2/r_0^2)$

Eq. 4.3: $Q = AV = \int u\,dA = \int_0^{r_0} V_c(1 - r^2/r_0^2)2\pi r\,dr = 2\pi V_c\int_0^{r_0}(r - r^3/r_0^2)dr = 2\pi V_c(r_0^2/4)$

Eq. 4.7: $V = Q/A = [2\pi V_c(r_0^2/4)]/(\pi r_0^2) = (1/2)V_c$ Q.E.D. ◄

8.9 *For laminar flow between two parallel, flat plates a small distance d apart, at what distance from the centerline (in terms of d) will the velocity be equal to the mean velocity?*

N

Let y_m = distance from centerline where $u = V_{mean}$. From solution to Exer. 8.6.3, $V_{mean} = (2/3)V_c$.

So when $y = y_m$, $u = (2/3)V_c$, i.e. $(2/3)V_c = V_c(1 - y_m^2/y_0^2)$; $y_m^2 = (1/3)y_0^2 = (1/3)(d/2)^2$;

$y_m = 0.289d$ ◄

8.10 *Oil with a viscosity of 0.18 N·s/m² and a density of 915 kg/m³ is flowing in a 25-cm-diameter pipe at 0.75 L/s. How much power is lost per meter of pipe length?*

SI

Eq. 4.7: $V = 0.000\,75/(\pi 0.125^2) = 0.015\,28$ m/s

Eq. 8.1: $\mathbf{R} = [0.25(0.015\,28)915]/0.18 = 19.42$, flow is laminar. Eq. 8.20: $f = 64/19.42$

Eq. 8.10: $h_L = (64/19.42)(1/0.25)0.015\,28^2/[2(9.81)] = 0.000\,1569$ meter per meter

Eq. 5.33: Power loss = $(915 \times 9.81)(\pi 0.125^2)0.015\,28(0.000\,1569) = 0.001\,056$ watts per meter ◄

Sec. 8.7: Entrance Conditions in Laminar Flow -- *Exercises*

8.7.1 *In Exer. 8.2.3 what will be the approximate distance from the pipe entrance to the first point at which the flow is established?*

Exer. 8.2.3: Oil with a kinematic viscosity of 0.004 ft²/sec flows through a 4-in-diameter pipe with a velocity of 13.5 fps.

BG

Eq. 8.1: $\mathbf{R} = (4/12)13.5/0.004 = 1125$

Sec. 8.7: $L' = 0.058\mathbf{R}D = 0.058(1125)(4/12) = 21.8$ feet ◄

8.7.2 *In Exer. 8.5.2 what will be the approximate distance from the pipe entrance to the first point at which the flow is established?*

Exer. 8.5.2: Oil (s = 0.93) of viscosity of 0.00034 m²/s flows in a 10-cm-diameter pipe at a rate of 0.45 L/s.

SI

Eq. 4.7: $V = Q/A = 0.000\,45/(\pi 0.05^2) = 0.0573$ m/s

Eq. 8.1: $\mathbf{R} = (0.10)0.0573/0.000\,34 = 16.85$

Sec. 8.7: $L' = 0.058\mathbf{R}D = 0.058(16.85)0.10 = 0.0977$ m ◄

Sec. 8.8: Turbulent Flow -- *Exercise*

8.8.1 *Tests on 70°F water flowing through a 10-in-diameter pipe showed that when V = 13 fps, f = 0.0162. (a) If, at a distance of 3 in from the center of the pipe, τ = 0.398 psf and the velocity profile gives a value for du/dy of 6.77 per second, find the viscous shear and the turbulent shear at that radius. (b) What is the value of the mixing length l, and what is the value of the ratio l/r_0?*

BG

Table A.1 at 70°F: $\gamma = 62.30$ lb/ft³, $\mu = 0.000\,0205$ lb·sec/ft²

(*a*) Eq. 8.22: $\tau = \mu dy/dy + \eta du/dy = 0.398$ psf (given)

$\mu du/dy = 0.000\,0205(6.77) = 0.000\,1387$ psf, viscous shear ◄

$\eta du/dy = \tau - \mu dy/dy = 0.398 - 0.000\,1387 = 0.398$ psf, turbulent shear ◄

(*b*) Eq. 8.24: $\tau_{\text{turb}} = \rho l^2 (du/dy)^2 = 0.398$ psf

$l = (\tau/\rho)^{1/2}/(du/dy) = [0.398(32.2/62.30)]^{1/2}/6.77 = 0.0670$ ft or 0.804 inch ◄

$l/r_0 = 0.0670/(5/12) = 0.1608$ ◄

Sec. 8.9: Viscous Sublayer in Turbulent Flow -- *Exercises*

8.9.1 *Compute δ_l for the data of Sample Prob. 8.3.*

Sample Prob. 8.3: The head loss in 200 ft of 6-in-diameter pipe is known to be 25 ft·lb/lb when oil (s = 0.90) of viscosity 0.0008 lb·sec/ft² flows at 2.0 cfs. V = Q/A = 10.19 fps; $f = h_L D(2g)/(LV^2) = 0.0388$.

BG

Eq. 8.27: $\delta_l = 14.14\nu/(Vf^{1/2}) = [14.14(0.0008)/(0.9 \times 1.94)]/(10.19 \times 0.0388^{1/2}) = 0.003\,23$ feet

$= 0.387$ in ◄

8.9.2 *Water in a pipe (f = 0.018) is at a temperature of 70°F. (a) If the mean velocity is 14 fps, what is the nominal thickness δ_l of the viscous sublayer? (b) What will δ_l be if the velocity is increased to 24 fps and f does not change?*

BG

Table A.1 for water at 70°F: $\nu = 1.059 \times 10^{-5}$ ft²/sec

(*a*) Eq. 8.27: $\delta_l = 14.14\nu/(Vf^{1/2}) = 14.14(1.059 \times 10^{-5})/(14 \times 0.018^{1/2})$

$= 0.000\,0797$ ft $= 0.000\,957$ inch ◄

(*b*) For V = 24 fps and same f, $\delta_l' = (14/24)\delta_l = 0.000\,558$ inch ◄

8.9.3 *Water in a pipe (f = 0.015) is at a temperature of 15°C. (a) If the mean velocity is 3.5 m/s, what is the nominal thickness δ_l of the viscous sublayer? (b) What will δ_l be if the velocity is increased to 5.8 m/s and f does not change?*

SI

Table A.1 for water at 15°C: $\nu = 1.139 \times 10^{-6}$ m²/s

(*a*) Eq. 8.27: $\delta_l = 14.14\nu/(Vf^{1/2}) = 14.14(1.139 \times 10^{-6})/[3.5(0.015)^{1/2}] = 3.76 \times 10^{-5}$ m

$= 0.0376$ mm ◄

(*b*) For V = 5.8 m/s and same f, $\delta_l' = (3.5/5.8)\delta_l = 0.0227$ mm ◄

8.9.4 *For the data in Exer. 8.9.2(a), what is the distance from the wall to the assumed limit of the transition region where true turbulent flow begins?*
Exer. 8.9.2(a): Water at 70°F flows in a pipe (f = 0.018) with a mean velocity of 14 fps.

BG

Table A.1 for water at 70°F: $\nu = 1.059 \times 10^{-5}$ ft²/sec

Eq. 8.27: $\delta_l = 14.14\nu/(Vf^{1/2}) = 14.14(1.059 \times 10^{-5})/(14 \times 0.018^{1/2}) = 0.000\,0797$ ft $= 0.000\,957$ inch

Sec. 8.9: $y_t = 70\nu/u_* = 14\delta_l = 14(0.000\,957) = 0.013\,40$ inch ◄

8.9.5 *For the data in Exer. 8.9.3(a), what is the distance from the wall to the assumed limit of the transition region where true turbulent flow begins?*
Exer. 8.9.3(a): Water at 15°C flows in a pipe (f = 0.015) with a mean velocity of 3.5 m/s.

SI

Table A.1 for water at 15°C: $\nu = 1.139 \times 10^{-6}$ m²/s

Eq. 8.27: $\delta_l = 14.14\nu/(Vf^{1/2}) = 14.14(1.139 \times 10^{-6})/[3.5(0.015)^{1/2}] = 3.76 \times 10^{-5}$ m $= 0.0376$ mm

Sec. 8.9: $y_t = 70\nu/u_* = 14\delta_l = 14(0.0376) = 0.526$ mm ◄

8.9.6 *Water at 50°C flows in a 15-cm-diameter pipe with V = 6.5 m/s and e = 0.14 mm. Head loss measurements indicate that f = 0.020. (a) What is the thickness of the viscous sublayer? (b) Is the pipe behaving as a fully rough pipe?*

SI

(*a*) Eq. 8.27: $\delta_l = 14.14\nu/(Vf^{1/2})$; Table A.1 for water at 50°C: $\nu = 0.553 \times 10^{-6}$ m²/s

$\delta_l = 14.14(0.553 \times 10^{-6})/[6.5(0.020)^{1/2}] = 8.51 \times 10^{-6}$ m $= 8.51$ μm ◄

(*b*) $e/14 = 140$ μm$/14 = 10$ μm, so $\delta_l < e/14$; so, yes, pipe is behaving as fully rough ◄

Sec. 8.9: Viscous Sublayer in Turbulent Flow -- Problem

8.11 *Water at 50°F enters a pipe with a uniform velocity of 12 fps. (a) What is the distance at which the transition occurs from a laminar to a turbulent boundary layer? (b) If the thickness of this initial laminar boundary layer is given by* $4.91\sqrt{\nu x/U}$*, what is the thickness reached by it at the point of transition?*

BG

Table A.1 for water at 50°F: $\nu = 1.410 \times 10^{-5}$ ft²/sec

(*a*) Sec. 8.9: At transition point (turbulent boundary layer begins): For $\mathbf{R}_x = 500{,}000 = Ux/\nu$

$x = 500{,}000\nu/U = 500{,}000(0.000\,0141)/12 = 0.588$ ft $= 7.05$ inches ◄

(*b*) Given: $\delta = 4.91\sqrt{\nu x/U} = 4.91\sqrt{0.000\,0141 \times 0.588/12} = 0.004\,08$ feet or 0.0490 inches ◄

Sec. 8.10: Velocity Profile in Turbulent Flow -- ***Exercises***

8.10.1 *In a 110-cm-diameter pipe velocities are measured as 4.75 m/s on the centerline and 4.36 m/s at r = 8 cm. Approximately what is the flow rate?*

SI

Eq. 8.29: $4.36 = 4.75 - 5.75u_*\log(55/47)$ from which $u_* = 0.994$ m/s

Eq. 8.31: $V = 4.75 - 3.75(0.994) = 1.024$ m/s

$Q = AV = (\pi/4)1.10^2(1.024) = 0.973$ m³/s ◄

8.10.2 *Oil (s = 0.91) with a viscosity of 0.0007 lb·s/ft² flows at a rate of 9 cfs through a 5-in-diameter pipe having e = 0.047 in. Find the head loss. Determine the shear stress at the pipe wall and the velocity at 1.5 in from the centerline.*

BG

Eq. 4.7: $9/(\pi/4)(5/12)^2 = 66.0$ fps; Eq. 8.1: $\mathbf{R} = DV\rho/\mu = (5/12)(66)(1.94 \times 0.91)/0.0007 = 69{,}360$

$e/D = 0.047/5 = 0.0094$; from Fig. 8.11 or Eq. 8.40 or 8.41: $f = 0.0380$

Eq. 8.10: $h_L = 0.0380(12/5)66.0^2/[(2)32.2] = 6.17$ feet per foot ◄

Eq. 8.14: $\tau_0 = f\rho V^2/8 = 0.0380(0.91 \times 1.940)66.0^2/8 = 36.5$ psf ◄

From Eq. 8.32: $u_{max} = V(1 + 1.326\sqrt{f}) = 66.0(1 + 1.326\sqrt{0.0380}) = 83.1$ fps

Eq. 8.29: $u = u_{max} - 5.75\sqrt{\dfrac{\tau_0}{\rho}}\log\dfrac{r_0}{r_0 - r} = 83.1 - 5.76\sqrt{\dfrac{36.5}{0.91(1.940)}}\log\dfrac{2.5}{2.5-1.5} = 72.6$ fps ◄

8.10.3 *For turbulent flow in a circular pipe, find r/r_0 at the radial distance from the centerline where the mean velocity occurs.*

N

Turbulent flow. Eq. 8.33 for $u = V$: $V = (1 + 1.326f^{1/2})V - 2.04f^{1/2}V\log[r_0/(r_0 - r)]$

$0 = 1.326f^{1/2}V - 2.04f^{1/2}V\log\dfrac{r_0}{r_0 - r}$; $\log\dfrac{r_0}{r_0 - r} = \dfrac{1.326}{2.04} = 0.65$; $\dfrac{r_0}{r_0 - r} = 4.47$; $\dfrac{r}{r_0} = 0.776$ ◄

8.12 *Water at 130°F flows in an 0.5-in-diameter copper tube at 1.2 gpm. Find the head loss per 100 ft. What is the centerline velocity, and what is the value of δ_l?*

BG

Q = 1.2 gpm = 0.002 67 cfs; Eq. 4.7: $V = Q/A = 0.002\,67/[(\pi/4)(0.5/12)^2] = 1.961$ fps

Table A.1 for water at 130°F: $\nu = 5.58 \times 10^{-6}$ ft²/sec

Eq. 8.1: $\mathbf{R} = (0.5/12)1.963/(5.58 \times 10^{-6}) = 14{,}640$ /cont...

Table 8.1 for drawn copper tubing: e = 0.000 005 feet; $e/D = 0.000\,005/(0.5/12) = 0.000\,12$

So from Fig. 8.11: $f = 0.018$; Eq. 8.10: $h_L = 0.018(100 \times 12/0.5)1.961^2/(2g) = 2.58$ feet ◄

Eq. 8.32: $1.961/u_{max} = 1/[1 + 1.326(0.018)^{1/2}]$; $u_{max} = 0.849(1.961) = 1.665$ fps ◄

Eq. 8.27: $\delta_l = 14.14(5.58 \times 10^{-6})/[1.961(0.018)^{1/2})] = 0.300 \times 10^{-3}$ ft ◄

8.13 *Water at 60°C flows in a 1.5-cm-diameter copper tube at 0.06 L/s. Find the head loss per 10 m. What is the centerline velocity, and what is the value of δ_l?*

SI

Table A.1: At 60°C, $\nu = 0.474 \times 10^{-6}$ m²/s

Eq. 4.7: $V = Q/A = (0.06 \times 10^{-3})/[(\pi/4)(0.015)^2] = 0.340$ m/s

Eq. 8.1: $\mathbf{R} = VD/\nu = [0.340(0.015)]/(0.474 \times 10^{-6}) = 10\,740$

Table 8.1 for drawn copper tubing: e = 0.0015 mm; $e/D = 0.0015/15 = 0.0001$

So from Fig. 8.11: $f = 0.031$; Eq. 8.10: $h_L = 0.031(10/0.015)(0.340)^2/[2(9.81)] = 0.1214$ m ◄

Eq. 8.32: $0.340/u_{max} = 1/[1 + 1.326(0.031)^{1/2}]$; $u_{max} = 0.419$ m/s ◄

Eq. 8.27: $\delta_l = 14.14(0.474 \times 10^{-6})/[0.340(0.031)^{1/2}] = 0.000\,1121$ m ◄

8.14 *Repeat Prob. 8.12 for flow rates of 0.08 and 18 gpm.*
Prob. 8.12: Water at 130°F flows in an 0.5-in-diameter copper tube at 1.2 gpm. Find the head loss per 100 ft. What is the centerline velocity, and what is the value of δ_l?

BG

cfs = gpm/449. Table A.1 for water at 130°F: $\nu = 5.58 \times 10^{-6}$ ft²/sec

Table 8.1 for drawn copper tubing: e = 0.000 005 ft

Q (given)	0.08 gpm	18 gpm
Q	0.000 1782 cfs	0.0401 cfs
$V = Q/A$	0.1307 fps	29.4 fps
e/D	0.000 12	0.000 12
$\mathbf{R} = VD/\nu$	976	2.20×10^5
Flow regime	laminar	turbulent
f	0.0655 (Eq. 8.20)	0.0163 (Fig. 8.11 or Eq. 8.40* or 8.41)
h_L (Eq. 8.10)	0.0417 ft ◄	525 ft ◄
u_{max} (from Eq. 8.32)	0.1751 fps ◄	34.4 fps ◄
δ_l (Eq. 8.27)	2.36×10^{-3} ft ◄	2.10×10^{-5} ft ◄

*By root solver on programmable scientific calculator.

8.15 *Repeat Prob. 8.13 for flow rates of 0.004 and 0.9 L/s.*

Prob. 8.13: Water at 60°C flows in a 1.5-cm-diameter copper tube at 0.06 L/s. Find the head loss per 10 m. What is the centerline velocity, and what is the value of δ_l?

SI

Table A.1: At 60°C, $\nu = 0.474 \times 10^{-6}$ m²/s

Table 8.1 for drawn copper tubing: $e = 0.0015$ mm

Q (given)	0.004 L/s	0.9 L/s
Q	0.000 004 m³/s	0.0009 m³/s
$V = Q/A$	0.0226 m/s	5.09 m/s
e/D	0.0001	0.0001
$\mathbf{R} = VD/\nu$	716	161 200
Flow regime	Laminar	Turbulent
f	0.089 (Eq. 8.20)	0.0170 (Fig. 8.11 or Eq. 8.40* or 8.41)
h_L (Eq. 8.10)	0.001 549 m ◀	14.98 m ◀
u_{max} (from Eq. 8.32)	0.0316 m/s ◀	5.97 m/s ◀
δ_l (Eq. 8.27)	0.000 993 m ◀	1.009×10^{-5} m ◀

*By root solver on programmable scientific calculator.

8.16 *Oil (s = 0.85) with a viscosity of 0.0056 N·s/m² flows at a rate of 80 L/s in a 15-cm-diameter pipe having e = 0.90 mm. Find the head loss. Determine the shear stress at the pipe wall. Find the velocity 2.5 cm from the centerline. Under these conditions is the pipe behaving as a fully rough, transitional, or smooth pipe?*

SI

$V = Q/A = 0.08/[(\pi/4)0.15^2] = 4.53$ m/s; $\rho = 0.85(1000) = 850$ kg/m³

Eq. 8.1: $\mathbf{R} = DV\rho/\mu = (0.15\text{ m})(4.53\text{ m/s})(850\text{ kg/m}^3)/0.0056\text{ N·s/m}^2) = 103\,100$

$e/D = 0.9/150 = 0.006$; $f = 0.0329$ (from Fig. 8.11)

From Eq. 8.10: $h_L/L = (f/D)(V^2/2g) = (0.0329/0.15)(4.53^2/2(9.81) = 0.229$ m/m ◀

Eq. 8.14: $\tau_0 = f\rho V^2/8 = 0.0329(850)4.53^2/8 = 71.6$ N/m² ◀

Eq. 8.33: $u = (1 + 1.326\sqrt{0.0329})4.53 - 2.04\sqrt{0.0329}(4.53)\log[7.5/(7.5 - 2.5)] = 5.32$ m/s ◀

Eq. 8.27: $\delta_l = 14.14/(Vf^{1/2}) = 14.14\mu/(\rho Vf^{1/2}) = 14.14(0.0056)/[850(4.53)0.0329^{1/2}] = 0.000\,1134$ m

$= 0.1134$ mm

So $e/14 = 0.0643$ mm $< \delta_l < e$. Sec. 8.9: pipe is behaving as transitionally rough ◀

8.17 *In a 42-in-diameter pipe velocities are measured as 17.0 fps at r = 0 and 16.5 fps at r = 3.5 in. Approximately what is the flow rate?*

BG

Eq. 8.29: $16.5 = 17 - 5.76u_*\log[21/(21 - 3.5]$

from which $u_* = 1.096 = (fV^2/8)^{1/2}$ (from Eq. 8.26). Thus $f = 9.61/V^2$ (1)

Eq. 8.32: $V/17 = 1/(1 + 1.326f^{1/2})$ (2)

Solving (1) and (2) simultaneously yields: $f = 0.0579$, $V = 12.89$ fps

$Q = AV = (\pi/4)(42/12)^2 12.89 = 124.0$ cfs ◀

8.18 *The flow rate in a 10-in-diameter pipe is 7 cfs. The flow is known to be turbulent, and the centerline velocity is *15.2 fps. Plot the velocity profile, and determine the head loss per foot of pipe. [*Note this correction to the textbook.]*

BG

$V = Q/A = 7/[\pi(10/12)^2/4] = 12.83$ fps

Eq. 8.32: $V/u_{max} = 12.83/15.2 = 1/(1 + 1.326f^{1/2})$; $f = 0.019\,32$

Eq. 8.10: $h_L = 0.019\,32(12/10)12.83^2/(2 \times 32.2) = 0.0593$ feet per foot ◄

From Eq. 8.26: $u_* = V(f/8)^{1/2} = 12.83(0.019\,32/8)^{1/2} = 0.631$ fps

Eq. 8.29: $u = 15.2 - 5.76(0.631)\log[r_0/(r_0 - r)] = 15.2 - 3.63\log[r_0/(r_0 - r)]$

r (in)	u (fps)
0	15.20
2	14.39
3	13.75
4	12.66
4.5	11.57
4.9	9.03
4.99	5.39

▲

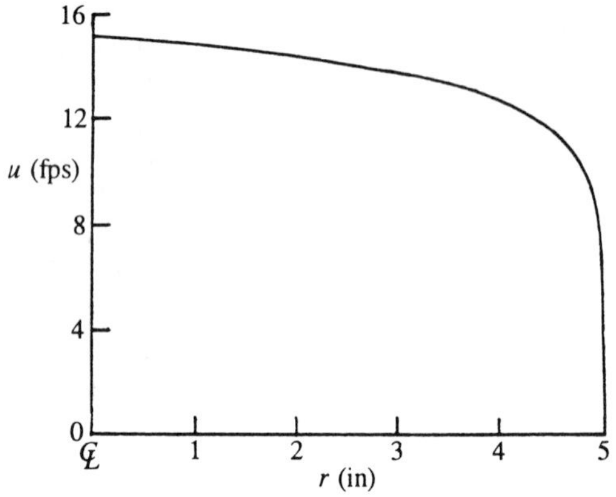

8.19 *The flow rate in a 25-cm-diameter pipe is 0.20 m³/s. The flow is know to be turbulent, and the centerline velocity is 4.75 m/s. Plot the velocity profile, and determine the head loss per meter of pipe.*

SI

$V = Q/A = 0.20/(\pi 0.25^2/4^2) = 4.07$ m/s

Eq. 8.32: $V/u_{max} = 4.07/4.75 = 1/(1 + 1.326f^{1/2})$; $f = 0.015\,64$

Eq. 8.10: $h_L = 0.015\,64(1/0.25)4.07^2/(2 \times 9.81) = 0.0529$ m per m ◄

From Eq. 8.26: $u_* = V(f/8)^{1/2} = 4.07(0.015\,64/8)^{1/2} = 0.1801$ m/s

Eq. 8.29: $u = 4.75 - 5.76(0.1801)\log[r_0/(r_0 - r)] = 4.75 - 1.038\log[r_0/(r_0 - r)]$

r (cm)	u (m/s)
0	4.75
5	4.52
8	4.29
10	4.02
12	3.30
12.25	2.99
12.49	1.537

▲

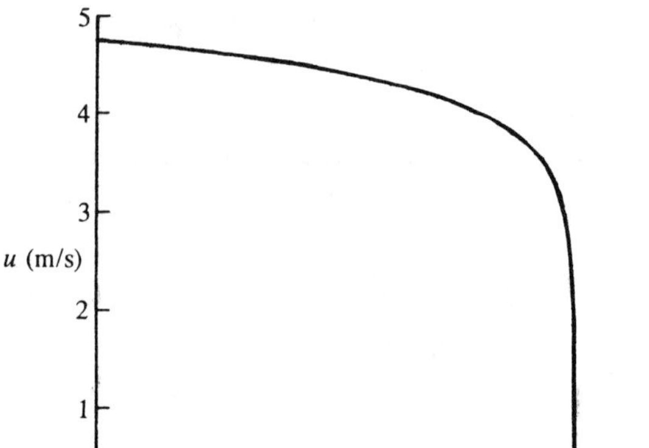

Sec. 8.11: Pipe Roughness -- ***Exercises***

8.11.1 *Using implicit Eq. (8.35), approximate Eq. (8.36), and Blasius' Eq. (8.37), solve for the smooth-pipe friction factor f using Reynolds numbers of (a) 4000, (b) 20,000, and (c) 10^5. (d) For which of these three values do the equations show the most variation in f?*

N

Part: **R**	(*a*) 4000	(*b*) 20,000	(c) 10^5
f values, computed from:			
Prandtl's Eq. 8.35:	0.0399	0.0259	0.017 99
Approx Eq. 8.36:	0.0404	0.0257	0.017 82
Blasius Eq. 8.37:	0.0397	0.0266	0.017 77
	▲	▲	▲
Average of 3:	0.0400	0.0261	0.017 86
Absolute variation	0.000 688	0.000 824	0.000 220
% variation	1.719	3.16 ◀	1.231

8.11.2 *Substitute into Eq. (8.40) the given and computed data of Sample Problem 8.5a. How well does the right-hand side of the equation agree with the left-hand side?*

Sample Prob. 8.5a: Water at 20°C flows in a 500-mm-diameter welded steel pipe, with a friction loss gradient of 0.006. e/D = 0.000 092, by trial and error **R** *= 1.059 × 10^6, and f = 0.0131.*

SI

$$\text{Eq. 8.40: Left side} = \frac{1}{0.0131^{1/2}}; \quad \text{Right side} = -2\log\left[\frac{0.000\,092}{3.7} + \frac{2.51}{(1.059 \times 10^6)0.0131^{1/2}}\right]$$

$$= 1/0.1145 \qquad\qquad = -2\log(0.000\,0248 + 0.000\,0207) = -2\log(0.000\,0455)$$

$$= 8.74 \;◀ \qquad\qquad = -2(-4.34) = 8.68 \;◀$$

Checks closely. The right side is 0.625% smaller than the left side. ◀

8.11.3 *Repeat Exer. 8.11.2 using Eq. (8.41).*

Exer. 8.11.2: Substitute in the given and computed data of Sample Prob. 8.5a. How well does the right side of the equation agree with the left side?

Sample Prob. 8.5a: Water at 20°C flows in a 500-mm-diameter welded steel pipe, with a friction loss gradient of 0.006. e/D = 0.000 092, by trial and error **R** *= 1.059 × 10^6, and f = 0.0131.*

SI

Left side Eq. 8.41: $1/(0.0131)^{1/2} = 1/0.1145 = 8.74$ ◀

Right side Eq. 8.41: $-1.8\log[(e/3.7D)^{1.11} + 6.9/\mathbf{R}] = -1.8\log[(0.000\,092/3.7)^{1.11} + 6.9/(1.059 \times 10^6)]$

$= -1.8\log[(0.000\,0248)^{1.11} + 0.000\,006\,52] = -1.8\log[0.000\,007\,75 + 0.000\,006\,52]$

$= -1.8\log(0.000\,014\,26) = -1.8(-4.85) = 8.72$ ◀

Checks closely. The right side is 0.167% smaller than the left side. ◀

8.20 *If the diameter of a pipe is doubled, what effect does this have on the flow rate for a given head loss? Consider (a) laminar flow; (b) turbulent flow.*

N

(*a*) Laminar flow, Eq. 8.22: $h_L = 32\nu LV/(gD^2) = \text{const.} \times V/D^2$

$V = kD^2$; $Q = AV = (\pi D^2/4)(kD^2) = k'D^4$

Thus doubling the diameter will increase the flow by a factor of $2^4 = 16$ ◄

(*b*) Turbulent flow, Eq. 8.13: $h_L = f(L/D)V^2/(2g)$

(*i*) If f = constant (complete turbulence), $h_L = k_1V^2/D$; $V = k_2D^{1/2}$ and $Q = AV = k_3D^{5/2}$

Thus doubling the diameter will increase the flow rate by a factor of $2^{5/2} = 5.66$ for turbulent flow with complete turbulence. ◄

(*ii*) For smooth pipe with $\mathbf{R} < 100{,}000$, Blasius Eq. 8.37: $f = 0.316/\mathbf{R}^{1/4} = k/(DV)^{1/4}$

$h_L = f(L/D)V^2/(2g) = k_1V^{7/4}/D^{5/4}$

$V = k_2D^{5/7}$ and $Q = AV = k_3D^{19/7}$

Thus doubling the diameter will increase the flow rate by a factor of $2^{19/7} = 6.56$ for turbulent flow in smooth pipes. ◄

8.22 *Refer to the data of Prob. 8.16. Above what flow rate will this pipe behave as a fully rough pipe? Below what flow rate will it behave as a smooth pipe?*

Prob. 8.16: Oil (s = 0.85) with a viscosity of 0.0056 N·s/m² flows at a rate of 80 L/s in a 15-cm-diameter pipe having e = 0.90 mm. V = 4.53 m/s, ρ = 850 kg/m³, $\mathbf{R}$ *=103 100,* $\delta_l = 0.1133$ *mm.*

SI

(*a*) From Sec. 8.9: Fully rough when $\delta_l < e/14$

In Eq. 8.27 assume f = constant; $\delta_l \propto 1/V$

So $\delta_l = 0.9/14 = 0.0643$ mm when $V = V_{\text{prev}}[(\delta_l)_{\text{prev}}/\delta_l] = 4.53(0.1133/0.0643) = 7.98$ m/s

and when $Q = 80(0.1133/0.0643) = 141$ L/s ◄

(*b*) From Fig. 8.9: Smooth when $\delta_l > e$. f varies rapidly with $\mathbf{R}$ for smooth pipe (Fig. 8.11).

Eqs. 8.27 and 8.36 to eliminate f:

$\delta_l = 14.14\nu/(V\sqrt{f}) = [14.14\mu/(\rho V)]1.8\log(\mathbf{R}/6.9) = [14.14(0.0056)1.8/(850V)]\log(\mathbf{R}/6.9)$ m

$= [0.1677/V]\log(\mathbf{R}/6.9)$ mm

where, from Prob. 8.16, $\mathbf{R} = 103\,100(V/4.53) = 22\,759V$,

so that $\delta_l = (0.1677/V)\log(22\,759V/6.9) = (0.1677/V)\log(3298V)$ mm

Solve by trials (or using an equation solver).

Try V m/s	$\mathbf{R}$	δ_e mm	*cf.* e mm
1	22 759	0.590	$< e = 0.9$
0.5	11 380	1.079	$> e$
0.7	15 932	0.806	$< e$
0.65	14 794	0.859	$< e$
0.616	14 020	0.9005	equal!

Then $Q = AV = (\pi/4)(0.15)^2 0.616 = 0.010\,88$ m³/s $= 10.88$ L/s ◄

8.23 *Kerosene (s = *0.813) at a temperature of 70°F flows in a 3-in-diameter smooth brass pipeline at a rate of 15 gpm. (a) Find the head loss per foot. (b) For the same head loss what would be the flow rate if the temperature of the kerosene were raised to 100°F? [*Note this correction to the textbook.]*

BG

15 gpm = 0.0334 cfs; $V = Q/A = 0.0334/[\pi(3/12)^2/4] = 0.681$ fps

(*a*) Fig. 2.4 for kerosene at 70°F: $\nu = 2.4 \times 10^{-5}$ ft²/sec.

Eq. 8.1: $\mathbf{R} = (3/12)0.681/(2.4 \times 10^{-5}) = 7090$

Eq. 8.37 for smooth pipe: $f = 0.316/7090^{0.25} = 0.0344$

Eq. 8.10: $h_L = 0.0344(12/3)0.681^2/[2(32.2)] = 0.000\,991$ ft per ft ◄

(*b*) Same h_L, raised temperature. Fig. 2.4 for kerosene at 100°F: $\nu = 1.8 \times 10^{-5}$ ft²/sec.

Substitute Eq. 8.37 into Eq. 8.10: $h_L = f\left(\frac{L}{D}\right)\left(\frac{V^2}{2g}\right) = \frac{0.316\nu^{0.25}}{D^{0.25}V^{0.25}}\left(\frac{L}{D}\right)\left(\frac{V^2}{2g}\right)$

i.e., $\frac{h_L}{L} = \frac{0.316\nu^{0.25}}{D^{1.25}2g}V^{1.75}$; $0.000\,991 = \frac{0.316(1.8 \times 10^{-5})^{0.25}V^{1.75}}{(3/12)^{1.25}2(32.2)}$; $V = 0.709$ fps

$Q = AV = (\pi/4)(3/12)^2 0.709 = 0.0348$ cfs = 15.63 gpm ◄

8.24 *Kerosene (s = 0.813) at a temperature of 20°C flows in a 7.5-cm-diameter smooth brass pipeline at a rate of 0.80 L/s. (a) Find the head loss per meter. (b) For the same head loss what would be the flow rate if the temperature of the kerosene were raised to 40°C?*

SI

$V = Q/A = 0.0008/(\pi 0.075^2/4) = 0.1811$ m/s;

(*a*) Fig. 2.3 for kerosene at 20°C: $\mu = 1.65 \times 10^{-3}$ N·s/m²

Eq. 8.1: $\mathbf{R} = 0.075(0.1811)(0.813 \times 1000)/(1.65 \times 10^{-3}) = 6690$

Eq. 8.37 for smooth pipe: $f = 0.316/(6690)^{0.25} = 0.0349$

Eq. 8.10: $h_L = 0.0349(1/0.075)0.1811^2/[(2)9.81] = 0.000\,779$ m/m ◄

(*b*) Same h_L, raised T°. Fig. 2.4 for kerosene at 40°C: $\nu = 1.7 \times 10^{-6}$ m²/s

Substitute Eq. 8.37 into Eq. 8.10 to get: $h_L = f\left(\frac{L}{D}\right)\left(\frac{V^2}{2g}\right) = \frac{0.316\nu^{0.25}}{D^{0.25}V^{0.25}}\left(\frac{L}{D}\right)\left(\frac{V^2}{2g}\right)$

$\frac{h_L}{L} = \frac{0.316\nu^{0.25}}{D^{1.25}2g}V^{1.75}$; i.e., $0.000\,779 = \frac{0.316(1.7\times10^{-6})^{0.25}V^{1.75}}{(0.075)^{1.25}(2)9.81}$; $V = 0.1857$ m/s

and $Q = 0.1857(\pi/4)(0.075)^2 = 0.000\,821$ m³/s ◄

/more...

8.25 *Water at 50°C flows in a 15-cm-diameter pipe with V = 7.5 m/s. Head loss measurements indicate that f = 0.020. Determine the value of e. Find the shear stress at the pipe wall and at r = 3 cm. What is the value of du/dy at r = 3 cm?*

SI

Table A.1 for water at 50°C: $\nu = 0.553 \times 10^{-6}$ m²/s, $\rho = 988$ kg/m³

Eq. 8.1: $\mathbf{R} = 0.15(7.5)/(0.553 \times 10^{-6}) = 2.03 \times 10^6$

Eq. 8.40, for $f = 0.020$ and $\mathbf{R} = 2.03 \times 10^6$: $e/D = 0.001\,046$

$e = 0.001046(15) = 0.01569$ cm $= 0.1569$ mm ◄

Eq. 8.14: $\tau_0 = f\rho V^2/8 = 0.020(988)7.5^2/8 = 138.9$ N/m² ◄ (using kg·m/s² = N)

Eq. 8.13 at $r = 3$ cm: $\tau = (3/7.5)138.9 = 55.6$ N/m² ◄

Sec. 8.10: $du/dy = (\tau_0/\rho)^{1/2}/(Ky) = [138.9\ (\text{kg/m}^3)(\text{m/s}^2)/(988\ \text{kg/m}^3)]^{1/2}/[0.40(0.075 - 0.03)]$

$= 20.8\ \text{sec}^{-1}$ ◄

Sec. 8.12: Chart for Friction Factor -- *Exercises*

8.12.1 *Water at 20°C flows through a 15-cm-diameter pipe with e = 0.015 mm. (a) If the mean velocity is 5 m/s, what is the nominal thickness δ_l of the viscous sublayer? (b) What will δ_l be if the velocity is increased to 6.2 m/s?*

SI

Table A.1 at 20°C: $\nu = 1.003 \times 10^{-6}$ m²/s

(*a*) Eq. 8.1: $\mathbf{R} = VD/\nu = 5(0.15)/(1.003 \times 10^{-6}) = 7.48 \times 10^5$

$e/D = 0.015/150 = 0.0001$; from Fig. 8.11 (or Eq. 8.40 or 8.41): $f = 0.0138$

Eq. 8.27: $$\delta_l = \frac{14.14\nu}{Vf^{1/2}} = \frac{14.14(1.003 \times 10^{-6})}{5(0.0138)^{1/2}} = 24.1 \times 10^{-6}\ \text{m} = 0.0241\ \text{mm}$$ ◄

(*b*) Eq. 8.1: $\mathbf{R} = VD/\nu = 6.2(0.15)/(1.003\times10^{-6}) = 9.27\times10^5$

$e/D = 0.0001$ as before ; from Fig 8.11 (or Eq. 8.40 or 8.41): $f = 0.0135$

Eq. 8.27: $$\delta_l = \frac{14.14(1.003\times10^{-6})}{6.2(0.0135)^{1/2}} = 19.67\times10^{-6} = 0.019\,67\ \text{mm}$$ ◄

8.12.2 *A straight, new, 48-in-diameter asphalted cast-iron pipe 700 ft long carries 78°F water at an average velocity of 12 fps. (a) Using the value of f as determined from Fig. 8.11, find the shear force on the pipe. (b) What will be the shear force if the average velocity is reduced to 4.2 fps?*

BG

Table 8.1: $e = 0.0004$ ft; $e/D = 0.004(12)/48 = 0.0001$

Interpolating in Table A.1: $\nu = 0.956 \times 10^{-5}$ ft²/sec

(*a*) Eq. 8.1: $\mathbf{R} = DV/\nu = (48/12)12/(0.956 \times 10^{-5}) = 5.02 \times 10^6$; Fig. 8.11: $f = 0.0123$

Eq. 8.17: $\tau_0 = f\rho V^2/8 = 0.0123(1.940)12^2/8 = 0.430$ lb/ft²

Shear force $= \pi D(L)\tau_0 = \pi(48/12)700(0.431) = 3780$ lb ◄

(*b*) If $V = 4.2$ fps, $\mathbf{R} = 1.76 \times 10^6$, $f = 0.0129$, $\tau_0 = 0.0552$ lb/ft²

Shear force $= \pi D(L)\tau_0 = \pi(48/12)700(0.0522) = 485$ lb ◄

8.26 *Water at 40°F flows in a 42-in-diameter concrete pipe (e = 0.022 in). Determine **R**, τ_0, u_{max}/V, δ_l, δ_l/e, and the flow regime (hydraulic smoothness) for flow rates of 250, 25, 2.5, 0.25, and 0.025 cfs.*

BG

E.g., for Q = 250 cfs: $V = Q/A = 250/[(\pi/4)(42/12)^2] = 26.0$ fps

Table A.1 for water at 40°F: $\nu = 1.664 \times 10^{-5}$ ft²/sec

Eq. 8.1: $\mathbf{R} = DV/\nu = (42/12)26.0/(1.664 \times 10^{-5}) = 5.47 \times 10^6$; $e/D = 0.022/42 = 0.000\,524$

Fig. 8.11: $f = 0.0170$; Eq. 8.17: $\tau_0 = f\rho V^2/8 = 0.0170(1.940)26.0^2/8 = 2.78$ psf

Determine u_{max}/V from Eq. 8.32 and δ_l from Eq. 8.27. For the other discharges:

Q cfs	V fps	**R**	f	τ_0 psf	u_{max}/V	δ_l ft	δ_l/e	Regime
250	26.0	5.47E6	0.0170	2.78	1.173	0.000 0694	0.0379	Rough
25	2.60	5.47E5	0.0177	0.0290	1.176	0.000 522	0.285	Trans
2.5	0.260	54,700	0.0223	3.54E−4	1.198	0.006 06	3.31	Smooth
0.25	0.0260	5470	0.0371	6.07E−6	1.255	0.0470	25.6	Smooth
0.025	0.00260	547	0.0787	1.29E−7	1.372	0.323	176	Laminar
		▲		▲	▲	▲	▲	▲

8.27 *For **R** ranging from 10^2 to 10^7, make a plot of the values of α and β versus **R** for smooth brass pipes. On the same plot also show values of u_{max}/V.*

N

Flow	f from	α from	β from	u_{max}/V from
Laminar	Eq. 8.20	Sec. 5.1	Sec. 6.3	Sec. 8.6
Turbulent	Fig. 8.11	Eq. 8.34*a*	Eq. 8.34*b*	Eq. 8.32

R	Flow	f	α	β	u_{max}/V
10^2	lam.	0.640	2.0	1.333	2.0
10^3	lam.	0.0640	2.0	1.333	2.0
10^4	turb.	0.0307	1.083	1.030	1.233
10^5	turb.	0.0178	1.048	1.017	1.177
10^6	turb.	0.0116	1.031	1.011	1.143
10^7	turb.	0.0081	1.022	1.008	1.120
▲			▲	▲	▲

8.28 *When water at 50°F flows at 2.5 cfs in a 20-in pipeline, the head loss is 0.0004 ft/ft. What will be the head loss when glycerin at 68°F flows through this same pipe at the same rate?*

BG

Table A.1 for water at 50°F: $\nu = 1.410 \times 10^{-5}$ ft²/sec

$V = Q/A = 2.5/[(\pi/4)(20/12)^2] = 1.146$ fps; $\mathbf{R} = (20/12)1.146/(1.410 \times 10^{-5}) = 1.355 \times 10^5$

Eq. 8.10: $0.0004 = f(12/20)1.146^2/2g$; $f = 0.0327$; Thus from Fig. 8.11: $e/D = 0.0060$

Table A.4 for glycerin at 68°F: $s = 1.26$, $\mu = 0.0312$ lb·sec/ft²

$\nu = 0.0312/(1.26 \times 1.940) = 0.012\,76$ ft²/sec; $\mathbf{R} = (20/12)1.146/0.012\,76 = 149.6$; flow is laminar.

Eq. 8.20: $f = 64/149.6 = 0.428$

Eq. 8.10: $h_L = 0.428(12/20)1.146^2/[2(32.2)] = 0.005\,23$ ft per ft ◄

8.29 *Oil (s = 0.92) with viscosity 1.8 × 10⁻⁴ lb·sec/ft² flows in a 3-in-diameter welded-steel pipe at 0.15 cfs. What is the head loss per foot of pipe?*

BG

$V = Q/A = 0.15/[\pi(3/12)^2/4] = 3.06$ fps; Table 8.1: $e = 0.000\,15$ ft

Eq. 8.1: $\mathbf{R} = (3/12)3.06(1.940 \times 0.92)/(1.8 \times 10^{-4}) = 7575$; $e/D = 0.000\,15/(3/12) = 0.0006$

Fig. 8.11: $f = 0.034$; Eq. 8.10: $h_L = 0.034(12/3)3.06^2/(2 \times 32.2) = 0.019\,72$ ft per ft ◄

8.30 *Oil (s = 0.91) with viscosity 0.0102 N·s/m² flows in a 7.5-cm-diameter welded-steel pipe at 4.9 L/s. What is the head loss per meter of pipe?*

SI

$V = Q/A = 0.0049/[(\pi/4)(0.075)^2] = 1.109$ m/s; Table 8.1: $e = 0.046$ mm

Eq. 8.1: $\mathbf{R} = 0.075(1.109)(1000 \times 0.91)/0.0102 = 7421$; $e/D = (0.046/1000)/0.075 = 0.000\,613$

Fig. 8.11: $f = 0.034$; Eq. 8.10: $h_L = 0.034(1/0.075)1.109^2/(2 \times 9.81) = 0.0284$ m/m ◄

8.31 *Crude oil (s = 0.855) at 50°C flows at 0.30 m³/s through a 45-cm-diameter pipe (e = 0.054mm) 1500 m long. Find the kilowatt loss.*

SI

Fig. 2.4 for crude oil ($s = 0.855$) at 50°C: $\nu = 3.8 \times 10^{-6}$ m²/s

$V = 0.30/(\pi 0.225^2) = 1.886$ m/s; Eq. 8.1: $\mathbf{R} = DV/\nu = 0.45(1.886)/(3.8 \times 10^{-6}) = 2.23 \times 10^5$

$e/D = 0.054/450 = 0.000\,12$; Fig. 8.11: $f = 0.0163$

Eq. 8.10: $h_L = f(L/D)(V^2/2g) = 0.0163(1500/0.45)1.886^2/(2 \times 9.81) = 9.85$ m

Eq. 5.33: Power loss $= \gamma Q h_L = (0.885 \times 9807)0.30(9.85)/1000 = 25.7$ kW ◄

8.32 *Air at 90 psia and 70°F flows through a 4.5-in-diameter welded-steel pipe at 60 lb/min. Find the head loss and pressure drop in 200 ft of this pipe. Assume the air to be of constant density.*

BG

Table A.5 for air: $R = 1715$ ft²/(s²·°R)

Eq. 2.5: $\gamma = \rho g = pg/RT = 90(144)32.2/[1715(460 + 70)] = 0.459$ lb/ft³

Eq. 4.5: $G = 60/60 = 1.00$ lb/sec $= \gamma A V = 0.459(\pi/4)(4.5/12)^2 V$; $V = 19.72$ ft/sec;

End of Sec. 2.11: ν depends strongly on density so we must look up μ.

Table A.2 for air at 70°F: $\mu = 0.382 \times 10^{-6}$ lb·sec/ft²

Eq. 8.1: $$\mathbf{R} = \frac{DV\rho}{\mu} = \frac{DQ\gamma}{(\pi D^2/4)g\mu} = \frac{4G}{\pi D g \mu} = \frac{4(1)}{\pi(4.5/12)32.2(0.382 \times 10^{-6})} = 276{,}000$$

Table 8.1 for welded steel: $e = 0.000\,15$ ft; $e/D = 0.000\,15/(4.5/12) = 0.000\,400$

Fig. 8.11: $f = 0.0177$; Eq. 8.10: $h_L = 0.0177(200 \times 12/4.5)19.72^2/[2(32.2)] = 57.0$ feet of air ◄

Per Eq. 3.4: $\Delta p = h_L(\gamma) = 57.0(0.459) = 26.2$ lb/ft² $= 0.1818$ psi ◄

8.33 *Air flows at an average velocity of 0.7 m/s through a long 3.8-m-diameter circular tunnel (e = 1.5 mm). Find the head-loss gradient at a point where the air temperature and pressure are 20°C and 102 kN/m² abs respectively. Find also the shear stress at the pipe wall and the thickness δ_l of the viscous sublayer.*

SI

Table A.2 for air at 20°C: $\mu = 18.1 \times 10^{-6}$ N·s/m², $\gamma = 11.82$ N/m³

Eq. 8.1: $\mathbf{R} = DV\rho/\mu = DV\gamma/(\mu g) = 3.8(0.7)11.82/(18.1 \times 10^{-6} \times 9.81) = 177\,100$

$e/D = 1.5/3800 = 0.000\,395$; Fig. 8.11: $f = 0.0185$

From Eq. 8.10: $h_L/L = (0.0185/3.8)0.7^2/[2(9.81)] = 0.000\,1216$ m/m ◄

Eq. 8.14: $\tau_0 = f\rho V^2/8 = 0.0182(11.82/9.81)0.7^2/8 = 0.001\,365$ N/m² ◄

Eq. 8.27: $\delta_l = \dfrac{14.14\nu}{Vf^{1/2}} = \dfrac{14.14\mu g}{Vf^{1/2}\gamma} = \dfrac{14.14(18.1 \times 10^{-6})9.81}{0.7(0.0185)^{1/2}11.82} = 0.002\,23 \text{ m} = 2.23 \text{ mm}$ ◄

8.34 *Repeat Prob. 8.33 for the case where the average velocity is 7.0 m/s.*

Prob. 8.33: Air flows through a long 3.8-m-diameter circular tunnel (e = 1.5 mm). Find the head-loss gradient at a point where the air temperature and pressure are 20°C and 102 kN/m² abs respectively. Find also the shear stress at the pipe wall and the thickness δ_l of the viscous sublayer.

SI

Table A.2 for air at 20°C: $\gamma = 11.82$ N/m³, $\mu = 18.1 \times 10^{-6}$ N·s/m²,

Eq. 8.1: $\mathbf{R} = DV\rho/\mu = DV\gamma/(\mu g) = 3.8(7.0)11.82/[18.1 \times 10^{-6}(9.81)] = 1\,771\,000$

$e/D = 1.5/3800 = 0.000\,395$; Fig. 8.11: $f = 0.0161$

From Eq. 8.10: $h_L/L = (0.0161/3.8)7.0^2/[2(9.81)] = 0.010\,58$ m/m ◄

Eq. 8.14: $\tau_0 = 0.0161(11.82/9.81)7.0^2/8 = 0.1188$ N/m² ◄

Eq. 8.27: $\delta_l = \dfrac{14.14\nu}{Vf^{1/2}} = \dfrac{14.14\mu g}{Vf^{1/2}\gamma} = \dfrac{14.14(18.1 \times 10^{-6})9.81}{7.0(0.0161)^{1/2}11.82} = 0.000\,239 \text{ m} = 0.239 \text{ mm}$ ◄

8.35 *Air at 70°F and atmospheric pressure flows with a velocity of 22 fps through a 3-in-diameter pipe (e = 0.00015 in). Find the head loss in 75 ft of pipe.*

BG

Table A.2 for air at 70°F: $\nu = 0.164 \times 10^{-3}$ ft²/sec

Eq. 8.1: $\mathbf{R} = DV/\nu = (3/12)22/(0.164 \times 10^{-3}) = 33{,}537$; $e/D = 0.000\,15/3 = 0.000\,05$

Fig. 8.11: $f = 0.0230$; Eq. 8.10: $h_L = 0.0230[75/(3/12)]22^2/[2(32.2)] = 51.9$ ft ◄

8.36 *Air at 30°C and atmospheric pressure flows with a velocity of 6.5 m/s through a 75-mm-diameter pipe (e = 0.002 mm). Find the head loss in 30 m of pipe.*

SI

Table A.2 for air at 30°C: $\nu = 16.0 \times 10^{-6}$ m²/s

Eq. 8.1: $\mathbf{R} = DV/\nu = (0.075)6.5/(16.0 \times 10^{-6}) = 3.05 \times 10^4$; $e/D = 0.002/75 = 0.000\,0267$

Fig. 8.11: $f = 0.0235$; Eq. 8.10: $h_L = 0.0235(30/0.075)6.5^2/[(2)9.81] = 20.2$ m ◄

8.13.1 *Compute the friction head per 100 ft of 3-in-diameter pipe for a Reynolds number of 50,000 if (a) the flow is laminar (achievable with great care); (b) the flow is turbulent in a smooth pipe; and (c) the flow is turbulent in a rough pipe with e/D = 0.05. Consider two situations, one where the fluid is 70°F water, the other where the fluid is SAE 10 (Western) oil at 150°F.*

BG

These are Type 1 problems, to find h_L. Table A.1 for water at 70°F: $\nu = 1.059 \times 10^{-5}$ ft²/sec

Eq. 8.1: $\mathbf{R} = (3/12)V/(1.059 \times 10^{-5}) = 50{,}000$; $V = 2.12$ fps

(*a*) For laminar flow, $f = 64/\mathbf{R}$ (Eq. 8.20).

Eq. 8.10: $h_L = (64/50{,}000)[100/(3/12)]2.12^2/[2(32.2)] = 0.0357$ ft ◄

(*b*) For turbulent flow in smooth pipe, Fig. 8.11 (smooth pipe curve): $f = 0.0209$

Eq. 8.10: $h_L = 0.0209[100/(3/12)]2.12^2/[2(32.2)] = 0.582$ ft ◄

(*c*) For turbulent flow in rough pipe ($e/D = 0.05$), Fig. 8.11: $f = 0.0720$

Eq. 8.10: $h_L = 0.0720[100/(3/12)]2.12^2/[2(32.2)] = 2.01$ ft ◄

Fig. 2.4 for SAE 10 (Western) oil at 150°F: $\nu = 0.000\,16$ ft²/sec

Eq. 8.1: $\mathbf{R} = (3/12)V/0.000\,16 = 50{,}000$; $V = 32.0$ fps

Friction factors are the same as before. Eq. 8.10:

(*a*) $h_L = 8.15$ ft ◄ (*b*) 133 ft ◄ (*c*) 459 ft ◄

8.13.2 *California crude oil, warmed until its kinematic viscosity is 0.0004 ft²/s and its specific weight is 53.45 lb/ft³, is pumped through a 3-in pipe (e = 0.001 in). (a) For laminar flow with* $\mathbf{R} = \mathbf{R}_{crit} = 2000$, *what would be the loss in energy head in pounds per square inch per 1000 ft of pipe? (b) What would be the loss in head per 1000 ft if the velocity were three times the value in (a)?*

BG

These are Type 1 problems, to find h_L.

(*a*) Eq. 8.1: $\mathbf{R}_{crit} = 2000 = (2/12)V/0.0004$; $V = 3.2$ fps

Eq. 8.20: $f = 64/2000 = 0.032$; $h_L = 0.032[1000/(2/12)]3.2^2/[2(32.2)] = 20.3$ ft per 1000 ft

Pressure loss $= \gamma h_L = 53.45(20.3)/144 = 7.55$ psi per 1000 ft ◄

(*b*) $V = 3(3.2) = 9.6$ fps; $\mathbf{R} = 6000$, $e/D = 0.001/3 = 0.000\,333$; Fig. 8.11: $f = 0.0359$

$h_L = 0.359[1000/(3/12)]9.6^2/[2(32.2)] = 205$ ft per 1000 ft ◄

8.13.3 *Water at 50°F flowing through 80 ft of 4-in-diameter average cast-iron pipe causes a head loss of 0.27 ft. Find the flow rate.*

BG

This is a Type 2 problem to find Q.

Table 8.1 for average cast-iron: $e = 0.000\,85$, $e/D = 0.00\,85/(4/12) = 0.002\,55$,

for which (Fig. 8.11) $f_{min} \approx 0.025$

Table A.1 at 50°F: $\nu = 1.410 \times 10^{-5}$ ft²/s. $h_L = 0.27$ ft in 80 ft, given.

From Eq. 8.10: $h_L = 0.27 = f[80/(4/12)]V^2/[2(32.2)]$; i.e. $V = 0.269/f^{1/2}$

Eq. 8.1: $\mathbf{R} = DV/\nu = (4/12)V/(1.410 \times 10^{-5}) = 23{,}640V$ /cont...

Try $f = 0.025 \approx f_{min}$. Then $V = 0.269/(0.025)^{1/2} = 1.703$ fps, $\mathbf{R} = 23{,}640(1.703) = 40{,}260$

Eq. 8.40 or Fig. 8.11 with $e/D = 0.002\,55$ and $\mathbf{R} = 4.03 \times 10^4$: $f = 0.0283$

This and subsequent trials:

Try f	V fps	**R**	Obtained f	
0.025	1.703	4.02×10^4	0.0283	Try again
0.0283	1.601	3.79×10^4	0.0284	Try again
0.0284	1.597	3.78×10^4	0.0284	Converged!

$Q = (\pi/4)(4/12)^2 1.597 = 0.1394$ cfs ◄

8.13.4 *When gasoline with a kinematic viscosity of 5×10^{-7} m²/s flows in a 20-cm-diameter smooth pipe the head loss is 0.43 m per 100 m. Find the flow rate.*

SI

This is a Type 2 problem to find Q.

For smooth pipe $e = 0$, for which (Fig. 8.11) initial assumptions: $f \approx 0.02$ (mid-range)

From Eq. 8.10: $h_L = 0.43 = f(100/0.2)V^2/[2(9.81)]$; i.e. $V = 0.1299/f^{1/2}$

Eq. 8.1: $\mathbf{R} = DV/\nu = 0.2V/(5 \times 10^{-7}) = 400\,000V$

Try $f = 0.02$. Then $V = 0.1299/(0.02)^{1/2} = 0.918$ m/s, $\mathbf{R} = 400\,000(0.918) = 3.67 \times 10^5$

Eq. 8.36 or Fig. 8.11 with for smooth pipe ($e = 0$) and $\mathbf{R} = 3.67 \times 10^5$: $f = 0.014$

This and subsequent trials:

Try f	V fps	**R**	Obtained f	
0.02	0.918	3.67×10^5	0.014	Try again
0.014	1.101	4.4×10^5	0.0134	Try again
0.0134	1.12	4.48×10^5	0.0134	Converged!

$Q = (\pi/4)(0.2)^2 1.12 = 0.0352$ m³/s $= 35.2$ L/s ◄

Sec. 8.13: Solution of Pipe Flow Problems by Trials -- *Problems*

8.37 *Water at 15°C flowing through 25 m of 10-cm-diameter galvanized iron pipe causes a head loss of 7.5 cm. Find the flow rate.*

SI

This is a Type 2 problem, to find Q.

Table 8.1 for galvanized iron: $e = 0.15$ mm $= 0.0150$ cm, $e/D = 0.015/10 = 0.0015$,

for which (Fig. 8.11) $f_{min} \approx 0.0215$

Eq. 8.10: $h_L = 0.075$ m $= f(25/0.10)V^2/[2(9.81)]$, i.e. $V = 0.0767/f^{1/2}$

Table A.1 for water at 15°C: $\nu = 1.139 \times 10^{-6}$ m²/s

Try f	V m/s	$\mathbf{R} = DV/\nu$	Chart f	
0.021	0.529	4.65×10^4	0.026	Try again
0.026	0.476	4.18×10^4	0.026	Converged!

$Q = (\pi/4)(0.1)^2 0.476 = 0.003\,74$ m³/s $= 3.74$ L/s ◄

8.38 *When gasoline with a kinematic viscosity of 6×10^{-6} ft²/sec flows in a 10-in-diameter smooth pipe the head loss is 0.36 ft per 100 ft. Find the flow rate.*

BG

This is a Type 2 problem to find Q.

For smooth pipe $e = 0$, for which (Fig. 8.11) initial assumptions: $f \approx 0.02$ (mid-range)

$\nu = 6 \times 10^{-6}$ ft²/sec (given); $h_L = 0.36$ ft per 100 ft (given).

From Eq. 8.10: $h_L = 0.36 = f\left(\frac{100}{10/12}\right)\frac{V^2}{2(32.2)}$; i.e. $V = \frac{0.44}{f^{1/2}}$; Eq. 8.1: $\mathbf{R} = \frac{(10/12)V}{6 \times 10^{-6}} = 138{,}900V$

Try $f = 0.02$. Then $V = 0.44/(0.02)^{1/2} = 3.11$ fps, $\mathbf{R} = 138{,}900(3.11) = 4.32 \times 10^5$

Eq. 8.36 or Fig. 8.11 with for smooth pipe ($e = 0$) and $\mathbf{R} = 4.32 \times 10^5$: $f = 0.0135$

This and subsequent trials:

Try f	V fps	$\mathbf{R}$	Obtained f	
0.02	3.11	4.32×10^5	0.0135	Try again
0.0135	3.78	5.25×10^5	0.0130	Try again
0.0130	3.85	5.35×10^5	0.0130	Converged!

$Q = (\pi/4)(10/12)^2 3.85 = 2.10$ cfs ◄

8.39 *Oil with a kinematic viscosity of 0.000 22 ft²/sec is to flow at 7.4 cfs with a head loss of 0.37 ft·lb/lb per 100 ft of pipe length. What size pipe (e = 0.000 13 ft) is required?*

BG

This is a Type 3 problem, to find D. Assume laminar flow.

Eq. 8.19: $h_L = 32\nu LV/(gD^2) = 32\nu LQ(4)/(g\pi D^4) = 32(0.000\,22)100(7.4)4/(32.2\pi D^4)$

But $h_L = 0.37$ ft·lb/lb (given); Solving for D, $D = 1.158$ ft if flow is laminar.

Check for laminar flow: Eq. 8.1: $\mathbf{R} = DV/\nu = DQ(4)/(\nu\pi D^2) = 7.4(4)/(0.000\,22\pi 1.158) = 36{,}900$

Flow is not laminar, so use Eq. 8.10: $h_L = 0.37 = f[100/(2 \times 32.2D)][7.4/(\pi D^2/4)]^2$

from which $D^5 = 373f$; $e/D = 0.000\,13/D$; $V = Q/A = 7.4/(\pi D^2/4) = 9.42/D^2$

Eq. 8.1: $\mathbf{R} = DV/\nu = D(9.42/D^2)/0.000\,22 = 42{,}800/D$. Start by assuming a mid-range value of f:

Try f	D ft	e/D	$\mathbf{R}$	Obtained f	
0.0300	1.621	0.000 0802	2.64×10^4	0.0245	Try again
0.0245	1.556	0.000 0835	2.75×10^4	0.0242	Close
0.0242	1.552	0.000 0837	2.76×10^4	0.0242	Converged

$D = 1.552$ ft $= 18.62$ in ◄

8.40 *Oil with a kinematic viscosity of 2.0×10^{-5} m²/s is to flow at 0.21 m³/s with a head loss of 0.42 N·m/N per 100 m of pipe length. What size pipe (e = 0.038 mm) is theoretically required?*

SI

This is a Type 3 problem, to find D. $e = 0.000\,038$ m (given)

Eq. 8.10: $h_L/L = 0.42/100 = [f/(2 \times 9.81D)][0.21/(\pi D^2/4)]^2$, i.e. $D^5 = 0.868f$

Try f	D m	$V = 4Q/(\pi D^2)$ m/s	$\mathbf{R} = DV/\nu$	e/D	Obtained f	
0.030	0.482	1.151	2.77×10^4	0.000 0788	0.0235	Try again
0.235	0.459	1.269	2.91×10^4	0.000 0827	0.0235	Converged

$D = 0.459$ m ◄

8.41 *Water at 150°F flows in a 350-ft-long 0.863-in-diameter iron pipe (e = 0.000 15 ft) between points A and B. At A the elevation of the pipe is 112.0 ft and the pressure is 8.35 psi. At B the elevation of the pipe is 105 ft and the pressure is 8.80 psi. Compute the flow rate as accurately as you can.*

BG

This is a Type 2 problem, to find Q.

e = 0.000 15 ft (given); e/D = 0.000 15/(0.863/12) = 0.002 09 for which (Fig. 8.11) $f_{min} \approx 0.024$.

Table A.1 at 150°F: γ = 61.2 pcf, $\nu = 0.476 \times 10^{-5}$ ft²/sec

Sec. 8.4: $h_L = (z_A + p_A/\gamma) - (z_B + p_B/\gamma)$ = [112 + 8.35(144)/61.2] − [105 + 8.8(144)/61.2] = 5.94 ft

Eq. 8.10: $h_L = 5.94 = f[350/(0.863/12)]V^2/[2(32.2)]$, i.e. $V = 0.280/f^{1/2}$

Try $f = f_{min} \approx 0.024$. Then $V = 0.280/(0.024)^{1/2}$ = 1.810 fps

Eq. 8.1: $\mathbf{R} = DV/\nu = (0.863/12)1.810/(0.476 \times 10^{-5}) = 2.73 \times 10^5$

Eq. 8.40 or Fig. 8.11 with e/D = 0.000 092 and $\mathbf{R} = 2.73 \times 10^5$: f = 0.286

This and subsequent trials:

Try f	V fps	$\mathbf{R}$	Obtained f	
0.024	1.81	2.73×10^5	0.0286	Try again
0.286	1.66	2.5×10^5	0.029	Try again
0.029	1.65	2.49×10^5	0.029	Converged!

$Q = (\pi/4)(0.863/12)^2 1.65$ = 0.006 69 cfs ◄

8.42 *Water at 60°C flows in a 100-m-long 2-cm-diameter pipe (e = 0.060 mm) between points A and B. At A the elevation of the pipe is 54.1 m and the pressure is 88.7 kPa. At B the elevation of the pipe is 52.0 m and the pressure is 91.8 kPa. Compute the flow rate as accurately as you can.*

SI

This is a Type 2 problem, to find Q.

e/D = 0.060/20 = 0.003, for which (Fig. 8.11) $f_{min} \approx 0.026$

Table A.1 for water at 60°C: $\nu = 0.474 \times 10^{-6}$ m²/s, γ = 9.642 kN/m³

Sec 8.4: $h_L = (z + p/\gamma)_A - (z + p/\gamma)_B$ = [54.1 + (88.7/9.642)] − [52.0 + (91.8/9.642)] = 1.778 m

Eq. 8.10: $h_L = 1.778 = f(100/0.02)V^2/[2(9.81)]$; i.e. $V = 0.0835/f^{1/2}$

Try f	V m/s	$\mathbf{R} = DV/\nu$	Obtained f	
0.026	0.518	2.19×10^4	0.0307	Try again
0.0307	0.477	2.01×10^4	0.0315	Close
0.0315	0.471	1.99×10^4	0.0315	Converged!

$Q = (\pi/4)(0.02)^2 0.471$ = 0.000 1479 m³/s = 0.1479 L/s ◄

/more...

8.43 *A steel pipe (e = 0.0002 ft) 13,450 feet long is to convey oil (ν = 0.000 54 ft²/sec) at 13 cfs from a reservoir with surface elevation 705 ft to one with surface elevation 390 ft. Theoretically, what pipe size is required?*

BG

Eq. 8.10: $h_L = (fL/D)(V^2/2g) = (fL/D)[(Q/A)^2/(2g)] = [fL/(2gD)][Q/(\pi D^2/4)]^2 = 8fLQ^2/(\pi^2 g D^5)$

$705 - 390 = 315 = 8f(13{,}450)13^2/[\pi^2(32.2)D^5]$; $D^5 = 181.6f$ (1)

$V = Q/A = Q/(\pi D^2/4) = 4Q/(\pi D^2) = 4(13)/(\pi D^2) = 16.55/D^2$ (2)

$\mathbf{R} = DV/\nu = DV/0.000\,54$ (3)

Try f	D ft (from 1)	V fps (from 2)	**R** (from 3)	e/D	Chart f	
0.030	1.404	8.40	21,837	0.000 142	≈ 0.0257	Try again
0.026	1.364	8.89	22,468	0.000 147	≈ 0.0257	Converged

$D = 1.364$ ft $= 16.37$ in ◄

8.44 *A steel pipe (e = 0.065 mm) 4200 m long is to convey oil (ν = 5.2 × 10⁻⁵ m²/s) at 0.3 m³/s from a reservoir with surface elevation 247 m to one with surface elevation 156 m. Theoretically what pipe size is required?*

SI

This is a Type 3 problem, to find D. $e = 0.000\,065$ m (given).

Eq. 8.10: $h_L = 247 - 156 = 91$ m $= f[4200/(2 \times 9.81D)][0.3/(\pi D^2/4)]^2$, i.e. $D^5 = 0.343f$

Try f	D m	$V = 4Q/(\pi D^2)$ m/s	$\mathbf{R} = DV/\nu$	e/D	Obtained f	
0.030	0.400	2.382	1.83×10^4	0.000 162	0.027	Try again
0.027	0.392	2.484	1.87×10^4	0.000 166	0.027	Converged

$D = 0.392$ m ◄

Sec. 8.14: Rigorous Solution of Pipe Flow Problems -- *Exercises*

8.14.1 *Solve Sample Prob. 8.5 using (a) an equation solver on a programmable scientific calculator, (b) Mathcad or similar software.*

Sample Prob. 8.5: Water at 20°C flows in a 500-mm-diameter welded steel pipe. If the friction loss gradient is 0.006, determine the flow rate.

SI

(*a*) Given: $D = 0.5$ m, $h_L = 0.006$ m, $L = 1$ m, $g = 9.81$ m/s², $e = 0.000\,046$ m,

$\nu = 1.003 \times 10^{-6}$ m²/s. Estimate $Q = 1$ m³/s.

Solve Colebrook-based combination Eq. 8.42*b* for Q, then continuity equation for V, then Eq. 8.1 for Reynolds number **R**, and then Colebrook Eq. 8.40 for f if desired.

Results are: $Q = 0.414$ m³/s ◄ $V = 2.11$ m/s, $\mathbf{R} = 1\,049\,989$, $f = 0.013\,27$

(*b*) Use the same given and estimated data as for part (*a*). Place in a Solve Block Eqs. 8.10, continuity, 8.1, and 8.40, and solve them simultaneously. Results are the same as for part (*a*).

8.14.2 *Solve Exer. 8.13.3 using (a) a programmable scientific calculator with an equation solver, (b) Mathcad or similar mathematics software, and (c) only a basic scientific calculator, without trial and error.*

Exer. 8.13.3: Water at 50°F flowing through 80 ft of 4-in-diameter average cast-iron pipe causes a head loss of 0.27 ft. Find the flow rate.

A basic scientific calculator is here defined to be one that is not programmable and does not have automatic equation solving capabilities.

BG

(*a*) Given: $h_L = 0.27$ ft, $L = 80$ ft, $D = 4/12$ ft, $g = 32.2$ ft/s^2, $e = 0.000\,85$ ft,

$\nu = 1.410 \times 10^{-5}$ ft^2/s. Estimate $Q = 1$ cfs.

Solve Colebrook-based combination Eq. 8.42*b* for Q, then continuity equation for V, then Eq. 8.1 for Reynolds number **R**, and then Colebrook Eq. 8.40 for f if desired.

Results are: $Q = 0.1393$ cfs ◄ $V = 1.596$ fps, $\mathbf{R} = 37{,}736$, $f = 0.0284$.

(*b*) Use the same given and estimated data as for part (*a*). Place in a Solve Block Eqs. 8.10, continuity, 8.1, and 8.40, and solve them simultaneously. Results are the same as for part (*a*). ◄

(*c*) For Eq. 8.42*a*, the quantity $\sqrt{2gDh_L/L} = \sqrt{2(32.2)4(0.27)/(12 \times 80)} = 0.269$ fps

Assuming turbulent flow:

$$\text{Eq. 8.42}a\text{:}\quad V = -2(0.269)\log\left[\frac{0.000\,85}{(4/12)3.7} + \frac{2.51(1.41 \times 10^{-5})}{(4/12)0.269}\right] = 1.596 \text{ fps}$$

$$\mathbf{R} = \frac{DV}{\nu} = \frac{(4/12)1.596}{1.41 \times 10^{-5}} = 37{,}736,$$ so flow is turbulent and use of Eq. 8.42*a* is appropriate.

$Q = AV = (\pi/4)(4/12)^2 1.596 = 0.1393$ cfs. ◄

8.14.3 *Solve Exer. 8.13.4 using (a) a programmable scientific calculator with an equation solver, (b) Mathcad or similar mathematics software, and (c) only a basic scientific calculator [see Exer. 8.14.2], without trial and error.*

Exer. 8.13.4: When gasoline with a kinematic viscosity of 5×10^{-7} m^2/s flows in a 20-cm-diameter smooth pipe the head loss is 0.43 m per 100 m. Find the flow rate.

SI

(*a*) Given: $h_L = 0.43$ m, $L = 100$ m, $D = 0.2$ m, $g = 9.81$ m/s^2, $e = 0$, $\nu = 5 \times 10^{-7}$ m^2/s.

Estimate $Q = 0.1$ m^3/s.

Solve Colebrook-based combination Eq. 8.42*b* for Q, then continuity equation for V, then Eq. 8.1 for Reynolds number **R**, and then Colebrook Eq. 8.40 for f if desired.

Results are: $Q = 0.0352$ m^3/s ◄ $V = 1.121$ m/s, $\mathbf{R} = 448\,506$, $f = 0.013\,42$.

(*b*) Use the same given and estimated data as for part (*a*). Place in a Solve Block Eqs. 8.10, continuity, 8.1, and 8.40, and solve them simultaneously. Results are the same as for part (*a*). ◄

(*c*) For Eq. 8.42*a*, the quantity $\sqrt{2gDh_L/L} = \sqrt{2(9.81)0.2(0.43)/100} = 0.1299$ m/s

Assuming turbulent flow:

Eq. 8.42*a*: $V = -2(0.1299)\log[0 + 2.51(5 \times 10^{-7})/[(0.2)0.1299]] = 1.121$ m/s

$\mathbf{R} = DV/\nu = (0.2)1.121/(5 \times 10^{-7}) = 448\,506$, so flow is turbulent and use of Eq. 8.42*a* is appropriate.

$Q = AV = (\pi/4)(0.2)^2 1.121 = 0.0352$ m^3/s $= 35.2$ L/s. ◄

8.45 *Using only a basic scientific calculator, solve Prob. 8.37 without trial and error.*

Prob. 8.37: L = 25 m, D = 0.1 m, h_L = 0.075 m, e/D = 0.0015, ν = 1.139 × 10^{-6} m²/s, find Q.

A basic scientific calculator is here defined to be one that is not programmable and does not have automatic equation solving capabilities.

SI

This is a type 2 problem to find Q.

For Eq. 8.24a, the quantity $\sqrt{2gDh_L/L} = \sqrt{(2)9.81(0.1)0.075/25} = 0.0767$ m/s

Assuming turbulent flow:

Eq. 8.42a: $V = -2(0.0767)\log\left[\dfrac{0.0015}{3.7} + \dfrac{2.51(1.139 \times 10^{-6})}{(0.1)0.0767}\right] = 0.477$ m/s

$\mathbf{R} = DV/\nu = 0.1(0.477)/(1.139 \times 10^{-6}) = 4.19 \times 10^4$. ∴ the use of turbulent Eq. 8.42a was valid.

$Q = AV = (\pi/4)(0.1)^2 0.477 = 0.003\,75$ m³/s = 3.75 L/s ◀

8.46 *Using only a basic scientific calculator [see Prob. 8.45], solve Prob. 8.40 without trial and error.*

Prob. 8.40: Q = 0.21 m³/s, h_L/L = 0.42/100 m, e = 0.000 038 m, ν = 2.0 × 10^{-5} m²/s, find D.

SI

This is a Type 3 problem, to find D.

Eq. 8.43: $\mathbf{N}_1 = \dfrac{128gh_LQ^3}{\pi^3L\nu^5} = \dfrac{128(9.81)0.42}{(2.0 \times 10^{-5})100}\left(\dfrac{0.21}{\pi}\right)^3 = 4.92 \times 10^{20}$

Eq. 8.44: $\mathbf{N}_2 = \dfrac{\pi e\nu}{4Q} = \left(\dfrac{\pi}{4}\right)\dfrac{0.000\,038}{0.21}2.0 \times 10^{-5} = 2.84 \times 10^{-9}$

If laminar, Eq. 8.45: $\mathbf{R} = (4.92 \times 10^{20}/64)^{0.25} = 5.27 \times 10^4$ (not laminar!)

If turbulent, Eq. 8.47: $\mathbf{R}^{2.5} = -2\sqrt{4.92 \times 10^{20}}\,\log\left[\dfrac{2.84 \times 10^{-9}(4.92 \times 10^{20})^{0.208}}{2.59} + \dfrac{4.29}{(4.92 \times 10^{20})^{0.188}}\right]$

$= 1.438 \times 10^{11}$ so $\mathbf{R} = 2.90 \times 10^4$

Comparing both $\mathbf{R}$ values with $\mathbf{R}_{crit}$ = 2000, the flow is turbulent, not laminar.

Eq. 8.48: $D = 4Q/(\pi\nu\mathbf{R}) = 4(0.21)/[\pi(2.0 \times 10^{-5})2.90 \times 10^4] = 0.461$ m ◀

8.47 *Using only a basic scientific calculator [see Prob. 8.45], solve Prob. 8.42 without trial and error.*

Prob. 8.42: Water at 60°C flows in a 100-m-long 2-cm-diameter pipe (e = 0.060 mm) between points A and B. At A the elevation of the pipe is 54.1 m and the pressure is 88.7 kPa. At B the elevation of the pipe is 52.0 m and the pressure is 91.8 kPa. Compute the flow rate as accurately as you can.

SI

This is a type 2 problem to find Q.

Table A.1 for water at 60°C: ν = 0.474 × 10^{-6} m²/s, γ = 9.642 kN/m³. e/D = 0.060/20 = 0.003.

Sec. 8.4: $h_L = (z + p/\gamma)_A - (z + p/\gamma)_B = (54.1 + 88.7/9.64) - (52.0 + 91.8/9.64) = 1.778$ m

For Eq. 8.42a, the quantity $(2gDh_L/L)^{1/2} = [(2)9.81(0.02)1.778/100]^{1/2} = 0.0835$ m/s

Assuming turbulent flow: Eq. 8.42a: $V = -2(0.0835)\log\left[\dfrac{0.003}{3.7} + \dfrac{2.51(0.474 \times 10^{-6})}{(0.020)0.0835}\right] = 0.471$ m/s

$\mathbf{R} = DV/\nu = 0.020(0.471)/(0.474 \times 10^{-6}) = 1.99 \times 10^4$. ∴ the use of turbulent Eq. 8.42a was valid.

$Q = AV = (\pi/4)(0.020)^2 0.471 = 0.000\,1478$ m³/s = 0.1478 L/s ◀

8.48 *Using only a basic scientific calculator [see Prob. 8.45], solve Prob. 8.44 without trial and error.*

Prob. 8.44: A steel pipe (e = 0.065 mm) 4200 m long is to convey oil ($\nu = 5.2 \times 10^{-5}$ m²/s) at 0.3 m³/s from a reservoir with surface elevation 247 m to one with surface elevation 156 m. Theoretically what pipe size is required?

SI

This is a Type 3 problem, to find D.

$h_L = 247 - 156 = 91$ m, $e = 0.000\,065$ m.

Eq. 8.43: $$\mathbf{N}_1 = \frac{128(9.81)91}{(5.2 \times 10^{-5})^5 4200}\left(\frac{0.3}{\pi}\right)^3 = 6.23 \times 10^{19}$$

Eq. 8.44: $$\mathbf{N}_2 = \left(\frac{\pi}{4}\right)\frac{0.000\,065}{0.3}5.2 \times 10^{-5} = 8.85 \times 10^{-9}$$

If laminar, Eq. 8.45: $\mathbf{R} = (6.23 \times 10^{19}/64)^{0.25} = 3.14 \times 10^4$ (not laminar!)

If turbulent,

Eq. 8.47: $$\mathbf{R}^{2.5} = -2\sqrt{6.23 \times 10^{19}}\log\left[\frac{8.85 \times 10^{-9}(6.23 \times 10^{19})^{0.208}}{2.59} + \frac{4.29}{(6.23 \times 10^{19})^{0.188}}\right] = 4.84 \times 10^{10}$$

so $\mathbf{R} = 1.87 \times 10^4$

Comparing both $\mathbf{R}$ values with $\mathbf{R}_{crit} = 2000$, the flow is turbulent, not laminar.

Eq. 8.48: $D = 4Q/(\pi \nu \mathbf{R}) = 4(0.3)/[\pi(5.2 \times 10^{-5})1.87 \times 10^4] = 0.393$ m ◀

8.49 *Using an equation solver on a programmable scientific calculator, or in mathematics software such as Mathcad, solve the following without manual trial and error: (a) Prob. 8.37; (b) Prob. 8.38; (c) Prob. 8.41; (d) Prob. 8.42.*

B

These are Type 2 problems, to find Q.

(*i*) Using an equation solver on a programmable scientific calculator, first solve Colebrook-based Eq. 8.42*b* for Q, then the continuity equation for V, then Eq. 8.1 for Reynold's number $\mathbf{R}$, and then Colebrook Eq. 8.40 for f if desired.

(*ii*) Using mathematics software such as Mathcad, place in a solve block Eqs. 8.10, continuity, 8.1, and 8.40, and solve them simultaneously.

Part: Solve Prob:	(*a*) 8.37	(*b*) 8.38	(*c*) 8.41	(*d*) 8.42
Input data				
D	0.1 m	10/12 ft	0.863/12 ft	0.02 m
L	25 m	100 ft	350 ft	100 m
h_L	0.075 m	0.36 ft	5.94 ft	1.778 m
e	0.000 15 m	0	0.000 15 ft	0.000 060 m
ν	6×10^{-6} m²/s	6×10^{-6} ft²/s	0.476×10^{-5} ft²/s	0.474×10^{-6} m²/s
Results				
Q	0.003 75 m³/s	2.10 cfs	0.006 69 cfs	0.000 15 m³/s ◀◀
V	0.477 m/s	3.86 fps	1.647 fps	0.471 m/s
$\mathbf{R}$	41 883	535,512	24,881	19 861
f	0.0259	0.013	0.0290	0.0315

8.50 *Using an equation solver on a programmable scientific calculator, or in mathematics software such as Mathcad, solve the following without manual trial and error: (a) Prob. 8.39; (b) Prob. 8.40; (c) Prob. *8.43; (d) Prob. 8.44. [*Note this correction to the textbook.]*

B

These are Type 3 problems, to find D.

(*i*) Using an equation solver on a programmable scientific calculator, first solve Colebrook-based Eq. 8.42*b* for D, then the continuity equation for V, then Eq. 8.1 for Reynold's number $\mathbf{R}$, and then Colebrook Eq. 8.40 for f if desired.

(*ii*) Using mathematics software such as Mathcad, place in a solve block Eqs. 8.10, continuity, 8.1, and 8.40, and solve them simultaneously.

Part:	(*a*)	(*b*)	(*c*)	(*d*)	
Solve Prob:	8.39	8.40	8.43	8.44	
Input data					
Q	7.4 cfs	0.21 m^3/s	13 cfs	0.3 m^3/s	
L	100 ft	100 m	13,450 ft	4200 m	
h_L	0.37 ft	0.42 m	315 ft	91 m	
e	0.000 13 ft	0.000 038 m	0.0002 ft	0.000 065 m	
ν	0.000 22 ft^2/s	2.0×10^{-5} m^2/s	0.000 54 ft^2/s	5.2×10^{-5} m^2/s	
Results					
D	1.552 ft	0.461 m	1.359 ft	0.391 m	◄◄
V	3.91 fps	1.261 m/s	8.97 fps	2.50 m/s	
$\mathbf{R}$	27,595	29 028	22,563	18 786	
f	0.0242	0.0239	0.0255	0.0266	

Sec. 8.15: Empirical Equations for Pipe Flow -- *Exercises*

8.15.1 *When water flows at 2.5 cfs through a 20-in-diameter pipeline, the head loss is 0.0004 ft/ft. Find the value of the Hazen-Williams coefficient.*

BG

Eq. 8.49*a*: $V = 1.318 C_{HW} R_h^{0.63} S^{0.54}$; Sec. 8.3: $R_h = (20/12)/4 = 0.417$ ft

$2.5/[\pi(10/12)^2] = 1.318 C_{HW}(0.417)^{0.63}(0.0004)^{0.54}$; $C_{HW} = 103.2$ ◄

8.15.2 *Water flows at 0.20 m^3/s through a 50-cm-diameter pipeline with a head loss of 0.0032 m/m. Find the value of the Hazen-Williams coefficient.*

SI

Eq. 8.49*a*: $V = 0.849 C_{HW} R_h^{0.63} S^{0.54}$; Sec. 8.3: $R_h = D/4 = 0.125$ m

$0.20/(\pi 0.25^2) = 0.849 C_{HW}(0.125)^{0.63}(0.0032)^{0.54}$; $C_{HW} = 98.9$ ◄

Sec. 8.15: Empirical Equations for Pipe Flow -- *Problems*

8.51 *In a field test of the 16-ft-diameter Colorado River aqueduct flowing full, Manning's n was found to have a value of 0.0132 when 50°F water was flowing at a Reynolds number of 10.5×10^6. Determine the average value of e for this conduit.*

BG

Write Eqs. 8.10 and 8.50*a* in terms of S: $S = \dfrac{h_L}{L} = \dfrac{f}{D}\dfrac{V^2}{2g} = \dfrac{n^2V^2}{1.486^2 R_h^{4/3}}$ where $R_h = \dfrac{D}{4}$ (Sec. 8.3)

Thus $f = (n/1.486)^2 D2g/(D/4)^{4/3} = (0.0132/1.486)^2 16(2)32.2/(16/4)^{4/3} = 0.012\,80$

Fig. 8.11 or Eq. 8.40 or 8.41 for $\mathbf{R} = 10.5 \times 10^6$ (given) and $f = 0.012\,80$: $e/D = 0.000\,133$

$e = 0.000\,133(16) = 0.002\,13$ ft ◄

[Note: Solutions for V (= 9.25 fps) and S (= 0.001 06) are unnecessary as they can be cancelled out.]

8.52 *Measurements were taken on pressure flow through a 2.5-m-diameter aqueduct built of concrete. When the water temperature was 10°C and the Reynolds number was 2.0×10^6, Manning's n was determined to be 0.0140. Find the average value of e for this aqueduct.*

SI

Write Eqs. 8.10 and 8.50*b* in terms of *S*: $S = \frac{h_L}{L} = \frac{f}{D}\frac{V^2}{2g} = \frac{n^2 V^2}{R_h^{4/3}}$ where $R_h = \frac{D}{4}$ (Sec. 8.3)

Thus $f = n^2 D 2g/(D/4)^{4/3} = (0.014)^2 2.5(2)9.81/(2.5/4)^{4/3} = 0.017\,99$

Fig. 8.11 or Eq. 8.40 or 8.41 for **R** $= 2 \times 10^6$ (given) and $f = 0.017\,99$: $e/D = 0.000\,658$

$e = 0.000\,658(2.5) = 0.001\,645$ m $= 1.645$ mm ◄

[Note: Solutions for V (= 1.045 m/s) and S (= 0.000 400) are unnecessary as they can be cancelled out.]

Sec. 8.16: Fluid Friction in Noncircular Conduits -- *Exercises*

8.16.1 *When fluid of specific weight 40 lb/ft³ flows in a 9-in-diameter pipe, the frictional stress between the fluid and the pipe wall is 0.6 lb/ft². Calculate the head loss per foot of pipe. If the flow rate is 2.4 cfs, how much power is lost per foot of pipe?*

BG

Eq. 8.51: $R_h = \frac{D}{4} = \frac{0.75}{4} = 0.1875$ ft. Eq. 8.6: $h_L = \frac{\bar{\tau}_0 L}{R_h \gamma} = \frac{(0.6)1}{0.1875(40)} = 0.0800$ ft per ft ◄

Eq. 5.32: $HP_{loss} = 40(2.4)0.0800/550 = 0.013\,96$ hp per ft ◄

8.16.2 *When fluid of specific weight 8.2 kN/m³ flows in a 15-cm-diameter pipe, the frictional stress between the fluid and the pipe wall is 25 N/m². Calculate the head loss per meter of pipe. If the flow rate is 42 L/s, how much power is lost per meter of pipe?*

SI

Eq. 8.51: $R_h = D/4 = 0.15/4 = 0.0375$ m

Eq. 8.6: $h_L = \bar{\tau}_0 L/(R_h \gamma) = (25\text{ N/m}^2)(1\text{ m})/[(0.0375\text{ m})8200\text{ N/m}^3] = 0.813$ m/m ◄

From Eq. 5.33: Power loss $= \gamma Q h_L = (8200\text{ N/m}^3)(0.042\text{ m}^3/\text{s})(0.0813\text{ m}) = 28.0$ W/m ◄

Sec. 8.16: Fluid Friction in Noncircular Conduits -- *Problems*

8.53 *Air at 160°F and standard sea-level atmospheric pressure flows in a 15-in by 21-in rectangular air duct (e = 0.0007 in) at the rate of 1.2 lb/min. Find the head loss per 100 ft of duct. Express the answer in feet of air flowing and in pounds per square inch.*

BG

Eq. 8.51: $R_h = A/P = 1.25(1.75)/6 = 0.365$ ft $= 4.38$ inches

Using Eq. 8.52: $e/D = e/(4R_h) = 0.0007/(4 \times 4.38) = 0.000\,0400$

Table A.5 for air: $R = 1715$ ft²/(sec²·°R). Eq. 2.5: $\gamma = \frac{pg}{RT} = \frac{60(144)32.2}{1715(460 + 160)} = 0.262$ lb/ft³

$Q = (1.2\text{ lb/min})(\text{ft}^3/0.262\text{ lb})(\text{min}/60\text{ sec}) = 0.0764$ ft³/s; $V = Q/A = 0.0764/2.1875 = 0.0350$ fps

Table A.2 for air at 160°F: $\nu = 0.212 \times 10^{-3}$ ft²/sec

From Eq. 8.54: **R** $= 4R_h V/\nu = 4(0.365)0.0350/(0.212 \times 10^{-3}) = 240$

Flow is laminar, so *e/D* is insignificant and from Eq. 8.20: $f = 64/\mathbf{R} = 64/240 = 0.267$

Eq. 8.53: $h_L = 0.267[100/(4 \times 0.365)]0.0350^2/[(2(32.2)] = 0.000\,347$ ft ◄

$\Delta p = h_L \gamma = 0.000\,347(0.262)/144 = 6.32 \times 10^{-7}$ psi ◄

8.54 *Water at 80°F flows in a conduit with a cross section shaped in the form of an equilateral triangle. The cross-sectional area of the duct is 160 in² and e = 0.0018 in. If the head loss is 3 ft in 150 ft, find the approximate flow rate.*

BG

$A = 3^{1/2}x^2 = 160\ \text{in}^2,\ x = 9.61\ \text{in} = 0.801\ \text{ft}$

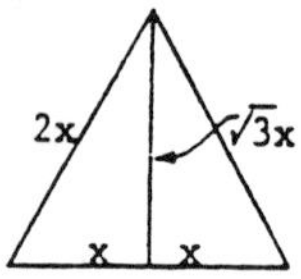

Eq. 8.51: $R_h = A/P = 160/(3 \times 2 \times 9.61) = 2.77\ \text{in} = 0.231\ \text{ft}$

Eq. 8.52: $D = 4R_h = 0.925\ \text{ft};\ e/4R_h = 0.0018/(4 \times 2.77) = 0.000\,1622$

Table A.1 for water at 80°F: $\nu = 0.930 \times 10^{-5}\ \text{ft}^2/\text{sec}$

From Eq. 8.54: $\mathbf{R} = 0.925V/(0.930 \times 10^{-5}) = 9.945 \times 10^4 V$ (1)

Eq. 8.53: $h_L = 3 = f(150/0.925)V^2/(2 \times 32.2);\ 1.191 = f\,V^2$ (2)

Solve by T and E to satisfy (1), (2), and Fig. 8.11

If $f = 0.015$, $V = 8.91$ fps, $\mathbf{R} = 8.86 \times 10^5$; Fig. 8.11: $f = 0.0143$

If $f = 0.0140$, $V = 9.22$ fps, $\mathbf{R} = 9.17 \times 10^5$; Fig. 8.11: $f = 0.0140$

Thus $V = 9.22$ fps and $Q = 9.22(160/144) = 10.25$ cfs ◄

8.55 *Water at 20°C flows in a conduit with a cross section shaped in the form of an equilateral triangle. The cross-sectional area of the duct is 1000 cm² and e = 0.045 mm. If the head loss is 1 m in 50 m, find the approximate flow rate.*

SI

$A = 3^{1/2}x^2 = 1000\ \text{cm}^2,\ x = 24.03\ \text{cm} = 0.2403\ \text{m}$

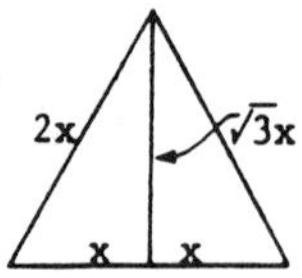

Eq. 8.51: $R_h = A/P = 1000/(6 \times 24.03) = 6.94\ \text{cm} = 0.0694\ \text{m}$

Eq. 8.52: $D = 4R_h = 0.278\ \text{m};\ e/(4R_h) = 0.045/(4 \times 69.4) = 0.000\,1622$

Table A.1 for water at 20°C: $\nu = 1.003 \times 10^{-6}\ \text{m}^2/\text{s}$

From Eq. 8.54: $\mathbf{R} = 0.278V/(1.003 \times 10^{-6}) = 2.77 \times 10^5 V$ (1)

Eq. 8.53: $h_L = 1 = f(50/0.278)V^2/(2 \times 9.81);\ 0.1091 = fV^2$ (2)

Solve by T and E to satisfy (1), (2), and Fig. 8.11

If $f = 0.015$, $V = 2.70$ m/s, $\mathbf{R} = 7.47 \times 10^5$; Fig. 8.11: $f = 0.0143$

If $f = 0.014$, $V = 2.79$ m/s, $\mathbf{R} = 7.73 \times 10^5$; Fig. 8.11: $f = 0.014$

Thus $V = 2.79$ m/s and $Q = AV = (1000 \times 10^{-4})2.79 = 0.279\ \text{m}^3/\text{s}$ ◄

Sec. 8.19: Loss of Head at Submerged Discharge -- *Exercises*

8.19.1 *A 12-in-diameter pipe (f = 0.02) carries fluid between two tanks at 8 fps. The entrance and exit conditions to and from the pipe are square-edged and flush with the wall of the tank. Find the ratio of the minor losses divided by the pipe friction loss if the length of the pipe is (a) 4 ft; (b) 80 ft; (c) 1600 ft.*

BG

Pipe friction loss: $0.02(L/D)(V^2/2g)$

Minor losses: $0.5(V^2/2g) + (V^2/2g) = 1.5(V^2/2g)$

L	Minor	Fiction	Minor/friction
4 ft	1.491	0.0795	18.75:1
80 ft	1.491	1.590	0.938:1
1600 ft	1.491	31.8	0.0469:1

▲

8.19.2 *A 30-cm-diameter pipe (f = 0.015) carries fluid between two tanks at 2.4 m/s. The entrance and exit conditions to and from the pipe are reentrant. Find the ratio of the minor losses divided by the pipe friction loss if the length of the pipe is (a) 2 m; (b) 50 m; (c) 1000 m.*

SI

Pipe friction loss: $0.015(L/D)(V^2/2g) = 0.015(L/0.3)(2.4)^2/(2 \times 9.81) = 0.014\,68L$

Minor losses: $0.5(V^2/2g) + (V^2/2g) = 1.5(V^2/2g) = 1.5(2.4)^2/(2 \times 9.81)$

L	Minor	Fiction	Minor/friction
2 m	0.440	0.0294	15.00:1
50 m	0.440	0.734	0.600:1
1000 m	0.440	14.68	0.0300:1

▲

8.19.3 *A smooth 15-in-diameter pipe is 400 ft long and has a flush entrance and a submerged discharge. It carries 70°F water at a velocity of 8 fps. What is the total head loss?*

BG

Eqs. 8.10, 8.55, 8.56: $\Sigma h_L = h_{Lf} + h_e' + h_d' = [f(L/D) + k_e + k_d]V^2/2g$

Table A.1 for water at 70°F: $\nu = 1.059 \times 10^{-5}$ ft²/sec

Eq. 8.1: $\mathbf{R} = DV/\nu = 1.25(8)/(1.059 \times 10^{-5}) = 9.44 \times 10^5$

Fig. 8.11 or Eq. 8.35 or Eq. 8.36 for smooth pipe: $f = 0.011\,76$; Fig. 8.13: $k_e = 0.5$

$\Sigma h_L = [0.011\,76(400/1.25) + 0.5 + 1.0]8^2/[2(32.2)] = (5.26)0.994 = 5.23$ ft ◄

8.19.4 *A smooth 30-cm-diameter pipe is 90 m long and has a flush entrance and a submerged discharge. It carries 15°C water at a velocity of 3 m/s. What is the total head loss?*

SI

Eqs. 8.10, 8.55, 8.56: $\Sigma h_L = h_{Lf} + h_e' + h_d' = [f(L/D) + k_e + k_d]V^2/2g$

Table A.1 for water at 20°C: $\nu = 1.003 \times 10^{-6}$ m²/s

Eq. 8.1: $\mathbf{R} = DV/\nu = 0.4(2.4)/(1.003 \times 10^{-6}) = 9.57 \times 10^5$

Fig. 8.11 or Eq. 8.35 or Eq. 8.36 for smooth pipe: $f = 0.011\,73$; Fig. 8.13: $k_e = 0.5$

$\Sigma h_L = [0.011\,73(120/0.4) + 0.5 + 1.0]2.4^2/[2(9.81)] = (5.02)0.294 = 1.474$ m ◄

8.19.5 *Suppose the fluid in Exer. 8.19.3 is oil with a kinematic viscosity of 0.001 ft²/s and a specific gravity of 0.92. What would be the head loss in feet of oil and in pounds per square inch?*

Exer. 8.19.3: The fluid flows at 8 fps through a smooth 15-in-diameter pipe which is 400 ft long with a flush entrance and submerged discharge.

BG

Eq. 8.1 for oil: $\mathbf{R} = DV/\nu = 1.25(8)/0.001 = 1 \times 10^4$

Fig. 8.11 or Eq. 8.35 or Eq. 8.36 for smooth pipe: $f = 0.0309$; Fig. 8.13: $k_e = 0.5$

$\Sigma h_L = h_{Lf} + h_e' + h_d' = [0.0309(400/1.25) + 0.5 + 1.0]8^2/[2(32.2)] = (11.38)0.994 = 11.31$ ft oil ◄

From Eq. 3.4: $\Delta p = \gamma \Sigma h_L = 0.92(62.4/144)11.31 = 4.51$ psi ◄

8.19.6 *Suppose the fluid in Exer. 8.19.4 is oil with a kinematic viscosity of 9.7 x 10^{-5} m^2/s and a specific gravity of 0.94. What would be the head loss in meters of oil and in kN/m^2?*

Exer. 8.19.4: The fluid flows at 3 m/s through a smooth 30-cm-diameter pipe which is 90 m long with a flush entrance and submerged discharge.

SI

Eq. 8.1 for oil: $\mathbf{R} = DV/\nu = 0.40(2.4)/(9.7 \times 10^{-5} = 9897$

Fig. 8.11 or Eq. 8.35 or Eq. 8.36 for smooth pipe: $f = 0.0310$; Fig. 8.13: $k_e = 0.5$

$\Sigma h_L = h_{Lf} + h_e' + h_d' = [0.0310(120/0.40) + 0.5 + 1.0]2.4^2/[2(9.81)] = (10.79)0.294 = 3.17$ m oil ◀

From Eq. 3.4: $\Delta p = \gamma \Sigma h_L = 0.94(9.81)3.17 = 29.2\ \text{kN/m}^2$ ◀

Sec. 8.21: Loss Due to Expansion -- ***Exercises***

8.21.1 *Two pipes with a diameter ratio of 1:2 are connected in series. With a velocity of 7.5 m/s in the smaller pipe, find the loss of head due to (a) sudden contraction; (b) sudden enlargement; (c) expansion in a conical diffuser with a total angle of 30°, and of 10°.*

SI

(*a*) Table 8.2: For $D_2/D_1 = 0.5$, $k_c = 0.33$

Eq. 8.57: $h_c = k_c V_2^2/2g = 0.33(7.5)^2/[2(9.81)] = 0.946$ m ◀

(*b*) Eq. 8.58: $h_x' = (V_1 - V_2)^2/2g$ where $V_2 = (0.5)^2 V_1 = (0.5)^2 7.5 = 1.875$ m/s

$h_x' = (7.5 - 1.875)^2/2g = 5.625^2/(2 \times 9.81) = 1.613$ m ◀

(*c*) Eq. 8.59: $h' = k'(V_1 - V_2)^2/2g = k'(1.613\text{ m})$

Fig. 8.19 for $\alpha = 30°$: $k' = 0.65$, $\therefore h' = 0.65(1.613) = 1.048$ m ◀

Fig. 8.19 for $\alpha = 10°$: $k' = 0.175$, $\therefore h' = 0.175(1.613) = 0.282$ m ◀

8.21.2 *A 5-in-diameter pipe ($f = 0.033$) 110 ft long connects two reservoirs whose water-surface elevations differ by 12 ft. The pipe entrance is flush, and the discharge is submerged. (a) Compute the flow rate. (b) If the last 10 ft of pipe were replaced with a conical diffuser with a cone angle of 10°, compute the flow rate.*

BG

(*a*) Energy: $12 - 0.5V^2/2g - 0.033[110/(5/12)](V^2/2g) - (V^2/2g) = 0$

$12 = V^2/2g(0.5 + 8.71 + 1.0) = 10.21V^2/2g$; $V^2/2g = 1.175$ ft, $V = 8.70$ fps

$Q = AV = (\pi/4)(5/12)^2 8.70 = 1.186$ cfs ◀

(*b*) Fig. 8.19 for $\alpha = 10°$: $k' = 0.175$

$D_d = D + 2L\tan(10°/2) = (5/12) + 2(10)\tan 5° = 2.17$ ft

Energy: $12 - 0.5V^2/2g - 0.033[100/(5/12)](V^2/2g) - 0.175(V - V_d)^2/2g = 0$

$V_d = V(D/D_d)^2 = V[(5/12)/2.17]^2 = 0.0370V$; $V - V_d = 0.963V$

$12 = (V^2/2g)(0.5 + 7.92 + 0.1623) = 8.58V^2/2g$; $V = 9.49$ fps

$Q = AV = (\pi/4)(5/12)^2 9.49 = 1.294$ cfs ◀

8.56 *A smooth pipe consists of 52 ft of 9-in pipe followed by 310 ft of 18-in pipe with an abrupt change of cross section at the junction. The entrance is flush and the discharge is submerged. If it carries water at 60°F, with a velocity of 19 fps in the smaller pipe, what is the total head loss?*

BG

We note (Eq. 8.58) that: $h_x' = \dfrac{(V_1 - V_2)^2}{2g} = \left(1 - \dfrac{V_2}{V_1}\right)^2 \dfrac{V_1^2}{2g} = \left(1 - \dfrac{A_1}{A_2}\right)^2 \dfrac{V_1^2}{2g}$

$$\Sigma h_L = h_e' + h_{Lf1} + h_x' + h_{Lf2} + h_d' = \left[k_e + f_1\frac{L_1}{D_1} + \left(1 - \frac{A_1}{A_2}\right)^2\right]\frac{V_1^2}{2g} + \left[f_2\frac{L_2}{D_2} + k_d\right]\frac{V_2^2}{2g}$$

Table A.1 for water at 60°F: $v = 1.217 \times 10^{-5}$ ft²/sec; Fig. 8.13: $k_e = 0.5$

Eq. 8.1 for 9-inch pipe: $\mathbf{R} = DV/v = (9/12)19/(1.217 \times 10^{-5}) = 1.171 \times 10^6$

Fig. 8.11 or Eq. 8.35 or Eq. 8.36 for smooth pipe: $f_1 = 0.011\,34$

Eq. 8.1 for 18-inch pipe: $\mathbf{R} = DV/v = (18/12)19(9/12)^2/(1.217 \times 10^{-5}) = 3.29 \times 10^5$

Fig. 8.11 or Eq. 8.35 or Eq. 8.36 for smooth pipe: $f_2 = 0.014\,21$

$$\Sigma h_L = \left\{0.5 + 0.011\,34\frac{52(12)}{9} + \left[1 - \left(\frac{9}{24}\right)^2\right]^2\right\}\frac{19^2}{2(32.2)} + \left(0.014\,21\frac{310(12)}{18} + 1.0\right)\left(\frac{9}{24}\right)^4\frac{19^2}{2(32.2)}$$

$= 12.03 + 0.436 = 12.46$ ft ◀

8.57 *A smooth pipe consists of 12 m of 18-cm pipe followed by 75 m of 55-cm pipe with an abrupt change of cross section at the junction. The entrance is flush and the discharge is submerged. If it carries water at 15°C, with a velocity of 5.7 m/s in the smaller pipe, what is the total head loss?*

SI

Eq. 4.3: $Q = A_1V_1 = (\pi/4)D_1^2V_1 = (\pi/4)(0.18)^2 5.7 = 0.1451$ m³/s

$Q = A_2V_2 = (\pi/4)(0.55)^2V_2;\ V_2 = (0.18/0.55)^2 5.7 = 0.611$ m/s

Table A.1 for water at 15°C: $v = 1.139 \times 10^{-6}$ m²/s

Eq. 8.1: $\mathbf{R}_1 = D_1V_1/v = 0.18(5.7)/(1.139 \times 10^{-6}) = 9.01 \times 10^5$

$\mathbf{R}_2 = D_2V_2/v = 0.55(0.611)/(1.139 \times 10^{-6}) = 2.95 \times 10^5$

Eq. 8.35 by T and E (or approx from Fig. 8.11 or Eq. 8.36): $f_1 = 0.011\,86$, $f_2 = 0.014\,51$

$$\Sigma h_L = [k_e + (f_1L_1/D_1)]V_1^2/(2g) + (V_1 - V_2)^2/(2g) + [(f_2L_2/D_2) + 1]V_2^2/(2g)$$

$$= \left[0.5 + \frac{0.011\,86(12)}{0.18}\right]\frac{5.7^2}{2(9.81)} + \frac{(5.7 - 0.611)^2}{2(9.81)} + \left[\frac{0.014\,51(75)}{0.55} + 1\right]\frac{0.611^2}{2(9.81)}$$

$= 2.14 + 1.320 + 0.0566 = 3.51$ m ◀

/more...

Sec. 8.23: Loss in Bends and Elbows -- *Exercises*

8.23.1 *Water at 72°F flows through a 110-ft-long, 5-in-diameter wrought-iron pipe that contains the following fittings: one open globe valve, one medium-radius elbow, and one 90° pipe bend (k_b = 0.13) with a radius of curvature of 45 in. The length of the bend is not included in the 110 ft. There are no entrance or discharge losses. Find the total head loss if the flow velocity is 4.8 fps.*

BG

From Table A.1 for water at 72°F, by linear interpolation: $\nu = 1.033 \times 10^{-5}$ ft²/s

Eq. 8.1: $\mathbf{R} = (5/12)4.8/(1.033 \times 10^{-5}) = 1.936 \times 10^5$; Table 8.1: $e = 0.000\,15$ ft

$e/D = 0.000\,15/(5/12) = 0.000\,360$; Fig. 8.11 or Eq. 8.35 or Eq. 8.36: $f = 0.018\,13$

$\Sigma h_L = [f(L/D) + k_v + k_{b1} + k_{b2}]V^2/(2g)$ where L = pipe + bend (Sec. 8.23)

Table 8.3: $$\Sigma h_L = \left(0.018\,13\frac{[(100)12 + (\pi/2)45]}{4} + 10 + 0.75 + 0.13\right)\frac{4.8^2}{2(32.2)} = 6.15 \text{ ft}$$ ◄

8.23.2 *Water at 18°C flows through a 25-m-long, 7.5-cm-diameter commercial steel pipe that contains the following fittings: one open angle valve, one short-radius elbow, and one 90° pipe bend (k_b = 0.12) with a radius of curvature of 60 cm. The length of the bend is not included in the 25 m. There are no entrance or discharge losses. Find the total head loss if the flow velocity is 2 m/s.*

SI

From Table A.1 for water at 18°C, by linear interpolation: $\nu = 1.057 \times 10^{-6}$ m²/s

Eq. 8.1: $\mathbf{R} = 0.075(2)/(1.057 \times 10^{-6}) = 141\,857$

Table 8.1 for commercial steel: $e = 0.046$ mm; $e/D = 0.000\,046/0.075 = 0.000\,613$

Fig. 8.11 or Eq. 8.40 or Eq. 8.41: $f = 0.200$

$$\Sigma h_L = \left(0.0200\frac{[25 + (\pi/2)0.6]}{0.075} + 5 + 0.9 + 0.12\right)\frac{2^2}{2(9.81)} = 2.64 \text{ m}$$ ◄

Sec. 8.24: Solution of Single-Pipe Flow Problems with Minor Losses -- *Exercises*

8.24.1 *An 8-in-diameter pipeline (f = 0.028) 500 ft long discharges a 3-in-diameter jet of water into the atmosphere at a point which is 250 ft below the water surface at intake. The entrance to the pipe is reentrant with k_e = 0.9, and the nozzle loss coefficient is 0.045. Find the flow rate and the pressure head at the base of the nozzle.*

BG

$250 - 0.9(V^2/2g) - 0.028(500 \times 12/8)(V^2/2g) - 0.045(V_j^2/2g) = V_j^2/2g$

$V_j = V(8/3)^2 = 7.11V$

$250 - (V^2/2g)(0.9 + 21 + 2.28) = 50.6(V^2/2g)$; $250 = (V^2/2g)74.7$; $V = 14.68$ fps

$Q = AV = 0.267(14.68) = 5.12$ cfs ◄

$(V^2/2g) = p/\gamma - 2.28(V^2/2g) = 50.6(V^2/2g)$

$p/\gamma = 51.9(V^2/2g) = 51.9(14.68)^2/(2 \times 32.2) = 173.6$ ft; $p = 173.6(62.4)/144 = 75.3$ psi ◄

8.24.2 *An 18-cm-diameter pipeline (f = 0.032) 150 m long discharges a 6-cm-diameter jet of water into the atmosphere at a point which is 80 m below the water surface at intake. The entrance to the pipe is reentrant with k_e = 0.9, and the nozzle loss coefficient is 0.055. Find the flow rate and the pressure head at the base of the nozzle.*

SI

$$80 - 0.9V^2/[(2(9.81)] - 0.032(150/0.18)V^2/[2(9.81)] - 0.055V_j^2/[2(9.81)] = V_j^2/[2(9.81)]$$

$V_j = 9V$; $80 = [V^2/(2 \times 9.81)]113.0$; $V = 3.73$ m/s

$Q = (\pi 0.09^2)3.73 = 0.0948\ \text{m}^3/\text{s} = 94.8$ L/s ◀

$(V^2/2g) + p/\gamma - 4.46(V^2/2g) = 81(V^2/2g)$

$p/\gamma = 84.5(V^2/2g) = 84.5(3.73^2)/(2 \times 9.81) = 59.8$ m; $p = 59.8(9.81) = 586\ \text{kN/m}^2$ ◀

8.24.3 *A horizontal 4-in-diameter pipe (f = 0.028) projects into a body of water 2.5 ft below the surface. Considering all losses, find the pressure at a point 15 ft from the end of the pipe if the velocity is 12 fps and the flow is (a) into the body of water; (b) out of the body of water.*

BG

(*a*) Outflow from pipe. Sec. 8.19: $k_d = 1.0$

$$h_L = \left(k_d + f\frac{L}{D}\right)\frac{V^2}{2g} = \left(1 + 0.028\frac{15}{4/12}\right)\frac{12^2}{2(32.2)} = 2.26(2.236) = 5.05 \text{ ft}$$

Energy Eq: $p/\gamma + 0 + 2.236 = 2.5 + 5.05$; $p/\gamma = 5.31$ ft, $p = 2.30$ psi ◀

(*b*) Inflow into pipe. Fig. 8.13: $k_e = 0.8$

$$h_L = \left(k_e + f\frac{L}{D}\right)\frac{V^2}{2g} = \left(0.8 + 0.028\frac{15}{4/12}\right)\frac{12^2}{2g} = 2.06(2.236) = 4.61 \text{ ft}$$

Energy Eq.: $p/\gamma + 0 + 2.236 = 2.5 - 4.61$; $p/\gamma = -4.34$ ft, $p = -1.882$ psi ◀

8.24.4 *A horizontal 10-cm diameter pipe (f = 0.027) projects into a body of water 1 m below the surface. Considering all losses, find the pressure at a point 5 m from the end of the pipe if the velocity is 4 m/s and the flow is (a) into the body of water; (b) out of the body of water.*

SI

(*a*) Outflow from pipe. Sec. 8.19: $k_d = 1.0$

$$h_L = \left(k_d + f\frac{L}{D}\right)\frac{V^2}{2g} = \left(1 + 0.027\frac{5}{0.10}\right)\frac{4^2}{2(9.81)} = 2.35(0.815) = 1.916 \text{ m}$$

Energy Eq: $p/\gamma + 0 + 0.815 = 1 + 1.916$; $p/\gamma = 2.10$ m, $p = 20.6\ \text{kN/m}^2$ ◀

(*b*) Inflow into pipe. Fig. 8.13: $k_e = 0.8$

$$h_L = \left(k_e + f\frac{L}{D}\right)\frac{V^2}{2g} = \left(0.8 + 0.027\frac{5}{0.10}\right)\frac{4^2}{2(9.81)} = 2.15(0.815) = 1.753 \text{ m}$$

Energy Eq: $p/\gamma + 0 + 0.815 = 1 - 1.753$; $p/\gamma = -1.569$ m, $p = -15.39\ \text{kN/m}^2$ ◀

8.24.5 *A 450-ft-long pipeline runs between two reservoirs, both ends being under water, and the intake end is square-edged and nonprojecting. The difference between the water levels in the two reservoirs is 150 ft. (What is the discharge if the pipe diameter is 12 in and f = 0.028? (b) When this same pipe is old, assu that the growth of tubercles has reduced the diameter to 11.25 in and that f = 0.06. What will the rate discharge be then?*

BG

(*a*) $150 = [1.5 + 0.028(450/1)]V^2/(2g) = (1.5 + 12.60)V^2/(2g)$; $V^2/(2g) = 150/14.1 = 10.64$

$V = [2(32.2)10.64]^{1/2} = 26.2$ fps ; $Q = AV = (0.7854)26.2 = 20.6$ cfs ◀

(*b*) $150 = [1.5 + 0.06(450)12/11.25]V^2/(2g) = (1.5 + 28.8)V^2/(2g)$; $V^2/(2g) = 150/30.3 = 4.95$

$V = [2(32.2)4.95]^{1/2} = 17.86$ fps ; $Q = AV = (0.690)17.86 = 12.33$ cfs ◀

8.24.6 *Solve Exer. 8.24.5a when f is unknown but given that e = 0.005 ft and the water temperature is 60°F. What, then, is the value of f?*

Exer. 8.24.5a: L = 450 ft, D = 12 in, f = 0.028, both pipe ends are submerged, k_e = 0.5. Find Q when h_L = 150 ft.

BG

Table A.1 for water at 60°F: $\nu = 1.217 \times 10^{-5}$ ft²/s. $e/D = 0.005/1 = 0.005$.

Energy: $h_L = [\Sigma k + (fL/D)]V^2/(2g)$, i.e. $150 = [1.5 + (450/1)f]V^2/(2 \times 32.2)$ (1)

Continuity: $Q = AV = (\pi/4)D^2V$, i.e. $Q = (\pi/4)(1^2)V$ (2)

Eq. 8.1: $\mathbf{R} = DV/\nu$, i.e. $\mathbf{R} = 1V/(1.217 \times 10^{-5})$ (3)

Eq. 8.40: $\dfrac{1}{\sqrt{f}} = -2\log\left[\dfrac{e/D}{3.7} + \dfrac{2.51}{\mathbf{R}\sqrt{f}}\right]$, i.e. $\dfrac{1}{\sqrt{f}} = -2\log\left[\dfrac{0.005}{3.7} + \dfrac{2.51}{\mathbf{R}\sqrt{f}}\right]$ (4)

In these four equations the four unkowns are *f*, *V*, *Q*, and **R**. Solve for them as in Sample Prob. 8.9, (*a*) by making trials for *f* in (1) and solving for *V*, then with this *V* solving for **R** in (3) and *f* in (4) or more conveniently in Eq. 8.41, and repeating these steps until the resulting *f* equals the assumed *f*, when *Q* may be found from (2); or (*b*) by using an equation solver on a programmable calculator to solve Eq. 8.61 for *Q* and then find *V*, **R**, and *f* from (2), (3), and Eq. 8.41; or (*c*) by using Mathcad to solve (1) - (4) simultaneously.

Results: $Q = 19.81$ cfs ◀ $V = 25.2$ fps, $\mathbf{R} = 2{,}072{,}372$, $f = 0.0304$ ◀

Sec. 8.24: Solution of Single-Pipe Flow Problems with Minor Losses -- *Problems*

8.58 *Water flows at 12 fps through a vertical 4-in-diameter pipe standing in a body of water with its lower end 5 ft below the surface. Considering all losses and with f = 0.026, find the pressure in the pipe at a point 1 ft above the surface of the water when the flow is (a) downward; (b) upward.*

BG

(*a*) Downward flow.

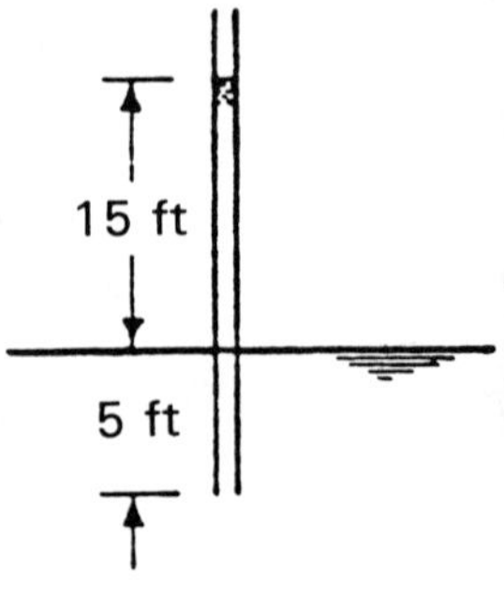

$$h_L = h_d' + f\frac{L}{D}\frac{V^2}{2g} = \left(1 + 0.026\frac{(15 + 5)}{4/12}\right)\frac{12^2}{2(32.2)} = (2.56)2.24 = 5.72 \text{ ft}$$

Energy: $(p/\gamma + 15 + 2.24) - 0 = 5.72$ ft

$p/\gamma = -11.51$ ft; $p = -11.52(62.4/144) = -4.99$ psi ◀

(*b*) Upward flow. Fig. 8.13: $k_e = 0.8$

$$h_L = \left(k_e + f\frac{L}{D}\right)\frac{V^2}{2g} = \left(0.8 + 0.026\frac{(15 + 5)}{4/12}\right)\frac{12^2}{2(32.2)} = (2.36)2.24 = 5.28 \text{ ft}$$

Energy: $0 - (p/\gamma + 15 + 2.24) = 5.28$; $p/\gamma = -22.5$ ft; $p = -22.5(62.4/14.4) = -9.76$ psi ◀

8.59 *Water flows at 2.5 m/s through a vertical 10-cm-diameter pipe standing in a body of water with its lower end 1 m below the surface. Considering all losses and with f = 0.024, find the pressure in the pipe at a point 4 m above the surface of the water when the flow is (a) downward; (b) upward.*

SI

(*a*) Downward flow.

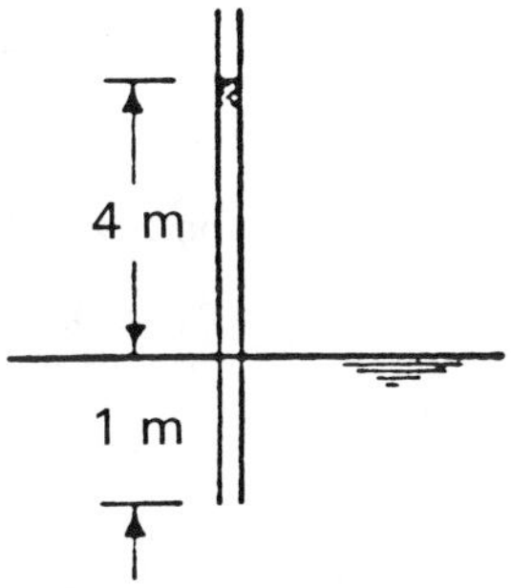

$$h_L = \left(k_d + f\frac{L}{D}\right)\frac{V^2}{2g} = \left(1 + 0.024\frac{(4+1)}{0.10}\right)\frac{2.5^2}{2(9.81)} = (2.20)0.319 = 0.701 \text{ m}$$

Energy: $(p/\gamma + 4 + 0.319) - 0 = 0.701$ m

$p/\gamma = -3.62$ m; $p = -3.62(9.81) = -35.5$ kN/m^2 ◄

(*b*) Upward flow. Fig. 8.13: $k_e = 0.8$

$$h_L = \left(k_e + f\frac{L}{D}\right)\frac{V^2}{2g} = \left(0.8 + 0.024\frac{(46+1)}{0.10}\right)\frac{2.5^2}{2(9.81)}$$

$= (2.0)0.319 = 0.637$ m

Energy: $0 - (p/\gamma + 4 + 0.319) = 0.637$; $p/\gamma = -4.96$ m

$p = -4.96(9.81) = -48.6$ kN/m^2 ◄

8.60 *A 12-in-diameter pipe (f = 0.028) 450 ft long runs from one reservoir to another, both ends of the pipe being under water. The intake is square-edged. The difference between the water levels in the two reservoirs is 150 ft. Find (a) the flow rate, and (b) the pressure in the pipe at a point 320 ft from the intake, where the elevation is 135 ft lower than the surface of the water in the upper reservoir.*

BG

(*a*) $h_L = 150 = [0.5 + 1 + 0.028(450/1)]V^2/(2g) = 14.10V^2/2g$

$V^2/2g = 150/14.10 = 10.64$ ft; $V = 26.2$ fps; $Q = 20.6$ cfs ◄

(*b*) $h_L' = [0.5 + 0.028(320/1)]10.64 = 100.6$ ft

Energy: $135 - (p/\gamma + V^2/2g) = 100.6$; $p/\gamma = 23.7$ ft; $p = 23.7(62.4/144) = 10.28$ psi ◄

8.61 *A 32-cm-diameter pipe (f = 0.025) 140 m long runs from one reservoir to another, both ends of the pipe being under water. The intake is square-edged. The difference between the water levels in the two reservoirs is 36 m. Find (a) the flow rate, and (b) the pressure in the pipe at a point 95 m from the intake, where the elevation is 39 m lower than the surface of the water in the upper reservoir.*

SI

(*a*) $h_L = 36$ m $= [0.5 + 1 + 0.025(140/0.32)]V^2/(2g) = 12.44V^2/2g$

$V^2/2g = 36/12.44 = 2.89$ m; $V = 7.42$ m/s; $Q = 0.606$ m^3/s ◄

(*b*) $h_L' = [0.5 + 0.025(95/0.32)]2.89 = 22.9$ m

Energy: $39 - (p/\gamma + V^2/2g) = 22.9$; $p/\gamma = 13.18$ m; $p = 13.18(9.81) = 129.3$ kN/m^2 ◄

8.62 *A pump delivers water through 300 ft of 4-in fire hose ($f = 0.025$) to a nozzle which throws a 1-in-diameter jet. The loss coefficient of the nozzle is 0.04. The nozzle is 20 ft higher than the pump, and a jet velocity of 70 fps is required. What must be the pressure in the hose at the pump?*

BG

Nozzle loss $= 0.04V_j^2/(2g)$; continuity: $V_j = (D_1/D_j)^2V_1 = 16V_1$

Hose friction $= 0.025(300 \times 12/4)V_1^2/(2g) = 22.5V_1^2/(2g) = 0.0879V_j^2/(2g)$

Energy: $p_1/\gamma + V_1^2/(2g) = 20 + V_j^2/(2g) + (0.04 + 0.0879)V_j^2/(2g)$

where $V_j^2/(2g) = 70^2/(2 \times 32.2) = 76.1$ ft

$p_1/\gamma = 20 + 76.1 + 0.1279(76.1) - 76.1/16^2 = 105.5$ ft; $p_1 = 105.5(62.4/144) = 45.7$ psi ◀

8.63 *A pump delivers water through 100 m of 10 cm fire hose ($f = 0.025$) to a nozzle which throws a 2.5-cm-diameter jet. The loss coefficient of the nozzle is 0.04. The nozzle is 6 m higher than the pump, and a jet velocity of 20 m/s is required. What must be the pressure in the hose at the pump?*

SI

Nozzle loss $= 0.04V_j^2/(2g)$; continuity: $V_j = (D_1/D_j)^2V_1 = 16V_1$

Hose friction $= 0.025(100/0.10)V_1^2/(2g) = 25V_1^2/(2g) = 0.0977V_j^2/(2g)$

Energy: $p_1/\gamma + V_1^2/(2g) = 6 + V_j^2/(2g) + (0.04 + 0.0977)V_j^2/(2g)$

where $V_j^2/(2g) = 20^2/(2 \times 9.81) = 20.4$ m

$p_1/\gamma = 6 + 20.4 + 0.1377(20.4) - 20.4/16^2 = 29.1$ m; $p_1 = 29.1(9.81) = 286$ kN/m^2 ◀

8.64 *A 6-in-diameter horizontal jet of water is discharged into air through a nozzle (loss coefficient 0.15) at a point 150 ft lower than the water level above the intake. The 12-in-diameter pipeline ($f = 0.015$) is 600 ft long, with a square-edged nonprojecting entrance. What is (a) the jet velocity, and (b) the pressure at the base of the nozzle?*

BG

(*a*) Nozzle loss $= 0.15V_j^2/(2g)$; continuity: $V_j = (D/D_j)^2V = 4V$

$h_L = [0.5 + 0.015(600/1)]V^2/(2g) + 0.15V_j^2/(2g) = 9.5V^2/(2g) + 0.15V_j^2/(2g)$

$= (9.5/4^2)V_j^2/(2g) + 0.15V_j^2/(2g) = 0.744V_j^2/(2g)$

Energy: $150 - 0.744V_j^2/(2g) = V_j^2/(2g)$; $V_j^2/(2g) = 86.0$ ft

$V_j = [2(32.2)86.0]^{1/2} = 74.4$ fps ◀

(*b*) (*i*) Energy, from water surface to nozzle base:

$(0 + 150 + 0) - 9.5V^2/(2g) = [p_{base}/\gamma + 0 + V^2/(2g)]$

$p_{base}/\gamma = 150 - 10.5V^2/2g = 150 - (10.5/4^2)V_j^2/2g = 150 - (10.5/16)86.0 = 93.5$ ft.

(*ii*) Energy, from nozzle base to jet:

$[p_{base}/\gamma + z + V^2/(2g)] - 0.15V_j^2/(2g) = [0 + z + V_j^2/(2g)]$

$p_{base}/\gamma = 1.15V_j^2/(2g) - V^2/(2g) = 1.15V_j^2/(2g) - (1/4^2)V_j^2/(2g) = 1.088V_j^2/(2g)$ ft

$= 1.088(86.0) = 93.5$ ft ; $p_{base} = 93.5(62.4/144) = 40.5$ psi ◀

8.65 *Refer to Fig. 8.23b. Suppose Δz = 50 ft, the line is 600 ft of 8-in-diameter pipe (f = 0.025), and the nozzle loss coefficient is 0.05. Find the jet diameter that will result in the greatest jet horsepower.*

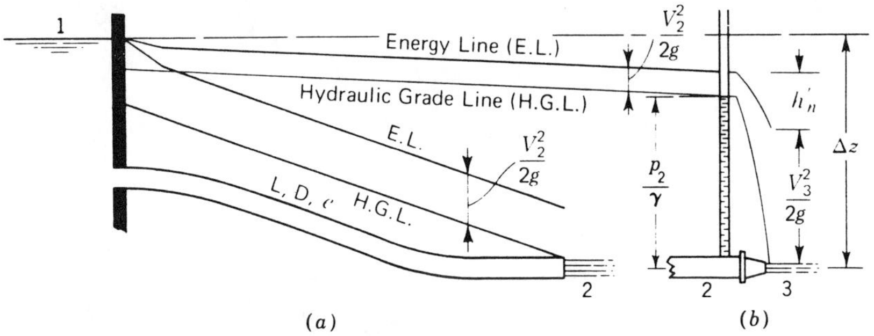

Figure 8.23

BG

Energy: $50 - 0.5V_8^2/(2g) - 0.025(600 \times 8/12)V_8^2/(2g) - 0.05V_j^2/(2g) = V_j^2/(2g)$

$50 - 23(V_8^2/(2g) = 1.05(V_j^2/2g)$

Continuity: $D_j^2V_j = D_8^2V_8$; $V_8 = V_j(D_j/D_8)^2$; D_8 = 8 inches

$50 = 1.05V_j^2/(2g) + 23(D_j/D_8)^4(V_j^2/2g)$; $V_j^2/(2g) = 50/[1.05 + 23(D_j/D_8)^4]$

Eq. 5.32: $hp = \gamma Q(V_j^2/2g)/550 = \gamma A_jV_j(V_j^2/2g)/550 = 0.001\,762A_jV_j^3$

D_j (in)	A_j (ft^2)	$V_j^2/2g$	V_j	hp
0	0	--	--	0
1.0	0.005 45	47.6	55.3	1.62
2.0	0.0218	44.0	53.0	5.70
3.0	0.0491	33.2	46.2	8.50
4.0	0.0873	20.1	36.0	7.15
6.0	0.196	6.04	19.7	2.61
8.0	0.349	2.08	11.5	0.93

Maximum power occurs with a jet diameter of approximately 3.0 inches. ◄

(An exact answer may be obtained by using differential calculus.)

8.66 *Refer to Fig. 8.23b. Suppose Δz = 15 m, the line is 180 m of 20-cm-diameter pipe (f = 0.025), and the nozzle loss coefficient is 0.05. Find the jet diameter that will result in the greatest jet power.*

See Fig. 8.23b accompanying Solution 8.65.

SI

Energy: $15 - 0.5V_8^2/(2g) - 0.025(180/0.20)V^2/(2g) - 0.05V_j^2/(2g) = V_j^2/(2g)$

$15 - 23(V^2/(2g) = 1.05(V_j^2/2g)$; continuity: $D_j^2V_j = D^2V$; $V = V_j(D_j/D)^2$; D = 20 cm

$15 = 1.05V_j^2/(2g) + 23(D_j/D)^4(V_j^2/2g)$; $V_j^2/(2g) = 15/[1.05 + 23(D_j/D)^4]$

Eq. 5.33: $kW = \gamma Q(V_j^2/2g)/1000 = \gamma A_jV_j(V_j^2/2g)/1000 = 0.5A_jV_j^3$

/cont...

D_j (cm)	A_j (m^2)	$V_j^2/2g$ (m)	V_j (m/s)	kW
0	0	--	--	--
2	0.000 314	14.25	16.72	0.734
4	0.001 257	13.80	16.46	2.80
6	0.002 83	12.13	15.43	5.20
8	0.005 03	9.15	13.40	6.05
10	0.007 85	6.03	10.88	5.05
12	0.011 31	3.72	8.54	3.53
14	0.015 39	2.28	6.69	2.31
16	0.0201	1.433	5.30	1.498

Maximum power occurs with a jet diameter of approximately 8 cm. ◄

(An exact answer may be obtained by using differential calculus.)

8.67 *Water at 60°F flows through 800 ft of 12-in-diameter pipe between two reservoirs whose water-surface elevation difference is 16 ft. The pipe entrance is flush and square-edged, and there is a half-open gate valve in the line. Using only a basic scientific calculator, find the flow rate (a) if e = 0.0018 in, and (b) if e is twenty times larger.*

A basic scientific calculator is here defined to be one that is not programmable and does not have automatic equation solving capabilities.

BG

$L/D = 800 < 1000$, so minor losses are not negligible.

Energy: $\Delta z = 16 \text{ ft} = \Sigma h_L = [0.5 + 2.06 + 1.0 + f(800/1)]V^2/2g$

from which $V = [32(32.2)/(800f + 3.56)]^{1/2}$ (1)

Table A.1 for water at 60°F: $\nu = 1.217 \times 10^{-5}$ ft²/sec

Eq. 8.1: $\mathbf{R} = DV/\nu = (1)V/(1.217 \times 10^{-5}) = 8.22 \times 10^4 V$ (2)

(*a*) $e/D = 0.0018/12 = 0.000\,15$

Solve by trial satisfying (1), (2), and Fig. 8.11 or Eq. 8.41:

Try f	V from (1)	**R** from (2)	Obtained f	
0.03	6.11 fps	5.02×10^5	0.0149	Try again
0.0149	8.15	6.70×10^5	0.0145	Close enough
0.0145	8.23			

$Q = (\pi/4)(1^2)8.23 = 6.47$ cfs ◄

(*b*) $e = 20(0.0018) = 0.036$ in; $e/D = 0.036/12 = 0.003$

Try f	V from (1)	**R** from (2)	Obtained f	
0.03	6.11 fps	5.02×10^5	0.0264	Try again
0.0264	6.46	5.31×10^5	0.0264	Converged.

$Q = (\pi/4)(1^2)6.46 = 5.07$ cfs ◄

Check minor losses: Minor losses / friction losses $= 3.56/(fL/D)$

(*a*) $3.56/(0.0145 \times 800/1) = 0.307 = 30.7\%$

(*b*) $3.56/(0.0264 \times 800/1) = 0.1686 = 16.9\%$

These minor losses are large fractions of the friction loss, so may not be neglected.

8.68 *Solve Prob. 8.67 without manual trial and error, by using an equation solver (i) on a programmable calculator or (ii) in mathematics software like Mathcad.*

Prob. 8.67: 60°F water flows through a pipeline, L = 800 ft, D = 12 in, Σk = 3.56, h_L = 16 ft. Find Q (a) if e = 0.0018 in, and (b) if e is 20 times larger.

BG

$L/D = 800 < 1000$, so minor losses are not negligible. $\Sigma k = 0.5 + 2.06 + 1.0 = 3.56$

Table A.1 for water at 60°F: $\nu = 1.217 \times 10^{-5}$ ft²/sec

Energy: $h_L = [\Sigma k + (fL/D)]V^2/2g$, i.e. $16 = [3.56 + (800/1)f]V^2/(2 \times 32.2)$ (1)

Continuity: $Q = AV = (\pi/4)D^2V$, i.e. $Q = (\pi/4)(1^2)V$ (2)

Eq. 8.1: $\mathbf{R} = DV/\nu$, i.e. $\mathbf{R} = (1)V/(1.217 \times 10^{-5})$ (3)

Eq. 8.40: $1/\sqrt{f} = -2\log[(e/D)/3.7 + 2.51/(\mathbf{R}\sqrt{f})]$ (4)

(*a*) $e/D = 0.0018/12 = 0.000\,15$; Substitute this into Eq. (4)

(*i*) Use an equation solver on a programmable calculator to solve Eq. 8.61 for Q.
Then find V, $\mathbf{R}$, and f from (2), (3) and Eq. 8.41 respectively.

(*ii*) Use Mathcad to solve Eqs. (1) – (4) simultaneously for Q, V, $\mathbf{R}$, f.

Results: $Q = 6.47$ cfs ◄, $V = 8.24$ fps, $\mathbf{R} = 676{,}757$, $f = 0.014\,54$.

(*b*) $e/D = 20(0.0018/12) = 0.003$; Substitute this into Eq. (4). Repeat (*i*) and (*ii*) as for (*a*).

Results: $Q = 5.07$ cfs ◄, $V = 6.46$ fps, $\mathbf{R} = 530{,}709$, $f = 0.0264$.

See solution of Prob. 8.67 for a check of minor losses.

8.69 *A 150-m-long commercial steel pipe is to convey 30 L/s of oil (s = 0.9, μ = 0.038 N·s/m²) from one tank to another where the difference in elevation of the free liquid surfaces is 2 m. The pipe entrance is flush and square-edged, and there is a fully-open gate valve in the line. Using only a basic scientific calculator, find the diameter theoretically required.*

A basic scientific calculator is here defined to be one that is not programmable and does not have automatic equation solving capabilities.

SI

Eq. 2.11: $\nu = \mu/\rho = 0.038/(1000 \times 0.9) = 0.000\,0422$ m²/s

Eq. 8.60: $h_L = (fL/D + \Sigma k)V^2/(2g) = (fL/D + \Sigma k)[4Q/(\pi D^2)]^2/2g$

from which $f = [2gh_L(\pi D^2/4Q)^2 - \Sigma k](D/L)$ (1)

Eq. 8.1: $\mathbf{R} = DV/\nu = (D/\nu)[4Q/(\pi D^2)] = 4Q/(\pi \nu D)$ (2)

Sec. 8.18 for flush entrance: $k_e = 0.5$; Sec. 8.19 for submerged discharge: $k_d = 1.0$

Table 8.3 for fully open gate valve: $k_v = 0.19$. ∴ $\Sigma k = 0.5 + 1.0 + 0.19 = 1.69$

Substituting this and the given values into (1) and (2) we get $f = 179.3D^5 - D/88.8$ (3); $\mathbf{R} = 905/D$ (4)

From Table 8.1 for commercial steel pipe: $e = 0.000\,046$ m

Because of the form of (3), it is easier in this case to try values of D (than f); but we must check for f_{min}.

Try D (m)	e/D	f_{min} (Eq.8.39)	f (Eq. 3)	$\mathbf{R}$ (Eq. 4)	f (Fig. 8.11 or Eq. 8.41)
1.0	0.000 046	0.010 39	179.3	905 = lam.	--
0.5	0.000 092	0.011 79	5.60	1810 = lam.	--
0.1	0.000 46	0.016 39	0.000 666	9052	$f < f_{min}$ is imposs.
0.2	0.000 23	0.014 13	0.0551	4526	0.0387
0.19	0.000 242	0.014 28	0.0423	4764	0.0382
0.187	0.000 246	0.014 33	0.0389	4840	0.0380
0.186			0.0378	4866	0.0380 Close enough

Eq. 8.41 is valid because $\mathbf{R} > 2000$.

$L/D = 150/0.186 = 806 < 1000$, so minor losses must be included. $D = 0.186$ m ◄

8.70 *Solve Prob. 8.69 without manual trial and error, by using an equation solver (a) on a programmable calculator or (b) in mathematics software like Mathcad.*

Prob. 8.69: Oil (s = 0.9, μ = 0.038 N·s/m²) is to flow at 30 L/s through a commercial steel pipe, L = 150 m, Σk = 1.69 with h_L = 2 m. Find the D theoretically required.

SI

$\Sigma k = 0.5 + 0.19 + 1.0 = 1.69$

Eq. 2.11: $\nu = \mu/\rho = 0.038/(1000 \times 0.9) = 0.000\,0422\ \text{m}^2/\text{s}$

Table 8.1 for comercial steel: $e = 0.046$ mm $= 0.000\,046$ m

Knowns: $L = 150$ m, $g = 9.81$ m²/s, $h_L = 2$ m, $Q = 0.03$ m³/s, $\Sigma k = 1.69$, $e = 0.000\,046$ m,

$\nu = 0.000\,0422$ m²/s

Find: D, V, $\mathbf{R}$, f.

Energy: $h_L = [\Sigma k + (fL/D)]V^2/(2g) = 2 = [1.69 + f(150/D)]V^2/(2 \times 9.81)$ (1)

Continuity: $Q = AV = (\pi/4)D^2V = 0.03$ (2)

Eq. 8.1: $\mathbf{R} = DV/\nu = DV/0.000\,0422$ (3)

Eq. 8.40: $1/\sqrt{f} = -2\log[0.000\,046/(3.7D) + 2.51/(\mathbf{R}\sqrt{f})]$ (4)

(*a*) Use an equation solver on a programmable calculator to solve Eq. 8.61 for D.

Then find V, $\mathbf{R}$, and f from (2), (3) and Eq. 8.41 respectively.

(*b*) Use Mathcad to solve Eqs. (1) - (4) simultaneously for D, V, $\mathbf{R}$, f.

Results: $D = 0.1861$ m ◄ $V = 1.102$ m/s, $\mathbf{R} = 4863$, $f = 0.0380$.

Eqs. 8.40, 8.41, and 8.61 are valid because $\mathbf{R} > 2000$.

$L/D = 150/0.1861 = 806 < 1000$, so minor losses must be included.

Sec. 8.25: Pipeline with Pump or Turbine -- ***Exercises***

8.25.1 *An 80-in-diameter pipe (f = 0.025) 7252 ft long delivers water to a powerhouse at a point 1500 ft lower in elevation than the water surface at intake. When the pipe delivers 450 cfs, what is the horsepower delivered to the plant?*

BG

$A = (\pi/4)(80/12)^2 = 34.9$ ft²; $V = 450/34.9 = 12.89$ fps; $V^2/(2g) = 2.58$ ft

$h_L = [0.5 + 0.025(7252 \times 12/80)]2.58 = 71.5$ ft

Head delivered to plant, $h = \Delta z - h_L = 1500 - 71.5 = 1429$ ft

Eq. 5.32: Power delivered $= 62.4(450)1429/550 = 72{,}900$ hp ◄

8.25.2 *A 10-in pipeline (f = 0.020) is 3 miles long. If 4 cfs of water is to be pumped through it, with a total actual lift of 25 ft, what will be the power required? The pump efficiency is 72 percent.*

BG

$V = Q/A = 4/[\pi(10/12)^2/4] = 7.33$ fps; $V^2/(2g) = 0.835$ ft

Eq. 8.10: $h_{Lf} = 0.020[3(5280)12/10]0.835 = 318$ ft; $h_p = 25 + 318 = 343$ ft

Using Eq. 5.32: Power required $= 62.4(4)343/(550 \times 0.72) = 216$ hp ◄

8.25.3 *A 25-cm pipeline (f = 0.025) is 4.7 km long. If 0.1 m^3/s of water is to be pumped through it, with a total actual lift of 10.5 m, what will be the power required? The pump efficiency is 75 percent.*

SI

$V = Q/A = 0.4/[0.25^2\pi/4)] = 2.04$ m/s; $V^2/(2g) = 0.212$ m

Eq. 8.10: $h_{Lf} = 0.025(3)(4700/0.25)0.212 = 99.4$ m; $h_p = 10.5 + 99.4 = 109.9$ m

Using Eq. 5.33: Power required = $9810(0.1)109.9/(1000 \times 0.75) = 143.8$ kW ◄

8.25.4 *A 12-in pipe 9200 ft long (f = 0.024) discharges freely into the air at an elevation 18 ft below the surface of the water at intake. It is necessary that the flow be doubled by inserting a pump. If the efficiency of the pump is 73 per cent, what will be the power required?*

BG

Eq. 8.10: $18 = 0.024(9200/1)V^2/(2g) + V^2/(2g)$; $V^2/(2g) = 0.0812$ ft

$V = 2.29$ fps; $Q_1 = AV = (\pi 1^2/4)2.29 = 1.796$ cfs

If the flow is doubled, $Q_2 = 2(1.796) = 3.59$ cfs

In the above $h_L = 18$ ft, and for double the flow (constant f), $h_L = 4(18) = 72$ ft

However gravity supplies 18 ft and so the pump must provide 54 ft

Eq. 5.32: Power required = $62.4(3.59)54/(550 \times 0.73) = 30.1$ hp ◄

8.25.5 *A 30-cm pipe 3400 m long (f = 0.022) discharges freely into the air at an elevation 5.6 m below the surface of the water at intake. It is necessary that the flow be doubled by inserting a pump. If the efficiency of the pump is 76 per cent, what will be the power required?*

SI

Eq. 8.10: $5.6 = 0.022(3400/0.3)V^2/(2g) + V^2/(2g)$; $V^2/(2g) = 0.0224$ m

$V = [2(9.81)0.0224]^{1/2} = 0.662$ m/s; $Q_1 = AV = (\pi 0.15^2)0.662 = 0.0468$ m^3/s

$Q_2 = 0.0937$ m^3/s; $h_L \approx 2^2(5.6) = 22.4$ m; $h_p = 22.4 - 5.6 = 16.8$ m

Eq. 5.33: Power required = $9810(0.0937)16.8/(1000 \times 0.76) = 20.3$ kW ◄

8.25.6 *Refer to Fig. 8.25. When the pump is delivering 1.2 cfs of water, a pressure gage at D reads 25 psi, while a vacuum gage at C reads 10 in Hg. The pressure gage at D is 2 ft higher than the vacuum gage at C. The pipe diameters are 4 in for the suction pipe and 3 in for the discharge pipe. Find the power delivered to the water.*

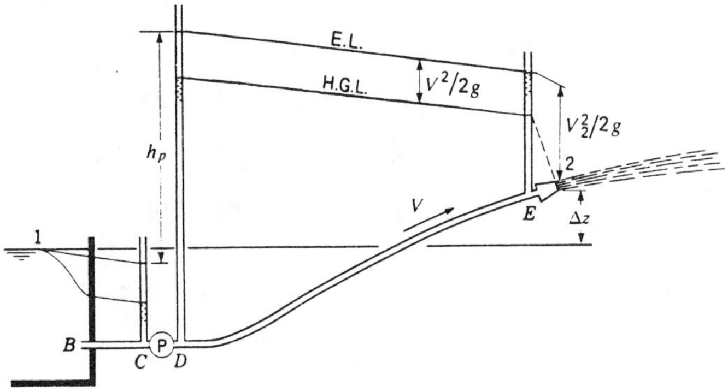

Figure 8.25

BG

$V_s = Q/A = 1.2/0.0873 = 13.75$ fps

$V_s^2/(2g) = 2.94$ ft

25 psi = 57.7 ft water

$V_d = Q/A = 1.2/0.491 = 24.4$ fps; $V_d^2/(2g) = 9.28$ ft; 10 inHg = 11.3 ft water

$h_p = (57.7 + 2 + 9.28) - (-11.3 + 2.94) = 77.3$ ft

Eq. 5.32: Power delivered = $62.4(1.2)77.3/550 = 10.53$ hp ◄

8.25.7 *Refer to Fig. 8.25. When the pump is delivering 35 L/s of water, a pressure gage at D reads 175 kN/m², while a vacuum gage at C reads 25 cmHg. The pressure gage at D is 60 cm higher than the vacuum gage at C. The pipe diameters are 10 cm for the suction pipe and 7.5 cm for the discharge pipe. Find the power delivered to the water.*

See Solution 8.25.6 for Fig. 8.25.

SI

$V_s = Q/A = 0.035/(0.10^2\pi/4) = 4.46$ m/s; $V_s^2/(2g) = 1.012$ m

$V_d = Q/A = 0.035/(0.075^2\pi/4) = 7.92$ m/s; $V_d^2/(2g) = 3.20$ m

175 kN/m² = 17.84 m water; 25 cmHg = 3.39 m water.

$h_p = (17.84 + 0.60 + 3.20) - (-3.39 + 1.012) = 24.0$ m

Eq. 5.33: Power delivered = 9810(0.035)24.0/1000 = 8.25 kW ◄

8.25.8 *If in Sample Problem 8.11 the vapor pressure of the liquid is 1.9 psia and the atmospheric pressure is 14.5 psia, what is the maximum theoretical flow rate?*

Sample Prob. 8.11: Liquid with γ = 52 lb/ft³ is in a closed tank, with gas at 5 psi in the space above it. It flows through 50 ft of 6-in-diameter suction line (f = 0.025), through a pump (elevation = 15 ft above tank liquid surface) which adds 2.0 hp to the liquid, and through 200 ft of 8-in-diameter line (f = 0.030). All flow is turbulent. The discharge is submerged, into liquid with its surface 10 ft below that in the tank.

BG

Eq. 5.36: $p_c/\gamma = -(p_{at}/\gamma - p_v/\gamma) = -(144/52)(14.5 - 1.9) = -34.9$ ft

Energy, tank to pump inlet: $\left(\frac{5(144)}{52} + 0 + 0\right) - 0.5\frac{V_6^2}{2g} - 0.025\left(\frac{50}{6/12}\right)\frac{V_6^2}{2g} = \left(-34.9 + 15 + \frac{V_6^2}{2g}\right)$

$V_6^2/(2g) = 8.43$, $V_6 = 23.3$ fps; $Q_{max} = (0.1964)23.3 = 4.58$ cfs ◄

8.25.9 *In Fig. 8.26 assume the pipe diameter is 12 in, f = 0.021, BC = 200 ft, and Δz = 120 ft. The entrance to the pipe at the intake is flush with the wall, and discharge losses are negligible. (a) If Q = 8 cfs of water, what is the head supplied to the turbine? (b) What is the power delivered by the turbine if its efficiency is 75 per cent?*

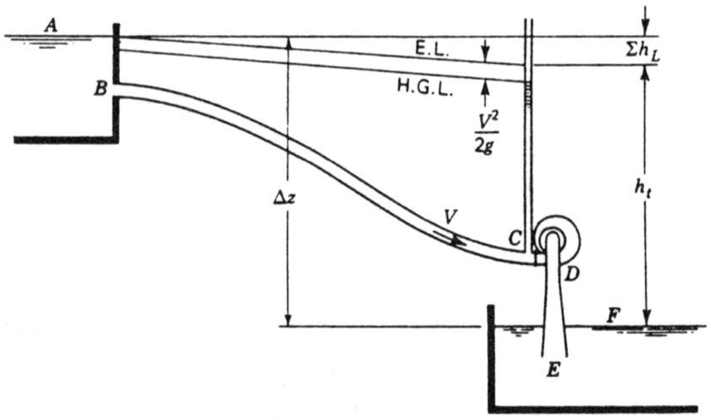

Figure 8.26

BG

(*a*) $V = Q/A = 8/0.7854 = 10.19$ fps; $V^2/2g = 1.611$ ft

Fig. 8.13: $k_e = 0.5$

$\therefore h_L = [0.5 + 0.021(200/1)]1.611 = 7.57$ ft

Eq. 8.65: $h_t = 120 - 7.57 = 112.4$ ft ◄

(*b*) Eq. 5.32: Power delivered by turbine = (0.75)62.4(8)112.4/550 = 76.5 hp ◄

8.25.10 *In Fig. 8.26 assume the pipe diameter is 30 cm, f = 0.021, BC = 60 m, and Δz = 36.5 m. The entrance to the pipe at the intake is flush with the wall, and discharge losses are negligible. (a) If Q = 225 L/s of water, what is the head supplied to the turbine? (b) What is the power delivered by the turbine if its efficiency is 75 per cent?*

See Solution 8.25.9 for Fig. 8.26.

SI

(*a*) $V = Q/A = 0.225/(\pi 0.15^2) = 3.18$ m/s; $V^2/(2g) = 0.516$ m

Fig. 8.13: $k_e = 0.5$, $\therefore h_L = [0.5 + 0.021(60/0.3)]0.516 = 2.43$ m

Eq. 8.65: $h_t = 36.5 - 2.43 = 34.1$ m ◄

(*b*) Eq. 5.33: Power delivered by turbine = (0.75)9810(0.225)34.1/1000 = 56.4 kW ◄

8.71 *In Fig. 8.24 assume the pipe diameter is 9 in, $f = 0.025$, $BC = 20$ ft, $DE = 2800$ ft, and $\Delta z = 125$ ft. (a) If $Q =$ *5.5 cfs of water and the pump efficiency is 78 per cent, what horsepower is required? (b) If the elevation of C above the lower water surface is 12 ft, that of D is 15 ft, and that of E is 100 ft, compute the pressure heads at C, D, and E. (c) Sketch the energy line and the hydraulic grade line. [*Note this correction to the textbook.]*

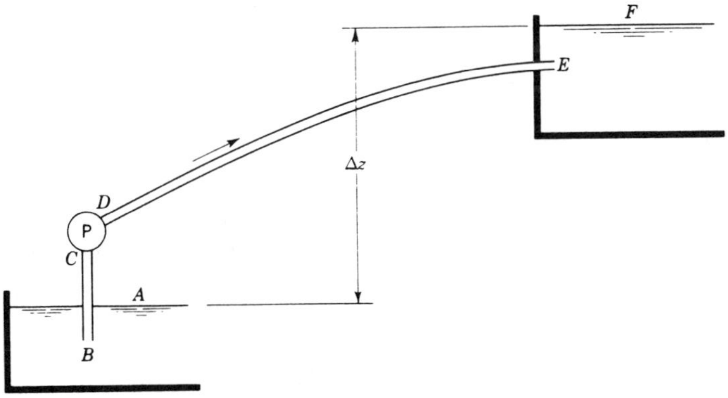

Figure 8.24

BG

(*a*) $V = Q/A = 5.5/0.442 = 12.45$ fps; $V^2/(2g) = 2.41$ ft

Suction pipe, $h_L = [0.8 + 0.025(20)12/9]2.41 = 3.53$ ft

Discharge pipe, $h_L = [0.025(2800)12/9 + 1]2.41 = 227$ ft

Total $h_L = 3.53 + 227 = 231$ ft; Pump head, $h_p = 125 + 231 = 356$ ft

Eq. 5.32: hp delivered to the water $= 62.4(5.5)356/550 = 222$ hp

From Eq. 5.34: hp delivered to the pump (excluding motor) $= 222/0.78 = 284$ hp ◀

(*b*) Eq. 5.28: $H_A = 0 + 0 + 0 = 0$; $H_B = p_B/\gamma - 8 + 2.41$; $H_C = p_C/\gamma + 12 + 2.41$

$H_D = p_D/\gamma + 15 + 2.41$; $H_E = p_E/\gamma + 100 + 2.41$; $H_F = 0 + 125 + 0$

Energy, *AB*: $H_A - H_B = 0 - (p_B/\gamma - 5.59) = 0$; $\therefore p_B/\gamma = 5.59$ ft

Energy, *AC*: $H_A - H_C = 0 - (p_C/\gamma + 12 + 2.41) = 3.53$; $\therefore p_C/\gamma = -17.94$ ft ◀

Energy, *DF*: $H_D - H_F = (p_D/\gamma + 17.41) - 125 = 227$ ft; $\therefore p_D/\gamma = 335$ ft ◀

Energy, *EF*: $H_E - H_F = (p_E/\gamma + 102.41) - 125 = 2.41$ or $p_E/\gamma = 25.0$ ft ◀ (either inside pipe or just outside.)

(*c*) For sketch: Positions of lines relative to *A* (datum, assumed at elevation 0 ft):

Point:	*A*	*B*	*C*	*D*	*E*	*F*	
p/γ, ft	0	5.59	−17.94	334.62	25.00	0	
z, ft	0	−8.00	12.00	15.0	100.00	125.00	
$V^2/(2g)$, ft	0	2.41	2.41	2.41	2.41	0	
HGL, ft	0	−2.41	−5.94	349.62	125.00	125.00	◀◀
EL, ft	0	0	−3.53	352.03	127.41	125.00	◀◀

8.72 *In Fig. 8.24 assume a pipe diameter of 35 cm, $f = 0.016$, $BC = 12$ m, $DE = 920$ m, and $\Delta z = 48$ m. Find the maximum theoretical flow rate if 15°C water is being pumped at an altitude of 1000 m above sea level. Point C is 6.0 m above the lower water surface.*

See Solution 8.71 for Fig. 8.24.

SI

Table A.1 for 15°C water: $p_v/\gamma = 0.17$ m abs

Table A.3 for 1000 m elevation: $p = 89.876$ kN/m^2; $p_{atm}/\gamma = 89.9/9.81 = 9.16$ m of water

Energy equation from water surface A to pump intake C using absolute pressure heads, when $p_C = p_v$:

$$9.16 - 0.8(V^2/2g) - 0.016(12/0.35)V^2/2g = 0.17 + 6.0 + V^2/2g$$

$$2.99 = (V^2/2g)(0.8 + 0.549 + 1) = 2.35(V^2/2g)$$

$V^2/2g = 2.99/2.35 = 1.274$; $V = 5.00$ m/s, $Q_{max} = 0.481$ m^3/s ◄

8.73 *In Fig. 8.24 assume that the pipe diameter is 4 in, $f = 0.035$, $BC = 15$ ft, $DE = 180$ ft, and $\Delta z = 60$ ft. The elevation of C is 10 ft above the lower water surface. (a) If the pressure head at C is to be no less than −20 ft, what is the maximum rate at which the water can be pumped? (b) If the efficiency of the pump is 65 percent, what is the horsepower required?*

See Solution 8.71 for Fig. 8.24.

BG

$$h_L = [0.8 + 0.035(15)12/4]V^2/2g = (0.8 + 1.575)V^2/2g = 2.38V^2/2g$$

Energy from A to C: $(0 + 0 + 0) = (-20 + 10 + V^2/2g) + 2.38V^2/2g$

$V^2/2g = 10/3.38 = 2.96$ ft; $V = [2(32.2)2.96]^{1/2} = 13.81$ fps

(*a*) $Q_{max} = AV = 0.0491(13.81) = 1.205$ cfs

(*b*) $h_p = 60 + [0.8 + 1.575 + 0.035(180)12/4]2.96 = 123.0$ fps

Eq. 5.32: hp delivered to the water = 62.4(1.205)123.0/500 = 16.83 ◄

From Eq. 5.34: hp delivered to the pump (excluding motor) = 16.83/0.65 = 25.9 ◄

8.74 *Refer to Fig. 8.25. Suppose that the water-surface elevation, elevation of the pump, and elevation of the nozzle tip are 100, 90, and 120 ft, respectively. Pipe BC is 40 ft long, has a diameter of 8 in, with $f = 0.025$; pipe DE is 200 ft long, has a diameter of 10 in, with $f = 0.030$; the jet diameter is 6 in, and the nozzle loss coefficient is 0.04. Assume the pump is 80 percent efficient under all conditions of operation. Make a plot of flow rate and p_C/γ versus pump horsepower input. At what flow rate will cavitation occur in the pipe at C if the water temperature is 50°F and the atmospheric pressure is 13.9 psia?*

See Solution 8.25.6 above for Fig. 8.25.

BG

Table A.1 for water at 50°F: $p_v/\gamma = 0.41$ ft; Cavitation begins (Eq. 5.36) when

$$p_c/\gamma = -(p_{at}/\gamma - p_v/\gamma) = -[33.9(13.9/14.7) - 0.41] = -31.6 \text{ ft}$$

Continuity: $V_8 = (6/8)^2V_6 = 0.563V_6$; $V_{10} = (6/10)^2V_6 = 0.360V_6$

Energy equation from 1 to 2:

$$100 - \left[0.8 + 0.025\frac{40}{8/12}\right]\frac{(0.563V_6)^2}{2g} + h_p - 0.030\frac{200}{10/12}\frac{(0.360V_6)^2}{2g} - 0.04\frac{V_6^2}{2g} = 120 + \frac{V_6^2}{2g}$$

i.e., $h_p = 20 + 2.70(V_6^2/2g)$. But $V_6 = Q/0.1963 = 5.09Q$; $V_6^2 = 25.9Q^2$; so $h_p = 20 + 1.088Q^2$

Eq. 5.32: $\text{HP}_{input} = \gamma Q h_p/[550(0.8)] = 0.1418Q(20 + 1.088Q^2) = 2.84Q + 0.1543Q^3$

Energy Eq. from 1 to C: $100 - [0.8 + 0.025(40)/(8/12)](V_8^2/2g) = 90 + p_C/\gamma + (V_8^2/2g)$

/cont...

i.e., $p_C/\gamma = 10 - 3.3(V_8^2/2g)$. But $V_8 = Q/0.349 = 2.86Q$; $V_8^2 = 8.21Q^2$; so $p_C/\gamma = 10 - 0.421Q^2$

Let $p_C/\gamma = -31.6 = 10 - 0.421Q^2$; $Q = 9.95$ cfs when cavitation begins ◄

For plot:

Q (cfs)	h_p (ft)	HP_{input}	p_C/γ (ft)
0	20	0	10.0
2	24.4	5.91	8.32
4	37.4	21.2	3.27
6	59.2	50.3	−5.14
8	89.6	101.7	−16.91
10	128.8	182.6	−32.1
▲		▲	▲

8.75 *In a testing laboratory, a certain turbine has been found to discharge 10 cfs under a head h_t of 45 ft. In the field, it is to be installed near the end of a pipe 360 ft long. The supply line (flush entrance) and discharge line (submerged exit) will both be 10 inches in diameter with $f = 0.025$. The total fall from the surface of the water at intake to the surface of the tailwater will be 52 ft. What will be the head on the turbine, the rate of discharge, and the power delivered to the flow? Note that for turbines, $Q \propto \sqrt{h_t}$ (Eq. 16.17).*

BG

$V = Q/A = Q/[(\pi/4)(10/12)^2] = 1.833Q$

$h_L = [0.5 + 0.025(360)(12/10) + 1]V^2/2g = 12.30V^2/2g$

Substituting for V: $h_L = 12.30(1.833Q)^2/(2 \times 32.2) = 0.642Q^2$

Thus, in Eq. 8.65: $h_t = 52 - 0.642Q^2$

Also, from Eq. 16.17: $Q_1/Q = 10/Q = [45/(52 - 0.642Q^2)]^{1/2}$, i.e. $Q^2/100 = (52 - 0.642Q^2)/45$

from which $Q = 6.90$ cfs ◄ so that $h_t = 52 - 0.642(6.90)^2 = 21.4$ ft ◄

Eq. 5.32: Power $= 62.4(6.90)21.4/500 = 16.78$ HP ◄

8.76 *In a testing laboratory, a certain turbine has been found to discharge 285 L/s under a head h_t of 13.5 m. In the field, it is to be installed near the end of a pipe 110 m long. The supply line (flush entrance) and discharge line (submerged exit) will both be 25 cm in diameter with $f = 0.024$. The total fall from the surface of the water at intake to the surface of the tailwater will be 15.5 m. What will be the head on the turbine, the rate of discharge, and the power delivered to the flow? Note that for turbines, $Q \propto \sqrt{h_t}$ (Eq.16.17).*

SI

$V = Q/A = 4Q/(\pi D^2) = 4Q/(\pi 0.25^2) = 20.4Q$

$h_L = [0.5 + 0.024(110/0.25) + 1]V^2/2g = 12.06V^2/2g$

Substituting for V: $h_L = 12.06(20.4Q)^2/(2 \times 9.81) = 255Q^2$

Thus, in Eq. 8.65: $h_t = 15.5 - 255Q^2$

Also, from Eq. 16.17: $Q_1/Q = 0.285/Q = [13.5/(15.5 - 255Q^2)]^{1/2}$

i.e. $Q^2/0.285^2 = (15.5 - 255Q^2)/13.5$ from which $Q = 0.1918$ m³/s $= 191.8$ L/s ◄

so that $h_t = 15.5 - 255(0.1918)^2 = 6.11$ m ◄

Eq. 5.32: Power $= 9810(0.1918)6.11/1000 = 11.51$ kW ◄

8.77 *Assume the total fall from one body of water to another is 130 ft. The water is conveyed by 250 ft of 15-in pipe (f = 0.020) which has its entrance flush with the wall. At the end of the pipe is a turbine and draft tube which discharged 6 cfs of water when tested under a head of 63.1 ft in another location. Discharge losses are negligible. What would be the rate of discharge through the turbine and the head on it under the present conditions? Note that for turbines, $Q \propto \sqrt{h_t}$ (Eq. 16.17).*

BG

$V = Q/A = Q/[(\pi/4)(15/12)^2] = 0.815Q$; $h_L = [0.5 + 0.020(250)(12/15) + 0]V^2/2g = 4.50V^2/2g$

Substituting for V: $h_L = 4.50(0.815Q)^2/(2 \times 32.2) = 0.0464Q^2$; $\therefore$ Eq. 8.65: $h_t = 130 - 0.0464Q^2$

Also, from Eq. 16.17: $Q_1/Q = 6/Q = [63.1/(130 - 0.0464Q^2)]^{1/2}$

i.e. $Q^2/36 = (130 - 0.0464Q^2)/63.1$ from which $Q = 8.50$ cfs ◄

so that $h_t = 130 - 0.0464(8.50)^2 = 126.6$ ft ◄

8.78 *A pump is installed to deliver water from a reservoir of surface elevation zero to another of elevation 200 ft. The 12-in-diameter suction pipe (f = 0.020) is 40 ft long, and the 10-in-diameter discharge pipe (f = 0.032) is 4500 ft long. The pump head may be defined as $h_p = 300 - 20Q^2$, where the pump head h_p is in feet and Q is in cubic feet per second. Compute the rate at which this pump will deliver the water. Also, what is the horsepower input to the water?*

BG

Continuity: $V_{10} = (12/10)^2V_{12} = 1.440V_{12}$ and $Q = (\pi/4)1^2V_{12}$

Eq. 8.63: $h_p = 200 + 0.02(40/1)V_{12}^2/2g + 0.032(4500)/(10/12)V_{10}^2/2g$

$= 200 + 0.8V_{12}^2/2g + 172.8(1.440V_{12})^2/2g = 200 + 359V_{12}^2/2g = 200 + 9.04Q^2$

But also $h_p = 300 - 20Q^2$ (given). Equating these, $300 - 20Q^2 = 200 + 9.04Q^2$; $Q = 1.856$ cfs ◄

$h_p = 300 - 20(1.856)^2 = 231$ ft. Eq. 5.32: Power input = 62.4(1.856)231/550 = 48.7 HP ◄

NOTE: Problems 8.112 - 114 also involve pumps.

Sec. 8.26: Branching Pipes -- *Problems*

8.79 *In Fig. 8.27, suppose that pipe 1 is 36-in smooth concrete, 5000 ft long; pipe 2 is 24-in cast iron, 3000 ft long; and pipe 3 is 20-in cast iron, 1300 ft long. The elevations of the water surfaces in reservoirs A and B are 225 and 200 ft, respectively, and the discharge through pipe 1 is 42 cfs. The water temperature is 60°F. Using a basic scientific calculator only, find the elevation of the surface of reservoir C. Neglect minor losses and assume the energy line and hydraulic grade line are coincident.*

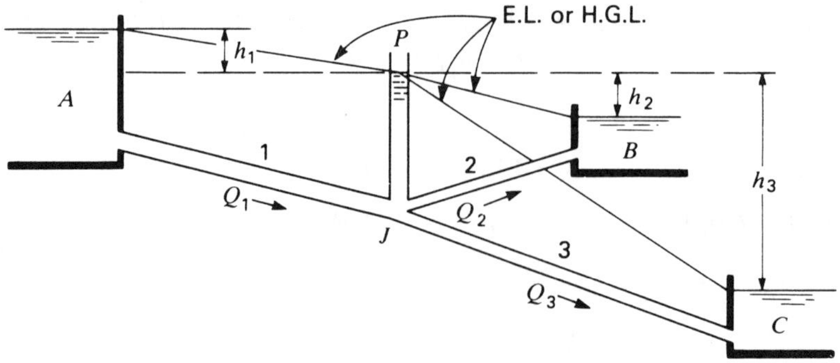

Figure 8.27

BG

This is an example of Case 1. Table A.1 for water at 60°F: $\nu = 1.217 \times 10^{-5}$ ft^2/sec

Pipe 1: $V_1 = Q_1/A_1 = 42/(\pi 3^2/4) = 5.94$ fps

Table 8.1 for smooth concrete: $e_1 = 0.001$ ft; $e/D = 0.001/3 = 0.000\,333$ /cont...

$\mathbf{R}_1 = D_1V_1/\nu = 3(5.94)/(1.217 \times 10^{-5}) = 1{,}464{,}700$

Haaland Eq. 8.41: $1/\sqrt{f_1} = -1.8\log[(0.000\,333/3.7)^{1.11} + 6.9/1{,}464{,}700]$; $f_1 = 0.015\,72$

Eq. 8.10: $h_1 = 0.015\,72(5000/3)5.94^2/(2 \times 32.2) = 14.36$ ft

So Elev P = Elev $A - h_1 = 225 - 14.36 = 210.64$ ft

Pipe 2: h_2 = Elev P − Elev B = 210.64 - 200 = 10.64 ft

Table 8.1 for cast iron: $e_2 = 0.000\,85$ ft; $e/D = 0.000\,85/2 = 0.000\,425$

In Eq. 8.42*b* for Pipe 2: The quantity $\sqrt{2gDh_L/L} = \sqrt{2(32.2)2(10.64)/3000} = 0.676$ fps

Eq. 8.42*b*: $Q_2 = -2(\pi 2^2/4)0.676\log[0.000\,425/3.7 + (2.51/2)1.217 \times 10^{-5}(0.676)] = 16.40$ cfs

Continuity at J: $Q_3 = Q_1 - Q_2 = 42 - 16.40 = 25.60$ cfs

Pipe 3: $V_3 = Q_3/A_3 = 25.6/[(\pi/4)(20/12)^2] = 11.73$ fps

$\mathbf{R}_3 = (20/12)11.73(1.217 \times 10^{-5}) = 1{,}607{,}100$; $e/D = 0.000\,85/(20/12) = 0.000\,51$

Haaland Eq. 8.41: $f_3 = 0.017\,08$; Eq. 8.10: $h_3 = 0.017\,08(1300)(12/20)11.73^2/(2 \times 32.2) = 28.49$ ft

∴ Elev C = Elev $P - h_3 = 210.64 - 28.49 = 182.15$ ft ◀

8.80 *In Fig. 8.27, suppose that pipe 1 is 90-cm smooth concrete, 1500 m long; pipe 2 is 60-cm cast iron, 900 m long; and pipe 3 is 50-cm cast iron, 400 m long. The elevations of the water surfaces in reservoirs A and B are 75 and 67 m, respectively, and the discharge through pipe 1 is 1.2 m³/s. The water temperature is 15°C. Using a basic scientific calculator only, find the elevation of the surface of reservoir C. Neglect minor losses and assume the energy line and hydraulic grade line are coincident.*

See Solution 8.79 for Fig. 8.27.

SI

This is an example of Case 1.

Table A.1 for water at 15°C: $\nu = 1.139 \times 10^{-6}$ m²/s

Pipe 1: $V_1 = Q_1/A_1 = 1.2/(\pi 0.9^2/4) = 1.886$ m/s

Table 8.1 for smooth concrete: $e_1 = 0.0003$ m; $e/D = 0.0003/0.9 = 0.000\,333$

$\mathbf{R}_1 = D_1V_1/\nu = 0.9(1.886)/(1.139 \times 10^{-6}) = 1\,490\,500$

Haaland Eq. 8.41: $1/\sqrt{f_1} = -1.8\log[(0.000\,333/3.7)^{1.11} + 6.9/1\,490\,500]$; $f_1 = 0.015\,71$

Eq. 8.10: $h_1 = 0.015\,71(1500/0.9)1.886^2/(2 \times 9.81) = 4.75$ m

So Elev P = Elev $A - h_1 = 75 - 4.75 = 70.25$ m

Pipe 2: h_2 = Elev P − Elev B = 70.25 - 67 = 3.25 m

Table 8.1 for cast iron: $e_2 = 0.000\,25$ m; $e/D = 0.000\,25/0.6 = 0.000\,417$

In Eq. 8.42*b* for Pipe 2: The quantity $\sqrt{2gDh_L/L} = \sqrt{2(9.81)0.6(3.25)/900} = 0.206$ m/s

Eq. 8.42*b*: $Q_2 = 2(\pi 0.6^2/4)0.206\log[0.000\,417/3.7 + (2.51/0.6)1.139 \times 10^{-6}(0.206)] = 460$ m³/s

Continuity at J: $Q_3 = Q_1 - Q_2 = 1.2 - 0.460 = 0.740$ m³/s

Pipe 3: $V_3 = Q_3/A_3 = 0.740/[(\pi/4)(0.5)^2] = 3.77$ m/s

$\mathbf{R}_3 = (0.5)3.77(1.139 \times 10^{-6}) = 1654\,400$; $e/D = 0.000\,25/0.5 = 0.0005$

Haaland Eq. 8.41: $f_3 = 0.017\,00$; Eq. 8.10: $h_3 = 0.017\,00(400/0.5)3.77^2/(2 \times 9.81) = 9.85$ m

∴ Elev C = Elev $P - h_3 = 70.25 - 9.85 = 60.40$ m ◀

☐ 8.81 *Solve Prob. 8.80 without manual trial and error, by using an equation solver in mathematics software like Mathcad.*

SI

Table A.1 for water at 15°C: $\nu = 1.139 \times 10^{-6}$ m²/s

Open a Mathcad worksheet and enter the following 12 known values, without units:

$g = 9.81$	$D1 = 0.90$	$L1 = 1500$	$e1 = 0.0003$
$\nu = 1.139 \cdot 10^{-6}$	$D2 = 0.60$	$L2 = 900$	$e2 = 0.000\,25$
$Q1 = 1.2$	$D3 = 0.50$	$L3 = 400$	$e3 = 0.000\,25$

Also enter guessed values for the following 14 unkowns:

$Q2 = 2$	$V1 = 3$	$h1 = 10$	$\mathrm{R1} = 800\,000$	$f1 = 0.01$
$Q3 = 2$	$V2 = 3$	$h2 = 2$	$\mathrm{R2} = 800\,000$	$f2 = 0.01$
	$V3 = 3$	$h3 = 10$	$\mathrm{R3} = 800\,000$	$f3 = 0.01$

In a "solve block" which begins with "Given" enter the following four governing equations for Pipe 1:

$$\frac{1}{\sqrt{f1}} = -2 \cdot \log\left(\frac{e1}{3.7 \cdot D1} + \frac{2.51}{\mathrm{R1} \cdot \sqrt{f1}}\right); \quad \mathrm{R1} = \frac{D1 \cdot V1}{\nu}$$

$$h1 = \frac{f1 \cdot L1 \cdot (V1)^2}{D1 \cdot 2 \cdot g}; \quad Q1 = \frac{V1 \cdot \pi \cdot (D1)^2}{4}$$

Enter four similar equations for Pipe 2, and four for Pipe 3, replacing the 1's with 2's and 3's. Also enter the equations $h1 + h2 = 8$ $\quad Q1 = Q2 + Q3$

Thus there are 14 simultaneous equations, corresponding to the 14 unknowns.

End the solve block with "Find($V1$, $V2$, $V3$, $h1$, $h2$, $h3$, R1, R2, R3, $f1$, $f2$, $f3$, $Q2$, $Q3$) ="

Mathcad obtains the following results:

	V	h	R	f	Q
Pipe 1:	1.886	4.76	1490 500	0.015 74	(given)
Pipe 2:	1.593	3.24	839 000	0.016 72	0.450
Pipe 3:	3.818	10.11	1676 100	0.017 01	0.750

So Elev C = Elev $A - h_1 - h_3 = 75 - 4.76 - 10.11 = 60.13$ m ◄

8.82 *With the sizes and lengths and materials of pipes given in Prob. 8.79, suppose that the surface elevations of reservoirs A and C are 250 and 180 ft, respectively, and the discharge through pipe 2 is 10 cfs into reservoir B. Using a basic scientific calculator only, find the surface elevation of reservoir B.*

See Solution 8.79 for Fig. 8.27. A basic scientific calculator is here defined to be one that is not programmable and does not have automatic equation solving capabilities.

BG

This is a Case 2 problem. Table A.1 for water at 60°F: $\nu = 1.217 \times 10^{-5}$ ft²/sec

	Pipe:	1	2	3
Given:	L, ft	5000	3000	1300
	D, ft	3	2	20/12
	e, ft (Table 8.1)	0.001	0.000 85	0.000 85
Calc:	L/D	1667	1500	780
	A, ft²	7.07	3.14	2.18
	e/D	0.000 333	0.000 425	0.000 51

Find the elevation of P (Fig. 8.27) by trial and error. /cont...

Elev. P lies between 180 and 250 ft. Calculate V from Eq. 8.42a. Trials:

El. P	h_1	h_3	V_1	V_3	Q_1	Q_3	ΣQ	Move P ?
200	50	20	11.15	9.81	78.8	19.97	+38.9	Up
	20	50	7.02	15.58	49.6	34.0	+5.64	Up
Extrapolation (Fig. 8.28): (El. P − 230)/(230 − 200) = 5.64/(38.9 − 5.64); El. P = 235.09								
235	15	55	6.07	16.34	42.9	35.7	−2.76	Down
Interpolation: (235 − El. P)/(235 − 230) = 2.76/(2.76 + 5.64); Elev P = 233.36, close enough!								

$\mathbf{R}_1 = D_1V_1/\nu = 2.75 \times 10^6$, $\mathbf{R}_3 = 2.13 \times 10^6$, both turbulent, OK.

$V_2 = Q_2/A_2 = 10/(\pi 2^2/4) = 3.18$ fps; $\mathbf{R}_2 = D_2V_2/\nu = 2(3.18)/(1.217 \times 10^{-5}) = 523{,}100$

Eq. 8.41: $f_2 = 0.017\,04$; Eq. 8.10: $h_2 = 4.02$ ft

Elev. B = Elev. $P - h_2 = 233.36 - 4.02 = 229.34$ ft ◄

8.83 *Solve Prob. 8.82 without manual trial and error, by using an equation solver in mathematics software like Mathcad.*

See Solution 8.79 for Fig. 8.27.

BG

Table A.1 for water at 60°F: $\nu = 1.217 \times 10^{-5}$ ft²/sec

Open a Mathcad worksheet and enter the following 12 known values, without units:

$g = 32.2$	$D1 = 36/12$	$L1 = 5000$	$e1 = 0.001$
$\nu = 1.217 \cdot 10^{-5}$	$D2 = 24/12$	$L2 = 3000$	$e2 = 0.00085$
$Q2 = 10$	$D3 = 20/12$	$L3 = 1300$	$e3 = 0.00085$

Also enter guessed values for the following 14 unkowns:

$Q1 = 20$	$V1 = 10$	$h1 = 10$	R1 = 800000	$f1 = 0.01$
$Q3 = 20$	$V2 = 10$	$h2 = 10$	R2 = 800000	$f2 = 0.01$
	$V3 = 10$	$h3 = 30$	R3 = 800000	$f3 = 0.01$

In a "solve block" which begins with "Given" enter the following four governing equations for Pipe 1:

$$\frac{1}{\sqrt{f1}} = -2\cdot\log\left(\frac{e1}{3.7\cdot D1} + \frac{2.51}{\mathrm{R1}\cdot\sqrt{f1}}\right); \quad \mathrm{R1} = \frac{D1\cdot V1}{\nu}; \quad h1 = \frac{f1\cdot L1\cdot (V1)^2}{D1\cdot 2\cdot g}; \quad Q1 = \frac{V1\cdot\pi\cdot(D1)^2}{4}$$

Enter four similar equations for Pipe 2, and four for Pipe 3, replacing the 1's with 2's and 3's. Also enter the equations $h1 + h3 = 70$ $\quad Q1 = Q2 + Q3$

Thus these are 14 simultaneous equations, corresponding to the 14 unknowns.

End the solve block with "Find($Q1$, $Q3$, $V1$, $V2$, $V3$, $h1$, $h2$, $h3$, R1, R2, R3, $f1$, $f2$, $f3$) ="

Mathcad obtains the following results:

	Q	V	h	$\mathbf{R}$	f
Pipe 1:	45.1	6.38	16.58	1,573,900	0.015 72
Pipe 2:	(given)	3.18	4.04	523,100	0.017 14
Pipe 3:	35.1	16.10	53.42	2,205,300	0.017 01

So Elev B = Elev $A - h_1 - h_2 = 250 - 16.58 - 4.04 = 229.37$ ft ◄

8.84 *With the sizes and lengths and materials of pipes given in Prob. 8.80, suppose that the surface elevations of reservoirs A and C are 60 and 38 m, respectively, and the discharge through pipe 2 is 0.3 m^3/s into reservoir B. Using a basic scientific calculator only [see Prob. 8.82], find the surface elevation of reservoir B.*

See Solution 8.79 for Figure 8.27.

SI

This is a Case 2 problem.

Table A.1 for water at 15°C: $\nu = 1.139 \times 10^{-6}\ m^2/s$

	Pipe:	1	2	3
Given:	L, m	1500	900	400
	D, m	0.90	0.60	0.50
	e, m (Table 8.1)	0.0003	0.000 25	0.000 25
Calc:	L/D	1667	1500	800
	A, m^2	0.636	0.283	0.1963
	e/D	0.000 333	0.000 417	0.0005

Find the elevation of P (Fig. 8.27) by trial and error.

Elev. P lies between 38 and 60 m. Calculate V from Eq. 8.42a. Trials:

El. P	h_1	h_3	V_1	V_3	Q_1	Q_3	ΣQ	Move P ?
50	10	12	2.75	4.16	1.747	0.817	+0.630	Up
55	5	17	1.934	4.96	1.231	0.974	−0.044	Down

Interpolation (Fig. 8.28): $\dfrac{55 - \text{El. } P}{55 - 50} = \dfrac{0.044}{0.044 + 0.630}$; El. P = 54.37, close enough!

$\mathbf{R}_1 = D_1V_1/\nu = 1.53 \times 10^6$, $\mathbf{R}_3 = 2.18 \times 10^6$, both turbulent, OK.

$V_2 = Q_2/A_2 = 0.3/(\pi 0.6^2/4) = 1.061$ m/s ; $\mathbf{R}_2 = D_2V_2/\nu = 0.6(1.061)/(1.139 \times 10^{-6}) = 558\,900$

Eq. 8.41: $f_2 = 0.016\,93$; Eq. 8.10: $h_2 = 1.457$ m

Elev. B = Elev. $P - h_2 = 54.67 - 1.46 = 53.21$ m ◄

8.85 *Repeat Prob. 8.82, except that the 10 cfs discharge through pipe 2 is now from (not into) reservoir B.*

See Solution 8.79 for Fig. 8.27. Elev. A = 250 ft, Elev. C = 180 ft, Q_2 = 10 cfs. Using only a basic scientific calculator (see Prob. 8.82), find Elev. B.

BG

This is a Case 2 problem.

Table A.1 for water at 60°F: $\nu = 1.217 \times 10^{-5}\ ft^2/sec$

	Pipe:	1	2	3
Given:	L, ft	5000	3000	1300
	D, ft	3	2	20/12
	e, ft (Table 8.1)	0.001	0.000 85	0.000 85
Calc:	L/D	1667	1500	780
	A, ft^2	7.07	3.14	2.18
	e/D	0.000 333	0.000 425	0.000 51

Find the elevation of P (Fig. 8.27) by trial and error.

/cont...

Elev. P lies between 180 and 250 ft. Calculate V from Eq. 8.42a. Trials:

El. P	h_1	h_3	V_1	V_3	Q_1	Q_3	ΣQ	Move P ?
230	20	50	7.02	15.58	49.6	34.0	+25.6	Up
245	5	65	3.47	17.77	24.5	38.8	−4.26	Down

Interpolation (Fig. 8.28): (245 − El. P)/(245 − 230) = 4.26/(4.26 − 25.6); El. P = 242.86

El. P	h_1	h_3	V_1	V_3	Q_1	Q_3	ΣQ	Move P ?
243	7	63	4.12	17.50	29.1	38.2	+0.945	Up

Interpolation: (245 − El. P)/(245 − 243) = 4.26/(4.26 + 0.945); Elev P = 243.36, close enough!

$\mathbf{R}_1 = D_1V_1/\nu = 1.02 \times 10^6$, $\mathbf{R}_3 = 2.40 \times 10^6$, both turbulent, OK.

$V_2 = Q_2/A_2 = 10/(\pi 2^2/4) = 3.18$ fps; $\mathbf{R}_2 = D_2V_2/\nu = 2(3.18)/(1.217 \times 10^{-5}) = 523{,}100$

Eq. 8.41: $f_2 = 0.017\,04$; Eq. 8.10: $h_2 = 4.02$ ft

Elev. B = Elev. $P + h_2 = 243.36 + 4.02 = 247.38$ ft ◀

8.86 *Repeat Prob. 8.84, except that the 0.3 m^3/s discharge through pipe 2 is now from (not into) reservoir B.*

See Solution 8.79 for Fig. 8.27. Elev. A = 60 m, Elev. C = 38 m, $Q_2 = 0.3\ m^3/s$. Using only a basic scientific calculator (see Prob. 8.82), find Elev. B.

SI

This is a Case 2 problem.

Table A.1 for water at 15°C: $\nu = 1.139 \times 10^{-6}\ m^2/s$

	Pipe:	1	2	3
Given:	L, m	1500	900	400
	D, m	0.90	0.60	0.50
	e, m (Table 8.1)	0.0003	0.000 25	0.000 25
Calc:	L/D	1667	1500	800
	A, m^2	0.636	0.283	0.1963
	e/D	0.000 333	0.000 417	0.0005

Find the elevation of P (Fig. 8.27) by trial and error.

Elev. P lies between 38 and 60 m. Calculate V from Eq. 8.42a. Trials:

El. P	h_1	h_3	V_1	V_3	Q_1	Q_3	ΣQ	Move P ?
55	5	17	1.934	4.96	1.231	0.974	+0.556	Up
58	2	20	1.214	5.38	0.772	1.057	+0.015	Up (close)

Extrapolation (Fig. 8.28): (El. P − 58)/(58 − 55) = 0.015/(0.556 − 0.015);

El. P = 58.08, close enough!

$\mathbf{R}_1 = D_1V_1/\nu = 9.59 \times 10^5$, $\mathbf{R}_3 = 2.36 \times 10^6$, both turbulent, OK.

$V_2 = Q_2/A_2 = 0.3/(\pi 0.6^2/4) = 1.061$ m/s; $\mathbf{R}_2 = D_2V_2/\nu = 0.6(1.061)/(1.139 \times 10^{-6}) = 558\,900$

Eq. 8.41: $f_2 = 0.016\,93$; Eq. 8.10: $h_2 = 1.457$ m

Elev. B = Elev. $P - h_2 = 58.08 + 1.46 = 59.54$ m ◀

8.87 *Suppose, in Fig. 8.27, that pipes 1, 2, and 3 are 900 m of 60 cm, 300 m of 45 cm, and 1200 m of 40 cm, respectively, of new welded-steel pipe. The surface elevations of reservoirs A, B, and C are 36, 22 and 0 m, respectively. The water temperature is 15°C. Using a basic scientific calculator only [see Prob. 8.82], find the approximate water flow in all pipes.*

See Solution 8.79 for Fig 8.27.

SI

This is a Case 3 problem. Solve by T and E.

Table A.1 for water at 15°C: $\nu = 1.139 \times 10^{-6}$ m/s

Pipe:	1	2	3
L, m (given)	900	300	1200
D m (given)	0.60	0.45	0.40
e, m (Table 8.1)	0.000 046	0.000 046	0.000 046
Trial 1: Try P at El of reservoir surface B = 22 m			
h, m	14	0	22
$\sqrt{2gDh_L/L}$, m/s	0.428	0	0.379
Q, m³/s (Eq. 8.42b)	1.088	0	0.410

At J, ΣQ = in − out = 1.088 − 0.410 = 0.678 m³/s. This must be zero, so P must be raised; then water will flow into reservoir B. Mid way between El A and El B = (36 + 22)/2 = 29 m.

Trial 2: Try P at El. 29 m.

h, m	7	7	29
Q, m³/s	0.759	0.632	0.473

At J, ΣQ = in − out = 0.759 − 0.632 − 0.473 = −0.346 m³/s

By interpolation (Fig. 8.28): (29 − El. P)/(29 − 22) = 0.346/(0.346 + 0.678); El. P = 26.63 m.

Trial 3: Try P at El. 26.6 m.

h, m	9.4	4.6	26.6
Q, m³/s	0.885	0.509	0.453

At J, ΣQ = 0.885 - 0.509 - 0.453 = −0.077 m³/s. Error = 0.077/0.885 = 8.7% (so try again).

By extrapolation: (26.6 − El. P)/(29 − 26.6) = 0.077/(0.346 − 0.077); El. P = 25.91 m

Trial 4: Try P at El. 25.9 m.

h, m	10.1	3.9	25.9
V, m/s (Eq. 8.42a)	3.25	2.94	3.55
Q, m³/s = AV = $\pi D^2 V/4$	0.919	0.467	0.446

At J, ΣQ = 0.919 - 0.467 - 0.446 = +005 57 m³/s. Error = 0.005 57/0.919 = 0.61%, close enough!

Check Reynolds numbers:

$\mathbf{R} = DV/\nu$	1.71×10^6	1.16×10^6	1.25×10^6

All three flows are turbulent, so use of Eq. 8.42 and results are valid.

Q_1 = 0.919 m³/s, Q_2 = 0.467 m³/s, Q_3 = 0.446 m³/s ◀◀

8.88 *Solve Prob. 8.87 accurately without manual trial and error, by using an equation solver in mathematics software like Mathcad.*

See Solution 8.79 for Fig. 8.27. Elev. A = 36 m, Elev. B = 22 m, Elev. C = 0 m, T = 15°C. Find the 3 Q's.

SI

Table A.1 for water at 15°C: $\nu = 1.139 \times 10^{-6}$ m²/sec

Open a Mathcad worksheet and enter the following 11 known values, without units:

$g = 9.81$	$D1 = 0.60$	$L1 = 900$	$e1 = 0.000\,046$
$\nu = 1.139 \cdot 10^{-6}$	$D2 = 0.45$	$L2 = 300$	$e2 = 0.000\,046$
	$D3 = 0.40$	$L3 = 1200$	$e3 = 0.000\,046$

Also enter guessed values for the following 15 unkowns:

$Q1 = 2$	$V1 = 3$	$h1 = 10$	$R1 = 800\,000$	$f1 = 0.01$
$Q2 = 2$	$V2 = 3$	$h2 = 5$	$R2 = 800\,000$	$f2 = 0.01$
$Q3 = 2$	$V3 = 3$	$h3 = 10$	$R3 = 800\,000$	$f3 = 0.01$

In a "solve block" which begins with "Given" enter the following four governing equations for Pipe 1:

$$\frac{1}{\sqrt{f1}} = -2 \cdot \log\left(\frac{e1}{3.7 \cdot D1} + \frac{2.51}{R1 \cdot \sqrt{f1}}\right); \quad R1 = \frac{D1 \cdot V1}{\nu}$$

$$h1 = \frac{f1 \cdot L1 \cdot (V1)^2}{D1 \cdot 2 \cdot g}; \quad Q1 = \frac{V1 \cdot \pi \cdot (D1)^2}{4}$$

Enter four similar equations for Pipe 2, and four for Pipe 3, replacing the 1's by 2's and 3's. Also enter the equations $h1 + h2 = 14$ $\quad h1 + h3 = 36$ and $Q1 = Q2 + Q3$

Thus there are 15 simultaneous equations, corresponding to the 15 unknowns.

End the solve block with "Find($Q1, Q2, Q3, V1, V2, V3, h1, h2, h3,$ R1, R2, R3, $f1, f2, f3$) ="

Mathcad obtains the following results:

	Q	V	h	R	f
Pipe 1:	0.917	3.24	10.05	1707 800	0.012 51
Pipe 2:	0.470	2.95	3.95	1167 400	0.013 31
Pipe 3:	0.447	3.55	25.95	1248 400	0.013 43

▲

/more...

8.89 *Suppose, in Fig. 8.27, that pipes 1, 2, and 3 are 3000 ft of 24 in, 1000 ft of 18 in, and 4000 ft of 16 in, respectively, of new welded-steel pipe. The surface elevations of reservoirs A, B, and C are 120, 75 and 0 ft, respectively. The water temperature is 60°F. Using a basic scientific calculator only, find the approximate water flow in all pipes.*

See Solution 8.79 for Fig. 8.27.

BG

Water surface Els.: A = 120 ft, B = 75 ft, C = 0 ft

To find the 3 Q's this is Case 3. Solve by T and E.

Table A.1 for water at 15°C: $\nu = 1.217 \times 10^{-5}$ ft²/sec.

Pipe:	1	2	3
L, ft (given)	3000	1000	4000
D ft (given)	2.0	1.5	16/12
e, ft (Table 8.1)	0.000 15	0.000 15	0.000 15

Trial 1: Try P at El of reservoir surface B = 75 ft

h, ft	45	0	75
$\sqrt{2gDh_L/L}$, fps	1.390	0	1.269
Q, cfs (Eq. 8.42b)	39.3	0	15.29

At J, ΣQ = in − out = 39.3 − 15.29 = 24.1 cfs. This must be zero, so P must be raised; then water will flow into reservoir B. Mid way between El A and El B = (120 + 75)/2 = 97.5 ft.

Trial 2: Try P at El. 100 ft.

h, ft	20	25	100
Q, cfs	25.8	24.1	17.74

At J, ΣQ = in − out = 25.8 − 24.1 − 17.74 = −16.07 cfs.

By interpolation (Fig. 8.28): (100 − El. P)/(100 − 75) = 16.07/(16.07 + 24.1); El. P = 90.0 ft.

Trial 3: Try P at El. 90.0 ft

h, ft	30	15	90
Q, cfs	31.89	18.54	16.80

At J, ΣQ = in - out = = 31.89 - 18.54 - 16.80 = -3.45 cfs.

Error = 3.45/31.89 = 10.8% (so try again).

By extrapolation: (90 − El. P)/(100 − 90) = 3.45/(16.07 − 3.45); El. P = 87.3 ft.

Trial 4: Try P at El. 87.3 ft.

h, ft	32.7	12.3	87.3
V, fps (Eq. 8.42a)	10.61	9.35	11.85
Q, cfs = $AV = \pi D^2 V/4$	33.348	16.523	16.540

At J, ΣQ = in - out = 33.348 - 16.523 - 16.540 = 0.285 cfs

Error = 0.285/33.348 = 0.85%, close enough! Check Reynolds numbers:

$\mathbf{R} = DV/\nu$	1.744×10^6	1.152×10^6	1.298×10^6

All 3 are turbulent, so use of Eq. 8.42 and results are valid. Q_1 = 33.3 cfs, $Q_2 = Q_3$ = 16.5 cfs ◄◄

☐ 8.90 *Solve Prob. 8.89 accurately without manual trial and error, by using an equation solver in mathematics software like Mathcad.*

See Solution 8.79 for Fig. 8.27. Elev. A = 120 ft, Elev. B = 75 ft, Elev. C = 0 ft, T = 60°F. Find the 3 Q's.

BG

Table A.1 for water at 60°F: $\nu = 1.217 \times 10^{-5}$ ft²/sec

Open a Mathcad worksheet and enter the following 12 known values, without units:

$g = 32.2$	$D1 = 24/12$	$L1 = 3000$	$e1 = 0.000\,15$
$\nu = 1.217 \cdot 10^{-5}$	$D2 = 18/12$	$L2 = 1000$	$e2 = 0.000\,15$
	$D3 = 16/12$	$L3 = 4000$	$e3 = 0.000\,15$

Also enter guessed values for the following 15 unkowns:

$Q1 = 70$	$V1 = 10$	$h1 = 30$	$R1 = 800\,000$	$f1 = 0.01$
$Q2 = 70$	$V2 = 10$	$h2 = 15$	$R2 = 800\,000$	$f2 = 0.01$
$Q3 = 70$	$V3 = 10$	$h3 = 30$	$R3 = 800\,000$	$f3 = 0.01$

In a "solve block" which begins with "Given" enter the following four governing equations for Pipe 1:

$$\frac{1}{\sqrt{f1}} = -2 \cdot \log\left(\frac{e1}{3.7 \cdot D1} + \frac{2.51}{R1 \cdot \sqrt{f1}}\right) ; \quad R1 = \frac{D1 \cdot V1}{\nu}$$

$$h1 = \frac{f1 \cdot L1 \cdot (V1)^2}{D1 \cdot 2 \cdot g} ; \quad Q1 = \frac{V1 \cdot \pi \cdot (D1)^2}{4}$$

Enter four similar equations for Pipe 2, and four for Pipe 3, replacing the 1's with 2's and 3's. Also enter the equations $h1 + h2 = 45$ $\quad h1 + h3 = 120$ $\quad$ and $\quad Q1 = Q2 + Q3$

Thus there are 15 simultaneous equations, corresponding to the 15 unknowns.

End the solve block with "Find($Q1, Q2, Q3, V1, V2, V3, h1, h2, h3,$ R1, R2, R3, $f1, f2, f3$) ="

Mathcad obtains the following results:

	Q	*V*	*h*	**R**	*f*
Pipe 1:	33.31	10.60	32.63	1,742,600	0.012 46
Pipe 2:	16.77	9.49	12.37	1,169,400	0.013 27
Pipe 3:	16.55	11.85	87.37	1,298,300	0.013 35

▲

/more...

8.91 *Suppose that, in Fig. 8.27, pipe 1 is 1500 ft of 12-in new cast-iron pipe, pipe 2 is 800 ft of 6-in wrought-iron pipe, and pipe 3 is 1200 ft of 8-in wrought-iron pipe. The water surface of reservoir B is 20 ft below that of A, while the junction J is 35 ft below the surface of A. In place of reservoir C, pipe 3 leads away to some other destination but its elevation at C is 60 ft below A. Find the flow in each pipe and the pressure head at C when the pressure head at J is 25 ft.*

See Solution 8.79 for Fig. 8.27.

BG

Table A.1 for water at 60°F: $\nu = 1.217 \times 10^{-5}$ ft²/sec

	Pipe:	1	2	3
Given:	L, ft	1500	800	1200
	D, ft	1	0.5	8/12
	e, ft	0.000 85	0.000 15	0.000 15
Calc:	h_L, ft	$35 - 25 = 10$	$20 - h_1 = 10$	
	$\sqrt{2gDh_L/L}$, fps	0.655	0.634	
	V, fps (Eq. 8.42*a*)	4.66	4.76	
	Q, cfs (Eq. 8.42*b*)	3.66 ◀	0.934 ◀	

$Q_3 = Q_1 - Q_2 = 3.66 - 0.934 = 2.73$ cfs ◀ $V_3 = Q_3/A_3 = 2.73/[(\pi/4)(8/12)^2] = 7.82$ fps

$\mathbf{R}_1 = D_1V_1/\nu = 383{,}100,\ \mathbf{R}_2 = 195{,}500,\ \mathbf{R}_3 = 428{,}100$

All the flows are turbulent, so the equations used and answers obtained are valid.

Haaland Eq. 8.41: $f_3 = 0.015\,72$; Eq. 8.10: $h_3 = 26.8$ ft

$H_J - H_C = h_3$. Neglecting velocity heads: $(p/\gamma + z)_J - (p/\gamma + z)_C = h_3$

$(25 - 35) - (p_C/\gamma - 60) = 26.8;\ p_C/\gamma = 23.2$ ft ◀

Sec. 8.27: Pipes in Series -- *Problems*

8.92 *Suppose that in Fig. 8.29 pipes 1, 2, and 3 are 750 ft of 4-in, 250 ft of 2-in, and 300 ft of 3-in asphalt-dipped cast-iron pipe. With a total head loss of 25 ft between A and B, find the flow of 60°F water.*

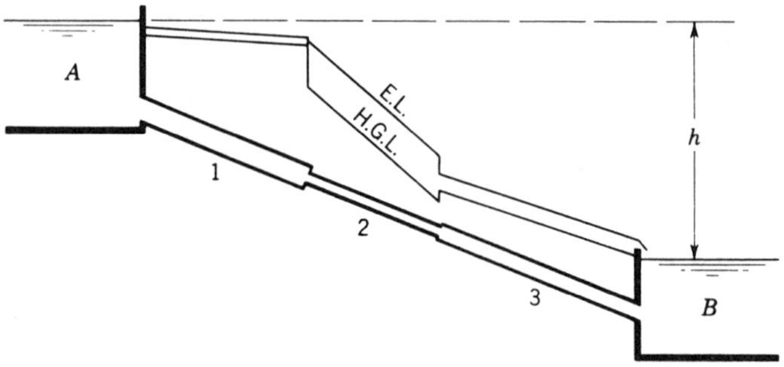

Figure 8.29

BG

Table 8.1: $e = 0.0004$ ft.

* Initially assume $f = f_{min}$ (fully rough value) from Fig. 8.11.

Pipe	L, ft	D, in	L/D	e/D	*Try f	D_1/D	$(D_1/D)^4$
1	750	4	2250	0.0012	0.0205	1.00	1.00
2	250	2	1500	0.0024	0.0246	2.00	16.00
3	300	3	1200	0.0016	0.0221	1.333	3.16

/cont...

Eq. 8.71: $25 = [0.0205(2250) + 0.0246(1500)16 + 0.0221(1200)3.16]V_1^2/2g$

$V_1^2/2g = 25/720 = 0.0347$ ft; $V_1 = 1.495$ fps

$D_1''V_1 = 4(1.495) = 5.98$; $\therefore f_1 = 0.0249$ (Fig. 8.11)

$D_2''V_2 = 2(1.495)2^2 = 11.96$; $\therefore f_2 = 0.0263$

$D_3''V_3 = 3(1.495)1.333^2 = 7.97$; $\therefore f_3 = 0.0251$

Repeating Eq. 8.71 with improved f values: $25 = 782V_1^2/2g$, $V_1 = 1.434$ fps

$D_1''V_1 = 4(1.434) = 5.74$; $\therefore f_1 = 0.0251$

$D_2''V_2 = 2(1.434)2^2 = 11.48$; $\therefore f_2 = 0.0263$

$D_3''V_3 = 3(1.434)1.333^2 = 7.65$; $\therefore f_3 = 0.0252$

Repeating Eq. 8.71: $25 = 783V_1^2/2g$, $V_1 = 1.434$ fps (no change)

$Q = A_1V_1 = (\pi/4)(4/12)^2 1.434 = 0.1251$ cfs ◀

Note: f values may instead be obtained from Haaland Eq. 8.41.

<u>Alternative solution, by Mathcad</u>:

Establish the four standard equations for each pipe (see Sample Prob. 8.12*b*) plus Eq. 8.69, for a total of 13 governing equations. Assign known and guessed values to the variables (see, e.g., the solution to Prob. 8.81). Solve for the following 13 unknowns: Three values each of V, h, **R**, and f, and one value of Q.

☐ 8.93 *Suppose that in Fig. 8.29 pipes 1, 2, and 3 are 150 m of 80-mm, 60 m of 50-mm, and 120 m of 60 mm wrought-iron pipe. With a total head loss of 6 m between A and B, find the flow of 15°C water.*

See Solution 8.92 for Fig. 8.29

SI

Table 8.1: $e = 0.000\,046$ m; Table A.1 at 15°C: $\nu = 1.139 \times 10^{-6}$ m²/s.

*Trial 1: assume $f = f_{\min}$ (fully rough value) from Eq. 8.39.

Pipe	L, m	D, m	L/D	e/D	*Try f	D_1/D	$(D_1/D)^4$
1	150	0.08	1875	0.000 575	0.017 24	1.00	1.00
2	60	0.05	1200	0.000 920	0.019 24	1.60	6.55
3	120	0.06	2000	0.000 767	0.018 43	1.333	3.16

Eq. 8.71: $6 = [0.017\,24(1875) + 0.019\,24(1200)6.55 + 0.018\,43(200)3.16]V_1^2/2g$

$V_1^2/2g = 6/300 = 0.0200$ m; $V_1 = 0.626$ m/s

$\mathbf{R}_1 = D_1V_1/\nu = 43\,988$; $\therefore f_1 = 0.022\,97$ (from Eq. 8.41)

$\mathbf{R}_2 = D_2V_2/\nu = D_2V_1(D_1/D_2)^2/\nu = D_1^2V_1/(D_2\nu) = \mathbf{R}_1(D_1/D_2) = 70\,381$; $\therefore f_2 = 0.022\,50$

$\mathbf{R}_3 = \mathbf{R}_1(D_1/D_3) = 58\,651$; $\therefore f_3 = 0.022\,54$

Trial 3 (Eq. 8.71): $6 = 367V_1^2/2g$; $V_1 = 0.566$ m/s (little change)

$Q = A_1V_1 = (\pi/4)(0.08^2)0.566 = 0.002\,85$ m³/s = 2.85 L/s ◀

Note: See solution to Prob. 8.92 for alternative solution by Mathcad.

8.94 *Two pipes connected in series are respectively 150 ft of 2-in (e = 0.000 006 ft) and 450 ft of 8-in (e = 0.0009 ft). With a total head loss of 30 ft, find the flow of 60°F water.*

BG

Start with minimum (fully turbulent) trial values of f:

Pipe	D, in	L, ft	e, ft	L/D	e/D	D_2/D	$(D_2/D)^4$	f_{min}
1	2	150	0.000 006	900	0.000 036	4	256	0.009 95
2	8	450	0.0009	675	0.001 35	1	1	0.0211

Trial 1: Eq. 8.71: $30 = (V_2^2/2g)[0.009\,95(900)256 + 0.0211(675)] = 2310V_2^2/2g$; $V_2 = 0.915$ fps

$D_1''V_1 = 2(0.915)4^2 = 29.3$; Fig. 8.11: $f_1 = 0.0165$ (40% change).

$D_2''V_2 = 8(0.915) = 7.32$; $f_2 = 0.025$ (15% change)

Trial 2: $30 = (V_2^2/2g)[0.0165(900)256 + 0.025(675]$; $V_2 = 0.711$ fps

$D_1''V_1 = 22.8$; $f_1 = 0.017$ (3% change)

$D_2''V_2 = 5.69$; $f_2 = 0.026$ (2% change)

Both of the $D''V$ (and so **R**) values are in the turbulent range.

Trial 3: $V_2 = 0.701$ fps (1.4% change, close enough).

$Q = A_2V_2 = (\pi/4)(8/12)^2 0.701 = 0.245$ cfs ◄

8.95 *Repeat Prob. 8.92 for the case where the fluid is an oil with s = 0.92, μ = *0.000 96 lb·sec/ft². [*Note this correction to the textbook.]*

See Solution 8.92 for Fig. 8.29. Pipe material = asphalt-dipped cast iron, h = 25 ft. Find Q.

BG

$\nu = \mu/\rho = 0.000\,96/(0.92 \times 1.940) = 5.38 \times 10^{-4}$ ft²/sec. Table 8.1: $e = 0.0004$ ft.

Initially assume $f = f_{min}$ (fully rough value) from Eq. 8.39.

Pipe	L, ft	D, in	L/D	e/D	Try f	D_1/D	$(D_1/D)^4$
1	750	4	2250	0.0012	0.020 54	1.00	1.00
2	250	2	1500	0.0024	0.024 60	2.00	16.00
3	300	3	1200	0.0016	0.022 09	1.333	3.16

Eq. 8.71: $25 = [0.020\,54(2250) + 0.0246(1500)16 + 0.022\,09(1200)3.16]V_1^2/2g$

$V_1^2/2g = 25/720 = 0.0347$ ft; $V_1 = 1.495$ fps

$\mathbf{R}_1 = D_1V_1/\nu = 927$

$\mathbf{R}_2 = D_2V_2/\nu = D_2V_1(D_1/D_2)^2/\nu = D_1^2V_1/(D_2\nu) = \mathbf{R}_1(D_1/D_2) = 1853$

$\mathbf{R}_3 = \mathbf{R}_1(D_1/D_3) = 1236$

All three flows may be laminar in which case a trial solution is unnecessary.

Use Eq. 8.19: $h_L = (32\nu/g)[(L_1V_1/D_1^2) + (L_2V_2/D_2^2) + L_3V_3/D_3^2)]$

i.e. $25 = (32/32.2)(5.38 \times 10^{-4})[(L_1V_1/D_1^2) + (L_2V_1/D_2^2)(D_1/D_2)^2 + (L_3V_1/D_3^2)(D_1/D_3)^2]$

Substituting for the L's and D's we find: $25 = 27.4V_1$; $V_1 = 0.912$ fps

$Q = A_1V_1 = (\pi/4)(4/12)^2 0.912 = 0.0796$ cfs ◄

Check: $\mathbf{R}_1 = 565$, $\mathbf{R}_2 = 1131$, $\mathbf{R}_3 = 754$ (all laminar, OK).

8.96 *Repeat Prob. 8.93 for the case where the fluid is an oil with $s = 0.94$, $\mu = 0.04\ N{\cdot}s/m^2$. See Solution 8.92 for Fig. 8.29. Pipe material = wrought iron, $h = 6$ m. Find Q.*

SI

$\nu = \mu/\rho = 0.04/(0.94 \times 999) = 4.26 \times 10^{-5}\ m^2/s$. Table 8.1: $e = 0.000\,046$ m

Trial 1: assume $f = f_{min}$ (fully rough value) from Eq. 8.39.

Pipe	L, m	D, m	L/D	e/D	Try f	D_1/D	$(D_1/D)^4$
1	150	0.08	1875	0.000 575	0.017 24	1.00	1.00
2	60	0.05	1200	0.000 920	0.019 24	1.60	6.55
3	120	0.06	2000	0.000 767	0.018 43	1.333	3.16

Eq. 8.17: $6 = [0.017\,24(1875) + 0.019\,24(1200)6.55 + 0.018\,43(200)3.16]V_1^2/2g$

$V_1^2/2g = 6/300 = 0.0200$ m; $V_1 = 0.626$ m/s

$\mathbf{R}_1 = D_1V_1/\nu = 1176$

$\mathbf{R}_2 = D_2V_2/\nu = D_2V_1(D_1/D_2)^2/\nu = D_1^2V_1/(D_2\nu) = \mathbf{R}_1(D_1/D_2) = 1882$

$\mathbf{R}_3 = \mathbf{R}_1(D_1/D_3) = 1568$

All three flows may be laminar in which case a trial solution is unnecessary.

Use Eq. 8.19: $h_L = (32\nu/g)[(L_1V_1/D_1^2) + (L_2V_2/D_2^2) + (L_3V_3/D_3^2)]$

i.e. $6 = (32/9.81)(4.26 \times 10^{-4})[(L_1V_1/D_1^2) + (L_2V_1/D_2^2)(D_1/D_2)^2 + (L_3V_1/D_3^2)(D_1/D_3)^2]$

Substituting for the L's and D's we find: $6 = 20.0V_1$; $V_1 = 0.300$ m/s

$Q = A_1V_1 = (\pi/4)(0.08^2)0.300 = 0.001\,506\ m^3/s = 1.506$ L/s ◀

Check: $\mathbf{R}_1 = 563$, $\mathbf{R}_2 = 900$, $\mathbf{R}_3 = 750$ (all laminar, OK).

8.97 *A pipeline 900 ft long discharges freely at a point 200 ft below the water level at intake. The pipe intake projects into the reservoir. The first 600 ft is of 10-in-diameter, and the remaining 300 ft is of 6-in diameter. Find the rate of discharge, assuming $f = 0.06$. If the junction point C of the two sizes of pipe is 150 ft below the intake water surface level, find the pressure head just upstream of C and just downstream of C. Assume a sudden contraction at C.*

BG

Sudden contraction at C, $D_2/D_1 = 0.6$; Table 8.2: $k_c = 0.28$

Contraction loss $= 0.28V_2^2/2g$. Fig. 8.13: $k_e = 0.8$ for reentrant entrance.

$h_L = [0.8 + 0.06(600)(12/10)]V_1^2/2g + [0.28 + 0.06(300/0.50)]V_2^2/2g = 44.0V_1^2/2g + 36.3V_2^2/2g$.

Continuity: $V_1 = (6/10)^2V_2 = 0.36V_2$. $\therefore\ h_L = 44.0(0.36V_2)^2/2g + 36.3V_2^2/2g = 42.0V_2^2/2g$

Energy: $200 - h_L = V_2^2/2g$, $\therefore\ 200 = (42.0 + 1)V_2^2/2g$, $V_2^2/2g = 4.65$ ft

$V_2 = \sqrt{2(32.2)4.65} = 17.31$ fps; $Q = A_2V_2 = (\pi/4)(0.5^2)17.31 = 3.40$ cfs ◀

Energy from water surface to point 1 just upstream of contraction:

$(0 + 150 + 0) - 44.0(0.36V_2)^2/2g = [p_2/\gamma + 0 + (0.36V_2)^2/2g]$; $p_1/\gamma = 122.86$ ft ◀

Energy from point 2 just downstream of contraction to jet:

$(p_2/\gamma + 50 + V_2^2/2g) - 0.06(300/0.5)V_2^2/2g = (0 + 0 + V_2^2/2g)$; $p_2/\gamma = 117.51$ ft ◀

Check head loss across contraction:

(*a*) $h_L = [122.86 + z + (0.36V_2)^2/2g] - [117.51 + z + V_2^2/2g] = 1.303$ ft

(*b*) $h_L = 0.28V_2^2/2g = 0.28(4.65) = 1.303$ ft. Agrees!

8.98 *A pipeline 300 m long discharges freely at a point 50 m below the water level at intake. The pipe intake projects into the reservoir. The first 200 m is of 35 cm diameter, and the remaining 100 m is of 25 cm diameter. Find the rate of discharge, assuming $f = 0.06$. If the junction point C of the two sizes of pipe is 40 m below the intake water surface level, find the pressure head just upstream of C and just downstream of C. Assume a sudden contraction at C.*

SI

Sudden contraction at C, $D_2/D_1 = 25/35 = 0.714$. Interpolating in Table 8.2: $k_c = 0.21$

Contraction loss $= 0.21V_2^2/2g$. Fig. 8.13: $k_e = 0.8$ for reentrant entrance.

$h_L = [0.8 + 0.06(600)(200/0.35)]V_1^2/2g + [0.21 + 0.06(100/0.25)]V_2^2/2g = 35.1V_1^2/2g + 24.2V_2^2/2g$.

Continuity: $V_1 = (25/35)^2V_2 = 0.510V_2$. $\therefore\ h_L = 35.1(0.510V_2)^2/2g + 24.2V_2^2/2g = 33.3V_2^2/2g$

Energy: $50 - h_L = V_2^2/2g$, $\therefore\ 50 = (33.3 + 1)V_2^2/2g$, $V_2^2/2g = 1.456$ m

$V_2 = \sqrt{2(9.81)1.456} = 5.34$ m/s; $Q = A_2V_2 = (\pi/4)(0.25^2)5.34 = 0.262$ m^3/s ◀

Energy from water surface to point 1 just upstream of contraction:

$(0 + 150 + 0) - 35.1(0.510V_2)^2/2g = [p_1/\gamma + 0 + (0.510V_2)^2/2g]$; $p_1/\gamma = 26.32$ m ◀

Energy from point 2 just downstream of contraction to jet:

$(p_2/\gamma + 10 + V_2^2/2g) - 0.06(100/0.25)V_2^2/2g = (0 + 0 + V_2^2/2g)$; $p_2/\gamma = 24.94$ m ◀

Check head loss across contraction:

(*a*) $h_L = [26.32 + z + (0.510V_2)^2/2g] - [24.94 + z + V_2^2/2g] = 0.306$ m

(*b*) $h_L = 0.21V_2^2/2g = 0.21(1.456) = 0.306$ m. Agrees!

8.99 *Repeat Prob. 8.97 neglecting minor losses.*

Prob. 8.97: $\Delta z = 200$ *ft*, $L_1 = 600$ *ft*, $D_1 = 10$ *in*, $L_2 = 300$ *ft*, $D_2 = 6$ *in*, $f = 0.06$. *Find Q. Also find p/γ just upstream and downstream of the junction point C, which is 150 ft below the water surface level.*

BG

$h_L = 0.06(600)(12/10)V_1^2/2g + 0.06(300/0.5)V_2^2/2g = 43.2V_1^2/2g + 36V_2^2/2g$

Continuity: $V_1 = (6/10)^2V_2 = 0.36V_2$. $\therefore\ h_L = 43.2(0.36V_2)^2/2g + 36V_2^2/2g = 41.6V_2^2/2g$

Energy: $200 - h_L = V_2^2/2g$, $\therefore\ 200 = (41.6 + 1)V_2^2/2g$, $V_2^2/2g = 4.69$ ft

$V_2 = \sqrt{2(32.2)4.69} = 17.39$ fps; $Q = A_2V_2 = (\pi/4)(0.5^2)17.39 = 3.41$ cfs ◀

Energy from water surface to point 1 just upstream of contraction:

$(0 + 150 + 0) - 43.2(0.36V_2)^2/2g = [p_1/\gamma + 0 + (0.36V_2)^2/2g]$; $p_1/\gamma = 123.11$ ft ◀

Energy from point 2 just downstream of contraction to jet:

$(p_2/\gamma + 50 + V_2^2/2g) - 0.06(300/0.5)V_2^2 = (0 + 0 + V_2^2/2g)$; $p_2/\gamma = 119.02$ ft ◀

8.100 *Repeat Prob. 8.98 neglecting minor losses.*

Prob. 8.98: $\Delta z = 50$ *m,* $L_1 = 200$ *m,* $D_1 = 35$ *cm,* $L_2 = 100$ *m,* $D_2 = 25$ *cm,* $f = 0.06$. *Find Q. Also find* p/γ *just upstream and downstream of the junction point C, which is 40 m below the water surface level.*

SI

$h_L = 0.06(200/0.35)V_1^2/2g + 0.06(100/0.25)V_2^2/2g = 34.3V_1^2/2g + 24V_2^2/2g$

Continuity: $V_1 = (25/35)^2 V_2 = 0.510V_2$. $\therefore h_L = 34.3(0.510V_2)^2/2g + 24V_2^2/2g = 32.9V_2^2/2g$

Energy: $50 - h_L = V_2^2/2g$, $\therefore 50 = (32.9 + 1)V_2^2/2g$, $V_2^2/2g = 1.474$ m

$V_2 = \sqrt{2(9.81)1.474} = 5.38$ m/s; $Q = A_2V_2 = (\pi/4)(0.25^2)5.38 = 0.264$ m³/s ◀

Energy from water surface to point 1 just upstream of contraction:

$(0 + 40 + 0) - 34.3(0.510V_2)^2/2g = [p_1/\gamma + 0 + (0.510V_2)^2/2g]$; $p_1/\gamma = 26.46$ m ◀

Energy from point 2 just downstream of contraction to jet:

$(p_2/\gamma + 10 + V_2^2/2g) - 0.06(100/0.25)V_2^2/2g = (0 + 0 + V_2^2/2g)$; $p_2/\gamma = 25.37$ m ◀

8.101 *Three new cast-iron pipes, having diameters of 24, 21, and 18 in, respectively, each 450 ft long, are connected in series. The 24-in pipe leads from a reservoir (flush entrance), and the 18-in pipe discharges into the air at a point 15 ft below the reservoir water surface level. Assuming all changes in section to be abrupt, find the rate of discharge of water at 60°F.*

BG

Table 8.1: for new cast iron: $e = 0.000\,85$ ft.

Pipe	D (in)	L/D	e/D	Trial f	D_1/D	$(D_1/D)^4$
1	24	225	0.000 425	0.016	1.00	1.00
2	21	257	0.000 486	0.017	1.143	1.706
3	18	300	0.000 567	0.018	1.333	3.16
Total		782				

$\Sigma(L/D) = 782 < 1000$, so include minor losses.

Flush entrance, Fig. 8.13: $k_e = 0.5$

Sudden contractions, Table 8.2: $k_c = 0.0825$ for $D_2/D_1 = 0.875$; $k_c = 0.0986$ for $D_3/D_2 = 0.857$

Discharge: $k_d = 1.0$ or $h_L = V_3^2/2g$

Energy: $(0 + 15 + 0) = (0 + 0 + V_3^2/2g) + \Sigma h_L$, i.e.,

$15 = [0.5 + 0.016(225) + (0.0825 + 0.017\times 257)1.706 + (0.0986 + 0.018\times 300)3.16 + 1(3.16)]V_1^2/2g$

$V_1^2/2g = 15/32.2 = 0.465$ ft; $V_1 = \sqrt{2(32.2)0.465} = 5.47$ fps

$D_1''V_1 = 24(5.47) = 131.4$; $\therefore f_1 = 0.0167$ (from Fig. 8.11)

$D_2''V_2 = 21(5.47)1.143^2 = 150.1$; $\therefore f_2 = 0.0171$

$D_3''V_3 = 18(5.47)1.133^2 = 175.2$; $\therefore f_3 = 0.0175$

$15 = [0.5 + 0.0167(225) + (0.0825 + 0.0171\times 257)1.706 + (0.0986 + 0.0175\times 300)3.16 + 1(3.16)]V_1^2/2g$

$V_1^2/2g = 15/32.0 = 0.469$; $V_1 = \sqrt{2(32.2)0.469} = 5.50$ fps (little change)

$Q = A_1V_1 = (\pi/4)(2^2)5.50 = 17.27$ cfs ◀

Sec. 8.28: Pipes in Parallel -- *Problems*

8.102 *Suppose that in Fig. 8.30 pipes 1, 2, and 3 are of smooth brass as follows: 500 ft of 2-in, 350 ft of 3-in; and *750 ft of 4-in respectively. When the total flow of 70°F crude oil (s = 0.855) is 0.7 cfs, find the head loss from A to B and the flow in each pipe. Note: This problem can be solved without trial and error, using a basic scientific calculator only. [*Note this correction to the textbook.]*

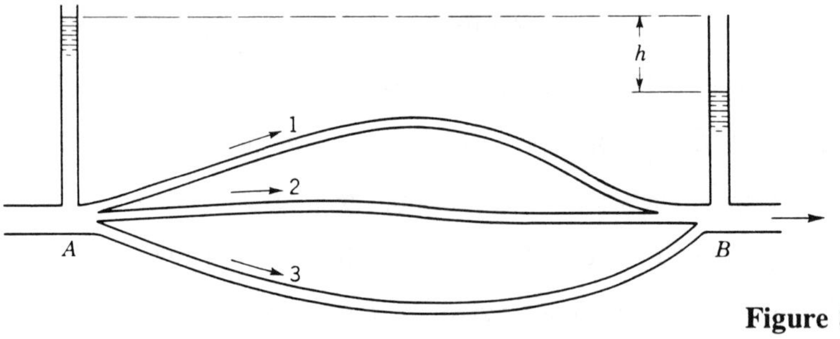

Figure 8.30

BG

Fig. 2.4 for crude oil (s = 0.855) at 70°F: $\nu = 7.6 \times 10^{-5}$ ft²/sec

Assume three Reynolds numbers are in the range ($3000 < \mathbf{R} < 10^5$) where Blasius' Eq. 8.37 for smooth pipe may be used. Eq. 8.37: $f = 0.316/\mathbf{R}^{1/4} = 0.316\nu^{1/4}/(D^{1/4}V^{1/4})$

Eq. 8.10: $h_L = f(L/D)(V^2/2g) = [0.316\nu^{1/4}/(D^{1/4}V^{1/4})](L/D)V^2/2g = (\text{const})LV^{7/4}/D^{5/4}$,

where const $= 0.316\nu^{1/4}/2g$

Hence Eq. 8.73: $(h_L)_1 = (h_L)_2 = (h_L)_3$ yields $L_1V_1^{7/4}/D_1^{5/4} = L_2V_2^{7/4}/D_2^{5/4} = L_3V_3^{7/4}/D_3^{5/4}$

Substituting for the L's and D's: $210V_1^{7/4} = 88.6V_2^{7/4} = 132.6V_3^{7/4}$; so $V_2 = 1.638V_1$, $V_3 = 1.301V_1$

Eq. 8.72: $Q = 0.70 = 0.0218V_1 + 0.0491(1.638V_1) + 0.0873(1.301V_1) = 0.216V_1$; $V_1 = 3.24$ fps.

$\mathbf{R}_1 = 7114$, $\mathbf{R}_2 = 17{,}478$, $\mathbf{R}_3 = 18{,}516$, all in the required range, so the assumed Eq. 8.37 is valid.

$Q_1 = (0.0218)3.24 = 0.0707$ cfs ◄

$Q_2 = (0.0491)1.638V_1 = 0.261$ cfs ◄

$Q_3 = (0.0873)1.301V_1 = 0.368$ cfs ◄ Check: $\Sigma Q = 0.700$ cfs, right on!

$h_L = h_{L1} = (0.316\nu^{1/4}/2g)500(3.24)^{7/4}/(2/12)^{5/4} = 16.87$ ft ◄

/more...

8.103 *Repeat Prob. 8.102 for the case where the total flow rate is 0.07 cfs.*

Prob. 8.102: In Fig. 8.30 (see Solution 8.102), pipe 1 is 500 ft of 2-in, pipe 2 is 350 ft of 3-in, and pipe 3 is 750 ft of 4-in diameter. Crude oil (s = 0.855) is flowing at 70°F. Find the head loss from A to B, and the flow in each pipe.

BG

Fig. 2.4 for crude oil (s = 0.855) at 70°F: $\nu = 7.6 \times 10^{-5}$ ft²/sec

$\bar{V} = \Sigma Q/\Sigma A = 0.07/(0.0218 + 0.0491 + 0.0873) = 0.442$ fps

With this $\bar{V}$ in the largest pipe, $\mathbf{R}_{max} = (4/12)\bar{V}/\nu = 1940$

Thus all three flows may be laminar, so use Eq. 8.19 with Eq. 8.73.

(Dividing the three **R**'s of Prob. 8.102 by 10 shows they are laminar.)

$gh_L/(32\nu) = L_1V_1/D_1^2 = L_2V_2/D_2^2 = L_3V_3/D_3^2 = \text{const}$

$500V_1/2^2 = 350V_2/3^2 = 750V_3/4^2$; $V_2 = 3.21V_1$, $V_3 = 2.67V_1$

Eq. 8.72: $0.07 = 0.0218V_1 + 0.0491(3.21V_1) + 0.0873(2.67V_1)$; $V_1 = 0.1698$ fps;

$\mathbf{R}_1 = 372$, $\mathbf{R}_2 = 1795$, $\mathbf{R}_3 = 1986$. All the flows are laminar, so the equations used are valid.

$Q_1 = A_1V_1 = 0.003\,70$ cfs ◄ $Q_2 = 0.0268$ cfs ◄ $Q_3 = 0.0395$ cfs ◄

Eq. 8.19: $h_L = 32(7.6 \times 10^{-5})500(0.1698)/[32.2(2/12)^2] = 0.231$ ft ◄

□ 8.104 *Repeat Prob. 8.102 for the case where all the pipes are of galvanized iron. Does the "Note" still apply?*

Prob. 8.102: In Fig. 8.30 (see Solution 8.102), pipe 1 is 500 ft of 2-in, pipe 2 is 350 ft of 3-in, and pipe 3 is 750 ft of 4-in diameter, all of smooth brass. Crude oil (s = 0.855) at 70°F is flowing at 0.7 cfs. Find the head loss from A to B, and the flow in each pipe. Note: This problem can be solved without trial and error, using a basic scientific calculator only.

BG

Fig. 2.4 for crude oil (s = 0.855) at 70°F: $\nu = 7.6 \times 10^{-5}$ ft²/sec. Table 8.1: $e = 0.0005$ ft.

As $e > 0$, Blasius Eq. 8.37 may not be used, so the "Note" does not apply. ◄

Eq. 8.73 using Eq. 8.10: $f_1(L_1/D_1)V_1^2/2g = f_2(L_2/D_2)V_2^2/2g = f_3(L_3/D_3)V_3^2/2g$

For the first trial assume the f's are the same. Substituting for L's and D's:

$3000V_1^2 = 1400V_2^2 = 2250V_3^2$; so $V_2 = 1.464V_1$, $V_3 = 1.155V_1$

Eq. 8.72: $Q = 0.7 = 0.0218V_1 + 0.0491(1.464V_1) + 0.0873(1.155V_1) = 0.1944V_1$; $V_1 = 3.60$ fps

$\mathbf{R}_1 = D_1V_1/\nu = 7895$, $\mathbf{R}_2 = 17{,}336$, $\mathbf{R}_3 = 18{,}233$

For 2nd trial: $f_1 = 0.036\,57$, $f_2 = 0.030\,18$, $f_3 = 0.029\,05$ (Eq. 8.41)

Eq. 8.73: $0.036\,57(3000)V_1^2 = 0.030\,18(1400)V_2^2 = 0.029\,05(2250)V_3^2$; so $V_2 = 1.611V_1$, $V_3 = 1.296V_1$

Eq. 8.72: $0.7 = 0.218V_1 + 0.0491(1.611V_1) + 0.0873(1.296V_1) = 0.214V_1$

$V_1 = 3.27$ fps (10.0% change); $\mathbf{R}_1 = 7174$, $\mathbf{R}_2 = 17{,}341$, $\mathbf{R}_3 = 18{,}589$

For 3rd trial: $f_1 = 0.037\,33$, $f_2 = 0.030\,18$, $f_3 = 0.028\,95$ (Eq. 8.41)

Eq. 8.73: $0.037\,33(3000)V_1^2 = 0.030\,18(1400)V_2^2 = 0.028\,95(2250)V_3^2$; so $V_2 = 1.628V_1$, $V_3 = 1.311V_1$

Eq. 8.72: $0.7 = 0.0218V_1 + 0.0491(1.628V_1) + 0.0873(1.311V_1) = 0.216V_1$

$V_1 = 3.24$ fps (1.02% change, close enough!). $Q_1 = A_1V_1 = 0.0218(3.24) = 0.0707$ cfs ◄

$Q_2 = 0.0491(1.628 \times 3.24) = 0.259$ cfs ◄ $Q_3 = 0.0873(1.311 \times 3.24) = 0.371$ cfs ◄

Eq. 8.10: $h_L = h_{L1} = 0.037\,33(3000)(3.24)^2/2g = 18.24$ ft ◄

8.105 *Suppose that in Fig. 8.30 [see Solution 8.102] pipes 1, 2, and 3 are of drawn copper tubing as follows: 90 m of 2 cm, 150 m of 4 cm and 80 m of 6 cm, respectively. When the total flow of 50°C crude oil (s = 0.855) is 7 L/s, find the head loss from A to B and the flow in each pipe. Note: This problem can be solved without trial and error, using a basic scientific calculator only.*

SI

Fig. 2.4 for crude oil ($s = 0.855$) at 50°C: $\nu = 3.8 \times 10^{-6}$ m²/s

Assume Blasius' Eq. 8.37 applies for smooth (drawn copper) pipe and then check **R**

(need $3000 < \mathbf{R} < 10^5$).

Eq. 8.10 with Eq. 8.37: $h_L = f(L/D)V^2/2g = 0.316\nu^{0.25}LV^2/[(DV)^{0.25}D2g] = 0.316\nu^{0.25}LV^{1.75}/(2gD^{1.25})$

i.e., $h_L \propto LV^{1.75}/D^{1.25}$. So in Eq. 8.73: $90V_1^{1.75}/2^{1.75} = 150V_2^{1.75}/4^{1.25} = 80V_3^{1.75}/6^{1.25}$

$37.8V_1^{1.75} = 26.5V_2^{1.75} = 8.52V_3^{1.75}$; so $V_1 = 0.816V_2$, $V_3 = 1.913V_2$

Eq. 8.72: $Q = 0.007 = (\pi/4)[0.02^2(0.816V_2) + 0.04^2V_2 + 0.06^2(1.913V_2)]$; $V_2 = 1.011$ m/s

Check: $\mathbf{R}_2 = 0.04(1.011)/(3.8 \times 10^{-6}) = 10\,644$; $\mathbf{R}_1 = 4343$; $\mathbf{R}_3 = 30\,547$

These **R** are all in the required range, so assumed Eq. 8.37 <u>does</u> apply.

Thus $V_1 = 0.825$ m/s, $V_2 = 1.011$ m/s, $V_3 = 1.935$ m/s

$Q_1 = 0.259$ L/s ◄ $Q_2 = 1.271$ L/s ◄ $Q_3 = 5.47$ L/s ◄

$h_L = h_{L1} = 0.316(3.8 \times 10^{-6})^{0.25}150(1.011)^{1.75}/[(0.04)^{1.25}2(9.81)] = 6.08$ m ◄

8.106 *Repeat Prob. 8.105 for the case where the total flow rate is *0.35 L/s.*

*Prob. 8.105: In Fig. 8.30 (see Solution 8.102) pipes 1,2, and 3 are of drawn copper tubing as follows: 90 m of 2 cm, 150 m of 4 cm, and 80 m of 6 cm, respectively. When crude oil (s = 0.855) at 50°C is flowing, find the head loss from A to B and the flow in each pipe. Note: This problem can be solved without trial and error, using a basic scientific calculator only. [*Note this correction to the textbook.]*

SI

$\bar{V} = Q/\Sigma A = (0.000\,35 \text{ m}^3/\text{s})/[(\pi/4)(0.02^2 + 0.04^2 + 0.06^2) \text{ m}^2] = 0.0796$ m/s (low)

With this $\bar{V}$ in the largest pipe, $\mathbf{R}_{max} = 0.06(0.0796)/\nu = 1256$

So assume all the flows are laminar, use Eq. 8.19, and then check **R** values.

Eq. 8.19: $h_L = 32\nu(L/gD^2)V$; $h_L \propto LV/D^2$

From Eq. 8.73: $90V_1/2^2 = 150V_2/4^2 = 80V_3/6^2$; $V_1 = 0.417V_2$, $V_3 = 4.22V_2$

Eq. 8.72: $Q = 0.000\,35 = \Sigma(AV) = (\pi/4)[0.02^2(0.417V_2) + 0.04^2V_2 + 0.06(4.22V_2)] = (\pi/4)0.016\,95V_2$

$V_2 = 0.0263$ m/s, $V_1 = 0.010\,95$ m/s, $V_3 = 0.1109$ m/s

Fig. 2.4 for crude oil ($s = 0.855$) at 50°C: $\nu = 3.8 \times 10^{-6}$ m²/s

Eq. 8.1: $\mathbf{R} = DV/\nu$: $\mathbf{R}_1 = 57.7$, $\mathbf{R}_2 = 277$, $\mathbf{R}_3 = 1751$

All 3 flows are laminar, so Eq. 8.19 and the results are valid.

$Q_1 = 0.003\,44$ L/s ◄ $Q_2 = 0.0330$ L/s ◄ $Q_3 = 0.314$ L/s ◄

Check: $\Sigma Q = 0.350$ L/s (as required)

Eq. 8.19: $h_L = h_{L1} = 32(3.8 \times 10^{-6})90(0.010\,95)/(9.81 \times 0.02^2) = 0.0305$ m ◄

8.107 *Repeat Prob. 8.105 for the case where all the pipes are of galvanized iron. Does the "Note" still apply?*

Prob. 8.105: In Fig. 8.30 (see Solution 8.102) pipe 1 is 90 m of 2-cm, pipe 2 is 150 m of 4-cm, and pipe 3 is 80 m of 6-cm diameter, all of drawn copper tubing. Crude oil (s = 0.855) at 50°C is flowing at 7 L/s. Find the head loss from A to B, and the flow in each pipe. Note: This problem can be solved without trial and error, using a basic scientific calculator only.

SI

Fig. 2.4 for crude oil (s = 0.855) at 50°C: $\nu = 3.8 \times 10^{-6}$ m²/s. Table 8.1: e = 0.000 15 m

As $e > 0$, Blasius Eq. 8.37 may not be used, so the "Note" does not apply. ◀

Eq. 8.73 using Eq. 8.10: $f_1(L_1/D_1)V_1^2/2g = f_2(L_2/D_2)V_2^2/2g = f_3(L_3/D_3)V_3^2/2g$

For the first trial assume the f's are the same. Substituting for L's and D's:

$4500V_1^2 = 3750V_2^2 = 1333V_3^2$; so $V_2 = 1.095V_1$, $V_3 = 1.837V_1$

Eq. 8.72: $Q = 0.007 = (\pi/4)[0.02^2V_1 + 0.04^2(1.095V_1) + 0.06^2(1.837V_1) = 0.006\,89V_1$

$V_1 = 1.017$ m/s; $\mathbf{R}_1 = D_1V_1/\nu = 5351$, $\mathbf{R}_2 = 11\,723$, $\mathbf{R}_3 = 29\,491$

For 2nd trial: $f_1 = 0.044\,43$, $f_2 = 0.034\,91$, $f_3 = 0.028\,86$ (Eq. 8.41)

Eq. 8.73: $0.044\,43(4500)V_1^2 = 0.034\,91(3750)V_2^2 = 0.028\,86(1333)V_3^2$; so $V_2 = 1.236V_1$, $V_3 = 2.28V_1$

Eq. 8.72: $0.007 = (\pi/4)[0.02^2V_1 + 0.04^2(1.236V_1) + 0.06^2(2.28V_1)] = 0.008\,31V_1$

$V_1 = 0.842$ m/s (21% change); $\mathbf{R}_1 = 4432$, $\mathbf{R}_2 = 10\,955$, $\mathbf{R}_3 = 30\,310$

For 3rd trial: $f_1 = 0.046\,08$, $f_2 = 0.035\,30$, $f_3 = 0.028\,77$ (Eq. 8.41)

Eq. 8.73: $0.046\,08(4500)V_1^2 = 0.035\,30(3750)V_2^2 = 0.028\,77(1333)V_3^2$; so $V_2 = 1.252V_1$, $V_3 = 2.33V_1$

Eq. 8.72: $0.007 = (\pi/4)[0.02^2V_1 + 0.04^2(1.252V_1) + 0.06^2(2.33V_1) = 0.008\,46V_1$

$V_1 = 0.827$ m/s (1.8% change, close enough!)

$Q_1 = A_1V_1 = (\pi/4)0.02^2(0.827) = 0.000\,260$ m³s = 0.260 L/s ◀

$Q_2 = (\pi/4)0.04^2(1.252 \times 0.827) = 0.001\,301$ m³/s = 1.301 L/s ◀

$Q_3 = (\pi/4)0.06^2(2.33 \times 0.827) = 0.005\,44$ m³/s = 5.44 L/s ◀

Eq. 8.10: $h_L = h_{L1} = 0.046\,08(4500)(0.827)^2/2g = 7.23$ m ◀

/more...

8.108 *The pipes in the system shown in the figure are all galvanized iron. (a) With a flow of 15 cfs, find the head loss from A to D. (b) What should be the diameter of a single pipe from B to C such that it replaces pipes 2, 3, and 4 without altering the capacity for the same head loss from A to D?*

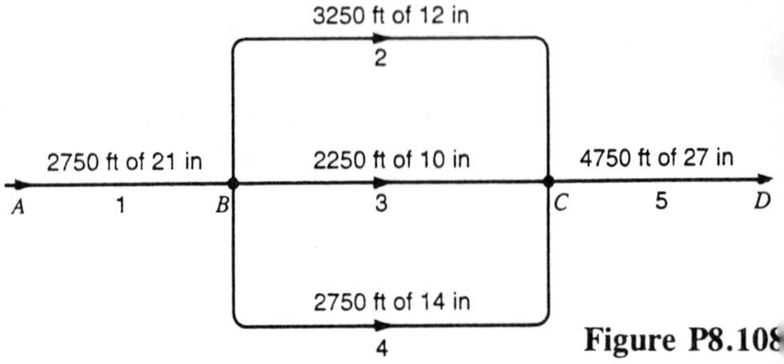

Figure P8.108

BG

Table 8.1: $e = 0.0005$ ft.

(*a*) Trial 1: Assume $f = f_{min}$ (complete turbulence) from Eq. 8.39.

Pipe	D (in)	L/D	e/D	Try f	A (ft²)
2	12	3250	0.0005	0.0167	0.785
3	10	2700	0.0004	0.0174	0.545
4	14	2357	0.000 429	0.0161	1.069

All L/D are > 1000, so neglect minor losses. Then in Eq. 8.75, $C = A\sqrt{2gD/fL}$

Pipe	C	% discharge	Q cfs	V fps	D''V	f (Fig. 8.11 or Eq. 8.41)	New C
2	0.856	29.6	4.45	5.66	67.9	0.017 64	0.832
3	0.639	22.1	3.32	6.09	60.9	0.018 32	0.622
4	1.393	48.2	7.24	6.77	94.8	0.016 91	1.359
Σ	2.888	99.9	15.01				2.814

ΣC has changed by only 2.6% so these values are very close to correct.

Eq. 8.75: $Q = \Sigma C h_L^{1/2}$; $15 = 2.814 h_L^{1/2}$; $\therefore h_L = 28.4$ ft between B and C

Pipe	D (in)	L/D	e/D	V	D''V	f	h_L ft (Eq. 8.10)
1	21	1571	0.000 286	6.24	131.0	0.015 54	14.75
5	27	2111	0.000 222	3.77	101.9	0.015 14	7.06

Then h_L (A to D) = 14.75 + 28.4 + 7.06 = 50.2 ft ◄

(*b*) This is a Type 3 problem, to find D. Given $L = 2250$ ft.

Eq. 8.43: $\mathbf{N}_1 = 2.12 \times 10^{28}$; Eq. 8.44: $\mathbf{N}_2 = 3.19 \times 10^{-10}$

Eq. 8.47: $\mathbf{R}^{2.5} = 1.146 \times 10^{15}$, so $\mathbf{R} = 1.056 \times 10^{6}$

Eq. 8.48: $D = 1.486$ ft $= 17.83$ in ◄

Alternative: Using a solver with Eq. 8.42*b*: $D = 1.482$ ft $= 17.79$ in

/more...

8.109 (a) *With the same pipe lengths, sizes, and connections as in Prob. 8.108, find the flow in each pipe if the head loss from A to D is 150 ft and if all pipes have $f = 0.020$. Also find the head losses from A to B, B to C, and C to D. (b) Find the new head loss distributions and the percentage increase in the capacity of the system achieved by adding another 12-in pipe 3250 ft long between B and C.*

See Fig. P8.108 with Solution 8.108.

BG

(a) Eq. 8.10: $h_L = f(L/D)V^2/2g = f(L/D)Q^2/(2gA^2) = (fL/2gD)[Q/(\pi D^2/4)]^2 = (8f/\pi^2 g)(LQ^2/D^5)$

Eq. 8.73: $h_{L2} = h_{L3} = h_{L4}$; f = const, so 3 values of LQ^2/D^5 are equal

i.e., $3250Q_2^2/1^5 = 2250Q_3^2/(10/12)^5 = 2750Q_4^2/(14/12)^5$ from which $Q_2 = 1.313Q_3$, $Q_4 = 2.10Q_3$

Eq. 8.72: $Q_1 = Q_5 = Q_2 + Q_3 + Q_4 = (1.313 + 1 + 2.10)Q_3 = 4.41Q_3$

Eq. 8.69: $(h_L)_{AD} = 150 = h_{L1} + h_{L3} + h_{L5} = 8(0.020/\pi^2 g)(L_1Q_1^2/D_1^5 + L_3Q_3^2/D_3^5 + L_5Q_5^2/D_5^5)$

$= 0.000\,503[2750(4.41Q_3)^2/(21/12)^5 + 2250Q_3^2/(10/12)^5 + 4750(4.41Q_3)^2/(27/12)^5] = 5.27Q_3^2$

So $Q_3 = 5.34$ cfs ◄ $Q_1 = Q_5 = 23.5$ cfs ◄ $Q_2 = 7.00$ cfs ◄ $Q_4 = 11.20$ cfs ◄

$(h_L)_{AB} = h_{L1} = 0.000\,503(2750)23.5^2/(21/12)^5 = 46.73$ ft ◄

$(h_L)_{BC} = h_{L3} = 0.000\,503(2250)5.34^2/(10/12)^5 = 80.29$ ft ◄

$(h_L)_{CD} = h_{L5} = 0.000\,503(4750)23.5^2/(27/12)^5 = 22.98$ ft ◄ Check: $\Sigma h_L = 100.00$ ft -- correct!

(b) The additional pipe from B to C is identical to pipe 2.

As above, $Q_2 = 1.313Q_3$, $Q_4 = 2.10Q_3$. But now $Q_1 = Q_5 = 2Q_2 + Q_3 + Q_4 = 5.72Q_3$, so now

$$(h_L)_{AD} = 150 = 0.000\,503\left[\frac{2750(5.72Q_3)^2}{(21/12)^5} + \frac{2250Q_3^2}{(10/12)^2} + \frac{4750(5.72Q_3)^2}{(27/12)^5}\right] = 6.94Q_3^2$$

So $Q_3 = 4.65$ cfs, $Q_1 = Q_5 = 26.6$ cfs

$(h_L)_{AB} = h_{L1} = 0.000\,503(2750)26.6^2/(21/12)^5 = 59.71$ ft ◄

$(h_L)_{BC} = h_{L3} = 0.000\,503(2250)4.65^2/(10/12)^5 = 60.93$ ft ◄

$(h_L)_{CD} = h_{L5} = 0.000\,503(4750)26.6^2/(27/12)^5 = 29.36$ ft ◄ Check: $\Sigma h_L = 100.00$ ft -- correct!

$Q_b/Q_a = 26.6/23.5 = 1.130$, or 13.0% increase in capacity ◄

8.110 *In the figure, pipe AB is 1200 ft long, of 8 in diameter, with $f = 0.035$; pipe BC (upper) is 800 ft long, of 6 in diameter, with $f = 0.025$; pipe BC (lower) is 900 ft long, of 4 in diameter, with $f = 0.045$; and pipe CD is 500 ft long, of 6 in diameter, with $f = 0.025$. The elevations are reservoir water surface = 150 ft, A = 120 ft, B = 70 ft, C = 60 ft, and D = 30 ft. There is free discharge to the atmosphere at D. Neglecting velocity heads, (a) compute the flow in each pipe and (b) determine the pressures at B and C.*

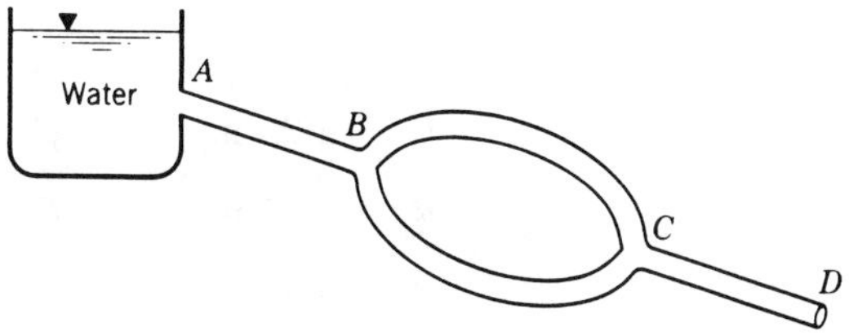

Figure P8.110

BG

(a) B to C, Eq. 8.73: $0.025(1600)V_u^2/2g = 0.045(2700)V_L^2/2g$; $V_u = 1.743V_L$

Continuity: $Q = 0.349V_1 = 0.1963(1.743V_L) + 0.0873V_L = 0.1963V_2$

from which $V_1 = 1.230V_L$, $V_2 = 2.19V_L$

All $L/D > 1000$, therefore neglect minor losses

/cont...

Energy Eq. (neglecting velocity heads) from water surface to D:

$(0 + 150) = (0 + 30) + h_{L1} + h_{LL} + h_{L2}$

$\therefore\ h_L = 150 - 30 = f_1(L_1/D_1)V_1^2/2g + f_L(L_L/D_L)V_L^2/2g + f_2(L_2/D_2)V_2^2/2g$

$120 = 0.035(1800)(1.230V_L)^2/2g + 0.045(2700)V_L^2/2g + 0.025(1100)(2.19V_L)^2/2g = 348V_L^2/2g$

$V_L = 4.71$ fps, $Q_L = 0.411$ cfs ◀

$V_u = 8.21$ fps, $Q_u = 1.612$ cfs ◀

$V_1 = 5.79$ fps, $Q_1 = 2.02$ cfs ◀

$V_2 = 10.30$ fps, $Q_2 = 2.02$ cfs ◀

(*b*) Energy, water surface to B: $(0 + 150) = (p_B/\gamma + 70) + (h_L)_{AB}$

$p_B/\gamma = 80 - 0.035(1800)(5.79)^2/2g = 47.16$ ft; $p_B = 47.16(62.4/144) = 20.4$ psi ◀

Energy, C to D: $(p_C/\gamma + 60) = (0 + 30) + (h_L)_{CD}$

$p_C/\gamma = 0.025(1100)(10.30^2/2g) - 30 = 15.31$ ft; $p_C = 15.31(62.4/144) = 6.64$ psi ◀

☐ 8.111 *In Fig. P8.110 [see Solution 8.110], pipe AB is 600 m long, of 18 cm diameter, with f = 0.035; pipe BC (upper) is 500 m long, of 12 cm diameter, with f = 0.025; pipe BC (lower) is 400 m long, of 16 cm diameter, with f = 0.030; and pipe CD is 900 m long, of 32 cm diameter, with f = 0.020. The elevations are: reservoir water surface = 150 m, A = 100 m, B = 60 m, C = 50 m, D = 20 m. There is free discharge to the atmosphere at D. Neglecting velocity heads, (a) compute the flow in each pipe and (b) determine the pressures at B and C. Comment on the practicality of this system.*

SI

(*a*) B to C, Eq. 8.73: $0.025(4167)V_u^2/2g = 0.030(2500)V_L^2/2g$; $V_u = 0.849V_L$

Continuity: $Q = 0.0254V_1 = 0.011\,31(0.849V_L) + 0.0201V_L = 0.0804V_2$

from which $V_1 = 1.167V_L$, $V_2 = 0.369V_L$

Energy Eq. (neglecting velocity heads) from water surface to D:

All $L/D > 1000$, therefore neglect minor losses

$(0 + 150) = (0 + 20) + h_{L1} + h_{LL} + h_{L2}$

$\therefore\ h_L = 150 - 20 = f_1(L_1/D_1)V_1^2/2g + f_L(L_L/D_L)V_L^2/2g + f_2(L_2/D_2)V_2^2/2g$

$130 = 0.035(3333)(1.167V_L)^2/2g + 0.030(2500)V_L^2/2g + 0.020(2813)(0.369V_L)^2/2g = 242V_L^2/2g$

$V_L = 3.25$ m/s, $Q_L = 0.0653\ m^3/s = 65.3$ L/s ◀

$V_u = 2.76$ m/s, $Q_u = 0.0312\ m^3/s = 31.2$ L/s ◀

$V_1 = 3.79$ m/s, $Q_1 = 0.0965\ m^3/s = 96.5$ L/s ◀

$V_2 = 1.200$ m/s, $Q_2 = 0.0965\ m^3/s = 96.5$ L/s ◀

(*b*) Energy, water surface to B: $(0 + 150) = (p_B/\gamma + 60) + (h_L)_{AB}$

$p_B/\gamma = 90 - 0.035(3333)(3.79)^2/2g = 4.48$ m; $p_B = 4.48(9.81) = 43.9$ kPa ◀

Energy, C to D: $(p_C/\gamma + 50) = (0 + 20) + (h_L)_{CD}$

$p_C/\gamma = 0.020(2813)(1.200^2/2g) - 30 = -25.9$ m (impossible) ◀

Comment: The head loss in the parallel pipes is so great that it wants to pull the HGL down below zero absolute pressure. ∴ The system will not function. ◀

8.112 *Referring to the figure, the pipe characteristics are as follows: pipe A is 5000 ft long, of 3 ft diameter, with $f = 0.035$; pipe B is 4500 ft long, of 2 ft diameter, with $f = 0.035$; and pipe C is 5000 ft long, of 3 ft diameter, with $f = 0.025$. When the pump develops 30 ft of head, the velocity of flow in pipe C is 5 fps. Neglecting minor losses, find (a) the flow rates in cubic feet per second in pipes A and B under these conditions, and (b) the elevation of the discharge end of pipe B.*

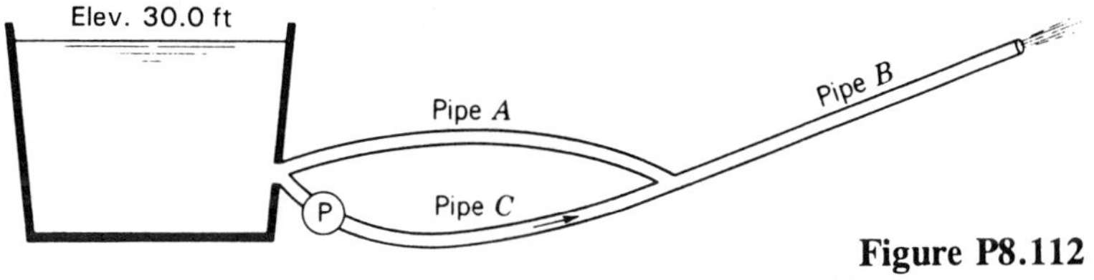

Figure P8.112

BG

(a) Energy Drop across A, Eq. 8.10: $(h_L)_A = 0.035(5000/3)V_A^2/2g = 58.3V_A^2/2g$

Given $V_C = 5$ fps, energy drop across $C = (h_L)_C - h_p = 0.025(5000/3)5^2/2g - 30 = -13.83$ ft

Hence energy is greater at right end of pipe C and flow will be to the left in pipe A.

$58.3(V_A^2/2g) = 13.83$; $V_A = 3.91$ fps; $Q_A = (\pi 3^2/4)3.91 = 27.6$ cfs to left ◄

$Q_C = (\pi 3^2/4)5 = 35.3$ cfs to right; $Q_C - Q_A = Q_B = 7.73$ cfs ◄

(b) $V_B = Q_B/A_B = 7.73/(\pi 2^2/4) = 2.46$ fps; $V_B^2/2g = 0.0940$ ft

$h_{LB} = f(L/D)V^2/2g = 0.035(4500/2)0.0940 = 7.40$ ft

Energy, water surface to jet J: $H_S + h_{LA} - h_{LB} = H_J$

$(0 + 30 + 0) + 13.83 - 7.40 = (0 + z_J + 0.0940)$; $z_J = 36.3$ ft ◄

8.113 *Repeat Prob. 8.112 for the case where the velocity in pipe C is 6 fps with all other data remaining the same.*

BG

(a) Eq. 8.10: $(h_L)_A = 58.3V_A^2/2g$; $(\Delta E)_C = (h_L)_C - h_p = (41.7)6^2/2g - 30 = -6.71$ ft

Energy is greater at right end of pipe C thus flow will be to the left in pipe A

$58.3(V_A^2/2g) = 6.71$; $V_A = 2.72$ fps, $Q_A = 19.24$ cfs to the left ◄

$V_C = 6.0$ fps (given), $Q_C = 42.4$ cfs to the right

$Q_B = Q_C - Q_A = 42.4 - 19.24 = 23.2$ cfs ◄

(b) $V_B = Q_B/A_B = 23.2/(\pi 2^2/4) = 7.38$ fps; $V_B^2/2g = 0.845$ ft

$h_{LB} = f(L/D)V^2/2g = 0.035(4500/2)0.845 = 66.5$ ft

Energy, water surface to jet J: $H_S + h_{LA} - h_{LB} = H_J$

$(0 + 30 + 0) + 6.71 - 66.5 = (0 + z_J + 0.845)$; $z_J = -30.7$ ft ◄

8.114 *Refer to Fig. P8.112 [with Solution 8.112]. Assume the water surface in the reservoir is instead at elevation 200 m. Pipes A, B, and C are all 600 m long, and they all have a diameter of 50 cm with $f = 0.030$. When the pump develops 15 m of head, the velocity in pipe C is 5.0 m/s. Neglecting minor losses, find (a) the flow rate in all pipes and (b) the elevation of the discharge end of pipe B.*

BG

(*a*) Δ Energy head across pipe $C = h_p - h_{L_C} = 15 - 0.030(600/0.50)5^2/(2 \times 9.81) = -30.9$ m

Thus flow in pipe A is to the right.

$h_{LA} = 30.9 \text{ m} = 0.030(600/0.50)V_A^2/(2 \times 9.81); \quad V_A = 4.10$ m/s

$Q_A = \pi 0.25^2(4.10) = 0.805 \text{ m}^3\text{/s}$ ◀

$Q_C = \pi 0.25^2(5) = 0.982 \text{ m}^3\text{/s}$ ◀

$Q_B = Q_A + Q_C = 1.787 \text{ m}^3\text{/s}$ ◀

(*b*) $V_B = Q_B/A_B = 1.787/(\pi 0.5^2/4) = 9.10 \text{ m/s}; \quad V_B^2/2g = 4.22$ m

$h_{LB} = f(L/D)V^2/2g = 0.030(600/0.50)4.22 = 152.0$ m

Energy, water surface to jet J: $H_S - h_{LA} - h_{LB} = H_J$

$(0 + 200 + 0) - 30.9 - 152.0 = (0 + z_J + 4.22); \quad z_J = 12.90$ m ◀

8.115 *A 10-in cast-iron pipe 1500 ft long forms one link in a pipe network. If the velocities to be encountered are assumed to fall within the range of 4 to 10 fps, derive an equation for the flow of water at 60°F in this pipe in the form $h_L = KQ^n$.* Hint: *Using information from Fig. 8.11 and Table 8.1, set up two simultaneous equations corresponding to the ends of the desired velocity range, then solve for the unknowns K and n.*

BG

Table 8.1: $e = 0.000\,85$ ft; $e/D = 0.000\,85(12/10) = 0.001\,02$

(*i*) At 4 fps: $D''V = 40$. Fig. 8.11 or Eq. 8.41: $f = 0.020\,77$

Eq. 8.10: $h_L = 0.020\,77[1500/(10/12)]4^2/(2 \times 32.2) = 9.29$ ft; $\quad Q = AV = 2.18$ cfs

Given $h_L = KQ^n$. $\therefore\ 9.29 = K(2.18)^n; \quad \log 9.29 = \log K + n \log 2.18$ (1)

(*ii*) At 10 fps: $D''V = 100$. Fig. 8.11 or Eq. 8.41: $f = 0.020\,17$

Eq. 8.10: $h_L = 0.020\,17[1500/(10/12)]10^2/(2 \times 32.2) = 56.4$ ft; $\quad Q = 5.45$ cfs

$\therefore\ 56.4 = K(5.45)^n; \quad \log 56.4 = \log K + n \log 5.45$ (2)

(2) − (1): $\log(56.4/9.29) = n \log(5.45/2.18); \quad n = 1.968; \quad K = 56.4/5.45^{1.968} = 2.00$

So required equation is $h_L = 2.00Q^{1.968}$ ◀

/more...

8.116 *A 25-cm cast-iron pipe 400 m long forms one link in a pipe network. If the velocities to be encountered are assumed to fall within the range of 0.75 to 3 m/s, derive an equation for the flow of water at 15°C in this pipe in the form $h_L = KQ^n$.* Hint: *See Prob. 8.115.*

SI

Table A.1 for water at 15°C: $\nu = 1.139 \times 10^{-6}$ m²/s

Table 8.1: $e = 0.25$ mm; $e/D = 0.25/250 = 0.001$

(*i*) At 0.75 m/s: $\mathbf{R} = DV/\nu = 164\,600$. Fig. 8.11 or Eq. 8.41: $f = 0.021\,30$

Eq. 8.10: $h_L = 0.021\,30(400/0.25)0.75^2/(2 \times 9.81) = 0.977$ m; $Q = AV = 0.0368$ m³/s

Given $h_L = KQ^n$. $\therefore$ $0.977 = K(0.0368)^n$; $\log 0.977 = \log K + n \log 0.0368$ (1)

(*ii*) At 2.3 m/s: $\mathbf{R} = 504\,800$. Fig. 8.11 or Eq. 8.41: $f = 0.020\,23$

Eq. 8.10: $h_L = 0.020\,23(400/0.25)2.3^2/(2 \times 9.81) = 8.73$ m; $Q = 0.1129$ m³/s

$\therefore$ $8.73 = K(0.1129)^n$; $\log 8.73 = \log K + n \log 0.1129$ (2)

(2) − (1): $\log(8.73/0.977) = n \log(0.1129/0.0368)$; $n = 1.954$; $K = 8.73/0.1129^{1.954} = 619$

So the required equation is $h_L = 619Q^{1.954}$ ◀

8.117 *Find the magnitude and direction of the flow in network lines ab and bc after making two sets of corrections. The numbers on the figure are the K values of each line; take n = 2.0. Start by assuming initial flows as follows: 9 cfs in lines ab and cd, 6 cfs in lines ac and bd, and 3 cfs in line bc.*

BG

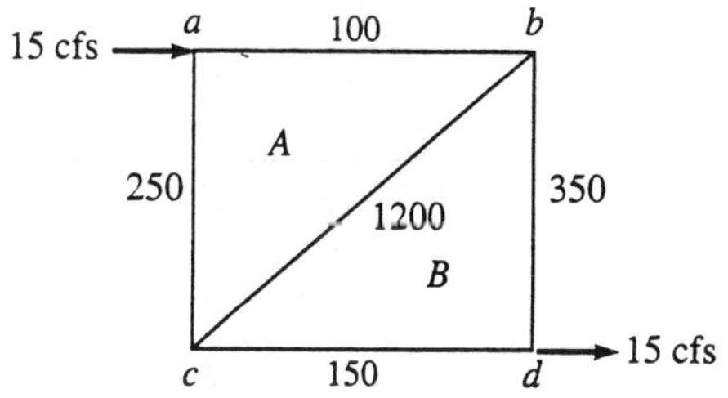

Figure P8.117

Given initial assumptions:

First approx:

Loop A

Pipe	K	Q	KQ^2	$\|2KQ\|$
ab	100	+9	+8100	1800
bc	1200	+3	+10,800	7200
ac	250	−6	−9000	3000
			+9900	12,000

$\Delta Q = -(+9900/12{,}000) = -0.825$

Loop B

Pipe	K	Q	KQ^2	$\|2KQ\|$
bd	350	+6	+12,600	4200
bc	1200	−3	−10,800	7200
cd	150	−9	−12,150	2700
			−10,350	14,100

$\Delta Q = -(-10{,}350)/14{,}100) = +0.734$

Second approx. (after first corrections):

Loop A

Pipe	K	Q	KQ^2	$\|2KQ\|$
ab	100	+8.175	+6683	1635
bc	1200	+1.441	+2492	3458
ac	250	−6.825	−11,645	3413
			−2470	8506

$\Delta Q = -(-2470/8506) = +0.290$

Loop B

Pipe	K	Q	KQ^2	$\|2KQ\|$
bd	350	+6.734	+15,871	4714
bc	1200	−1.441	−2492	3458
cd	150	−8.266	−10,249	2480
			+3130	10,652

$\Delta Q = -(+3130/10{,}652) = -0.294$

After second corrections:

Flow in line *ab* = +8.175 + 0.290 = 8.465 cfs from *a* to *b* ◀

Flow in line *bc* = +1.441 + 0.290 − (−0.294) = 2.025 cfs from *b* to *c* ◀

8.118 *Find the magnitude and direction of the flow in network lines ab and bc after making two sets of corrections. The numbers on the figure are the K values of each line; take n = 2.0. Start by assuming initial flows as follows: 0.3 m^3/s in lines ab and cd, 0.2 m^3/s in lines ac and bd, and 0.1 m^3/s in line bc.*

SI

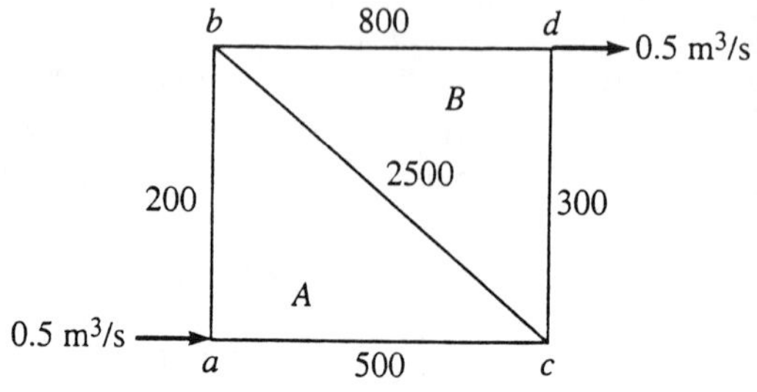

Figure P8.118

Given initial assumptions:

First approx:

Loop A

Pipe	K	Q	KQ^2	$\lvert 2KQ \rvert$
ab	200	+0.3	+18.0	120
bc	2500	+0.1	+25.0	500
ac	500	−0.2	−20.0	200
			+23.0	820

$\Delta Q = -(+23/820) = -0.028$

Loop B

Pipe	K	Q	KQ^2	$\lvert 2KQ \rvert$
bd	800	+0.2	+32.0	320
bc	2500	−0.1	−25.0	500
cd	300	−0.3	−27.0	180
			−20.0	1000

$\Delta Q = -(-20.0/1000) = +0.020$

Second approx. (after first corrections):

ab	200	+0.272	+14.80	109
bc	2500	+0.052	+6.76	260
ac	500	−0.228	−25.99	228
			−4.43	597

$\Delta Q = -(-4.43/597) = +0.007$

bd	800	+0.22	+38.72	352
bc	2500	−0.052	−6.76	260
cd	300	−0.28	−23.52	168
			+8.44	780

$\Delta Q = -(+8.44/780) = -0.011$

After second corrections: Flow in line *ab* = +0.272 + 0.007 = 0.279 m^3/s from *a* to *b* ◀

Flow in line *bc* = +0.052 + 0.007 − (−0.011) = 0.070 m^3/s from *b* to *c* ◀

8.119 *Solve the pipe network shown using four approximations, to find the flow in each pipe. For simplicity, take n = 2.0 and use the value of f for complete turbulence, as given in Fig. 8.11. All pipe is cast iron. For initial flows, assume only values of 30, 15, and 0 L/s. If the pressure head at a is 40 m, find the pressure head at d (which might represent a fire demand, for example) neglecting velocity heads.*

SI

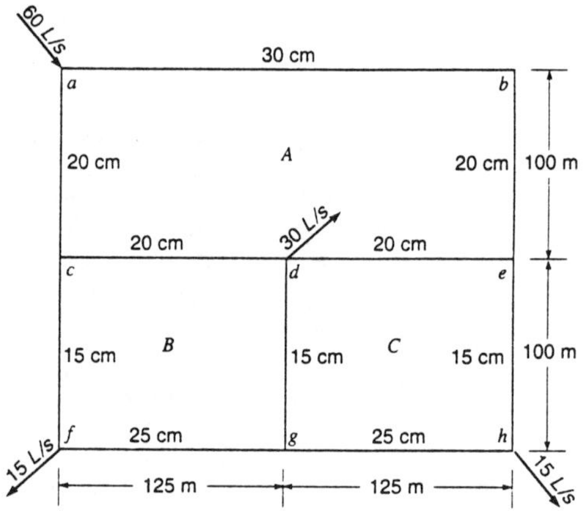

Figure. P8.119

Per Eq. 8.66 and Sec. 8.26 (Approximate Solutions):

$h_L = KQ^2$ with $K = 8fL/(\pi^2 g D^5)$.

Table 8.1 for cast iron: $e = 0.000\,25$ m

Using $f = f_{min}$ for complete turbulence from Fig. 8.11 or Eq. 8.39, the pipe characteristics are:

Pipes	L, m	D, m	e/D	f_{min}	K
ab	250	0.30	0.000 833	0.018 79	160
ac, be	100	0.20	0.001 25	0.020 75	536
cd, ed	125	0.20	0.001 25	0.020 75	670
cf, dg, eh	100	0.15	0.001 667	0.022 33	2430
gf, gh	125	0.25	0.001 00	0.019 64	208

Initial flow assumptions (L/s):

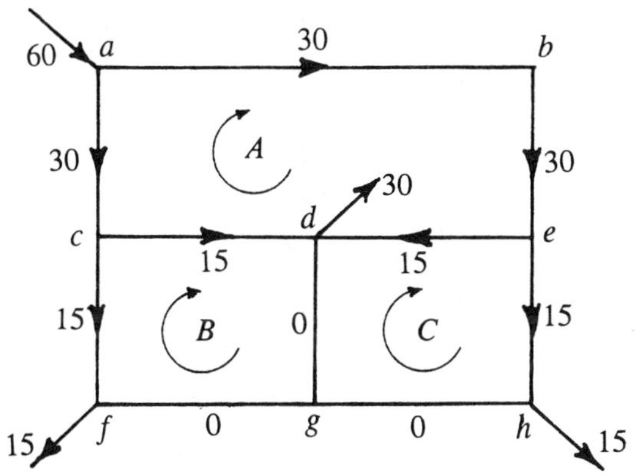

First approximation and correction

Note: +/− signs are used to indicate clockwise/counter-clockwise flows in the loops indicated.

Loop A

Pipe	Q	K	KQ^n	$\lvert KnQ^{n-1}\rvert$
ab	+30	160	+144 000	9600
be	+30	536	+482 400	32 160
ed	+15	670	+150 750	20 100
cd	−15	670	-150 750	20 100
ac	−30	536	-482 400	32 160
			+144 000	114 120

ΔQ = -(+144 000)/114 120 = -1.26 L/s

Loop B

Pipe	Q	K	KQ^n	$\lvert KnQ^{n-1}\rvert$
cd	+15	670	+150 750	20 100
dg	0	2430	0	0
gf	0	208	0	0
cf	−15	2430	-546 750	72 900
			-396 000	93 000

ΔQ = -(-396 000)/93 000 = +4.26 L/s

Loop C

Pipe	Q	K	KQ^n	$\lvert KnQ^{n-1}\rvert$
ed	−15	670	-150 750	20 100
eh	+15	2430	+546 750	72 900
gh	0	208	0	0
dg	0	2430	0	0
			+396 000	93 000

ΔQ = -(+396 000)/93 000 = -4.26 L/s

Pipe	In Loop	Q, L/s	(2nd Approximation)			
ab	A	+30	−1.26			= +28.74
be	A	+30	−1.26			= +28.74
ed	A	+15	−1.26		+4.26	= +18.00
ed	C	−15	+1.26		-4.26	= −18.00
cd	A	−15	−1.26	−4.26		= −20.52
cd	B	+15	+1.26	+4.26		= +20.52
ac	A	−30	−1.26			= −31.26
dg	B	0		+4.26	+4.26	= +8.52
dg	C	0		−4.26	−4.26	= −8.52
gf	B	0		+4.26		= +4.26
cf	B	−15		+4.26		= −10.74
eh	C	+15			−4.26	= +10.74
gh	C	0			−4.26	= −4.26

Repeating the same procedure, the subsequent loop corrections ΔQ in L/s are as follows:

Loop	1st	2nd	3rd correction
A	−1.26	+0.11	−0.13
B	+4.26	−1.48	+0.28
C	+4.26	+0.98	−0.41

The loop corrections are quite rapidly becoming small.

/cont...

The approximations of the flows Q, in L/s, were as follows:

Pipe	In Loop	1st	2nd	3rd	4th
ab	A	+30	+28.74	+28.85	+28.72
be	A	+30	+28.74	+28.85	+28.72
ed	A	+15	+18.00	+17.13	+17.41
ed	C	−15	−18.00	−17.13	−17.41
cd	A	−15	−20.52	−18.93	−19.34
cd	B	+15	+20.52	+18.93	+19.34
ac	A	−30	−31.26	−31.15	−31.28
dg	B	0	+8.52	+6.06	+6.75
dg	C	0	−8.52	−6.06	−6.75
gf	B	0	+4.26	+2.78	+3.06
cf	B	−15	−10.74	−12.22	−11.94
eh	C	+15	+10.74	+11.72	+11.31
gh	C	0	−4.26	−3.28	−3.69

▲

This table shows the process of converging, which occurs more slowly in pipes that are common to two loops. Resulting final flows, L/s:

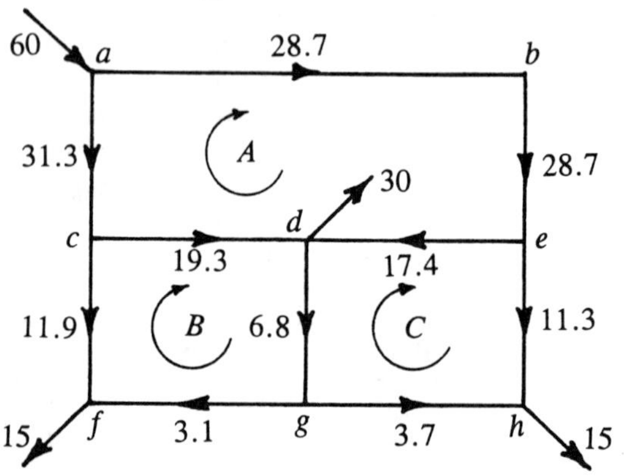

Assume all pipes (junctions) are at the same elevation. Final approximated flows from a to d are: 31.26 L/s from a to c, and 19.34 L/s from c to d.

The head losses are given by $h_L = KQ^2$.

$\therefore$ neglecting velocity heads:

$$p_d/\gamma = p_a/\gamma - (h_L)_{ac} - (h_L)_{cd} = 40 - 536(31.26/1000)^2 - 670(19.34/1000)^2 = 40 - 0.523 - 0.251$$

$$= 39.23 \text{ m} \quad ◄$$

/more...

☐ 8.120 *Solve the pipe network shown using five trials. The 12-in and 16-in pipes are of average cast iron, while the 18-in and 24-in sizes are of average concrete (e = 0.003 ft). Assume n = 2.0, and use the values of f from Fig. 8.11 for complete turbulence. If the pressure at h is 80 psi, find the pressure at f.*

BG

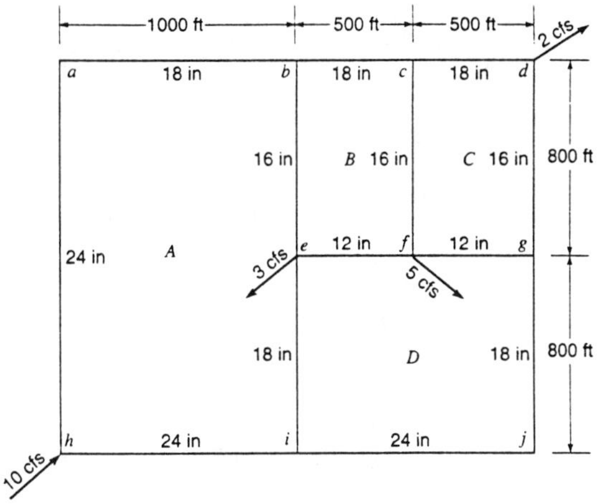

Figure P8.120

Per Eq. 8.66 and Sec. 8.26 (Approximate Solutions):

$h_L = KQ^2$ with $K = 8fL/(\pi^2 g D^5) = fL/(39.7D^5)$

Table 8.1 for average cast iron: e = 0.000 85 ft

Using $f = f_{min}$ for complete turbulence from Fig. 8.11 or Eq. 8.39, the pipe characteristics are:

Pipe	Mat'l	L, ft	D, ft	e/D	f	D^5	$K = \dfrac{fL}{39.7D^5}$
ab	avg conc	1000	1.50	0.002 00	0.0233	7.59	0.0772
bc	avg conc	500	1.50	0.002 00	0.0233	7.59	0.0386
cd	avg conc	500	1.50	0.002 00	0.0233	7.59	0.0386
ef	avg C.I.	500	1.00	0.000 85	0.0188	1.00	0.237
fg	avg C.I.	500	1.00	0.000 85	0.0188	1.00	0.237
hi	avg conc	1000	2.00	0.001 50	0.0218	32.00	0.017 15
ij	avg conc	1000	2.00	0.001 50	0.0218	32.00	0.017 15
ah	avg conc	1600	2.00	0.001 50	0.0218	32.00	0.0274
be	avg C.I.	800	1.33	0.000 62	0.0175	4.21	0.0836
ei	avg conc	800	1.50	0.002 00	0.0233	7.59	0.0618
cf	avg C.I.	800	1.33	0.000 62	0.0175	4.21	0.0836
dg	avg C.I.	800	1.33	0.000 62	0.0175	4.21	0.0836
gi	avg conc	800	1.50	0.002 00	0.0233	7.59	0.0618

Initial flow assumptions (cfs):

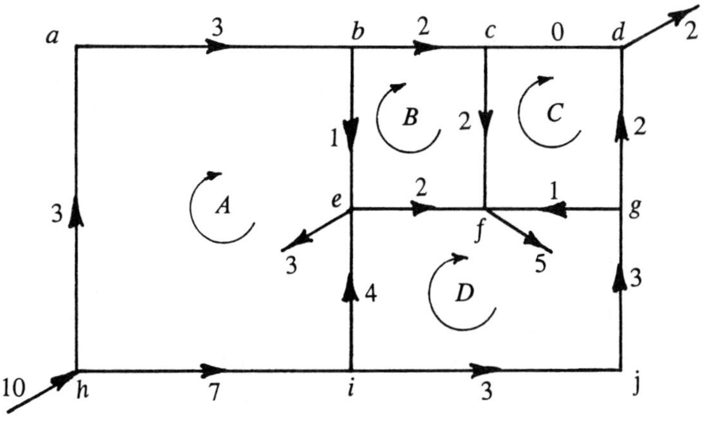

/cont...

First approximation (note that +/− signs are used to indicate clockwise/counter-clockwise flows in the loops indicated):

Loop A

Pipe	Q_0	KQ_0^2	$\lvert h_L/Q_0 \rvert$
ha	+3	+0.247	0.0823
ab	+3	+0.695	0.232
be	+1	+0.0836	0.0836
hi	−7	−0.840	0.1200
ie	−4	−0.989	0.247
Σ		−0.803	0.765

ΔQ = -(-0.803)/(2 × 0.765) = +0.525 cfs

Loop B

Pipe	Q_0	KQ_0^2	$\lvert h_L/Q_0 \rvert$
bc	+2	+0.1545	0.0772
cf	+2	+0.335	0.1673
be	−1	−0.0836	0.0836
ef	−2	−0.947	0.473
Σ		−0.541	0.801

ΔQ = -(-0.541)/(2 × 0.801) = +0.338 cfs

Loop C

Pipe	Q_0	KQ_0^2	$\lvert h_L/Q_0 \rvert$
cd	0	0.000	0.000
gf	+1	+0.237	0.237
cf	−2	−0.335	0.1673
gd	−2	−0.335	0.1673
Σ		−0.571	0.571

ΔQ = -(-0.571)/(2 × 0.571) = +0.379 cfs

Loop D

Pipe	Q_0	KQ_{02}	$\lvert h_L/Q_0 \rvert$
ie	+4	+0.989	0.247
ef	+2	+0.947	0.473
ij	−3	−0.1543	0.0514
jg	−3	−0.556	0.1854
gf	−1	−0.237	0.237
Σ		+0.989	1.194

ΔQ = -(+0.989)/(2 × 1.194) = -0.414 cfs

After 5 trials, the flows (cfs) within approx 1% are:

ab = 3.54; *bc* = 2.48; *cd* = 0.40; *ef* = 1.47; *gf* = 1.45; *hi* = 6.46; *ij* = 3.05; *ha* = 3.54

be = 1.06; *ie* = 3.41; *cf* = 2.08; *gd* = 1.60; *jg* = 3.05 ◄◄

Head loss (Eq. 8.63) from *h* to *f* (any path): *hi* = 0.716 ft; *ie* = 0.719 ft; *ef* = 0.511 ft;

Total = 1.945 ft. p_f = 80 - 1.945(62.4)/144 = 80 - 0.843 = 79.2 psi ◄

Chapter 9

Forces on Immersed Bodies

PROBLEM SELECTION GUIDE

Sec	Exer/Prob	Units	Difficulty	Length	Parts	Similar	Special features
9.3	*Laminar Boundary Layer for Incompressible Flow Along a Smooth Flat Plate*						
	X[1]9.3.1	BG	Easy	Short	1	P9.3	*
	9.3.2	SI	Medium	Medium	3	P9.4	
	9.3.3	N	Easy	Short	1		
	P 9.1	N	Medium	Medium	1		Derivation; differentiation.
	9.2	N	Medium	Medium	1		Proof; integration, differentiation.
	9.3	SI	Easy	Short	1	X9.3.1	*
	9.4	BG	Medium	Medium	3	X9.3.2	
9.4	*Turbulent Boundary Layer for Incompressible Flow Along a Smooth Flat Plate*						
	X 9.4.1	BG	Easy	Short	3	S9.2	
	9.4.2	SI	Medium	Medium	3	P9.8	Must solve, or have Ans to, Exer 9.3.2
	9.4.3	BG	Medium	Medium	1	P9.9	Power
	9.4.4	BG	V Easy	V Short	1	P9.10	
	9.4.5	N	V Easy	V Short	1		
	P 9.5	N	Medium	Short	1		Derivation; integration.
	9.6	N	Medium	Short	1		Derivation; integration.
	9.7	N	Medium	Short	1		Derivation.
	9.8	BG	Medium	Medium	3	X9.4.2	Must solve, or have Ans to, Prob 9.4
	9.9	SI	Medium	Medium	1	X9.4.3	Power
	9.10	SI	V Easy	V Short	1	X9.4.4	
9.5	*Friction Drag for Incompressible Flow Along a Smooth Flat Plate with a Transition Regime*						
	X 9.5.1	BG	Easy	Short	1		
	9.5.2	BG	Medium	Medium	2	P9.12	
	9.5.3	SI	Medium	Medium	2	P9.13	
	P 9.11	N	V Easy	V Short	1		Proof
	9.12	SI	Medium	Medium	2	X9.5.2	
	9.13	BG	Medium	Medium	2	X9.5.3	
	9.14	BG	Medium	Long	1		Plots
	9.15	BG	Medium	Medium	1	9.14	Plots
	9.16	BG	Medium	V Long	2	9.17	See note with solution. Plots
	9.17	SI	Medium	V Long	2	9.16	See note with solution; interpol'n. Plots

/cont...

[1] For all Exercises (marked "X"), answers are given in Appendix D of the textbook.

* Note a correction made to early printings of the textbook.

† Answers are sensitive to values (C_d, C_f, τ, σ) which must be read from a graph.

X = Exercise, P = (end-of-chapter) Problem, S = Sample Problem.

Sec	Exer/Prob	Units	Difficulty	Length	Parts	Similar	Special features
9.7	*Drag on Three-Dimensional Bodies (Incompressible Flow)*						
	X 9.7.1	BG	Easy	Medium	1	9.7.2	Assume **R** < 1 and confirm
	9.7.2	SI	Easy	Medium	1	9.7.1	Assume **R** < 1 and confirm
	9.7.3	SI	Medium	Long	1	P9.19	†
	9.7.4	BG	Medium	Medium	1	P9.23	† Make assumption. Power
	9.7.5	BG	Easy	Short	1	P9.24	†
	9.7.6	BG	Easy	Medium	1	9.7.7	Assume **R** > 1000 and confirm
	9.7.7	SI	Easy	Medium	1	9.7.6	Assume **R** > 1000 and confirm
	9.7.8	SI	V Easy	Short	1		
	P 9.18	BG	Medium	Long	1		† T & E
	9.19	BG	Medium	Long	1	9.20,X9.7.3	† Plot
	9.20	BG	Medium	Long	1	9.19	† Plot
	9.21	BG	Medium	Medium	1	9.22	† T & E
	9.22	SI	Medium	Medium	1	9.21	† T & E
	9.23	SI	Medium	Medium	1	X9.7.4	† Make assumption. Power
	9.24	SI	V Easy	V Short	1	X9.7.5	†
	9.25	BG	Medium	Medium	2	9.26	† Uses Fig. 9.13 of (next) Sec. 9.8
	9.26	SI	Medium	Medium	2	9.25	† Uses Fig. 9.13 of (next) Sec. 9.8
	9.27	BG	Medium	Long	1	9.28	† T & E
	9.28	SI	Medium	Long	1	9.27	† T & E
	9.29	SI	Medium	Long	3		† T & E
	9.30	BG	Medium	Medium	2		† Uses $pv = RT$ (Sec. 2.7)
	9.31	BG	Hard	Long	3		†
9.8	*Drag on Two-Dimensional Bodies (Incompressible Flow)*						
	X 9.8.1	BG	Easy	Short	1		†
	9.8.2	BG	Easy	Short	2		
	P 9.32	SI	Medium	Medium	2		†
	9.33	SI	Medium	Medium	2		†
9.9	*Lift and Circulation*						
	X 9.9.1	BG	Medium	Medium	1		
9.10	*Ideal Flow About a Cylinder*						
	X 9.10.1	BG	V Easy	V Short	1	9.10.2	
	9.10.2	SI	V Easy	V Short	1	9.10.1	
	P 9.34	N	Medium	Medium	1		Proof; integration.
	9.35	BG	Medium	Medium	1		† Interpolation
	9.36	BG	Medium	Medium	1		
	9.37	BG	Medium	Medium	2	9.38	Plot
	9.38	SI	Medium	Medium	2	9.37	Plot
	9.39	BG	Easy	Short	2	9.40	
	9.40	SI	Easy	Short	2	9.39	
	9.41	BG	Medium	Medium	1	9.42	
	9.42	SI	Medium	Medium	1	9.41	

/cont...

Sec	Exer/Prob	Units	Difficulty	Length	Parts	Similar	Special features
9.12	*Induced Drag on Airfoil of Finite Length*						
	X 9.12.1	SI	Medium	Medium	3	P9.43	
	P 9.43	BG	Medium	Medium	3	X9.12.1	
9.13	*Lift and Drag Diagrams*						
	X 9.13.1	BG	Medium	Medium	4		Power
	9.13.2	BG	Easy	V Short	1	9.13.3	
	9.13.3	SI	Easy	V Short	1	9.13.2	
	P 9.44	SI	Medium	Medium	1	9.45	†
	9.45	BG	Medium	Medium	1	9.44	†
	9.46	N	Medium	Medium	1		†
	9.47	BG	Medium	Long	1		† Interpolation
	9.48	SI	Medium	Medium	3		†
	9.49	BG	Medium	Long	1		Interpolation, T & E
9.14	*Effects of Compressibility on Drag and Lift*						
	X 9.14.1	BG	Medium	Medium	2	9.14.2	†
	9.14.2	SI	Medium	Medium	2	9.14.1	†
	P 9.50	BG	Hard	Long	1	9.51	
	9.51	SI	Hard	Long	1	9.50	
	9.52	BG	Medium	Long	1	9.53	† T & E
	9.53	SI	Medium	Long	1	9.52	† T & E

Chapter 9
FORCES ON IMMERSED BODIES

Sec. 9.3: Laminar Boundary Layer for Incompressible Flow Along a Smooth Flat Plate -- *Exercises*

9.3.1 *Determine the shear stress at 9 in and 18 in back from the leading edge of the plate in Sample Prob. 9.1.*
*Sample Prob. 9.1: The plate is 6 in wide × 18 in long, placed longitudinally in oil (s = 0.925, ν = *0.001 05 ft²/sec) flowing with undisturbed U = 2 fps. [*Note this correction to the textbook.]*

BG

Eq. 9.11: $\tau_0 = 0.332(\mu U/x)(\mathbf{R}_x)^{0.5}$

where (Sec. 9.3) $\mathbf{R}_x = xU/\nu$, which increases linearly in the downstream direction.

$\therefore\ \mu = \nu\rho = 0.001\,05(0.925)(62.4/32.2) = 0.001\,882\ \text{lb·sec/ft}^2$

x	$\mathbf{R}_x$	τ_0
0.75 ft	1429	0.0630 lb/ft²
1.5 ft	2860	0.0445 lb/ft²

▲

9.3.2 *Find the shear stress and the thickness of the boundary layer (a) at the center and (b) at the trailing edge of a smooth, flat plate 3.0 m wide and 0.6 m long parallel to the flow, immersed in 15°C water flowing at an undisturbed velocity of 0.9 m/s. Assume a laminar boundary layer over the whole plate. Also, (c) find the total friction drag on one side of the plate.*

SI

Table A.1 for water at 15°C: $\rho = 999.1\ \text{kg/m}^3$, $\nu = 1.139\times10^{-6}\ \text{m}^2/\text{s}$, $\mu = 0.001\,139\ \text{N·s/m}^2$

(*a*) At center, $x = 0.3$ m, $\mathbf{R}_x = \dfrac{xU}{\nu} = \dfrac{0.3(0.9)}{1.139\times10^{-6}} = 237\,000$

For $\mathbf{R}_x < 500\,000$ the boundary layer will be laminar if undisturbed.

From Eq. 9.10: $\delta = \dfrac{4.91(0.3)}{237\,000^{1/2}} = 0.003\,03\ \text{m} = 3.03\ \text{mm}$ ◀

Eq. 9.11: $\tau_0 = 0.332\dfrac{(0.001\,139)0.9}{0.3}(237\,000)^{1/2} = 0.552\ \text{N/m}^2$ ◀

(*b*) At trailing edge $x = 0.6$ m, $\mathbf{R}_x = 474\,000$, $\delta = 0.004\,28\ \text{m} = 4.28\ \text{mm}$ ◀

$\tau_0 = 0.391\ \text{N/m}^2$ ◀

(*c*) Drag on one side of plate: Eq. 9.14: $C_f = 1.328/474\,000^{1/2} = 0.001\,929$

Eq. 9.2: $F_f = (0.001\,929)999.1(0.9^2/2)3.0(0.6) = 1.405\ \text{kg·m/s}^2 = 1.405\ \text{N}$ ◀

9.3.3 *For the critical Reynolds number of 500,000 for transition from laminar to turbulent flow in the boundary layer, find the corresponding critical Reynolds number for flow in a circular pipe. How does this compare with the value given in Chap. 8? Hint: Let the boundary layer thickness correspond to the radius of the pipe in laminar flow, and let the undisturbed velocity U of the boundary layer theory represent the centerline velocity u_{max} of the pipe flow.*

N

Eq. 9.10: $\dfrac{\delta}{x} = \dfrac{4.91}{500{,}000^{1/2}} = 0.006\,94$, $\therefore\ \mathbf{R} = 500{,}000 = \dfrac{Ux}{\nu} = \dfrac{U\delta}{0.006\,94\,\nu}$

Since: δ corresponds to pipe radius, $D/2$, U corresponds to $U_{\max} = 2V$ for laminar flow.

Then for pipe flow, $500{,}000 = \dfrac{2V(D/2)}{0.006\,94\,\nu}$ so that $\dfrac{DV}{\nu} = \mathbf{R}_{\text{pipe}} = 500{,}000(0.006\,94) = 3470$ ◀

This is a fair check with the value of 2000 given in Chapter 8. ◀

9.1 *Given the general equation for a parabola, $u = ay^2 + by + c$, derive the dimensionless velocity distribution of Eq. (9.9).*

N

(*a*) $u = ay^2 + by + c$; differentiating: $du/dy = 2ay + b$

Boundary conditions: (1) At $y = 0,\ u = 0$

(2) At $y = \delta,\ u = U$

(3) At $y = \delta,\ du/dy = 0$

(*b*) From (1): $c = 0$

From (2): $U = a\delta^2 + b\delta$

From (3): $(du/dy)_{y=\delta} = 2a\delta + b = 0; \ \therefore\ b = -2a\delta$

Substitute for b into (2): $U = a\delta^2 - 2a\delta^2 = -a\delta^2$

Then, $a = -U/\delta^2$; $b = 2U/\delta$, and $\therefore\ u = -(U/\delta^2)y^2 + (2U/\delta)y$

With $\eta = y/\delta$, $u/U = f(\eta) = -\eta^2 + 2\eta$ which is Eq. (9.9) Q.E.D. ◀

9.2 *Given the parabolic velocity distribution of Eq. (9.9), prove that $\alpha = 0.1333$ [in Eq. (9.5)] and $\beta = 2.0$ [in Eq. (9.7)].*

Eq. (9.9): $u/U = f(\eta) = 2\eta - \eta^2$

N

From Eq. 9.5: $\alpha = \int_0^1 f(\eta)[1 - f(\eta)]d\eta$; substituting for $f(\eta)$ from Eq. 9.9:

$$\alpha = \int_0^1 (2\eta - \eta^2)[1 - 2\eta + \eta^2]d\eta = \int_0^1 (2\eta - 5\eta^2 + 4\eta^3 - \eta^4)d\eta$$

$$= [\eta^2 - (5/3)\eta^3 + \eta^4 - (1/5)\eta^5]_0^1 = 0.1333$$ ◀

From Eq. 9.7: $\beta = \dfrac{\tau_0\delta}{\mu U} = \left[\dfrac{df(\eta)}{d\eta}\right]_{\eta=0}$ Substituting for $f(\eta)$ from Eq. 9.9:

$$\beta = \frac{d}{d\eta}(2\eta - \eta^2)_{\eta=0} = [2 - 2\eta]_{\eta=0} = 2.0$$ ◀

9.3 *Determine the shear stress at 15 cm and 30 cm back from the leading edge of the plate in Sample Prob. 9.1.*

*Sample Prob. 9.1: The plate is 15 cm wide × 50 cm long, placed longitudinally in oil ($s = 0.925$, $\nu = {}^*0.73 \times 10^{-4}$ m²/s) flowing with undisturbed $U = 0.60$ m/s. [*Note this correction to the textbook.]*

SI

Eq. 9.11: $\tau_0 = 0.332(\mu U/x)(\mathbf{R}_x)^{0.5}$

where (Sec. 9.3) $\mathbf{R}_x = xU/\nu$, which increases linearly in the downstream direction.

$\therefore\ \mu = \nu\rho = 0.73 \times 10^{-4}(0.925)1000\ \text{N·s}^2/\text{m}^2 = 0.0675\ \text{N·s/m}^2$

x	$\mathbf{R}_x$	τ_0
0.15 m	1233	3.15 N/m²
0.30 m	2470	2.23 N/m²

▲

9.4 *Find the shear stress and the thickness of the boundary layer (a) at the center and (b) at the trailing edge of a smooth, flat plate 10 ft wide and 2 ft long parallel to the flow, immersed in 60°F water flowing at an undisturbed velocity of 3 fps. Assume a laminar boundary layer over the whole plate. Also, (c) find the total friction drag on one side of the plate.*

BG

Table A.1 for water at 60°F: $\rho = 1.938$ slug/ft³, $\nu = 1.217 \times 10^{-5}$ ft²/s, $\mu = 2.359 \times 10^{-5}$ lb·sec/ft²

(*a*) At center, $x = 1$ ft, $\mathbf{R}_x = \dfrac{xU}{\nu} = \dfrac{1(3)}{1.217 \times 10^{-5}} = 247{,}000$

For $\mathbf{R}_x < 500{,}000$ boundary layer will be laminar if undisturbed.

Eq. 9.10: $\delta = \dfrac{4.91(1)}{247{,}000^{1/2}} = 0.009\,89$ ft $= 0.1187$ inches ◀

Eq. 9.11: $\tau_0 = 0.332\dfrac{(2.359 \times 10^{-5})3}{1}(247{,}000)^{1/2} = 0.011\,67$ lb/ft² ◀

(*b*) At trailing edge $x = 2$ ft, $\mathbf{R}_x = \dfrac{2(3)}{1.217 \times 10^{-5}} = 493{,}000$

$\delta = \dfrac{4.91(2)}{493{,}000^{1/2}} = 0.013\,99$ ft or 0.1678 inches ◀

$\tau_0 = 0.332(2.359 \times 10^{-5})(3/2)493{,}000^{1/2} = 0.008\,25$ lb/ft² ◀

(*c*) Drag on one side of plate: Eq. 9.14: $C_f = \dfrac{1.328}{493{,}000^{1/2}} = 0.001\,891$

Eq. 9.2: $F_f = 0.001\,891(1.938)(3^2/2)(10)2 = 0.330$ lb ◀

Sec. 9.4: Turbulent Boundary Layer for Incompressible Flow Along a Smooth Flat Plate -- *Exercises*

9.4.1 *Find the shear stress on the sides of the van in Sample Prob. 9.2 at (a) 2 ft, (b) 12 ft, and (c) 22 ft back from the leading edge of the sides.*

Sample Prob. 9.2: $U = 88$ *fps*, $\nu = 0.000\,152$ *ft²/sec*, $\rho = 0.002\,42$ *slug/ft³*.

BG

Eq. 9.22: $\tau_0 = 0.0587\rho\dfrac{U^2/2}{\mathbf{R}_x^{1/5}}$; Sec. 9.3: $\mathbf{R}_x = \dfrac{xU}{\nu}$

Given: $U = 88$ fps, $\nu = 0.000\,152$ ft²/sec, $\rho = 0.002\,42$ lb·sec²/ft⁴

$\therefore\ 0.0587\rho(U^2/2) = 0.0587(0.002\,42\ \text{lb·sec}^2/\text{ft}^4)(88^2/2)\ \text{ft}^2/\text{sec}^2 = 0.550\ \text{lb/ft}^2$ and $\tau_0 = 0.550/\mathbf{R}_x^{1/5}$ lb/ft²

	x	$\mathbf{R}_x = \dfrac{xU}{\nu}$	$\mathbf{R}_x^{1/5}$	τ_0
(*a*)	2 ft	1,158,000	16.32	0.0337 lb/ft²
(*b*)	12 ft	6,950,000	23.35	0.0236 lb/ft²
(*c*)	22 ft	12,740,000	26.36	0.0209 lb/ft²

▲

9.4.2 *Assume that the boundary layer of Exer. 9.3.2 is disturbed near the leading edge. Compute the corresponding quantities for the turbulent boundary layer covering the entire plate, and compare the results.*

Exer. 9.3.2: A smooth, flat plate 3.0 m wide × 0.6 m long parallel to the flow is immersed in 15°C water (ρ = 999.1 kg/m^3) flowing at undisturbed U = 0.9 m/s. Find τ_0 and δ at (a) the center (where x = 0.3 m and $\mathbf{R}_x$ = 237 000) and (b) the trailing edge (where x = 0.6 m and $\mathbf{R}_x$ = 474 000). Also (c) find F_f on one side.

SI

(*a*) At center of plate: x = 0.3 m, $\mathbf{R}_x$ = 237 000 ; Eq. 9.21: $\dfrac{\delta}{x} = \dfrac{0.377}{237\,000^{1/5}} = \dfrac{0.377}{11.88} = 0.0317$

$\therefore\ \delta = 0.3(0.0317) = 0.009\,52 \text{ m} = 9.52 \text{ mm}$ ◀

Eq. 9.22: $\tau_0 = 0.0587(999.1)\left(\dfrac{0.9^2}{2}\right)\left(\dfrac{1}{11.88}\right) = 1.999 \text{ N/m}^2$ ◀

(*b*) At trailing edge: x = 0.6 m, $\mathbf{R}_x$ = 474 000

$\delta = 0.016\,57 \text{ m} = 16.57 \text{ mm}$ ◀ $\tau_0 = 1.740 \text{ N/m}^2$ ◀

(*c*) Total drag on one side of plate, for $\mathbf{R} < 10^7$: Eq. 9.25: $C_f = \dfrac{0.0735}{474\,000^{1/5}} = \dfrac{0.0735}{13.65} = 0.005\,38$

Eq. 9.2: $F_f = 0.005\,38(999.1)(0.9^2/2)(3.0)0.6 = 3.92 \text{ N}$ ◀

The shear stresses, boundary-layer thicknesses, and total friction drag for this turbulent boundary layer are 2.8−4.5 times greater than those for a laminar boundary layer (Exer. 9.3.2). ◀

9.4.3 *A 280-ft-long streamlined train has 8.5-ft-high sides and an 8-ft-wide top. Compute the power required to overcome the skin-friction drag when the train is traveling at 90 mph through the ICAO standard atmosphere at sea level, assuming the drag on the sides and top to be equal to that on one side of a flat plate 25 ft wide and 280 ft long.*

BG

Table A.3 for air at sea level: ρ = 0.002 377 slug/ft^3, ν = 0.000 1572 ft^2/sec.

With V = 90(44/30) = 132.0 fps, Eq. 7.6: $\mathbf{R}$ = 280(132.0)/0.000 1572 = 2.35 × 10^8

$\mathbf{R}$ > 500,000, so boundary layer is turbulent.

As $\mathbf{R} > 10^7$, Eq. 9.26: $C_f = \dfrac{0.455}{[\log(2.35 \times 10^8)]^{2.58}} = 0.001\,893$

Eq. 9.2: $F_f = C_f\rho(V^2/2)BL = 0.001\,893(0.002\,377)(132^2/2)(25)280 = 274 \text{ lb}$

Power = $FV/550$ = 274(132)/550 = 65.9 hp ◀

9.4.4 *A lifeguard determines the wind velocity 6 ft above the beach to be 25 fps. If one wishes to get out of the wind by lying down, what would be the velocity at (a) 0.5 ft, and (b) at 1.0 ft above beach level?*

BG

Use the seventh-root law to approximate the velocity profile. Eq. 9.20 (and 8.38): $u = 25(y/6)^{1/7}$

(*a*) $u_{0.5} = 25(0.5/6)^{1/7} = 17.53$ fps ◀ (*b*) $u_{1.0} = 25(1/6)^{1/7} = 19.35$ fps ◀

9.4.5 *Compute C_f for $\mathbf{R} = 10^7$ using Eqs. (9.25) and (9.26), and compare the two values.*

N

Eq. 9.25: $C_f = \dfrac{0.0735}{\mathbf{R}^{1/5}} = \dfrac{0.0735}{(10^7)^{1/5}} = 0.002\,93$ ◀

Eq. 9.26: $C_f = \dfrac{0.455}{(\log 10^7)^{2.58}} = \dfrac{0.455}{7^{2.58}} = 0.003\,00$ ◀

The two values of C_f agree closely (within 2.6%). ◀

Sec. 9.4: Turbulent Boundary Layer for Incompressible Flow Along a Smooth Flat Plate -- ***Problems***

9.5 *Derive Eq. (9.19) along the lines suggested in the text.*

N

Rearranging Eq. 9.18: $\delta^{1/4}\, d\delta = \dfrac{0.023\, dx}{\alpha(U/\nu)^{1/4}}$

Integrating, with $\delta = 0$ at $x = 0$: $\dfrac{4}{5}\delta^{5/4} = \dfrac{0.023x}{\alpha(U/\nu)^{1/4}}$

Multiply by $\dfrac{5}{4}$ and take $\dfrac{4}{5}$ root: $\delta = \left(\dfrac{0.0288}{\alpha}\right)^{4/5}\left(\dfrac{\nu}{Ux}\right)^{1/5} x$ which is Eq. (9.19) Q.E.D. ◀

9.6 *For the turbulent boundary layer, derive the value of $\alpha = 0.0972$ from the seventh-root law given in Eq. (9.20).*

N

Eq. 9.20: $u = U\eta^{1/7}$ or $f(\eta) = \eta^{1/7}$

Then $\alpha = \int_0^1 f(\eta)[1 - f(\eta)]d\eta = \int_0^1 \eta^{1/7}(1 - \eta^{1/7})d\eta = \left[\dfrac{7}{8}\eta^{8/7} - \dfrac{7}{9}\eta^{9/7}\right]_0^1 = 0.0972$ Q.E.D. ◀

9.7 *From the information given, derive Eq. (9.22).*

N

Eq. 9.17: $\tau_0 = 0.023\rho U^2[\nu/(\delta U)]^{1/4}$

Eq. 9.19 with $\alpha = 0.0972$: $\delta = 0.377[\nu/(Ux)]^{1/5}x$ which is Eq. 9.21.

Substituting this δ into Eq. 9.17: $\tau_0 = 0.023\rho U^2\left(\dfrac{\nu}{U}\right)^{1/4}\left[\dfrac{1}{0.377x}\left(\dfrac{\nu}{Ux}\right)^{-1/5}\right]^{1/4}$

$= \dfrac{0.023}{0.377^{1/4}}\rho U^2 \dfrac{\nu^{1/4-1/20}}{(Ux)^{1/4-1/20}} = 0.0587\rho(U^2/2)\left(\dfrac{\nu}{Ux}\right)^{1/5} = 0.0587\rho\dfrac{U^2/2}{\mathbf{R}_x^{1/5}}$ Q.E.D. ◀

9.8 *Assume that the boundary layer of Prob. 9.4 is disturbed near the leading edge. Compute the corresponding quantities for the turbulent boundary layer covering the entire plate, and compare the results.*

Prob. 9.4: A smooth flat plate 10 ft wide × 2 ft long parallel to the flow is immersed in 60°F water (ρ = 1.938 slug/ft^3) flowing at undisturbed U = 3 fps. Find τ_0 and δ at (a) the center (where x = 1 ft and $\mathbf{R}_x$ = 247,000) and (b) the trailing edge (where x = 2 ft and $\mathbf{R}_x$ = 493,000). Also (c) find F_f on one side.

BG

(*a*) At center of plate: $x = 1.0$ ft, $\mathbf{R}_x = 247{,}000$; Eq. 9.21: $\dfrac{\delta}{x} = \dfrac{0.377}{247{,}000^{1/5}} = \dfrac{0.377}{11.98} = 0.0315$

$\therefore\ \delta = 1.0(0.0315) = 0.0315$ ft $= 0.378$ inches ◀

/cont...

Eq. 9.22: $\tau_0 = 0.0587(1.938)(3^2/2)(1/11.98) = 0.0427$ lb/ft^2 ◄

(*b*) At trailing edge: $x = 2.0$ ft, $\mathbf{R}_x = 493{,}000$

Eq. 9.21: $\delta = 2.0\dfrac{0.377}{493{,}000^{1/5}} = \dfrac{(2.0)0.377}{13.76} = 0.0548$ ft $= 0.658$ in ◄

Eq. 9.22: $\tau_0 = 0.0587(1.938)(3^2/2)(1/13.76) = 0.0372$ lb/ft^2 ◄

(*c*) Total drag on one side, for $\mathbf{R} < 10^7$: Eq. 9.25: $C_f = \dfrac{0.0735}{493{,}000^{0.2}} = \dfrac{0.0735}{13.76} = 0.005\,34$

Eq. 9.2: $F_f = 0.005\,34(1.938)(3^2/2)(10)2.0 = 0.932$ lb ◄

The shear stresses, boundary-layer thicknesses, and total friction drag for this turbulent boundary layer are 2.8−4.5 times greater than those for a laminar boundary layer (Prob. 9.4).

9.9 *A 85-m-long streamlined train has 2.5-m-high sides and a 2.5-m-wide top. Compute the power required to overcome the skin-friction drag when the train is traveling at 40 m/s through the ICAO standard atmosphere at sea level, assuming the drag on the sides and top to be equal to that on one side of a flat plate 7.5 m wide and 85 m long.*

SI

Table A.3 for air at sea level: $\rho = 1.225$ kg/m^3, $\nu = 14.61 \times 10^{-6}$ m^2/s

Eq. 7.6: $\mathbf{R} = \dfrac{40(85)}{14.61 \times 10^{-6}} = 2.33 \times 10^8$; Eq. 9.26: $C_f = \dfrac{0.455}{[\log(2.33 \times 10^8)]^{2.58}} = 0.001\,896$

Eq. 9.2: $F_f = 0.001\,896(1.225)(40^2/2)(7.5)85 = 1184$ N

Power $= 1184(40) = 47\,400$ W $= 47.4$ kW ◄

9.10 *A lifeguard determines the wind velocity 2 m above the beach to be 8 m/s. If one wishes to get out of the wind by lying down, what would be the velocity at (a) 0.15 m, and (b) at 0.3 m above beach level?*

SI

Use the seventh-root law to approximate the velocity profile. Eq. 9.20 (and 8.38): $u = 8(y/2)^{1/7}$

(*a*) $u_{0.15} = 8(0.15/2)^{1/7} = 5.53$ m/s ◄ (*b*) $u_{0.30} = 8(0.3/2)^{1/7} = 6.10$ m/s ◄

Sec. 9.5: Friction Drag for Incompressible Flow Along a Smooth Flat Plate with a Transition Regime -- *Exercises*

9.5.1 *A 7.5-ft by 1.5-ft smooth, thin, flat plate with sharpened edges is submerged in 60°F water moving with a velocity of 1.4 fps in the direction of the 7.5 ft length. What is the total drag?*

BG

Table A.1 for 60°F water: $\rho = 1.938$ slugs/ft^3; $\nu = 1.217 \times 10^{-5}$ ft^2/sec

Eq. 7.6: $\mathbf{R} = \dfrac{7.5(1.4)}{1.217 \times 10^{-5}} = 863{,}000$

Eq. 9.27: $C_f = \dfrac{0.455}{(\log 863{,}000)^{2.58}} - \dfrac{1700}{863{,}000} = \dfrac{0.455}{99.0} - 0.001\,970 = 0.002\,63$

Eq. 9.2: $F_f = 2 \times 0.002\,63(1.938)(1.4^2/2)(1.5)7.5 = 0.1124$ lb for 2 sides ◄

9.5.2 *An 1-in-diameter harpoon 6 ft long, with a sharp tip, is launched at 20 fps into 60°F water. Find (a) the friction drag, and (b) the maximum thickness of the boundary layer.*

BG

(*a*) Table A.1 for 60°F water: ρ = 1.938 slugs/ft^3; ν = 1.217 × 10^{-5} ft^2/sec

Secs. 9.3 and 9.5: $x_c = \dfrac{\mathbf{R}_c \nu}{V} = \dfrac{500{,}000(1.217 \times 10^{-5})}{20} = 0.304$ ft

$\therefore\ x_c < L = 6$ ft, and a transition region must exist.

Eq. 7.6: $\mathbf{R} = \dfrac{VL}{\nu} = \dfrac{20(6)}{1.217 \times 10^{-5}} = 9.86 \times 10^6$

Eq. 9.27: $C_f = \dfrac{0.455}{[\log(9.86 \times 10^6)]^{2.58}} - \dfrac{1700}{9.86 \times 10^6} = 0.002\,84$

Eq. 9.2: $F_f = 0.002\,84(1.938)(20^2/2)[\pi(1)/12]6 = 1.728$ lb ◄

(*b*) From Eq. 9.21: $\delta = \dfrac{0.377 \times 6}{(9.86 \times 10^6)^{1/5}} = 0.0903$ ft or 1.084 in ◄

9.5.3 *An airplane wing with a chord length of 3 m parallel to the flow moves through standard atmospheric air at an altitude of 6 km with a speed of 350 km/h. Find (a) the critical roughness for a point one-tenth the chord length back from the leading edge, and (b) the surface drag on an 8-m-span section of this wing.*

SI

Table A.3 for air at 6 km altitude: ρ = 0.660 kg/m^3, ν = 24.16 × 10^{-6} m^2/s

(*a*) For $x = c/10 = 3/10 = 0.3$ m, and U = 350 km/h = 350(1000/3600) m/s = 97.2 m/s,

Eq. 7.6: $\mathbf{R}_x = \dfrac{97.2(0.3)}{2.416 \times 10^{-5}} = 1.207 \times 10^6$

Eq. 9.28: $e_c = 26.0(2.416 \times 10^{-5}/97.2)(1.207 \times 10^6)^{1/4} = 0.000\,214$ m = 0.214 mm ◄

(*b*) For entire wing: $\mathbf{R} = \dfrac{97.2(3)}{2.416 \times 10^{-5}} = 1.207 \times 10^7$

Eq. 9.27: $C_f = \dfrac{0.455}{[\log(1.207 \times 10^7)]^{2.58}} - \dfrac{1700}{1.207 \times 10^7} = 0.002\,77$

Eq. 9.2 for both upper and lower surfaces: $F_f = 0.002\,77(0.660)(97.2^2/2)(8)3(2) = 415$ N ◄

Sec. 9.5: Friction Drag for Incompressible Flow Along a Smooth Flat Plate with a Transition Regime -- *Problems*

9.11 *Verify that the two expressions of Eq. (9.28) are equal.*

N

Eq. 9.11: $\tau_0 = 0.332(\mu U/x)(Ux/\nu)^{1/2}$

Substituting into the first expression of Eq. 9.28: $e_c = \dfrac{15\nu}{(\tau_0/\rho)^{1/2}} = \dfrac{15\nu}{[0.322(\mu U/\rho x)(Ux/\nu)^{1/2}]^{1/2}}$

$= \dfrac{15\nu x^{1/2}\nu^{1/4}}{0.332^{1/2}\nu^{1/2}U^{1/2}U^{1/4}x^{1/4}} = 26.0\dfrac{\nu}{U}\left(\dfrac{Ux}{\nu}\right)^{1/4} = 26.0\dfrac{\nu}{U}\mathbf{R}_x^{1/4}$ Q.E.D. ◄

9.12 *A 20-mm-diameter harpoon 1.8 m long, with a sharp tip, is launched at 6 m/s into 15°C water. Find (a) the friction drag, and (b) the maximum thickness of the boundary layer.*

SI

(*a*) Table A.1 for 15°C water: $\nu = 1.139 \times 10^{-6}$ m²/s, $\rho = 999$ kg/m³

$$x_c = \frac{\mathbf{R}_c \nu}{V} = \frac{500\,000(1.139 \times 10^{-6})}{6} = 0.0949 \text{ m}$$

$\therefore\ x_c < L = 1.8$ m, and a transition region must exist.

$$\text{Eq. 7.6:}\quad \mathbf{R} = \frac{VL}{\nu} = \frac{6(1.8)}{1.139 \times 10^{-6}} = 9.48 \times 10^6$$

$$\text{Eq. 9.27:}\quad C_f = \frac{0.455}{[\log(9.48 \times 10^6)]^{2.58}} - \frac{1700}{9.48 \times 10^6} = 0.002\,85$$

$$\text{Eq. 9.2:}\quad F_f = 0.002\,85(999)(6^2/2)(0.020\pi)1.8 = 5.80 \text{ N} \quad ◄$$

$$(b)\ \text{Eq. 9.21:}\quad \delta = \frac{0.377 \times 1.8}{(9.48 \times 10^6)^{1/5}} = 0.0273 \text{ m or } 27.3 \text{ mm} \quad ◄$$

9.13 *An airplane wing with a chord length of 7 ft parallel to the flow moves through standard atmospheric air at an altitude of 15,000 ft with a speed of 300 mph. Find (a) the critical roughness for a point one-tenth the chord length back from the leading edge, and (b) the surface drag on a 25-ft-span section of this wing.*

BG

Table A.3 for air at 15,000 ft altitude: $\rho = 0.001\,496$ slug/ft³, $\nu = 0.229 \times 10^{-3}$ ft²/sec

(*a*) For $x = c/10 = 7/10 = 0.70$ ft, and $U = 300(44/30) = 440$ fps,

$$\text{Eq. 7.6:}\quad \mathbf{R}_x = \frac{440(0.70)}{2.29 \times 10^{-4}} = 1.345 \times 10^6$$

$$\text{Eq. 9.28:}\quad e_c = 26.0(2.29 \times 10^{-4}/440)(1.345 \times 10^6)^{1/4} = 0.000\,461 \text{ ft or } 0.005\,53 \text{ in} \quad ◄$$

$$(b)\ \text{For entire wing:}\quad \mathbf{R} = \frac{440(7)}{2.29 \times 10^{-4}} = 1.345 \times 10^7$$

$$\text{Eq. 9.27:}\quad C_f = \frac{0.455}{[\log(1.345 \times 10^7)]^{2.58}} - \frac{1700}{1.345 \times 10^7} = 0.002\,74$$

Eq. 9.2 for both upper and lower surfaces: $F_f = 0.002\,74(0.001\,496)(440^2/2)25(7)2 = 138.8$ ◄

9.14 *A flat plate 25 ft long and 1.5 ft wide is towed at 8 fps through a liquid ($\gamma = 50$ lb/ft³, $\mu = 0.000\,26$ lb·sec/ft²). Determine the total drag on the plate. Plot the thickness of the boundary layer, and the local shear stress τ_0, as functions of x along the plate. Determine the area under the stress curve, and compare it with the previously computed value of the drag. Assume that the boundary layer changes from laminar to turbulent at a Reynolds number of 300,000.*

BG

$$\rho = 1.940(50/62.4) = 1.554 \text{ slug/ft}^3\ ;\quad \text{Given:}\quad 300{,}000 = \mathbf{R}_c = x_c\frac{U\rho}{\mu} = x_c\frac{(8)1.554}{0.000\,26}$$

$\therefore\ x_c = 6.27$ ft, where the turbulent boundary layer begins.

$x_c < L = 25$ ft, so a transition region must exist.

/cont...

For entire plate: $\mathbf{R} = LU\rho/\mu = (25)8(1.554)/0.000\,26 = 1.196 \times 10^6$

Sec. 9.5: $F_f = \rho\dfrac{U^2}{2}B\left[\dfrac{1.328x_c}{\mathbf{R}_c^{1/2}} + \dfrac{0.455L}{(\log\,\mathbf{R})^{2.58}} - \dfrac{0.0735x_c}{\mathbf{R}_c^{1/5}}\right]$

$F_f = 1.554(8^2/2)1.5[0.0152 + 0.1081 - 0.0370] = 6.44$ lb/side

$\therefore$ Total drag $= 2 \times 6.44 = 12.88$ lb ◀

For $x < 6.27$ ft (laminar B.L.), Eqs. 9.10 and 9.11 are applicable:

$\delta = 4.91\dfrac{x}{\mathbf{R}_x^{1/2}}$; $\tau_0 = 0.332\dfrac{\mu U}{x}\mathbf{R}_x^{1/2} = 0.000\,691\dfrac{\mathbf{R}_x^{1/2}}{x}$ where $\mathbf{R}_x = \dfrac{x(8)1.554}{0.000\,26} = 47{,}800x$

x (ft)	2	4	6	6.27	
$\mathbf{R}_x$	95,660	191,320	287,000	300,000	
δ (ft)	0.0318	0.0449	0.0550	0.0562	◀◀
τ_0 (psf)	0.1068	0.0755	0.0617	0.0603	◀◀

For $x > 6.27$ ft (turbulent B.L.), Eqs. 9.21 and 9.22 apply: $\delta = 0.377\dfrac{x}{\mathbf{R}_x^{1/5}}$; $\tau_0 = 0.0587\dfrac{\rho U^2}{2\mathbf{R}_x^{1/5}} = \dfrac{2.92}{\mathbf{R}_x^{1/5}}$

x (ft)	6.27	12	16	25	
$\mathbf{R}_x$	300,000	574,000	765,300	1,195,800	
δ (ft)	0.1898	0.319	0.402	0.574	◀◀
τ_0 (psf)	0.234	0.206	0.1944	0.1778	◀◀

Area A under stress curve $= \int \tau_0\,dx = 4.47$ lb/ft (by integration),

so drag $= AB = 4.47(1.5) = 6.71$ lb/side ◀

This compares very well with the previously above computed drag of $F_f = 6.44$ lb/side. ◀
Actually the δ's and the τ_0's in the second table are too large because they represent values for the case of a turbulent boundary layer over the entire length.

9.15 *For velocities of 0, 10, 20, 30, 40, 50, and 60 fps, make the necessary calculations to plot drag vs. velocity for the plate and data of Prob. 9.14.*

Prob. 9.14: $L = 25$ ft, $B = 1.5$ ft, $\rho = 1.554$ slug/ft^3, $\mu = 0.000\,26$ lb·sec/ft^2, $\mathbf{R}_c = 300{,}000$.

BG

Sec. 9.5: $F_f = \rho\dfrac{U^2}{2}B\left[\dfrac{1.328x_c}{\mathbf{R}_c^{1/2}} + \dfrac{0.455L}{(\log\,\mathbf{R})^{2.58}} - \dfrac{0.0735x_c}{\mathbf{R}_c^{1/5}}\right]$

Given $\mathbf{R}_c = \dfrac{U\rho x_c}{\mu} = 300{,}000$, $\therefore\ x_c = \dfrac{50.2}{U}$, $\mathbf{R}_c^{1/2} = 548$, $\mathbf{R}_c^{1/5} = 12.46$

$\mathbf{R} = LU\rho/\mu = 25(U)1.554/0.000\,26 = 149{,}500U$; $\rho B/2 = 1.166$

$\therefore\ F_f = 1.166U^2\left(\dfrac{1.328x_c}{548} + \dfrac{0.455(25)}{(\log\,\mathbf{R})^{2.58}} - \dfrac{0.0735x_c}{12.46}\right) = 1.166U^2\left(\dfrac{11.38}{(\log\,\mathbf{R})^{2.58}} - 0.003\,48x_c\right)$

U	0	10	20	30	40	50	60 (fps)	
x_c	—	5.02	2.51	1.673	1.254	1.004	0.836 (ft)	
$\mathbf{R}$	0	1.495	2.99	4.48	5.98	7.47	8.97 ($\times 10^6$)	
F_f	—	10.07	38.7	83.8	144.2	219	309 (lb/side)	◀◀

9.16 *Determine the drag on the harpoon of Exer. 9.5.2 for velocities of 0, 10, 30, and 50 fps, through (a) 60°F water and (b) 60°F air at standard sea-level pressure. Calculate the length x_c of the laminar zone and plot curves of drag vs. velocity for each case.*

Exer. 9.5.2: Harpoon diameter = 1 in, L = 6 ft.

BG

*Note to instructor: At the lowest V in air, $\mathbf{R} < 500{,}000$, so the boundary layer is only laminar.

Eq. 9.2: $F_f = C_f \rho \dfrac{V^2}{2} BL$

With transition regime (i.e., $x_c < L$), Eq. 9.27: $C_f = \dfrac{0.455}{(\log \mathbf{R})^{2.58}} - \dfrac{1700}{\mathbf{R}}$

With laminar flow only (i.e., $x_c > L$), Eq. 9.14: $C_f = 1.318/\mathbf{R}^{1/2}$

(*a*) Table A.1 for 60°F water: $\rho = 1.938$ slugs/ft^3, $\nu = 1.217 \times 10^{-5}$ ft^2/sec

$$x_c = \frac{\mathbf{R}_c \nu}{V} = \frac{500{,}000}{V}(1.217 \times 10^{-5}) = \frac{6.09}{V}\text{ ft}$$

Eq. 7.6: $\mathbf{R} = \dfrac{LV}{\nu} = \dfrac{6V}{1.217 \times 10^{-5}} = 493{,}000V$; Eq. 9.2: $F_f = C_f 1.938 \dfrac{V^2}{2}\left(\dfrac{1\pi}{12}\right)6 = 1.522 C_f V^2$

V (fps)	0	10	30	50	
x_c (ft)	–	0.609	0.203	0.1217	◄◄
R	0	4.93×10^6	1.479×10^7	2.47×10^7	
Eq. for C_f	–	9.27	9.27	9.27	
C_f	–	0.003 03	0.002 71	0.002 54	
F_f (lb)	0	0.461	3.71	9.67	◄◄

(*b*) Table A.2 for air at 60°F: $\rho = 0.002\,37$ slugs/ft^3, $\nu = 0.000\,158$ ft^2/sec

$$x_c = \frac{\mathbf{R}_c \nu}{V} = \frac{500{,}000}{V}(0.000\,158) = \frac{79.0}{V}\text{ ft}$$

Eq. 7.6: $\mathbf{R} = \dfrac{LV}{\nu} = \dfrac{6V}{0.000\,158} = 38{,}000V$; Eq. 9.2: $F_f = C_f 0.002\,37 \dfrac{V^2}{2}\left(\dfrac{1\pi}{12}\right)6 = 0.001\,861 C_f V^2$

V (fps)	0	10	30	50	
x_c (ft)	–	7.90 > L	2.63	1.580	◄◄
R	0	$*3.80 \times 10^5$	1.139×10^6	1.899×10^6	
Eq. for C_f	–	9.14	9.27	9.27	
C_f	–	0.002 14	0.002 87	0.003 08	
F_f (lb)	0	0.000 398	0.004 81	0.014 34	◄◄

9.17 *Determine the drag on the harpoon of Prob. 9.12 for velocities of 0, 3, 9, and 15 m/s, through (a) 15°C water and (b) 15°C air at standard sea-level pressure. Calculate the length x_c of the laminar zone and plot curves of drag vs. velocity for each case.*

Prob. 9.12: Harpoon diameter = 20 mm, L = 1.8 m.

SI

*Note to instructor: At the lowest V in air, $\mathbf{R} < 500{,}000$, so the boundary layer is only laminar.

Eq. 9.2: $F_f = C_f \rho \dfrac{V^2}{2} BL$

With transition regime (i.e., $x_c < L$), Eq. 9.27: $C_f = \dfrac{0.455}{(\log \mathbf{R})^{2.58}} - \dfrac{1700}{\mathbf{R}}$

With laminar flow only (i.e., $x_c > L$), Eq. 9.14: $C_f = 1.318/\mathbf{R}^{1/2}$

/cont...

(*a*) Table A.1 for 15°C water: ρ = 999.1 kg/m³, ν = 1.139 × 10⁻⁶ m²/s

$$x_c = \frac{\mathbf{R}_c \nu}{V} = \frac{500\,000}{V}(1.139 \times 10^{-6}) = \frac{0.570}{V}\ \text{m}$$

$$\text{Eq. 7.6:}\quad \mathbf{R} = \frac{LV}{\nu} = \frac{1.8V}{1.139 \times 10^{-6}} = 1580\,000V$$

$$\text{Eq. 9.2:}\quad F_f = C_f 999.1(V^2/2)(0.020\pi)1.8 = 56.5C_f V^2$$

V (m/s)	0	3	9	15	
x_c (m)	–	0.1898	0.0633	0.0380	◄◄
R	0	4.74×10^6	1.422×10^7	2.37×10^7	
Eq. for C_f	–	9.27	9.27	9.27	
C_f	–	0.003 04	0.002 72	0.002 55	
F_f (N)	0	1.544	12.45	32.5	◄◄

(*b*) Table A.2 for air at 15°C, by interpolation: ρ = 1.227 kg/m³, ν = 1.455 × 10⁻⁵ m²/s

$$x_c = \frac{\mathbf{R}_c \nu}{V} = \frac{500\,000}{V}(1.455 \times 10^{-5}) = \frac{7.28}{V}\ \text{N};\quad \text{Eq. 7.6:}\quad \mathbf{R} = \frac{LV}{\nu} = \frac{1.5V}{1.455 \times 10^{-5}} = 103\,100V$$

$$\text{Eq. 9.2:}\quad F_f = C_f 1.227(V^2/2)(0.020\pi)1.8 = 0.0694C_f V^2$$

V (m/s)	0	3	9	15	
x_c (m)	–	2.43 > L	0.808	0.485	◄◄
R	0	*3.09×10^5	9.28×10^5	1.546×10^6	
Eq. for C_f	–	9.14	9.27	9.27	
C_f	–	0.002 37	0.002 70	0.003 03	
F_f (N)	0	0.001 480	0.015 18	0.0473	◄◄

Sec. 9.7: Drag on Three-Dimensional Bodies (Incompressible Flow) -- *Exercises*

9.7.1 *An 0.3-in-diameter steel sphere (s = 8.0) is released in a large tank of oil (s = 0.80). The terminal velocity of the sphere is determined to be 2.5 fpm. Calculate the viscosity of the oil?*

BG

Assume **R** < 1, ∴ Use Stoke's Law (Eq. 9.33): $F_D = 3\pi\mu VD$. Table A.8: $\forall_{\text{sphere}} = \pi D^3/6$

At terminal velocity $\sum F_z = 0 = W - F_B - F_D$; ∴ $(\gamma_s - \gamma_l)\pi D^3/6 = 3\pi\mu VD$

$$\text{from which}\quad \mu = (\gamma_s - \gamma_l)(D^2/18V) = (8 - 0.8)62.4\frac{(0.3/12)^2}{18(2.5/60)} = 0.374\ \text{lb·sec/ft}^2 \quad ◄$$

$$\text{Check:}\quad \mathbf{R} = \frac{VD\rho}{\mu} = \frac{2.5}{60}\left(\frac{0.3}{12}\right)\frac{1.940 \times 0.8}{0.374} = 0.004\,32 < 1,$$ so the use of Stoke's Law here is valid.

9.7.2 *A 5-mm-diameter steel sphere (s = 8.0) is released in a large tank of oil (s = 0.80). The terminal velocity of the sphere is determined to be 0.7 m/min. Calculate the viscosity of the oil?*

SI

Assume **R** < 1, ∴ Use Stoke's Law (Eq. 9.33): $F_D = 3\pi\mu VD$. Table A.8: $\forall_{\text{sphere}} = \pi D^3/6$

At terminal velocity $\sum F_z = 0 = W - F_B - F_D$; ∴ $(\gamma_s - \gamma_l)\pi D^3/6 = 3\pi\mu VD$

$$\text{from which}\quad \mu = (\gamma_s - \gamma_l)(D^2/18V) = (8 - 0.8)9810\frac{0.005^2}{18(0.7/60)} = 8.41\ \text{N·s/m}^2 \quad ◄$$

Check: $\mathbf{R} = VD\rho/\mu = (0.7/60)(0.005)(1000 \times 0.8)/8.41 = 0.005\,55$

As **R** < 1, the use of Stoke's Law here is valid.

9.7.3 *For a 38-cm-diameter sphere, compute the drag from wind under sea level conditions in a standard ICAO atmosphere. Plot drag vs. velocity for 0, 10, 20, and 30 m/s.*

SI

Table A.3 for air at sea level: $T = 15°C$, $\rho = 1.225$ kg/m³, $\nu = 1.461 \times 10^{-5}$ m²/s

$A = \pi r^2 = \pi 0.19^2 = 0.1134$ m². Eq. 9.32: $F_D = C_D 1.225(V^2/2)0.1134 = 0.0695 V^2 C_D$

Eq. 7.6: $\mathbf{R} = \dfrac{DV}{\nu} = \dfrac{0.38V}{1.461 \times 10^{-5}} = 26\,000V$. Solve for F_D by finding C_D from Fig. 9.10

V (m/s)	0	10	20	30
R	—	260 000	520 000	780 000
C_D	0	0.18	0.19	0.20
F_D (N)	0	1.250	5.28	12.50

◄◄

9.7.4 *A poorly-streamlined automobile has a body form corresponding roughly to the 1 : 0.75 oblate ellipsoid of Fig. 9.10, while a well-streamlined car has a body approximating the streamlined body in the same figure, each with a diameter of 5 ft. If the velocity is 50 mph through standard air at sea level (Appendix A, Table A.3), find the horsepower required to overcome air resistance in each case.*

BG

Table A.3: $\rho = 0.002\,377$ slug/ft³, $\nu = 0.000\,1572$ ft²/sec. $V = 50(44/30) = 73.3$ fps.

$\mathbf{R} = \dfrac{5(73.3)}{0.000\,1572} = 2.33 \times 10^6$. Assuming in Fig. 9.10 that C_D remains constant for $\mathbf{R} > 10^6$:

(*a*) C_D (poorly-streamlined = 1 : 0.75 oblate ellipsoid) = 0.20 (*b*) C_D (streamlined body) = 0.04

(*a*) Eq. 9.32: $F_D = 0.2 \times 0.002\,377(73.3^2/2)(\pi/4)5^2 = 25.1$ lb (*b*) $F_D = (0.04/0.2)25.1 = 5.02$ lb

Power (*a*) $F_D V/550 = (25.1)73.3/550 = 3.35$ hp ◄

(*b*) $F_D V/550 = (5.02)73.3/550 = 0.669$ hp ◄

9.7.5 *For the streamlined train in Exer. 9.4.3, calculate the pressure drag. As a rough approximation, assume that the nose and tail of the train are of the shape of the two halves of the prolate ellipsoid of Fig. 9.10, of 8.5 ft diameter. Find the drag on the ellipsoid (pressure drag on the train), and compare this with the skin-friction drag on the train determined earlier.*

Exer. 9.4.3: $\rho = 0.002\,377$ *slug/ft³*, $\nu = 0.000\,1572$ *ft²/sec*, $V = 132.0$ *fps, skin-friction drag* $F_f = 274$ *lb.*

BG

Eq. 7.6: $\mathbf{R} = \dfrac{8.5(132.0)}{0.000\,1572} = 7.14 \times 10^6$; Fig. 9.10 (prolate ellipsoid): $C_D = 0.08$

Eq. 9.1: $F_p = (0.08)0.002\,377(132^2/2)(\pi/4)8.5^2 = 94.0$ lb ◄

This is about 1/3 of the friction drag of 274 lb determined in Exer. 9.4.3 ◄

9.7.6 *A hemispherical shell with its concave side upstream has a drag coefficient of approximately 1.33 if* $\mathbf{R} > 10^3$. *If the total load is 250 lb, find the diameter of a hemispherical parachute needed to provide a fall velocity no greater than that caused by jumping from a height of 10 ft. Assume standard air at sea level.*

BG

In jump from 10 ft height, final $V = \sqrt{2gh} = \sqrt{(2)32.2(10)} = 25.4$ fps.

Table A.3 for standard atmosphere at sea level: $\rho = 0.002\,377$ slug/ft³, $\nu = 0.000\,1572$ ft²/s.

Then, by Eq. 9.32, if $\mathbf{R} > 1000$: $250 = (1.33)0.002\,377(25.4^2/2)(\pi/4)D^2$; $D = 17.68$ ft ◄

Check: $\mathbf{R} = \dfrac{17.68(25.4)}{0.000\,1572} = 2.85 \times 10^6 > 1000$, $\therefore$ O.K.

9.7.7 *A hemispherical shell with its concave side upstream has a drag coefficient of approximately 1.33 if* **R** > *10^3. If the total load is 1000 N, find the diameter of a hemispherical parachute needed to provide a fall velocity no greater than that caused by jumping from a height of 3 m. Assume standard air at sea level.*

SI

In jump from 3 m height, final $V = \sqrt{2gh} = \sqrt{(2)9.81(3)} = 7.67$ m/s.

Table A.3 for standard atmosphere at sea level: $\rho = 1.225$ kg/m³, $\nu = 1.461 \times 10^{-5}$ m²/s.

Then, by Eq. 9.32, if **R** > 1000: $1000 = (1.33)1.225(7.67^2/2)(\pi/4)D^2$; $D = 5.15$ m ◄

Check: $\mathbf{R} = \dfrac{5.15(7.67)}{1.461 \times 10^{-5}} = 2.71 \times 10^6 > 1000$, ∴ O.K.

9.7.8 *Assuming $C_D = 1.25$ and a speed of 30 m/s, find the drag force exerted by a 3.5-m-diameter braking parachute at sea level? At what speed will the same braking force be exerted by this parachute at an elevation of 2000 m? Assume C_D remains constant.*

SI

Table A.3 for air: at sea level, $\rho = 1.225$ kg/m³; at 2000 m, $\rho = 1.007$ kg/m³

Eq. 9.32 at sea level: $F_D = C_D A \rho V^2/2 = 1.25\pi(1.75)^2 1.225(30^2/2) = 6630$ N ◄

At 2000 m: $6630 = 1.25\pi(1.75)^2 1.007(V^2/2)$; $V = 33.1$ m/s ◄

Sec. 9.7: Drag on Three-Dimensional Bodies (Incompressible Flow) -- *Problems*

9.18 *What will be the terminal velocity of the sphere of Exer. 9.7.1 in 120°F water?*
Exer. 9.7.1: $D = 0.3$ in, $s = 8.0$.

BG

Table A.1 for 120°F water: $\rho = 1.918$ slug/ft³, $\gamma = 61.71$ pcf, $\nu = 6.09 \times 10^{-6}$ ft²/sec

Select a value of C_D from Fig. 9.10. After several trials, assume $C_D = 0.43$

At terminal V, $W - F_B = F_D$ using Eqs. 9.32 and 9.33: $(\gamma_s - \gamma)(\pi D^3/6) = C_D \rho (A/2) V^2$

$(8 \times 62.4 - 61.71)(\pi/6)(0.30/12)^3 = 0.43(1.918)(\pi/2)(0.15/12)^2 V^2$; $V = 4.20$ fps

Eq. 7.6: $\mathbf{R} = \dfrac{(0.3/12)4.20}{6.09 \times 10^{-6}} = 17{,}300$

This yields $C_D = 0.43$ for a sphere in Fig. 9.10, therefore $V = 4.20$ fps ◄

9.19 *For a 18-in-diameter sphere, compute the drag from wind under sea level conditions in a standard ICAO atmosphere. Plot drag vs. velocity for 0, 25, 50, and 100 fps.*

BG

Table A.3 for air at sea level: $T = 59.0$°F, $\rho = 0.002\,377$ slug/ft³, $\nu = 0.000\,1572$ ft²/sec,

$A = \pi r^2 = \pi(9/12)^2 = 1.767$ ft²

Eq. 9.32: $F_D = C_D \rho (A/2) V^2 = C_D 0.002\,377(1.767/2) V^2 = 0.002\,10 V^2 C_D$

Eq. 7.6: $\mathbf{R} = \dfrac{DV}{\nu} = \dfrac{(18/12)V}{0.000\,1572} = 9540V$. Solve for F_D by finding C_D from Fig. 9.10

V (fps)	0	25	50	100
R	–	239,000	477,000	954,000
C_D	0	0.17	0.19	0.20
F_D (lb)	0	0.223	0.998	4.20

◄◄

9.20 *Repeat Prob. 9.19 for wind at a 10,000 ft elevation in a standard ICAO atmosphere.*
Prob. 9.19: Sphere D = 18 in. Plot drag vs. velocity for 0, 25, 50, and 100 fps.

BG

Table A.3 for air at 10,000 ft altitude: ρ = 0.001 756 slug/ft^3, ν = 0.000 201 ft^2/sec.

$A = \pi r^2 = \pi(9/12)^2 = 1.767$ ft^2 ; Eq. 9.32: $F_D = C_D 0.001\,756(1.767/2)V^2 = 0.001\,552 V^2 C_D$

Eq. 7.6: $\mathbf{R} = \dfrac{DV}{\nu} = \dfrac{(18/12)V}{0.000\,201} = 7460V$; Solve by finding C_D from Fig. 9.10

V (fps)	0	25	50	100
R	–	186,600	373,000	746,000
C_D	0	0.44	0.185	0.195
F_D (lb)	0	0.427	0.718	3.03

◄◄

9.21 *A 1.5-ft-diameter metal ball weighing 150 lb is dropped into the ocean from a boat. Determine the maximum velocity of the ball as it falls through sea water with* ρ = *2.0 slugs/ft*^{3} *and* $\mu = 3.3 \times 10^{-5}$ *lb·sec/ft*^{2}.

BG

$W - F_B = F_D = C_D \rho A V^2/2; \quad W - \gamma(\pi D^3/6) = C_D \rho(\pi D^2/4)V^2/2$

$$150 - \frac{(62.4)2.0}{1.940}\left(\frac{\pi}{6}\right)1.5^3 = C_D(2.0)\frac{\pi}{8}1.5^2V^2; \quad 20.6 = C_D V^2 \qquad (1)$$

$$\text{Eq. 7.6:} \quad \mathbf{R} = \frac{DV\rho}{\mu} = \frac{1.5V2.0}{3.3\times 10^{-5}} = 90{,}900V \qquad (2)$$

Solve (1) and (2) by trial and error:

(*i*) Try $C_D = 0.4$: $V^2 = 20.6/0.4 = 51.4$

$V = 7.17$ fps, $\mathbf{R}$ = 652,000; from Fig. 9.10, $C_D = 0.19$

(*ii*) Try $C_D = 0.19$: $V^2 = 20.6/0.19 = 108.2$

$V = 10.40$ fps, $\mathbf{R}$ = 946,000; from Fig. 9.10, $C_D = 0.20$

Hence $V = (20.6/0.20)^{1/2} = 10.14$ fps ◄

9.22 *A 40-cm-diameter metal ball weighing 500 N is dropped into the ocean from a boat. Determine the maximum velocity of the ball as it falls through sea water with* ρ = *1030 kg/m*^{3} *and* μ = *0.0016 N·s/m*^{2}.

SI

$500 - 9810(1030/1000)(\pi/6)0.4^3 = C_D 1030(\pi/8)0.4^2V^2; \quad 2.49 = C_D V^2 \qquad (1)$

$\mathbf{R} = DV\rho/\mu = 0.4V(1030)/0.0016 = 257\,500V \qquad (2)$

Solve (1) and (2) by T and E:

(*i*) Try $C_D = 0.4$: $V = (2.49/C_D)^{1/2} = 2.497$ m/s ; $\mathbf{R}$ = 643 000; from Fig. 9.10: $C_D = 0.19$

(*ii*) Try $C_D = 0.2$: $V = 3.62$ m/s; $\mathbf{R}$ = 933 000 ; From Fig. 9.10, $C_D = 0.20$

Hence $V = (2.49/0.20)^{1/2} = 3.53$ m/s ◄

9.23 *A poorly-streamlined automobile has a body form corresponding roughly to the 1:0.75 oblate ellipsoid of Fig. 9.10, while a well-streamlined car has a body approximating the streamlined body in the same figure, each with a diameter of 1.5 m. If the velocity is 30 m/s through standard air at sea level (Appendix A, Table A.3), find the kW required to overcome air resistance in each case.*

SI

Table A.3 for air at sea level: $\rho = 1.225$ kg/m^3, $\nu = 1.461 \times 10^{-5}$ m^2/s.

$\mathbf{R} = 1.5(30)/(1.461 \times 10^{-5}) = 3.08 \times 10^6$. Assuming in Fig. 9.10 that C_D remains constant for $\mathbf{R} > 10^6$:

(*a*) C_D (poorly-streamlined = 1:0.75 oblate ellipsoid) = 0.20

(*b*) C_D (streamlined body) = 0.04

(*a*) Using Eq. 9.32: Power = $F_D V = C_D \rho A(V^3/2) = 0.20(1.225)(\pi 1.5^2/4)30^3/2 = 5.84$ kW ◀

(*b*) Power = (0.04/0.20)5.84 = 1.169 kW ◀

9.24 *For the streamlined train in Prob. 9.9, calculate the pressure drag. As a rough approximation, assume that the nose and tail of the train are of the shape of the two halves of the prolate ellipsoid of Fig. 9.10, of 2.5 m diameter. Find the drag on the ellipsoid (pressure drag on the train), and compare this with the skin-friction drag on the train determined earlier.*

Prob. 9.9: $\rho = 1.225$ *kg/m*^{3}, $\nu = 14.61 \times 10^{-6}$ *m*^{2}*/s,* $V = 40$ *m/s, skin-friction drag* $F_f = 1184$ *N.*

SI

Eq. 7.6: $\mathbf{R} = (2.5)40/(14.61 \times 10^{-6}) = 6.84 \times 10^6$; Fig. 9.10 (prolate ellipsoid): $C_D = 0.08$

Eq. 9.1: $F_p = (0.08)1.225(40^2/2)(\pi/4)2.5^2 = 385$ N ◀

This is about 1/3 of the friction drag of 1184 N determined in Prob. 9.9. ◀

9.25 *A submerged 7.5-ft by 1.5-ft flat plate is dragged through 60°F water at 1.4 fps with the flat side normal to the direction of motion. (a) What is the approximate drag force; (b) how does this compare with the drag force of 0.1124 lb when pulled in the direction of the 7.5 ft length? Hint: Assume that the drag coefficient for the plate of finite length is in the same ratio to the coefficient for the infinite plate as is the ratio of coefficients for the finite cylinder (L/D = 5) and the infinite cylinder of Fig. 9.13, for the same Reynolds number.*

BG

(*a*) Table A.1 for 60°F water: $\rho = 1.938$ slugs/ft^3, $\nu = 1.217 \times 10^{-5}$ ft^2/sec

$\mathbf{R} = 1.5(1.4)/(1.217 \times 10^{-5}) = 172{,}600$. From Fig. 9.13, for infinite plate, $C_D = 1.76$.

But (Fig. 9.13) for cylinders at $\mathbf{R} = 172{,}600$, $\dfrac{C_D \text{ for } L/D = 5}{C_D \text{ for infinite cyl}} = \dfrac{0.70}{1.15}$

Thus, for the given finite plate ($L/D = 5$), approximately $C_D = 1.76(0.70/1.15) = 1.071$

Eq. 9.32: $F_D = 1.071(1.938)(1.4^2/2)(7.5)(1.5) = 22.9$ lb ◀

(*b*) This is 22.9/0.1124 = 204 times the friction drag when pulled in the direction of the 7.5 ft length (see Exer. 9.5.1). ◀

9.26 *A submerged 2.5-m by 0.5-m flat plate is dragged through 15°C water at 0.5 m/s with the flat side normal to the direction of motion. (a) What is the approximate drag force; (b) how does this compare with the drag force of 0.888 N when pulled in the direction of the 2.5 m length? Hint: See Prob. 9.25.*

SI

(*a*) Table A.1 for 15°C water: $\rho = 999.1$ kg/m³, $\nu = 1.139 \times 10^{-6}$ m²/s.

$\mathbf{R} = 0.5(0.5)/(1.139 \times 10^{-6}) = 219\,000$. From Fig. 9.13, for infinite plate, $C_D = 1.76$.

But for cylinders at $\mathbf{R} = 439\,000$, $\dfrac{C_D \text{ for } L/D = 5}{C_D \text{ for } L/D = \infty} = \dfrac{0.68}{1.1}$

Thus, for the given finite plate, approximately $C_D = 1.76(0.68/1.1) = 1.088$

Eq. 9.32: $F_D = 1.088(999.1)(0.5^2/2)2.5(0.5) = 169.8$ N ◀

(*b*) This is 169.8/0.888 = 191 times the friction drag when pulled in the direction of the 2.5 m length. ◀

9.27 *Compare the velocity of an 0.15-in-diameter drop of water falling through standard air at sea level, with that of a spherical bubble of air of the same size rising through water at the same temperature.*

BG

(*a*) For air bubble rising through water:

Table A.1 for water at 60°F: $\rho = 1.938$ slugs/ft³, $\nu = 0.000\,0127$ ft²/sec.

At terminal velocity, neglecting weight of air, $F_D = F_B$

$\therefore$ Using Eq. 9.32: $F_D = C_D(1.938)\dfrac{(\pi/4)(0.15/12)^2}{2}V^2 = F_B = 62.4\dfrac{\pi}{6}\left(\dfrac{0.15}{12}\right)^3 = 63.8 \times 10^{-6}$ lb.

From which $V = 0.733/C_D^{1/2}$

Eq. 7.6: $\mathbf{R} = \dfrac{DV}{\nu} = \dfrac{(0.15/12)V}{0.000\,012\,17} = 1033V$

By trial, with Fig. 9.10: $C_D = 0.47$, $V = 1.069$ ft/sec ◀ and $\mathbf{R} = 1104$

(*b*) For water drop falling through air:

Table A.2 for air at 60°F: $\rho = 0.002\,374$ slugs/ft³, $\nu = 0.000\,158$ ft²/sec.

At terminal velocity, neglecting weight of air, $W = F_D$

$\therefore$ Using Eq. 9.32: $W = 63.8 \times 10^{-6}$ lb $= F_D = C_D(0.002\,374)(V^2/2)(\pi/4)(0.15/12)^2$

from which $V = 20.9/C_D^{1/2}$; also $\mathbf{R} = \dfrac{(0.15/12)V}{0.000\,158} = 79.1V$

By trial, with Fig. 9.10: $C_D = 0.40$, $V = 33.1$ fps ◀ and $\mathbf{R} = 2618$

Velocity ratio = $V_{\text{drop}}/V_{\text{bubble}}$ = 33.1/1.1069 = 31.0 ◀

9.28 *Compare the velocity of an 3-mm-diameter drop of water falling through standard air at sea level, with that of a spherical bubble of air of the same size rising through water at the same temperature.*

SI

(*a*) For air bubble rising through water:

Table A.1 for water at 15°C: $\rho = 999.1$ kg/m³, $\nu = 1.139 \times 10^{-6}$ m²/s.

At terminal velocity, neglecting weight of air, $F_D = F_B$

Eq. 9.32: $F_D = C_D(999.1)(V^2/2)(\pi/4)(0.003)^2 = F_B = 9810(\pi/6)0.003^3 = 1.387 \times 10^{-4}$ N

from which $V = 0.1982/C_D^{1/2}$; Eq. 7.6: $\mathbf{R} = \dfrac{DV}{\nu} = \dfrac{0.003V}{1.139 \times 10^{-6}} = 2634V$

By trial, with Fig. 9.10: $C_D = 0.49$, $V = 0.283$ m/s ◄ and $\mathbf{R} = 746$

(*b*) For water drop falling through the air:

Table A.2 for air at 15°C, by interpolation: $\rho = 1.227$ kg/m³, $\nu = 1.455 \times 10^{-5}$ m²/s.

At terminal velocity, neglecting weight of air, $W = F_D$

∴ Using Eq. 9.32: $W = 1.387 \times 10^{-4}$ N $= F_D = C_D(1.227)(V^2/2)(\pi/4)(0.003)^2$

from which $V = 5.66/C_D^{1/2}$; also $\mathbf{R} = \dfrac{0.003V}{1.455 \times 10^{-5}} = 206V$

By trial, with Fig. 9.10: $C_D = 0.41$, $V = 8.83$ m/s ◄ and $\mathbf{R} = 1821$

Velocity ratio $= V_{drop}/V_{bubble} = 8.83/0.283 = 31.2$ ◄

9.29 *Find the rate of fall of a spherical particle of sand (s = 2.65) in 15°C water if the diameter is (a) 0.1 mm; (b) 1.0 mm; (c) 10 mm. Express answers in millimeters per second.*

SI

Table A.1 for water at 15°C: $\gamma = 9798$ N/m³, $\rho = 999.1$ kg/m³, $\nu = 1.139 \times 10^{-6}$ m²/s.

At terminal velocity, net weight in water $= W - F_B = F_D$

So using Eq. 9.32: $(\pi/6)D^3(2.65 - 1)9798 = C_D(999.1)(V^2/2)\pi D^2/4$

which gives $V = 4.64(D/C_D)^{1/2}$. Also $\mathbf{R} = DV/\nu = DV/(1.139 \times 10^{-6}) = 878\,000DV$. From which:

	(*a*)	(*b*)	(*c*)	
D (m)	0.0001	0.001	0.01	
V	$0.0464/C_D^{1/2}$	$0.1469/C_D^{1/2}$	$0.464/C_D^{1/2}$	
$\mathbf{R}$	$87.8V$	$878V$	$8780V$	
C_D (by trial, using Fig. 9.10 sphere)	34.5*	0.82	0.37	
V (m/s)	0.00791	0.1622	0.764	◄◄
$\mathbf{R}$	0.694	142.4	6700	

*Stokes' Law, $C_D = 24/\mathbf{R}$, for $\mathbf{R} < 1$.

9.30 *A regulation football has a shape very similar to the prolate ellipsoid of Fig. 9.10, with a diameter of 6.78 in and a weight of 14.5 oz. (a) Find the resistance when the ball is passed through still air (14.5 psia and 80°F) at a velocity of 45 fps. Neglect the effect of spin about the longitudinal axis. What is the deceleration at the beginning of the trajectory? (b) Find the percentage change in resistance if the air temperature is 30°F rather than 80°F, and the drag coefficient is unchanged.*

BG

Table A.5 for air: $R = 1715 \text{ ft}^2/(\text{sec}^2\cdot°\text{R})$

(*a*) At 80°F, Table A.2 for air: $\nu = 0.000\,169 \text{ ft}^2/\text{sec}$

$$\text{Eq. 2.5:}\quad \gamma = \frac{gp}{RT} = \frac{32.2(14.7 \times 144)}{1715(460 + 80)} = 0.0736 \text{ lb/ft}^3$$

$$\mathbf{R} = \frac{(6.78/12)45}{0.000\,169} = 1.504 \times 10^5;\ \text{from Fig. 9.10:}\ C_D = 0.056$$

$$\text{Eq. 9.32:}\quad F_D = (0.056)\frac{0.0736}{32.2}\left(\frac{45^2}{2}\right)\frac{\pi}{4}\left(\frac{6.78}{12}\right)^2 = 0.0325 \text{ lb} \quad ◀$$

$$\text{At start of trajectory,}\quad a = \frac{F_D}{m} = \frac{0.0325(32.2)}{14.5/16} = 1.155 \text{ ft/sec}^2 \quad ◀$$

(*b*) At 30°F, Table A.2 for air: $\nu = 0.000\,142 \text{ ft}^2/\text{sec}$.

$$\text{Eq. 2.5:}\quad \gamma = \frac{32.2(14.7 \times 144)}{1715(460 + 30)} = 0.0811 \text{ lb/ft}^3$$

$\mathbf{R} = (6.78/12)45/0.000\,142 = 1.790 \times 10^5$; from Fig. 9.10: $C_D = 0.055$

$$\text{Eq. 9.32:}\quad F_D = (0.055)\frac{0.0811}{32.2}\left(\frac{45^2}{2}\right)\frac{\pi}{4}\left(\frac{6.78}{12}\right)^2 = 0.0352 \text{ lb or } 8.24\% \text{ increase} \quad ◀$$

9.31 *An eight-oar racing shell is traveling through 60°F water at a mean velocity of 12 mph. Each oar is 9 ft long, with a length of 6 ft from the oarlock to the center of the "spoon", which has a projected area of 120 in². Assume that all drag is due to the spoon, and that this drag is equal to that of a disk of equal area. (a) If the "stroke" is 32 per min, and if each oarsman sweeps a right angle in one-fourth of his rowing cycle, what is the maximum thrust of the oars? It must be assumed that the shell moves at something less (say, 20 percent) than its mean velocity when the oars are in the water. (b) The maximum velocity occurs during the backstroke when the oarsmen shift their weight toward the stern. Why? (c) The oarsman on his backstroke moves at half the angular velocity of his forward stroke, while the shell moves at perhaps 10 percent above its mean velocity. Find the drag in 60°F air resulting from a "feathered" oar (turbulent boundary layer, Fig 9.7) and an unfeathered one, in percentage of the forward thrust from part (a). What is the ratio of these two drag forces?*

BG

(*a*) Period of rowing cycle = 60/32 = 1.875 sec

Period of 1/4 cycle = 0.469 sec; angular sweep = $\pi/2$ radians

Velocity of oar relative to boat = $r\omega = 6(\pi/2)(1/0.469) = 20.1$ fps

Forward velocity of boat = (1 − 0.2)12(44/30) = 14.08 fps

Velocity of oar relative to water = 20.1 − 14.08 = 6.02 fps

Table A.1 for water at 60°F: $\rho = 1.938 \text{ slugs/ft}^3$, $\nu = 0.000\,012\,17 \text{ ft}^2/\text{sec}$.

$$\mathbf{R} = \frac{[(4/\pi)120]^{1/2}(6.02)}{12(0.000\,012\,17)} = 5.10 \times 10^5;\ \text{Fig. 9.10 for perpendicular disk:}\ C_D = 1.1$$

/cont...

Eq. 9.32: $F_D = (1.1)1.938(6.02^2/2)(120/144) = 32.3$ lb ; Total thrust $= 8(32.3) = 258$ lb ◀

(*b*) $V = V_{max}$ when oarsmen's weight is shifted toward the stern because of conservation of momentum. The reduced forward momentum of the oarsmen increases the forward momentum and velocity of the shell. ◀

(*c*) Backstroke: $r\omega = 20.1/2 = 10.05$ fps

$V_{boat} = (1.1)12(44/30) = 19.36$ fps ; $V_{oar\ rel\ to\ air} = 10.05 + 19.36 = 29.4$ fps

Table A.2 for air at 60°F: $\rho = 0.002\,37$ slug/ft³, $\nu = 0.000\,158$ ft²/sec.

Unfeathered oar: $\mathbf{R} = \dfrac{[(4/\pi)120]^{1/2}(29.4)}{12(0.000\,158)} = 1.917 \times 10^5$

Fig. 9.10 for perpendicular disk: $C_D = 1.1$

Eq. 9.32: $F_D = 1.1(0.002\,37)(29.4^2/2)(120/144) = 0.940$ lb

or $0.940/32.3 = 2.91\%$ of forward thrust ◀

Feathered oar (two surfaces): Fig. 9.7 (turbulent) for $\mathbf{R} = 1.917 \times 10^5$: $C_f = 0.0063$

Eq. 9.2 assuming $BL = A$: $F_f = 2(0.0063)0.002\,37(29.4^2/2)(120/144) = 0.010\,76$ lb

or $0.010\,76/32.3 = 0.0334\%$ of forward thrust. ◀

$0.940/0.010\,76 = 87.3$ ◀

So an unfeathered oar causes about 87 times the resistance of a feathered oar.

Sec. 9.8: Drag on Two-Dimensional Bodies (Incompressible Flow) -- *Exercises*

9.8.1 *What is the bending moment at the base of a vertical 0.3-in-diameter cylindrical radio antenna 6 ft tall on an automobile traveling at 80 mph through standard air at sea level?*

BG

Table A.3 for air at sea level: $\mu = 3.737 \times 10^{-7}$ lb·sec/ft²; $\rho = 0.002\,377$ slug/ft³

$V = 80(44/30) = 117.3$ fps

Eq. 7.6: $\mathbf{R} = \dfrac{0.30(117.3)0.002\,377}{12(3.737 \times 10^{-7})} = 18{,}660$; from Fig. 9.13 (infinite cylinder), $C_D = 1.15$

Eq. 9.32: $F_D = 1.15(0.002\,377)(117.3^2/2)(0.3/12)6 = 2.82$ lb

$M = F_D(L/2) = 2.82(6/2) = 8.47$ ft·lb ◀

9.8.2 *A 70 mph wind blows across an 0.1-in-diameter wire. Find the frequency of oscillation which results in standard atmosphere (a) at sea level, and (b) at 10,000 ft elevation.*

BG

$V = 70$ mph $= 70(44/30) = 102.7$ fps;

(*a*) Table A.3 for air at sea level: $\rho = 0.002\,377$ slug/ft³; $\nu = 0.000\,1572$ ft²/sec

$\mathbf{R} = \left(\dfrac{0.1}{12}\right)\dfrac{102.7}{0.000\,1572} = 5442$; Eq. 9.34: $f \approx 0.20\dfrac{102.7}{(0.1/12)}\left(1 - \dfrac{20}{5442}\right) = 2455$ Hz ◀

(*b*) Table A.3 for air at 10,000 ft: $\rho = 0.001\,756$ slug/ft³; $\nu = 0.000\,2013$ ft²/sec

$\mathbf{R} = \left(\dfrac{0.1}{12}\right)\dfrac{102.7}{0.000\,2013} = 4250$; Eq. 9.34: $f \approx 0.20\dfrac{102.7}{(0.1/12)}\left(1 - \dfrac{20}{4250}\right) = 2452$ Hz ◀

Sec. 9.8: Drag on Two-Dimensional Bodies (Incompressible Flow) -- *Problems*

9.32 *(a) Find the bending moment at the base of a 30-cm-diameter cylindrical light post 12 m high when it is subject to a uniform wind velocity of 30 m/s at standard sea level. Neglect end effects. (b) Discuss the consequences of considering the atmospheric boundary layer above the surface of the earth.*

SI

(*a*) Table A.3 for air at sea level: $\rho = 1.225 \text{ kg/m}^3$, $\nu = 14.61 \times 10^{-6} \text{ m}^2/\text{s}$

Eq. 7.6: $\mathbf{R} = \dfrac{0.30(30)}{14.61 \times 10^{-6}} = 6.16 \times 10^5$; from Fig. 9.13, $C_D = 0.31$

Eq. 9.32: $F_D = 0.31(1.225)(30^2/2)(12)0.30 = 615 \text{ N}$

$M = F_D(L/2) = 615(15/2) = 3690 \text{ N·m}$ ◄

(*b*) In the atmospheric boundary layer wind velocities are reduced close to the surface of the earth. For velocities < 24.3 m/s ($\mathbf{R}$ < 500 000) Fig. 9.13 indicates that C_D increases suddenly, by up to four times. But the resulting increased drag forces have small moment arms about the base. So the net effect of the atmospheric boundary layer is probably a small increase of bending moment. ◄

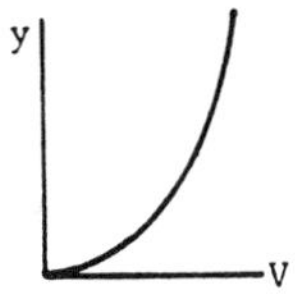

9.33 *(a) Repeat Prob. 9.32(a) for the case where the pole has a uniform taper from 35 cm diameter at the base to 25 cm at its top. (b) Determine if the drag force would be larger or smaller if the pole had instead been tapered from 35 cm down to 15 cm. Why?*

Prob. 9.32 (a): Find the bending moment at the base of a 12-m-high light pole subject to a uniform wind velocity of 30 m/s at standard sea level (neglect end effects).

SI

Table A.3 for air at sea level: $\rho = 1.225 \text{ kg/m}^3$, $\nu = 14.61 \times 10^{-6} \text{ m}^2/\text{s}$

(*a*) Given the wind velocity is uniform everywhere 30 m/s (actually it falls to zero at ground level).

Position on pole	Base	Top
D (m)	0.35	0.25
$\mathbf{R} = D(30)/(14.61 \times 10^{-6})$	7.19×10^5	5.13×10^5
C_D (Fig. 9.13)	0.32	0.30
Drag force per unit length (Eq. 9.32)		
$F_D/L = C_D\rho(V^2/2)D$ (N/m)	61.7	41.3

In this diameter $\mathbf{R}$ range C_D varies little (see Fig. 9.13). Thus the drag force depends entirely on D, and so its variation is almost linear.

Thus total drag force = ½(61.7 + 41.3)12 = 619 N ◄

$M = 41.3(12)6 + (61.7 - 41.3)(12/2)4 = 3470 \text{ N·m}$ ◄

(*b*) For $D = 0.15$ m, $\mathbf{R} = 3.08 \times 10^5$, $C_D = 0.90$, $F_D = 74.4$ N/m

Thus if the pole had been tapered from 35 cm at the base down to 15 cm (instead of 25 cm) at the top, the drag force per unit length on the upper portion of the pole would have been considerably larger (by about 80%) than for the 25 cm diameter, because of the abrupt rise in C_D as the Reynolds number is reduced (see Fig. 9.13). ◄

Sec. 9.9: Lift and Circulation -- *Exercise*

9.9.1 *Estimate the lift per unit length of span if the mean velocity along the top of a wing having a 10 ft chord is 100 mph and that along the bottom of the wing is 80 mph when the wing moves at 85 mph through still air (γ = 0.072 lb/ft^3).*

BG

Velocities (100, 80, and 85 mph) are 146.7, 117.3, and 124.7 fps.

At a distance where the velocity is undisturbed and uniform (= U), let the pressure = p_u (as in Sec. 9.10).

Then $\frac{U^2}{2g} + \frac{p_u}{\gamma} = \frac{V_{top}^2}{2g} + \frac{p_{top}}{\gamma}$; $\frac{U^2}{2g} + \frac{p_u}{\gamma} = \frac{V_{bot}^2}{2g} + \frac{p_{bot}}{\gamma}$

Subtracting: $\frac{p_{top} - p_{bot}}{\gamma} = \frac{V_{bot}^2 - V_{top}^2}{2g} = \frac{117.3^2 - 146.7^2}{2(32.2)} = -120.2$ ft of air

$\therefore$ Δp = 120.2(0.072) = 8.66 lb/ft^2 ; lift per unit length of span = $\Delta p(c)$ = 8.66(10) = 86.6 lb/ft ◀

Sec. 9.10: Ideal Flow About a Cylinder -- *Exercises*

9.10.1 *Imagine that a circulation of 25 ft^2/s is superimposed around a 1-ft-diameter cylinder immersed in 60°F water flowing at 10 fps perpendicular to the cylinder axis. Find the location of the stagnation points, and the lift on a 20-ft length of cylinder.*

BG

Table A.1 for water at 60°F: ρ = 1.938 slug/ft^3

Eq. 9.41: $\Gamma = 25 = -4\pi(0.5)10\sin\theta_0$; $\sin\theta_0 = -0.398$; $\theta_0 = -23.4°$ ◀

Eq. 9.40: F_L = 1.938(20)10(25) = 9690 lb ◀

9.10.2 *Imagine that a circulation of 2.4 m^2/s is superimposed around a 30-cm-diameter cylinder immersed in 15°C water flowing at 3.5 m/s perpendicular to the cylinder axis. Find the location of the stagnation points, and the lift on a 10-m length of cylinder.*

SI

Table A.1 for water at 15°C: ρ = 999.1 kg/m^3

Eq. 9.41: $\Gamma = 2.4 = -4\pi(0.3/2)3.5\sin\theta_0$; $\sin\theta_0 = -0.364$; $\theta_0 = -21.3°$ ◀

Eq. 9.40: F_L = 999.1(10)3.5(2.4) = 83 900 N ◀

Sec. 9.10: Ideal Flow About a Cylinder -- *Problems*

9.34 *Beginning with the expression above Eq. (9.40), fill in the steps leading to Eq. (9.40). Take care to account for all changes in sign.*

N

Expression above Eq. 9.40: $F_L = -B\int_0^{2\pi}(p - p_u)R\sin\theta\, d\theta$; substituting for $p - p_u$ from Eq. 9.39:

$$F_L = -B\int_0^{2\pi} \rho/2[U^2 - (2U\sin\theta + \Gamma/2\pi R)^2]R\sin\theta\, d\theta$$

$$= -\rho/2BR\int_0^{2\pi}[U^2\sin\theta - 4U^2\sin^3\theta - (2\Gamma U/\pi R)\sin^2\theta + \Gamma^2\sin\theta/(4\pi^2R^2)]d\theta$$

Noting: $\int_0^{2\pi}\sin\theta\, d\theta = 0$, $\int_0^{2\pi}\sin^3\theta\, d\theta = [-⅓\cos\theta(\sin^2\theta + 2)]_0^{2\pi} = 0$

and $\int_0^{2\pi}\sin^2\theta\, d\theta = [½\,\theta - ¼\sin 2\theta]_0^{2\pi} = \pi$

Thus, $F_L = -(\rho/2)BR[0 - 0 - (2\Gamma U/\pi R)\pi - 0] = \rho BU\Gamma$ Q.E.D. ◀

9.35 *For the rotating cylinder in Sample Prob. 9.7, calculate the value of the lift coefficient. Assume the effective circulation is half the theoretical. Assuming the drag coefficient to be unchanged by the rotation of the cylinder, find the magnitude and direction of the total wind force on the rotor.*

Sample Prob. 9.7: U = 120 fps, D = 4 ft, γ of air = 0.0765 lb/ft^3, theoretical Γ = 237 ft^2/sec, theoretical F_L = 1688 lb.

BG

Given effective = ½(theoretical). ∴ effective Γ = 118.4 ft^2/sec, effective F_L = 844 lb.

From Eqs. 9.40 and 9.44: $F_L = C_L\rho(U^2/2)BD = \rho BU\Gamma$

∴ $C_L = 2\Gamma/UD$ = [2(118.4)]/(120 × 4) = 0.493 ◄

Table A.2 for air with γ = 0.0765 lb/ft^3 (given), by interpolation: ν = 0.000 1564 ft^2/sec

∴ $\mathbf{R}$ = 4(120)/0.000 1576 = 3.05 × 10^6; From Fig. 9.13: C_D = 0.35

∴ From Eqs. 9.32 and 9.44: $F_D = (C_D/C_L)F_L$ = (0.35/0.493)844 = 599 lb

Total force = (844^2 + 599^2)$^{1/2}$ = 1035 lb ◄

Direction = tan^{-1}(844/599) = 54.7° from wind direction ◄

9.36 *Assume the rotor in Sample Problem 9.7 to be installed upright on a ship traveling due north at 20 knots. The wind has an absolute velocity of 35 mph due east. If the drag coefficient of the cylinder is 1.0 and the "lines" of stagnation are separated by 120° on the rotor, find approximately the component of the total air force on the rotor in the direction of the ship's motion. Assume standard air.*

Sample Prob. 9.7: D = 4 ft, B = 25 ft

BG

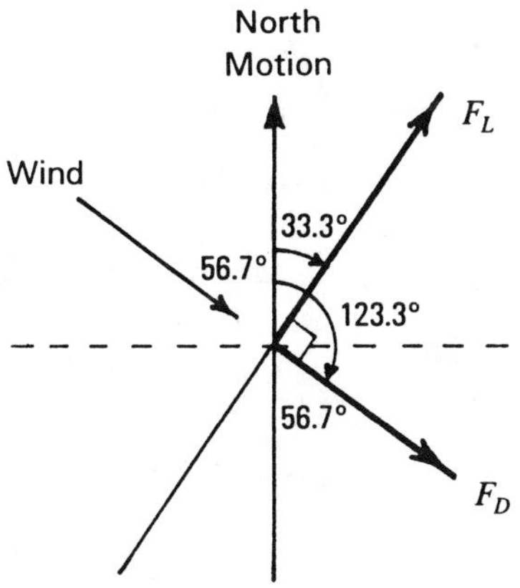

Ship velocity = 20 knots = 20(1.688) fps = 33.8 fps;

wind velocity = 35 mph = 35(44/30) fps = 51.3 fps

Wind relative to ship: $V = \sqrt{51.2^2 + 33.8^2}$ = 61.4 fps at

tan^{-1}(51.3/33.8) = 56.7° E of S.

Eq. 9.41: $\Gamma = -4\pi(2)61.4\sin(-30°)$ = 772 ft^2/sec

Table A.2 for standard air at 60°F: ρ = 0.002 37 slug/ft^3.

Eq. 9.40: F_L = 0.002 37(25)61.4(772) = 2810 lb

Given C_D = 1.00.

Eq. 9.32: F_D = (1.0)0.002 37(61.4^2/2)4(25) = 447 lb

Angle of F_D to ship's forward axis is 180° − 56.7° = 123.3°;

angle of F_L to ship's forward axis is 90° - 56.7° = 33.3°

Net forward component, F_{north} = 2810 cos 33.3° + 447 cos 123.3° = 2350 - 246 = 2100 lb ◄

9.37 *Consider a cylinder of radius a in a stream of ideal fluid in which the undisturbed velocity and pressure are U and p_u and the density is ρ. (a) Using Eq. (9.36) and the Bernoulli theorem, evaluate the dimensionless pressure coefficient $(p - p_u)/(\rho U^2/2)$ for every 10° over the surface of one quadrant of the cylinder, measuring the angle from the forward stagnation point. Plot the pressure coefficient to scale, plotting it radially from the cylinder surface. (b) What is the actual pressure in pounds per square inch on the surface of a 1-ft-diameter cylinder, 70° from the forward stagnation point, if the cylinder is 15 ft below the free surface of a stream of water at 60°F, flowing at 12 fps?*

BG

(*a*) Bernoulli theorem from point distant from cylinder (U, p_u) to point on surface (v, p):

$$\frac{p_u}{\gamma} + \frac{U^2}{2g} = \frac{p}{\gamma} + \frac{v^2}{2g}; \quad \text{so} \quad \frac{p - p_u}{\rho U^2/2} = 1 - \frac{v^2}{U^2} \quad \text{and by substituting Eq. 9.36:}$$

$$= 1 - \frac{(2U\sin\theta)^2}{U^2} = 1 - 4\sin^2\theta$$

$\theta°$	0	10	20	30	40	50	60	70	80	90
$\frac{p-p_u}{\rho U^2/2}$	1	0.879	0.532	0	−0.653	−1.347	−2	−2.53	−2.88	−3 ◄

(*b*) Table A.1 for water at 60°F: $\rho = 1.938\ \text{slug/ft}^3$, $\gamma = 62.37\ \text{lb/ft}^3$

$$\text{At } \theta = 70°, \quad \frac{p - p_u}{\rho U^2/2} = 1 - (4 \times 0.940^2) = -2.53$$

$$p = p_u - 2.53\rho U^2/2 = 15(62.37) - 2.53(1.938)12^2/2 = 936 - 353 = 582 \text{ psf or } 4.04 \text{ psi} \quad ◄$$

9.38 *Consider a cylinder of radius a in a stream of ideal fluid in which the undisturbed velocity and pressure are U and p_u and the density is ρ. (a) Using Eq. (9.36) and the Bernoulli theorem, evaluate the dimensionless pressure coefficient $(p - p_u)/(\rho U^2/2)$ for every 10° over the surface of one quadrant of the cylinder, measuring the angle from the forward stagnation point. Plot the pressure coefficient to scale, plotting it radially from the cylinder surface. (b) What is the actual pressure in newtons per square meter on the surface of a 30-cm-diameter cylinder, 70° from the forward stagnation point, if the cylinder is 5 m below the free surface of a stream of water at 15°C, flowing at 4 m/s?*

SI

(*a*) Solution is the same as for Prob. 9.37(*a*)

(*b*) Table A.1 for water at 15°C: $\rho = 999.1\ \text{kg/m}^3$, $\gamma = 9.798\ \text{kN/m}^3 = 9798\ \text{N/m}^3$

$$\text{At } \theta = 70°, \quad \frac{p - p_u}{\rho U^2/2} = 1 - (4 \times 0.940^2) = -2.53$$

$$p = p_u - 2.53\rho U^2/2 = 5(9798) - 2.53(999.1)4^2/2 = 49\,000 - 20\,200 = 28\,800\ \text{N/m}^2 \quad ◄$$

9.39 *A double stagnation point is observed to occur on a 4-ft-diameter cylinder rotating in an air stream moving at 60 fps. Given standard atmospheric conditions at sea level, find (a) the lift force per foot length of the cylinder, and (b) the lift coefficient.*

BG

(*a*) Eq. 9.41 with $\theta_0 = -90°$: $\Gamma = -4\pi(2)60\sin(-90°) = 1508\ \text{ft}^2/\text{sec}$

Table A.3 for standard air at sea level: $\rho = 0.002\,38\ \text{slug/ft}^3$

Eq. 9.40: $F_L = (0.002\,38)1(60)1508 = 215$ lb/ft of length ◄

(*b*) Eq. 9.44: $215 = C_L(0.002\,38)(60^2/2)4(1)$; $C_L = 12.57$ ◄

9.40 *A double stagnation point is observed to occur on a 1.2-m-diameter cylinder rotating in an air stream moving at 18 m/s. Given standard atmospheric conditions at sea level, find (a) the lift force per meter length of the cylinder, and (b) the lift coefficient.*

SI

(*a*) Eq. 9.41 with $\theta_0 = -90°$: $\Gamma = -4\pi(1.2/2)18\sin(-90°) = 135.7\ \text{m}^2/\text{s}$

Table A.3 for standard air at sea level: $\rho = 1.225\ \text{kg/m}^3$

Eq. 9.40: $F_L = (1.225)1(18)135.7 = 2990$ N/m of length ◀

(*b*) Eq. 9.44: $2990 = C_L(1.225)(18^2/2)1.2(1)$; $C_L = 12.57$ ◀

9.41 *There have been many arguments over the validity and extent of the curve of a pitched baseball. According to tests (Life, July 27, 1953), a pitched baseball was found to rotate 1400 rpm while traveling at 43 mph. The horizontal projection of the trajectory revealed a smooth curve of about 800 ft radius. If the ball had a circumference of 9 in and weighed 5 oz, find the transverse force required to produce the observed curvature. Assume the shape of the ball to be roughly that of a cylinder having a diameter equal to the ball's diameter and a length of two-thirds its diameter. Find the circulation that would be required to produce the transverse force. Compare this with that obtained by assuming no slip at the equator of the ball. Assume standard air at sea level.*

BG

$U = 43(44/30) = 63.1$ fps ; $F_L = ma_n = m\dfrac{U^2}{r} = \left(\dfrac{5/16}{32.2}\right)\dfrac{63.1^2}{800} = 0.0483$ lb ◀

$D = (9/\pi)/12 = 0.239$ ft; given $B = (2/3)0.239 = 0.1592$ ft

Table A.3 for standard air at sea level: $\rho = 0.002\,38\ \text{slug/ft}^3$

Eq. 9.40: $0.0483 = 0.002\,38(0.1592)63.1\Gamma$; $\Gamma = 2.02\ \text{ft}^2/\text{sec}$ ◀

Comparison: From rotation at $\omega = 1400(2\pi/60) = 146.6$ rad/sec,

$\upsilon = r\omega = (0.239/2)146.6 = 17.50$ fps ; Eq. 9.37: $17.50 = \Gamma/[2\pi(0.239/2)]$; $\Gamma = 13.12\ \text{ft}^2/\text{sec}$ ◀

Note: Even with 50% slip, it is evident that only the middle 1/4 of the ball is effective in producing circulation.

9.42 *Solve Prob. 9.41 for a speed of 69 km/h and a trajectory radius of 245 m, with a ball weighing 1.4 N and having a circumference of 23 cm.*

Prob. 9.41: Find the transverse force which produces the observed curved path of the pitched baseball, rotating at 1400 rpm. Assume the shape of the ball to be roughly that of a cylinder having a diameter equal to the ball's diameter and a length of two-thirds its diameter. Find the circulation that would be required to produce the transverse force. Compare this with that obtained by assuming no slip at the equator of the ball. Assume standard air at sea level.

SI

$U = 69(1000/3600) = 19.17$ m/s ; $F_L = ma_n = m\dfrac{U^2}{r} = \left(\dfrac{1.4}{9.81}\right)\dfrac{19.17^2}{245} = 0.214$ N ◀

$D = 0.23/\pi = 0.0732$ m; $B = (2/3)0.0732 = 0.0488$ m

Eq. 9.40: $0.214 = 1.225(0.0488)19.17\Gamma$; $\Gamma = 0.1867\ \text{m}^2/\text{s}$ ◀

Comparison: From rotation at $\omega = 1400(2\pi/60) = 146.6$ rad/s,

$\upsilon = r\omega = (0.0732/2)146.6 = 5.37$ m/s ; Eq. 9.37: $5.37 = \Gamma/[2\pi(0.0732/2)]$; $\Gamma = 1.234\ \text{m}^2/\text{s}$ ◀

Note: Even with 50% slip, it is evident that only the middle 1/4 of the ball is effective in producing circulation.

Sec. 9.12: Induced Drag on Airfoil of Finite Length -- *Exercise*

9.12.1 *A wing with a 15-m span and a 45-m^2 "planform" area moves horizontally through the standard atmosphere at 6000 m with a velocity of 750 km/hr. If the wing supports 180 kN, find (a) the required value of the lift coefficient; (b) the downwash velocity, assuming an elliptical distribution of lift over the span; and (c) the induced drag.*

SI

Table A.3 for standard air at 6 km altitude: $\rho = 0.660$; $V = 750(1000/3600) = 208$ m/s

(*a*) Eq. 9.44: $F_L = 180\,000 = C_L(0.660)(208^2/2)45$; so $C_L = 0.279$ ◄

(*b*) $\dfrac{B^2}{A} = \dfrac{15^2}{45} = 5$; $\dfrac{B}{C} = \dfrac{15}{3} = 5$; Eq. 9.45: $\dfrac{V_i}{V} = \dfrac{C_L}{\pi(B^2/A)} = \dfrac{0.279}{\pi(5)} = 0.017\,78$

$\therefore\ V_i = 0.017\,78(208) = 3.70$ m/s ◄

(*c*) Eq. 9.46: $C_{Di} = (0.279)^2/[\pi(15^2/45)] = 0.004\,96$

From Eqs. 9.42 and 9.44: $F_{Di} = (0.004\,96/0.279)180\,000 = 3200$ N ◄

Sec. 9.12: Induced Drag on Airfoil of Finite Length -- ***Problem***

9.43 *A wing of 44-ft span and a 350-ft^2 "planform" area moves horizontally through standard atmosphere at 20,000 ft with a velocity of 250 mph. If the wing supports 9000 lb, find (a) the required value of the lift coefficient; (b) the downwash velocity, assuming an elliptical distribution of lift over the span; and (c) the induced drag.*

BG

Table A.3 for standard air at 20,000 ft altitude: $\rho = 0.001\,267$ slug/ft^3

$V = 250(5280/3600) = 367$ fps

(*a*) Eq. 9.44: $F_L = 9000 = C_L(0.001\,267)(367^2/2)350$; so $C_L = 0.302$ ◄

(*b*) Eq. 9.45: $\dfrac{V_i}{V} = \dfrac{C_L}{\pi(B^2/A)} = \dfrac{0.302}{\pi(44^2/350)} = 0.017\,37$; $V_i = 0.017\,37(367) = 6.37$ fps ◄

(*c*) Eq. 9.46: $C_{Di} = (0.302)^2/[\pi(44^2/350)] = 0.005\,25$

From Eqs. 9.42 and 9.44: $F_{Di} = (0.005\,25/0.302)9000 = 156.4$ lb ◄

Sec. 9.13: Lift and Drag Diagrams -- ***Exercises***

9.13.1 *A 1400-lb airplane has a Clark Y airfoil wing of 6-ft chord by 36-ft span, with polar diagram given in Fig. 9.24. Find (a) the speed required to get the plane off the ground; (b) the horsepower required; (c) the circulation about the wing; and (d) the strength of the starting vortex. Assume standard air at sea level and the angle of attack for maximum ratio of lift to drag. Neglect aerodynamic forces on the fuselage and tail.*

BG

(*a*) Fig. 9.24: For max ratio C_L/C_D, $\alpha = -0.1°$, $C_L = 0.4$, $C_D = 0.018$

Table A.3 for standard air at sea level: $\rho = 0.002\,38$ slug/ft^3

Eq. 9.44: $1400 = (0.4)0.002\,38(V^2/2)6(36)$; $V = 116.7$ fps ◄

(*b*) From Eqs. 9.32 and 9.44: $F_D = (C_D/C_L)F_L = (0.018/0.40)1400 = 63.0$ lb

Power = (63.0)116.7/550 = 13.37 hp ◄

(*c*) Eq. 9.40: $1400 = 0.002\,38(36)116.7\Gamma$; $\Gamma = 140.0$ ft^2/sec ◄

(*d*) Sec. 9.9, vortex strength: $C = \Gamma/(2\pi) = 140.0/(2\pi) = 22.3$ ft^2/sec ◄

9.13.2 *A sailplane including its load weighs 300 lb. It has a Clark Y section wing with a 3.5-ft chord by 21-ft span. Given that it has the same characteristics as the larger wing of the same aspect ratio shown in Fig. 9.24, find the angle of glide through standard air at 2000 ft which will produce the greatest horizontal distance range. Neglect air forces on the fuselage and the tail. (Note: The aspect ratio of 6 is here chosen to facilitate working the problem with the available data. Actually, the sailplane may be constructed with an aspect ratio of about twice this, in order to reduce the drag to a minimum.)*

BG

Fig. 9.24 for max C_L/C_D: $\alpha = -0.1°$, $C_L = 0.4$, $C_D = 0.018$

For max range, glide angle β is minimum. $\beta = \tan^{-1}(F_D/F_L) = \tan^{-1}(0.018/0.4) = 2.58°$ ◀

9.13.3 *Solve Exer. 9.13.2 for a sailplane with a 1-m chord by 6-m span wing, flying at 600 m altitude with a total weight of 1800 N.*

SI

See solution to Exer. 9.13.2

Sec. 9.13: Lift and Drag Diagrams -- *Problems*

9.44 *Determine the induced angle of attack and the induced drag if the plan form of the wing in Exer. 9.12.1 were rectangular.*

Exer. 9.12.1: The wing with span B = 15 m and planform area A = 45 m² supports 180 kN when moving horizontally at 750 km/h through the standard atmosphere at an elevation of 6000 m.

SI

Table A.3 for standard air at 6 km altitude: $\rho = 0.660$; $V = 750(1000/3600) = 208$ m/s

Eq. 9.44: $F_L = 180\,000 = C_L(0.660)(208^2/2)45$; so $C_L = 0.279$ ◀

Fig. 9.25 for $B/c = B^2/A = 15^2/45 = 5.00$: $\tau = 0.154$, $\sigma = 0.041$

Eq. 9.49: $\alpha_i = \dfrac{0.279}{\pi(5)}(1 + 0.154) = 0.0205$ radian; $\therefore\ \alpha_i = 0.0205(180/\pi)° = 1.175°$ ◀

Eq. 9.50: $C_{Di} = \dfrac{0.279^2}{\pi(5)}(1 + 0.041) = 0.005\,17$

From Eqs. 9.42 and 9.44: $F_{Di} = (C_{Di}/C_L)F_L = (0.005\,17/0.279)180 = 3.33$ kN ◀

9.45 *Determine the induced angle of attack and the induced drag if the plan form of the wing in Prob. 9.43 were rectangular.*

Prob. 9.43: The wing with span B = 44 ft and planform area A = 350 ft² supports 9000 lb when moving horizontally at 250 mph through the standard atmosphere at an elevation of 20,000 ft.

BG

Table A.3 for standard air at 20,000 ft altitude: $\rho = 0.001\,267$ slug/ft³ ; $V = 250(5280/3600) = 367$ fps

Eq. 9.44: $F_L = 9000 = C_L(0.001\,267)(367^2/2)350$; so $C_L = 0.302$ ◀

Fig. 9.25 for $B/c = B^2/A = 44^2/350 = 5.53$: $\tau = 0.166$, $\sigma = 0.046$

Eq. 9.49: $\alpha_i = \dfrac{0.302}{\pi(5.53)}(1 + 0.166) = 0.0203$ radian; $\therefore\ \alpha_i = 0.0203(180/\pi)° = 1.161°$ ◀

Eq. 9.50: $C_{Di} = \dfrac{0.302^2}{\pi(5.53)}(1 + 0.046) = 0.005\,49$

From Eqs. 9.42 and 9.44: $F_{Di} = (C_{Di}/C_L)F_L = (0.005\,49/0.302)9000 = 163.6$ lb ◀

9.46 *For the Clark Y airfoil of Fig. 9.24, evaluate the friction coefficient η of Eq. (9.48) for values of C_L of 0.6, 1.0, and 1.4.*

N

Fig. 9.25 for $B/c = 36/6 = 6$: $\tau = 0.175$;

At $C_L = 1.0$, Eq. 9.49: $\alpha_i = [1.0/\pi(6)](1 + 0.175) = 0.0623$ radian;

$\therefore\ \alpha_i = 0.0623(180/\pi)° = 3.57°$ ◀

Fig. 9.24 for $C_L = 1.0$: $\alpha = 8.0°$; for $C_L = 0$, $\alpha = -5.6°$

Fig. 9.22: $\alpha_0 = \alpha - \alpha_i = 8.0 - 3.57 = 4.43°$

From zero angle of attack (Sec. 9.13), $\alpha_0' = 4.43 + 5.6 = 10.03° = 0.1750$ radian

Eq. 9.48: $1.0 = 2\pi\eta(0.1750)$; $\eta = 0.909$ (at $C_L = 1.0$) ◀

Similarly				
	C_L	0.6	1.0	1.4
	α_i	2.14°	3.57°	5.00°
	α	2.6°	8.0°	14.3°
	α_0	0.457°	4.43°	9.30°
	α_0'	6.06°	10.03°	14.90°
	η	0.903	0.909	0.857 ◀◀

9.47 *A kite has the shape of a rectangular airfoil with a chord length of 2.5 ft and a span of 4.5 ft. When rigged and oriented as shown in the figure, the guideline exerts a tension T of 14 lb when the wind velocity V is 28 mph in standard air at 1500 ft altitude. Find C_L, C_{DO}, and the friction coefficient η.*

BG

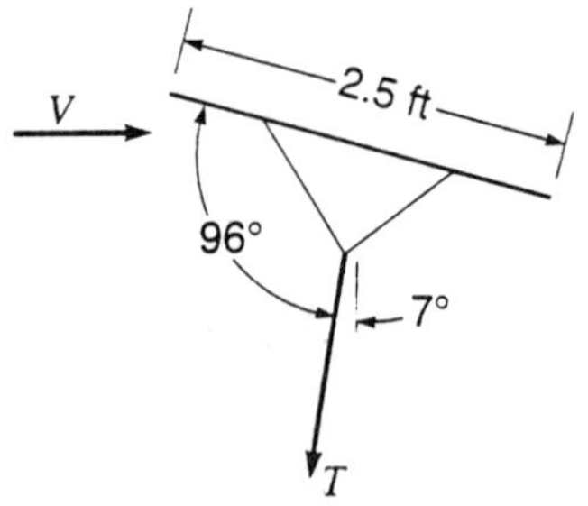

Figure P9.47

From Fig. P9.47: $F_L = 14\cos 7° = 13.90$ lb; $F_D = 14\sin 7° = 1.706$ lb

Table A.3 for standard air at 1500 ft by interpolation:

$\rho = 0.002\,28$ slug/ft^3; $V = 28(44/30) = 41.1$ fps

Eq. 9.44: $13.90 = C_L(0.002\,28)(41.1^2/2)2.5(4.5)$; $\therefore\ C_L = 0.642$

Eqs. 9.32 and 9.44: $C_D = 0.642(1.706/13.90) = 0.0789$

Fig. 9.25 for $B/c = 4.5/2.5 = 1.80$: $\tau \approx 0.072$, $\sigma \approx 0.007$

Eq. 9.50: $C_{Di} = \dfrac{0.642^2}{\pi(1.80)}(1 + 0.007) = 0.0735$

From Eq. 9.43: $0.0789 = C_{DO} + 0.0735$; $C_{DO} = 0.005\,38$ ◀

Eq. 9.49: $\alpha_i = [0.642/\pi(1.80)](1 + 0.072) = 0.1218$ radian or $\alpha_i = 0.1218(180/\pi)° = 6.98°$ ◀

Fig. P9.47: $\alpha = 96° + 7° - 90° = 13°$; Fig. 9.22: $\alpha_0 = \alpha - \alpha_i = 13 - 6.98 = 6.02°$

Since the angle of zero lift is undoubtedly zero, per Sec. 9.13: $\alpha_0' = \alpha_0 + 0 = 6.02° = 0.1051$ radian

Eq. 9.48: $0.642 = 2\pi\eta(0.1051)$; $\therefore\ \eta = 0.973$ ◀

9.48 *A rectangular airfoil with a 2.5-m chord and 15-m span has a drag coefficient of 0.062 and a lift coefficient of 0.95 at an angle of attack of 6.7°. What would be the corresponding (a) lift coefficient, (b) drag coefficient, and (c) angle of attack for a wing having the same profile but with an aspect ratio of 7.6?*

SI

(*a*) $C_L = 0.95$ ◀ irrespective of aspect ratio B/c (Sec. 9.13).

(*b*) Fig. 9.25 for $B/c = 15/2.5 = 6$: $\tau = 0.175$, $\sigma = 0.0505$

Fig. 9.25 for $B/c = 7.6$, $\tau = 0.204$, $\sigma = 0.066$

Eq. 9.47 with $B^2/A = B/c = 6$: $0.062 = C_{D0} + [0.95^2/(\pi 6)](1 + 0.0505)$; $C_{D0} = 0.011\,70$

Eq. 9.47 with $B^2/A = B/c = 7.6$: $C_D = 0.011\,76 + [0.95^2/(\pi 7.6)](1 + 0.066) = 0.0520$ ◀

(*c*) Eq. 9.49 with $B/c = 6$: $\alpha_i = [0.95/(\pi 6)](1 + 0.175) = 0.0592$ radian $= 0.0592(180/\pi)° = 3.39°$

Thus (Fig. 9.22): $\alpha_0 = \alpha - \alpha_i = 6.7 - 3.39 = 3.31°$

and for $B/c = 7.6$: $\alpha = \alpha_0 + \alpha_i = 3.31° + [0.95/(\pi 7.6)](1 + 0.204) = 3.31° + 0.0479$ radian

$= 6.05°$ ◀

9.49 *Determine the angle of glide that will allow the sailplane in Exer. 9.13.2 to remain in the air for the longest time? (Note: A trial and error solution will be required here.)*

Exer. 9.13.2: The sailplane has a Clark Y section wing with c = 3.5 ft, B = 21 ft, and the same characteristics as the larger wing of the same aspect ratio in Fig. 9.24. Its total weight is 300 lb, as it glides through standard air at 2000 ft altitude.

BG

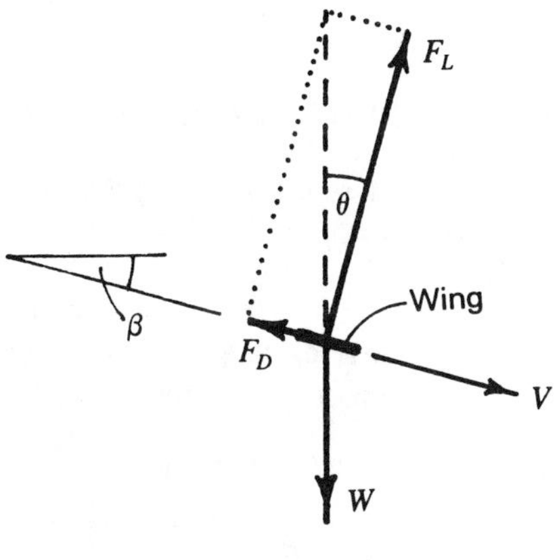

Table A.3 by linear interpolation for standard air at 2000 ft:

$\rho = 0.002\,25$ slug/ft^3

For the maximum time in the air, we need a minimum velocity of descent, V_z. This will occur when the largest upward force on the wing, the resultant of F_L and F_D, is vertical (see figure), i.e., when the glide angle

$$\beta = \theta = \tan^{-1}(F_D/F_L) = \tan^{-1}(C_D/C_L)$$

Then $W\cos\beta = F_L$, and using Eq. 9.44

$$300\cos\beta = C_L(0.002\,25)(V^2/2)3.5(21)$$

from which $V = \sqrt{3628\cos\beta/C_L}$ and $V_z = V\sin\beta$

Trying various values of α given in Fig. 9.24:

$\alpha°$	C_L	C_D	$\beta°$	V (fps)	V_z (fps)	
2.6	0.60	0.0309	2.95	77.7	4.00	
3.9	0.70	0.0375	3.07	71.9	3.85	min
5.4	0.80	0.0473	3.38	67.3	3.97	

The minimum V_z occurs at a glide angle $\beta = 3.07°$ ◀

(Compare with Exer. 9.13.2 for maximum range, where $\alpha = -0.1°$, $C_L = 0.4$, $\beta = 2.58°$, $V = 95.2$ fps, $V_z = 4.28$ fps.)

Sec. 9.14: Effects of Compressibility on Drag and Lift -- *Exercises*

9.14.1 *The blunt-nosed projectile (cylinder) of Fig. 9.27 has a diameter of 16 in and weighs 500 lb. Find its rate of deceleration when moving through standard sea level atmosphere (a) horizontally at 1200 mph; (b) upward at an angle of 40° with the horizontal at a velocity of 1200 mph.*

BG

1200 mph = 1200(5280/3600) fps = 1760 fps

Table A.3 for standard air at sea level: $\rho = 0.002\,38$ slug/ft^3, $c = 1116$ fps

$V/c = 1760/1117 = 1.577$, for which (Fig. 9.27, cylinder) $C_D = 1.36$

Eq. 9.32: $F_D = C_D\rho A(V^2/2) = 1.36(0.002\,38)[\pi(8/12)^2](1760^2/2) = 7000$ lb

(*a*) $F_D = 7000 = -(W/g)a = -(500/32.2)a; \quad a = -451$ ft/sec^2 ◄

(*b*) $F_D + W\sin\theta = (W/g)a; \quad 7000 + 500\sin 40° = -(500/32.2)a; \quad a = -471$ ft/sec^2 ◄

9.14.2 *The blunt-nosed projectile (cylinder) of Fig. 9.27 has a diameter of 0.45 m and weighs 2500 N. Find its rate of deceleration when moving through standard sea level atmosphere (a) horizontally at 450 m/s; (b) upward at an angle of 45° with the horizontal at a velocity of 450 m/s.*

SI

Table A.3 for standard air at sea level: $\rho = 1.225$ kg/m^3, $c = 340.3$ m/s

$V/c = 450/340.3 = 1.322$, for which (Fig. 9.27, cylinder) $C_D = 1.24$

Eq. 9.32: $F_D = 1.24(1.225)[\pi(0.45/2)^2]450^2/2 = 24\,500$ N

(*a*) $F_D = 24\,500 = -(W/g)a = -(2500/9.81)a; \quad a = -96.0$ m/s^2 ◄

(*b*) $F_D + W\sin\theta = (W/g)a; \quad 24\,500 + 2500\sin 45° = -(2500/9.81)a; \quad a = -102.9$ m/s^2 ◄

Sec. 9.14: Effects of Compressibility on Drag and Lift -- *Problems*

9.50 *A supersonic jet plane is flying horizontally at 1500 mph. How soon after it passes overhead at an elevation of 5000 ft will the shock wave will be felt at sea level?*

BG

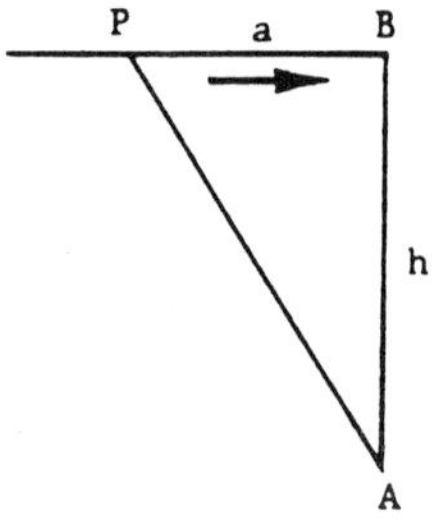

Let V = speed of aircraft, c = speed of shock wave. The first shock that arrives at A will have emanated from some point P. Let t = time of travel of the airplane from P to B. Thus $a = Vt$.

Time after plane passes B until shock is felt at A is

$$T = \frac{PA}{c} - t = \frac{\sqrt{h^2 + a^2}}{c} - t = \frac{(h^2 + V^2t^2)^{0.5}}{c} - t$$

T is minimum when $\dfrac{\partial T}{\partial t} = 0$, i.e., $\dfrac{\partial T}{\partial t} = \dfrac{0.5(h^2 + V^2t^2)^{-0.5}}{c}2V^2t - 1 = 0$

$$\therefore\ V^2t = c\sqrt{h^2 + V^2t^2}; \quad V^4t^2 = c^2h^2 + c^2V^2t^2; \quad c^2h^2 = t^2V^2(V^2 - c^2); \quad t = ch/\left(V\sqrt{V^2 - c^2}\right)$$

$$\therefore\ T = \frac{1}{c}\sqrt{h^2 + V^2\frac{c^2h^2}{V^2(V^2 - c^2)}} - \frac{ch}{V\sqrt{V^2 - c^2}} = h\left(\frac{V}{c} - \frac{c}{V}\right)/\sqrt{V^2 - c^2} \qquad (1)$$

$V = 1500$ mph $= 1500(5280/3600)$ fps $= 2200$ fps

Table A.3, at 5000 ft: $c = 1097$ fps; at sea level: $c = 1116$ fps. So average $c = 1107$ fps

Eq. (1): $T = 5000\left(\dfrac{2200}{1107} - \dfrac{1107}{2200}\right)/\sqrt{2200^2 - 1107^2} = 3.91$ sec ◄

9.51 *A supersonic jet plane is flying horizontally at 2800 km/h. How soon after it passes overhead at an elevation of 2 km will the shock wave will be felt at sea level?*

SI

See Solution to Prob. 9.50 for developement of equation (1) for T.

$V = 2800$ kph $= 2800(1000/3600) = 778$ m/s

Table A.3 for air at 2 km altitude: $c = 333$ m/s; at sea level: $c = 340$ m/s. So average $c = 336$ m/s

Eq. (1): $T = 2000\left(\frac{778}{336} - \frac{336}{778}\right)/\sqrt{778^2 - 336^2} = 5.37$ s ◄

9.52 *If the round-nosed and sharp-nosed projectiles of Fig. 9.27 each represent a 850 lb bomb with a diameter of 18 in, and assuming they travel vertically downward nose first, find their terminal velocities in standard air at sea level.*

BG

At terminal velocity $W = F_D$. $\therefore$ using Eq. 9.32: 850 lb $= C_D \rho A(V^2/2)$; $A = \pi(9/12)^2 = 1.767$ ft^2

Table A.3 for standard air at sea level: $\rho = 0.002\,38$ slug/ft^3, $c = 1116$ fps

$\therefore$ $850 = C_D(0.002\,38)(1.767)(V^2/2)$; $C_D V^2 = 404{,}000$; $V = \sqrt{404{,}000/C_D}$ (1)

(*a*) For round nose: Solve by trial

Try C_D	V [Eq. (1)]	V/c	C_D (Fig. 9.27)
0.50	899 fps	0.806	0.24
0.24	1298	1.163	0.56
0.37	1045	0.937	0.29
0.29	1181	1.058	0.40
0.33	1107	0.992	0.35
0.34	1090	0.977	0.33

Thus $V_{\text{terminal}} \approx 1090$ fps (round nose) ◄

(*b*) For sharp nose: Solve by trial

0.40	1005 fps	0.901	0.23
0.23	1326	1.188	0.35
0.30	1161	1.040	0.29
0.29	1181	1.058	0.29

Thus $V_{\text{terminal}} \approx 1181$ fps (sharp nose) ◄

9.53 *If the round-nosed and sharp-nosed projectiles of Fig. 9.27 each represent a 3500-N bomb with a diameter of 45 cm, and assuming they travel vertically downward nose first, find their terminal velocities in standard air at sea level.*

SI

At terminal velocity $W = F_D$. ∴ using Eq. 9.32: $3500\ \text{N} = C_D \rho A(V^2/2)$; $A = \pi(0.225)^2 = 0.1590\ \text{m}^2$

Table A.3 for standard air at sea level: $\rho = 1.225\ \text{kg/m}^3$, $c = 340\ \text{m/s}$

$\therefore\ 3500 = C_D(1.225)(0.1590)(V^2/2)$; $C_D V^2 = 35\,900$; $V = \sqrt{35\,900/C_D}$ (1)

(*a*) For round nose: Solve by trial

Try C_D	V [Eq. (1)]	V/c	C_D (Fig. 9.27)
0.50	268 m/s	0.788	0.23
0.23	395	1.162	0.54
0.37	312	0.917	0.28
0.28	358	1.054	0.41
0.33	330	0.970	0.33

Thus $V_{\text{terminal}} \approx 330$ m/s (round nose) ◀

(*b*) For sharp nose: Solve by trial

0.40	300 m/s	0.881	0.21
0.21	414	1.217	0.37
0.30	346	1.018	0.28
0.28	358	1.054	0.29
0.29	352	1.035	0.29

Thus $V_{\text{terminal}} \approx 352$ m/s (sharp nose) ◀

Chapter 10

Steady Flow in Open Channels

PROBLEM SELECTION GUIDE

Sec.	Exer/Prob	Units	Difficulty	Length	Parts	Similar	Special features
10.1	*Open Channels*						
	P 10.1	BG	Medium	Medium	1		Uses Secs 8.12 & 8.16
10.3	*Solution of Uniform Flow Problems*						
	X[1]10.3.1	BG	Easy	Medium	1		Plot; Manning
	10.3.2	BG	Easy	Short	1	10.3.3	Manning
	10.3.3	SI	Easy	Short	1	10.3.2	Manning
	10.3.4	SI	Easy	Medium	1		☐ T & E (Trial & Error), Manning
	10.3.5	BG	Easy	Short	1		Manning
	10.3.6	SI	Easy	Short	1		Manning
	10.3.7	BG	V Easy	V Short	1	10.3.8	Manning
	10.3.8	SI	V Easy	V Short	1	10.3.7	Manning
	10.3.9	BG	Easy	Short	1	10.3.10	Composite section, Manning
	10.3.10	SI	Easy	Short	1	10.3.9	Composite section, Manning
	P 10.2	BG	Medium	Medium	1		Uses Sec 8.12, T & E
	10.3	BG	Medium	Medium	2	10.5	☐ T & E for (*b*), Manning
	10.4	BG	Medium	Medium	1	10.6	Uses Sec 8.12, Manning
	10.5	SI	Medium	Medium	2	10.3	☐ T & E for (*b*), Manning
	10.6	SI	Medium	Medium	1	10.4	Uses Sec 8.12, Manning
	10.7	BG	Medium	Medium	2	10.8	Uses Sec 8.12, Manning
	10.8	SI	Medium	Medium	2	10.7	Uses Sec 8.12, Manning
	10.9	BG	Medium	Medium	1		Manning
	10.10	BG	Medium	Medium	1		Composite section, Manning
10.4	*Velocity Distribution in Open Channels*						
	X 10.4.1	BG	Medium	Medium	1		Uses (next) Sec. 10.5
	10.4.2	SI	Easy	Medium	1		Plot
	P 10.11	BG	Medium	Medium	1		Uses (next) Sec. 10.5
10.5	*"Wide and Shallow" Open Channels*						
	X 10.5.1	BG	V Easy	V Short	1	10.5.2	
	10.5.2	SI	V Easy	V Short	1	10.5.1	
	10.5.3	BG	Easy	Short	1		

/cont...

[1] For all Exercises (marked "X"), answers are given in Appendix D of the textbook.
* Note a correction made to early printings of the textbook.
† Answers are sensitive to values (y/D, Q/Q_{full}, V/V_{full}, y/E, A, P) which are or may be read from a figure.
X = Exercise, P = (end-of-chapter) Problem, S = Sample Problem. ☐ = could use computing aids.

Sec.	Exer/Prob	Units	Difficulty	Length	Parts	Similar	Special features
10.6	***Most Efficient Cross Section***						
	X 10.6.1	BG	Easy	Short	1		
	10.6.2	BG	Easy	Medium	3	10.6.3	
	10.6.3	SI	Easy	Medium	3	10.6.2	
	P 10.12	N	Medium	Medium	1		
	10.13	BG	Easy	Medium	1		Plot
	10.14	N	Hard	Medium	1		Differentiation
	10.15	N	Hard	Medium	1		Differentiation
	10.16	BG	Medium	Medium	2	10.17	* ☐ T & E
	10.17	SI	Medium	Medium	2	10.16	* ☐ T & E
	10.18	BG	Medium	Medium	1	10.19	*
	10.19	SI	Medium	Medium	1	10.18	*
10.7	***Circular Sections Not Flowing Full***						
	X 10.7.1	BG	Easy	Short	1		†
	10.7.2	SI	Easy	Short	2		†
	P 10.20	N	Hard	Long	1		☐ Differentiation, T & E (Alt: Plot)
10.8	***Laminar Flow in Open Channels***						
	X 10.8.1	BG	Easy	Short	2	10.8.2	Uses Sec 2.11
	10.8.2	SI	Easy	Short	2	10.8.1	Uses Sec 2.11
	10.8.3	BG	Medium	Medium	1	10.8.4	
	10.8.4	SI	Medium	Medium	1	10.8.3	
	P 10.21	N	Medium	Medium	1		Uses Secs 8.6 & 8.16
	10.22	N	Medium	Medium	1		Integration; uses Sec 5.1
10.9	***Specific Energy and Alternate Depths of Flow in Rectangular Channels***						
	X 10.9.1	BG	Easy	Short	1	10.9.2	Depth = critical
	10.9.2	SI	Easy	Short	1	10.9.1	Depth = critical
	10.9.3	SI	Medium	Short	1		Wide and shallow
	P 10.23	N	Medium	Short	1		Wide and shallow
	10.24	N	Medium	Short	1		Differentiation
	10.25	BG	Medium	Medium	1		
	10.26	BG	Easy	Short	1		
	10.27	BG	Medium	Medium	2		†
	10.28	BG	Medium	Medium	3	10.29	†
	10.29	SI	Medium	Medium	3	10.28	☐ T & E
10.10	***Subcritical and Supercritical Flow***						
	X 10.10.1	BG	Easy	Short	1		Critical flow, wide and shallow

/cont...

Sec.	Exer/Prob	Units	Difficulty	Length	Parts	Similar	Special features
10.11	***Critical Depth in Nonrectangular Channels***						
	X 10.11.1	BG	Easy	Short	1	10.11.2	
	10.11.2	SI	Easy	Short	1	10.11.1	
	10.11.3	BG	Easy	Medium	2		
	10.11.4	SI	Medium	Medium	1		□ T & E
	P 10.30	BG	Medium	Medium	2	10.31	□ T & E
	10.31	SI	Medium	Medium	2	10.30	□ T & E
	10.32	BG	Medium	Long	1		□ Plot?
	10.33	BG	Medium	Long	2	10.34	† □ Plot?
	10.34	SI	Medium	Long	2	10.33	† □ Plot?
10.13	***Humps and Contractions***						
	X 10.13.1	SI	Easy	Short	1		
	10.13.2	BG	Easy	Medium	1	10.13.3	□ T & E
	10.13.3	SI	Easy	Medium	1	10.13.2	□ T & E
	10.13.4	BG	Medium	Medium	1	P10.40	* □ T & E
	P 10.35	BG	Hard	Long	4	S10.8	□ Damming action; T & E
	10.36	SI	Medium	Medium	3		Damming action
	10.37	BG	Hard	Long	3	S10.8	□ T & E
	10.38	BG	Medium	Medium	1	10.39	□ T & E
	10.39	SI	Medium	Medium	1	10.38	□ T & E
	10.40	BG	Medium	Medium	1	X10.13.4	□ T & E
	10.41	BG	Medium	Medium	1	10.42	□ T & E
	10.42	SI	Medium	Medium	1	10.41	□ T & E
10.15	***Energy Equation for Gradually Varied Flow***						
	X 10.15.1	BG	Medium	Medium	1	10.15.2	Assume/confirm dir'n of depth decrease
	10.15.2	SI	Medium	Medium	1	10.15.1	Assume/confirm dir'n of depth decrease
	10.15.3	BG	Medium	Medium	1	10.15.4	
	10.15.4	SI	Medium	Medium	1	10.15.3	
	P 10.43	BG	Medium	Medium	1	10.44	Sketch
	10.44	SI	Medium	Medium	1	10.43	Sketch
	10.45	BG	Medium	Long	1	10.46	□ T & E
	10.46	SI	Medium	Long	1	10.45	□ T & E
	10.47	BG	Medium	Long	1	10.48	□ T & E; uses (next) Sec. 10.16
	10.48	SI	Medium	Long	1	10.47	□ T & E; uses (next) Sec. 10.16
	10.49	BG	Hard	Long	2	10.50	□
	10.50	SI	Hard	Long	2	10.49	□
	10.51	BG	Hard	V Long	2		□ T & E; wide and shallow
	10.52	BG	Hard	Long	2		□ Chézy, T & E
	10.53	BG	V Hard	Long	3		† GVF in circular section; sketch
	10.54	BG	Medium	Long	1		† Ass./confirm dir'n of depth decrease
	10.55	BG	Medium	Medium	1		Uses Chézy *C*

/cont...

Sec.	Exer/Prob	Units	Difficulty	Length	Parts	Similar	Special features
10.17	*Examples of Water Surface Profiles*						
	X 10.17.1	N	Easy	Short	1		
	10.17.2	BG	Medium	Medium	1	10.17.3	☐ T & E
	10.17.3	SI	Medium	Medium	1	10.17.2	☐ T & E
	10.17.4	BG	Easy	Short	1	10.17.5	
	10.17.5	SI	Easy	Short	1	10.17.4	
	10.17.6	BG	Easy	Medium	1		Sketch
	10.17.7	BG	Medium	Medium	1	10.17.8	☐ T & E; sketch
	10.17.8	SI	Medium	Medium	1	10.17.7	☐ T & E; sketch
	10.17.9	BG	Medium	Medium	1		☐ T & E; damming action
	P 10.56	BG	Medium	Medium	1	10.57	☐ T & E; sketch
	10.57	SI	Medium	Medium	1	10.56	☐ T & E; sketch
	10.58	BG	Medium	Long	1		☐
	10.59	BG	Hard	Long	2		☐ Damming action, T & E; sketch
	10.60	BG	Medium	Medium	1		
	10.61	BG	Medium	Medium	1	10.62	* ☐ Break in slopes, T & E; sketch
	10.62	BG	Medium	Medium	1	10.61	* ☐ Break in slopes, T & E; sketch
	10.63	BG	Hard	Long	1		☐ Break in slopes, T & E; sketch
	10.64	BG	Hard	Long	3		☐ Damming action, T & E
10.18	*Hydraulic Jump*						
	X 10.18.1	BG	Medium	Medium	2	10.18.2	
	10.18.2	SI	Medium	Medium	2	10.18.1	
	10.18.3	BG	Medium	Medium	1		
	P 10.65	N	Medium	Medium	1		Derivation
	10.66	BG	Medium	Medium	2	10.67	
	10.67	SI	Medium	Medium	2	10.66	
	10.68	BG	Hard	Medium	2	10.69	☐ Jump on slope, T & E
	10.69	SI	Hard	Medium	2	10.68	☐ Jump on slope, T & E
	10.70	BG	Hard	Medium	1		Moving jump
	10.71	BG	Hard	Medium	1	10.72	☐ T & E
	10.72	SI	Hard	Medium	1	10.71	☐ T & E
10.19	*Location of Hydraulic Jump*						
	P 10.73	BG	Hard	Long	1		Wide and shallow
10.20	*Velocity of Gravity Wave*						
	X 10.20.1	BG	Easy	V Short	1		
	P 10.74	BG	Easy	Medium	1	10.75	
	10.75	SI	Easy	Medium	1	10.74	
10.21	*Flow Around Channel Bends*						
	X 10.21.1	SI	Easy	Short	1		Subcritical
	P 10.76	BG	Medium	Medium	2		Sub and supercritical

/cont...

Sec.	Exer/Prob	Units	Difficulty	Length	Parts	Similar	Special features
10.22	*Transitions*						
	X 10.22.1	BG	Medium	Medium	1		☐ T & E
	10.22.2	SI	Medium	Long	1		☐ Assume/confirm flow condit's; T&E
	P 10.77	BG	Medium	Medium	2		☐ Assume/confirm flow condit's; T&E
	10.78	SI	Medium	Long	3		☐ T & E
	10.79	BG	Medium	Long	2		☐ Assume/confirm flow condit's; T&E
10.23	*Hydraulics of Culverts*						
	X 10.23.1	SI	Medium	Long	1	S10.13	☐ Assume/confirm flow condit's; T&E
	P 10.80	BG	Hard	Long	3		☐ Assume/confirm flow condit's; T&E
	10.81	BG	Hard	Long	2	10.80*a*	☐ Assume/confirm flow condit's; T&E
	10.82	SI	Hard	Long	2		☐ Assume/confirm flow condit's; T&E
	10.83	BG	Hard	Long	1		

Chapter 10
STEADY FLOW IN OPEN CHANNELS

Sec. 10.1: Open Channels -- *Problem*

10.1 *For the channel of Sample Prob. 10.1 compute the "open-channel Reynolds number" assuming that water at 50°F is flowing. Refer to Fig. 8.11 to verify whether or not the flow is fully rough. Determine e from Fig. 8.11 and compare it with the value computed in the example.*

BG

Per Sample Prob. 10.1: R_h = 2.27 ft, A = 57.1 ft², Q = 225 cfs, f = 0.0226, e = 0.015 85

∴ $V = Q/A = 225/57.1 = 3.94$ ft/s; Table A.1 at 50°F: $\nu = 1.410 \times 10^{-5}$ ft²/s

Open channel Sec. 10.1: $\mathbf{R} = \dfrac{R_h V \rho}{\mu} = \dfrac{R_h V}{\nu} = \dfrac{2.27 \times 3.94}{1.410 \times 10^{-5}} = 6.34 \times 10^5$ ◄

Equivalent pipe Eq. 8.54: $\mathbf{R} = 4(6.34 \times 10^5) = 2.53 \times 10^6$

Fig. 8.11 for $\mathbf{R} = 2.53 \times 10^6$ and $f = 0.0226$: $e/D = 0.0018$ and the flow is fully rough ◄

$e = 0.0018D = 0.0018(4R_h) = 0.0018(4)2.27 = 0.016\ 27$ ft ◄

This e from Fig. 8.11 is very close to 0.015 85 ft as computed in Sample Prob. 10.1. ◄

Sec. 10.3: Solution of Uniform Flow Problems -- *Exercises*

10.3.1 *For the channel of Sample Prob. 10.1, compute the flow rate for depths of 2, 4, 6, and 8 ft. Plot a curve of Q versus y.*

BG

Manning Eq. 10.8*a* with $n = 0.016$ and $S = 0.0006$: $Q = (1.486/0.016)AR^{2/3}(0.0006)^{1/2} = 2.28AR^{2/3}$

y ft	A ft²	P ft	R_h ft	$R_h^{2/3}$	Q cfs
2	28	18.94	1.478	1.300	82.7
4	72	27.9	2.58	1.881	308
6	132	36.8	3.58	2.34	703
8	208	45.8	4.54	2.74	1298

▲

10.3.2 *Water flows uniformly in a trapezoidal channel with a 15-ft-wide bed and 1.8:1 (H:V) side slopes. If the bed slope is 1.25 ft/mile and n = 0.016, find the flow rate when the depth is 8 ft.*

BG

$A = [15 + 1.8(8)]8 = 235$ ft²; $P = 15 + (2)8\sqrt{1 + 1.8^2} = 47.9$ ft; $R_h = A/P = 4.91$ ft.

Eq. 10.8*a*: $Q = (1.486/0.016)235(4.91)^{2/3}(1.25/5280)^{1/2} = 970$ cfs ◄

10.3.3 *Water flows uniformly in a 3-m-wide rectangular channel at a depth of 40 cm. The channel slope is 0.003 and n = 0.014. Find the flow rate in m³/s.*

SI

$A = 3(0.40) = 1.2$ m², $P = 2(0.40) + 3 = 3.8$ m, $R_h = A/P = 0.316$ m

Eq. 10.8*b*: $Q = (1/0.014)1.2(0.316)^{2/3}(0.003)^{1/2} = 2.18$ m³/s ◄

10.3.4 *At what depth will water flow in a 4-m-wide rectangular channel if n = 0.018, S = 0.0009, and Q = 7 m^3/s?*

SI

Depth = y, $A = 4y$, $P = 4 + 2y$, $R_h = A/P$

Eq. 10.8*b*: $7 = (1/0.018)4y[4y/(4 + 2y)]^{2/3}(0.0009)^{1/2}$

Solve by trial or by equation solver per Sample Prob. 10.1 to obtain $y = 1.251$ m ◄

10.3.5 *Figure X10.3.5 shows a tunnel section on the Colorado River Aqueduct. The area of the water cross section is 191 ft^2, and the wetted perimeter is 39.1 ft. The flow is 1600 cfs. If n = 0.013 for its concrete lining, find the slope.*

BG

Eq. 8.2: $R_h = A/P = 191/39.1 = 4.88$ ft

Eq. 10.8*a*: $1600 = (1.486/0.013)191(4.88)^{2/3}S^{1/2}$

$S = 0.000\,648$ or 3.42 ft/mile ◄

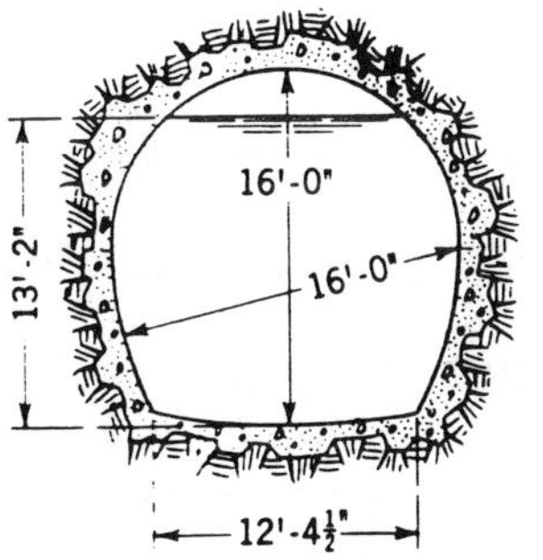

Figure X10.3.5

10.3.6 *If a = 1 m, b = 4 m, d = 2 m, w = 20 m, and n = 0.016, what slope is required so that the flow will be 25 m^3/s when the depth of flow is 1.50 m?*

SI

$A = 1.5(4) + 0.5(8)/2 = 8\ m^2$;

$P = 1.5 + 4.0 + 1.0 + \sqrt{0.5^2 + 8^2} = 14.52$ m;

$R_h = A/P = 0.551$ m

Eq. 10.8*a*: $25 = (1/0.016)8(0.551)^{2/3}S^{1/2}$; $S = 0.005\,53$ ◄

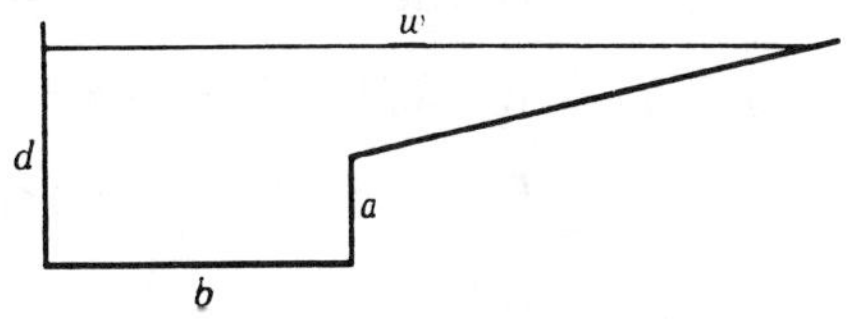

Figure X10.3.6

10.3.7 *Water flows in a rectangular flume 6 ft wide made of unplaned timber (n = 0.013). Find the necessary channel slope if the water flows uniformly at a depth of 3 ft and at 20 fps.*

BG

Eq. 10.7*b*: $20 = (1.486/0.013)(3 \times 6/12)^{2/3}S^{1/2}$; $S = 0.017\,83$ ◄

10.3.8 *Water flows in a rectangular flume 1.5 m wide made of unplaned timber (n = 0.013). Find the necessary channel slope if the water flows uniformly at a depth of 0.6 m and at 4.5 m/s.*

SI

Eq. 10.7*a*: $4.5 = (1/0.013)(0.6 \times 1.5/2.7)^{2/3}S^{1/2}$; $S = 0.014\,81$ ◄

10.3.9 *Refer to Fig. 10.5. Suppose the widths of A_1, A_2, and A_3 are 120, 50, and 240 ft and the total depths are 3, 12, and 4 ft. Compute the flow rate if S = 0.0018, $n_1 = n_3 = 0.06$, and $n_2 = 0.030$.*

BG

Eq. 10.11: $Q = Q_1 + Q_2 + Q_3 = \frac{1.486}{0.06}(120 \times 3)\left(\frac{120 \times 3}{120 + 3}\right)^{2/3}(0.0018)^{1/2}$

$$+ \frac{1.486}{0.030}(50 \times 12)\left(\frac{50 \times 12}{9 + 50 + 8}\right)^{2/3}(0.0018)^{1/2} + \frac{1.486}{0.06}(240 \times 4)\left(\frac{240 \times 4}{240 + 4}\right)^{2/3}(0.0018)^{1/2}$$

$Q = 774 + 5438 + 2514 = 8730$ cfs ◄

10.3.10 *Refer to Fig. 10.5. Suppose the widths of A_1, A_2, and A_3 are 30, 10, and 60 m and the total depths are 0.5, 3.0, and 1.0 m. Compute the flow rate if $S = 0.0016$, $n_1 = n_3 = 0.04$, and $n_2 = 0.025$.*

SI

Applying Eq. 10.8*b* to each section:

$$Q = Q_1 + Q_2 + Q_3 = \frac{30 \times 0.5}{0.04}\left(\frac{30 \times 0.5}{30 + 0.5}\right)^{2/3}(0.0016)^{1/2}$$

$$+ \frac{10 \times 3.0}{0.025}\left(\frac{10 \times 3.0}{2.5 + 10 + 2}\right)^{2/3}(0.0016)^{1/2} + \frac{60 \times 1.0}{0.04}\left(\frac{60 \times 1.0}{60 + 1.0}\right)^{2/3}(0.0016)^{1/2}$$

$Q = 9.35 + 77.9 + 59.3 = 146.6\ \text{m}^3/\text{s}$ ◄

Sec. 10.3: Solution of Uniform Flow Problems -- *Problems*

10.2 *Assuming the values of f versus* **R** *and e/D given for pipes in Fig. 8.11 also apply to open channels, find the rate of discharge of water at 60°F in a 100-in-diameter smooth concrete pipe flowing half full, if the pipe is laid on a grade of 2 ft/mi. Why must D be replaced by $4R_h$?*

BG

Flow is uniform. From Table 8.1 (concrete, smooth), choose $e = 0.001$ ft $= 0.012$ inches;

$e/D = 0.000\ 12$; From Fig. 8.11, try complete turbulent flow $f = 0.0125$

D must be replaced by $4R_h$ per Sec. 8.16 because it is open channel flow, not pipe flow. ◄

Eq. 8.39: $R_h = D/4 = (100''/12)/4 = 2.08$ ft; $S = 2/5280 = 0.000\ 379$

Eq. 10.5: $V = [(8g/f)R_hS]^{1/2} = [(8 \times 32.2/0.0125)2.08 \times 0.000\ 379]^{1/2} = 4.03$ fps

$D''V = 4R''V = 4(2.08)12(4.03) = 403$ and $f = 0.0130$. Try this.

Eq. 10.5: $V = [(8 \times 32.2/0.0130)2.08 \times 0.000\ 379]^{1/2} = 3.95$ fps

$D''V = 4(2.08)12(3.95) = 395$ and $f = 0.0130$ which checks.

$\therefore\ Q = AV = (1/2)(\pi/4)(100/12)^2(3.95) = 107.8$ cfs ◄

10.3 *Figure P10.3 shows a cross section of a canal that is to carry 2000 cfs. The canal is lined with concrete, for which n is 0.014. (a) What is the slope of this canal, and what is the drop in elevation per mile? (b) If the flow in the canal were to decrease to 1000 cfs, all other data, including the slope and n, being the same, what would be the depth of the water?*

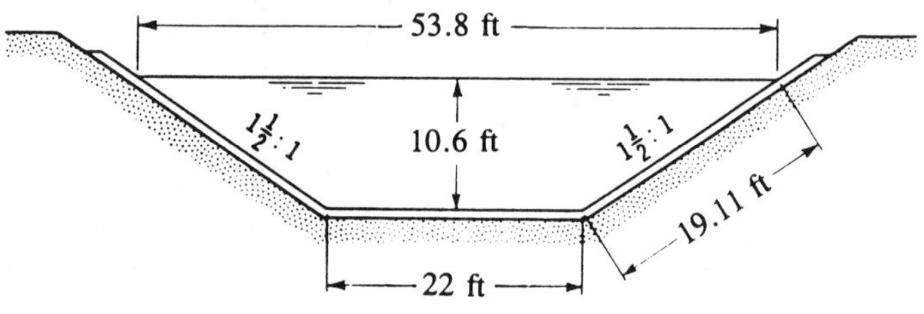

Figure P10.3

BG

(*a*) $A = [(53.8 + 22)/2]10.6 = 37.9(10.6) = 402\ \text{ft}^2$; $V = Q/A = 2000/402 = 4.98$ fps

$P = 22 + 2(19.11) = 60.2$ ft; $R_h = A/P = 402/60.2 = 6.67$ ft

Eq. 10.7*b*: $4.98 = (1.486/0.014)(6.68)^{2/3}S^{1/2}$; $S^{1/2} = 0.013\,23$

$S = 0.000\ 1752$ ◄ or 0.925 ft per mile ◄

Note: A very small drop per mile will produce a substantial velocity.

/cont...

(b) Eq. 10.8a: $1000 = (1.486/0.014)AR_h^{2/3}(0.000\ 1752)^{1/2} = 1.405AR_h^{2/3}$

i.e. $712 = AR_h^{2/3} = A(A^{2/3}/P^{2/3}) = A^{5/3}/P^{2/3}$

where $A = 22y + 1.5y^2$ and $P = 22 + 2y(19.11/10.6) = 22 + 3.61y$

$\therefore\ (22y + 1.5y^2)^{5/3} = 712(22 + 3.61y)^{2/3}$.

Solve by trial or by equation solver per Sample Prob. 10.1 to obtain $y = 7.36$ ft ◄

10.4 *In Prob. 10.3 find the corresponding value of e and compare it with values previously given for concrete pipe. Does it fall in the range given?*

BG

From Eqs. 10.9 and 10.10a, for $R_h = 6.67$ ft:

$1/f^{1/2} = 2\log[14.8(6.67/e)] = 0.0928(6.67)^{1/6}/0.014 = 9.12;\quad e = 0.002\ 80$ ft ◄

Table 8.1 gives $e = 0.001$ to 0.01 ft for concrete ◄

So the computed value of e does fall in the range given ◄

10.5 *Figure P10.5 shows a cross section of a canal forming a portion of the Colorado River Aqueduct, which is to carry 44 m³/s. The canal is lined with concrete, for which n is 0.014. (a) What is the slope of this canal, and what is the drop in elevation per kilometer? (b) If the flow in the canal were to decrease to 22 m³/s, all other data, including the slope and n, being the same, what would be the depth of the water?*

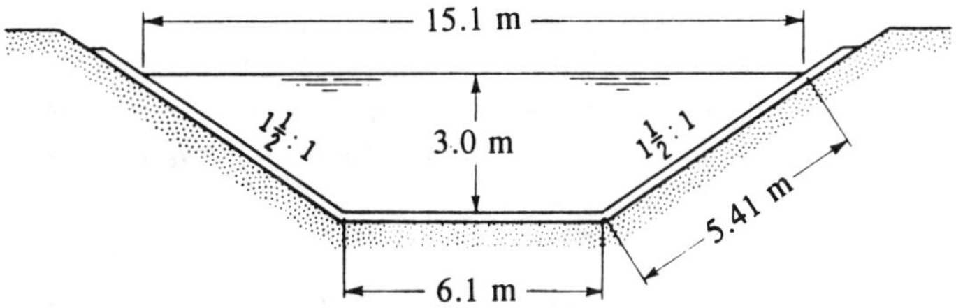

Figure P10.5

SI

(a) $A = [(15.1 + 6.1)/2]3 = 31.8\ \text{m}^2;\ V = Q/A = 44/31.8 = 1.384$ m/s

$P = 6.1 + 2(5.41) = 16.92$ m; $R_h = A/P = 31.8/16.92 = 1.879$ m

Eq. 10.7a: $1.384 = (1/0.014)(1.879)^{2/3}S^{1/2};\quad S^{1/2} = 0.012\ 72$

$S = 0.000\ 1618$ ◄ or 0.1618 m/km ◄

(b) Eq. 10.8b: $22 = (A/0.014)R_h^{2/3}(0.000\ 1618)^{1/2} = 0.909AR_h^{2/3}$

i.e. $24.2 = AR_h^{2/3} = A(A^{2/3}/P^{2/3}) = A^{5/3}/P^{2/3}$

where $A = 6.1y + 1.5y^2$ and $P = 6.1 + 2y(5.41/3) = 6.1 + 3.61y$

$\therefore\ (6.1y + 1.5y^2)^{5/3} = 24.2(6.1 + 3.61y)^{2/3}$.

Solve by trial or by equation solver per Sample Prob. 10.1 to obtain $y = 2.08$ m ◄

10.6 *In Prob. 10.5 find the corresponding value of e and compare it with values previously given for concrete pipe. Does it fall in the range given?*

SI

From Eqs. 10.9 and 10.10b for $R_h = 1.879$ m: $\dfrac{1}{f^{1/2}} = 2\log\left(14.8\dfrac{1.879}{e}\right) = \dfrac{0.113(1.879)^{1/6}}{0.014} = 8.97$

So $e = 0.000\ 914$ m $= 0.914$ mm ◄ Table 8.1 gives $e = 0.3$ to 3.0 mm for concrete ◄

So the computed value of e does fall in the range given ◄

10.7 *A monolithic concrete inverted siphon is circular in cross section and is 20 ft in diameter. Obviously, it is completely filled with water. (a) If n = 0.013, find the slope of the hydraulic grade line for a flow of 2100 cfs. (b) Solve the same problem using the methods of Chap. 8 and assuming a mean value of e from Table 8.1 for concrete pipe. Compare the result with that of part (a).*

BG

(*a*) $A = (\pi/4)20^2 = 314\ \text{ft}^2$; $R_h = D/4 = 20/4 = 5$ ft

Eq. 10.8*a*: $2100 = (1.486/0.013)(314)5^{2/3}S^{1/2}$

$S = 0.000\,400$ or 2.11 ft per mile ◀

(*b*) Table 8.1: Mean value of *e* for concrete = (0.01 + 0.001)/2 = 0.0055 ft

Then $e/D = 0.0055/20 = 0.000\,275$; $V = Q/A = 2100/314 = 6.68$ fps

$D''V = 20(12)6.68 = 1604$; Fig. 8.11 with $e/D = 0.000\,275$, $D''V = 1604$: $f = 0.0147$

Eq. 8.10: $h_L = 0.0147(L/20)6.68^2/2g$; $h_L/L = S = 0.000\,510$ ◀

This is within 28% of the value found in Part (*a*) ◀

10.8 *A monolithic concrete inverted siphon on the Colorado River Aqueduct is circular in cross section and is 4.88 m in diameter. Obviously, it is completely filled with water. (a) If n = 0.013, find the slope of the hydraulic grade line for a flow of 45 m³/s. (b) Solve the same problem using the methods of Chap. 8 and assuming a mean value of e from Table 8.1 for concrete pipe. Compare the result with that of part (a).*

SI

(*a*) $A = (\pi/4)4.88^2 = 18.70\ \text{m}^2$; $R_h = D/4 = 4.88/4 = 1.22$ m

Eq. 10.8*b*: $45 = (18.70/0.013)(1.22)^{2/3}S^{1/2}$; $S = 0.000\,750$ or 0.750 m/km ◀

(*b*) Table 8.1: Mean value of *e* for concrete = (0.3 + 3.0)/2 = 1.65 mm

So $e/D = 1.65/4880 = 0.000\,338$; $V = Q/A = 45/18.70 = 2.41$ m/s

Table A.1 at standard 15°C: $\nu = 1.139 \times 10^{-6}\ \text{m}^2/\text{s}$

$\mathbf{R} = DV/\nu = 4.88(2.41)/(1.139 \times 10^{-6}) = 1.031 \times 10^7$

Fig. 8.11 (or Eq. 8.40) with $e/D = 0.000\,338$ and $\mathbf{R} = 1.031 \times 10^7$: $f = 0.0154$

Eq. 8.13: $h_L = 0.0154(L/4.88)2.41^2/(2 \times 9.81)$; $h_L/L = S = 0.000\,931$ ◀

This is within 24% of the value found in Part (*a*). ◀

10.9 *A 36-in-diameter pipe is known to have a Manning's n of 0.024. What is Manning's n for a 108-in-diameter pipe having exactly the same e as the smaller pipe?*

BG

Equate $1/f^{1/2}$ from Eqs. 10.9 and 10.10*a*: $2\log(14.8R_h/e) = 0.0928R_h^{1/6}/n$ (1)

36 inch dia pipe with $n = 0.024$: $R_h = A/P = D/4 = (1/4)30/12 = 0.75$ ft

Substituting into Eq. (1): $2\log(14.8 \times 0.75/e) = 3.69$; $e = 0.1594$ ft

96 inch dia pipe with same *e*: $R_h = (1/4)108/12 = 2.25$ ft

Substituting into Eq. (1): $2\log(14.8 \times 2.25/0.1594) = 0.0928(2.25)^{1/6}/n$; $n = 0.0229$ ◀

10.10 *Refer to Fig. X10.3.6. Find the flow rate at water depths of 1, 2.5, 4, 5.5, and 7 ft if n = 0.025 and S = 0.0020. The dimensions are as follows: a = 4 ft, b = 5 ft, d = 7 ft, and w = 35 ft.*

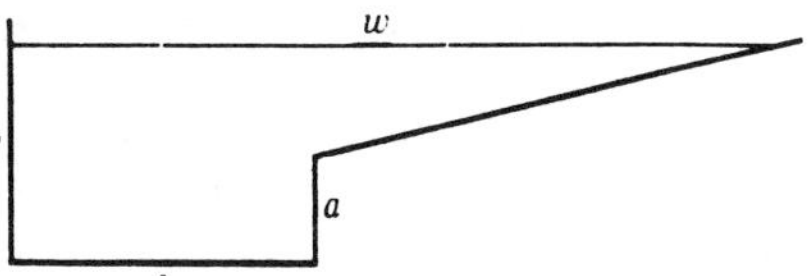

Figure X10.3.6

BG

Given $n = 0.025$, $S = 0.0020$.

Eq. 10.8*a*: $Q = (1.486/0.025)AR_h^{2/3}0.0020^{1/2} = 2.66AR_h^{2/3}$

y ft	A ft^2	P ft	R_h ft	$R_h^{2/3}$	Q cfs
1	5	7	0.714	0.799	10.62
2.5	12.5	10	1.250	1.160	38.56
4	20	13	1.538	1.333	70.85

▲

For depths greater than 4 ft the flow should be computed in parts as shown below, i.e., using Eq. 10.11.

$y = 5.5$ ft: $Q_1 = (1.486/0.025)27.5(27.5/14.5)^{2/3}(0.0020)^{1/2} = 112.0$ cfs

$Q_2 = (1.486/0.025)11.25(11.25/15.07)^{2/3}(0.0020)^{1/2} = 24.6$ cfs; $Q = Q_1 + Q_2 = 136.6$ cfs ◄

$y = 7$ ft: $Q_1 = (1.486/0.025)35(35/16)^{2/3}(0.0020)^{1/2} = 156.8$ cfs

$Q_2 = (1.486/0.025)45(45/30.1)^{2/3}(0.0020)^{1/2} = 156.2$ cfs; $Q = Q_1 + Q_2 = 313$ cfs ◄

Sec. 10.4: Velocity Distribution in Open Channels -- *Exercises*

10.4.1 *In Fig. X10.4.1 with uniform flow in the wide open channel, a = 3.20 ft. Find V and b if n = 0.015.*

BG

For a "wide" open channel, assume $R_h = y_0 = 9$ ft (Sec. 10.5).

Eq. 10.8*a*: $q = 9 \times V = (1.486/0.015)(9 \times 1)(9)^{2/3}S^{1/2}$

thus $S^{1/2} = 0.002\,33V$

Velocity head $u_a^2/2g = a = 3.2$ ft, $\therefore u_a = 14.36$ fps

Eq. 10.12: $u_a = V + \dfrac{1}{0.4}[32.2(9)]^{1/2}(0.002\,33V)\left(1 + 2.3\log_{10}\dfrac{6}{9}\right)$

$= 14.36$

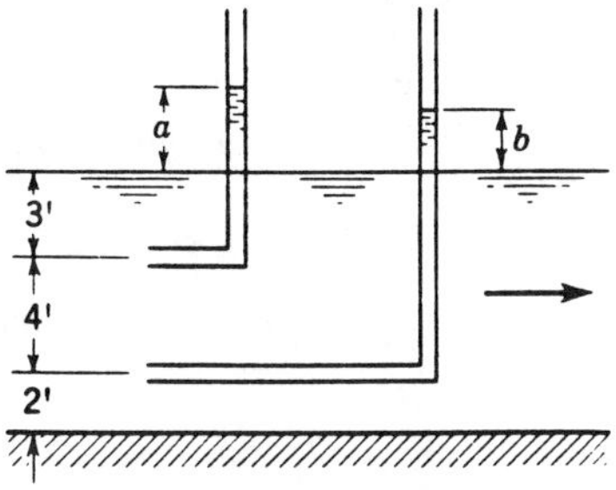

Figure X10.4.1

From which mean velocity, $V = 13.55$ fps ◄

Eq. 10.12: $u_b = 13.55 + (1/0.4)(32.2 \times 9)^{1/2}(0.002\,33 \times 13.55)(1 + 2.3\log_{10}2/9)$

$u_b = 12.99$ fps; $b = (12.88^2/2g) = 2.58$ ft ◄

10.4.2 *Water flows uniformly in a very wide rectangular channel ($R_h = y$) at a depth y_0 of 2.0 m. S = 0.005 and n = 0.018. Calculate the velocities at y-values of 0.2, 0.4, 0.8, 1.2, and 2.0 m above the bed and plot the velocity profile. Note the value of the maximum velocity at the water surface.*

BG

For a very wide channel, $R_h = y_0 = 2.0$ m (given)

Eq. 10.7*a*: $V = (1/0.018)2.0^{2/3}(0.005)^{1/2} = 6.24$ m/s

Eq. 10.12: $u = 6.24 + (1/0.40)[9.81(2.0)0.005]^{1/2}[1 + 2.3\log(y/2.0)]$

y m:	0.20	0.40	0.80	1.20	2.00
u m/s:	5.22	5.76	6.30	6.62	7.02 = u_{max} ◄◄

Sec. 10.4: Velocity Distribution in Open Channels -- ***Problem***

10.11 *Using Eq. (10.12), determine the depth below the surface at which the velocity is equal to the mean velocity. Also find the average of the velocities at 0.2 and 0.8 depths. Let y = 6 ft, S = 0.0005, and n = 0.024.*

BG

Let $y = y'$ when $u = V$. Then Eq. 10.12: $u = V + (1/K)(gy_0S)^{1/2}(1 + 2.3 \log y'/y_0)$

where $u = V$, so $V = V + (1/K)(gy_0S)^{1/2}(1 + 2.3 \log y'/y_0)$

$\therefore\ 1 + 2.3 \log y'/y_0 = 0;\ \log y'/y_0 = -(1/2.3) = 0.435$

$\log y_0/y' = 0.435,\ y_0/y' = 10^{0.435} = 2.72;\ y' = y_0/2.72 = 0.367y_0$

$\therefore$ The depth below the surface at which $u = V$ is equal to $y_0 - y'$ or $y_0 - 0.368y_0 = 0.633y_0$, where y_0 is the depth of water in the channel. ◀

Given to use Eq. 10.12, which is for "wide and shallow" flow, so assume $R_h = y_0$ (Sec. 10.5).

Eq. 10.7*b*: Mean velocity, $V = (1.486/0.024)6^{2/3}(0.0005)^{1/2} = 4.57$ fps

From Eq. 10.12 with $K = 0.4$: At 0.2 depth, $y/y_0 = 0.8$, $u = 5.18$ fps
At 0.8 depth, $y/y_0 = 0.2$, $u = 4.10$ fps

Average of these two u's $= (5.18 + 4.10)/2 = 4.64$ fps ◀

Therefore the average of the velocities at 0.2 depth and 0.8 depth is almost identical to V.

Sec. 10.5: "Wide and Shallow" Open Channels -- ***Exercises***

10.5.1 *Water flows 1/8 in deep in a 9-in-wide rectangular flume. What would be the percentage error in the hydraulic radius if this were assumed to be wide and shallow open channel flow?*

BG

$b/y = 9/(1/8) = 72$

Error percent = 200/72 = 2.78% ◀

10.5.2 *Water flows 6.9 mm deep in a 0.4-m-wide rectangular flume. What would be the percentage error in the hydraulic radius if this were assumed to be wide and shallow open channel flow?*

SI

$b/y = 400/6.9 = 58.0$

Error percent = 200/58.0 = 3.45% ◀

10.5.3 *At a location where the cross section of a stream is reasonably rectangular, the width is 13 ft and the average depth is computed to be 18.4 in. What would be the percentage error in the hydraulic radius if this stream were assumed to be "wide and shallow"? Would it be reasonable to make this assumption?*

BG

$b/y = 13(12)/18.4 = 8.48$

Error percent = 200/8.48 = 23.6% ◀

This error is large, and $b/y < 10$, so the assumption would not be reasonable. ◀

Sec. 10.6: Most Efficient Cross Section -- *Exercises*

10.6.1 *What diameter of semicircular channel will have the same capacity as a 4-ft-deep rectangular channel 10 ft wide. Assume S and n are the same for both channels. Compare the lengths of the wetted perimeters.*

BG

$Q_1 = Q_2$; $\therefore$ From Eq. 10.8*a* with $(1.486/n)S^{1/2}$ = constant, $A_1 R_1^{2/3} = A_2 R_2^{2/3}$

Substituting (from Eq. 8.2) for $R_h = A/P$: $A_1^{5/3}/P_1^{2/3} = A_2^{5/3}/P_2^{2/3}$

$\therefore (\pi D_1^2/8)^{5/3}/(\pi D_1/2)^{2/3} = (10 \times 4)^{5/3}/18^{2/3}$; $D_1 = 9.78$ ft ◀

$P_1 = \pi D/2 = 15.36$ ft, compared with $P_2 = 18$ ft ◀

10.6.2 *A rectangular flume of planed timber (n = 0.012) slopes 0.5 ft per 1000 ft. (a) Compute the rate of discharge if the width is 7 ft and the depth of water is 3.5 ft? (b) What would be the rate of discharge if the width were 3.5 ft and the depth of water 7 ft? (c) Which of the two forms would have the greater capacity and which would require less lumber?*

BG

(*a*) $b = 7$, $y = 3.5$, $\therefore A = 24.5$, $P = 14$, $R_h = 24.5/14 = 1.75$ ft; $n = 0.012$, $S = 0.0005$.

Eq. 10.7*b*: $V = (1.486/0.012)(1.75)^{2/3}(0.0005)^{1/2} = 4.02$ fps, $\therefore Q = 24.5(4.02) = 98.5$ cfs ◀

(*b*) $b = 3.5$, $y = 7$, $\therefore A = 24.5$, $P = 17.5$, $R_h = 1.4$ ft; $n = 0.012$, $S = 0.0005$.

$V = (1.486/0.012)(1.4)^{2/3}(0.0005)^{1/2} = 3.47$ fps, $\therefore Q = 24.5(3.47) = 84.9$ cfs ◀

(*c*) Lumber ratio = 14/17.5 = 0.80; flow ratio = 98.5/84.9 = 1.160
The first design provides 16% more flow capacity while requiring only 80% as much lumber. ◀

10.6.3 *A rectangular flume of planed timber (n = 0.012) slopes 1 m per km. (a) Compute the rate of discharge if the width is 2 m and the depth of water is 1 m? (b) What would be the rate of discharge if the width were 1 m and the depth of water 2 m? (c) Which of the two forms would have the greater capacity and which would require less lumber?*

SI

(*a*) $b = 2$, $y = 1$, $\therefore A = 2$, $P = 4$, $R_h = 2/4 = 0.5$ m; $n = 0.012$, $S = 0.001$.

Eq. 10.7*a*: $V = (1/0.012)(0.5)^{2/3}(0.001)^{1/2} = 1.660$ m/s $\therefore Q = 2(1.660) = 3.32\ m^3/s$ ◀

(*b*) $b = 1$, $y = 2$, $\therefore A = 2$, $P = 5$, $R_h = 2/5 = 0.4$ m; $n = 0.012$, $S = 0.001$.

Eq. 10.7*a*: $V = (1/0.012)(0.4)^{2/3}(0.001)^{1/2} = 1.431$ m/s $\therefore Q = 2(1.431) = 2.86\ m^3/s$ ◀

(*c*) Lumber ratio = 4/5 = 0.8; flow ratio = 3.32/2.86 = 1.160
The first design provides 16% more flow capacity while requiring only 80% as much lumber. ◀

Sec. 10.6: Most Efficient Cross Section -- *Problems*

10.12 *Prove that the most efficient hydraulic trapezoidal cross section discussed in Sec. 10.6 envelopes a semi-circle with its center in the water surface.*

N

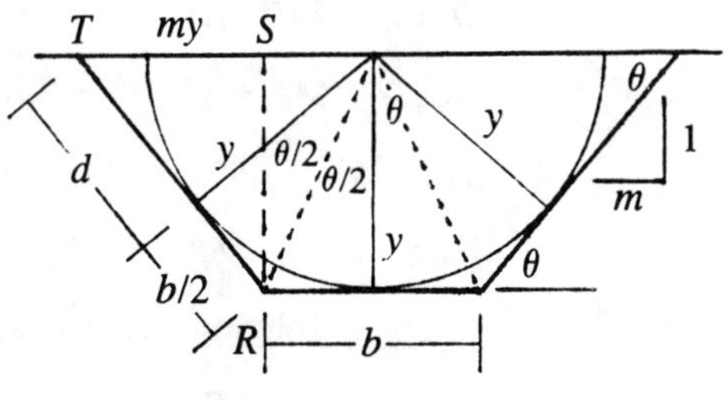

For such an enveloping trapezoid $\tan\theta = 1/m = y/d$, so $d = my$

ΔRST, Pythagoras: $(d + b/2)^2 = y^2 + m^2y^2 = y^2(1 + m^2)$

from which $b = 2y\left(\sqrt{1 + m^2} - m\right)$. Also

$P = 2(b + d) = 2\left[2y\left(\sqrt{1 + m^2} - m\right) + my\right] = 2y\left(2\sqrt{1 + m^2} - m\right)$

and $A = 2[(1/2)dy] + 4[(1/2)y(b/2)] = y(b + d)$, so $R_h = A/P = [y(b + d)]/[2(b + d)] = y/2$

These results for P, A, and R_h are the same as those obtained in Sec. 10.6 for the optimum trapezoid by differentiation. Therefore the hydraulically optimal trapezoid must envelope such a semicircle. Q.E.D.

10.13 *Consider a variety of rectangular sections all of which have a cross-sectional area of 20 ft². Plot the hydraulic radii versus channel widths for a range of channel widths from 2 to 20 ft and note the depth:width ratio when R_h is maximum.*

BG

$A = bh = 20\ \text{ft}^2$, $P = b + 2h$. Sample calculations are:

b ft	$h = A/b$ ft	P ft	h/b	$R_h = A/P$ ft
2	10.0	22	5.0	0.909
4	5.0	14	1.250	1.429
6	3.33	12.67	0.556	1.579
6.324	3.163	12.649	0.500	1.581 ◄
8	2.5	13	0.313	1.538
10	2	14	0.20	1.429
20	1	22	0.05	0.909

▲

10.14 *Set up a general expression for the wetted perimeter P of a trapezoidal channel in terms of the cross-sectional area A, depth y, and angle of side slope ϕ. Then differentiate P with respect to y with A and ϕ held constant. From this prove that $R_h = y/2$ for the section of greatest hydraulic efficiency (i.e., smallest P for a given A).*

N

Let b = bottom width, y = depth, ϕ = angle of side slope

$A = by + y(y\tan\phi) = by + y^2\tan\phi$; $b = (A/y) - y\tan\phi$

$P = b + 2y\sec\phi = (A/y) - y\tan\phi + 2y\sec\phi$. Differentiate with respect to y:

$dP/dy = -A/y^2 - \tan\phi - 2\sec\phi$. Substituting for A: $dP/dy = -(by + y^2\tan\phi)/y^2 - \tan\phi + 2\sec\phi$

Equating to 0: $(by + y^2\tan\phi)/y^2 = 2\sec\phi - \tan\phi$; $b = 2y\sec\phi - 2y\tan\phi = 2y(\sec\phi - \tan\phi)$ (1)

Substituting for b from (1) into $R_h = A/P = (by + y^2\tan\phi)/(b + 2y\sec\phi)$ gives $R_h = y/2$ Q.E.D. ◄

10.15 *Using the results of Prob. 10.14, prove that the most efficient triangular section is the one with a 90° vertex angle.*

N

Prob. 10.14 with $B = 0$: $A = y^2\tan\phi$, $P = 2y\sec\phi = 2y/\cos\phi$

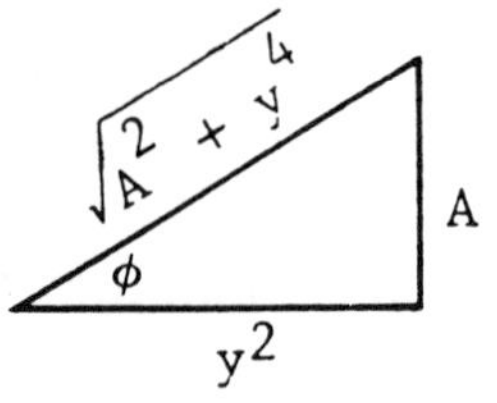

$\tan\phi = A/y^2$, $\therefore \cos\phi = \dfrac{y^2}{\sqrt{A^2 + y^4}}$; $P = \dfrac{2y\sqrt{A^2 + y^4}}{y^2} = 2(A^2 + y^4)^{1/2}y^{-1}$

For a given A, P is minimum when

$dP/dy = 2y^{-1}(1/2)(A^2 + y^4)^{-1/2}(4y^3) + 2(A^2 + y^4)^{1/2}(-y^{-2}) = 0$

i.e., $4y^2(A^2 + y^4)^{-1/2} - 2(A^2 + y^4)^{1/2}y^{-2} = 0$; i.e., $4y^4 - 2(A^2 + y^4) = 0$

i.e., $2y^4 = 2A^2$, or $y = A^{1/2}$ when P is min

Then $\tan\phi = A/y^2 = y^2/y^2 = 1$, $\phi = 45°$ = side slope, and vertex angle = 90° Q.E.D. ◄

<u>Alternate solution</u>: $A = a^2\sin\phi\cos\phi$; $P = 2a$; $R_h = (a/2)\sin\phi\cos\phi$

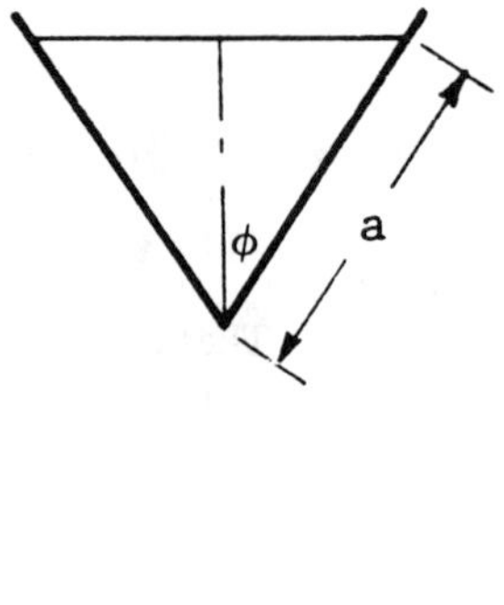

For a given A, R_h is minimum when

$dR_h/d\phi = (a/2)(\cos^2\phi - \sin^2\phi) = 0$; i.e., $\cos\phi = \sin\phi$

i.e., $\phi = 45°$, i.e., vertex angle $= 2\phi = 90°$ Q.E.D. ◄

10.16 *The amount of water to be carried by a canal excavated in smooth earth (n = 0.030) is 370 cfs. It has side slopes of 2:1 (see figure), and a bed slope of 2.5 ft/mi. (a) If the depth of water y is to be 5 ft, what must be the bottom width b? (b) How does this compare with the most efficient trapezoidal section for the given conditions? [*Note wording corrections]*

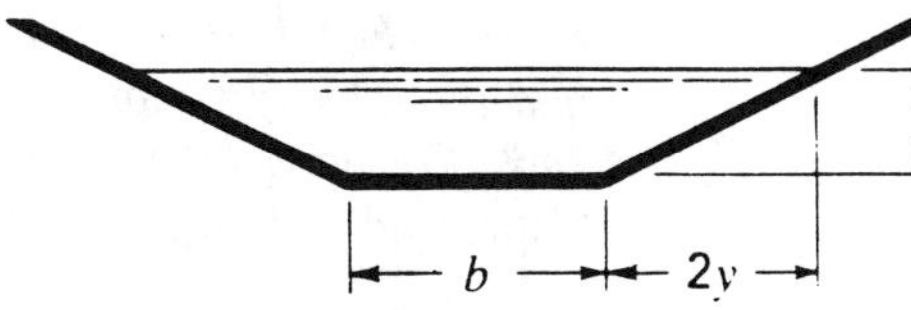

Figure P10.16

BG

(*a*) $A = 5(b + 10)$; $P = b + 2(5^2 + 10^2)^{1/2} = b + 22.36$

$R_h = A/P = 5(b + 10)/(b + 22.36)$; $S = 2.5/5280 = 0.000\ 473$

Eq. 10.8*a*: $370 = \dfrac{1.486}{0.03}\dfrac{[5(b + 10)]^{5/3}}{(b + 22.36)^{2/3}}\sqrt{0.000\,473}$

After several trials, or by equation solver per Sample Prob. 10.1*d*, we find $b = 19.63$ ft. ◄

(*b*) Eq. 10.8*a*: $370 = \dfrac{1.486}{0.030}AR_{h_{max}}^{2/3}\sqrt{0.000\,473}$ with (Sec. 10.6) $A = y_{opt}^2(2\sqrt{5}-2)$ and $R_{h_{max}} = \dfrac{y_{opt}}{2}$

from which $y_{opt} = 7.56$ ft ◄ $b_{opt} = 2(\sqrt{5}-2)7.56 = 3.57$ ft ◄

10.17 *The amount of water to be carried by a canal excavated in smooth earth (n = 0.030) is 12 m³/s. It has side slopes of 2:1 (see Fig. P10.16), and a bed slope of 60 cm/km. (a) If the depth of water y is to be 2.0 m, what must be the bottom width b? (b) How does this compare with the most efficient trapezoidal section for the given conditions? [*Note corrections to wording of textbook.]*

SI

(*a*) $A = 2.0(b + 4)$; $P = b + 2(2.0^2 + 4^2)^{1/2} = b + 8.944$

$R_h = A/P = 2.0(b + 4)/(b + 8.944)$; $S = 0.60/1000 = 0.000\ 60$

Eq. 10.8*b*: $12 = \dfrac{1}{0.03}\dfrac{[2.0(b + 4)]^{5/3}}{(b + 8.944)^{2/3}}\sqrt{0.000\,60}$

After several trials, or by equation solver per Sample Prob. 10.1*d*, we find $b = 2.69$ m ◄

(*b*) Eq. 10.8*b*: $12 = \dfrac{1}{0.030}AR_{h_{max}}^{2/3}\sqrt{0.000\,60}$ with (Sec. 10.6) $A = y_{opt}^2(2\sqrt{5}-2)$ and $R_{h_{max}} = \dfrac{y_{opt}}{2}$

from which $y_{opt} = 2.32$ m ◄ $b_{opt} = 2(\sqrt{5}-2)2.32 = 1.096$ m ◄

10.18 *Refer to Fig. P10.16. If the discharge in the canal (n = 0.030) is to be 200 cfs while the depth y is 5 ft and the velocity is not to exceed 150 ft/min, what must be the bottom width b and the drop in elevation per mile? Compare this with the bottom width for maximum efficiency *when using the same side slopes, bed slope, and discharge. [*Note this correction to the textbook.]*

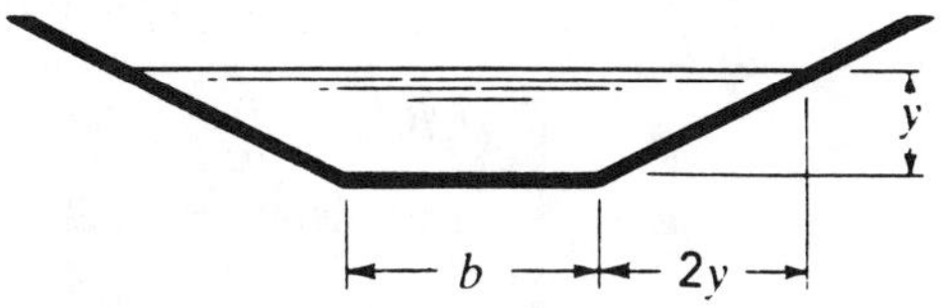

Figure P10.16

BG

$V_{max} = 150/60 = 2.5$ fps; $A_{min} = Q/V_{max} = 200/2.5 = 80$ ft²

$A_{min} = 5(b_{min} + 10) = 80$, $\therefore\ b_{min} = 6$ ft ◄

$P_{min} = 6 + 2(5^2 + 10^2)^{1/2} = 28.36$ ft; min $R_h = 80/28.36 = 2.82$ ft; min $R_h^{2/3} = 1.996$

$\therefore$ from Eq. 10.7*b*: max $S^{1/2} = 2.5(0.03)/[1.486(1.996)] = 0.0253$; $S_{max} = 0.000\,639$

$\therefore$ max drop $= 0.000\,639(5280) = 3.37$ ft per mile ◄

Eq. 10.8*a*: $200 = \dfrac{1.486}{0.030}AR_{h_{max}}^{2/3}\sqrt{0.000\,639}$ with (Sec. 10.6) $A = y_{opt}^2(2\sqrt{5}-2)$ and $R_{h_{max}} = \dfrac{y_{opt}}{2}$

from which $y_{opt} = 5.68$ ft ◄ $b_{opt} = 2(\sqrt{5}-2)5.68 = 2.68$ ft ◄

10.19 *Refer to Fig. P10.16. If the discharge in the canal (n = 0.030) is to be 6 m³/s while the depth y is 1.8 m and the velocity is not to exceed 50 m/min, what must be the bottom width b and the drop in elevation per km? Compare this with the bottom width for maximum efficiency *when using the same side slopes, bed slope, and discharge. [*Note this correction to the textbook.]*

SI

$V_{max} = 50/60 = 0.833$ m/s; $A_{min} = Q/V_{max} = 6/0.833 = 7.20$ m²

$A_{min} = 1.8(b_{min} + 3.6) = 7.20, \therefore b_{min} = 0.400$ m ◄

$P_{min} = 0.400 + 2(1.8^2 + 3.6^2)^{1/2} = 8.450$ m; min $R_h = 7.20/8.450 = 0.852$ m; min $R_h^{2/3} = 0.899$

$\therefore$ from Eq. 10.7*a*: max $S^{1/2} = 0.833(0.03)/0.899 = 0.0278$; $S_{max} = 0.000\,774$

$\therefore$ max drop $= 0.000\,774(1000) = 0.774$ m/km ◄

Eq. 10.8*b*: $6 = \dfrac{1}{0.030}AR_{h_{max}}^{2/3}\sqrt{0.000\,774}$ with (Sec. 10.6) $A = y_{opt}^2(2\sqrt{5}-2)$ and $R_{h_{max}} = \dfrac{y_{opt}}{2}$

from which $y_{opt} = 1.706$ m ◄ $b_{opt} = 2(\sqrt{5}-2)1.706 = 0.805$ m ◄

Sec. 10.7: Circular Sections Not Flowing Full -- *Exercises*

10.7.1 *Water flows uniformly in a circular concrete pipe (n = 0.014) of diameter 14 ft at a depth of 6 ft. Using Fig. 10.10, determine the flow rate and the average velocity of flow. S = 0.0004.*

BG

Eq. 10.8*a*: $Q_{full} = (1.486/0.014)\pi 7^2(\pi 7^2/14\pi)^{2/3}(0.0004)^{1/2} = 753$ cfs; $V_{full} = 753/(\pi 7^2) = 4.89$ fps

Fig. 10.10 for $y/D = 6/14 = 0.429$: $Q/Q_{full} \approx 0.315$, $V/V_{full} \approx 0.725$

$\therefore Q \approx 0.315(753) = 237$ cfs ◄ $V \approx 0.725(4.89) = 3.55$ fps ◄

10.7.2 *At what depth will water flow at 0.30 m³/s in a 120-cm-diameter concrete pipe on a slope of 0.003? (a) Assume n = 0.013. (b) Repeat with n = 0.015 and compare results.*

SI

(*a*) Eq. 10.8*b*: $Q_{full} = (1/0.013)(\pi \times 0.6^2)[(\pi \times 0.6^2)/(\pi \times 1.2)]^{2/3}(0.003)^{1/2} = 2.14$ m³/s

$Q/Q_{full} = (0.30/2.14) = 0.1405$; $\therefore$ from Fig. 10.10, $y/D \approx 0.275$

Thus depth $\approx 0.275(120) = 33.0$ cm ◄

(*b*) With *n* changed to 0.015, $Q_{full} = 2.14(0.013/0.015) = 1.851$ m³/s

$Q/Q_{full} = 0.30/1.851 = 0.1621$; $\therefore$ from Fig. 10.10, $y/D \approx 0.295$

Thus depth $\approx 0.295(120) = 35.4$ cm ◄

Increasing the *n* value increased the flow depth. ◄

Sec. 10.7: Circular Sections Not Flowing Full -- *Problem*

10.20 *Prove that the value of θ given in Sec. 10.7 for the point of maximum discharge (assuming constant n) is correct. After differentiating, a trial-and-error type of solution will be found most practical here.*

N

Eq. 10.8: $Q \propto AR_h^{2/3}$ for variable θ

Sec. 10.7: $A = (D^2/4)(\theta - 0.5\sin 2\theta)$ and $R_h = (D/4)[1 - (\sin 2\theta)/2\theta]$

$\therefore$ Write an expression for $AR_h^{2/3}$, differentiate it with respect to θ, and equate to zero.

By trial and error or equation solver solve the resulting expression for θ; $\theta = 151.2°$ ◄

Alternate solution: Plot values of $AR_h^{2/3}$ vs θ to find the approximate value of θ for which $AR_h^{2/3}$ is a maximum.

Sec. 10.8: Laminar Flow in Open Channels -- *Exercises*

10.8.1 *Eastern lubricating oil (SAE 30) at 100°F flows down a flat plate 12 ft wide. (a) What is the maximum rate of discharge at which laminar flow may be ensured, assuming that the critical Reynolds number is 500? (b) What should be the slope of the plate to secure a depth of 8 in at this flow rate?*

BG

$\mathbf{R} = R_h V/\nu = y_0 q/\nu y_0 = q/\nu = 500$

Fig. 2.4 for SAE 30 Eastern lubricating oil at 100°F: $\nu = 0.0011$ ft²/sec

$\therefore\ q = 500\nu = 500(0.0011) = 0.55$ cfs per ft

(*a*) For $B = 12$ ft: $Q = Bq = 12(0.55) = 6.60$ cfs ◄

(*b*) From Eq. 10.15: $S = \dfrac{3q\nu}{gy_0^3} = \dfrac{3(0.55)0.0011}{32.2(8/12)^3} = 0.000\,1902$ ◄

10.8.2 *Eastern lubricating oil (SAE 30) at 30°C flows down a flat plate 3 m wide. (a) What is the maximum rate of discharge at which laminar flow may be ensured, assuming that the critical Reynolds number is 500? (b) What should be the slope of the plate to secure a depth of 15 cm at this flow rate?*

SI

$\mathbf{R} = R_h V/\nu = y_0 q/\nu y_0 = q/\nu = 500$

Fig. 2.4 for SAE 30 Eastern lubricating oil at 30°C: $\nu = 0.000\,15$ m²/s

$\therefore\ q = 500\nu = 500(0.000\,15) = 0.075$ m³/s per m

(*a*) For $B = 3$ m: $Q = Ba = 3(0.075) = 0.225$ m³/s ◄

(*b*) From Eq. 10.15: $S = \dfrac{3q\nu}{gy_0^3} = \dfrac{3(0.075)0.000\,15}{9.81(0.15)^3} = 0.001\,019$ ◄

10.8.3 *At what rate (cfs/ft of width) will 70°F water flow in a wide, smooth, rectangular channel on a slope of 0.000 10, if the depth is 0.02 ft? Assume laminar flow and justify this assumption by computing Reynolds number.*

BG

Table A.1 for water at 70°F: $\nu = 1.059 \times 10^{-5}$ ft²/sec

Assuming the flow is laminar, Eq. 10.15: $q = \dfrac{32.2(0.02)^3 \times 0.0001}{3(1.059 \times 10^{-5})} = 8.11 \times 10^{-4}$ cfs/ft ◄

Sec. 10.1: $\mathbf{R} = \dfrac{R_h V}{\nu} = \dfrac{y_0 V}{\nu} = \dfrac{y_0(q/y_0)}{\nu} = \dfrac{q}{\nu} = \dfrac{8.11 \times 10^{-4}}{1.059 \times 10^{-5}} = 76.6$ ◄

Thus **R** is < 500, and the assumption that the flow is laminar is justified.

10.8.4 *At what rate (L/s per meter of width) will water at 20°C flow in a wide, smooth, rectangular channel on a slope of 0.0004, if the depth is 6.0 mm? Assume laminar flow and justify this assumption by computing Reynolds number.*

SI

Table A.1 for water at 20°C: $\nu = 1.003 \times 10^{-6}$ m²/s

Assuming the flow is laminar, Eq. 10.15: $q = \dfrac{(9.81\ \text{m/s}^2)(0.006\ \text{m})^3 0.0004}{3(1.003 \times 10^{-6}\ \text{m}^2/\text{s})}$

$q = 0.000\,282$ m²/s $= 0.282$ L/s per m of width ◄

Sec. 10.1: $\mathbf{R} = R_h V/\nu = y_0 V/\nu = q/\nu = 0.000\,282/(1.003 \times 10^{-6}) = 281$ ◄

Thus **R** is < 500, therefore the flow is laminar and the answer is valid.

Sec. 10.8: Laminar Flow in Open Channels -- *Problems*

10.21 *Evaluate the friction factor f for laminar flow in a wide and shallow open channel in terms of the Reynolds number, and compare it with f for pipe flow using Eq. (8.20). (Note: Recall that for a wide and shallow channel the hydraulic radius is approximately equal to the depth.)*

N

Sec. 10.1: $\mathbf{R}_{oc} = R_h V/\nu$; Eq. 8.54: $\mathbf{R}_{pipe} = 4R_h V/\nu$. $\therefore\ \mathbf{R}_{oc} \equiv \mathbf{R}_{pipe}/4$

For laminar open channel flow, Eq. 10.15: $V = Q/A = q/y_0 = gSy_0^2/(3\nu)$; Eq. 10.5: $V = \sqrt{(8g/f)R_h S}$

$\therefore\ V^2 = (8g/f)R_h S = VgSy_0^2/(3\nu)$, i.e., $f_{oc} = 24\nu R_h/(Vy_0^2)$

But $R_h = y_0$ for a wide and shallow channel, so: $f_{oc} = 24\nu/(VR_h) = 24/\mathbf{R}_{oc}$ ◄

For laminar pipe flow; Eq. 8.20: $f_{pipe} = 64/\mathbf{R}_{pipe}$

$\therefore\ f_{oc} = 24/\mathbf{R}_{oc} \equiv 24(4)/\mathbf{R}_{pipe} = 96/\mathbf{R}_{pipe} = 96(f_{pipe}/64) = 1.5f_{pipe}$ ◄

10.22 *In Sec. 10.8 the velocity distribution in laminar sheet flow was found to be given by* $u = (gS/2\nu)y_0^2\left[1 - (y/y_0)^2\right]$, *where y in this case is the variable distance downward from the surface. Evaluate* α, *and compare it with the result of Sample Prob. 5.1.*

N

$$\text{Eq. 5.4:}\quad \alpha = \frac{1}{V^3 y_0}\int_0^{y_0}\left\{\frac{gS}{2\nu}y_0^2\left[1 - \left(\frac{y}{y_0}\right)^2\right]\right\}^3 dy = \frac{16}{35V^3}\left\{\frac{gS}{2\nu}y_0^2\right\}^3$$

With $V = (2/3)u_{max} = (2/3)(gS/2\nu)y_0^2(1 - 0/y_0)$; $\alpha = 1.542$ ◄

In comparison, $\alpha = 2$ for flow in a circular pipe (Sample Prob. 5.1) ◄

Sec. 10.9: Specific Energy and Alternate Depths of Flow in Rectangular Channels -- *Exercises*

10.9.1 *A long straight rectangular channel 10 ft wide is observed to have a wavy water surface at a depth of about 6 ft. Estimate the rate of discharge.*

BG

Sec. 10.9: Wavy surface means the depth is near critical.

If $y_c = 6$ ft, by Eq. 10.29: $q = \sqrt{32.2(6)^3} = 83.4$ cfs/ft; so $Q = bq = 10(83.4) = 834$ cfs ◄

10.9.2 *A long straight rectangular channel 5 m wide is observed to have a wavy water surface at a depth of about 3 m. Estimate the rate of discharge.*

SI

Sec. 10.9: A wavy surface means the depth is near critical.

If $y_c = 3$ m, by Eq. 10.29: $q = \sqrt{9.81(3)^3} = 16.27$ m^3/s per m; $Q = Bq = 5(16.27) = 81.4$ m^3/s ◄

10.9.3 *Water flows down a wide and shallow rectangular channel of concrete (n = 0.015) laid on a slope of 0.004. Find the depth and rate of flow in SI units for critical conditions in this channel.*

SI

For wide and shallow rectangular channel (Sec. 10.5), $R_h = y$ and $V = q/y$.

Equating V's from Eq. 10.22 and from Eq. 10.7*a* (with $R_h = y$):

$(9.81y_c)^{1/2} = (1/0.015)y_c^{2/3}(0.004)^{1/2}$; $y_c^{1/6} = 0.743$; $y_c = 0.168$ m ◄

Eq. 10.22: $V_c = (gy_c)^{1/2} = (9.81 \times 0.168)^{1/2} = 1.284$ m/s

$\therefore\ q_c = V_c y_c = 1.284(0.168) = 0.216$ m^3/s per m ◄

10.23 *Consider a wide rectangular channel on a given slope. With what power of the discharge does the depth vary? With what power of the discharge does the critical depth vary? As the flow increases, does the Froude number increase or decrease? Assume Manning's equation, with a constant value of n.*

N

Assuming Manning's Eq. with constant "n" and $R_h = y$ (wide and shallow):

From Eq. 10.7*b*: $q = yV = y(1.486/n)y^{2/3}S^{1/2}$ or $q \propto y^{5/3}$, or $y \propto q^{3/5}$ ◄

Eq. 10.23: $y_c = (q^2/g)^{1/3}$, or $y_c \propto q^{2/3}$ ◄

$$\text{Eq. 7.8:}\quad \mathbf{F} = \frac{V}{\sqrt{gL}} = \frac{V}{\sqrt{gy}} = \frac{q}{y\sqrt{gy}} = \frac{q}{g^{1/2}y^{3/2}} \propto \frac{q}{(q^{3/5})^{3/2}} = q^{1/10}$$

So as the flow, q, increases, the Froude number <u>in</u>creases. ◄

10.24 *Differentiate Eq. (10.17) to obtain the expression for y_c given in Eq. (10.23).*

N

Eq. 10.17: $E = y + (q^2/2gy^2) = y + (q^2/2g)y^{-2}$; $y = y_c$ when $dE/dy = 0$

i.e., $1 + (q^2/2g)(-2y^{-3}) = 0$, i.e., $y_c = (q^2/g)^{1/3}$ Q.E.D. ◄

10.25 *A flow of 140 cfs is carried in a rectangular channel 12 ft wide at a depth of 1.8 ft. If the channel is made of smooth concrete (n = 0.012), find the slope necessary to sustain uniform flow at this depth. What roughness coefficient would be required to produce uniform critical flow for the given rate of discharge on this slope?*

BG

$V = 140/(12 \times 1.8) = 6.48$ fps; $R_h = A/P = 1.8(12)/[(12 + 2(1.8)] = 1.385$ ft

Eq. 10.7*b*: $6.48 = (1.486/0.012)1.385^{2/3}S^{1/2}$; $S^{1/2} = 0.0421$; $S = 0.001\,775$ ◄

For $q = 140/12 = 11.67$ cfs/ft, Eq. 10.23: $y_c = (11.67^2/32.2)^{1/3} = 1.617$ ft

To find n required, $V_c = 11.67/1.617 = 7.22$ fps; $R_h = 12(1.617)/[12 + 2(1.617)] = 1.274$ ft

Eq. 10.7*b*: $7.22 = (1.486/n)(1.274)^{2/3}(0.0421)$; $n = 0.010\,20$ ◄

10.26 *A rectangular channel 12 ft wide carries a flow of 240 cfs. Find the critical depth and the critical velocity for this flow. Find also the critical slope if n = 0.015.*

BG

$q = Q/b = 240/12 = 20$ cfs/ft; Eq. 10.23: $y_c = (20^2/32.2)^{1/3} = 2.32$ ft ◄

Eq. 10.22: $V_c = (32.2 \times 2.32)^{1/2} = 8.64$ fps ◄

Sec. 10.10: When $y = y_c$, then $S = S_c$.

$$\text{Eq. 10.8}a\text{:}\quad 240 = \frac{1.486}{0.015}(12 \times 2.32)\left(\frac{12(2.32)}{12 + 2(2.32)}\right)^{2/3} S_c^{1/2};\quad S_c = 0.003\,79$$ ◄

Note that Eq. 10.30 (which yields $S_c = 0.002\,46$) is not accurate because the channel is not sufficiently wide and shallow.

10.27 *Water flows with a velocity of 6 fps and at a depth of 3 ft in a wide rectangular channel. Is the flow subcritical or supercritical? Find the alternate depth for the same discharge and specific energy by two methods; (a) by direct solution of Eq. (10.17); (b) by use of Fig. 10.13.*

BG

$V^2/2g = 6^2/(2 \times 32.2) = 0.559$ ft and $y/2 = 1.5$ ft; Sec. 10.10: $0.559 < 1.5$ so flow is subcritical ◀

To find alternate (supercritical) depth:

(*a*) Eq. 10.16 with $\alpha = 1$: $E = 3 + 0.559 = 3.559$ ft; $q = 6(3) = 18$ cfs/ft

Eq. 10.17: $3.559 = y + 18^2/[2(32.2)y^2]$ or $y^3 - 3.559y^2 + 5.031 = 0$

Since $y = 3$ is one known solution, divide by $(y - 3)$ to yield $y^2 - 0.559y - 1.677 = 0$

from which $y = 1.604$ ft ◀ (the other root, $y = -1.045$ ft, is meaningless)

(*b*) $y/E = 3/3.559 = 0.843$ ($> 2/3$, $\therefore$ subcritical)

So in Fig. 10.13: $q/q_{max} \approx 0.865$, alternate $y/E \approx 0.45$ (supercritical)

$\therefore$ $y \approx 3.559(0.45) = 1.602$ ft ◀

(*b*) Alt: Eq. 10.28: $y_c = (2/3)3.559 = 2.37$ ft ; Eq. 10.29: $q_{max} = (32.2 \times 2.37^3)^{1/2} = 20.70$ cfs/ft

Thus, abscissa of Fig. 10.13 is $q/q_{max} = 18/20.70 = 0.869$

For supercritical limb, $y/E \approx 0.44$, or $y \approx 0.44(3.559) = 1.566$ ft ◀

10.28 *Water is released from a sluice gate in a rectangular channel 6 ft wide such that the depth is 3 ft and the velocity is 18 fps. Find (a) the critical depth for this specific energy; (b) the critical depth for this rate of discharge; (c) the type of flow and the alternate depth by either direct solution or the discharge curve.*

BG

(*a*) Eq. 10.16 with $\alpha = 1$: $E = 3 + 18^2/[2(32.2)] = 8.03$ ft

For this given (constant) E (Fig. 10.13*b*, q may vary), Eq. 10.28: $y_c = (2/3)8.03 = 5.35$ ft ◀

(*b*) $q = yV = 3(18) = 54$ cfs/ft. For this given (constant) q (Fig. 10.12, E may vary),

Eq. 10.23: $y_c = (54^2/32.2)^{1/3} = 4.49$ ft ◀

(*c*) $y < y_c$, $\therefore$ the flow is supercritical ◀

By direct solution of Eq. 10.17: $8.03 = y + 54^2/[2(32.2)y^2]$ or $y^3 - 8.03y^2 + 45.28 = 0$

As $y = 3$ ft is one solution, divide by $(y - 3)$ to give $y^2 - 5.03y - 15.09 = 0$

yielding $y = -2.11$ ft (meaningless), and $y = 7.14$ ft, the alternate depth ◀

10.29 *Water is released from a sluice gate in a rectangular channel 1.5 m wide such that the depth is 0.6 m and the velocity is 4.5 m/s. Find (a) the critical depth for this specific energy; (b) the critical depth for this rate of discharge; (c) the type of flow and the alternate depth by either direct solution or the discharge curve.*

SI

(*a*) Eq. 10.16 with $\alpha = 1$: $E = 0.6 + 4.5^2/[2(9.81)] = 1.632$ m

Eq. 10.28: $y_c = (2/3)1.632 = 1.088$ m ◀

(*b*) $q = yV = 0.6(4.5) = 2.7$ m³/s per m

Eq. 10.23: $y_c = (2.7^2/9.81)^{1/3} = 0.906$ m ◀

(*c*) $y < y_c$, $\therefore$ the flow is supercritical ◀

Eq. 10.17: $E = 1.632 = y + 2.7^2/[2(9.81)y^2]$; by trial, the alternate depth $y = 1.457$ m ◀

Sec. 10.10: Subcritical and Supercritical Flow -- *Exercise*

10.10.1 *A rectangular channel 50 ft wide (n = 0.015) carries a flow of 375 cfs. For this flow, find the critical depth, the critical velocity, and the critical bed slope.*

BG

$q = Q/b = 375/50 = 7.5$ cfs/ft; Eq. 10.23: $y_c = (7.5^2/32.2)^{1/3} = 1.204$ ft ◄

Eq. 10.22: $V_c = \sqrt{32.2(1.204)} = 6.23$ fps ◄

$b/y_c = 50/1.204 = 41.5$, ∴ (Sec. 10.5) critical flow is wide and shallow, $R_h \approx y_c$.

Eq. 10.30: $S_c = \left(\dfrac{0.015}{1.486}\right)^2 \dfrac{32.2}{1.204^{1/3}} = 0.003\,08$ ◄

Sec. 10.11: Critical Depth in Nonrectangular Channels -- *Exercises*

10.11.1 *A flow of 12 cfs of water is carried in a 90° triangular flume built of planed timber (n = 0.011). Find the critical depth and the critical slope.*

BG

$A = y^2$, $B = 2y$; Eq. 10.33: $12^2/32.2 = y_c^6/2y_c$; $y_c^5 = 8.94$; $y_c = 1.550$ ft ◄

Then: $A = (1.55)^2 = 2.40$ ft²; $V_c = 12/2.40 = 5.00$ fps

$P = 2(2^{1/2})1.55 = 4.38$ ft; $R_h = 2.40/4.38 = 0.548$ ft

Eq. 10.7*b*: $5.00 = (1.486/0.011)(0.548)^{2/3}S_c^{1/2}$; $S_c^{1/2} = 0.0552$; $S_c = 0.003\,05$ ◄

10.11.2 *A flow of 0.28 m³/s of water is carried in a 90° triangular flume built of planed timber (n = 0.011). Find the critical depth and the critical slope.*

SI

$A = y^2$, $B = 2y$; Eq. 10.33: $0.28^2/9.81 = y_c^6/2y_c$; $y_c^5 = 0.015\,98$; $y_c = 0.437$ m ◄

Eq. 10.8*b*: $0.28 = \dfrac{1}{0.011}(0.437^2)\left[\dfrac{0.437^2}{2(2^{1/2})0.437}\right]^{2/3} S_c^{1/2}$

$S_c^{1/2} = 0.0559$, $S_c = 0.003\,13$ ◄

10.11.3 *A circular conduit of well-laid (smooth) brickwork when flowing half full is to carry 320 cfs at a velocity of 8 fps. (a) What will be the necessary fall per mile? (b) Will the flow be subcritical or supercritical?*

BG

(*a*) Table 11.1, p. 425: For well-laid brickwork, minimum $n = 0.012$

$A = 320/8 = 40$ ft² $= 0.5(\pi/4)D^2$; $D = 10.09$ ft; $R_h = 10.09/4 = 2.52$ ft

Eq. 10.7*b*: $8 = (1.486/0.012)(2.52)^{2/3}S^{1/2}$; $S = 0.001\,215 = 6.42$ ft/mile ◄

(*b*) For flow half full, $B = D$.

Sec. 10.11: $\mathbf{F} = V/\sqrt{gy_h} = V\sqrt{B/gA} = 8\sqrt{10.09/(32.2 \times 40)} = 0.708$

Sec. 10.10: $\mathbf{F} < 1$, so the flow is subcritical. ◄

10.11.4 *A trapezoidal channel has a 4-m-wide bed and m = 2 (i.e., 2H:1V). What is the critical flow depth when it is carrying 85 m³/s?*

SI

Equation preceding Sample Prob. 10.7: $\dfrac{Q^2}{g} = \dfrac{85^2}{9.81} = \dfrac{(4y_c + 2y_c^2)^3}{4 + 4y_c} = \dfrac{8(2y_c + y_c^2)^3}{4(1 + y_c)}$

i.e., $368 = (2y_c + y_c^2)^3/(1 + y_c)$; by T & E, or by equation solver: $y_c = 2.44$ m. ◄

10.30 *A trapezoidal canal with side slopes of 2:1 (H:V) has a bottom width of 10 ft and carries a flow of 600 cfs. (a) Find the critical depth and critical velocity. (b) If the canal is lined with brick (n = 0.015), find the critical slope for the same rate of discharge.*

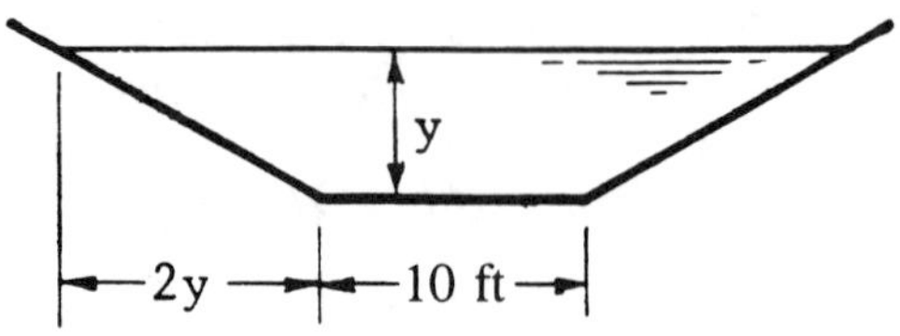

BG

(*a*) $A = 10y + 2y^2 = 2y(5 + y)$

$B = 10 + 4y = 4(2.5 + y)$

When $y = y_c$, from Eq. 10.33: $\dfrac{600^2}{32.2} = \dfrac{8y_c^3(5 + y_c)^3}{4(2.5 + y_c)}$; i.e., $11{,}180 = \dfrac{2y_c^3(5 + y_c)^3}{2.5 + y_c}$

By trial and error, or by equation solver per Sample Prob. 10.1: $y_c = 3.74$ ft ◀

From above: $A_c = 65.4\ \text{ft}^2$; $B_c = 25.0$ ft; $V_c = Q/A_c = 9.18$ fps ◀

(*b*) $P_c = 10 + 2(3.74)5^{1/2} = 26.7$ ft; $R_h = A_c/P_c = 2.45$ ft

Eq. 10.7*b*: $9.18 = (1.486/0.015)2.45^{2/3}S_c^{1/2}$; $S_c = 0.002\,59$ ◀

Note: Eq. 10.30 (yielding $S_c = 0.002\,11$) is not valid (not "wide and shallow").

10.31 *A trapezoidal canal with side slopes of 1:1 has a bottom width of 4 m and carries a flow of 25 m³/s. (a) Find the critical depth and critical velocity. (b) If the canal is lined with brick (n = 0.015), find the critical slope for the same rate of discharge.*

SI

(*a*) $A = y(4 + y)$; $B = 4 + 2y$

When $y = y_c$, Eq. 10.33: $25^2/9.81 = 63.71 = y_c^3(4 + y_c)^3/(4 + 2y_c)$

By trial and error, or by equation solver per Sample Prob. 10.1: $y_c = 1.401$ m ◀

From above, $A_c = 7.57\ \text{m}^2$; $B_c = 6.80$ m; $V_c = Q/A_c = 3.30$ m/s ◀

(*b*) $P_c = 4 + 2(1.401)2^{1/2} = 7.96$ m; $R_h = A_c/P_c = 0.950$ m

Eq. 10.7*a*: $3.30 = (1/0.015)(0.950)^{2/3}S_c^{1/2}$; $S_c = 0.002\,63$ ◀

Note: Eq. 10.30 (yielding $S_c = 0.001\,973$) is not valid (not "wide and shallow").

10.32 *For a circular conduit with a diameter of 12 ft, compute the specific energy for a flow of 120 cfs at depths of 2, 4, 6, and 10 ft assuming α = 1.0. At what depth is E the least? Check to see if Eq. (10.33) is satisfied at this depth.*

BG

Given: $D = 12$ ft, $Q = 120$ cfs, $\alpha = 1$. Sec. 10.7: $A = (D^2/4)(\theta - 0.5\sin 2\theta)$

Compute $A(y)$, or use Tables such as Table 7.4 in Brater et al's "Handbook of Hydraulics," 7th ed, McGraw-Hill, 1996. A spreadsheet could be helpful with this problem.

y ft	θ deg	A ft²	V fps (= Q/A)	$V^2/2g$ ft	E ft (Eq. 10.16)
2	48.2	12.4	9.69	1.46	3.46
2.5	54.3	17.1	7.03	0.77	3.27
3	60.0	22.1	5.43	0.46	3.46
4	70.5	33.0	3.64	0.21	4.21
6	90.0	56.6	2.12	0.07	6.07
10	131.8	100.7	1.19	0.02	10.02

▲

/cont...

Plot the above and more data and note that $E_{min} \approx 3.26$ ft at $y \approx 2.42$ feet ◄

Check: When $y = 2.42$ ft, $A = 16.29$ ft^2, $B = 9.63$ ft, and so in Eq. 10.33:

$Q^2/g = 120^2/32.2 = 447$ and $A^3/B = 16.29^3/9.63 = 449$

∴ Eq. 10.33 is satisfied well at this depth. ◄

10.33 *Figure P10.33 describes the cross section of an open channel for which $S_0 = 0.03$ and $n = 0.020$. The sketch is drawn to the scale shown. When the flow rate is 120 cfs, find (a) the depth for uniform flow and (b) the critical depth. Hint: The cross-sectional area may be found by planimetry or counting squares and the wetted perimeter by use of dividers.*

y ft	x_L ft	x_R ft
0	0	0
1	1.5	1.1
2	2.6	2.0
3	3.25	3.3
4	3.6	5.5
5	3.85	8.2

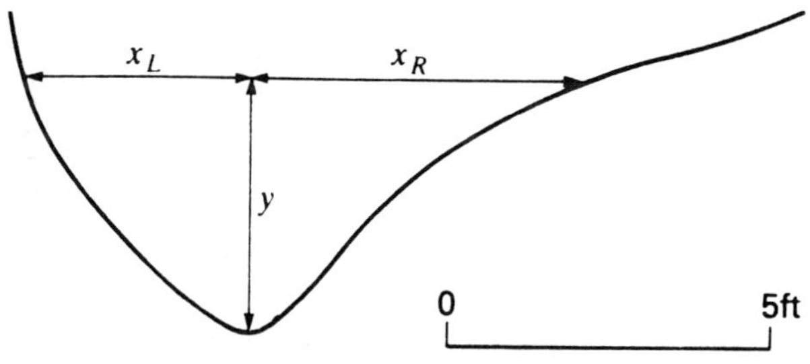

Figure P10.33

BG

Given: $Q = 120$ cfs, $S_0 = 0.03$, $n = 0.020$

(*a*)

y_0 ft	A^* ft^2	P^* ft	$R_h = A/P$ ft	$R_h^{2/3}$	Q cfs Eq. 10.8a
1	1.30	3.29	0.395	0.538	9.01
2	4.90	6.12	0.800	0.862	54.4
3	10.48	8.96	1.170	1.110	149.7
4	18.30	12.43	1.472	1.294	304.8
5	28.88	16.34	1.767	1.462	543.1

* From Fig. P10.33 (measurements may vary)

Plot Q vs y_0; from the plot, for $Q = 120$ cfs, find that $y_0 = 2.70$ ft ◄

(*b*)

y_c ft	A^* ft^2	$B = x_L + x_R$ ft	gA^3/B	$Q = \sqrt{gA^3/B}$ cfs Eq. 10.33
1	1.30	2.60	27.2	5.22
2	4.90	4.60	823	28.7
3	10.48	6.55	5,650	75.2
4	18.30	9.10	21,700	147.3
5	28.88	12.05	64,000	254

* From Fig. P10.33 (measurements may vary)

Plot Q vs y_c; from the plot, for $Q = 120$ cfs, find that $y_c = 3.75$ ft ◄

10.34 *Refer to the figure, table, and hint for Prob. 10.33, replacing feet dimensions with meters. Let the slope be 0.009 with $n = 0.020$. When the flow rate is 60 m^3/s, find (a) depth for uniform flow; (b) the critical depth.*

SI

Given: $Q = 60\ m^3/s$, $S_0 = 0.009$, $n = 0.020$

(*a*)

y_0 m	A^* m^2	P^* m	$R_h = A/P$ m	$R_h^{2/3}$	Q m^3/s Eq. 10.8b
1	1.30	3.29	0.395	0.538	3.32
2	4.90	6.12	0.800	0.862	20.0
3	10.48	8.96	1.170	1.110	55.2
4	18.30	12.43	1.472	1.294	112.3
5	28.88	16.34	1.767	1.462	200.2

* From Fig. P10.33 (measurements may vary)

Plot Q vs y_0; from the plot, for $Q = 60\ m^3/s$, find that $y_0 = 3.10$ m ◄

(*b*)

y_c m	A^* m^2	$B = x_L + x_R$ m	gA^3/B	$Q = \sqrt{gA^3/B}$ m^3/s Eq. 10.33
1	1.30	2.60	8.29	2.88
2	4.90	4.60	250	15.84
3	10.48	6.55	1 721	41.5
4	18.30	9.10	6 610	81.3
5	28.88	12.05	19 600	140.0

Plot Q vs y_c; from the plot, for $Q = 60\ m^3/s$, find that $y_c = 3.50$ m ◄

Sec. 10.13: Humps and Contractions -- *Exercises*

10.13.1 *A rectangular channel 3 m wide carries 4 m^3/s of water in subcritical uniform flow at a depth of 1.2 m. A frictionless hump is to be installed across the bed. Find the critical hump height (i.e., the minimum hump which causes y_c on it).*

SI

$V_0 = Q/A_0 = 4/(3 \times 1.2) = 1.111$ m/s

Eq. 10.16: $E_0 = 1.2 + 1.111^2/[2(9.81)] = 1.263$ m; $q = Q/b = 4/3 = 1.333\ m^3/s$ per m

Eq. 10.23: $y_c = (1.333^2/9.81)^{1/3} = 0.566$ m; Eq. 10.25: $E_{min} = (3/2)0.566 = 0.849$ m

Sec 10.13: $(\Delta z)_{crit} = E_0 - E_{min} = 1.263 - 0.849 = 0.414$ m ◄

10.13.2 *Suppose that the depth of uniform flow in a 4-ft-wide rectangular channel is 1.10 ft. Find the change in water-surface elevation caused by a 1-ft-wide bridge pier placed in the middle of the channel. The flow rate is 50 cfs.*

BG

$V_0 = Q/(by_0) = 50/(4 \times 1.10) = 11.36$ fps ; Eq. 10.16: $E_0 = 1.10 + 11.36^2/(2 \times 32.2) = 3.11$ ft

With a critical constriction, $E_{min} = 3.11$ ft $= (3/2)y_c$, $y_c = 2.07$ ft

and $q = \sqrt{gy_c^3} = \sqrt{32.2(2.07)^3} = 16.90$ cfs/ft, $b_{crit} = Q/q = 50/16.90 = 2.96$ ft.

At the given constriction, $b = 4 - 1 = 3$ ft $> b_{crit}$, so $V = (50/3)/y = 16.67/y$

and $3.11 = y + (16.67/y)^2/(2 \times 32.2)$

By trial and error or by equation solver, $y = 1.869$ ft and there is a 0.769-ft rise in the water surface. ◄

10.13.3 *Suppose that the depth of uniform flow in a 1.2-m-wide rectangular channel is 27 cm. Find the change in water-surface elevation caused by a 30-cm-wide bridge pier placed in the middle of the channel. The flow rate is 1.1 m³/s.*

SI

$V_0 = Q/(by_0) = 1.1/(1.2 \times 0.27) = 3.40$ m/s ; Eq. 10.16: $E_0 = 0.27 + 3.40^2/(2 \times 9.81) = 0.857$ m

With a critical constriction, $E_{min} = 0.857$ m $= (3/2)y_c$, $y_c = 0.572$ m

and $q = \sqrt{gy_c^3} = \sqrt{9.81(0.572)^3} = 1.354$ m³/s per m, $b_{crit} = Q/q = 1.1/1.354 = 0.813$ m.

At the given constriction, $b = 1.2 - 0.3 > b_{crit}$, so $V = 1.1/(0.9y) = 1.222/y$

and $0.857 = y + (1.222/y)^2/(2 \times 9.81)$

By trial and error or by equation solver, $y = 0.415$ m and there is a 0.145-m rise in the water surface. ◄

10.13.4 *A rectangular channel 12 ft wide carries 24 cfs in uniform flow at a depth of 0.80 ft. Find the local change in water-surface elevation caused by a hump *0.12 ft high across the floor of the channel. [*Note this correction to the textbook.]*

BG

$q = Q/b = 24/12 = 2$ cfs/ft. Eq. 10.23: $y_c = (2^2/32.2)^{1/3} = 0.499$ ft.

Eq. 10.25: $E_{min} = 1.5y_c = 0.748$ ft

Given $y_1 = 0.8$: $y_1 > y_c$ so flow is subcritical; $V_1 = q/y_1 = 2/0.8 = 2.50$ fps; $V_1^2/2g = 0.0970$ ft

Eq. 10.16: $E_1 = 0.8 + 0.0970 = 0.897$ ft; $\Delta z_{crit} = E_1 - E_{min} = 0.897 - 0.748 = 0.1486$ ft

$\therefore\ \Delta z < \Delta z_{crit}$. At 0.12-ft rise in bed, $E_2 = E_1 - \Delta z = 0.897 - 0.12 = 0.777$ ft

Eq. 10.17: $E_2 - y_2 + 2^2/[2(32.3)y_2^2] = 0.777$ ft. By T & E or by equation solver, $y_2 = 0.610$ ft.

Then change in water surface $= y_1 - (\Delta z + y_2) = 0.800 - (0.120 + 0.610) = 0.070$ ft drop ◄

Sec. 10.13: Humps and Contractions -- *Problems*

10.35 *Work Sample Prob. 10.8 ($y_0 = 2.0$ ft) for the case where the flow rate is 275 cfs.*

Sample Prob. 10.8: The flow is uniform in a 20-ft-wide rectangular channel. (a) Is it subcritical or supercritical? (b) If a hump of height $\Delta z = 0.30$ ft is placed in the bottom of the channel, calculate the water depth on the hump, and the change in the water surface level at the hump. (c) If the hump height is raised to $\Delta z = 0.60$ ft, what then are the water depths upstream and downstream of the hump? (d) If the 0.30-ft hump is accompanied by a local contraction to 18 ft, find the water depth on the hump. In all cases neglect head losses over the hump and through the contraction.

BG

Given: $y_0 = 2.00$ ft, $Q = 275$ cfs. *Note* the needs for extra significant figures.

(*a*) $q = Q/b = 275/20 = 13.75$ cfs/ft; Eq. 10.23: $y_c = (13.75^2/32.2)^{1/3} = 1.804$ ft

$y_0 > y_c$, so flow is subcritical (and slope is mild) ◄

(*b*) Instal hump, height $\Delta z = 0.30$ ft. First find critical hump height.

Eq. 10.17: $E_0 = 2.00 + [1/(2 \times 32.2)](13.75/2.00)^2 = 2.7339$ ft

Eq. 10.25: $E_{min} = 1.5(1.804) = 2.7061$ ft

Sec 10.13: $E_0 = \Delta z_{crit} + E_{min}$, so $\Delta z_{crit} = 2.7339 - 2.7061 = 0.0279$ ft

$\Delta z > \Delta z_{crit}$, so damming action occurs and y_c occurs on the hump. On hump $y = 1.804$ ft ◄

(*c*) Increase hump height Δz to 0.60 ft. Still $\Delta z > \Delta z_{crit}$, so y_c occurs on the hump.

/cont...

Let y = upstream or downstream depth, with specific energy E. Then, Sec. 10.13: $E = \Delta z + E_{\min}$

Eq. 10.17: $y + [1/(2 \times 32.2)](13.75/y)^2 = 0.60 + 2.706$ i.e., $y + (2.94/y^2) = 3.306$

By trial and error or by equation solver: y = 2.97, 1.17, or a negative root which has no physical meaning.

This hump causes damming action, and the depths just upstream and downstream of the hump are 2.97 ft (subcritical) and 1.17 ft (supercritical) respectively. ◄

See Sample Prob. 10.8(*c*) for an explanation of which depth occurs where.

(*d*) At section 2 with 0.30 ft hump and contraction to 18 ft:

$q_2 = 275/18 = 15.28$ cfs/ft; $y_{c2} = (15.28^2/32.2)^{1/3} = 1.935$ ft

$(E_{\min})_2 = 1.5(1.935) = 2.903$ ft ; $\Delta z_{\text{crit}} = E_1 - (E_{\min})_2 = 2.734 - 2.903 = -0.169$ ft

So $\Delta z > \Delta z_{\text{crit}}$, damming action will occur with critical depth on the hump.

$\therefore\ y_2 = y_{c2} = 1.935$ ft ◄

10.36 *A flow of 2.0 m³/s is carried in a rectangular channel 1.8 m wide at a depth of 1.0 m. Will critical depth occur at a section where (a) a frictionless hump 18 cm high is installed across the bed, (b) a frictionless sidewall constriction (with no hump) reduces the channel width to 1.4 m, and (c) the hump and the sidewall constriction are installed together? Show calculations.*

SI

Given: $Q = 2.0\ \text{m}^3/\text{s}$, channel $b = 1.8$ m, $y_0 = 1.0$ m

(*a*) With hump 0.18 m high: $q = Q/b = 2.0/1.8 = 1.111\ \text{m}^3/\text{s}$ per m

Eq. 10.17: $E_0 = 1.0 + [1/(2 \times 9.81)](1.111/1.0)^2 = 1.063$ m

Eq. 10.23: $y_c = (1.111^2/9.81)^{1/3} = 0.501$ m

Eq. 10.25: $E_{\min} = 1.5(0.501) = 0.752$ m; Sec. 10.13: $\Delta z_{\text{crit}} = E_0 - E_{\min} = 0.311$ m

$\therefore\ \Delta z = 0.18\ \text{m} < \Delta z_{\text{crit}}$, so y_c does not occur at the hump. ◄

Alternative solution:

$E_h = E_0 - \Delta z = 1.063 - 0.180 = 0.883$ m

$E_h > E_{\min} = 0.752$, so y_c does not occur at the hump. ◄

(*b*) With constricted $b = 1.4$ m, and no hump: $q = 2.0/1.4 = 1.429\ \text{m}^3/\text{s}$ per m

Eq. 10.23: $y_c = (1.429^2/9.81)^{1/3} = 0.593$ m; Eq. 10.25: $E_{\min} = 1.5(0.593) = 0.889$ m

$\therefore\ E_{\min} < E_0 = 1.063$, so y_c does not occur at constriction. ◄

(*c*) With both 0.18-m hump and constricted $b = 1.4$ m:

From part (*b*): $q_{\text{constriction}} = 1.429\ \text{m}^3/\text{s}$ per m, $y_c = 0.593$ m, $E_{\min} = 0.889$ m

$\Delta z_{\text{crit}} = E_0 - E_{\min} = 1.063 - 0.889 = 0.1741$ m

So hump height = 0.18 m > Δz_{crit}, so y_c does occur at hump + constriction ◄

Also therefore, the flow must back up, and so increase the y and E just upstream of the hump.

Alternative solution:

$E_h = E_0 - \Delta z = 0.883$; $E_{\min} = 0.889$ m

$E_h < E_{\min}$ is impossible, so y_c does occur and flow must back up, etc. ◄

10.37 *Work Sample Prob. 10.8 (y_0 = 2.0 ft) for the case where the flow rate is 90 cfs.*

Sample Prob. 10.8: The flow is uniform in a 20-ft-wide rectangular channel. (a) Is it subcritical or supercritical? (b) If a hump of height Δz = 0.30 ft is placed in the bottom of the channel, calculate the water depth on the hump, and the change in the water surface level at the hump. (c) If the hump height is raised to Δz = 0.60 ft, what then are the water depths upstream and downstream of the hump? (d) If the 0.30-ft hump is accompanied by a local contraction to 18 ft, find the water depth on the hump. In all cases neglect head losses over the hump and through the contraction.

BG

Given: $y_0 = 2.00$ ft, $Q = 90$ cfs

(*a*) $q = Q/b = 90/20 = 4.5$ cfs/ft; Eq. 10.23: $y_c = (4.5^2/32.2)^{1/3} = 0.857$ ft

$y_0 > y_c$, so flow is subcritical (and slope is mild) ◀

(*b*) Instal hump, height $\Delta z = 0.30$ ft. First find critical hump height.

Eq. 10.17: $E_0 = 2.00 + [1/(2 \times 32.2)](4.5/2.00)^2 = 2.079$ ft

Eq. 10.25: $E_{min} = 1.5(0.857) = 1.286$ ft

Sec 10.13: $E_0 = \Delta z_{crit} + E_{min}$, so $\Delta z_{crit} = 2.079 - 1.286 = 0.793$ ft

$\Delta z < \Delta z_{crit}$, so y_c does not occur on the hump, and damming action does not result.

Sec. 10.13: $E_h = E_0 - \Delta z$

∴ Eq. 10.17: $y_h + [1/(2 \times 32.2)](4.5/y_h)^2 = 2.079 - 0.30$ i.e., $y_h + 0.314/y_h^2 = 1.779$

By trial and error or by equation solver: $y_h = 1.6652$ (subcritical) or 0.495 (supercritical) or a negative root which has no physical meaning. The flow does not pass through y_c so it cannot become supercritical.

∴ Water depth on 0.30-ft hump = 1.665 ft ◀

Change in water surface level at hump = 2.00 − (1.16652 + 0.30) = 0.0348 ft drop ◀

(*c*) Increase hump height Δz to 0.60 ft

Still $\Delta z < \Delta z_{crit}$, so y_c does not occur on the hump and damming action does not result.

Sec 10.13: $E_h = E_0 - \Delta z$

∴ using Eq. 10.17: $y_h + [1/(2 \times 32.2)](4.5/y_h)^2 = 2.079 - 0.60$ i.e., $y_h + 0.314/y_h^2 = 1.479$

By trial and error or by equation solver: y = 1.290 ft (subcritical) or 0.597 ft (supercritical) or a negative root which has no physical meaning. The flow does not pass through y_c so it cannot become supercritical. So water depth on the 0.60-ft hump is 1.290 ft.

With no damming action, water depths upstream and downstream of the hump = y_0 = 2.00 ft ◀

(*d*) At section 2 with 0.3-ft hump and contraction to 18 ft:

$q_2 = 90/18 = 5$ cfs/ft; $y_{c2} = (5^2/32.2^2)^{1/3} = 0.919$ ft.

$(E_{min})_2 = 1.5(0.919) = 1.379$ ft

$\Delta z_{crit} = E_1 - (E_{min})_2 = 2.079 - 1.379 = 0.700$ ft.

So still $\Delta z < \Delta z_{crit}$, and damming action does not occur, y_c does not occur on the hump.

$E_2 = E_1 - \Delta z = 2.079 - 0.3 = 1.779$ ft.

Eq. 10.17: $1.779 = y_2 + (5/y_2)^2/(2 \times 32.2)$

By trial and error, or by equation solver, $y_2 = 1.633$ ft. ◀

10.38 *A rectangular channel 5 ft wide carries 50 cfs of water in uniform flow at a depth of 3.20 ft. If a bridge pier 1.5 ft wide is placed in the middle of this channel, find the local change in the water-surface elevation. What is the minimum width of constricted channel which will not cause a rise in water surface upstream?*

BG

Given y_0 = 3.20 ft, from Sec. 10.9: $V_0 = q/y_0 = (Q/b)/y_0 = (50/5)/3.2 = 3.125$ fps

At constriction, $V = (50/3.5)/y = 14.29/y$

Eq. 10.16: $E_0 = 3.20 + 3.125^2/(2 \times 32.2) = 3.35 \text{ ft} = y + (14.29/y)^2/(2 \times 32.2)$

By trial and error or by equation solver: y = 2.999 ft or 1.218 ft. Supercritical conditions are impossible at the pier. $\therefore$ y = 3.00 ft.

Change in water depth = 3.20 − 2.999 = 0.201 ft drop ◀

The minimum width of constricted channel which would produce critical flow is found by getting q_{max} for E = 3.35 ft.

From Eqs. 10.28 and 10.29: $q_{max} = g^{1/2}(2E/3)^{3/2} = 32.2^{1/2}(2/3 \times 3.35)^{3/2} = 18.95$ cfs/ft

Minimum channel width = 50/18.95 = 2.64 ft ◀

(A width less than this would produce backwater/damming action.)

10.39 *A rectangular channel 1.2 m wide carries 1.1 m³/s of water in uniform flow at a depth of 0.85 m. If a bridge pier 0.3 m wide is placed in the middle of this channel, find the local change in the water-surface elevation. What is the minimum width of constricted channel which will not cause a rise in water surface upstream?*

SI

Given y_0 = 0.85 m, $V_0 = q/y_0 = (Q/b)/y_0 = (1.1/1.2)/0.85 = 1.078$ m/s

At constriction, $V = (1.1/0.9)/y = 1.222/y$

Eq. 10.16: $E_0 = 0.85 + 1.078^2/(2 \times 9.81) = 0.909 \text{ m} = y + (1.222/y)^2/(2 \times 9.81)$

By trial and error or be equation solver: y = 0.7861 m or 0.379 m. Supercritical conditions are impossible at the pier so y = 0.786 ft.

Change in water depth = 0.850 − 0.7861 = 0.0639 m drop ◀

From Eqs. 10.28 and 10.29: $q_{max} = g^{1/2}(2E/3)^{3/2} = 9.81^{1/2}(2 \times 0.909/3)^{3/2} = 1.478$ m³/s per m

Minimum channel width = 1.1/1.478 = 0.744 m ◀

(A width less than this would produce backwater/damming action.)

10.40 *A rectangular channel 12 ft wide carries 24 cfs in uniform flow at a depth of 0.32 ft. Find the local change in water-surface elevation caused by a 0.15-ft-high obstruction across the channel bed.*

BG

$q = Q/b = 24/12 = 2$ cfs/ft. Eq. 10.23: $y_c = (2^2/32.2)^{1/3} = 0.499$ ft. Eq. 10.25: $E_{min} = 1.5y_c = 0.748$ ft

Given y_1 = 0.32 ft: $\therefore$ $y_1 < y_c$, flow is supercritical, and $V_1 = 2/0.32 = 6.25$ fps ; $V_1^2/2g = 0.607$ ft

$E_1 = 0.320 + 0.607 = 0.927$ ft ; $\Delta z_{crit} = E_1 - E_{min} = 0.927 - 0.748 = 0.1781$ ft

So $\Delta z < \Delta z_{crit}$, $\therefore$ at 0.15-ft rise in bed, $E_2 = E_1 - \Delta z = 0.927 - 0.150 = 0.777$ ft

Also, Eq. 10.17: $E_2 = y_2 + (2/y_2)^2/(2 \times 32.2) = 0.777$ ft

or $y_2^3 - 0.777y_2^2 + 0.0621 = 0$; By trial or by equation solver: $y_2 = 0.414$ ft.

Hence water surface change = (0.150 + 0.414) − 0.320 = 0.244 ft rise ◀

☐ 10.41 *Fifty cubic feet per second of water flows uniformly in a 6-ft-wide channel at a depth of 2.5 ft. What is the change in water-surface elevation at a section contracted to a 4-ft width with a 0.2-ft depression in the bottom?*

BG

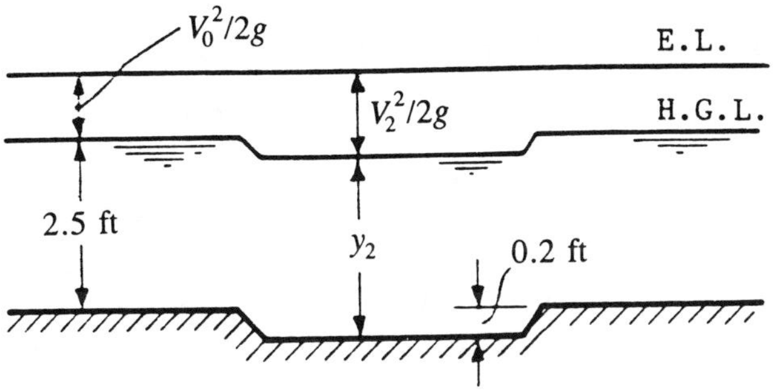

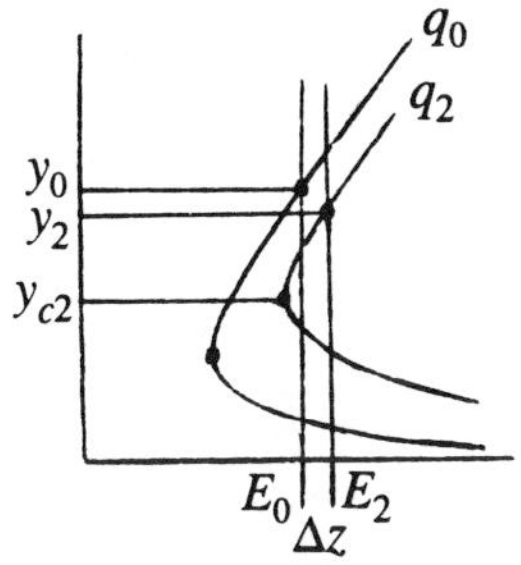

$q_2 = Q/b = 50/4 = 12.5$ cfs/ft; Eq. 10.23: $y_{c2} = (12.5^2/32.2)^{1/3} = 1.693$ ft

Sec. 10.13 and Eq. 10.16: $2.5 + V_0^2/2g = -0.2 + y_2 + V_2^2/2g$

$V_0 = 50/(6 \times 2.5) = 3.33$ fps, $V_2 = 50/(4 \times y_2)$

$\therefore\ 2.5 + 3.33^2/(2 \times 32.2) = 2.673$ ft $= -0.2 + y + (12.5/y)^2/(2 \times 32.2)$

By trial or by equation solver: $y = 2.477$ (subcritical) or 1.207 ft (supercritical).

y_2 cannot be less than y_{c2}, therefore $y_2 = 2.48$ ft.

Water-surface change $= 2.50 - (2.477 - 0.20) = 0.223$ ft drop ◄

☐ 10.42 *Water flows uniformly in a 2.0-m-wide channel at a rate of 1.6 m³/s and a depth of 0.75 m. What is the change in water-surface elevation at a section contracted to a 1.4 m width with an 8-cm depression in the bottom?*

SI

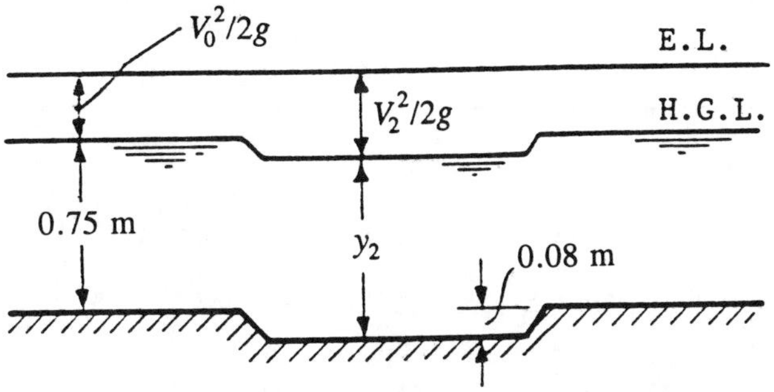

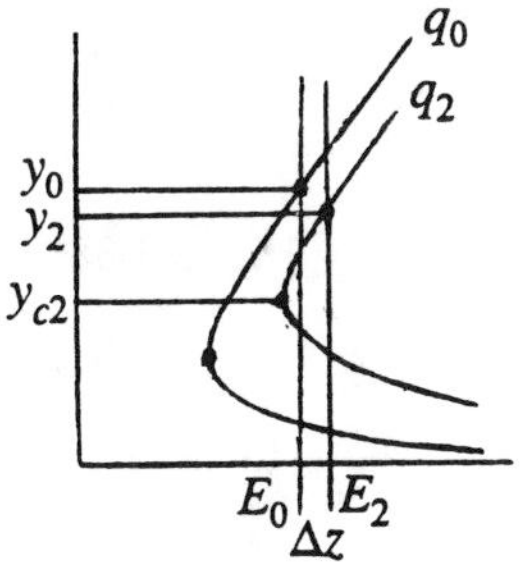

$q_2 = Q/b = 1.6/1.4 = 1.143$ m³/s per m; Eq. 10.23: $y_{c2} = (1.143^2/9.81)^{1/3} = 0.511$ m

Sec. 10.13 and Eq. 10.16: $E_0 = E_2 - \Delta z$, $\therefore\ 0.75 + V_0^2/2g = -0.08 + y_2 + V_2^2/2g$

$V_0 = 1.6/(2.0 \times 0.75) = 1.067$ m/s, $V_2 = 1.6/(1.4 y_2) = 1.143/y_2$

$\therefore\ 0.75 + 1.067^2/(2 \times 9.81) = 0.808$ m $= y_2 + (1.143/y_2)^2/(2 \times 9.81) - 0.080$

By trial and error or by equation solver, $y_2 = 0.7780$ m (subcritical) or 0.353 m (supercritical).

y_2 cannot be less than $y_{c2} = 0.511$ m, therefore $y_2 = 0.778$ m.

Water-surface change $= 0.750 - (0.7780 - 0.080) = 0.0520$ m drop ◄

10.15.1 *A rectangular flume of planed timber (n = 0.012) is 5 ft wide and carries 60 cfs of water. The bed slope is 0.0006, and at a certain section the depth is 3 ft. Find the distance (in one reach) to the section where the depth is 2.5 ft. Is the distance upstream or downstream?*

BG

Assume the 2.5 ft depth is downstream. Then (for a single reach):

$A = by = 5y;\ P = b + 2y = 5 + 2y;\ R = A/P;\ V = Q/A = 60/A;\ E = y + V^2/2g$

y ft	V fps	$V^2/2g$ ft	E ft	A ft^2	P ft	R ft	R_m ft	V_m fps
3.0	4.0	0.248	3.248	15.0	11	1.364		
							1.3068	4.4
2.5	4.8	0.358	2.858	12.5	10	1.250		

Eq. 10.38a: $S = [0.012 \times 4.40/(1.486 \times 1.3068^{2/3})]^2 = 0.000\,884$

Eq. 10.37: $3 + 0.248 = 2.5 + 0.358 + (0.000\,884 - 0.0006)\Delta x;\quad \Delta x = 1377$ ft ◄

Since Δx is positive, the 2.5-ft depth is downstream, as assumed. ◄

10.15.2 *A rectangular flume of planed timber (n = 0.013) is 1.6 m wide and carries 1.7 m³/s of water. The bed slope is 0.0005, and at a certain section the depth is 1.0 m. Find the distance (in one reach) to the section where the depth is 0.85 m. Is the distance upstream or downstream?*

SI

Assume the 0.85 m depth is downstream. Then (for a single reach):

$A = by = 1.6y;\ P = b + 2y = 1.6 + 2y;\ R = A/P;\ V = Q/A = 1.9/A;\ E = y + V^2/2g$

y m	V m/s	$V^2/2g$ m	E m	A m^2	P m	R m	R_m m	V_m m/s
1.00	1.063	0.0575	1.0575	1.60	3.6	0.444		
							0.428	1.156
0.85	1.250	0.0796	0.9296	1.36	3.3	0.412		

Eq. 10.38b: $S = [0.013 \times 1.156/0.428^{2/3}]^2 = 0.000\,700$

Eq. 10.37: $1.0 + 0.0575 = 0.85 + 0.0796 + (0.000\,700 - 0.0005)\Delta x;\quad \Delta x = 640$ m ◄

Since Δx is positive, the 0.85-m depth is downstream, as assumed. ◄

10.15.3 *A test on a rectangular glass flume 9 in wide yielded the following data on a reach of 40 ft length: with still water, $z_1 - z_2 = 0.009$ ft; with a measured flow of 0.1446 cfs, $y_1 = 0.381$ ft, $y_2 = 0.386$ ft. Find the value of the roughness coefficient n using only one reach.*

BG

$S_0 = 0.009/40 = 0.000\,225;\quad B_0 = 9 \text{ in} = 0.750$ ft

$A_1 = 0.286$ ft^2; $P_1 = 1.512$ ft; $R_1 = A_1/P_1 = 0.1890$ ft; $V_1 = Q/A_1 = 0.506$ fps; $V_1^2/2g = 0.003\,98$ ft

$A_2 = 0.290$ ft^2; $P_2 = 1.522$ ft; $R_2 = 0.1902$ ft; $V_2 = Q/A_2 = 0.499$ fps; $V_2^2/2g = 0.003\,87$ ft;

$\therefore R_m = 0.1896$ ft; $V_m = 0.503$ fps

Eq. 10.37: $0.381 + 0.003\,98 = 0.386 + 0.003\,87 + (S - 0.000\,225)40;\quad S = 0.000\,1026$

With these values of S, R_m, and V_m in Eq. 10.38*a*: $n = 0.009\,88$ ◄

10.15.4 *A test on a rectangular glass flume 25 cm wide yielded the following data on a reach of 9-m length: with still water, $z_1 - z_2 = 2.7$ mm; with a measured flow of 4.3 L/s, $y_1 = 11.02$ cm, $y_2 = 11.17$ cm. Find the value of the roughness coefficient n using only one reach.*

SI

$S_0 = 0.0027/9 = 0.0003$; $B_0 = 0.25$ m

$A_1 = 0.0276$ m^2; $P_1 = 0.470$ m; $R_1 = A_1/P_1 = 0.0586$ m; $V_1 = Q/A_1 = 0.1561$ m/s;

$V_1^2/2g = 0.001\ 242$ m; $A_2 = 0.0279$ m^2; $P_2 = 0.473$ m; $R_2 = 0.0590$ m;

$V_2 = Q/A_2 = 0.1540$ m/s; $V_2^2/2g = 0.001\,209$;

$\therefore\ R_m = 0.0588$; $V_m = 0.1550$ m/s

Eq. 10.37: $0.1102 + 0.001\,242 = 0.1117 + 0.001\,209 + (S - 0.0003)9$; $S = 0.000\,1370$

With these values of S, R_m, and V_m in Eq. 11.38*b*: $n = 0.011\ 41$ ◄

Sec. 10.15: Energy Equation for Gradually Varied Flow -- *Problems*

10.43 *A 5-ft-wide rectangular flume (n = 0.012) carries 60 cfs of water. At one point the water depth is found to be 4 ft; 1000 ft downstream is is measured at 3 ft. Calculate the bed slope of the flume, using one reach. Sketch the flume bed, water surface, and energy line, to check that the answer is reasonable.*

BG

$y_1 = 4$ ft; $A_1 = 20$ ft^2; $P_1 = 13$ ft; $R_1 = A_1/P_1 = 1.538$ ft; $V_1 = Q/A_1 = 3$ fps; $V_1^2/2g = 0.1398$ ft;

$y_2 = 3$ ft; $A_2 = 15$ ft^2; $P_2 = 11$ ft; $R_2 = 1.364$ ft; $V_2 = 4$ fps; $V_2^2/2g = 0.248$ ft;

$R_m = (R_1 + R_2)/2 = 1.451$ ft and $V_m = (V_1 + V_2)/2 = 3.5$ fps

Eq. 10.38*a*: $S = [0.012 \times 3.5/(1.486 \times 1.451^{2/3})]^2 = 0.000\,486$

Eq. 10.37: $4 + 0.1398 = 3 + 0.248 + (0.000\,486 - S_0)1000$; $S_0 = -0.000\,405$ ◄

The minus sign indicates the slope is adverse. Sketch:

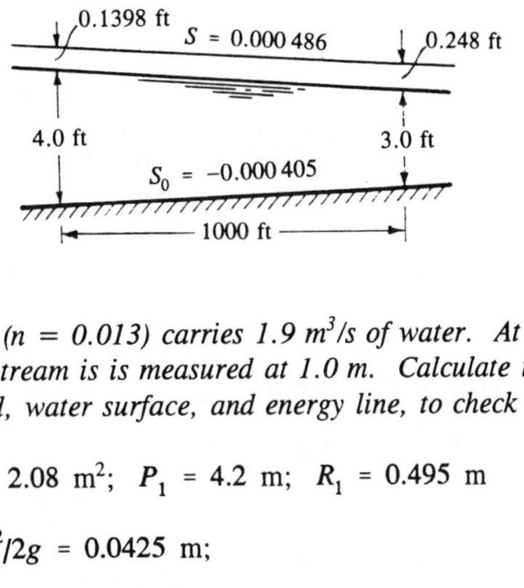

◄

10.44 *A 1.6-m-wide rectangular flume (n = 0.013) carries 1.9 m^3/s of water. At one point the water depth is found to be 1.3 m; 320 m downstream is is measured at 1.0 m. Calculate the bed slope of the flume, using one reach. Sketch the flume bed, water surface, and energy line, to check that the answer is reasonable.*

SI

$y_1 = 1.3$ m; $A_1 = 1.3 \times 1.6 = 2.08$ m^2; $P_1 = 4.2$ m; $R_1 = 0.495$ m

$V_1 = 1.9/2.08 = 0.913$ m/s; $V_1^2/2g = 0.0425$ m;

$y_2 = 1.0$ m; $A_2 = 1.0 \times 1.6 = 1.60$ m^2; $P_2 = 3.6$ m; $R_2 = 0.444$ m

$V_2 = 1.9/1.60 = 1.1875$ m/s; $V_2^2/2g = 0.0719$ m; $R_m = 0.470$ m; $V_m = 1.050$ m/s

Eq. 10.38*b*: $S = [0.013 \times 1.050/0.470^{2/3}]^2 = 0.000\,511$

/cont...

Eq. 10.37: $1.3 + 0.0425 = 1.0 + 0.0719 + (0.000\,511 - S_0)320$; $S_0 = -000\,335$ ◀

The minus sign indicates the slope is adverse. Sketch:

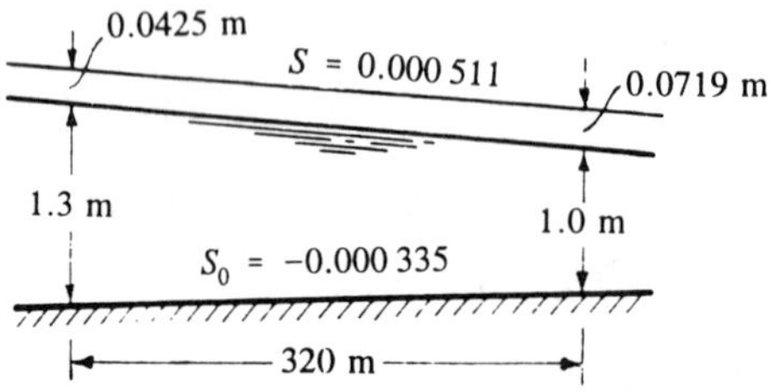

◀

10.45 *Suppose that the slope of the flume in Prob. 10.43 is now 0.01. With the same flow as before, find the depth 50 ft downstream from a section where the flow is 1.5 ft deep. Use only one reach. Is the flow subcritical or supercritical? [Note: In this case it will not be possible to make a direct solution from Eq. (10.37). A trial-and-error solution may best be set in the form of a table with the headings y_2, V_2, V_m, A_2, P_2, R_2, R_m, $R_m^{2/3}$, S, etc.]*

Prob. 10.43: Rectangular flume, b = 5 ft, n = 0.012, Q = 60 cfs.

BG

$q = Q/b = 60/5 = 12$ cfs/ft; Eq. 10.23: $y_c = (12^2/32.2)^{1/3} = 1.648$ ft

As $y_1 = 1.5 < y_c$, the flow is <u>supercritical</u> ◀

$A_1 = 7.5\ \text{ft}^2$; $P_1 = 5 + 2(1.5) = 8$ ft; $R_1 = A_1/P_1 = 0.938$ ft;

$V_1 = Q/A_1 = 8$ fps; $V_1^2/2g = 0.994$ ft

Eq. 10.37: $1.5 + 0.994 = y_2 + (12/y_2)^2/2g + (S - 0.01)50$

where, by Eq. 10.38*a*: $S = (0.012V_m/1.486R_m^{2/3})^2$

This expression for S may be substituted into the equation above it and by successive trials of y_2 or by equation solver yields the result: $y_2 = 1.241$ ft. ◀

10.46 *Suppose that the slope of the flume in Prob. 10.44 is now 0.008. With the same flow as before, find the depth 17 m downstream from a section where the flow is 50 cm deep. Use only one reach. Is the flow subcritical or supercritical? [Note: In this case it will not be possible to make a direct solution from Eq. (10.37). A trial-and-error solution may best be set in the form of a table with the headings y_2, V_2, V_m, A_2, P_2, R_2, R_m, $R_m^{2/3}$, S, etc.]*

Prob. 10.44: Rectangular flume, b = 1.6 m, n = 0.013, Q = 1.9 m³/s.

SI

$q = Q/b = 1.9/1.6 = 1.1875\ \text{m}^3/\text{s}$ per m

Eq. 10.23: $y_c = (1.1875^2/9.81)^{1/3} = 0.524$ m

As $y_1 = 0.50\ \text{m} < y_c$, the flow is <u>supercritical</u> ◀

$A_1 = 0.800\ \text{m}^2$; $P_1 = 1.6 + 2(0.50) = 2.6$ m; $R_1 = A_1/P_1 = 0.308$ m;

$V_1 = Q/A_1 = 2.375$ m/s; $V_1^2/2g = 0.2875$ m

Eq. 10.37: $0.50 + 0.2875 = y_2 + (1.1875/y_2)^2/2g + (S - 0.008)17$

where, by Eq. 10.38*b*: $S = (0.013V_m/R_m^{2/3})^2$

This expression for S may be substituted into the equation above it and by successive trials of y_2 or by equation solver yields the result: $y_2 = 0.423$ m. ◀

10.47 *Repeat Prob. 10.45, but increase the 50-ft distance to 500 ft.*

Prob. 10.45: Rectangular flume, b = 5 ft, S = 0.01, n = 0.012, Q = 60 cfs, y_1 = 1.5 ft. Find the downstream y_2 using only one reach. Is the flow subcritical or supercritical?

BG

$q = Q/b = 12$ cfs/ft; Eq. 10.23: $y_c = (12^2/32.2)^{1/3} = 1.648$ ft

As $y_1 = 1.5 < y_c$, the flow is supercritical ◀

To find y_2 at $\Delta x = 500$ ft downstream of y_1. First find y_0:

From Eq. 10.8*a*: $60 = \dfrac{1.486}{0.012}\dfrac{(5y_0)^{5/3}}{(5 + 2y_0)^{2/3}}0.01^{1/2}$;

By trial and error or by equation solver, per Sample Prob. 10.1, $y_0 = 1.140$ ft

Find the distance to $y = y_0$. Conditions at Section 1 are as for Prob. 10.45. At the section where $y = y_0$:

$A_0 = 5.70$ ft^2; $P_0 = 7.28$ ft; $R_0 = A_0/P_0 = 0.783$ ft

$V_0 = Q/A_0 = 10.53$ fps; $V_0^2/2g = 1.721$ ft; $R_m = 0.860$ ft; $V_m = 9.26$ fps

Eq. 10.38*a*: $S = [0.012 \times 9.26/(1.486 \times 0.860^{2/3})]^2 = 0.006\,84$

Eq. 10.39: $\Delta x = [(1.5 + 0.994) - (1.140 + 1.721)]/(0.006\,84 - 0.01) = 116.0$ ft

Downstream, where $\Delta x > 116$ ft, y remains at y_0

$\therefore$ At $\Delta x = 500$ ft, $y_2 = y_0 = 1.140$ ft ◀

Note: Calculation of y_2 using Eq. 10.37 with $\Delta x = 500$ ft would give an erroneous (impossible) result with $y_2 < y_0$.

10.48 *Repeat Prob. 10.46, but increase the 17-m distance to 170 m.*

Prob. 10.46: Rectangular flume, b = 1.6 m, S = 0.008, n = 0.013, Q = 1.9 m^3/s, y_1 = 0.50 m. Find the downstream y_2 using only one reach. Is the flow subcritical or supercritical?

SI

$q = Q/b = 1.9/1.6 = 1.1875$ m^3/s per m; Eq. 10.23: $y_c = (1.1875^2/9.81)^{1/3} = 0.524$ m

As $y_1 = 0.50$ m $< y_c$, the flow is supercritical ◀

To find y_2 at $\Delta x = 150$ m downstream of y_1. First find y_0:

From Eq. 10.8*b*: $1.9 = \dfrac{1}{0.013}\dfrac{(1.6y_0)^{5/3}}{(1.6 + 2y_0)^{2/3}}(0.008)^{1/2}$

By trial and error or by equation solver per Sample Prob. 10.1, $y_0 = 0.411$ m

Find the distance to $y = y_0$. Conditions at Section 1 are as for Prob. 10.46. At the section where $y = y_0$:

$A_0 = 0.658$ m^2; $P_0 = 2.42$ m; $R_0 = A_0/P_0 = 0.272$ m;

$V_0 = Q/A_0 = 2.89$ m/s; $V_0^2/2g = 0.425$ m; $R_m = 0.290$ m; $V_m = 2.63$ m/s

Eq. 10.38*b*: $S = [0.013 \times 2.63/0.290^{2/3}]^2 = 0.006\,11$

Eq. 10.39: $\Delta x = [(0.50 + 0.287) - (0.411 + 0.425)]/(0.006\,11 - 0.008) = 25.9$ m

Downstream, where $\Delta x > 26$ m, y remains at y_0

$\therefore$ At $\Delta x = 170$ m, $y_2 = y_0 = 0.411$ m ◀

Note: Calculation of y_2 using Eq. 10.37 with $\Delta x = 170$ ft would give an erroneous (impossible) result with $y_2 < y_0$.

□ 10.49 *A rectangular flume 9 ft wide is built of planed timber ($n = 0.012$) on a bed slope of 0.3 ft per 1000 ft, ending in a free overfall. If the measured depth at the fall is 1.67 ft, find (a) the rate of flow; (b) the distance upstream from the fall to where the depth is 3.6 ft. [Note: Assume that critical depth occurs at a distance of $4y_c$ upstream from the fall, and employ reaches with end depths of 2.5, 2.8, 3.2, and 3.6 ft].*

BG

Given: $b = 9$ ft, $n = 0.012$, $S_0 = 0.3/1000 = 0.0003$

Sec. 10.12: $y_b \approx 0.72y_c = 1.67$ ft; $\therefore\ y_c \approx 1.67/0.72 = 2.32$ ft

At $y = y_c$, by Eq. 10.29: $q = (32.2 \times 2.32^3)^{1/2} = 20.04$ cfs/ft

$Q = bq = 9(20.05) = 180.40$ cfs ◄

Note: Between y_b and y_c the flow is rapidly varying. So Manning's equation is not valid there, and cannot be used there to determine Q.

y ft	V fps	$V^2/2g$ ft	E ft	A ft^2	P ft	R ft	R_m ft	V_m fps	S Eq. 10.38*a*	Δx ft Eq. 10.39
3.6	5.57	0.481	4.081	32.4	16.2	2.00				
							1.94	5.92	0.000 941	424
3.2	6.26	0.609	3.809	28.8	15.4	1.87				
							1.798	6.71	0.001 336	206
2.8	7.16	0.796	3.596	25.2	14.6	1.726				
							1.667	7.59	0.001 890	61
2.5	8.02	0.998	3.498	22.5	14.0	1.607				
							1.569	8.33	0.002 47	9
2.32	8.64	1.159	3.479	20.88	13.64	1.531				
										$4y_c = 9$
y_b									$\Sigma(\Delta x)$	$= 710$ ◄

Note: A spreadsheet could be helpful on this problem.

□ 10.50 *A rectangular flume 3 m wide is built of planed timber ($n = 0.012$) on a bed slope of 20 cm per km, ending in a free overfall. If the measured depth at the fall is 0.55 m, find (a) the rate of flow; (b) the distance upstream from the fall to where the depth is 1.2 m. [Note: Assume that critical depth occurs at a distance of $4y_c$ upstream from the fall, and employ reaches with end depths of 0.82, 0.9, 1.0, and 1.2 m.]*

SI

Given: $b = 3$ m, $n = 0.012$, $S_0 = 0.2/1000 = 0.0002$

Sec. 10.12: $y_b \approx 0.72y_c = 0.55$ m; $y_c \approx 0.55/0.72 = 0.764$ m

At $y = y_c$, by Eq. 10.29: $q = \sqrt{9.81(0.764)^3} = 2.09$ m^3/s per m; $Q = bq = 3(2.09) = 6.27$ m^3/s ◄

Note: Between y_b and y_c the flow is rapidly varying. So Manning's equation is not valid there, and cannot be used there to determine Q.

/cont...

y m	V m/s	$V^2/2g$ m	E m	A m^2	P m	R m	R_m m	V_m m/s	S Eq. 10.38*b*	Δx m Eq. 10.39
1.2	1.743	0.1548	1.355	3.6	5.40	0.667				
							0.633	1.917	0.000 973	171
1.0	2.09	0.223	1.223	3.0	5.00	0.600				
							0.581	2.21	0.001 446	38
0.9	2.32	0.275	1.175	2.7	4.80	0.563				
							0.546	2.44	0.001 915	14
0.82	2.55	0.331	1.151	2.46	4.64	0.530				
							0.518	2.64	0.002 42	3
0.764	2.74	0.382	1.146	2.29	4.53	0.506				
										$4y_c$ = 3
y_b									$\Sigma(\Delta x)$	= 228 ◄

Note: A spreadsheet could be helpful on this problem.

10.51 *A wide and shallow rectangular channel dredged in earth (n = 0.035) is laid on a slope of 9 ft/mi and carries a flow of 90 cfs/ft of width. (a) Find the water depth 2 mi upstream of a section where the depth is 26.1 ft, using a single reach. (b) Compare the result with that obtained using three reaches of equal length.*

BG

Given: q = 90 cfs/ft, channel is "wide and shallow," ∴ (Sec. 10.5) $R_h = y$.

$n = 0.035$, $S_0 = 9/5280 = 0.001\ 704$, $y_2 = 26.1$ ft; find y, at $\Sigma(\Delta x)$ = 2 mi = 10,560 ft upstream.

(*a*) Using a single reach:
$$10{,}560 = \frac{\left[y_1 + \frac{1}{2g}\left(\frac{90}{y_1}\right)^2\right] - \left[26.1 + \frac{1}{2g}\left(\frac{90}{26.1}\right)^2\right]}{S - 0.001\ 704}$$

where, by Eq. 10.38*a*: $S = \left[0.035 V_m/(1.486 R_m^{2/3})\right]^2$ and $V_m = (90/26.1 + 90/y_1)/2$ and

$R_m = (26.1 + y_1)/2$

Solving the above equations by trial and error or by equation solver, y_1 = 11.191 ft ◄

(*b*) Using three equal reaches of Δx = 10,560/3 = 3520 ft each:

Reach	$y = R$ ft	$V = 90/y$ fps	$V^2/2g$ ft	E ft	$\Delta E = E_1 - E_2$ ft	V_m fps	R_m ft	S Eq. 10.38*a*	$\Delta x = \Delta E/(S - S_0)$ Want 3520 ft
	26.1	3.45	0.185	26.28					
3					−5.54	3.93	23.27	0.000 1287	3518
	20.44*	4.40	0.301	20.74					
2					−4.89	5.14	17.88	0.000 3136	3519
	15.31*	5.88	0.537	15.85					
1					−3.21	6.78	13.52	0.000 7919	3518
	11.72*	7.68	0.916	12.64					

*By equation solver or by trial and error, (4 - 6 trials) until Δx closely = 3520 ft.

Thus, using 3 reaches, y_1 = 11.72 ft. ◄ Note: A spreadsheet could be helpful on this problem.

10.52 *The slope of a stream of rectangular cross section is $S_0 = 0.0003$, the width is 170 ft, and the value of the Chézy C is 78.3 $ft^{1/2}/s$. (a) Find the depth for a uniform flow of 101.54 cfs per foot of width of the stream. (b) If a dam raises the water level so that at a certain distance upstream the increase is 5 ft, how far from this latter section will the increase by only 1 ft? Use reaches with 1-ft depth increments.*

BG

Given: $q = 101.54$ cfs/ft, $b = 170$ ft, Chézy $C = 78.3$, $S_0 = 0.0003$

(*a*) Uniform flow: $V_0 = q/y_0$ and Eq. 10.6: $V_0 = C(RS)^{1/2}$, where $R = A/P = 170y_0/(170 + 2y_0)$

Equating V_0's: $101.54/y_0 = 78.3\sqrt{170y_0 \times 0.0003/(170 + 2y_0)}$

Solve by trial and error or by equation solver, to obtain $y_0 = 19.00$ ft ◀

(*b*) GVF with 4 reaches, from $y_1 = 20$ ft to $y_2 = 24$ ft

y ft	A $170y$ ft²	P ft	R A/P ft	V q/y fps	$V^2/2g$ ft	E ft	ΔE $E_1 - E_2$ ft	V_m fps	R_m ft	S Eq. 10.6	Δx Eq. 10.39 ft
24	4080	218	18.72	4.23	0.278	24.278					
							−0.975	4.32	18.41	0.000 1656	7,255
23	3910	216	18.10	4.41	0.303	23.303					
							−0.972	4.52	17.79	0.000 1869	8,595
22	3740	214	17.48	4.62	0.331	22.331					
							−0.968	4.73	17.16	0.000 2123	11,030
21	3570	212	16.84	4.84	0.363	21.363					
							−0.963	4.96	16.52	0.000 2426	16,772
20	3400	210	16.19	5.08	0.400	20.400					
										$\Sigma(\Delta x)$	= 43,651

Note: A spreadsheet could be helpful on this problem. ▲

10.53 *A portion of an outfall sewer is approximately a circular conduit 5 ft in diameter and with a slope of 1 ft in 1100 ft. It is of brick, for which $n = 0.013$. (a) What would be its maximum capacity for uniform flow? (b) If it discharges 120 cfs with a depth at the end of 2.90 ft, how far back from the end must it become a pressure conduit? Proceeding from the mouth upstream, find by tabular solution the length of sewer that is not flowing full. Use three reaches with equal depth increments. (c) Find the pressure at the top and bottom of the pipe at the section that is 600 ft upstream from the point at which the sewer no longer flows full. Sketch the energy line and hydraulic grade line from this point to the end of the sewer.*

BG

Given: $D = 5$ ft, $S_0 = 1/1100 = 0.000\,909$, $n = 0.013$

(*a*) For uniform flow: Eq. 10.8*a*: $Q_{full} = (1.486/0.013)(\pi/4)5^2(5/4)^{2/3}0.000\,909^{1/2} = 78.5$ cfs

Fig. 10.10: $Q_{max}/Q_{full} \approx 1.025$; $\therefore$ $Q_{max} \approx 1.025(78.5) = 80.5$ cfs ◀

(*b*) GVF profile for part-full flow with $Q = 120$ cfs:

/cont...

y ft	θ deg	A ft^2 (Sec.	P ft 10.7)	R A/P ft	V Q/A fps	$V^2/2g$ ft	E ft	ΔE $E_1 - E_2$ ft	V_m fps	R_m ft	S Eq 10.38a	Δx ft Eq 10.39
2.9	99.2	11.81	8.66	1.364	10.16	1.603	4.503					
								0.073	9.05	1.429	0.003 89	24
3.6	116.1	15.13	10.13	1.494	7.93	0.976	4.576					
								0.417	7.30	1.503	0.002 37	285
4.3	136.1	17.96	11.87	1.513	6.68	0.693	4.993					
								0.587	6.40	1.382	0.002 03	522
5.0	180.0	19.63	15.71	1.250	6.11	0.580	5.580					
											$\Sigma(\Delta x)$	831

The sewer must become a pressure conduit 831 ft back from the end. ◀

Note: A spreadsheet could be helpful with the above.

(*c*) First 600 ft of pipe flowing full: Eq. 10.8*a*: $120 = (1.486/0.013)(\pi/4)5^2(5/4)^{2/3}S^{1/2}$; $S = 0.002\ 12$

For 600 ft length AB: $h_L = SL = 0.002\ 12(600) = 1.274$ ft water

Eq. 5.14: $(p/\gamma + z + V^2/2g)_A = (p/\gamma + z + V^2/2g)_B + h_L$; $V_A = V_B$

$\therefore (p_A - p_B)/\gamma = h_L - (z_A - z_B) = 1.274 - 600(1/1100) = 0.728$ ft water

$\therefore p_A - p_B = 62.4(0.728) = 45.4$ psf $= 0.316$ psi

At the point B where the sewer no longer flows full, $(p_B)_{top} = p_{atmos} = 0$ psig

At the point A 600 ft upstream of B, $(p_A)_{top} = (p_B)_{top} + 0.316 - 0 + 0.316 = 0.316$ psig ◀

The vertical pressure distribution is hydrostatic for parallel streamlines.

$\therefore (p_{bottom} - p_{top})/\gamma = 5$ ft water; $(p_A)_{bottom} - (p_A)_{top} = 5(62.4) = 312$ psf $= 2.17$ psi

$(p_A)_{bottom} = (p_A)_{top} + 2.17 = 0.316 + 2.17 = 2.48$ psi ◀

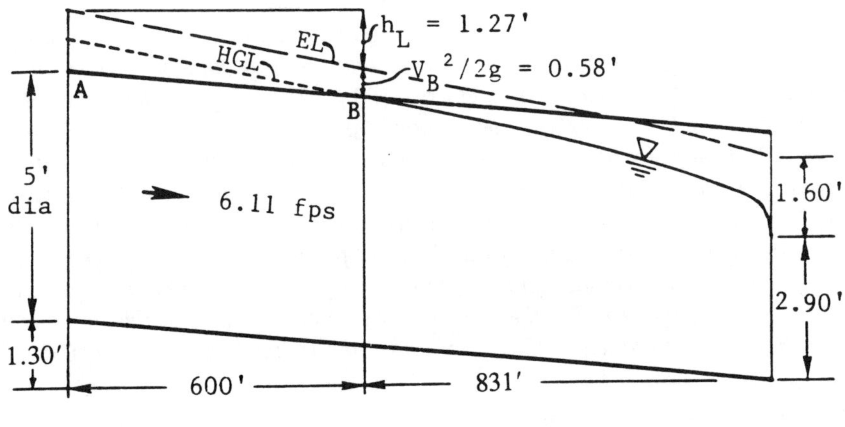

/more...

10.54 *For the channel of Prob. 10.33 find the distance between a section where the depth is 3.5 ft to another where the depth is 3.0 ft. Which section is upstream? Use only one reach. See the hint of Prob. 10.33.*

Prob. 10.33: $S_0 = 0.03$, $n = 0.020$, and $Q = 120$ cfs. The sketch is drawn to the scale shown. Hint: The cross-sectional area may be found by planimetry or counting squares and the wetted perimeter by use of dividers.

y ft	x_L ft	x_R ft
0	0	0
1	1.5	1.1
2	2.6	2.0
3	3.25	3.3
4	3.6	5.5
5	3.85	8.2

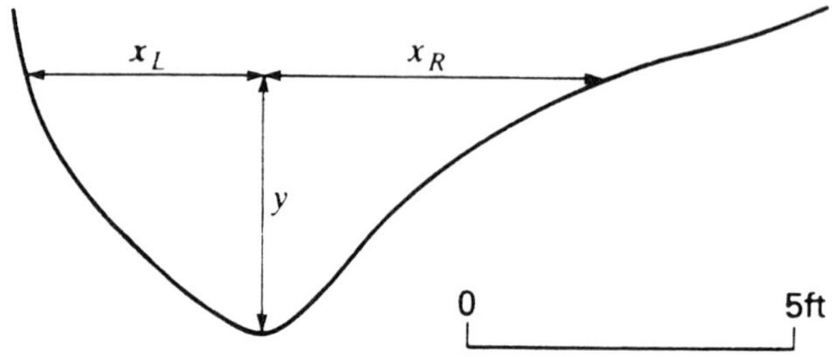

Figure P10.33

BG

Assume the 3.0-ft depth is upstream: $y_1 = 3.0$ ft, $y_2 = 3.5$ ft.

y ft	A^* ft^2	P^* ft	R ft	V fps	$V^2/2g$ ft	E ft	V_m fps	R_m ft
3.0	10.48	8.96	1.170	11.45	2.04	5.036		
							9.99	1.243
3.5	14.07	10.695	1.316	8.53	1.130	4.630		

*From Fig. P10.33 (measurements may vary).

Eq. 10.38*a*: $S = [0.020 \times 9.99/(1.486 \times 1.243^{2/3})]^2 = 0.013\,53$

Eq. 10.39: $\Delta x = (E_1 - E_2)/(S - S_0) = (5.036 - 4.630)/(0.013\,53 - 0.03) = -24.7$ ft

Distance Δx between sections = 24.7 ft ◄

Negative Δx indicates the assumption was incorrect; the 3.5-ft depth is upstream ◄

10.55 *When the flow in a certain natural stream is 7600 cfs, it is required to find the elevation of the water surface at different sections upstream from a certain initial point. A survey of the channel shows that conditions are fairly similar for a length of 1500 ft upstream from the initial point, and then beyond that there is another stretch of 2200 ft, and so on. Assuming a rise in the water surface in the distance of 1500 ft to be 0.20 ft, a study of the stream bed shows the average values of the area and wetted perimeter to be as given in the table below. Using Eq. (10.6), the computed head loss, based on average velocity and hydraulic radius, is seen to be 0.283 ft, which is greater than that assumed. Hence assume a larger value and repeat. Note that average areas and wetted perimeters have also been determined for alternative water-surface slopes. Complete the following table, and find the probable rise in elevation in the first 1500 ft. [In a similar manner, the rises in other lengths may be computed, and the sum of all of them up to the desired point will give the elevation at that point above the initial.] Assume $n = 0.036$.*

Assumed rise SL, ft	*A average, ft^2*	*P average, ft*	*R average, ft*	*V average, fps*	*$SL = LV^2/C^2R$, ft*
0.20	3100	350	8.86	2.45	0.283
0.25	3180	359	8.86		
0.26	3190	360	8.86		
0.27	3220	363	8.86		
0.28	3230	364	8.86		

/cont...

BG

Given: $Q = 7600$ cfs, $n = 0.036$

Comparing Eqs. 10.6 and 10.7*b*: Chézy $C = (1.486/n)R_h^{1/6}$

Here R_h = constant, $\therefore$ $C = (1.486/0.036)8.86^{1/6} = 59.4$

Completing the indicated spaces in the given table, the best agreement for the first 1500 ft is found with an assumed rise of $SL = 0.27$ ft.

Then $V = Q/A = 7600/3220 = 2.36$ fps and $LV^2/C^2R = 1500(2.36)^2/(59.4^2 \times 8.86) = 0.268$ ft

Thus the probable rise in the water surface over this 1500-ft reach is 0.27 ft ◀

Sec. 10.17: Examples of Water Surface Profiles -- *Exercises*

10.17.1 *Under what different conditions can supercritical flow leaving a hump ($\Delta z > \Delta z_{crit}$) return to normal depth without passing through a hydraulic jump?*

N

Supercritical flow can return to y_0 without a jump if $y_0 \le y_c$ (it must jump if $y_0 > y_c$).

Refer to Fig. 10.20: Return without a jump is possible only by S_2, S_3, or C_3 profiles. ◀

10.17.2 *Classify the water-surface profile of Exer. 10.15.1 as one of the forms shown in Fig. 10.20. Show all necessary calculations.*

Exer. 10.15.1: $n = 0.012$, $b = 5$ ft, $S_0 = 0.0006$, $Q = 60$ cfs, y varies from 2.5 to 3.0 ft.

BG

Eq. 10.8*a*: $60 = (1.486/0.012)(5y_0)[5y_0/(5 + 2y_0)]^{2/3}(0.0006)^{1/2}$;

By trial and error or by equation solver per Sample Prob. 10.1, $y_0 = 3.17$ ft

Eq. 10.23: Critical depth $y_c = (q^2/g)^{1/3} = [(60/5)^2/32.2]^{1/3} = 1.648$ ft

Fig. 10.20: $y_0 = 3.17$ ft $> y_c = 1.648$ ft, so the slope is Mild.

Given y varies from 2.5 to 3.0 ft. Fig. 10.20: Since $y_c < y < y_0$ this is an M_2 profile. ◀

10.17.3 *Classify the water-surface profile of Exer. 10.15.2 as one of the forms shown in Fig. 10.20. Show all necessary calculations.*

Exer. 10.15.2: Rectangular flume, $n = 0.013$, $b = 1.6$ m, $S_0 = 0.0005$, $Q = 1.7$ m³/s, y varies from 0.85 to 1.00 m.

SI

Eq. 10.8*b*: $1.7 = (1/0.013)(1.6y_0)[1.6y_0/(1.6 + 2y_0)]^{2/3}(0.0005)^{1/2}$

By trial and error or by equation solver per Sample Prob. 10.1, $y_0 = 1.047$ m

Eq. 10.23: Critical depth $y_c = [(1.7/1.6)^2/9.81]^{1/3} = 0.486$ m

Fig. 10.20: $y_0 = 1.047$ m $> y_c = 0.486$ m, so the slope is Mild.

Given y varies from 0.85 m to 1.00 m. Fig. 10.20: since $y_c < y < y_0$ this is an M_2 profile. ◀

10.17.4 *Repeat Exer. 10.17.2 when the channel slope is $-0.000\,40$.*

Exer. 10.17.2: $b = 5$ ft, $Q = 60$ cfs, y varies from 2.5 to 3.0 ft.

BG

Given: $S_0 = -0.000\,40$. Since $S_0 < 0$, the slope is Adverse, $y_0 = \infty$. y_c is independent of S_0.

Eq. 10.23: $y_c = (q^2/g)^{1/3} = [(60/5)^2/32.2]^{1/3} = 1.648$ ft.

Given y varies from 2.5 to 3.0 ft. Fig. 10.20: Since $y_c < y < y_0$, this is an A_2 profile ◀

10.17.5 *Repeat Exer. 10.17.3 when the channel slope is −0.000 40.*
Exer. 10.17.3: b = 1.6 m, Q = 1.7 m³/s, y varies from 0.85 to 1.00 m.

SI

Since $S_0 < 0$, the slope is Adverse, $y_0 = \infty$. y_c is independent of S_0.

Eq. 10.23: $y_c = (q^2/g)^{1/3} = [(1.7/1.6)^2/9.81]^{1/3} = 0.486$ m.

Given y varies from 0.85 m to 1.00 m. Fig. 10.20: Since $y_c < y < y_0$, this is an A_2 profile. ◄

10.17.6 *The flow in a 15-ft-wide rectangular channel that has a constant bottom slope is 1400 cfs. A computation using Manning's equation indicates that the normal depth is 6.0 ft. At a certain section the depth of flow in the channel is 2.8 ft. Does the depth increase, decrease, or remain the same as one proceeds downstream from this section? Sketch a physical situation where this type of flow will occur.*

BG

Eq. 10.23: $y_c = [(1400/15)^2/32.2]^{1/3} = 6.47$ ft; $y_0 = 6.0$ ft (given).

Fig. 10.20: $y_0 = 6.0$ ft $< y_c = 6.47$ ft, so the slope of the channel is Steep.

Fig. 10.20: Since the given $y = 2.8$ ft $< y_0 < y_c$, this is an S_3 curve, and the depth will increase (asymptotically to y_0) as one proceeds downstream. ◄

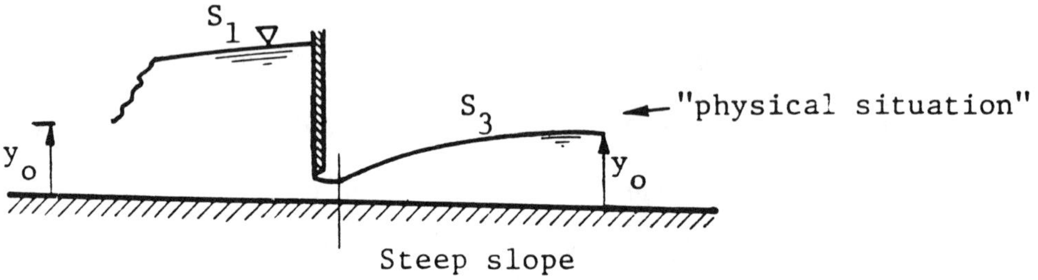

◄

10.17.7 *A laboratory flume (n = 0.012) is 10 inches wide and set on a slope of 0.0003. With a measured flow of 0.1516 cfs the depth is observed to vary between 0.361 and 0.366 ft. Classify the water-surface profile as one of the forms of Fig. 10.20. Show all necessary calculations, and sketch the profile.*

BG

$b = 10$ in $= 0.833$ ft, $Q = 0.1516$ cfs, $S_0 = 0.0003$

Eq. 10.8*a*: $0.1516 = (1.486/0.012)(0.833y_0)[0.833y_0/(0.833 + 2y_0)]^{2/3}(0.0003)^{1/2}$

By trial and error or by equation solver per Sample Prob. 10.1: $y_0 = 0.279$ ft

Eq. 10.23: $y_c = [(0.1516/0.833)^2/32.2]^{1/3} = 0.1009$ ft

Fig. 10.20 for $y_0 = 0.279$ ft $> y_c = 0.1009$ ft; the slope is Mild.

Given y varies in the range 0.361 to 0.366 ft

Fig. 10.20: Since given $y > y_0 > y_c$, this is an M_1 profile. ◄

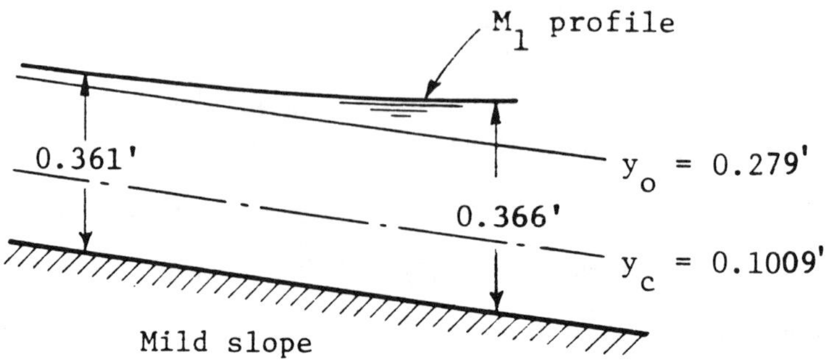

◄

10.17.8 *A laboratory flume (n = 0.012) is 25 cm wide and set on a slope of 0.0003. With a measured flow of 4.3 L/s, the depth is observed to vary between 11.02 and 11.17 cm. Classify the water-surface profile as one of the forms of Fig. 10.20. Show all necessary calculations, and sketch the profile.*

SI

$b = 0.25$ m, $Q = 0.0043$ m^3/s, $S_0 = 0.0003$

Eq. 10.8*b*: $0.0043 = (1/0.012)(0.25y_0)[0.25y_0/(0.25 + 2y_0)]^{2/3}(0.0003)^{1/2}$

By trial and error or by equation solver per Sample Prob. 10.1: $y_0 = 0.0865$ m;

Eq. 10.23: $y_c = [(0.0043/0.25)^2/9.81]^{1/3} = 0.0311$ m

Fig. 10.20: $y_0 = 0.0865$ m $> y_c = 0.0311$ m; the slope is Mild.

Given 0.1102 m $< y <$ 0.1117 m; Fig. 10.20: Since given $y > y_0 > y_c$, this is an M_1 profile. ◄

The sketch is the same as above (Exer. 10.17.7), but with the SI dimensions substituted. ◄

10.17.9 *In an 8-ft-wide rectangular channel (S_0 = 0.003, n = 0.015) water flows at 300 cfs. A low dam (broad-crested weir) placed in the channel raises the water to a depth of 8.0 ft. Analyze the water surface profile upstream from the dam.*

BG

Eq. 10.8*a*: $300 = (1.486/0.015)(8y_0)[8y_0/(8 + 2y_0)]^{2/3}(0.003)^{1/2}$

By trial and error or by equation solver per Sample Prob. 10.1: $y_0 = 4.26$ ft

Eq. 10.23: $y_c = [(300/8)^2/32.2]^{1/3} = 3.52$ ft

Fig. 10.20: $y_0 > y_c$ so the channel slope is Mild. $y = 8.0$ ft caused by the dam is $> y_0 > y_c$.

Therefore, per Fig. 10.20: Upstream of the dam is an M_1 profile of great length with the depth gradually decreasing upstream to $y_0 = 4.26$ ft. ◄

Note: As damming action has occurred (Sec. 10.13), critical depth $y_c = 3.52$ ft occurs on the dam.

Sec. 10.17: Examples of Water Surface Profiles -- *Problems*

10.56 *Classify the water-surface profile of Prob. 10.45 as one of the forms shown in Fig. 10.20. Show all necessary calculations. Sketch the profile.*

Prob. 10.45: Rectangular flume, b = 5 ft, n = 0.012, S_0 = 0.01, Q = 60 cfs, y_1 = 1.5 ft.

BG

Eq. 10.8*a*: $60 = (1.486/0.012)(5y_0)[5y_0/(5 + 2y_0)]^{2/3}(0.01)^{1/2}$

By trial and error or by equation solver per Sample Prob. 10.1: $y_0 = 1.140$ ft

Eq. 10.23: $y_c = (q^2/g)^{1/3} = [(60/5)^2/32.2]^{1/3} = 1.648$ ft.

Fig. 10.20: $y_0 = 1.140$ ft $< y_c = 1.648$ ft, so the slope is Steep.

Fig. 10.20: Since $y_0 <$ given $y_1 = 1.5$ ft $< y_c$, this is an S_2 profile. ◄

Sketch:

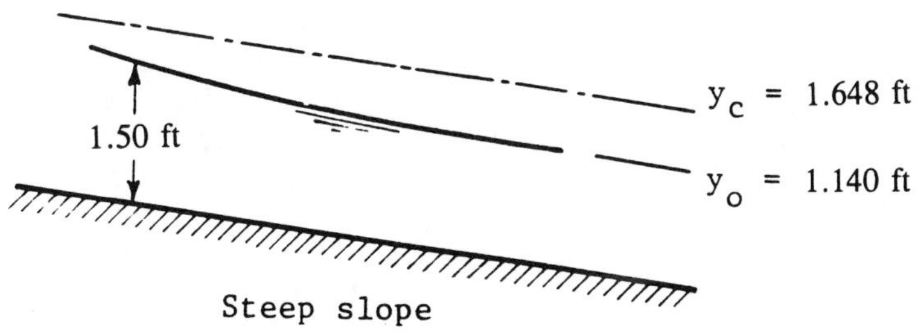

◄

10.57 *Classify the water-surface profile of Prob. 10.46 as one of the forms shown in Fig. 10.20. Show all necessary calculations. Sketch the profile.*

Prob. 10.46: Rectangular flume, $b = 1.6$ m, $n = 0.013$, $S_0 = 0.008$, $Q = 1.9\ m^3/s$, $y_1 = 0.50$ m.

SI

Eq. 10.8*b*: $1.9 = (1/0.013)(1.6y_0)[1.6y_0/(1.6 + 2y_0)]^{2/3}(0.008)^{1/2}$

By trial and error or by equation solver per Sample Prob. 10.1, $y_0 = 0.411$ m

Eq. 10.23: $y_c = (q^2/g)^{1/3} = [(1.9/1.6)^2/9.81]^{1/3} = 0.524$ m.

Fig. 10.20: $y_0 = 0.411$ m $< y_c = 0.524$ m, so slope is Steep.

Fig. 10.20: Since $y_0 <$ given $y_1 = 0.50$ m $< y_c$, this is an S_2 profile. ◀

Sketch:

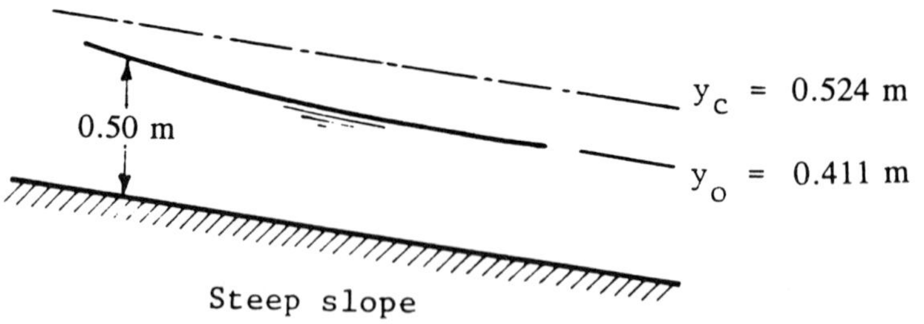

◀

10.58 *A trapezoidal canal dredged in smooth earth ($n = 0.030$) has a bottom width of 15 ft, side slopes of 1:1, and a bed slope $S_0 = 0.0003$. With a flow of 800 cfs, $y_c = 4.05$ ft, and $y_0 = 10.8$ ft. Find the length of M_2 curve extending from a free overfall back to where the depth is 10 ft. Use reaches with end depths of 6, 8, and 10 ft.*

BG

Given: $Q = 800$ cfs in trapezoidal channel with $b = 15$ ft and 1:1 side slopes, $n = 0.030$, $S_0 = 0.0003$, $y_0 = 10.8$ ft, $y_c = 4.05$ ft.

y ft	V fps	$V^2/2g$ ft	E ft	A ft²	P ft	$R = A/P$ ft	R_m ft	V_m fps	S Eq. 10.38*a*	Δx ft Eq. 10.39
10	3.20	0.159	10.159	250	43.3	5.78				
							5.33	3.77	0.000 623	5775
8	4.35	0.294	8.294	184	37.6	4.89				
							4.42	5.35	0.001 609	1273
6	6.35	0.626	6.626	126	32.0	3.94				
							3.43	8.36	0.005 51	174
4.05	10.37	1.670	5.720	77.2	26.5	2.92				
										Assume $4y_c = 16$
y_b										
										7239

▲

10.59 *A rectangular channel 10 ft wide carries 120 cfs in uniform flow at a depth of 2.00 ft. Suppose that an obstruction such as a submerged weir is placed across the channel, rising to a height of 8 in above the bottom. (a) Does this obstruction cause a hydraulic jump to form upstream? Why or why not? (b) Find the water depth over the obstruction, and just upstream of it. Classify the surface profile, if possible, to be found upstream from the weir. Sketch the resulting water surface profile and energy line, showing* y_c *and* y_0.

BG

(*a*) $q = Q/b = 120/10 = 12$ cfs/ft; Eq. 10.23: $y_c = (12^2/32.2)^{1/3} = 1.648$ ft

Fig. 10.20: The given $y_0 = 2.00$ ft $> y_c = 1.648$ ft, so the slope is Mild.

For $y = y_0$, the flow is subcritical, so no jump can occur ◄

(*b*) Eq. 10.17: $E_0 = 2.00 + (12/2.00)^2/(2 \times 32.2) = 2.56$ ft

Eq. 10.25: $E_{min} = 1.5(1.648) = 2.47$ ft

Sec. 10.13: $\Delta z_{crit} = E_0 - E_{min} = 2.56 - 2.47 = 0.0877$ ft

$\Delta z = 0.667$ ft $> \Delta z_{crit}$, so there is damming action, y_c occurs over the obstruction.

Water depth over obstruction = 1.648 ft ◄

Let y_1 = water depth just upstream of obstruction. Assuming no energy loss over obstruction,

$E_1 = E_{min} + \Delta z$, i.e., $y_1 + (12/y_1)^2/(2 \times 32.2) = 2.47 + 0.667 = 3.14$

By trial and error or by equation solver per Sample Prob. 10.1, $y_1 = 2.866$ ft, 1.030 ft, or a negative root which has no physical meaning.

Fig. 10.20: There is no *M* profile which enables normal flow to become supercritical further downstream, i.e., $y_1 = 1.030$ ft is not possible here. Thus $y_1 = 2.866$ ft. ◄

Fig. 10.20: Upstream of y_1, the water surface tends asymptotically to $y_0 = 2.00$ along an M_1 profile. ◄

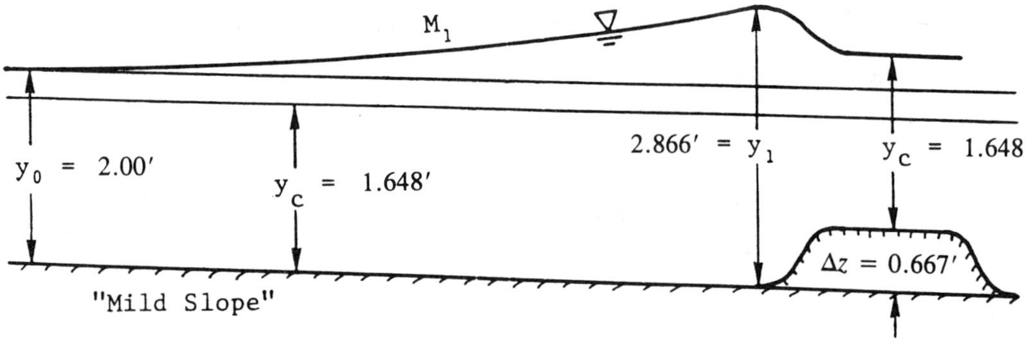

◄

10.60 *Suppose that the slope and roughness of the channel in Prob. 10.59 are such that uniform flow of 120 cfs occurs at a depth of 1.20 ft. Consider an obstruction rising 3 in above the bottom of the channel. Will a hydraulic jump form upstream? Classify the surface profile found just upstream from the obstruction.*
Prob. 10.59: Rectangular channel, b = 10 ft.

BG

$q = Q/b = 120/10 = 12$ cfs/ft; Eq. 10.23: $y_c = (12^2/32.2)^{1/3} = 1.648$ ft

Fig. 10.20: The given $y_0 = 1.20$ ft $< y_c = 1.648$ ft, so the slope is Steep.

For $y = y_0$, the flow is supercritical.

Eq. 10.17: $E_0 = 1.20 + (12/1.20)^2/(2 \times 32.2) = 2.75$ ft

Eq. 10.25: $E_{min} = 1.5(1.648) = 2.47$ ft

Sec. 10.13: $\Delta z_{crit} = E_0 - E_{min} = 2.75 - 2.47 = 0.281$ ft

$\Delta z = 3$ in $= 0.250$ ft $< \Delta z_{crit}$, so the obstruction is not sufficiently high to produce critical flow.

Upstream the flow is straight supercritical uniform flow, and a hydraulic jump cannot form. ◄

10.61 *Analyze the water-surface profile in a long rectangular channel (n = 0.014). The channel is 12 ft wide, the flow rate is 480 cfs, and there is an abrupt change in slope from *0.0019 to 0.018. Make a sketch showing normal depths, critical depths, and water surface profile types. [*Note this correction to the textbook.]*

BG

Eq. 10.23: $y_c = (q^2/g)^{1/3} = [(480/12)^2/32.2]^{1/3} = 3.68$ ft

On the upper slope ($S_{01} = 0.0019$), Eq. 10.8*a*: $480 = \dfrac{1.486}{0.014}(12y_{01})\left(\dfrac{12y_{01}}{12+2y_{01}}\right)^{2/3}(0.0019)^{1/2}$

By trial and error or by equation solver per Sample Prob. 10.1, $y_{01} = 4.58$ ft.

Using the same equation for the lower slope ($S_{02} = 0.018$), we find $y_{02} = 2.09$ ft.

Thus the flow is subcritical ($y_0 > y_c$) before the break and supercritical ($y_0 < y_c$) after the break.

Using Fig. 10.20:

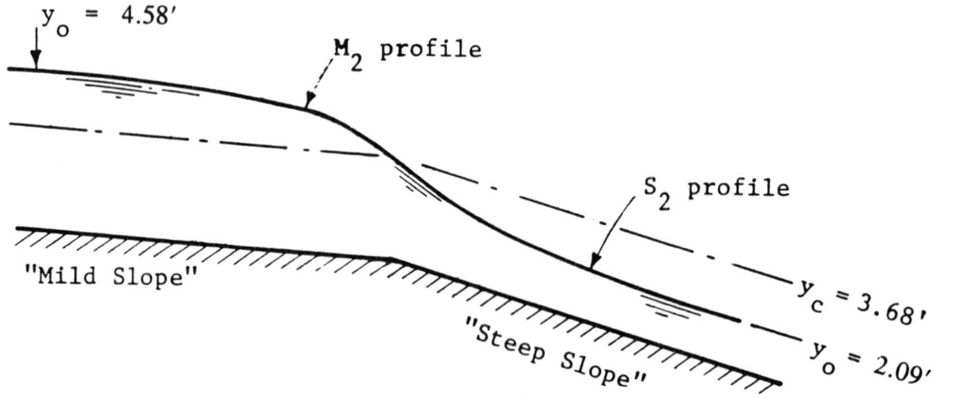

◄

Critical depth occurs in the vicinity of the break in slope. Depths at various distances upstream and downstream of the break may be found by use of Eq. 10.39.

10.62 *Repeat Prob. 10.61 for the case where the flow rate is 180 cfs.*

*Prob. 10.61: Rectangular 12-ft-wide channel, n = 0.014, abrupt change in slope from *0.0019 to 0.018. [*Note this correction to the textbook.]*

BG

Eq. 10.23: $y_c = [(180/12)^2/32.2]^{1/3} = 1.912$ ft

On the upper slope ($S_{01} = 0.0019$), Eq. 10.8*a*: $180 = (1.486/0.014)12y_0[12y_0/(12 + 2y_0)]^{2/3}0.0019^{1/2}$

By trial and error or by equation solver per Sample Prob. 10.1: $y_{01} = 2.31$ ft

Similarly, on the lower slope with $S_{02} = 0.0180$: $y_{02} = 1.104$ ft

$y_{01} > y_c > y_{02}$. The water surface profile has the same characteristics as that of Prob. 10.61. ◄

10.63 *Repeat Prob. 10.61 for the case where the slope change is from 0.0019 to 0.0007, and the flow rate is 180 cfs. Compute the approximate distance upstream from the break to the point where normal depth occurs, using one reach.*

Prob. 10.61: Rectangular 12-ft-wide channel, n = 0.014.

BG

Eq. 10.23: $y_c = (q^2/g)^{1/3} = [(180/12)^2/32.2]^{1/3} = 1.912$ ft

on the upper slope ($S_{01} = 0.0019$), Eq. 10.8*a*: $180 = \frac{1.486}{0.014}(12y_{01})\left(\frac{12y_{01}}{12 + 2y_{01}}\right)^{2/3}(0.0019)^{1/2}$

By trial and error or by equation solver per Sample Prob. 10.1, $y_{01} = 2.31$ ft.

Similarly, on the lower slope ($S_{02} = 0.0007$) we find: $y_{02} = 3.25$ ft

The flow is subcritical ($y_0 > y_c$) before the break in slope and is also subcritical ($y_0 > y_c$) after the break.

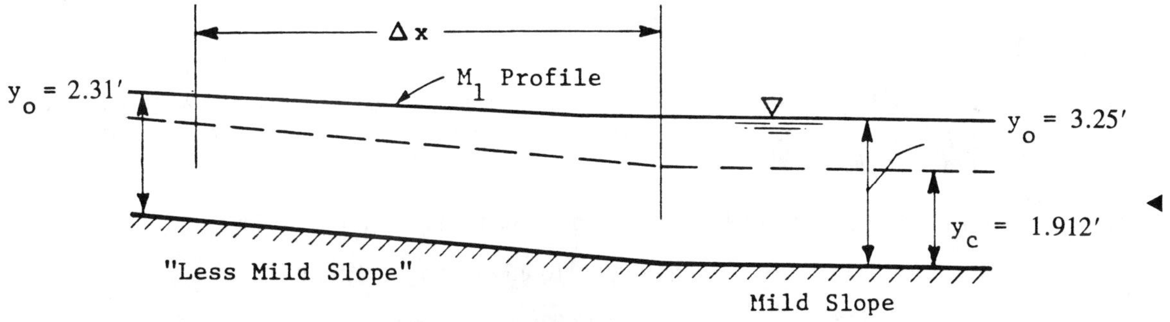

◄

Fig. 10.20: A water surface on a mild slope cannot diverge from y_0 in the upstream direction. Therefore the depth at the break in slope is 3.24 ft.

For the length of the M_1 profile (given $Q = 180$ cfs, $b = 12$ ft):

Sta	y ft	A ft^2 $10y$	P ft $12 + 2y$	R_h ft A/P	V fps $180/A$	$V^2/2g$ ft	E ft $y + V^2/2g$
1	2.31	27.7	16.61	1.668	6.49	0.655	2.965
2	3.25	39.0	18.50	2.108	4.62	0.331	3.581

$R_m = 1.888$ ft, $V_m = 5.55$ fps

Eq. 10.38*a*: $S = [0.014 \times 5.55/(1.486 \times 1.888^{2/3})]^2 = 0.001\,174$

Eq. 10.39: $\Delta x = (E_1 - E_2)/(S - S_0) = (2.965 - 3.581)/(0.001\,174 - 0.0019) = 848$ ft ◄

Greater accuracy can be achieved if more steps are taken.

10.64 *In an 8-ft-wide rectangular channel (n = 0.015) water flows at 300 cfs. A low dam (broad-crested weir) placed in the channel raises the water to a depth of 8.0 ft. Analyze the water-surface profile upstream from the dam if the channel slope is (a) 0.0006, (b) 0.0009, and (c) 0.006.*

BG

y_c is independent of slope. Eq. 10.23: $y_c = (q^2/g)^{1/3} = [(300/8)^2/32.2]^{1/3} = 3.52$ ft

(*a*) With $S_0 = 0.0006$: Eq. 10.8*a*: $300 = (1.486/0.015)\,8y_0\,[8y_0/(8+2y_0)]^{2/3}0.0006^{1/2}$

By trial and error or by equation solver per Sample Prob. 10.1, $y_0 = 8.03$ ft

Fig. 10.20: $y_0 = 8.03 > y_c = 3.52$, so the slope is Mild.

Insertion of a dam cannot lower the water surface below y_0, to $y = 8.0$ ft, so this situation is impossible. ◀

(*b*) With $S_0 = 0.0009$: Repeating the procedure of Part (*a*), by trial or by equation solver, $y_0 = 6.81$ ft

Fig. 10.20: $y_0 = 6.81 > y_c = 3.52$, so the slope is Mild.

$y = 8.0$ ft caused by the dam is $> y_0 > y_c$. Therefore, per Fig. 10.20: Upstream of the dam is a long M_1 profile with the depth gradually decreasing upstream to $y_0 = 6.81$ ft. As damming action has occurred (Sec. 10.13), depth $y_c = 3.52$ ft occurs on the dam. ◀

(*c*) With $S_0 = 0.006$: Repeating the procedure of Part (*a*), by trial or by equation solver, $y_0 = 3.29$ ft

Fig. 10.20: $y_0 = 3.29 < y_c = 3.52$, so the slope is Steep.

$y = 8.0$ ft caused by the dam is $> y_c > y_0$. Therefore, per Fig. 10.20: Upstream of the dam is an S_1 profile, preceded by a hydraulic jump. Upstream of the jump is straight supercritical uniform flow with depth $y_0 = 3.29$ ft. ◀

Sec. 10.18: Hydraulic Jump -- *Exercises*

10.18.1 *A hydraulic jump occurs in a 15-ft-wide rectangular channel carrying 200 cfs on a slope of 0.005. The depth after jump is 4.5 ft. (a) What must be the depth before jump? (b) What are the losses of energy and power in the jump?*

BG

(*a*) Neglecting the frictional effect and weight component (small slope),

$q = 200/15 = 13.33$ cfs/ft; Eq. 10.46*b* with $y_2 = 4.5$ ft:

$y_1 = (4.5/2)\left(-1 + \sqrt{1 + (8 \times 13.33^2)/(32.2 \times 4.5^3)}\right) = 0.492$ ft ◀

(*b*) Eq. 10.47: $h_L = \left[0.492 + (1/2g)(13.33/0.492)^2\right] - \left[4.5 + (1/2g)(13.33/4.5)^2\right] = 7.28$ ft ◀

Eq. 5.32: Power loss = 62.4(200)7.28/550 = 165.2 hp ◀

10.18.2 *A hydraulic jump occurs in a 6-m-wide rectangular channel carrying 8 m³/s on a slope of 0.005. The depth after jump is 1.5 m. (a) What must be the depth before jump? (b) What are the losses of energy and power in the jump?*

SI

(*a*) $q = 8/6 = 1.333$ m³/s per m; Eq. 10.46*b* with $y_2 = 1.5$ m:

$y_1 = (1.5/2)\left(-1 + \sqrt{1 + (8 \times 1.333^2)/(9.81 \times 1.5^3)}\right) = 0.1467$ m ◀

(*b*) $V = Q/A = q/y$;

Eq. 10.47: $h_L = \left[0.1467 + (1/2g)(1.333/0.1467)^2\right] - \left[1.5 + (1/2g)(1.333/1.5)^2\right] = 2.81$ m ◀

Eq. 5.33: Power loss = 9810(8)2.82/1000 = 221 kW ◀

10.18.3 *The hydraulic jump may be used as a crude flowmeter. Suppose that in a horizontal rectangular channel 6 ft wide the observed depths before and after a hydraulic jump are 0.80 and 3.60 ft, respectively. Find the rate of flow and the head loss.*

BG

From Eq. 10.45: $q = \sqrt{32.2(0.80)3.60(0.80 + 3.60)/2} = 14.28$ cfs/ft; $Q = bq = 6(14.28) = 85.7$ cfs ◄

(An alternative, but less convenient, solution for Q uses Eq. 10.42.)

$V_1 = 14.28/0.80 = 17.85$ fps, $V_1^2/2g = 4.95$ ft; $V_2 = 14.28/3.60 = 3.97$ fps, $V_2^2/2g = 0.244$ ft

$h_L = E_1 - E_2 = (0.80 + 4.95) - (3.60 + 0.244) = 1.906$ ft ◄

Sec. 10.18: Hydraulic Jump -- *Problems*

10.65 *Derive Eq. (10.48) in the manner suggested in Sec. 10.18.*

N

Eq. 10.47: $h_{Lj} = E_1 - E_2 = y_1 - y_2 + (V_1^2 - V_2^2)/2g$; writing $V = q/y$ this becomes:

$$h_{Lj} = (y_1 - y_2) + \frac{q^2}{2g}\left(\frac{1}{y_1^2} - \frac{1}{y_2^2}\right) = (y_1 - y_2) + \frac{q^2}{2g}\left(\frac{y_2^2 - y_1^2}{y_1^2 y_2^2}\right) = \frac{q^2}{2g}\frac{(y_2 - y_1)(y_2 + y_1)}{y_1^2 y_2^2} - (y_2 - y_1)$$

Also, Eq. 10.45 states that: $q^2/g = y_1 y_2 (y_1 + y_2)/2$; Substituting to eliminate q^2/g:

$$h_{Lj} = \frac{y_1 y_2 (y_2 + y_1)}{4}\frac{(y_2 - y_1)(y_2 + y_1)}{y_1^2 y_2^2} - (y_2 - y_1) = \left[\frac{(y_2 + y_1)^2}{4y_1 y_2} - 1\right](y_2 - y_1)$$

$$= \left[\frac{y_2^2 + 2y_1 y_2 + y_1^2 - 4y_1 y_2}{4y_1 y_2}\right](y_2 - y_1) = \left[\frac{(y_2 - y_1)^2}{4y_1 y_2}\right](y_2 - y_1) = \frac{(y_2 - y_1)^3}{4y_1 y_2} = \text{Eq. 10.48}$$ ◄

10.66 *In a rectangular channel 10 ft wide with a flow of 200 cfs the depth is 1 ft. If a hydraulic jump is produced, (a) what will be the depth after it? (b) What will be the loss of energy?*

BG

(*a*) Given: $Q = 200$ cfs, $b = 10$ ft, $y_1 = 1.00$ ft; $\therefore$ $q = 200/10 = 20$ cfs/ft

Eq. 10.46*a*: $y_2 = (1.00/2)(-1 + [1 + (8 \times 20^2)/(32.2 \times 1.00^3)]^{1/2}) = 4.51$ ft ◄

(*b*) $V_1 = 20/1.00 = 20.0$ fps, $V_1^2/2g = 20^2/(2 \times 32.2) = 6.21$ ft;

$V_2 = 20/4.51 = 4.44$ fps, $V_2^2/2g = 0.305$ ft

Eq. 10.47: $h_L = E_1 - E_2 = (1.00 + 6.21) - (4.51 + 0.305) = 2.40$ ft·lb/lb ◄

(*b*) Alt: Eq. 10.48: $h_L = (4.51 - 1.00)^3/[4(1.00)4.51] = 2.40$ ft·lb/lb ◄

10.67 *In a rectangular channel 4 m wide with a flow of 7.65 m^3/s the depth is 0.4 m. If a hydraulic jump is produced, (a) what will be the depth after it? (b) What will be the loss of energy?*

SI

(*a*) Given: $Q = 7.65$ m^3/s, $b = 4$ m, $y_1 = 0.4$ m; $\therefore$ $q = 7.65/4 = 1.9125$ m^3/s per m

Eq. 10.46*a*: $y_2 = (0.4/2)(-1 + [1 + (8 \times 1.9125^2)/(9.81 \times 0.4^3)]^{1/2}) = 1.180$ m ◄

(*b*) $V_1 = 1.9125/0.4 = 4.78$ m/s, $V_1^2/2g = 4.78^2/(2 \times 9.81) = 1.165$ m;

$V_2 = 1.9125/1.180 = 1.621$ m/s, $V_2^2/2g = 0.1339$ m

Eq. 10.47: $h_L = E_1 - E_2 = (0.4 + 1.165) - (1.180 + 0.1339) = 0.251$ N·m/N ◄

(*b*) Alt: Eq. 10.48: $h_L = (1.180 - 0.4)^3/[4(0.4)1.180] = 0.251$ N·m/N ◄

10.68 *Repeat Prob. 10.66 for the case where the channel bed slopes at 10°. For this slope, jump length $\approx 4y_2$. Assume friction force = 400 lb/ft of width. Also find the horsepower loss.*

Prob. 10.66: Rectangular channel, b = 10 ft, Q = 200 cfs, y = 1 ft. If a hydraulic jump is produced, (a) what will be the depth after it? (b) What will be the loss of energy?

BG

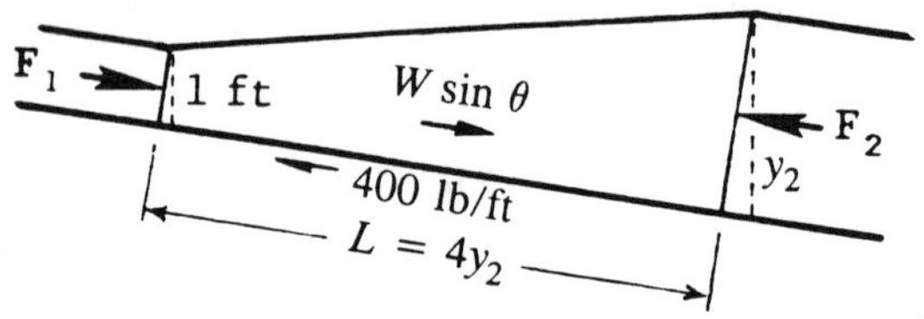

Note (per Secs. 10.1 and 10.2) that L is measured along the channel bed and depth is measured vertically.

(*a*) Impulse-momentum equation 6.7*a* parallel to the bed, per foot of channel width:

$F_1 - F_2 + W\sin\theta - 400 = \rho(Q/b)(V_2 - V_1)$ where $F_1 = (\gamma y_1/2)(y_1\cos 10°) = 30.7$ lb/ft,

$F_2 = \gamma y_2^2\cos 10°/2 = 30.7y_2^2$ lb/ft, and $W = 0.5(1 + y_2)\cos 10°(4y_2)62.4 = 122.9y_2(1 + y_2)$

Thus $30.7(1 - y_2^2) + 122.9y_2(1 + y_2)\sin 10° - 400 = 1.938(20)[(20/y_2) - 20]/\cos 10°$

This is a cubic equation in y_2. By trial and error or by equation solver, $y_2 = 6.95$ ft ◄

Note to instructor: In Chow's *Open Channel Hydraulics*, McGraw-Hill, 1959, a procedure for jumps in sloping channels is given in Sec. 15-16. By that method $y_2 = 6.3$ ft.

(*b*) $V_2 = Q/(by_2\cos 10°) = 200/(10 \times 6.95\cos 10°) = 2.92$ fps; similarly, $V_1 = 20.3$ fps

Eq. 10.47: $h_L = [1 + (20.3^2/2g)] - [6.95 + (2.92^2/2g)] = 7.40 - 7.08 = 0.325$ ft ◄

Eq. 5.32: hp loss $= \gamma Qh_L/550 = 62.4(200)0.325/550 = 7.38$ hp ◄

10.69 *Repeat Prob. 10.67 for the case where the channel bed slopes at 10°. For this slope, jump length $\approx 4y_2$. Assume friction force = 6 kN/m of width. Also find the power (kW) loss.*

Prob. 10.67: Rectangular channel, b = 4 m, Q = 7.65 m³/s, y = 0.4 m. If a hydraulic jump is produced, (a) what will be the depth after it? (b) what will be the loss of energy?

SI

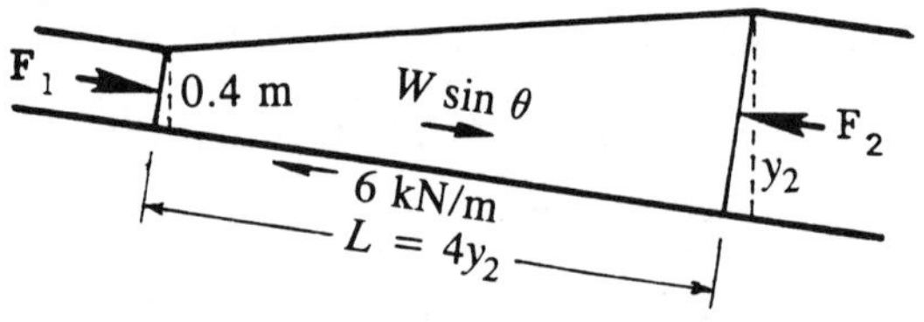

Note (per Secs. 10.1 and 10.2) that L is measured along the channel bed and depth is measured vertically.

(*a*) Impulse-momentum equation 6.7*a* parallel to the bed, per meter of channel width:

$F_1 - F_2 + W\sin\theta - 6000 = \rho(Q/b)(V_2 - V_1)$ where $F_1 = (\gamma y_1/2)(y_1\cos 10°) = 773$ N/m,

$F_2 = \gamma y_2^2\cos 10°/2 = 4830y_2^2$ N/m, and $W = 0.5(0.4 + y_2)\cos 10°(4y_2)9810 = 19\,320y_2(0.4 + y_2)$

Thus $773 - 4830y_2^2 + 19\,320y_2(0.4 + y_2)\sin 10° - 6000 = 1000(1.913)^2[(1/y_2) - 1/0.4]/\cos 10°$

This is a cubic equation in y_2. By trial and error or by equation solver, $y_2 = 0.918$ m ◄

See *Note to instructor* in solution to Prob. 10.68.

(*b*) Eq. 10.47: $h_L = [0.4 + (1.913/0.4\cos 10°)^2/2g] - [0.918 + (1.913/0.918\cos 10°)^2/2g] = 0.455$ m ◄

Eq. 5.33: Power loss $= \gamma Qh_L/1000 = 9810(7.65)0.455/1000 = 34.2$ kW ◄

10.70 *The tidal bore, which carries the tide into the estuary of a large river, is an example of an abrupt translatory wave, or moving hydraulic jump. Suppose such a bore is observed to rise to a height of 13 ft above the normal low-tide river depth of 10 ft. The speed of travel of the bore upstream is observed to be 13 mph. Find the approximate velocity of the undisturbed river. Does this represent subcritical or supercritical flow? [Note: The theory developed in Sec. 10.18 is based on the hydraulic jump in a fixed position. In the case of a moving jump, all kinematic terms must be considered relative to the moving wave as a frame of reference.]*

BG

$y_1 = 10$ ft; $y_2 = 10 + 13 = 23$ ft; Eq. 10.45: $q^2/32.2 = (23)10(23 + 10)/2 = 3795$;

from which $q = 349.6$ cfs/ft, and $V_1 = 349.6/10 = 34.96$ fps (relative to the jump)

Velocity of jump $= 13(44/30) = 19.07$ fps

The jump is moving upstream at a velocity of 19.07 fps while the river is moving into the jump at 34.96 fps. Therefore, the river is moving downstream at $(34.96 - 19.07) = 15.89$ fps. ◄

$V_2/2g = 15.89^2/(2 \times 32.2) = 3.92$ ft $< y/2 = 5$ ft; From Sec. 10.10: The flow is subcritical. ◄

10.71 *A hydraulic jump occurs in a triangular flume having side slopes at 1:1. The flow rate is 18 cfs and the depth before the jump is 1.2 ft. Find the depth after the jump and the power loss in the jump.*

BG

Divide Eq. 10.42 by γ and substitute $V_i = Q/A_i$: $Q^2/gA_1 + h_{c1}A_1 = Q^2/gA_2 + h_{c2}A_2$

For triangular section $h_c = y/3$, $A = y^2$; $y_1 = 1.2$ ft, $Q = 18$ cfs (given)

Thus $18^2/[32.2(1.2)^2] + (1.2/3)(1.2)^2 = 18^2/(32.2y_2^2) + (y_2/3)(y_2)^2$

or $22.7 = 30.2/y_2^2 + y_2^3$; by trial and error or by equation solver, $y_2 = 2.64$ ft ◄

Eq. 10.47: $h_L = [1.2 + (1/2g)(18/1.2^2)^2] - [2.64 + (1/2g)(18/2.64^2)^2] = 0.883$ ft

Eq. 5.32: Power loss $= 62.4(18)0.883/550 = 1.803$ hp ◄

10.72 *A hydraulic jump occurs in a triangular flume having side slopes at 1:1. The flow rate is 0.45 m³/s and the depth before the jump is 0.3 m. Find the depth after the jump and the power loss in the jump.*

SI

Divide Eq. 10.42 by γ and substitute $V_i = Q/A_i$: $Q^2/gA_1 + h_{c1}A_1 = Q^2/gA_2 + h_{c2}A_2$

For triangular section $h_c = y/3$, $A = y^2$; $y_1 = 0.3$ m, $Q = 0.45$ m^3/s (given)

Thus $0.45^2/[9.81(0.3)^2] + (0.3/3)(0.3)^2 = 0.45^2/(9.81y_2^2) + (y_2/3)(y_2)^2$

or $0.715 = 0.0619/y_2^2 + y_2^3$; by trial and error or by equation solver, $y_2 = 0.858$ m ◄

Eq. 10.47: $h_L = [0.3 + (1/2g)(0.45/0.3^2)^2] - [0.858 + (1/2g)(0.45/0.858^2)^2] = 0.697$ m

Eq. 5.33: Power loss $= 9810(0.45)0.697/1000 = 3.08$ kW ◄

Sec. 10.19: Location of Hydraulic Jump -- ***Problem***

10.73 *A wide and shallow rectangular channel with bed slope S_0 = 0.0004 and roughness n = 0.022 carries a steady flow of 65 cfs/ft of width. If a sluice gate (Fig. 11.34) is adjusted as to produce a minimum depth of 1.6 ft in the channel, determine whether a hydraulic jump will form downstream, and if so, find (using one reach) the distance from the gate to the jump.*

BG

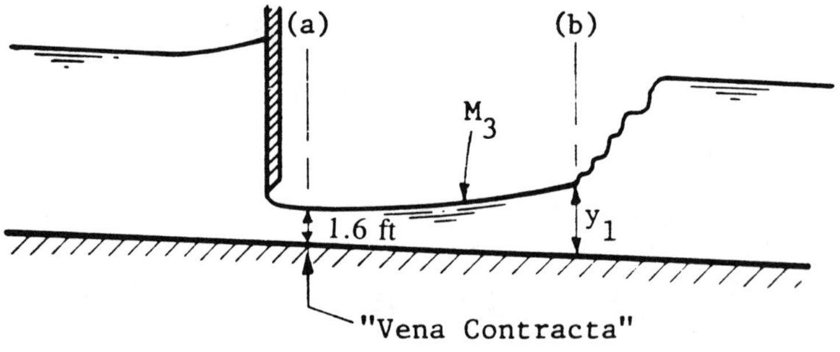

Given: q = 65 cfs/ft, n = 0.022, S_0 = 0.0004, and the channel is "wide and shallow" so that $R_h = y$

Eq. 10.8*a* for unit width: $65 = (1.486/0.022)(1y_0)(y_0)^{2/3}(0.0004)^{1/2}$; y_0 = 10.22 ft

Eq. 10.23: $y_c = (65^2/32.2)^{1/3}$ = 5.08 ft. $y_0 > y_c$, so (per Fig. 10.20) the channel slope is Mild.

Alt. from Eq. 10.30 or 31: S_c = 0.004 11. $\therefore S_0 < S_c$ and (Sec. 10.10) the bed slope is Mild.

y = 1.6 ft caused by the gate is $< y_c < y_0$. Therefore, per Fig. 10.20:

Downstream of the sluice gate is an M_3 profile, which must be followed by a hydraulic jump to enable the flow to return to normal depth. ◀

Sec. 10.18: The channel slope (0.0004) is much less than 3° (or S = 0.0524), so the weight component in the jump may be neglected.

Eq. 10.46*b* with $y_2 = y_0$ = 10.20: y_1 = 2.087 ft

For the distance along the M_3 profile from y = 1.6 ft to y_1:

Sta	$y = R_h$ ft	$V = q/y$ fps	$V^2/2g$ ft	$E = y + V^2/2g$ ft	R_m ft	V_m fps
(*a*)	1.6	40.6	25.63	27.227		
					1.844	35.88
(*b*)	2.087	31.1	15.06	17.148		

Eq. 10.38*a*: $S = [(0.022 \times 35.88)/(1.486 \times 1.844^{2/3})]^2 = 0.1249$

Eq. 10.39: $\Delta x = (E_1 - E_2)/(S - S_0) = (27.227 - 17.148)/(0.1249 - 0.0004) = 81.0$ ft ◀

Sec. 10.20: Velocity of Gravity Wave -- ***Exercise***

10.20.1 *A thin rod is placed vertically in a stream which is 4 ft deep, and the resulting small disturbance wave is observed to make an angle of about 60° with the axis of the stream. Find the approximate velocity of the stream.*

BG

Eq. 10.49: $c \approx (32.2 \times 4)^{1/2}$ = 11.35 fps; Eq. 10.51: $\sin\beta = \sin 60° \approx 11.35/V$

$\therefore V \approx 11.35/\sin 60° = 13.10$ fps ◀

Sec. 10.20: Velocity of Gravity Wave -- *Problems*

10.74 *At a point in a shallow lake, the wave from a passing boat is observed to rise 0.5 ft above the undisturbed water surface. The observed speed of the wave is 8 mph. Find the approximate depth of the lake at this point. Compute it three ways and comment on them.*

BG

c = 8 mph = 8(44/30) fps = 11.73 fps; Δy = 0.5 ft. Using Eq. 10.49 three ways:

(*i*) $c^2 = 137.7 = g(y + 0.5)(y + 0.25)/y$ (quadratic); y = 3.49 ft ◄

(*ii*) $137.7 \approx g(y + 0.5)$; $y \approx 3.78$ ft ◄ (*iii*) $137.7 \approx gy$; $y \approx 4.28$ ft ◄

y = 3.49 ft is accurate; the other two values are approximations with errors of 8% and 23% respectively, on the high side. ◄

10.75 *At a point in a shallow lake, the wave from a passing boat is observed to rise 30 cm above the undisturbed water surface. The observed speed of the wave is 16 km/h. Find the approximate depth of the lake at this point. Compute it three ways and comment on them.*

SI

c = 16 kph = 16(1000/3600) m/s = 4.44 m/s; Δy = 0.30 m. Using Eq. 10.49 three ways:

(*i*) $c^2 = 19.75 = g(y + 0.3)(y + 0.15)/y$ (quadratic); y = 1.534 m ◄

(*ii*) $19.75 \approx g(y + 0.3)$; $y \approx 1.714$ m ◄ (*iii*) $19.75 \approx gy$; $y \approx 2.01$ m ◄

y = 1.534 m is accurate; the other two values are approximations with errors of 12% and 31% respectively, on the high side. ◄

Sec. 10.21: Flow Around Channel Bends -- *Exercise*

10.21.1 *A rectangular channel 5 m wide carries 8 m³/s in uniform flow at a depth of 2 m. What will be the maximum difference in water-surface elevations between the inside and outside of a circular bend in this channel if the radius of the bend at the centerline is 20 m?*

SI

$V = Q/A = 8/(5 \times 2) = 0.80$ m/s; Sec. 10.10: $\mathbf{F} = 0.8/\sqrt{9.81(2.0)} = 0.181$, so flow is <u>sub</u>critical

Eq. 10.52: $\Delta y = 0.8^2(5)/(9.81 \times 20) = 0.016\,31$ m = 16.31 mm ◄

Sec. 10.21: Flow Around Channel Bends -- *Problem*

10.76 *A rectangular channel 12 ft wide carries 360 cfs under uniform flow conditions. By how much should the outside wall be elevated above the inside wall for a bend of 45-ft radius to the center line of the channel, if (a) the normal depth is 4.4 ft and (b) the normal depth is 2.2 ft.*

BG

(*a*) $V = Q/A = 360/(12 \times 4.4) = 6.82$ fps

Sec. 10.10: $\mathbf{F} = 6.82/\sqrt{32.2(4.4)} = 0.573 < 1$, so flow is <u>sub</u>critical.

Eq. 10.52: Difference in wall elevations = $\Delta y = 6.82^2(12)/(32.2 \times 45) = 0.385$ ft ◄

(*b*) $V = Q/A = 360/(12 \times 2.2) = 13.64$ fps

Sec. 10.10: $\mathbf{F} = 13.64/(32.2 \times 2.2)^{1/2} = 1.620 > 1$, so flow is <u>super</u>critical.

Eq. 10.52: $\Delta y = 13.64^2(12)/(32.2 \times 45) = 1.540$ ft

Because of (supercritical) wave action, maximum water depth at inside wall = y_0 and maximum water depth at outside wall = $(y_0 + \Delta y)$.

∴ Required difference in wall elevations = Δy = 1.540 ft ◄

10.22.1 *Water flows at 180 cfs from a trapezoidal channel (b = 10 ft, side slopes = 2.5 H:1V) into a rectangular channel with the same bed width. If the upstream depth is 3.5 ft and the transition is abrupt (k_t = 0.4), find the downstream depth.*

BG

Trapezoidal: $A_1 = [10 + (2.5)3.5]3.5 = 65.6 \text{ ft}^2$; rectangular: $A_2 = 10y_2$

$V_1 = Q/A_1 = 180/65.6 = 2.74$ fps; $V_2 = Q/A_2 = 180/(10y_2) = 18/y_2$

Using Eq. 10.53: $E_1 = E_2 + h_{Lt}$, i.e., $3.5 + \dfrac{2.74^2}{2(32.2)} = y_2 + \dfrac{V_2^2}{2(32.2)} + 0.4\left(\dfrac{V_2^2}{2(32.2)} - \dfrac{2.74^2}{2(32.2)}\right)$

from which $3.66 = y_2 + 0.217V_2^2$ and substituting for $V_2 = 18/y_2$: $3.66 = y_2 + 7.04/y_2^2$;

by trial and error or by equation solver, $y_2 = 2.18$ ft ◄

10.22.2 *A rectangular, 3-m-wide channel is connected to a trapezoidal channel (b = 1 m, side slopes = 1.5H:1V) by a warped transition. When 7.5 m³/s of water is flowing, the depth in the upstream channel is 1.5 m. (a) Find y_2. (b) Is the flow through the transition expanding or contracting?*

SI

Upstream rectangular: $A_1 = 3(1.5) = 4.5 \text{ m}^2$; downstream trapezoidal: $A_2 = (1 + 1.5y_2)y_2$

$V_1 = Q/A_1 = 7.5/4.5 = 1.667$ m/s; $V_2 = Q/A_2 = 7.5/[(1 + 1.5y_2)y_2]$

(*i*) Assume flow through transition is contracting ($A_2 < A_1$, $V_2 > V_1$).

Table 10.4 for contracting flow through warped transition: $k_t = 0.1$

Using Eq. 10.53: $E_1 = E_2 + h_{Lt}$, i.e., $1.5 + \dfrac{1.667^2}{2(9.81)} = y_2 + \dfrac{V_2^2}{2(9.81)} + 0.1\left(\dfrac{V_2^2}{2(9.81)} - \dfrac{1.667^2}{2(9.81)}\right)$

and substituting for V_2 this yields: $1.656 = y_2 + \dfrac{3.15}{[(1 + 1.5y_2)y_2]^2}$

By trial and error or by equation solver, $y_2 = 1.532$ m

$\therefore A_2 = (1 + 1.5 \times 1.532)1.532 = 5.05 \text{ m}^2$

$A_2 = 5.05 > A_1 = 4.5$, so the flow is <u>not</u> contracting, our assumption is not valid.

(*ii*) Assume flow is expanding ($A_2 > A_1$, $V_2 < V_1$). Table 10.4: $k_t = 0.3$

Using Eq. 10.53: $E_1 = E_2 + h_{Lt}$, i.e., $1.5 + \dfrac{1.667^2}{2(9.81)} = y_2 + \dfrac{V_2^2}{2(9.81)} + 0.3\left(\dfrac{1.667^2}{2(9.81)} - \dfrac{V_2^2}{2(9.81)}\right)$

and substituting for V_2 this yields: $1.599 = y_2 + \dfrac{2.01}{[(1 + 1.5y_2)y_2]^2}$

By trial and error or by equation solver, $y_2 = 1.518$ m

$\therefore A_2 = (1 + 1.5 \times 1.518)1.518 = 4.97 \text{ m}^2$

$A_2 = 4.97 > A_1 = 4.5$, so the flow <u>is</u> expanding, our assumption is valid

(*a*) $y_2 = 1.518$ m ◄ (*b*) The flow is expanding ◄

10.77 *Refer to Fig. 10.32. A rectangular channel changes in width from 5 to 7.5 ft. Measurements indicate that y_1 = 3.00 ft and Q = 62 cfs. Determine the depth y_2 by (a) neglecting head loss; (b) considering the head loss to be given as shown on the figure.*

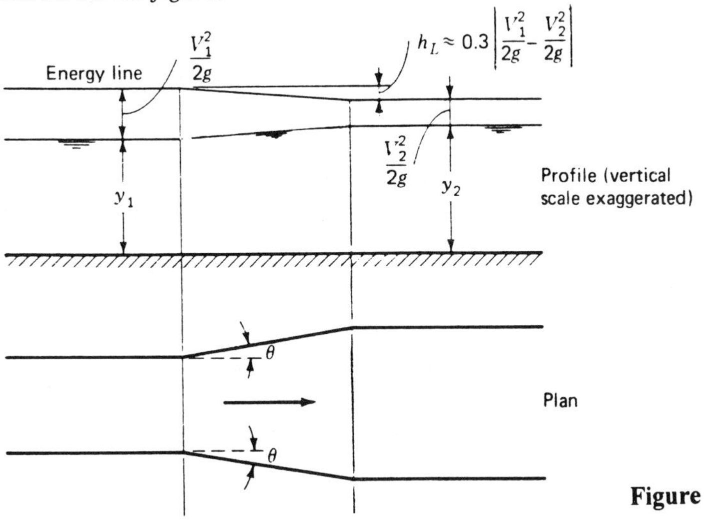

Figure 10.32

BG

$V_1 = Q/A_1 = 62/(3 \times 5) = 4.13$ fps; $V_2 = 62/7.5y_2 = 8.27/y_2$ fps

(*a*) $h_L = 0$, $\therefore E_1 = E_2$ i.e., $3.0 + 4.13^2/(2 \times 32.2) = y_2 + (8.27/y_2)^2/(2 \times 32.2)$

By trial and error or by equation solver, $y_2 = 3.16$ ft ◀

(*b*) Fig. 10.32 and assuming expanding flow: $E_1 = E_2 + h_L = E_2 + 0.3(V_1^2/2g - V_2^2/2g)$

$y_1 + V_1^2/2g = y_2 + V_2^2/2g + 0.3(V_1^2/2g - V_2^2/2g)$ i.e., $y_1 + 0.7V_1^2/2g = y_2 + 0.7V_2^2/2g$

i.e., $3.0 + 0.186 = 3.186 = y_2 + 0.743/y_2^2$; by trial and error or by equation solver, $y_2 = 3.11$ ft

Check: $A_1 = 5(3) = 15$ ft²; $A_2 = 7.5(3.11) = 23.3$ ft²

So flow is expanding, assumption and result are valid. $y_2 = 3.11$ ft ◀

10.78 *A rectangular channel changes in width from 1.35 m to 2.0 m with the wedge-shaped transition shown in Fig. 10.32. Given that y_1 = 0.75 m and Q = 1.5 m³/s, find the depth y_2 by (a) neglecting head loss and (b) using a loss coefficient k_t of 0.4. (c) Does the head loss increase or decrease the flow expansion?*

SI

See Fig. 10.32 with Solution 10.77 above.

$V_1 = Q/A_1 = 1.5/(1.35 \times 0.75) = 1.481$ m/s; $V_2 = 1.5/2y_2 = 0.75/y_2$ m/s

(*a*) $h_L = 0$, $\therefore E_1 = E_2$, i.e., $0.75 + 1.481^2/(2 \times 9.81) = y_2 + (0.75/y_2)^2/(2 \times 9.81)$

By trial and error or by equation solver, $y_2 = 0.819$ m ◀

(*b*) Table 10.4 for wedge transition (given in Fig. 10.32) with $k_t = 0.4$: the flow expands, so $V_1 > V_2$

Per Fig. 10.32: $E_1 = E_2 + 0.4(V_1^2/2g - V_2^2/2g)$; $y_1 + V_1^2/2g = y_2 + V_2^2/2g + 0.4(V_1^2 - V_2^2)/2g$

$y_1 + 0.6V_1^2/2g = y_2 + 0.6V_2^2/2g$; $0.75 + 0.0671 = 0.817 = y_2 + 0.01720/y_2^2$

By trial and error or by equation solver, $y_2 = 0.790$ m ◀

(*c*) Expansion (*a*) without head loss $= A_2/A_1 = 2(0.819)/(1.35 \times 0.75) = 1.618$

Expansion (*b*) with head loss $= 2(0.790)/(1.35 \times 0.75) = 1.560$, so head loss <u>reduces</u> expansion ◀

10.79 *Flow from a trapezoidal channel (20-ft bed width, 1:1 side slopes) enters a 20-ft-wide rectangular channel via a cylinder-quadrant transition. (a) If the upstream depth is 4 ft when the flow rate is 530 cfs, find the resulting downstream depth. (b) If the flow increases to 550 cfs, causing the upstream depth to increase to 4.1 ft, does the downstream depth increase or decrease? By how much?*

BG

(*a*) $A_1 = (20 + 1y_1)y_1 = (20 + 4)4 = 96\ \text{ft}^2$; $A_2 = 20y_2$

$V_1 = Q/A_1 = 530/96 = 5.52$ fps; $V_2 = 530/(20y_2) = 26.5/y_2$

The downstream rectangular channel is "narrower," so try contracting flow ($A_2 < A_1$, $V_2 > V_1$).

Table 10.4 for contracting flow through cylinder-quadrant transition: $k_t = 0.2$

Using Eq. 10.53: $E_1 = E_2 + h_L$, i.e., $4.0 + 5.52^2/2g = y_2 + V_2^2/2g + 0.2(V_2^2/2g - 5.52^2/2g)$

$4.57 = y_2 + 0.018\,63V_2^2$. Substituting for V_2: $4.57 = y_2 + 13.09/y_2^2$

By trial and error or by equation solver, $y_2 = 3.4994$ ft

$\therefore\ A_2 = 20(3.50) = 70.0\ \text{ft}^2 < A_1 = 96\ \text{ft}^2$, flow is contracting.

Assumption and result are valid, $y_2 = 3.50$ ft ◄

(*b*) $Q = 550$ cfs, $y_1 = 4.1$ ft. Repeating procedure of part (*a*), $y_2 = 3.5743$ ft.

$A_2 = 71.5\ \text{ft}^2 < A_1 = 98.8\ \text{ft}^2$ so flow is still contracting.

Downstream depth change = 3.5743 − 3.4994 = 0.0750 ft increase ◄

Sec. 10.23: Hydraulics of Culverts -- *Exercise*

10.23.1 *Repeat Sample Prob. 10.13 for the situation where the culvert must be 100 m long.*
Sample Prob. 10.13: $S_0 = 0.003$, $Q = 4.3\ m^3/s$, maximum headwater = 3.6 m above invert. Neglect velocity of approach, assume square-edged inlet with $k_e = 0.5$, $C_d = 0.65$, and free discharge at outlet. What size of corrugated pipe ($n = 0.025$) is needed?

SI

Sec. 10.23: Assume headwater depth/$D > 1.2$ so that entrance is submerged, per Fig. 10.33.

$\therefore\ D <$ headwater depth/1.2 = 3.6 m/1.2 = 3.0 m

Given discharge is free (not submerged), so flow is per Fig. 10.33*b* or 10.33*c*.

Assume the full flow of Fig. 10.33*b*: $V = Q/A = 4.3/[\pi D^2/4] = 5.47/D^2$; $R_h = D/4$

Fig. 10.33*b*: $\Delta h = (y_1 - y_3) + (z_1 - z_3) = y_1 - y_3 + S_0L = 3.6 - D + 0.003(100) = 3.9 - D$

Eq. 10.56: (in SI units) $$\Delta h = \left(0.5 + \frac{(2)9.81(0.025)^2 100}{(D/4)^{4/3}} + 1\right)\frac{5.47^2}{2(9.81)D^4} = \left(1.5 + \frac{7.79}{D^{4/3}}\right)\frac{1.528}{D^4}$$

Equating these two expressions for Δh and simplifying: $3.9 = D + \left(1.5 + \frac{7.79}{D^{4/3}}\right)\frac{1.528}{D^4}$

By trial and error or by equation solver: $D = 1.409$ m

Thus the first assumption ($D < 3.0$ m, i.e. submerged entrance) is OK.

To check assumption 2, find the d_0 which just flows full with uniform flow:

Eq. 10.8*b*: $4.3 = (1/0.025)(\pi d_0^2/4)(d_0/4)^{2/3}0.003^{1/2}$; $d_0 = 1.994$ m

$D < d_0$, so the culvert flows full, we do have the assumed flow of Fig. 10.33*b*.

The assumptions and results are valid, $D = 1.409$ m.

Standard sizes are 1.35 and 1.50 m. $\therefore$ Use $D = 1.50$ m ◄

10.80 *What is the capacity of a 5-ft by 5-ft concrete box culvert ($n = 0.013$) with a rounded entrance ($k_e = 0.05$, $C_d = 0.95$) if the culvert slope is 0.004, the length is 180 ft, and the headwater level is 7 ft above the culvert invert? Assume (a) free outlet conditions, (b) tailwater elevation 1 ft above top of box at outlet, and (c) tailwater elevation 2 ft above top of box at outlet. Neglect headwater and tailwater velocity heads.*

BG

(*a*) Find capacity with free outlet.

Sec. 10.23: Headwater/D = 7/5 = 1.4 > 1.2, so conditions are those of Fig. 10.33*b* or 10.33*c*.

Assume Case (*b*) of Fig. 10.33. $R = A/P = 5^2/(4 \times 5) = 1.25$ ft

$\Delta h = (y_1 - y_3) + (z_1 - z_3) = y_1 - y_3 + S_0 L$. Equating this to Δh in Eq. 10.56:

$7 - 5 + 0.004(180) = [0.05 + 29.2(0.013^2)180/1.25^{4/3} + 1]V^2/(2 \times 32.2) = 0.0265V^2$

from which $V = 10.12$ fps, $Q = AV = 5^2(10.12) = 253$ cfs

Now find the depth y_0 which occurs with normal/uniform flow at this Q:

Eq. 10.8*a*: $253 = (1.486/0.013)5y_0[5y_0/(5 + 2y_0)]^{2/3}0.004^{1/2}$

By trial and error or by equation solver per Sample Prob. 10.1, $y_0 = 4.983$ ft.

$y_0 < D = 5$ ft, so the culvert does <u>not</u> flow full; free discharge at the outlet is given; therefore the preceeding assumption and computations are not valid, and the flow conditions must be Case (*c*).

Case (*c*) of Fig. 10.33, Eq. 10.57: $Q = C_d A\sqrt{2gh} = 0.95(5^2)\sqrt{2(32.2)(7-2.5)} = 404$ cfs ◄

(*b*) Find capacity when tailwater elevation is 1 ft above top of box at outlet.

Sec. 10.23: Headwater/D = 7/5 = 1.4 > 1.2, so conditions are per Fig. 10.33*a*, with $y_2 = 6$ ft.

Using the same equations as Part (*a*) (neglecting headwater and tailwater velocity heads):

$7 - 6 + 0.004(180) = 0.0265V^2$ from which $V = 8.05$ fps, $Q = 5^2(8.05) = 201$ cfs ◄

(*c*) TW elevation 2 ft above top of box at outlet. Conditions are again per Fig. 10.33*a*, but with $y_2 = 7$ ft.

Using the same equations as Part (*a*): $7 - 7 + 0.004(180) = 0.0265V^2$

from which $V = 5.21$ fps, so $Q = 5^2(5.21) = 130.2$ cfs ◄

10.81 *Repeat Prob. 10.80(a) for the cases where the culvert slope is (a) 0.03 and (b) 0.07.*

Prob. 10.80(a): Find the capacity of a 5-ft by 5-ft box culvert ($n = 0.013$) 180 ft long with a rounded entrance ($k_e = 0.05$, $C_d = 0.95$). The headwater is 7 ft above invert, and the outlet discharge is free; neglect headwater and tailwater velocity heads.

BG

(*a*) Find capacity when $S_0 = 0.03$

Sec. 10.23: Headwater/D = 7/5 = 1.4 > 1.2, so conditions are per Fig. 10.33*b* or 10.33*c*.

Assume Case (*b*) of Fig. 10.33. $R = A/P = 5^2/(4 \times 5) = 1.24$ ft.

$\Delta h = (y_1 - y_3) + (z_1 - z_2) = y_1 - y_3 + S_0 L$. Equating this to Δh in Eq. 10.56:

$7 - 5 + 0.03(180) = [0.05 + 29.2(0.013^2)180/1.25^{4/3} + 1]V^2/(2 \times 32.2) = 0.0265V^2$

from which $V = 16.70$ fps, $Q = AV = 5^2(16.70) = 417$ cfs

Now find the depth y_0 which occurs with normal/uniform flow at this Q:

Eq. 10.8*a*: $417 = (1.486/0.013)5y_0[5y_0/(5 + 2y_0)]^{2/3}0.03^{1/2}$

/cont...

By trial and error or by equation solver per Sample Prob. 10.1, $y_0 = 3.33$ ft.

$y_0 < D = 5$ ft, so the culvert does not flow full and Case (*b*) cannot occur. Case (*c*) must occur, i.e., with entrance control:

Eq. 10.57: $Q = 0.95(5^2)\sqrt{2(32.2)(7 - 2.5)} = 404$ cfs ◄

(*b*) Find capacity when $S_0 = 0.07$. Conditions are again those of Fig. 10.33*b* or 10.33*c*.

Following the same procedure as in Part (*a*):

$7 - 5 + 0.07(180) = 0.0265V^2$ from which $V = 23.5$ fps, $Q = 586$ cfs, $y_0 = 3.12$ ft

$y_0 < D = 5$ ft, so the culvert does not flow full and Case (*b*) cannot occur. Case (*c*) must occur, i.e., with entrance control: As for Part (*a*): $Q = 404$ cfs ◄

10.82 *A culvert under a road must carry 4.5 m³/s. (a) If the culvert length is 32 m, the slope is 0.004, and the maximum permissible headwater level above the culvert invert is 3.8 m, what size of corrugated-pipe culvert (n = 0.025) would you select? The outlet will discharge freely. Neglect velocity of approach. Assume square-edged entrance with $k_e = 0.5$, $C_d = 0.65$. (b) Repeat for a culvert length of 110 m.*

SI

(*a*) Find diameter D when $L = 32$ m.

Sec. 10.23: Assume $D < 3.0$ m, so that headwater/$D > 3.8/3 = 1.267$ and the conditions are those of Fig. 10.33*b* or 10.33*c*.

Assume Case (*b*), Fig. 10.33. $V = Q/A = 4.5/(\pi D^2/4) = 5.73/D^2$; $R = D/4$

$\Delta h = (y_1 - y_3) + (z_1 - z_3) = y_1 - y_3 + S_0L$; Equating this to Δh in Eq. 10.56:

$$3.8 - D + 0.004(32) = \left[0.5 + \frac{2(9.81)0.025^2(32)}{(D/4)^{4/3}} + 1\right]\frac{5.73^2}{2(9.81)D^4}$$

i.e., $3.93 - D = [1.5 + (2.49/D^{4/3})]1.673/D^4$

By trial and error or by equation solver per Sample Prob. 10.1, $D = 1.206$ m

Now find the diameter d_0 which just flows full with normal/uniform flow:

Eq. 10.8*b*: $4.5 = (1/0.025)(\pi d_0^2/4)(d_0/4)^{2/3}0.004^{1/2}$ from which, by trial and error, $d_0 = 1.922$ m.

$d_0 > D$, so the culvert does flow full; free discharge at outlet is given;

Therefore the above assumptions and results are valid; use standard $D = 1.35$ m ◄

(*b*) Find diameter D when $L = 110$ m

Using the same procedures as Part (*a*):

$$3.8 - D + 0.004(110) = \left[0.5 + \frac{(2)9.81(0.025)^2 110}{(D/4)^{4/3}} + 1\right]\frac{5.73^2}{2(9.81)D^4}$$

i.e., $4.24 - D = [1.5 + (8.56/D^{4/3})]1.673/D^4$

By trial and error or by equation solver per Sample Prob. 10.1, $D = 1.421$ m

As in Part (*a*), the diameter which just flows full with normal/uniform flow is $d_0 = 1.922$ m.

$d_0 > D$, so the culvert does flow full; free discharge at outlet is given; therefore the above assumptions and analysis are valid.

Standard sizes are 1.35 and 1.50 m. Use $D = 1.50$ m ◄

10.83 *A 120-ft-long corrugated metal pipe ($n = 0.022$) of 30-in diameter is tested in a laboratory. The headwater is maintained at a level which is 5 ft above the pipe invert at entrance, and the outlet cannot be submerged. Assume a square-edged entrance with $k_e = 0.5$, $C_d = 0.68$, and neglect headwater and tailwater velocity heads. Compute values of Q for S_0 values of 0.0, 0.01, 0.03, and 0.08. As the slope is increased, at what slope does the flow change to condition (c) of Fig. 10.33?*

BG

Given: conditions of Fig. 10.33*b* prevail when $y_0 > D$.

Sec. 10.23: Headwater/D = 5/2.5 = 2 > 1.2, so conditions are those of Fig. 10.33*b* or 10.33*c*.

For Case (*b*) of Fig. 10.33: $R = D/4 = 2.5/4 = 0.625$ ft

$\Delta h = (y_1 - y_3) + (z_1 - z_3) = y_1 - y_3 + S_0 L$

Equating this to Δh in Eq. 10.56: $5 - 2.5 + 120S_0 = \left[0.5 + 29.2(0.022)^2 120/0.625^{4/3} + 1\right]V^2/(2 \times 32.2)$

i.e., $2.5 + 120S_0 = 0.0726V^2$ (1)

The change in conditions, from Case (*b*) to Case (*c*), occurs when $y_0 = D = 2.5$ ft.

Then, using Eq. 10.7*b*: $V_{\text{full}} = (1.486/0.022)0.625^{2/3}S_0^{1/2}$

i.e., $V_{\text{full}} = 49.4S_0^{1/2}$; $(V_{\text{full}})^2 = 2438S_0$

Substituting this into Eq. (1): $2.5 + 120S_0 = 0.0726(2438S_0)$; $S_0 = 0.0439$ ◄

So when $S_0 > 0.0439$ orifice flow (Condition 10.33*c*) occurs.

For the required slopes:

Slope S_0	Condition	Eq.	V fps	Q cfs
0.0	Fig. 10.33*b*	(1)	5.87	28.8
0.01	Fig. 10.33*b*	(1)	7.14	35.0
0.03	Fig. 10.33*b*	(1)	9.17	45.0
(0.0439)	Fig. 10.33*b*	(1)	(10.35)	(50.8)
0.08	Fig. 10.33*c*	10.57*	10.57	51.9

▲

*With $C_d = 0.68$, $h = 5 - D/2 = 3.75$ ft

Chapter 11

Fluid Measurements

PROBLEM SELECTION GUIDE

Sec	Exer/Prob	Units	Difficulty	Length	Parts	Similar	Special features
11.1	***Measurement of Fluid Properties***						
	X[1]11.1.1	BG	Easy	Short	1	11.1.2	Uses Sec. 2.3
	11.1.2	SI	Easy	Short	1	11.1.1	Uses Sec. 2.3
	11.1.3	BG	Easy	V Short	1	11.1.4	Uses Sec. 3.2
	11.1.4	SI	Easy	V Short	1	11.1.3	Uses Sec. 3.2
	11.1.5	BG	Medium	Short	1	11.1.6	Uses Sec. 2.11
	11.1.6	SI	Medium	Short	1	11.1.5	Uses Sec. 2.11
	11.1.7	B	Easy	V Short	1		
	11.1.8	BG	Medium	Short	1	11.1.9	Assume laminar flow
	11.1.9	SI	Medium	Short	1	11.1.8	Assume laminar flow
	11.1.10	BG	Easy	Short	1		Uses Sec. 7.4
	P 11.1	BG	Medium	Medium	1	11.2	
	11.2	SI	Medium	Medium	1	11.1	
	11.3	BG	Medium	Short	1		Assume laminar flow, uses Sec 2.11
	11.4	SI	Easy	Medium	1		Uses Secs 2.11, 7.4, & 8.6
	11.5	SI	Easy	Medium	1		Uses Secs 2.11 & 7.4
	11.6	BG	Medium	Short	1	11.7	Uses Sec 9.7. T & E (Trial and Error).
	11.7	SI	Medium	Short	1	11.6	Uses Sec 9.7. T & E
	11.8	BG	Easy	Medium	1		Plot
11.3	***Measurement of Velocity with Pitot Tubes***						
	X 11.3.1	SI	Easy	Short	1		
	11.3.2	BG	Easy	V Short	1	11.3.3	Uses Sec. 3.6
	11.3.3	SI	Easy	V Short	1	11.3.2	Uses Sec. 3.6
	11.3.4	BG	Easy	V Short	1	11.3.5	* Uses Sec. 3.6
	11.3.5	SI	Easy	V Short	1	11.3.4	* Uses Sec. 3.6
	11.3.6	BG	V Easy	V Short	1	11.3.7	Uses Sec. 3.6
	11.3.7	SI	V Easy	V Short	1	11.3.6	Uses Sec. 3.6
	P 11.9	BG	Medium	Short	1		Uses Sec. 2.7
	11.10	BG	Easy	Short	1		Uses Sec. 3.6
	11.11	BG	Easy	Short	1	11.12	Uses Sec. 3.6
	11.12	SI	Easy	Short	1	11.11	Uses Sec. 3.6
	11.13	BG	Easy	Medium	1		Plot
	11.14	BG	Medium	Medium	1		Uses Secs. 3.6, 8.10, & 8.12.
11.5	***Measurement of Discharge***						
	P 11.15	BG	Medium	Medium	1	11.16	Integration; uses Sec. 4.5
	11.16	SI	Medium	Medium	1	11.15	Integration; uses Sec. 4.5

/cont...

[1] For all Exercises (marked "X"), answers are given in Appendix D of the textbook.

* Note a correction made to early printings of the textbook.

† Answers are sensitive to values (C, C_d, K, k', Y) which are read from graphs.

X = Exercise, P = (end-of-chapter) Problem, S = Sample Problem. ☐ = could use computing aids.

Sec	Exer/Prob	Units	Difficulty	Length	Parts	Similar	Special features
11.6	*Orifices, Nozzles, and Tubes*						
	X 11.6.1	BG	Easy	Short	1		
	11.6.2	BG	Easy	Short	1	11.6.3	Uses Sec 5.16
	11.6.3	SI	Easy	Short	1	11.6.2	Uses Sec 5.16
	11.6.4	BG	Easy	Short	1		
	11.6.5	BG	Easy	Short	2		
	11.6.6	BG	Easy	Short	1		
	11.6.7	BG	Easy	Medium	1		Uses Sec 5.5
	11.6.8	SI	Easy	Short	1		
	11.6.9	SI	Easy	Short	1		* Interpolation
	11.6.10	SI	Easy	Short	1		
	P 11.17	BG	Medium	Medium	4	11.18	* Uses Sec 5.5
	11.18	SI	Medium	Medium	4	11.17	Uses Sec 5.5
	11.19	BG	Medium	Medium	4	11.20	Uses Secs 5.5 & 5.10
	11.20	SI	Medium	Medium	4	11.19	Uses Secs 5.5 & 5.10
	11.21	BG	Medium	Medium	2	11.22	
	11.22	SI	Medium	Medium	2	11.21	
	11.23	BG	Medium	Medium	1	11.24	Uses Secs 3.5 & 5.5
	11.24	SI	Medium	Medium	1	11.23	Uses Secs 3.5 & 5.5
	11.25	BG	Medium	Medium	1		
11.7	*Venturi Meter*						
	X 11.7.1	BG	Medium	Medium	1	11.7.2	† Assume/verify *C* value.
	11.7.2	SI	Medium	Medium	1	11.7.1	† Assume/verify *C* value.
	11.7.3	SI	Medium	Medium	1		† Assume/verify *C*; uses Sec 7.4
	11.7.4	BG	Medium	Medium	1		† Assume/verify *C*; uses Sec 7.4
	P 11.26	BG	Medium	Long	3	11.27	† Assume/verify *C*; uses Sec 8.21
	11.27	SI	Medium	Long	3	11.26	† Assume/verify *C*; uses Sec 8.21
	11.28	BG	Medium	Long	3	11.29	† Assume/verify *C*; uses Sec 8.21
	11.29	SI	Medium	Long	3	11.28	† Assume/verify *C*; uses Sec 8.21
11.8	*Flow Nozzle*						
	X 11.8.1	SI	Medium	Medium	3	P11.30	†
	P 11.30	SI	Medium	Medium	3	X11.8.1	†
	11.31	BG	Medium	Medium	1		† Assume/verify *K* value; uses Sec 7.4
11.9	*Orifice Meter*						
	X 11.9.1	SI	Medium	Medium	3	P11.32	†
	P 11.32	SI	Medium	Medium	4	X11.9.1	†
	11.33	BG	Medium	Medium	1		† Assume/verify *K* value; uses Sec 7.4
	11.34	BG	Medium	Long	3		† Interpolation

/cont...

Sec	Exer/Prob	Units	Difficulty	Length	Parts	Similar	Special features
11.10	*Flow Measurement of Compressible Fluids*						
	X 11.10.1	BG	Easy	Short	1		† Uses $pv = RT$ (Sec. 2.7)
	11.10.2	BG	Easy	Short	1		Uses $pv = RT$ (Sec. 2.7)
	11.10.3	BG	Easy	Short	1	11.10.4	† Uses $pv = RT$ (Sec. 2.7)
	11.10.4	SI	Easy	Short	1	11.10.3	† Uses $pv = RT$ (Sec. 2.7)
	11.10.5	BG	Easy	Short	1	11.10.6	† Uses $pv = RT$ (Sec. 2.7). Extrapol'n.
	11.10.6	SI	Easy	Short	1	11.10.5	† Uses $pv = RT$ (Sec. 2.7). Extrapol'n.
	11.10.7	BG	Medium	Short	2	11.10.8	Uses Secs. 13.3 & 13.5-6.
	11.10.8	SI	Medium	Short	2	11.10.7	Uses Secs. 13.3 & 13.5-6.
	P 11.35	BG	Medium	Medium	3	11.36	* † Uses $pv = RT$ (Sec. 2.7)
	11.36	SI	Medium	Medium	3	11.35	* † Uses $pv = RT$ (Sec. 2.7)
	11.37	BG	Easy	Short	1	11.38	†
	11.38	SI	Easy	Short	1	11.37	†
	11.39	SI	Medium	Long	1		† Uses Secs 2.7, 2.11, 4.7, interpolation
	11.40	SI	Hard	Long	4		† Uses Sec 13.8
11.11	*Thin Plate Weirs*						
	X 11.11.1	BG	V Easy	V Short	1	11.11.2	
	11.11.2	SI	V Easy	V Short	1	11.11.1	
	11.11.3	BG	V Easy	V Short	2	11.11.4	*
	11.11.4	SI	V Easy	V Short	2	11.11.3	*
	11.11.5	BG	Easy	Medium	3		† Extrapolation.
	P 11.41	BG	Easy	Medium	1		Plots
	11.42	N	Medium	Medium	1		
	11.43	N	Medium	Medium	1		
	(11.44-45 are in the next section)						
	11.46	BG	Medium	Medium	1	11.47	† Differentiation.
	11.47	SI	Medium	Medium	1	11.46	† Differentiation.
11.12	*Streamlined Weirs and Free Outfall*						
	X 11.12.1	BG	Easy	Medium	1	11.12.2	T & E
	11.12.2	SI	Easy	Medium	1	11.12.1	T & E
	11.12.3	BG	Medium	Medium	1	11.12.4	T & E
	11.12.4	SI	Medium	Medium	1	11.12.3	T & E
	11.12.5	BG	Medium	Short	1	11.12.6	Uses Sec. 10.9
	11.12.6	SI	Medium	Short	1	11.12.5	Uses Sec. 10.9
	P 11.44	BG	Hard	Long	1		□ Successive T & E; plots
	11.45	N	Hard	Medium	1		Integration.
11.14	*Sluice Gate*						
	X 11.14.1	BG	Medium	Short	1	11.14.2	
	11.14.2	SI	Medium	Short	1	11.14.1	
	P 11.48	BG	Medium	Medium	1	11.49	
	11.49	SI	Medium	Medium	1	11.48	
	11.50	SI	Medium	Medium	3		T & E for V

Chapter 11
FLUID MEASUREMENTS

Sec. 11.1: Measurement of Fluid Properties -- *Exercises*

11.1.1 *A small object has a volume of 0.0070 ft^3 and weighs 1.50 lb in air and 1.18 lb in a liquid. What is the density of the liquid?*

BG

Buoyant force = 1.50 - 1.18 = 0.32 lb

γ = 0.32/0.007 = 45.7 lb/ft^3 ; Eq. 2.1: $\rho = \gamma/g$ = 45.7/32.2 = 1.420 slugs/ft^3 ◄

11.1.2 *A small object has a volume of 300 cm^3 and weighs 15.50 N in air and 10.50 N in a liquid. What is the density of the liquid?*

SI

Buoyant force = 15.50 - 10.50 = 5.00 N; γ = 5.00/300 = 0.016 67 N/cm^3 = 0.016 67 kg·m/(s^2·cm^3)

Eq. 2.1: $\rho = \gamma/g$ = [0.016 67 kg·m/(s^2·cm^3)]/(9.81 m/s^2) = 0.001 699 kg/cm^3 = 1.699 g/cm^3 ◄

11.1.3 *Carbon tetrachloride (s = 1.59) is placed in an open U tube. A liquid is poured into one of the legs of the tube. A column of this liquid 19.4 in high balances a carbon tetrachloride column 15.0 in high. What is the specific weight of the liquid? In Sec. 11.1 it is mentioned that this method will give only approximate values. Why is this so?*

BG

Eq. 3.4: $p = \gamma_1 h_1 = \gamma_2 h_2$; $19.4\gamma = 15(1.59 \times 62.4)$; γ = 76.7 lb/ft^3 ◄

Results are only approximate because of difficulty with menisci. ◄

11.1.4 *Carbon tetrachloride (s = 1.59) is placed in an open U tube. A liquid is poured into one of the legs of the tube. A column of this liquid 39.2 cm high balances a carbon tetrachloride column 30.0 cm high. What is the specific weight of the liquid? In Sec. 11.1 it is mentioned that this method will give only approximate values. Why is this so?*

SI

Eq. 3.4: $p = \gamma_1 h_1 = \gamma_2 h_2$; $0.392\gamma = 0.30(1.59 \times 9810)$; γ = 11 940 N/m^3 ◄

Results are only approximate because of difficulty with menisci. ◄

11.1.5 *A rotational viscometer is constructed of two concentric cylinders of height 15.0 in. The OD of the inner cylinder is 4.00 in, and the ID of the outer cylinder is 4.10 in. When a torque of 6.0 ft·lb is applied to the outer cylinder, it was found to rotate at 1 revolution per 3.8 sec. Find the (absolute) viscosity of the fluid. Neglect mechanical friction.*

BG

Gap distance = (4.10 - 4.00)/2 = 0.05 inches = 0.004 17 ft

Peripheral velocity = ωr = [1(2π)/3.8]4.10/(2 × 12) = 0.282 fps ; du/dy = 0.282/0.004 17 = 67.8 fps/ft

Eq. 2.9: $\tau = \mu\, du/dy$; $F = \tau A = (\mu\, du/dy)\pi DL$; $T = FD/2 = (\mu\, dy/dy)\pi D^2 L/2$

$\therefore$ 6.0 = $\mu(67.8)\pi(4.10/12)^2(15/12)/2$ = 15.54μ ; μ = 0.386 lb·s/ft^2 ◄

11.1.6 *A rotational viscometer is constructed of two concentric cylinders of height 30.0 cm. The OD of the inner cylinder is 10.00 cm, and the ID of the outer cylinder is 10.20 cm. When a torque of 8.0 N·m is applied to the outer cylinder, it was found to rotate at 1 revolution per 4.0 s. Find the (absolute) viscosity of the fluid. Neglect mechanical friction.*

SI

Gap distance = (10.20 − 10.00)/2 = 0.10 cm = 0.001 00 m

Peripheral velocity = ωr = $[1(2\pi)/4.0]10.2/(2 \times 100)$ = 0.0801 m/s

du/dy = 0.0801/0.00100 = 80.1 m/s per m

Eq. 2.9: $\tau = \mu\, du/dy$; $F = \tau A = (\mu\, du/dy)\pi DL$; $T = FD/2 = (\mu\, dy/dy)\pi D^2L/2$

$\therefore\ 8.0 = \mu(80.1)\pi(0.102)^2 0.30/2$; $\mu = 20.4\ \text{N·s/m}^2$ ◄

11.1.7 *A tube viscometer similar to the one of Fig. 11.1 has a diameter of 0.0420 in and a length of 3.10 in. A volume of 60 cm³ flowed through the tube in 128.7 sec, causing the vertical distance from the liquid surface in the reservoir to the tube outlet to change from 10.00 to 9.50 in. Find the kinematic viscosity of the liquid.*

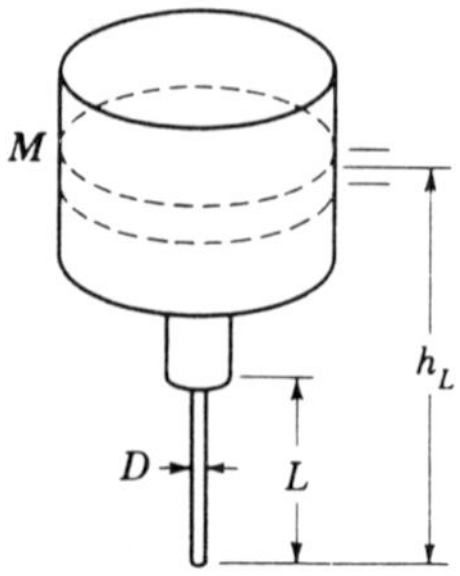

Figure 11.1

B

Eq. 11.2: $$\nu = \frac{\pi(0.0420/12)^4(9.75/12)}{128[60/(2.54 \times 12)^3](3.10/12)}(32.2)128.7 = 0.000\,0227\ \text{ft}^2/\text{sec}$$ ◄

Check to confirm that flow is laminar: V = 1.711 fps, **R** = 265, ∴ OK.

11.1.8 *If 60°F water flows through a tube-type viscometer in 120 sec, how long will it take the same volume of 100°F water to pass though the same viscometer?*

BG

Table A.1 for water: $\nu = 1.217 \times 10^{-5}\ \text{ft}^2/\text{sec}$ at 60°F, $0.739 \times 10^{-5}\ \text{ft}^2/\text{sec}$ at 100°F

Assume laminar flow (**R** < 2000) in both cases.

From Eq. 11.2, $\nu \propto t$: $t_{100°} = t_{60°}(\nu_{100°}/\nu_{60°})$ = 120(0.739/1.217) = 72.9 sec ◄

11.1.9 *If 15°C water flows through a tube-type viscometer in 85 s, how long will it take the same volume of 40°C water to pass though the same viscometer?*

SI

Table A.1 for water: $\nu = 1.139 \times 10^{-6}\ \text{m}^2/\text{s}$ at 15°C, $0.658 \times 10^{-6}\ \text{m}^2/\text{s}$ at 40°C

Assume laminar flow (**R** < 2000) in both cases.

From Eq. 11.2, $\nu \propto t$: $t_{40} = t_{15}(\nu_{40}/\nu_{15})$ = 85(0.658/1.139) = 49.1 s ◄

11.1.10 *A lead sphere (s = 11.4) with a diameter of 0.25 in falls through oil (s = 0.86) in a 2.25-in-diameter cylinder. If the sphere moves with a constant velocity of 0.150 fps, what is the absolute viscosity of the oil? Check* **R** *to see if it is less than 1.0.*

BG

Eq. 11.4: $$\frac{V}{V_t} = 1 + \frac{9(0.25)}{4(2.25)} + \left(\frac{1}{4}\right)^2 = 1.313\ ; \quad V = 0.150(1.313) = 0.1969\ \text{fps}$$

Eq. 11.3: $$\mu = \frac{(0.25/12)^2(11.4 - 0.86)62.4}{18(0.1969)} = 0.0806\ \text{lb·s/ft}^2$$ ◄

Check: Eq. 7.6: $$\mathbf{R} = \frac{(0.25/12)(0.150)(1.940 \times 0.86)}{0.0806} = 0.0647$$ ◄

Sec. 11.1: Measurement of Fluid Properties -- *Problems*

11.1 *A hydrometer (Figs. 3.21 and 3.22a) is made in the form of an 8-in-long cylinder with a diameter of 3/8 in. Attached to the end of the cylinder is a 1.2-in-diameter sphere. The entire device weighs 0.9 oz. What range of specific gravities can be measured with this device? When floating in a liquid with a density of 1.80 slugs/ft³, how much of the cylinder will project above the surface?*

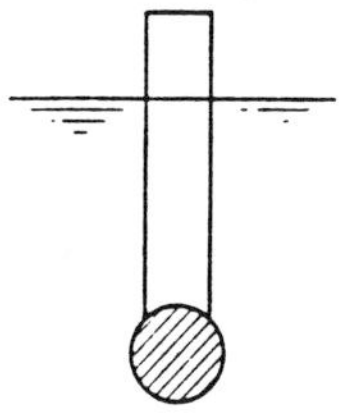

Figure 3.22*a*

BG

$$\mathcal{V}_{sphere} + \mathcal{V}_{rod} = \frac{\pi D^3}{6} + \frac{\pi}{4}d^2 h = \frac{\pi(1.2/12)^3}{6} + \frac{\pi}{4}\left(\frac{3/8}{12}\right)^2\left(\frac{8}{12}\right)$$

$$= 0.000\,524 + 0.000\,511 = 0.001\,035 \text{ ft}^3$$

$$\gamma_{min} = \frac{W}{\mathcal{V}_{tot}} = \frac{0.9/16 \text{ lb}}{0.001\,035 \text{ ft}^3} = 54.4 \text{ lb/ft}^3 \ ; \quad s_{min} = \frac{\gamma_{min}}{\gamma_{water}} = \frac{54.4}{62.4} = 0.871 \quad ◄$$

$$\gamma_{max} = \frac{W}{\mathcal{V}_{sphere}} = \frac{0.9/16}{0.000\,524} = 107.4 \ ; \quad s_{max} = \frac{\gamma_{max}}{\gamma_{water}} = \frac{107.4}{62.4} = 1.722 \quad ◄$$

$s = 1.80/1.940 = 0.928$; so $(s_{min} = 0.871) < (s = 0.928) < (s_{max} = 1.722)$

$\gamma = \rho g = 1.80(32.2) = 58.0$ pcf

Displaced $\mathcal{V} = W/\gamma = (0.9/16)/58.0 = 0.000\,970 \text{ ft}^3 = 0.000\,524 + (L/8)0.000\,511$

Immersed $L = 6.99$ in ; $\therefore$ Length projecting $= 8 - 6.99 = 1.01$ in ◄

11.2 *A hydrometer (Figs. 3.21 and 3.22a) is made in the form of a 20-cm-long cylinder with a diameter of 6 mm. Attached to the end of the cylinder is a 22-mm-diameter sphere. The mass of the entire device is 8 g. What range of specific gravities can be measured with this device? When floating in a liquid having a density of 900 kg/m³, how much of the cylinder will project above the surface?*

See Fig. 3.22a with Solution 11.1.

SI

sphere: $\mathcal{V}_s = \frac{\pi}{6}D^3 = \frac{\pi}{6}(22/1000)^3 = 5.58 \times 10^{-6} \text{ m}^3$

cylinder: $\mathcal{V}_c = \frac{\pi}{4}d^2 h = \frac{\pi}{4}(6/1000)^2(20/100) = 5.65 \times 10^{-6} \text{ m}^3$; $\mathcal{V}_{total} = 11.23 \times 10^{-6} \text{ m}^3$

Device mass $= m = 8 \text{ g} = 0.008$ kg

$$\rho_{min} = \frac{m}{\mathcal{V}_{total}} = \frac{0.008 \text{ kg}}{11.23 \times 10^{-6} \text{ m}^3} = 712 \text{ kg/m}^3 \ ; \quad s_{min} = \frac{\rho_{min}}{\rho_{water}} = \frac{712}{1000} = 0.712 \quad ◄$$

$$\rho_{max} = \frac{m}{\mathcal{V}_s} = \frac{0.008 \text{ kg}}{5.58 \times 10^{-6} \text{ m}^3} = 1435 \text{ kg/m}^3 \ ; \quad s_{max} = \frac{\rho_{max}}{\rho_{water}} = \frac{1435}{1000} = 1.435 \quad ◄$$

For $\gamma = 900 \text{ kg/m}^3$, displaced $\mathcal{V} = W/\gamma = 0.008/900 = 8.89 \times 10^{-6} \text{ m}^3$

$= \mathcal{V}_s + (L/h)\mathcal{V}_c = 5.58 \times 10^{-6} + (L/20)5.65 \times 10^{-6}$; immersed $L = 11.72$ cm

Length projecting above surface $= h - L = 20 - 11.72 = 8.28$ cm ◄

11.3 *Fifty cubic centimeters of 70°F water flows through a certain tube-type viscometer in 53.7 sec, whereas the same volume of 50°F oil (s = 0.90) flows through it in 850 sec. Find the absolute viscosity of the oil.*

BG

Table A.1 for water at 70° F: $\nu = 1.059 \times 10^{-5}$ ft²/sec

Assume laminar flow (**R** < 2000) in both cases. From Eq. 11.2, $\nu \propto t$ or $\nu = kt$

Oil at 50°F: $\nu_{oil} = (k)850$; water at 70°F: $1.059 \times 10^{-5} = (k)53.7$

Dividing to eliminate k: $\nu_{oil} = 1.059 \times 10^{-5}(850/53.7) = 1.676 \times 10^{-4}$ ft²/sec

Eq. 2.11 for the oil: $\mu = \rho\nu = (1.940 \times 0.90)1.676 \times 10^{-4} = 2.93 \times 10^{-4}$ lb·sec/ft² ◄

11.4 *A liquid (ρ = 889 kg/m³) under a head of 50 cm flows at a steady rate of 30 cm³/min through a glass tube of diameter 2.0 mm and length 4.5 m. What is the absolute viscosity and the kinematic viscosity of the liquid? Express the answers in stokes and poises. Confirm by calculation that the flow is laminar.*

SI

Sec 11.1: For such a small diameter tube, assume laminar flow.

$V = Q/A = Q/(\pi D^2/4) = [(30\text{ cm}^3/\text{min})/(60\text{ sec/min})]/[\pi(0.2\text{ cm})^2/4] = 15.92\text{ cm/s}$

Eq. 8.19: $50\text{ cm} = 32\nu\dfrac{450\text{ cm}}{981\text{ cm/s}^2(0.2)^2\text{ cm}^2}15.92\text{ cm/s}$; $\nu = 0.00856\text{ cm}^2/\text{s} = 0.00856$ St ◄

Eq. 2.11: $\mu = \rho\nu = 889(0.00856) = 7.61\dfrac{\text{kg}}{\text{m}^3}\dfrac{\text{cm}^2}{\text{s}} = 7.61\dfrac{\text{N}\cdot\text{s}^2}{\text{m}^4}\dfrac{\text{cm}^2}{\text{s}} = 7.61 \times 10^{-4}$ P ◄

Check assumption: $\mathbf{R} = \dfrac{DV}{\nu} = \dfrac{0.2\text{ cm}(15.92\text{ cm/s})}{0.00856\text{ cm}^2/\text{s}} = 372 < 2000$ so flow <u>is</u> laminar. ◄

11.5 *A glass bead (s = 2.60) with a diameter of 16 mm falls through a liquid (s = 1.59) in a 10-cm-diameter tube. If the bead moves with a constant velocity of 14.5 cm/min, what is the absolute viscosity and the kinematic viscosity of the liquid?*

SI

Eq. 11.4: $\dfrac{V}{V_t} = 1 + \dfrac{9 \times 16}{4 \times 100} + \left(\dfrac{9 \times 16}{4 \times 100}\right)^2 = 1.490$; $V = 1.490(14.5/60) = 0.360\text{ cm/s} = 0.003\,60\text{ m/s}$

Eq. 11.3: $\mu = \dfrac{(0.016)^2 9810(2.60 - 1.59)}{18(0.003\,60)} = 39.1\text{ N}\cdot\text{s/m}^2$ ◄

Eq. 2.11: $\nu = \mu/\rho = 39.1/(1.59 \times 1000) = 0.0247\text{ m}^2/\text{s}$ ◄

Check solution validity: Eq. 7.6: $\mathbf{R} = \dfrac{0.016(0.003\,60)}{0.0247} = 0.002\,34$

R < 1, so these answers are O.K. [If **R** > 1, data from Fig. 9.10 must be used.]

11.6 *A 3.6-in-diameter tube contains oil (s = 0.93) with a viscosity of 0.0065 lb·sec/ft². What is the maximum size of steel sphere (s = 7.84) that will satisfy Stokes' law when released into this oil. What will be the fall velocity of this sphere?*

BG

Sec 9.7: When D is max for Stokes' Law validity: $\mathbf{R} = \dfrac{DV_t(0.93 \times 1.94)}{0.0065} = 1$; $V_t = \dfrac{1}{278D}$

Eq. 11.3: $\mu = 0.0065 = \dfrac{D^2(7.84 - 0.93)62.4}{18V}$; $V = 3685D^2$

Eq. 11.4: $\dfrac{V}{V_t} = 1 + \dfrac{9D}{4(3.6/12)} + \left[\dfrac{9D}{4(3.6/12)}\right]^2 = 1 + 7.5D + (7.5D)^2$

Solution of these three equations (by eliminating V and V_t) yields $1.023 \times 10^6 D^3 = 1 + 7.5D + (7.5D)^2$

By trials: $D = 0.01019$ ft $= 0.1223$ inches ◄ ; $V = 0.383$ fps ◄

11.7 *A 10-cm-diameter tube contains oil (s = 0.9) with a viscosity of 0.25 N·s/m². What is the maximum size of steel sphere (s = 7.8) that will satisfy Stokes' law when released into this oil. What will be the fall velocity of this sphere?*

SI

Sec 9.7: When D is max for Stokes' Law validity: $\mathbf{R} = \dfrac{DV_t(0.9\times 1000)}{0.25} = 1$; $V_t = \dfrac{1}{3600D}$

Eq. 11.3: $\mu = 0.25 = \dfrac{D^2(7.8 - 0.9)9810}{18V}$; $V = 15\,040D^2$

Eq. 11.4: $\dfrac{V}{V_t} = 1 + \dfrac{9D}{4(0.1)} + \left[\dfrac{9D}{4(0.1)}\right]^2 = 1 + 22.5D + (22.5D)^2$

Solution of the above three equations (by eliminating V and V_t) yields

$54.2\times 10^6 D^3 = 1 + 22.5D + (22.5D)^2$

By trials: $D = 0.00270$ m $= 2.70$ mm ◄ ; $V = 0.1097$ m/s ◄

11.8 *An 0.10-in-diameter sphere has a velocity of 0.005 fps when falling through liquid in a 1.0-in-diameter tube. What will be its fall velocities through the same liquid in tubes of diameter 0.50, 2.0, 4.0, and 10.0 in. Plot fall velocity vs tube diameter.*

BG

Eq. 11.4: $\dfrac{V}{0.005} = 1 + \dfrac{9(0.1)}{4(1)} + \left[\dfrac{9(0.1)}{4(1)}\right]^2$; $V = 0.005(1 + 0.225 + 0.0506) = 0.006\,38$ fps

Eq. 11.4: $\dfrac{0.006\,38}{V_t} = 1 + \dfrac{9(0.1)}{4D_t} + \left[\dfrac{9(0.1)}{4D_t}\right]^2$

D_t (inches)	V_t (fps)	
0.5	0.003 87	◄
1.0	0.005 00	
2.0	0.005 66	◄
4.0	0.006 04	◄
10.0	0.006 25	◄
∞	0.006 38	

Sec. 11.3: Measurement of Velocity with Pitot Tubes -- *Exercises*

11.3.1 *In Fig. S11.1 kerosene (s = 0.81) is flowing. The pressure gages at A and B read 70 and 150 N/m². Find the velocity u assuming C_I = 1.0.*

SI

$\gamma = 0.81(9.81) = 7.95$ kN/m³

From Eq. 11.5: $\gamma(V^2/2g) = 150 - 70 = 80$ N/m²

$V^2 = 2(9.81)(80/7950) = 0.1974$; $V = \sqrt{0.1974} = 0.444$ m/s ◄

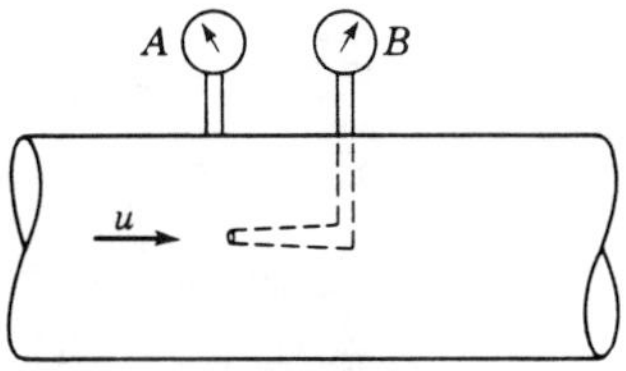

Figure S11.1

11.3.2 *A pitot tube and a wall piezometer tube are installed in a pipe carrying 60°F water. The tubes are connected to a water-mercury manometer which registers a differential of 3.4 in. Given that C_I = 0.98, find the flow velocity approaching the tube.*

BG

Eqs. 11.7 and 3.12: $u = C_I\sqrt{2g[(s_M/s_F) - 1]MR} = 0.98\sqrt{2(32.2)[(13.56/1.00) - 1]3.4/12} = 14.84$ fps ◄

11.3.3 *A pitot tube and a wall piezometer tube are installed in a pipe carrying 15°C water. The tubes are connected to a water-mercury manometer which registers a differential of 78 mm. Given that C_I = 0.97, find the flow velocity approaching the tube.*

SI

Eqs. 11.7 and 3.12: $u = C_I\sqrt{2g[(s_M/s_F) - 1]MR} = 0.97\sqrt{2(9.81)[(13.56/1.00) - 1]0.078} = 4.25$ m/s ◄

11.3.4 *The fluids of Exer. 11.3.2 are reversed so that mercury (s = *13.56) is flowing in the pipe and water is the gage fluid (with the manometer now inverted). With the same gage differential, find the flow velocity of the mercury. [*Note this correction to the textbook.]*

BG

Eqs. 11.7 and 3.13: $u = C_I\sqrt{2g[1 - (s_M/s_F)]MR} = 0.98\sqrt{2(32.2)[1 - (1.00/13.56)]3.4/12} = 4.03$ fps ◄

11.3.5 *The fluids of Exer. 11.3.3 are reversed so that mercury (s = *13.56) is flowing in the pipe and water is the gage fluid (with the manometer now inverted). With the same gage differential, find the flow velocity of the mercury. [*Note this correction to the textbook.]*

SI

Eqs. 11.7 and 3.13: $u = C_I\sqrt{2g[1 - (s_M/s_F)]MR} = 0.97\sqrt{2(9.81)[1 - (1.00/13.56)]0.078} = 1.156$ m/s ◄

11.3.6 *A pitot-static tube (C_I = 0.985) is connected to an inverted U tube containing oil (s = 0.875). Find the velocity of flowing water if the manometer reading is 3.6 in.*

BG

From Eq. 3.13: $\Delta p/\gamma = [1 - (0.875/1.00)]3.6/12 = 0.0375$ ft

Eq. 11.7: $u = 0.985\sqrt{2(32.2)0.0375} = 1.531$ fps ◄

11.3.7 *A pitot-static tube (C_I = 0.992) is connected to an inverted U tube containing oil (s = 0.91). Find the velocity of flowing water if the manometer reading is 127 mm.*

SI

From Eq. 3.13: $\Delta p/\gamma = [1 - (0.91/1.00)]0.127 = 0.011\,43$ m

Eq. 11.7: $u = 0.992\sqrt{2(9.81)0.011\,43)} = 0.470$ m/s ◄

Sec. 11.3: Measurement of Velocity with Pitot Tubes -- *Problems*

11.9 *In Fig. S11.1, pressure gages A and B read 11.0 psi and 12.0 psi, respectively. If 60°F air is flowing, what is the velocity u? Atmospheric pressure is 27.0 inHg. Assume C_I = 1.0 and neglect compressibility effects.*

See Fig. S11.1 with Solution 11.3.1 above.

BG

$p_{abs} = p_{atmos} + p_{gage} = 14.7(27.0/29.9) + 11.0 = 24.3$ psia

Eq. 2.5: $\gamma = gp/RT = 32.2(24.3\times 144)/(1715\times 520) = 0.1262$ lb/ft^3

$\Delta p = p_0 - p = (12.0 - 11.0)144 = 144$ psf

From Eq. 11.5: $u^2/2g = \Delta p/\gamma = 144/0.1262 = 1141$ ft ; $u = \sqrt{2(32.2)1141)} = 271$ fps ◄

11.10 *Suppose the two pressure gages of Prob. 11.9 are replaced by a differential manometer containing water. What is the reading on the manometer?*

Prob. 11.9: Air is flowing, γ = 0.1262 pcf; $\Delta p/\gamma$ = 1142 ft of air.

BG

From Eq. 3.12: $\dfrac{\Delta p}{\gamma} = \left(\dfrac{s_M}{s_F} - 1\right)\dfrac{MR(\text{in})}{12}$; $1141 = \left(\dfrac{62.4}{0.1262} - 1\right)\dfrac{MR}{12}$; $MR = 27.7$ inches ◄

11.11 *The pitometers in Fig. P11.11 are connected to a mercury manometer, which reads 5.0 in. The velocity of flowing carbon tetrachloride (s = 1.59) is known to be 9.37 fps. What is C_I for this instrument?*

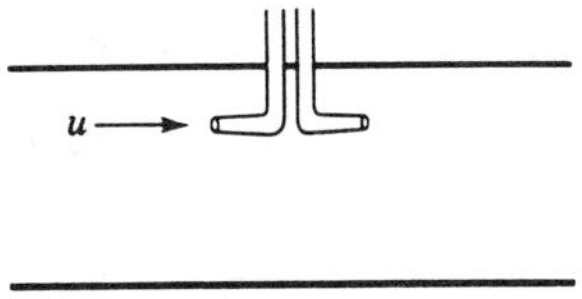

Figure P11.11

BG

Eq. 3.12: $\dfrac{\Delta p}{\gamma} = \left(\dfrac{13.56}{1.59} - 1\right)\dfrac{5.0}{12} = 3.14$ ft

Eq. 11.7: $9.37 = C_I\sqrt{2(32.2)3.14}$; $C_I = 0.659$ ◄

11.12 *The pitometers in Fig. P11.11 are connected to a mercury manometer, which reads 12.0 cm. The velocity of flowing carbon tetrachloride (s = 1.59) is known to be 2.76 m/s. What is C_I for this instrument?*

See Fig. P11.11 with Solution 11.11 above.

SI

Eq. 3.12: $\dfrac{\Delta p}{\gamma} = \left(\dfrac{13.56}{1.59} - 1\right)\dfrac{12}{100} = 0.903$ m

Eq. 11.7: $2.76 = C_I\sqrt{2(9.81)0.903}$; $C_I = 0.656$ ◄

11.13 *Given air at 60°F and standard atmospheric pressure is flowing in Fig. P11.11. The pitometers are attached to a manometer containing a liquid with s = 0.90. Plot the velocity u versus the manometer reading assuming $C_I = 0.93$.*

See Fig. P11.11 with Solution 11.11 above.

BG

Table A.2: Density of air $= 2.37\times10^{-3}$ slug/ft^3. Ratio of densities $= \dfrac{0.90(1.940)}{2.37\times10^{-3}} = 737$

Thus 1 foot of Manometer Reading (*MR*) is equivalent to 737 ft of air, and $\Delta p/\gamma = 737(MR)$

Eq. 11.7: $u = C_I\sqrt{2g(\Delta p/\gamma)} = 0.93\sqrt{2(32.2)737(MR)} = 203\sqrt{MR}$

MR (in)	*MR* (ft)	*u* (fps)
1.0	0.083	58.5
5.0	0.416	130.8
10.0	0.833	184.9
20.0	1.667	262

▲ ▲

At larger manometer readings the effect of compressibility must be considered, see Eq. 13.31.

11.14 *Water at 90°F flows through a smooth, 12-in-diameter pipe. A Prandtl tube is placed on the center line of the pipe. The reading on a differential manometer attached to this Prandtl tube is 8 in of carbon tetrachloride (s = 1.59). What is the flow rate?*

BG

Table A.1 for water at 90°F: $\nu = 0.826\times10^{-5}$ ft^2/s. Sec. 11.3 for Prandtl tube: $C_I = 1$

Eq. 3.12: $\Delta p/\gamma = [(1.59/1.00) - 1]8/12 = 0.393$ ft

Eq. 11.7 on centerline: $u_{max} = C_I\sqrt{2g(\Delta p/\gamma)} = 1\sqrt{2(32.2)0.393} = 5.03$ fps

Assume $V = 5$ fps, then $\mathbf{R} = 1(5)/(0.826\times10^{-5}) = 6.05\times10^5$

Fig. 8.11 for $e = 0$: $f = 0.0127$; Eq. 8.32: $\dfrac{V}{u_{max}} = \dfrac{1}{1 + 1.326\sqrt{f}} = \dfrac{1}{1.149}$

$V = u_{max}/1.149 = 5.03/1.149 = 4.38$ fps ($\mathbf{R} = 5.30\times10^5$, f is now 0.0130, little change)

$Q = (\pi/4)(1^2)4.38 = 3.44$ cfs ◄

11.15 *Assume the pipe diameter in Fig. 11.10 is 16 in, and the flow is laminar, i.e., $u = u_{max} - kr^2$, when $u_{max} = 0.1$ fps. Divide the circle into concentric rings with radii of 2, 4, 6, and 8 in, and compute the flow rate by the method of Fig. 11.10 by taking the velocities at radii of 1, 3, 5, and 7 in as representative of the rings. Compare the result with that obtained by integration.*

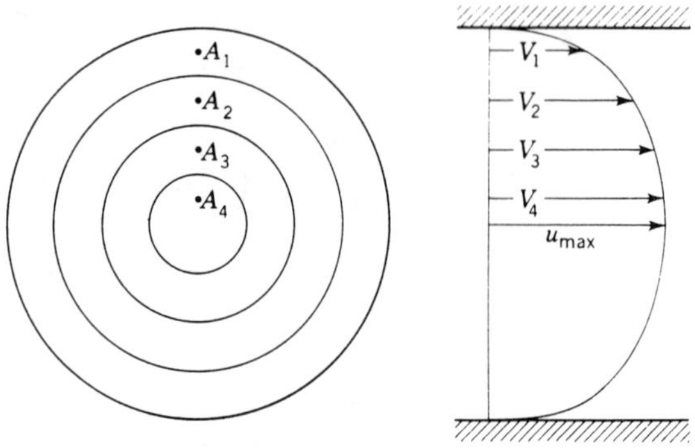

Figure 11.10

BG

$u = 0.1 - kr^2$; when $r = 8$ inches, $u = 0$ fps

$\therefore \quad 0 = 0.1 - k(8)^2, \quad k = 1/640$

Hence $u = 0.1 - r^2/640$

$\bar{r}$ (in)	A (in²)	u (fps)	$\Delta Q = Au$
1	$\pi 2^2 = 4\pi$	0.0984	1.237
3	$\pi(4^2 - 2^2) = 12\pi$	0.0859	3.24
5	$\pi(6^2 - 4^2) = 20\pi$	0.0609	3.83
7	$\pi(8^2 - 6^2) = 28\pi$	0.0234	2.06
			Σ = 10.37 ft·in²/sec = 0.0720 cfs ◄

By integration: Eq. 4.3: $Q = \int_0^8 u 2\pi r dr = 2\pi \int_0^8 (0.1r - r^3/640) dr$

$Q = 2\pi[0.1r^2/2 - r^4/2560]_0^8 = 3.2\pi$ ft·in²/sec $= 0.0698$ cfs ◄

11.16 *Assume the pipe diameter in Fig. 11.10 is 48 cm, and the flow is laminar, i.e., $u = u_{max} - kr^2$, when $u_{max} = 30$ mm/s. Divide the circle into concentric rings with radii of 6, 12, 18, and 24 cm, and compute the flow rate by the method of Fig. 11.10 by taking the velocities at radii of 3, 9, 15, and 21 cm as representative of the rings. Compare the result with that obtained by integration.*

See Fig. 11.10 with Solution 11.15 above.

SI

$u = 0.03 - kr^2$; when $r = 0.24$ m, $u = 0$ m/s ; $\therefore \quad 0 = 0.03 - k(0.24)^2, \quad k = 0.521$

Hence $u = 0.03 - 0.521r^2$

$\bar{r}$ (m)	A (m²)	u (m/s)	$\Delta Q = Au$
0.03	$\pi(0.06)^2 = 0.003\,60\,\pi$	0.0295	0.000 334
0.09	$\pi(0.12^2 - 0.06^2) = 0.010\,80\pi$	0.0258	0.000 825
0.15	$\pi(0.18^2 - 0.12^2) = 0.018\,00\pi$	0.0183	0.001 034
0.21	$\pi(0.24^2 - 0.18^2) = 0.0252\pi$	0.007 03	0.000 557
			Σ = 0.002 80 m³/s ◄

By integration: Eq. 4.3: $Q = \int_0^{0.24} u 2\pi r dr = 2\pi \int_0^{0.24} (0.03r - 0.521r^3) dr$

$Q = 2\pi[0.03r^2/2 - 0.521r^4]_0^{0.24} = 0.002\,71$ m³/s ◄

Sec. 11.6: Orifices, Nozzles, and Tubes -- *Exercises*

11.6.1 *Water issues from a circular 5-in-diameter orifice under a head of 45 ft. If 534 ft^3 are discharged in 2 min, what is the coefficient of discharge? If the diameter at the vena contracta is measured to be 3.93 in, what is the coefficient of contraction and what is the coefficient of velocity?*

BG

Eqs. 11.9 and 11.11: $Q = C_d A_o \sqrt{2gh}$; $534/(2\times 60) = C_d \frac{\pi}{4}\left(\frac{5}{12}\right)^2 \sqrt{2(32.2)45}$; $C_d = 0.606$ ◄

Sec 11.6: $C_c = A/A_o = (3.93/5.00)^2 = 0.618$ ◄

Eq. 11.11: $C_v = C_d/C_c = 0.606/0.618 = 0.981$ ◄

11.6.2 *A jet discharges 0.131 cfs from a 1.25-in-diameter orifice in a vertical plane under a head of 10 ft. The jet center line passes through the point 11.47 ft horizontally from the vena contracta and 3.5 ft below the center of the orifice. Find the coefficients of (a) discharge, (b) velocity, and (c) contraction.*

BG

Eq. 5.38: $V = x\sqrt{g/2z} = 11.47\sqrt{32.2/(2\times 3.5)} = 24.6$ fps $= C_v\sqrt{2gh} = C_v\sqrt{2(32.2)10}$; $C_v = 0.969$ ◄

Eq. 11.9: $0.131 = C_d A_o = C_d(\pi/4)(1.25/12)^2\sqrt{2(32.2)10} = C_d(0.008\,52)$; $C_d = 0.606$ ◄

From Eq. 11.11: $C_c = C_d/C_v = 0.606/0.969 = 0.625$ ◄

Alternatively: $A = Q/V = 0.131/24.6 = 0.005\,33$; $C_c = A/A_o = 0.005\,33/0.008\,52 = 0.625$ ◄

11.6.3 *A jet discharges 5.19 L/s from a 35-mm-diameter orifice in a vertical plane under a head of 4 m. The jet center line passes through the point 4.28 m horizontally from the vena contracta and 1.2 m below the center of the orifice. Find the coefficients of (a) discharge, (b) velocity, and (c) contraction.*

SI

Eq. 5.38: $V = x\sqrt{g/2z} = 4.28\sqrt{9.81/(2\times 1.2)} = 8.65$ m/s $= C_v\sqrt{2gh} = C_v\sqrt{2(9.81)4}$; $C_v = 0.977$ ◄

$A_o = (\pi/4)(3.5/100)^2 = 0.000\,962\ m^2$; Eq. 11.9: $0.005\,19 = C_d(0.000\,962)\sqrt{2(9.81)4}$; $C_d = 0.609$ ◄

From Eq. 11.11: $C_c = C_d/C_v = 0.609/0.977 = 0.623$ ◄

Alternatively: $A = Q/V = 0.005\,19/8.65 = 0.000\,600$; $C_c = A/A_o = 0.000\,600/0.000\,962 = 0.623$ ◄

11.6.4 *A 4-in-diameter water jet discharges from a nozzle whose velocity coefficient is 0.96. The pressure in the 9-in-diameter pipe is 11 psi. Assuming the jet does not contract, what is the velocity at the tip of the nozzle? What is the flow rate?*

BG

Eq. 11.12: $$V_{2i} = \frac{1}{\sqrt{1-(4/9)^4}}\sqrt{2(32.2)\frac{11(144)}{62.4}} = 41.2 \text{ fps}$$

$V_2 = C_v(V_{2i}) = 0.96(41.2) = 39.6$ fps ◄ $Q = A_2V_2 = (\pi/4)(4/12)^2 39.6 = 3.46$ cfs ◄

11.6.5 *In Fig X11.6.5 the area A is twice area A_o. If the diverging tube discharges water when h = 6 ft, find (a) the velocity at the throat; (b) the pressure head at the throat. Neglect all friction losses.*

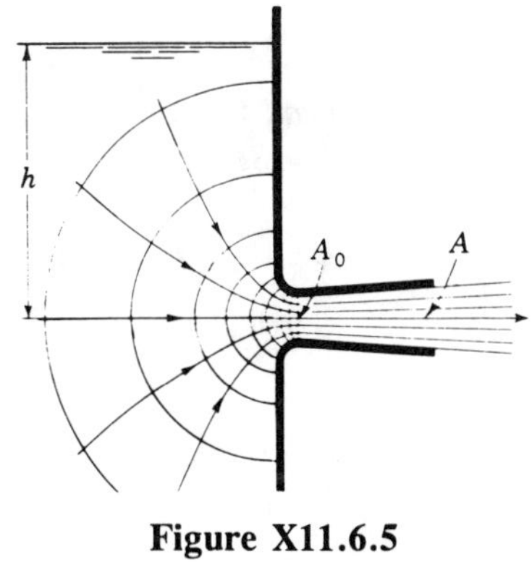

Figure X11.6.5

BG

$V = \sqrt{2gh} = \sqrt{2(32.2)6} = 19.66$ fps

(a) Continuity: $V_o = V(A/A_o) = 19.66(2) = 39.3$ fps ◄

(b) $$\frac{p_o}{\gamma} = \frac{V^2}{2g} - \frac{V_o^2}{2g} = 6.00 - \frac{39.3^2}{2(32.2)} = -18.00 \text{ ft}$$ ◄

11.6.6 *For the same data as Exer. 11.6.5, with a barometric pressure of 14.8 psia and a water temperature of 90°F, what is the maximum value of h at which the tube will flow full? What will happen if h is made greater than this?*

Exer. 11.6.5: $A = 2A_o$, neglect friction. See Fig. X11.6.5 with Solution 11.6.5 above.

BG

Table A.1 for water at 90°F: $\gamma = 62.11$ pcf, $p_v = 0.70$ psia

When $h = h_{max}$ for full flow, $p_o = (p_o)_{min} = p_v$

Given $h_L = 0$, then: $\dfrac{p_o}{\gamma} + \dfrac{V_o^2}{2g} = \dfrac{p_a}{\gamma} + \dfrac{V^2}{2g}$; $\dfrac{0.70(144)}{62.11} + \dfrac{4V^2}{2g} = \dfrac{14.8(144)}{62.11} + V^2/2g$

$3V^2/2g = 32.7$; $h_{max} = V^2/2g = 32.7/3 = 10.90$ ft ◄

If $h > 10.90$ ft, the jet will spring free from the diverging tube, so that $p_o = p_a$, and the discharge will decrease. ◄

11.6.7 *For the rounded entrance and tube flowing full as in Fig. X11.6.5, $C_c = 1.0$ for both the throat and the exit, and thus $C_v = C_d$ for both sections. For the throat, assume the value of C_v as given for Fig. 11.16a, and assume that for the tube as a whole the discharge coefficient applied to the exit end is 0.72. Find the velocity at the throat and the pressure head at the throat if h = 6 ft. Compare your answer with Exer. 11.6.5, which neglected friction.*

Fig. 11.16a: $C_v = 0.98$. Exer. 11.6.5 neglecting friction: Ans = $V_o = 39.3$ fps, $p_o/\gamma = -18.00$ ft. See Fig. X11.6.5 with Solution 11.6.5 above.

BG

From Eq. 11.9: $V = Q/A = 0.72A\sqrt{2g(6)}/A = 14.15$ fps. Given: $C_d = C_v = 0.98$

Continuity: $V_o = VA/A_o = 14.15(2) = 28.3$ fps ◄

From Eq. 11.15 (surface to throat): $h_L = \left(\dfrac{1 - C_v^2}{C_v^2}\right)\dfrac{(V_{2i} \times C_v)^2}{2g} = (1 - C_v^2)h = (1 - 0.98^2)6 = 0.238$ ft

Eq. 5.14, from surface to throat:

$0 + 6 + 0 = p_o/\gamma + 0 + 28.3^2/[2(32.2)] + 0.238$; $p_o/\gamma = -6.68$ ft ◄

At throat:	V_o	p_o/γ
Included friction (this Exer.)	28.3 fps	−6.68 ft
Neglecting friction (Exer. 11.6.5)	39.3 fps	−18.00 ft

Friction decreases velocity and increases pressure at the throat, considerably. ◄

11.6.8 *Given that the diverging tube shown in Fig. X11.6.5 has $A = 1.75A_o$. Find the throat velocity and pressure head when h = 3 m of water. Neglect all friction losses.*

See Fig. X11.6.5 with Solution 11.6.5 above.

SI

$V = \sqrt{2gh} = \sqrt{2(9.81)3} = 7.67$ m/s. $V_o = V(A/A_0) = 7.67(1.75) = 13.43$ m/s ◄

$p_o/\gamma = V^2/2g - V_o^2/2g = 3.0 - 13.43^2/(2 \times 9.81) = -6.19$ m ◄

11.6.9 *Given that the diverging tube shown in Fig. *X11.6.5 has $A = 1.75A_o$. If it is operating at 1000 m elevation under standard atmospheric conditions, what will be the maximum value of h for which the tube will flow full? [*Note this correction to the textbook.]*

See Fig. X11.6.5 with Solution 11.6.5 above.

SI

Table A.3 at 1000 m elevation: $p_{atm} = 89.9 \text{ kN/m}^2$ abs and $T = 8.501°C$

Table A.1 at 8.50°C by interpolation: $p_v/\gamma = 0.111$ m

$$\frac{p_o}{\gamma} + \frac{V_o^2}{2g} = \frac{p_a}{\gamma} + \frac{V^2}{2g}\ ; \quad 0.111 + \frac{(1.75V)^2}{2g} = \frac{89.9}{9.81} + \frac{V^2}{2g}$$

$0.111 + 3.06V^2/2g = 9.16 + V^2/2g$; $2.06(V^2/2g) = 9.05$ m ; $V^2/2g = 4.39$ m

The tube will not flow full if $h > 4.39$ m. ◄

11.6.10 *Water flows through a 60-mm-diameter sharp-edged orifice which connects two adjacent tanks. The head on one side of the orifice is 2.5 m and 0.5 m on the other. Given $C_c = 0.62$ and $C_v = 0.95$, calculate the flow rate.*

SI

$D = 60 \text{ mm} = 0.06$ m ; $A = \pi D^2/4 = \pi(0.06)^2/4 = 0.002\,83 \text{ m}^2$

Sec. 11.6 for submerged discharge: $Q = C_c C_v A\sqrt{2gh} = 0.62(0.95)0.002\,83\sqrt{(2)9.81(2.5 - 0.5)}$

$Q = 0.01043 \text{ m}^3/\text{s} = 10.43$ L/s ◄

Sec. 11.6: Orifices, Nozzles, and Tubes -- *Problems*

11.17 *Water flows at 12 fps through a horizontal 3-in-diameter pipe. A nozzle at the end of the pipe has a velocity coefficient of 0.98. If the pressure in the pipe is *7.5 psi, find (a) the velocity in the jet; (b) the rate of discharge; (c) the diameter of the jet; and (d) the head loss through the nozzle. [*Note this correction to the textbook.]*

BG

(*a*) Eq. 5.15: $V_1^2/2g + p_1/\gamma = (V_j)_i^2/2g$ since $p_j = p_{atm} = 0$. $\therefore$ $(V_j)_i = \sqrt{2g(p_1/\gamma + V_1^2/2g)}$

$$\text{Sec 11.6:}\quad V_j = C_v(V_j)_i = 0.98\sqrt{2(32.2)\left(\frac{7.5(144)}{62.4} + \frac{12^2}{2(32.2)}\right)} = 34.8 \text{ fps} \quad ◄$$

(*b*) $Q = A_1V_1 = (\pi/4)(3/12)^2 12 = 0.589$ cfs ◄

(*c*) $A_j = Q/V_j = 0.589/34.8 = 0.016\,94 \text{ ft}^2$; $D_j = \sqrt{(4/\pi)0.016\,94} = 0.1469 \text{ ft} = 1.762$ in ◄

$$(d)\quad h_L = \frac{V_1^2}{2g} + \frac{p_1}{\gamma} + z_1 - \left(\frac{V_j^2}{2g} + \frac{p_j}{\gamma} + z_j\right) = \frac{12^2}{2(32.2)} + \frac{7.5(144)}{62.4} + z_1 - \left(\frac{34.8^2}{2(32.2)} + 0 + z_1\right)$$

$= 2.24 + 17.31 - 18.77 = 0.774$ ft ◄

11.18 *Water flows at 3.5 m/s through a horizontal 12-cm-diameter pipe. A nozzle at the end of the pipe has a velocity coefficient of 0.98. If the pressure in the pipe is 45 kN/m², find (a) the initial velocity of the jet; (b) the rate of discharge; (c) the diameter of the jet; and (d) the head loss through the nozzle.*

SI

(*a*) Eq. 5.15: $V_1^2/2g + p_1/\gamma = (V_j)_i^2/2g$ since $p_j = p_{\text{atm}} = 0$. $\therefore$ $(V_j)_i = \sqrt{2g(p_1/\gamma + V_1^2/2g)}$

$$\text{Sec 11.6:}\quad V_j = C_v(V_j)_i = 0.98\sqrt{2(9.81)\left(\frac{45}{9.81} + \frac{3.5^2}{2(9.81)}\right)} = 9.91\ \text{m/s} \quad ◄$$

(*b*) $Q = \pi(0.12^2/4)3.5 = 0.0396\ \text{m}^3/\text{s}$ ◄

(*c*) $Q = A_jV_j = (\pi/4)D_j^2V_j$; $0.0396 = (\pi/4)D_j^2(9.91)$; $D_j = 0.005\,09\ \text{m} = 0.509\ \text{cm}$ ◄

$$(d)\quad h_L = \frac{V_1^2}{2g} + \frac{p_1}{\gamma} + z_1 - \left(\frac{V_j^2}{2g} + \frac{p_j}{\gamma} + z_j\right) = \frac{3.5^2}{2(9.81)} + \frac{45}{9.81} + z_1 - \left(\frac{9.91^2}{2(9.81)} + 0 + z_j\right)$$

$= 0.206$ m ◄

11.19 *In Fig. P11.19 the pressure at the base of the nozzle, point 1, is 21.0 psi, and the power available in the jet at point 2 is 3.42 hp. Find (a) the theoretical height to which the jet will rise; (b) the coefficient of velocity for this nozzle; (c) the head loss between points 1 and 2; and (d) the theoretical diameter of the jet at a point 20 ft above point 2.*

BG

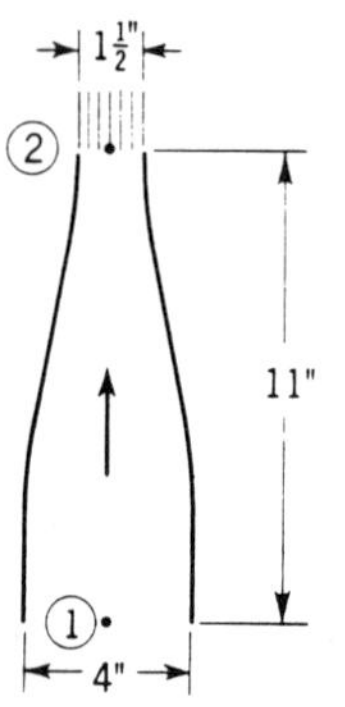

Figure P11.19

Eq. 5.31: $P = h\gamma Q = Q\gamma V_2^2/2g = A_2\gamma V_2^3/2g$ where $A_2 = \pi(1.5/12)^2/4 = 0.012\,27\ \text{ft}^2$

$3.42(550) = 0.021\,27(62.4)V_2^3/(2 \times 32.2)$; $V_2^3 = 158{,}200$; $V_2 = 54.1$ fps

(*a*) Theoretical height $= V_2^2/2g = 54.1^2/(2 \times 32.2) = 45.4$ ft ◄

(*b*) Then $V_1^2/2g = 45.4(1.5/4)^4 = 0.898$ ft

Ideal Eq. 5.15 from point 1 to 2, writing (from Sec. 11.6) $V_i = V/C_v$:

$$\frac{21.0(144)}{62.4} + 0 + \frac{0.898}{C_v^2} = 0 + \frac{11}{12} + \frac{45.4}{C_v^2}\ ; \quad C_v = 0.968 \quad ◄$$

$$(c)\quad \text{Eq. 11.14:}\quad h_L = \left(\frac{1}{0.968^2} - 1\right)\left[1 - \left(\frac{1.5}{4}\right)^4\right]45.4 = 3.02\ \text{ft} \quad ◄$$

The above answer is very sensitive to the value of C_v that is used.

(*d*) $Q = AV = (\pi/4)D^2V = \text{constant}$, $\therefore$ $D_2^2V_2 = D_3^2V_3$ or $(D_3/D_2)^2 = V_2/V_3$

or $(D_3/D_2)^4 = (V_2/V_3)^2 = (V_2^2/2g)/(V_3^2/2g)$ from which

$$D_3 = D_2\left(\frac{V_2^2/2g}{V_3^2/2g}\right)^{1/4} = 1.5\left(\frac{45.4}{45.4 - 20}\right)^{1/4} = 1.734\ \text{in} \quad ◄$$

11.20 *In Fig. P11.19 the pressure at the base of the nozzle, point 1, is 145 kPa, and the power available in the jet at point 2 is 2.55 kW. Find (a) the theoretical height to which the jet will rise; (b) the coefficient of velocity for this nozzle; (c) the head loss between points 1 and 2; and (d) the theoretical diameter of the jet at a point 6 m above point 2.*

See Fig. P11.19 with Solution 11.19 above.

SI

Eq. 5.31: $P = h\gamma Q = Q\gamma V_2^2/2g = A_2\gamma V_2^3/2g$

$2.55 \text{ kW} = 2.55 \text{ kN·m/s} = (\pi/4)(1.5\times0.0254)^2(9.81)V_2^3/(2\times9.81) = 0.000\,570\,V_2^3\ ;\quad V_2 = 16.48 \text{ m/s}$

(*a*) Theoretical height $= V_2^2/2g = 16.48^2/(2\times9.81) = 13.84$ m ◀

(*b*) Then $V_1^2/2g = 13.84(1.5/4)^4 = 0.274$ m

Ideal Eq. 5.15 from point 1 to 2, writing (from Sec. 11.6) $V_i = V/C_v$:

$$\frac{145}{9.81} + 0 + \frac{0.274}{C_v^2} = 0 + 11(0.0254) + \frac{13.84}{C_v^2}\ ;\quad C_v = 0.967 \quad ◀$$

$$(c)\ \text{Eq. 11.14:}\ h_L = \left(\frac{1}{0.967^2} - 1\right)\left[1 - \left(\frac{1.5}{4}\right)^4\right]13.84 = 0.938 \text{ m} \quad ◀$$

The above answer is very sensitive to the value of C_v that is used.

$$(d)\ D_3 = D_2\left(\frac{V_2^2/2g}{V_3^2/2g}\right)^{1/4} = 1.5\left(\frac{13.84}{13.84 - 6}\right)^{1/4} = 1.729 \text{ in} = 43.9 \text{ mm} \quad ◀$$

11.21 *The loss of head due to friction in a nozzle, tube, or orifice may be expressed as $h_L = kV^2/2g$, where V is the actual velocity of the jet. (a) What is k for the three tubes in Fig. 11.16? (b) If these tubes discharge water under a head of 6 ft, what is the loss of head for each?*

Fig. 11.16: The three tubes have $C_v = 0.98$, 0.82, and 0.74.

BG

$$(a)\ \text{Using Eq. 11.15:}\ h_L = k\frac{V^2}{2g} = \left[\frac{1}{C_v^2} - 1\right]\frac{V^2}{2g} \quad \therefore \quad k = \left[\frac{1}{C_v^2} - 1\right]$$

Fig 11.16: $C_v = 0.98$, ∴ $k = 0.0412$ ◀

$C_v = 0.82$, $k = 0.487$ ◀

$C_v = 0.74$, $k = 0.826$ ◀

(*b*) $V = C_v\sqrt{2gH}$ so $V^2 = C_v^2(2gH)$ and $V^2/2g = C_v^2 H$

$$\text{So using Eq. 11.15:}\ h_L = \left[\frac{1}{C_v^2} - 1\right]\frac{V^2}{2g} = \left[\frac{1}{C_v^2} - 1\right]C_v^2 H = (1 - C_v^2)H$$

The respective head losses are: 0.238 ft, 1.966 ft, and 2.71 ft ◀◀

11.22 *The loss of head due to friction in a nozzle, tube, or orifice may be expressed as $h_L = kV^2/2g$, where V is the actual velocity of the jet. (a) What is k for the three tubes in Fig. 11.16? (b) If these tubes discharge water under a head of 1.8 m, what is the loss of head for each?*

Fig. 11.16: The three tubes have C_v = 0.98 , 0.82, and 0.74.

SI

(*a*) Using Eq. 11.15: $h_L = k\frac{V^2}{2g} = \left(\frac{1}{C_v^2} - 1\right)\frac{V^2}{2g} \quad \therefore \quad k = \frac{1}{C_v^2} - 1$

Fig 11.16: $C_v = 0.98, \therefore \; k = 0.0412$ ◄
$C_v = 0.82, \; k = 0.487$ ◄
$C_v = 0.74, \; k = 0.826$ ◄

(*b*) $V = C_v\sqrt{2gH}$ so $V^2 = C_v^2(2gH)$ and $V^2/2g = C_v^2 H$

So using Eq. 11.15: $h_L = \left(\frac{1}{C_v^2} - 1\right)\frac{V^2}{2g} = \left(\frac{1}{C_v^2} - 1\right)C_v^2 H = (1 - C_v^2)H$

The respective head losses are: 0.713 m, 0.590 m, and 0.814 m ◄◄

11.23 *Find the maximum theoretical head h at which the Borda tube of Fig. 11.17 will flow full if the liquid is 90°F water and the barometer reads 29.2 inHg. Assume $C_d = 0.72$ when the tube is flowing full.*

BG

Flowing full: $V = 0.72\sqrt{2gh}$, thus $h = \frac{1}{(0.72)^2}\frac{V^2}{2g}$ or $\frac{V^2}{2g} = (0.72)^2 h$

If, while flowing full, $C_c = 0.52$ and $C_v = 0.98$ for the contracted throat section (per Sec. 11.6),

$V_o^2/(2g) = (1/0.52)^2(V^2/2g) = (0.72)^2 h/(0.52)^2 = 1.917h$

Eq. 5.14 from surface to throat: $p_a/\gamma + h + 0 = p_o/\gamma + 0 + V_o^2/2g + h_L$

Table A.1 for water at 90°F: $p_v/\gamma = 1.61$ ft

When $h = h_{max}$ for full flow, $p_o = (p_o)_{min} = p_v$, $\therefore \; (p_o/\gamma)_{min} = 1.61$ ft

Thus $(29.2/12)13.56 + h + 0 = 1.61 + 0 + 1.917h + [1 - (0.98)^2](h + 33.9)$

where (per Sec 3.5) the quantity $(h + 33.9)$ ft is the total absolute pressure head acting on the orifice.

$\therefore \; h_{max} = (33.0 - 1.61 - 1.342)/0.957 = 31.4$ ft ◄

11.24 *Find the maximum theoretical head h at which the Borda tube of Fig. 11.17 will flow full if the liquid is 30°C water and the barometer reads 735 mmHg. Assume $C_d = 0.72$ when the tube is flowing full.*

SI

Flowing full: $V = 0.72\sqrt{2gh}$, thus $h = (1/0.72)^2(V^2/2g)$ or $V^2/(2g) = (0.72)^2 h$

If while flowing full, $C_c = 0.52$ and $C_v = 0.98$ for the contracted throat section (per Sec. 11.6),

$V_o^2/(2g) = (1/0.52)^2(V^2/2g) = (0.72)^2 h/(0.52)^2 = 1.917h$

Eq. 5.14 from surface to throat: $p_a/\gamma + h + 0 = p_o/\gamma + 0 + V_o^2/2g + h_L$

Table A.1 at 30°C: $p_v/\gamma = 0.44$ m

When $h = h_{max}$ for full flow, $p_o = (p_o)_{min} = p_v$, $\therefore \; (p_o/\gamma)_{min} = 0.44$ m

Thus $0.735(13.56) + h + 0 = 0.44 + 0 + 1.917h + [1 - (0.98)^2](h + 10.3)$

where (per Sec 3.5) the quantity $(h + 10.3)$ m is the total absolute pressure head acting on the orifice.

$\therefore \; h_{max} = (9.97 - 0.44 - 0.408)/0.957 = 9.53$ m ◄

11.25 *In Fig. P11.25 the orifice at the bottom of the large open tank has a diameter of 1.26 in. If the flow rate is 0.19 cfs, and the pitot tube registers a pressure of 16.8 psi, find the C_c and the C_v of the orifice. Neglect air resistance.*

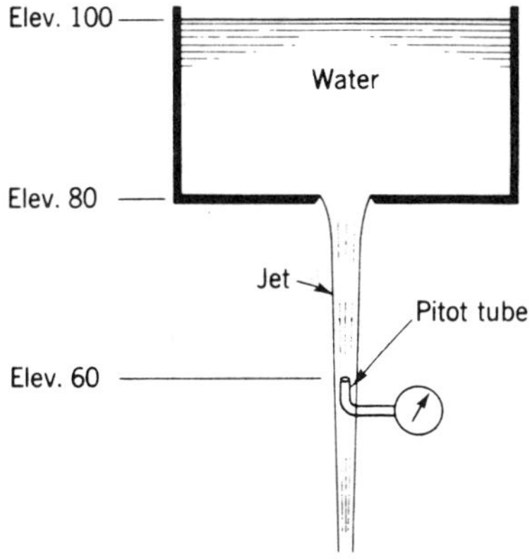

Figure P11.25

BG

Given: $Q = 0.19$ cfs

Let subscript 3 by at pitot tube, and j be at orifice.

$V_3^2/2g = 16.8(144)/62.4 = 38.8$ ft, $V_3 = 50.0$ fps

$V_j^2/2g + 80 = 16.8(144)/62.4 + 60$; $V_j = 34.8$ fps

$V_{\text{ideal}} = \sqrt{2g(20)} = 35.9$ fps ; $C_v = 34.8/35.9 = 0.969$ ◄

$$Q = A_j V_j = C_c A V_j \; ; \quad 0.19 = C_c\left(\frac{\pi}{4}\right)\left(\frac{1.26}{12}\right)^2 34.8 \; ; \quad C_c = 0.631$$ ◄

Sec. 11.7: Venturi Meter -- *Exercises*

11.7.1 *The venturi meter of Fig. S11.4 has $D_1 = 8$ in, $D_2 = 4$ in, and $\Delta z = 1.5$ ft. Assume the discharge coefficients of Fig. 11.19 are applicable. Find the flow rate of 72°F water when the manometer containing carbon tetrachloride ($s = 1.59$) reads $y = 4$ in.*

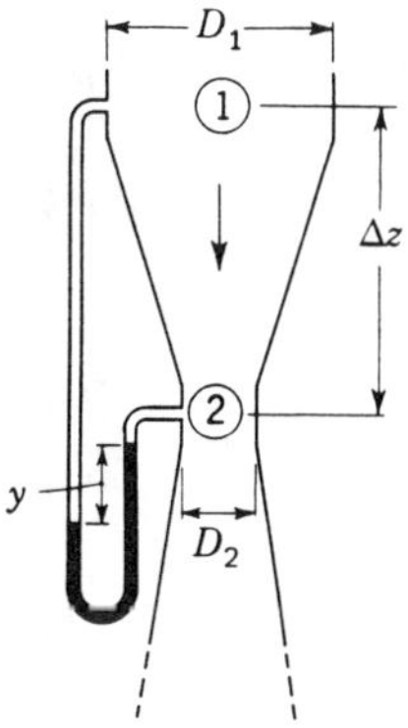

Figure S11.4

BG

Eq. 11.17: $\Delta(z + p/\gamma) = (4/12)[(1.59/1.00) - 1.0] = 0.1967$ ft

$A_2 = (\pi/4)(4/12)^2 = 0.0873$ ft²

From Fig. 11.19: As a first trial assume $C = 0.95$

$$\text{Eq. 11.16:} \quad Q = \frac{0.95(0.0873)}{\sqrt{1 - (4/8)^4}} \sqrt{2(32.2)0.1967} = 0.305 \text{ cfs}$$

$V_2 = Q/A_2 = 0.305/0.0873 = 3.49$ fps

$D_2'' V_2 = 13.88$, and $T = 72$°F (given) ; Fig. 11.19: $C = 0.985$

Thus $Q = (0.985/0.95)0.305 = 0.316$ cfs ◄

11.7.2 *The venturi meter of Fig. S11.4 has $D_1 = 20$ cm, $D_2 = 10$ cm, and $\Delta z = 0.45$ m. Assume the discharge coefficients of Fig. 11.19 are applicable. Find the flow rate of 20°C water when the manometer containing carbon tetrachloride ($s = 1.59$) reads $y = 10$ cm.*

See Fig. S11.4 with Solution 11.7.1 above.

SI

Eq. 11.17: $\Delta(z + p/\gamma) = 0.1[(1.59/1.00) - 1.0] = 0.0590$ m. $A_2 = (\pi/4)(0.1)^2 = 0.007\,85$ m²

$T = 20$°C $= 68$°F ; this is close enough to 72°F to use Fig. 11.22

Venturi meter: $D_1 \times D_2 = 20$ cm $\times$ 10 cm $= 7.87'' \times 3.94''$. This is close to $8'' \times 4''$

∴ From Fig. 11.19: As a trial assume C = 0.95

$$\text{Eq. 11.16:} \quad Q = \frac{0.95(0.00785)}{\sqrt{1 - (10/20)^4}} \sqrt{2(9.81)0.0590} = 0.008\,29 \text{ m}^3\text{/s}$$

$V_2 = Q/A_2 = 0.008\,29/0.007\,85 = 1.056$ m/s

Table A.1 at 20°C: $\nu = 1.003 \times 10^{-6}$ m²/s ; $\mathbf{R} = D_2 V_2/\nu = 0.1(1.056)/(1.003 \times 10^{-6}) = 1.052 \times 10^5$

Fig. 11.19: $C = 0.983$. Thus $Q = (0.983/0.95)0.00829 = 0.00858$ m³/s $= 8.58$ L/s ◄

11.7.3 *The venturi meter of Fig. S11.4 has D_1 = 90 cm, D_2 = 45 cm, and Δz = 2.25 m. Assume the discharge coefficients of Fig. 11.19 are applicable. Find the flow rate of 20°C water when the mercury (s = 13.55) manometer reads y = 16 cm.*

See Fig. S11.4 with Solution 11.7.1 above.

SI

Venturi meter: $D_1 \times D_2 = 90 \text{ cm} \times 45 \text{ cm} = 35.4'' \times 17.7''$.

This is about midway between the top two curves on Fig. 11.19. So assume $C = C_{max} = 0.988$.

Eq. 11.16: $$Q = \frac{0.988\,\pi\,0.225^2}{\sqrt{1-(45/90)^4}}\sqrt{2(9.81)0.16(13.55-1.00)} = 1.019 \text{ m}^3\text{/s}$$

$V_2 = Q/A_2 = 1.019/(\pi\,0.225^2) = 6.40$ m/s ; Table A.1 for water at 20°C: $\nu = 1.003\times10^{-6}$ m²/s

Eq. 7.6: $$\mathbf{R} = \frac{D_2V_2}{\nu} = \frac{0.45(6.40)}{1.003\times10^{-6}} = 2.87\times10^6$$

Fig. 11.19: $C = 0.988$ (as assumed); hence $Q = 1.019 \text{ m}^3\text{/s}$ ◄

11.7.4 *The venturi meter of Fig. S11.4 has D_1 = 4 in, D_2 = 2 in, and Δz = 12 in. Assume the discharge coefficients of Fig. 11.19 are applicable. Find the reading y of the mercury (s = 13.55) manometer when oil (s = 0.92) with a kinematic viscosity of 0.0005 ft²/s is flowing at a rate of 0.40 cfs.*

See Fig. S11.4 with Solution 11.7.1 above.

BG

$D_1 \times D_2 = 4'' \times 2''$ is about 3/4 of the way from $1/2'' \times 1/4''$ to $8'' \times 4''$ on Fig. 11.19.

So assume $C = C_{max} = 0.983$; $A_2 = \pi(1/12)^2 = 0.0218 \text{ ft}^2$

Eqs. 11.16 and 11.17: $$Q = \frac{0.983(0.0218)}{\sqrt{1-(2/4)^4}}\sqrt{2(32.2)\frac{y''}{12}\left(\frac{13.55}{0.92}-1\right)} = 0.40 \text{ cfs}$$

from which $y = 4.43$ inches ; $V_2 = Q/A_2 = 0.40/0.0218 = 18.33$ fps

Eq. 7.6: $$\mathbf{R} = \frac{D_2V_2}{\nu} = \frac{2/12)18.33}{0.0005} = 6110 \; ; \quad \text{Fig. 11.19: } C = 0.940$$

Thus, from Eqs 11.16 and 11.17: $y = 4.43(0.983/0.940)^2 = 4.84$ inches ◄

Sec. 11.7: Venturi Meter -- *Problems*

11.26 *The venturi meter of Fig. S11.4 has D_1 = 10 in, D_2 = 5 in, and Δz = 2 ft. Assume the discharge coefficients of Fig. 11.19 are applicable and the diverging cone angle is 12°. (a) Find the flow rate of 72°F water when the mercury (s = 13.55) manometer reads y = 5 in. Find also the head loss (b) from inlet to throat, and (c) across the entire meter.*

See Fig. S11.4 with Solution 11.7.1 above.

BG

(*a*) Eq. 11.17: $[(z_1 + p_1/\gamma) - (z_2 + p_2/\gamma)] = (5/12)[(13.55/1.00) - 1] = 5.23$ ft

$A_2 = (\pi/4)(5/12)^2 = 0.1363 \text{ ft}^2$. From Fig. 11.19: As a first trial assume $C = 0.95$

Eq. 11.16: $$Q = \frac{0.95(0.1363)}{\sqrt{1-(5/10)^4}}\sqrt{2(32.2)5.23} = 2.46 \text{ cfs}$$

$Q = A_2V_2 = 2.46 \text{ ft}^3\text{/s}$; $V_2 = 2.46/0.1363 = 18.01$ fps ; $D_2''V_2 = 5(18.01) = 90.0$

Fig. 11.19 for $18'' \times 5''$, $D_2''V_2 = 90.0$ and T = 72°F (given) : $C = 0.985$

Curve is flat in this area of Fig 11.19, so $C = 0.985$ is O.K.

Thus $Q = (0.985/0.95)2.46 = 2.55 \text{ ft}^3\text{/sec}$ ◄ /cont...

(*b*) $V_2 = Q/A_2 = 2.55/0.1363 = 18.67$ fps ; $V_1 = V_3 = V_2(5/10)^2 = 4.67$ fps

Eq. 11.14: $h_{L_{1\text{-}2}} = \left(\frac{1}{0.985^2} - 1\right)\left[1 - \left(\frac{5}{10}\right)^4\right]\frac{18.67^2}{2(32.2)} = 0.1557$ ft ◀

(*c*) Fig 8.19 for $\alpha = 12°$: $k' = 0.219$. Therefore

Eq. 8.59: $h_{L_{2\text{-}3}} = \frac{0.219(V_2 - V_3)^2}{2g} = \frac{0.219(18.67 - 4.67)^2}{2(32.2)} = 0.533$ ft

Total head loss = 0.1557 + 0.533 = 0.688 ft ◀

11.27 *The venturi meter of Fig. S11.4 has D_1 = 25 cm, D_2 = 12.5 cm, and Δz = 0.60 m. Assume the discharge coefficients of Fig. 11.19 are applicable and the diverging cone angle is 12°. (a) Find the flow rate of 20°C water when the mercury (s = 13.55) manometer reads y = 12.8 mm. Find also the head loss (b) from inlet to throat, and (c) across the entire meter.*
See Fig. S11.4 with Solution 11.7.1 above.

SI

(*a*) Eq. 11.17: $[(z_1 + p_1/\gamma) - (z_2 + p_2/\gamma)] = 0.0128[(13.55/1.00) - 1) = 0.1606$ m

$A_2 = \pi(0.125)^2/4 = 0.012\,27$ m^2. $T = 20°C = 68°F$ (close enough to 72°F for Fig. 11.19)

From Fig. 11.19: As a first trial assume $C = 0.95$

Eq. 11.16: $Q = \frac{0.95(0.012\,27)}{\sqrt{1 - (12.5/25)^4}}\sqrt{2(9.81)0.1606} = 0.0214$ m^3/s

$V_2 = Q/A_2 = 0.0214/0.012\,27 = 1.742$ m/s. Table A.1 for water at 20°C: $\nu = 1.003 \times 10^{-6}$ m^2/s

$\mathbf{R}_2 = (0.125)1.742/(1.003 \times 10^{-6}) = 2.17 \times 10^5$; Fig. 11.19: $C = 0.985$

Curve is flat in this area of Fig 11.19, so $C = 0.985$ is O.K.

Thus $Q = (0.985/0.95)0.0214 = 0.0222$ m^3/s ◀

(*b*) $V_2 = Q/A_2 = 0.0222/0.012\,27 = 1.806$ m/s ; $V_1 = V_3 = V_2(12.5/25)^2 = 0.452$ m/s

Eq. 11.14: $h_{L_{1\text{-}2}} = \left(\frac{1}{0.985^2} - 1\right)\left[1 - \left(\frac{12.5}{25}\right)^4\right]\frac{1.806^2}{2(9.81)} = 0.002\,34$ m ◀

(*c*) Fig 8.19 for $\alpha = 12°$: $k' = 0.219$ Therefore

Eq. 8.59: $h_{L_{2\text{-}3}} = \frac{0.219(V_2 - V_3)^2}{2g} = \frac{0.219(1.806 - 0.452)^2}{2(9.81)} = 0.0205$ m

Total head loss = 0.002 34 + 0.0205 = 0.0228 m ◀

11.28 *Repeat Prob. 11.26 when the venturi meter is set horizontally (i.e., Δz = 0), with all other data the same.*
BG

When Eq. 11.17 is substituted into Eq. 11.26 to eliminate $\Delta(p/\gamma + z)$, the result is the same regardless of meter orientation. ∴ Answers are the same as for Prob. 11.26. ◀

11.29 *Repeat Prob. 11.27 when the venturi meter is set horizontally (i.e., Δz = 0), with all other data the same.*
SI

When Eq. 11.17 is substituted into Eq. 11.27 to eliminate $\Delta(p/\gamma + z)$, the result is the same regardless of meter orientation. ∴ Answers are the same as for Prob. 11.27. ◀

Sec. 11.8: Flow Nozzle -- *Exercise*

11.8.1 *A 10 cm ISA flow nozzle (Figs. 11.21 and 11.22) is used to measure the flow of 40°C water through a 20-cm pipe. What would be the reading on a mercury manometer for the following flow rates: (a) 1.5 L/s; (b) 15 L/s; (c) 150 L/s?*

SI

Eq. 11.17: $\Delta(p/\gamma + z) = MR[(13.56/1.00) - 1] = 12.56MR$; Substitute into Eq. 11.18:

$Q = KA_2\sqrt{2(9.81)12.56MR} = K(\pi 0.05^2)\sqrt{246MR} = 0.1233K\sqrt{MR}$. $\therefore$ $MR = [Q/(0.1233K)]^2$

Table A.1 for 40°C water: $\nu = 0.658\times10^{-6}$ m²/s. $A_1 = (\pi/4)(0.2\text{ m})^2 = 0.0314\text{ m}^2$

Case	Q (m³/s)	$V_1 = Q/A_1$ (m/s)	$\mathbf{R} = D_1V_1/\nu$	K (Fig. 11.22*)	MR, m
(*a*)	0.0015	0.0477	1.45×10^4	0.978	0.000 1548
(*b*)	0.015	0.477	1.45×10^5	1.005	0.014 65
(*c*)	0.15	4.77	1.45×10^6	1.005	1.465

*For $D_2/D_1 = 10/20 = 0.5$ ▲

Sec. 11.8: Flow Nozzle -- *Problems*

11.30 *An 8 cm ISA flow nozzle (Figs. 11.21 and 11.22) is used to measure the flow of 40°C water through a 16-cm pipe. What would be the reading on a mercury manometer for the following flow rates: (a) 3.8 L/s; (b) 38 L/s; (c) 380 L/s. (d) For which of the preceding three flows is the mercury manometer practical?*

SI

Eq. 11.17: $\Delta(p/\gamma + z) = MR[(13.56/1.00) - 1] = 12.56MR$; Substitute into Eq. 11.18:

$Q = KA_2\sqrt{2(9.81)12.56MR} = K(\pi 0.04^2)\sqrt{246MR} = 0.0789K\sqrt{MR}$. $\therefore$ $MR = [Q/(0.0789K)]^2$

Table A.1 for 40°C water: $\nu = 0.658\times10^{-6}$ m²/s. $A_1 = (\pi/4)(0.16\text{ m})^2 = 0.0201\text{ m}^2$

Case	Q (m³/s)	$V_1 = Q/A_1$ (m/s)	$\mathbf{R} = D_1V_1/\nu$	K (Fig. 11.22*)	MR, m
(*a*)	0.0038	0.1890	4.60×10^4	1.001	0.00231
(*b*)	0.038	1.890	4.60×10^5	1.005	0.230
(*c*)	0.38	18.90	4.60×10^6	1.005	23.0

*For $D_2/D_1 = 8/16 = 0.5$ ▲

(*d*) The manometer is only practical for flow (*b*), because (*a*) is too small for good accuracy and (*c*) is too large to conveniently read. ◄

11.31 *A 7-in ISA flow nozzle is used to measure the flow of 20°F crude oil (s = 0.855) through a 10-in-diameter pipe. If a mercury manometer shows a reading of 6.9 in, find the flow rate.*

BG

Figure 11.22 for $D_2/D_1 = 7/10 = 0.70$: Assume $K = 1.08$. $A_2 = (\pi/4)(7/12)^2 = 0.267\text{ ft}^2$

Eqs. 11.17 and 11.18: $Q = 1.08(0.267)\sqrt{2(32.2)\frac{6.9}{12}\left(\frac{13.56}{0.855} - 1.0\right)} = 6.77\text{ cfs}$

Check $\mathbf{R}$ to confirm that the assumed K is correct.

$V_2 = Q/A_2 = 6.77/0.267 = 25.3$ fps ; Fig 2.4 at 20°F: $\nu = 0.00028$ ft²/s

Eq. 7.6: $\mathbf{R} = DV/\nu = (7/12)25.3/0.000\,28 = 5.28\times10^4$

Fig. 11.22 for $D_2/D_1 = 0.70$: $K = 1.062$; Thus $Q = (1.062/1.08)6.77 = 6.66$ cfs ◄

Sec. 11.9: Orifice Meter -- *Exercise*

11.9.1 *Repeat Exer. 11.8.1 for a VDI orifice.*

Exer. 11.8.1: A 10 cm flow nozzle is used to measure the flow of 40°C water through a 20-cm pipe. What would be the reading on a mercury manometer for the following flow rates: (a) 1.5 L/s; (b) 15 L/s; (c) 150 L/s?

SI

Eq. 11.17: $\Delta(p/\gamma + z) = MR[(13.55/1.00) - 1] = 12.56MR$; Substitute into Eq. 11.20:

$Q = KA_o\sqrt{2(9.81)12.56MR} = K(\pi 0.05^2)\sqrt{246MR} = 0.1233K\sqrt{MR}$. $\therefore$ $MR = [Q/(0.1233K)]^2$

Table A.1 for 40°C water: $\nu = 0.658\times10^{-6}$ m²/s ; $A_1 = (\pi/4)(0.2\text{ m})^2 = 0.0314\text{ m}^2$

Case	Q (m³/s)	$V_1 = Q/A_1$ (m/s)	$\mathbf{R} = D_1V_1/\nu$	K (Fig. 11.24*)	MR, m
(*a*)	0.0015	0.0477	1.45×10^4	0.647	0.000354
(*b*)	0.015	0.4775	1.45×10^5	0.622	0.0383
(*c*)	0.15	4.775	1.45×10^6	0.622	3.83

* For $D_o/D_1 = 10/20 = 0.5$ ▲

Sec. 11.9: Orifice Meter -- *Problems*

11.32 *Repeat Prob. 11.30 for a VDI orifice.*

Prob. 11.30: An 8 cm flow nozzle is used to measure the flow of 40°C water through a 16-cm pipe. What would be the reading on a mercury manometer for the following flow rates: (a) 3.8 L/s; (b) 38 L/s; (c) 380 L/s. (d) For which of the preceding three flows is the mercury manometer practical?

SI

Eq. 11.17: $\Delta(p/\gamma + z) = MR[(13.56/1.00) - 1.00] = 12.56MR$; Substitute into Eq. 11.20:

$Q = KA_o\sqrt{2(9.81)12.56MR} = K(\pi 0.04^2)\sqrt{246MR} = 0.0789K\sqrt{MR}$. $\therefore$ $MR = [Q/(0.0789K)]^2$

Table A.1 for 40°C water: $\nu = 0.658\times10^{-6}$ m²/s. $A_1 = (\pi/4)(0.16\text{ m})^2 = 0.0201\text{ m}^2$

Case	Q (m³/s)	$V_1 = Q/A_1$ (m/s)	$\mathbf{R} = D_1V_1/\nu$	K (Fig. 11.24*)	MR, m
(*a*)	0.0038	0.1890	4.60×10^4	0.630	0.005 86
(*b*)	0.038	1.890	4.60×10^5	0.623	0.599
(*c*)	0.38	18.90	4.60×10^6	0.623	59.9

* For $D_o/D_1 = 8/16 = 0.5$ ▲

(*d*) The manometer is only practical for flow (*b*), because (*a*) is too small for good good accuracy and (*c*) is too large to conveniently read. ◄

11.33 *Repeat Prob. 11.31 for a VDI orifice.*

Prob. 11.31: A 7-in flow nozzle is used to measure the flow of 20°F crude oil (s = 0.855) through a 10-in-diameter pipe. If a mercury manometer shows a reading of 6.9 in, find the flow rate.

BG

Fig. 11.24 for $D_o/D_1 = 7/10 = 0.70$: Assume $K = 0.70$. $A_o = (\pi/4)(7/12)^2 = 0.267$ ft²

Eqs. 11.17 and 11.20: $Q = 0.7(0.267)\sqrt{2(32.2)(6.9/12)[(13.56/0.855) - 1.0]} = 4.39$ cfs

Check $\mathbf{R}$ to confirm that the correct K is correct.

$V_o = Q/A_o = 4.39/0.267 = 16.42$ fps ; Fig. 2.4 at 20°F: $\nu = 0.00028$ ft²/s

Eq. 7.4: $\mathbf{R} = DV/\nu = (7/12)16.42/0.000\,28 = 3.42\times10^4$

Fig. 11.24 for $D_o/D_1 = 0.70$: $K = 0.716$; $Q = (0.716/0.70)4.39 = 4.49$ cfs ◄

11.34 *An 8-in-diameter pipe carries 2.7 cfs of water at 72°F. Find the differential head and head loss across the following types of meters: (a) an 8-in by 4-in VDI orifice; (b) an 8-in by 4-in ISA flow nozzle; (c) a 8-in by 4-in venturi meter.*

BG

$A_1 = (\pi/4)(8/12)^2 = 0.349 \text{ ft}^2$; $A_2 = A_1/4 = 0.0873 \text{ ft}^2$.

$V_1 = Q/A_1 = 2.7/0.349 = 7.73$ fps ; $V_2 = 4V_1 = 30.9$ fps.

Table A.1 for water at 72°F, by interpolation: $\nu = 10.33 \times 10^{-6} \text{ ft}^2/\text{sec}$

Differential heads:

(*a*) Orifice meter: Approach $\mathbf{R} = D_1V_1/\upsilon = (8/12)7.73/(10.33 \times 10^{-6}) = 4.99 \times 10^5$

$D_o/D_1 = 4/8 = 0.5$. Fig. 11.24: $K = 0.622$

Eq. 11.20: $2.7 = 0.622(0.0873)\sqrt{2(32.2)\Delta h}$; $\Delta h = 38.4$ ft ◄

(*b*) Flow nozzle: Approach $\mathbf{R} = 4.99 \times 10^5$

$D_2/D_1 = 4/8 = 0.5$. Fig. 11.22: $K = 1.005$

Eq. 11.18: $2.7 = 1.005(0.0873)\sqrt{2(32.2)\Delta h}$; $\Delta h = 14.72$ ft ◄

(*c*) Venturi meter: Throat $\mathbf{R} = D_2V_2/\upsilon = (4/12)30.9/(10.33 \times 10^{-6}) = 9.98 \times 10^5$

Fig. 11.19 for $8'' \times 4''$: $C = 0.985$

Eq. 11.16: $2.7 = \dfrac{0.985(0.0873)}{\sqrt{1 - (0.5)^4}}\sqrt{2(32.2)\Delta h}$; $\Delta h = 13.47$ ft ◄

Head losses:

Due to converging section $h_L \approx 0.04\,V_2^2/2g$

$V_2 = 30.9$ fps for gradually contracting nozzle and venturi

$V_2 \approx 30.9/0.62 = 49.9$ fps for suddenly contracting orifice

Thus head losses due to converging section are:

Orifice $\approx 0.04(49.9^2)/(2 \times 32.2) = 1.547$ ft

Nozzle and venturi $\approx 0.04(30.9^2)/(2 \times 32.2) = 0.595$ ft

Due to diverging sections: $h_L = K(V_1 - V_2)^2/2g$

Orifice $\approx 1.0(49.9 - 7.73)^2/(2 \times 32.2) = 27.6$ ft

Nozzle $\approx 1.0(30.9 - 7.73)^2/(2 \times 32.2) = 8.36$ ft

Venturi $\approx 0.2(30.9 - 7.73)^2/(2 \times 32.2) = 1.67$ ft

∴ Total (converging plus diverging) head losses are approximately:

(*a*) 29.2 ft for orifice meter ◄

(*b*) 8.96 ft for flow nozzle ◄

(*c*) 2.27 ft for Venturi meter ◄

Sec. 11.10: Flow Measurement of Compressible Fluids -- ***Exercises***

11.10.1 *Natural gas, for which $R = 3100$ ft·lb/(slug·°R) and $k = 1.3$, flows through a venturi tube. The pipe diameter is 18 in and the throat diameter is 9 in. The initial pressure of the gas is 160 psia at 64°F. What is the weight flow rate of the gas if the throat pressure is 104 psia and the meter coefficient is 0.98?*

BG

$p_2/p_1 = 104/160 = 0.65$; $A_2 = (\pi/4)(9/12)^2 = 0.442$ ft^2 ; $D_2/D_1 = 9/18 = 0.5$

Fig. 11.25 for $p_2/p_1 = 0.65$, and $D_2/D_1 = 0.5$: $Y = 0.778$

Eq. 2.5: $\gamma_1 = \dfrac{gp_1}{RT} = \dfrac{32.2(160)144}{3100(460 + 64)} = 0.457$ pcf

Eq. 11.23: $G = 0.98(0.778)0.442\sqrt{2(32.2)0.457\dfrac{(160 - 104)144}{1 - 0.5^4}} = 169.4$ lb/sec ◄

11.10.2 *Helium [$R = 12,420$ ft·lb/(slug·°R), $k = 1.66$] is stored in a tank at 60 psia and 76°F. The gas flows out through a ¾-in-diameter orifice for which $C_v = 0.98$ and $C_c = 0.62$ for liquids. Find the weight flow rate if $Y = 0.95$ and the pressure outside the tank is 44 psia.*

BG

Eq. 2.5: $\gamma_1 = \dfrac{gp_1}{RT_1} = \dfrac{32.2(60)144}{12,420(460 + 76)} = 0.0418$ pcf ; $A_2 = \pi(0.75/12)^2/4 = 0.003\,07$ ft^2

Sec. 11.10: Replacing C by $K\sqrt{1 - (D_o/D_1)^4}$, Eq. 11.23 becomes for an orifice meter

$G = KYA_2\sqrt{2g\gamma_1(p_1 - p_2)}$

where the flow (discharge) coefficient K is equivalent to $C_d = C_cC_v$ (Eq. 11.11) for liquids.

Thus $G = 0.98(0.62)0.95(0.003\,07)\sqrt{2(32.2)0.0418(60 - 44)144} = 0.1394$ lb/sec ◄

Note: Also, in this problem, we have $D_1 = \infty$ so that $\sqrt{1 - (D_o/D_1)^4} = 1$

11.10.3 *A tank contains air at 200 psia and 100°F. It flows out through an orifice with an area of 1.5 in^2, and enters a space where the pressure is 80 psia. Assuming $C_d = 0.60$, find the weight flow rate.*

BG

Table A.5 for air: $R = 1715$ ft^2/(sec^2·°R) . $p_2/p_1 = 80/200 = 0.40$; $D_o/D_1 = D_o/\infty = 0$

From Fig. 11.25: $Y = 0.825$. Eq. 2.5: $\gamma_1 = \dfrac{32.2(200)144}{1715(460 + 100)} = 0.966$ pcf

Eq. 11.23 for orifice: $G = 0.60(0.825)\dfrac{1.5}{144}\sqrt{2(32.2)0.966\dfrac{(200 - 80)144}{1 - 0}} = 5.34$ lb/sec ◄

11.10.4 *A tank contains air at 1460 kN/m^2 abs and 44°C. It flows out through an orifice with an area of 12.5 cm^2, and enters a space where the pressure is 713 kN/m^2 abs. Assuming $C_d = 0.62$, find the weight flow rate.*

SI

Table A.5 for air: $R = 287$ m^2/(s^2·K) . $p_2/p_1 = 713/1460 = 0.488$; $D_o/D_1 = D_o/\infty = 0$

From Fig. 11.25: $Y = 0.850$. Eq. 2.5: $\gamma_1 = \dfrac{9.81(1\,460\,000)}{287(273 + 44)} = 157.4$ N/m^3

Eq. 11.23 for orifice: $G = 0.62(0.850)\dfrac{12.5}{100^2}\sqrt{2(9.81)157.4\dfrac{(1460 - 713)1000}{1 - 0}} = 31.6$ N/s ◄

11.10.5 *Repeat Exer. 11.10.3, but discharging into a space where the pressure is 15 psia.*

Exer. 11.10.3: $p_1 = 200$ *psia,* $T_1 = 100°F$*,* $D_1 = \infty$*,* $A_o = 1.5\ in^2$*,* $C_d = 0.60$*; find G of air through orifice.*

BG

Table A.5 for air: $R = 1715\ \text{ft}^2/(\text{sec}^2\cdot°\text{R})$. $D_o/D_1 = D_o/\infty = 0$; $p_2/p_1 = 15/200 = 0.0750$

To extrapolate Fig. 11.25 for $D_o/D_1 = 0$:

line passes through $(p_2/p_1, Y) = (1, 1)$ and $(0.4, 0.825)$, so $Y = \frac{0.175}{0.6}\left(\frac{p_2}{p_1}\right) + \left(1 - \frac{0.175}{0.6}\right)$

Thus, when $p_2/p_1 = 0.0750$, $Y = 0.730$. Eq. 2.5: $\gamma_1 = \frac{32.2(200)144}{1715(460 + 100)} = 0.966$ pcf

Eq. 11.23 for orifice: $G = 0.60(0.730)\frac{1.5}{144}\sqrt{2(32.2)0.966\frac{(200 - 15)144}{1 - 0}} = 5.87$ lb/sec ◄

11.10.6 *Repeat Exer. 11.10.4, but discharging into a space where the pressure is 105 kN/m² abs.*

Exer. 11.10.4: $p_1 = 1460\ kN/m^2$ *abs,* $T_1 = 44°C$*,* $D_1 = \infty$*,* $A_o = 12.5\ cm^2$*,* $C_d = 0.62$*; find G of air through orifice.*

SI

Table A.5 for air: $R = 287\ \text{m}^2/(\text{s}^2\cdot\text{K})$. $D_o/D_1 = D_o/\infty = 0$; $p_2/p_1 = 105/1460 = 0.0719$

To extrapolate Fig. 11.25 for $D_o/D_1 = 0$:

line passes through $(p_2/p_1, Y) = (1, 1)$ and $(0.4, 0.825)$, so $Y = \frac{0.175}{0.6}\left(\frac{p_2}{p_1}\right) + \left(1 - \frac{0.175}{0.6}\right)$

Thus, when $p_2/p_1 = 0.0719$, $Y = 0.729$. Eq. 2.5: $\gamma_1 = \frac{9.81(1\,460\,000)}{287(273 + 44)} = 157.4\ \text{N/m}^3$

Eq. 11.23 for orifice: $G = 0.62(0.729)\frac{12.5}{100^2}\sqrt{2(9.81)157.4\frac{(1460 - 105)1000}{1 - 0}} = 36.6$ N/s ◄

11.10.7 *Find (a) the critical pressure and the corresponding throat velocity in a suitable nozzle, assuming air at 70°F and* $p_1 = 100$ *psia. Neglect the velocity of approach. Refer to Secs. 13.3 and 13.6. (b) What will the values be if* $D_2/D_1 = 0.80$*?*

BG

(*a*) Table A.5 for air: $R = 1715\ \text{ft}^2/(\text{sec}^2\cdot°\text{R})$, $k = 1.40$

Neglecting velocity of approach, $p_1 = p_o$

From Eq. 13.33: $p* = 0.528p_o = 0.528p_1 = 0.528(100) = 52.8$ psia ◄

From Eq. 13.36: $T* = T_1(p*/p_1)^{(k-1)k} = (460 + 70)0.528^{0.4/1.4} = 442°\text{R}$

Sec. 13.5 for critical conditions: $\mathbf{M} = 1$. ∴ (Eq. 7.10 and 13.10) $V = c$

Eq. 13.15: $c = \sqrt{kRT} = \sqrt{1.4(1715)442}$ so $V = 1030$ fps ◄

(*b*) If $D_2/D_1 = 0.80$, $p*$ will be the same and $V = c$ will be a little larger, perhaps about 1045 fps ◄◄

11.10.8 *Find (a) the critical pressure and the corresponding throat velocity in a suitable nozzle, assuming air at 20°C and p_1 = 700 kN/m² abs. Neglect the velocity of approach. Refer to Secs. 13.3 and 13.6. (b) What will the values be if D_2/D_1 = 0.80?*

SI

(*a*) Table A.5 for air: $R = 287\ \text{m}^2/(\text{s}^2\cdot\text{K})$, $k = 1.40$. Neglecting velocity of approach, $p_1 = p_o$

From Eq. 13.33: $p* = 0.528p_o = 0.528p_1 = 0.528(700) = 370\ \text{kN/m}^2$ abs ◄

From Eq. 13.36: $T* = T_1(p*/p_1)^{(k-1)k} = (273 + 20)0.528^{0.4/1.4} = 244\ \text{K}$

Sec. 13.5 for critical conditions: $\mathbf{M} = 1$. ∴ (Eq. 7.10 and 13.10) $V = c$

Eq. 13.15: $c = \sqrt{kRT} = \sqrt{1.4(287)244}$ so $V = 313$ m/s ◄

(*b*) If $D_2/D_1 = 0.80$, $p*$ will be the same and $V = c$ will be a little larger, perhaps about 318 m/s ◄◄

Sec. 11.10: Flow Measurement of Compressible Fluids -- *Problems*

11.35 *Air at 70°F and 104 psia flows through a venturi tube with a coefficient of 0.98. The pressure at the throat is 67.6 psia, the inlet area is 0.60 ft², and the throat area is 0.15 ft². Given k for the air is 1.40. (a) Using Eq. (11.22), find the *ideal weight flow rate. (b) Evaluate Y from Fig. 11.25, and use it to find the *actual weight flow rate. (c) What is the throat velocity? [*Note these corrections to the textbook.]*

BG

Table A.5 for air: $R = 1715\ \text{ft}^2/(\text{sec}^2\cdot°\text{R})$, $c_p = 6000\ \text{ft}^2/(\text{sec}^2\cdot°\text{R})$

Eq. 2.5: $\gamma_1 = 32.2(104)144/[1715(460 + 70)] = 0.531$ pcf ; $p_1 = 104(144) = 14{,}976$ psf abs

(*a*) Eq. 11.22:

$$G_{\text{ideal}} = 0.15\sqrt{2(32.2)\frac{1.4}{0.4}(14{,}976)0.531\left(\frac{67.6}{104}\right)^{\frac{2}{1.4}}\frac{1-\left(\frac{67.6}{104}\right)^{\frac{0.4}{1.4}}}{1-\left(\frac{0.15}{0.60}\right)^2\left(\frac{67.6}{104}\right)^{\frac{2}{1.4}}}} = 51.1\ \text{lb/sec} \quad ◄$$

(*b*) Fig. 11.25 for $D_2/D_1 = 0.5$ and $p_2/p_1 = 0.65$: $Y = 0.779$ ◄

Eq. 11.23: $G = 0.98(0.779)0.15\sqrt{2(32.2)0.531(104 - 67.6)144/(1 - 0.5^4)} = 50.0$ lb/sec ◄

(*c*) Eq. 11.21: $V^2/2 = 6000(460 + 70)\sqrt{1 - 0.65^{0.4/1.4}} = 3.68 \times 10^5$, $V = 858$ fps ◄

11.36 *Air at 20°C and 700 kN/m² abs flows through a venturi tube with a coefficient of 0.98. The pressure at the throat is 420 kN/m² abs, the inlet area is 0.060 m², and the throat area is 0.015 m². Given k for the air is 1.40. (a) Using Eq. (11.22), find the *ideal weight flow rate. (b) Evaluate Y from Fig. 11.25, and use it to find the *actual weight flow rate. (c) What is the throat velocity? [*Note these corrections to the textbook.]*

SI

Air, Table A.5: $R = 287\ \text{m}^2/(\text{s}^2\cdot\text{K})$, $c_p = 1003\ \text{m}^2/(\text{s}^2\cdot\text{K})$. Eq. 2.5: $\gamma_1 = \dfrac{9.81(700\,000)}{287(273 + 20)} = 81.7\ \text{N/m}^3$

(*a*) Eq. 11.22:

$$G_{\text{ideal}} = 0.015\sqrt{2(9.81)\frac{1.4}{0.4}(700\,000)81.7\left(\frac{420}{700}\right)^{\frac{2}{1.4}}\frac{1-\left(\frac{420}{700}\right)^{\frac{0.4}{1.4}}}{1-\left(\frac{0.015}{0.060}\right)^2\left(\frac{420}{700}\right)^{\frac{2}{1.4}}}} = 244\ \text{N/s} \quad ◄$$

(*b*) Fig. 11.25 for $D_2/D_1 = 0.5$ and $p_2/p_1 = 0.60$: $Y = 0.744$ ◄

$$\text{Eq. 11.23:}\quad G = 0.98(0.744)0.015\sqrt{2(9.81)81.7\frac{(700-420)1000}{1-0.5^4}} = 239\ \text{N} \quad ◄$$

(*c*) Eq. 11.21: $V^2/2 = 1003(273 + 20)(1 - 0.60^{0.4/1.4}) = 3.99 \times 10^4$; $V = 283$ m/s ◄

11.37 *Using the same data as Prob. 11.35, what would be the value of Y and the weight flow rate for a square-edged orifice with C = 0.59?*

Prob. 11.35: Air flows, $A_1 = 0.60\ ft^2$, $A_o = 0.15\ ft^2$, $T_1 = 70°F$, $p_1 = 104$ psia, $p_2 = p_o = 67.6$ psia.

BG

Table A.5 for air: $R = 1715\ ft^2/(sec^2 \cdot °R)$. Eq. 2.5: $\gamma_1 = \frac{32.2(104)144}{1715(460 + 70)} = 0.531$ pcf

Fig. 11.25 for square-edged orifice with $D_o/D_1 = 0.5$ and $p_2/p_1 = 0.65$: $Y = 0.891$ ◄

Eq. 11.23: $G = 0.59(0.891)0.15\sqrt{2(32.2)0.531\frac{(104 - 67.6)144}{1 - 0.5^4}} = 34.5$ lb/sec ◄

11.38 *Using the same data as Prob. 11.36, what would be the value of Y and the weight flow rate for a square-edged orifice with C = 0.62?*

Prob. 11.36: Air flows, $A_1 = 0.060\ m^2$, $A_o = 0.015\ m^2$, $T_1 = 20°C$, $p_1 = 700\ kN/m^2$ abs, $p_2 = p_o = 420\ kN/m^2$ abs.

SI

Table A.5 for air: $R = 287\ m^2/(s^2 \cdot K)$. Eq. 2.5: $\gamma_1 = \frac{9.81(700\,000)}{287(273 + 20)} = 81.7\ N/m^3$

Fig. 11.25 for square-edged orifice with $D_o/D_1 = 0.5$ and $p_2/p_1 = 0.60$: $Y = 0.876$ ◄

Eq. 11.23: $G = 0.62(0.876)0.015\sqrt{2(9.81)81.7\frac{(700 - 420)1000}{1 - 0.5^4}} = 178.2$ N/s ◄

11.39 *Air flows through a 20-cm by 10-cm venturi meter. At inlet the air temperature is 15°C and the gage pressure is 150 kN/m². Determine the flow rate if a mercury manometer reads 18 cm. Assume an atmospheric pressure of 101.3 kN/m² abs.*

SI

Venturi meter: $D_1 \times D_2 = 20\ cm \times 10\ cm = 7.9'' \times 3.9''$; $D_2/D_1 = 10/20 = 0.5$

$p_1 - p_2 = \gamma_M MR = 13.56(9807\ N/m^3)0.18\ m = 23\,900\ N/m^2 = 23.9\ kN/m^2$

$p_1 = p_{gage} + p_{at} = 150 + 101.3 = 251.3\ kN/m^2$ abs

$p_2 = p_1 - (p_1 - p_2) = 251.3 - 23.9 = 227.4\ kN/m^2$ abs ; $p_2/p_1 = 227.4/251.3 = 0.905$

Fig. 11.25 for $p_2/p_1 = 0.905$ and $D_2/D_1 = 0.5$: $Y = 0.940$; Table A.5 for air: $R = 287\ m^2/(s^2 \cdot K)$

Eq. 2.5: $\gamma_1 = \frac{gp_1}{RT_1} = \frac{(9.81)251\,300\ N/m^2\ abs}{287(273 + 15)} = 29.8\ N/m^3$

Fig. 11.19 for $D_2 \times D_1 = 7.9'' \times 3.9''$ and $\mathbf{R}_2$ unknown: Assume $C = C_{max} = 0.985$

Eq. 11.23: $G = 0.985(0.940)\frac{\pi(0.1)^2}{4}\sqrt{2(9.81)29.8\frac{(23.9)1000}{1 - 0.5^4}} = 28.1$ N/s

Check $\mathbf{R}_2$:

From Eqs. 2.2 and 2.6: $\frac{p_2}{p_1} = \left(\frac{v_1}{v_2}\right)^n = \left(\frac{\rho_2}{\rho_1}\right)^k \quad \therefore \quad \frac{\rho_2}{\rho_1} = \left(\frac{p_2}{p_1}\right)^{1/k} = 0.905^{1/1.4} = 0.931$

So $\rho_2 = 0.931\rho_1 = 0.931(\gamma_1/g) = 0.931(29.8/9.81) = 2.83\ kg/m^3$

Eq. 2.7: $T_2/T_1 = (p_2/p_1)^{(k-1)/k} = 0.905^{0.4/1.4} = 0.972$

So $T_2 = 0.972T_1 = 0.972(273 + 15) = 279.9\ K = 6.9°C$ /cont...

Table A.2 for air at 6.9°C (noting, per Sec. 2.11, that μ is practically independent of p), by interpolation:

$\mu_2 = 17.45 \times 10^{-6} = \text{N·s/m}^2$

From Eq. 4.16b: $V_2 = \dfrac{G}{\rho_2 g A_2} = \dfrac{28.1 \text{ N/s}}{(2.83 \text{ kg/m}^3)9.81 \text{ m/s}^2(\pi 0.1^2/4) \text{ m}^2} = 128.9 \text{ m/s}$

At throat: $\mathbf{R}_2 = \dfrac{D_2 V_2 \rho_2}{\mu_2} = \dfrac{0.1(128.9)2.83}{17.45 \times 10^{-6}} = 2.09 \times 10^6$

Fig. 11.19 for $\mathbf{R}_2 = 2.09 \times 10^6$ and $D_1 \times D_2 = 7.9'' \times 3.9''$: $C = 0.985$

$\therefore$ Assumed C was correct; $G = 28.1$ N/s ◀

11.40 *Air flows through a 5.0 cm diameter orifice from a tank at 1500 kN/m² abs and 40°C into a space where the pressure is 500 kN/m² abs. (a) Compute the weight flow rate assuming $C_d = 0.60$. Refer to Sec. 13.8. Repeat for external pressures of (b) 750, (c) 1000, and (d) 1250 kN/m² abs.*

SI

Table A.5 for air: $R = 287 \text{ m}^2/(\text{s}^2\text{·K})$.

(*a*) $p_2/p_1 = 500/1500 = 0.333$.

Sec. 13.8: Because $p_2/p_1 < 0.528 = p_2^*/p_1$, sonic velocity occurs at the throat, use Eq. 13.45.

From Eq. 13.45: $G = C_d \dot{m} g = 0.60(9.81)\dfrac{\pi(0.025)^2 1500}{\sqrt{273 + 40}}\sqrt{\dfrac{1.4}{287}\left(\dfrac{2}{2.4}\right)^{2.4/0.4}} = 0.0396 \text{ kN/s}$ ◀

(*b*) Sec. 13.8: If $p_2 < 0.528 p_1 = 0.528(1500) = 792 \text{ kN/m}^2$ abs, then sonic flow occurs at the throat and Eq. 13.45 applies (as above). Thus for $p_2 = 750 \text{ kN/m}^2$ abs: $G = 0.0396$ kN/s ◀

(*c*) If $p_2 > 792 \text{ kN/m}^2$ abs, then use Eq. 11.23, for which

Eq. 2.5: $\gamma_1 = \dfrac{g p_1}{R T_1} = \dfrac{9.81(1500)}{287(273 + 40)} = 0.1638 \text{ kN/m}^3$

So for $p_2 = 1000 \text{ kN/m}^2$ abs:

Fig. 11.25 for $D_o/D_1 = 0$ and $p_2/p_1 = 1000/1500 = 0.667$: $Y = 0.902$

Eq. 11.23: $G = 0.60(0.902)\pi 0.025^2\sqrt{2(9.81)0.1638(1500 - 1000)/(1 - 0)} = 0.0426 \text{ kN/s}$ ◀

(*d*) Similarly for $p_2 = 1250 \text{ kN/m}^2$ abs: $p_2/p_1 = 0.833$, $Y = 0.951$, $G = 0.0318$ kN/s ◀

Sec. 11.11: Thin-Plate Weirs -- *Exercises*

11.11.1 *A 2.5-ft-high rectangular sharp-crested weir extends across 6-ft-wide rectangular channel. When the head is 1.12 ft, determine the flow rate by neglecting the velocity of approach.*

BG

Eq. 11.27: $Q = 3.32 L H^{3/2} = 3.32(6)1.12^{3/2} = 23.6$ cfs ◀

11.11.2 *An 0.75-m-high rectangular sharp-crested weir extends across a 2.2-m-wide rectangular channel. When the head is 32 cm, determine the flow rate by neglecting the velocity of approach.*

SI

Eq. 11.27: $Q = 1.83 L H^{3/2} = 1.83(2.2)0.32^{3/2} = 0.729 \text{ m}^3/\text{s}$ ◀

11.11.3 *Suppose the rectangular weir of Exer. 11.11.1 is contracted at both ends. (a) Find the flow rate for a head of 1.12 ft by the *Francis formula. (b) What would be the maximum value of H for which the Francis formula could be used? [*Note: The results of the substitutions described at the top of textbook page 553 are called Francis formulas.]*

Exer. 11.11.1: P = 2.5 ft, L = 6 ft.

BG

(*a*) Eq. 11.27 with end contractions: $Q = 3.32[6.00 - 2(0.1)1.12]1.12^{3/2} = 22.7$ cfs ◄

(*b*) Fig. 11.27: For Francis formula, require $L > 3H$; $H_{max} = L/3 = 6/3 = 2.00$ ft ◄

11.11.4 *Suppose the rectangular weir of Exer. 11.11.2 is contracted at both ends. (a) Find the flow rate for a head of 32 cm by the *Francis formula. (b) What would be the maximum value of H for which the Francis formula could be used? [*Note: The results of the substitutions described at the top of textbook page 553 are called Francis formulas.]*

Exer. 11.11.2: P = 0.75 m, L = 2.2 m.

SI

(*a*) Eq. 11.27 with end contractions: $Q = 1.83[2.2 - 2(0.1)0.32]0.32^{3/2} = 0.708$ m^3/s ◄

(*b*) Fig. 11.27: For Francis formula, require $L > 3H$; $H_{max} = L/3 = 2.2/3 = 0.733$ m ◄

11.11.5 *(a) What is the rate of discharge of water over a 45° triangular weir when the head is 0.6 ft? (b) With the same head, what would be the increase in discharge obtained by doubling the notch angle, i.e., for a 90° weir? (c) What would be the head for a discharge of 1.5 cfs of water over a 60° triangular weir?*

BG

(*a*) Fig. 11.29 for $\theta = 45°$ and $H = 0.6$ ft: $C_d = 0.593$

Eq. 11.28: $Q = 0.593(8/12)\sqrt{2(32.2)}[\tan(45°/2)]0.6^{5/2} = 0.293$ cfs ◄

(*b*) Fig. 11.29 for $\theta = 90°$ and $H = 0.6$ ft: $C_d = 0.583$

Eq. 11.28: $Q = 0.583(8/15)\sqrt{2(32.2)}[\tan(90°/2)]0.6^{5/2} = 0.696$ cfs ◄

Discharge increase = (0.696 - 0.293)/0.293 = 1.375 = 137.5% ◄

(*c*) Fig. 11.29 for $\theta = 60°$ and unknown H: Assume $C_d = 0.581$ (the minimum).

Eq. 11.28: $1.5 = 0.581(8/15)\sqrt{2(32.2)}[\tan(60°/2)]H^{5/2}$; $H = 1.018$ ft

Fig. 11.29 for $\theta = 60°$ and $H = 1.02$ ft, by extrapolation: $C_d = 0.582$

Eq. 11.28 with $C_d = 0.582$: $H = 1.017$ ft ◄

Sec. 11.11: Thin-Plate Weirs -- *Problems*

11.41 *Using the Rehbock formula, plot a family of curves of C_d versus P/H with H as a parameter. These curves give a complete picture of the variation of C_d for sharp-crested rectangular weirs. Include P/H values of 0.5, 1.0, 2.0, 5.0, 10.0 and H values of 0.2, 1.0, and 5.0 ft.*

BG

Eq. 11.25*a*:

P/H	C_d		
	H = 0.2 ft	1.0 ft	5.0 ft
0.5	0.781	0.768	0.766
1.0	0.701	0.688	0.686
2.0	0.661	0.648	0.646
5.0	0.637	0.624	0.622
10.0	0.629	0.616	0.614
	▲	▲	▲

11.42 *For the Cipolletti weir of Sec. 11.11, confirm that the side slopes of the trapezoid are 4V:1H by setting the reduction in discharge due to contraction equal to the increase in discharge due to the triangular area added.*

N

Eq. 11.24: The decrease in flow due to two end contractions is $\Delta Q = C_d(2/3)\sqrt{2g}[2(0.1)H]H^{3/2}$

This deficiency is made up by the equivalent triangular weir of angle θ, per Eq. 11.28. Equating these:

$C_d(2/3)\sqrt{2g}0.2H^{5/2} = C_d(8/15)\sqrt{2g}[\tan(\theta/2)]H^{5/2}$

Assuming C_d is the same for both weirs: $\tan(\theta/2) = 0.25$, or side slope = 4V:1H QED ◀

11.43 *Develop in general terms an expression for the percent of error in Q over a triangular weir if there is a small error in the measurement of the vertex angle. Assume there is no error in the weir coefficient. Compute the percent error in Q if there is a 1° error in the measurement of the total vertex angle of a triangular weir having a total vertex angle of 60°.*

N

Eq. 11.28: $Q = C_d(8/15)\sqrt{2g}H^{5/2}\tan(\theta/2)$

$dQ/d\theta = C_d(8/15)\sqrt{2g}H^{5/2}d/d\theta[\tan(\theta/2)] = C_d(8/15)\sqrt{2g}H^{5/2}(1/2)\sec^2(\theta/2)$

Dividing: $\dfrac{dQ}{Q} = \dfrac{\sec^2(\theta/2)\,d\theta}{2\tan(\theta/2)} = \dfrac{d\theta}{2\sin(\theta/2)\cos(\theta/2)} = \dfrac{d\theta}{\sin\theta}$ ◀

For $\theta = 60°$ and $d\theta = 1°$: $dQ/Q = \dfrac{1°}{\sin 60°} = \dfrac{\pi/180}{0.866} = 0.0202 = 2.02\%$ ◀

11.44 - 45: See Sec. 11.12 below.

11.46 *A 60° V-notch weir and a rectangular weir with end contractions (crest length = 2 ft) are both used to measure a flow rate of approximately 0.25 cfs. Assuming C_d is known to be minimum for each weir, compute the percentage of error in Q that would result from an error of 0.01 ft in the respective head measurements.*

BG

For V-notch weir: Eq. 11.28: $Q = C_d(8/15)\sqrt{2g}[\tan(\theta/2)]H^{5/2}$

Differentiating: $dQ/dH = C_d(8/15)\sqrt{2g}[\tan(\theta/2)](5/2)H^{3/2}$

Dividing by Eq. 11.28: $dQ/Q = 2.5dH/H$

Fig. 11.29: Minimum C_d is 0.581. With Q = 0.25 cfs and $\theta = 60°$,

Eq. 11.28: $0.25 = 0.581(8/15)\sqrt{2(32.2)}[\tan(60°/2)]H^{5/2}$; H = 0.498 ft

For ΔH = 0.01 ft: $dQ/Q = 2.5(0.01/0.498) = 0.0502 = 5.02\%$ ◀

For rectangular weir: Eq. 11.27: $Q = 3.32LH^{3/2}$

Differentiating: $dQ/dH = 3.32L(3/2)H^{1/2}$; Dividing by Eq. 11.27: $dQ/Q = 1.5(dH/H)$

Sec. 11.11: Minimum C_d is about 0.62. With Q = 0.25 cfs and L = 2 ft,

Eq. 11.27: $0.25 = 3.32(2.0)H^{3/2}$; H = 0.1123 ft

For ΔH = 0.01 ft: $dQ/Q = 1.5(0.01/0.1123) = 0.1335 = 13.35\%$ ◀

11.47 *A 60° V-notch weir and a rectangular weir with end contractions (crest length = 0.6 m) are both used to measure a flow rate of approximately 7 L/s. Assuming C_d is known to be minimum for each weir, compute the percentage of error in Q that would result from an error of 5 mm in the respective head measurements.*

SI

For V-notch weir: Eq. 11.28: $Q = C_d(8/15)\sqrt{2g}[\tan(\theta/2)]H^{5/2}$

Differentiating: $dQ/dH = C_d(8/15)\sqrt{2g}[\tan(\theta/2)](5/2)H^{3/2}$; dividing by Eq. 11.28: $dQ/Q = 2.5dH/H$

Fig. 11.29: Minimum C_d is 0.581. With $Q = 0.007$ m³/s and $\theta = 60°$,

Eq. 11.28: $0.007 = 0.581(8/15)\sqrt{2(32.2)}[\tan(60°/2)]H^{5/2}$; $H = 0.1509$ m

For $\Delta H = 0.005$ m: $dQ/Q = 2.5(0.005/0.1509) = 0.0828 = 8.28\%$ ◄

For rectangular weir: Eq. 11.27: $Q = 1.83LH^{3/2}$

Differentiating: $dQ/dH = 1.83L(3/2)H^{1/2}$; Dividing by Eq. 11.27: $dQ/Q = 1.5(dH/H)$

Sec. 11.11: Minimum C_d is about 0.62. With $Q = 0.007$ m³/s and $L = 0.6$ m,

Eq. 11.27: $0.007 = 1.83(0.6)H^{3/2}$; $H = 0.0344$ m

For $\Delta H = 0.005$ m: $dQ/Q = 1.5(0.005/0.0344) = 0.218 = 21.8\%$ ◄

Sec. 11.12: Streamlined Weirs and Free Outfall -- *Exercises*

11.12.1 *What is the flow rate per unit width over a broad crested weir which rises 1.0 ft above the bed of a horizontal channel when the head is 3.0 ft above the crest?*

BG

Eq. 11.30: $Q = L\sqrt{g}(2/3)^{3/2}E^{3/2} = L\sqrt{32.2}(2/3)^{3/2}(3 + V_0^2/2g)^{3/2}$

$Q/L = q = 3.09(3 + V_0^2/2g)^{3/2}$, but $q = 4V_0$ or $V_0 = q/4$

$\therefore\ q = 3.09[3 + q^2/(32 \times 32.2)]^{3/2}$; by trial $q = 18.92$ cfs/ft ◄

11.12.2 *What is the flow rate per unit width over a broad crested weir which rises 0.3 m above the bed of a horizontal channel when the head is 0.9 m above the crest?*

SI

Eq. 11.30: $Q = L\sqrt{g}(2/3)^{3/2}E^{3/2} = L\sqrt{9.81}(2/3)^{3/2}(0.90 + V_0^2/2g)^{3/2}$

$Q/L = q = 1.705(0.90 + V_0^2/2g)^{3/2}$, but $q = 1.20V_0$ or $V_0 = q/1.20$

$\therefore\ q = 1.705[0.90 + q^2/(2.88 \times 9.81)]^{3/2}$; by trial $q = 1.715$ m³/s per m ◄

11.12.3 *Find the water depth just upstream of a 1.5-ft-high broad-crested weir in a channel 6 ft wide when the flow is 18 cfs.*

BG

$q = 18/6 = 3$ cfs/ft ; $y_2 = y_c = (q^2/g)^{1/3} = (3^2/32.2)^{1/3} = 0.654$ ft

$V_2 = q/y_2 = 3/0.654 = 4.59$ fps ; $y_0 + V_0^2/2g = 1.5 + 0.654 + (4.59^2/2g)$; but $V_0 = 3/y_0$

$\therefore\ y_0 + (3/y_0)^2/(2 \times 32.2) = 2.48$; by trial, $y_0 = 2.46$ ft ◄

11.12.4 *Find the water depth just upstream of an 0.4-m-high broad-crested weir in a channel 1.8 m wide when the flow is 0.54 m³/s.*

SI

$q = 0.54/1.8 = 0.30$ m³/s per m ; $y_2 = y_c = (q^2/g)^{1/3} = (0.30^2/9.81)^{1/3} = 0.209$ m

$V_2 = q/y_2 = 0.30/0.209 = 1.433$ m/s ; $y_0 + V_0^2/(2g) = 0.4 + 0.209 + 1.433^2/(2g)$; $V_0 = 0.3/y_0$

$\therefore\ y_0 + 1/(218y_0^2) = 0.714$; by trial, $y_0 = 0.705$ m ◄

11.12.5 *Subcritical flow in a rectangular open channel 6 ft wide ends in a free outfall. If the water depth at the brink measures 2.70 ft, estimate the flow rate.*

BG

Sec. 11.12: $y_c = y_b/0.72 = 2.7/0.72 = 3.75$ ft

From Eq. 10.23: $q = \sqrt{g y_c^3} = \sqrt{32.2(3.75)^3} = 41.2$ cfs/ft ; $Q = bq = 6(41.2) = 247$ cfs ◄

11.12.6 *Subcritical flow in a rectangular open channel 2 m wide ends in a free outfall. If the water depth at the brink measures 0.75 m, estimate the flow rate.*

SI

Sec. 11.12: $y_c = y_b/0.72 = 0.75/0.72 = 1.042$ m

From Eq. 10.23: $q = \sqrt{g y_c^3} = \sqrt{9.81(1.042)^3} = 3.33\ \text{m}^3/\text{s}$ per m; $Q = bq = 2(3.33) = 6.66\ \text{m}^3/\text{s}$ ◄

Sec. 11.12: Streamlined Weirs and Free Outfall -- *Problems*

11.44 *Plot C_d versus P/H for broad-crested weirs using Eq. (11.31). Include P/H values of 0.5, 1.0, 2.0, 5.0, 10.0 and H values of 0.2, 1.0, and 5.0 ft.*

BG

Successive trial and error solutions are required.

Given P/H and H: $P = H(P/H)$

Fig. 11.30: $y_0 + V_0^2/2g = P + y_c + V_c^2/(2g) = P + 1.5y_c$; $y_0 = P + H$

i.e., $P + H + [1/(2g)]\,[q/(P + H)]^?= 1.5(q^2/g)^{1/3}$; $E = H + V_0^2/(2g)$

i.e., $E = H + [1/(2g)][q/(P + H)]^2 = 1.5(q^2/g)^{1/3}$

Given P and H, solve this by trial for q, and hence find E.

Then find C_d using Eq. 11.31. Repeat for other P/H and H values.

Trial and error results:

P/H	C_d		
	H = 0.2 ft	1.0 ft	5.0 ft
0.5	0.656	0.656	0.652
1.0	0.614	0.614	0.614
2.0	0.594	0.593	0.592
5.0	0.581	0.581	0.581
10.0	0.579	0.578	0.579
	▲	▲	▲

These repetitive trial and error solutions can be performed effectively using a spreadsheet or Mathcad.

/more...

11.45 *All the weir crests discussed in this chapter produce flow rates which vary as the head to some power greater than 1. In certain cases, such as in the outlet of a constant-velocity sedimentation chamber, it is desirable to employ a weir form in which Q varies directly with H. The proportional-flow weir is set flush with the bottom of the channel, as shown in the figure, while the sides taper inward, following the hyperbola $x\sqrt{y} = k$, a constant. Commencing with the head $h = H - y$, on the element of area $dA = 2x\,dy = 2(k/\sqrt{y})dy$, prove that the discharge equation for such a weir may be written as $Q = C_d\pi k\sqrt{2g}H$, and evaluate k in terms of the width B and the velocity V in the rectangular approach channel.*

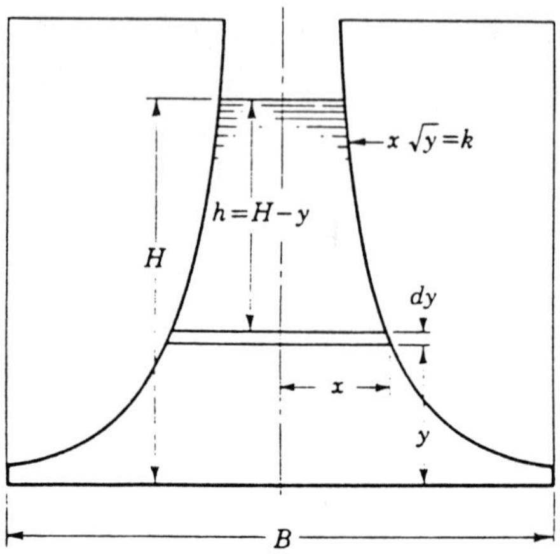

Figure P11.45

N

$$dQ_i = V dA = \sqrt{2gh}(2x dy) = \sqrt{2g}\sqrt{H-y}\,2k\,dy/\sqrt{y}$$

$$Q_i = 2k\sqrt{2g}\int_0^H \left[\frac{H-y}{y}\right]^{1/2} dy$$

To integrate, let $\sqrt{(H-y)/y} = \tan\theta$; Then,

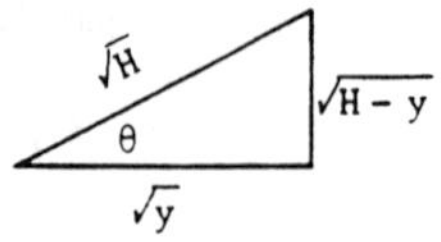

$\sqrt{y/H} = \cos\theta$; $y = H\cos^2\theta$; $dy = -2H\cos\theta\sin\theta\,d\theta$

At $y = 0$, $\tan\theta = \infty$, $\theta = \pi/2$; at $y = H$, $\tan\theta = 0$, $\theta = 0$

$$\text{Then, } Q_i = 2k\sqrt{2g}\int_{\pi/2}^{0} -2H\cos\theta\sin\theta\tan\theta\,d\theta = 2k\sqrt{2g}\int_{\pi/2}^{0} -2H\sin^2\theta\,d\theta$$

$$= 2k\sqrt{2g}(-2H)[\theta/2 - (1/4)\sin 2\theta]_{\pi/2}^{0} = \pi H k\sqrt{2g}$$

$$Q = C_d Q_i = C_d\pi k\sqrt{2g}H \qquad \text{Q.E.D.} \quad ◀$$

$$Q = C_d\pi k\sqrt{2g}H = VBH \;; \quad \text{so} \quad k = VB/(\pi C_d\sqrt{2g}) \quad ◀$$

Sec. 11.14: Sluice Gate -- *Exercises*

11.14.1 *A rectangular channel contains a sluice gate which extends across the full width of the channel with an opening of 0.24 ft. Given $C_d = 0.60$ and free flow, find the flow rate per unit width and the sluice coefficient when the upstream and downstream depths are 2.05 ft and 0.15 ft respectively.*

BG

$$\text{Eq. 11.35: } V_{2i} = \frac{1}{\sqrt{1-(0.15/2.10)^2}}\sqrt{2(32.2)(2.05-0.15)} = 11.09 \text{ fps}$$

$$q_i = aV_{2i} = 0.24(11.09) = 2.66 \text{ cfs/ft} \;; \quad q = C_d q_i = 0.6(2.66) = 1.597 \text{ cfs/ft} \quad ◀$$

$$\text{From Eq. 11.36: } K_s = \frac{q}{a\sqrt{2gy_1}} = \frac{1.597}{0.24\sqrt{2(32.2)2.10}} = 0.579 \quad ◀$$

11.14.2 *A rectangular channel contains a sluice gate which extends across the full width of the channel with an opening of 0.10 m. Given $C_d = 0.60$ and free flow, find the flow rate per unit width and the sluice coefficient when the upstream and downstream depths are 0.85 m and 0.06 m respectively.*

SI

$$\text{Eq. 11.35: } V_{2i} = \frac{1}{\sqrt{1-(0.06/0.85)^2}}\sqrt{2(9.81)(0.85-0.06)} = 3.96 \text{ m/s}$$

$$q_i = aV_{2i} = 0.10(3.96) = 0.396 \text{ m}^3\text{/s per m} \;; \quad q = C_d q_i = 0.6(0.396) = 0.237 \text{ m}^3\text{/s per m} \quad ◀$$

$$\text{From Eq. 11.36: } K_s = \frac{q}{a\sqrt{2gy_1}} = \frac{0.237}{0.10\sqrt{2(9.81)0.85}} = 0.581 \quad ◀$$

Sec. 11.14: Sluice Gate -- *Problems*

11.48 *An 8-ft-wide rectangular channel contains a sluice gate which extends across the full width of the channel with an opening of 0.6 ft. Assuming $C_d = 0.60$, $C_c = 0.62$, and free flow, find the flow rate and the sluice coefficient when the upstream depth is 5 ft.*

BG

$y_2 = C_c a = 0.62(0.6) = 0.372$ ft ; Eq. 11.35: $V_{2i} = \sqrt{\dfrac{2(32.2)(5 - 0.372)}{1 - (0.372/5)^2}} = 17.31$ fps

$Q = C_d Q_i = C_d A V_{2i} = C_d a V_{2i} = C_d a B V_{2i} = 0.6(0.6)8(17.31) = 49.9$ cfs ◄

From Eq. 11.36: $K_s = \dfrac{Q}{A\sqrt{2gy_1}} = \dfrac{49.9}{8(0.6)\sqrt{2(32.2)5}} = 0.579$ ◄

11.49 *A 2.0-m-wide rectangular channel contains a sluice gate which extends across the full width of the channel with an opening of 0.15 m. Assuming $C_d = 0.60$, $C_c = 0.62$, and free flow, find the flow rate and the sluice coefficient when the upstream depth is 1.25 m.*

SI

$y_2 = C_c a = 0.62(0.15) = 0.0930$ m ; Eq. 11.35: $V_{2i} = \sqrt{\dfrac{2(9.81)(1.25 - 0.0930)}{1 - (0.0930/1.25)^2}} = 4.78$ m/s

$Q = C_d Q_i = C_d A V_{2i} = C_d a B V_{2i} = 0.6(0.15)2.0(4.78) = 0.860$ m^3/s ◄

From Eq. 11.36: $K_s = \dfrac{0.860}{2.0(0.15)\sqrt{2(9.81)1.25}} = 0.579$ ◄

11.50 *Refer to Sample Problem 5.11. (a) If $C_v = 0.98$, what is the flow rate? (b) If $C_c = 0.62$, what is the height of the opening? (c) What is K_s?*

Sample Prob. 5.11: For no energy losses, $V_{1i} = 2.12$ m/s, $V_{2i} = 5.29$ m/s.

SI

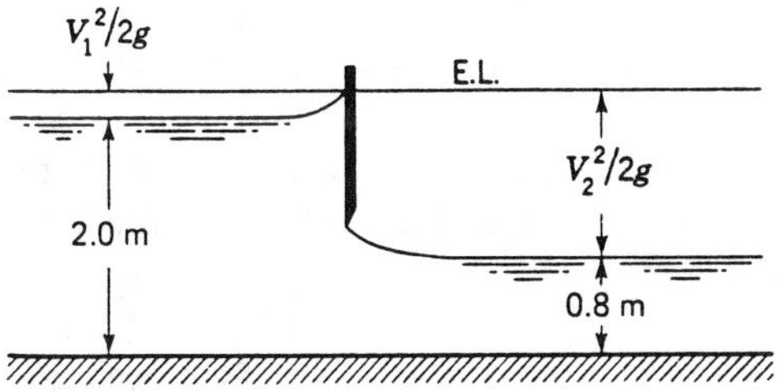

Figure S5.11

(*a*) First trial (using Sec. 11.6):

$V_2 = C_v V_{2i} = 0.98(5.29) = 5.18$ m/s

$V_1 = V_2(y_2/y_1) = 5.18(0.8/2.0) = 2.07$ m/s

$V_1^2/(2g) = 0.219$ m

Second trial: (using Fig. S5.11):

$V_{2i}^2/2g = 2.0 + 0.219 - 0.8 = 1.419$ m

$V_{2i} = 5.28$ m/s; $V_2 = 0.98(5.28) = 5.17$ m/s

This is little change from the first trial, so expect little further change. Close enough.

$q = y_2 V_2 = 0.8(5.17) = 4.14$ m^3/s per m of width ◄

(*b*) Height of gate opening, $a = y_2/C_c = 0.8/0.62 = 1.290$ m ◄

(*c*) From Eq. 11.36: $K_s = \dfrac{Q}{A\sqrt{2gy_1}} = \dfrac{q}{a\sqrt{2gy_1}} = \dfrac{4.14}{1.290\sqrt{2(9.81)2.0}} = 0.512$ ◄

Chapter 12

Unsteady Flow Problems

PROBLEM SELECTION GUIDE

Sec	Exer/Prob	Units	Difficulty	Length	Parts	Similar	Special features
12.2	*Discharge With Varying Head*						
	X[1] 12.2.1	BG	Medium	Medium	2	12.2.2	Integration
	12.2.2	SI	Medium	Medium	2	12.2.1	Integration
	12.2.3	BG	Medium	Medium	1	12.2.4	Integration
	12.2.4	SI	Medium	Medium	1	12.2.3	Integration
	12.2.5	BG	Medium	Medium	1	12.2.6	Integration
	12.2.6	SI	Medium	Medium	1	12.2.5	Integration
	12.2.7	BG	Medium	Short	1	12.2.8	Integration
	12.2.8	SI	Medium	Short	1	12.2.7	Integration
	P 12.1	BG	Medium	Medium	3	12.2	Integration
	12.2	SI	Medium	Medium	3	12.1	* Integration
	12.3	N	Medium	Medium	2		Integration
	12.4	BG	Medium	Medium	1	12.5	Integration; plot
	12.5	SI	Medium	Medium	1	12.4	Integration; plot
	12.6	BG	Hard	Long	1	12.7	Numerical integration
	12.7	SI	Hard	Long	1	12.6	Numerical integration
	12.8	BG	Hard	Long	1		Integration (using table of integrals)
12.3	*Unsteady Flow of Incompressible Fluids in Pipes*						
	X 12.3.1	BG	Medium	Medium	1		
	12.3.2	BG	Medium	Short	1		
	P 12.9	BG	Medium	Medium	2	12.10-12	Differentiation
	12.10	SI	Medium	Medium	2	12.9-12	Differentiation
	12.11	BG	Medium	Medium	2	12.9-12	Differentiation
	12.12	SI	Medium	Medium	2	12.9-11	Differentiation
	12.13	BG	Hard	Medium	1		Integration
	12.14	BG	Medium	Long	1	12.15	Assume/verify laminar flow; integration
	12.15	SI	Medium	Long	1	12.14	Assume/verify laminar flow; integration
	12.16	BG	Hard	Long	1	12.17	Integration
	12.17	SI	Hard	Long	1	12.16	Integration
	12.18	BG	Hard	Long	2	12.19	Integration
	12.19	SI	Hard	Long	2	12.18	Integration

/cont...

[1] For all Exercises (marked "X"), answers are given in Appendix D of the textbook.
* Note a correction made to early printings of the textbook.
† Answers are sensitive to values (f) if they are read from a graph.
X = Exercise, P = (end-of-chapter) Problem, S = Sample Problem. ☐ = could use computing aids.

Sec	Exer/Prob	Units	Difficulty	Length	Parts	Similar	Special features
12.4	*Establishment of Steady Flow*						
	X 12.4.1	BG	Easy	Short	1	12.4.2	Neglect minor losses (h', Sec 8.24)
	12.4.2	SI	Easy	Short	1	12.4.1	Neglect h' (Sec 8.24)
	12.4.3	SI	Easy	Medium	1	12.4.4,5,6	Neglect h' (Sec 8.24)
	12.4.4	BG	Easy	Medium	1	12.4.3,5,6	Include h' (Sec 8.24)
	12.4.5	BG	Easy	Short	1	12.4.3,4,6	Include h' (Sec 8.24)
	12.4.6	SI	Easy	Short	1	12.4.3,4,5	Include h' (Sec 8.24)
	P 12.20	BG	Medium	Medium	1	12.21,22	Neglect h' (Sec 8.24); plot
	12.21	BG	Medium	Medium	1	12.20,22	Include h' (Sec 8.24); plot
	12.22	SI	Medium	Medium	1	12.20,21	Neglect h' (Sec 8.24); plot
	12.23	BG	Medium	Long	1		Neglect h' (Sec 8.24)
	12.24	BG	Hard	Long	1	12.25	Integration; plot
	12.25	SI	Hard	Long	1	12.24	Integration; plot
	12.26	BG	Hard	Long	1		Integration; plot
	12.27	BG	Hard	Long	1	12.28	Neglect h' (Sec 8.24); assume/verify
	"						flow is laminar (Sec 8.6) and velocity
	"						head is negligible; integration; plot.
	12.28	SI	Hard	Long	1	12.27	Neglect h' (Sec 8.24); assume/verify
	"						flow is laminar (Sec 8.6) and velocity
	"						head is negligible; integration; plot.
	12.29	BG	V Hard	Long	3	12.30	Assume/verify laminar flow; integration.
	12.30	SI	V Hard	Long	3	12.29	Assume/verify laminar flow; integration.
12.5	*Velocity of Pressure Wave in Pipes*						
	X 12.5.1	BG	Easy	Short	1	12.5.2	
	12.5.2	SI	Easy	Short	1	12.5.1	
	12.5.3	BG	Easy	Short	1	12.5.4	
	12.5.4	SI	Easy	Short	1	12.5.3	
12.6	*Water Hammer*						
	X 12.6.1	BG	Easy	Short	2	12.6.2	Instantaneous closure
	12.6.2	SI	Easy	Short	2	12.6.1	Instantaneous closure
	12.6.3	BG	Medium	Medium	2	12.6.4	Rapid closure
	12.6.4	SI	Medium	Medium	2	12.6.3	Rapid closure
	12.6.5	BG	Medium	Medium	2	12.6.6	Slow closure
	12.6.6	SI	Medium	Medium	2	12.6.5	Slow closure
	P 12.31	BG	Medium	Medium	4	12.32	
	12.32	SI	Medium	Medium	4	12.31	
	12.33	BG	Medium	Medium	1		Uses Sec 8.5
	12.34	SI	Medium	Medium	2		† Slow closure; uses Sec 8.14 (or 8.13)
	12.35	BG	Medium	Medium	1		Slow closure
	12.36	BG	Medium	Medium	2	12.37	Rapid closure
	12.37	SI	Medium	Medium	2	12.36	Rapid closure

/cont...

Sec	Exer/Prob	Units	Difficulty	Length	Parts	Similar	Special features
12.7	***Surge Tanks***						
	X 12.7.1	BG	Easy	Short	1	12.7.2	Neglect h' & $V^2/2g$; find z_{max}
	12.7.2	SI	Easy	Short	1	12.7.1	Neglect h' & $V^2/2g$; find D_s
	P 12.38	N	Hard	Medium	1		Integration, differentiation.
	12.39	SI	Medium	Short	1	12.40	Include h' & $V^2/2g$; find Q_0
	12.40	SI	Medium	Short	1	12.39	Neglect h' & $V^2/2g$; find Q_0
	12.41	BG	Hard	Medium	1	12.42-3,46-7	* ☐ Incl. h' & $V^2/2g$, find z_{max}. T&E
	12.42	BG	Hard	Medium	1	12.41-3,46-7	☐ Negl. h' & $V^2/2g$, find z_{max}. T&E
	12.43	BG	Hard	Medium	1	12.41-2,46-7	☐ Incl. h' & $V^2/2g$, find z_{max}. T&E
	12.44	BG	Hard	Medium	1	12.45	☐ Include h' & $V^2/2g$, find D_s. T & E
	12.45	BG	Hard	Medium	1	12.44	☐ Neglect h' & $V^2/2g$, find D_s. T & E
	12.46	SI	Hard	Medium	1	12.41-3,47	☐ Include h' & $V^2/2g$, find z_{max}. T&E
	12.47	SI	Hard	Medium	1	12.41-3,46	☐ Neglect h' & $V^2/2g$, find z_{max}. T&E

Chapter 12
UNSTEADY-FLOW PROBLEMS

Sec. 12.2: Discharge With Varying Head -- *Exercises*

12.2.1 *(a) A ship lock has vertical sides. Water enters or leaves through a conduit area of cross-sectional area A. The flow through the conduit is given by $Q = C_d A(2gz)^{1/2}$, where z is the variable difference in elevation between the water surface in the lock and that outside. Prove that the time required for the water level in the lock to drop from z_1 to z_2 is*

$$t = \frac{2A_s}{C_d A\sqrt{2g}}(z_1^{1/2} - z_2^{1/2})$$

(Note: If the lock is being filled, the signs must be reversed.) (b) Suppose the lock is 400 ft long by 100 ft wide, and the discharge coefficient for the conduit is 0.50. If the water surface in the lock is initially 40 ft below the level of the surface of the water upstream, how large must the conduit be if the lock is to be filled in 10 min?

BG

(*a*) In Eq 12.2: $Q_i = 0$; $Q_o = C_d A(2gz)^{1/2}$

$$\text{thus } t = \int_{z_1}^{z_2} \frac{A_s dz}{-C_d A(2gz)^{1/2}} = \frac{A_s}{C_d A(2g)^{1/2}}\int_{z_1}^{z_2}\frac{dz}{-z^{1/2}}$$

$$= \frac{A_s}{C_d A(2g)^{1/2}}\left[-2z^{1/2}\right]_{z_1}^{z_2} = \frac{2A_s}{C_d A\sqrt{2g}}\left(z_1^{1/2} - z_2^{1/2}\right) \quad \text{Q.E.D.} \blacktriangleleft$$

(*b*) From the above Eq: $10(60) = \dfrac{2(400)100}{0.50A\sqrt{2(32.2)}}(40^{1/2} - 0^{1/2})$ from which $A = 210\text{ ft}^2$ ◄

12.2.2 *(a) Repeat Exer. 12.2.1(a) (b) Suppose the lock is 100 m long by 30 m wide, and the discharge coefficient for the conduit is 0.50. If the water surface in the lock is initially 12 m below the level of the surface of the water upstream, how large must the conduit be if the lock is to be filled in 10 min?*

SI

(*a*) See solution to Exer. 12.2.1(*a*).

(*b*) From the above Eq: $10(60) = \dfrac{2(100)30}{0.50A\sqrt{2(9.81)}}(12^{1/2} - 0^{1/2})$ from which $A = 15.64\text{ m}^2$ ◄

12.2.3 *The tank in Fig. X12.2.3 has the shape of the frustrum of a cone with a 2-ft² orifice ($C_d = 0.62$) in the bottom. Given $D_1 = 44$ ft, $D_2 = 20$ ft, $z_0 = 36$ ft, and that the water level outside the tank is constant at section 2, how long will it take the water level in the tank to drop from section 1 to section 2? (Note: The tank diameter = ky, and y = z + h_2, where z is the variable distance between surface levels.)*

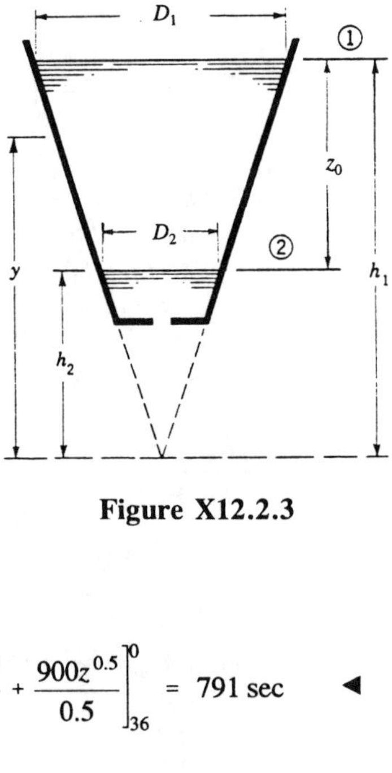

Figure X12.2.3

BG

$$\frac{h_2}{z_0} = \frac{D_2}{D_1 - D_2} \quad \text{so} \quad h_2 = \frac{D_2 z_0}{D_1 - D_2} = \frac{20(36)}{44 - 20} = 30\text{ ft}$$

$z = y - h_2$; $y = z + h_2 = z + 30$; $k = D/y = (44 - 20)/36 = 2/3$

$A_s = (\pi/4)(2y/3)^2 = (\pi/9)y^2 = (\pi/9)(z + 30)^2$

$Q_i = 0$; $Q_o = C_d A\sqrt{2gz} = 0.62(2)\sqrt{2gz}$

$$\text{Eq 12.2: } t = \int \frac{A_s dz}{Q_i - Q_o} = \int_{36}^{0} \frac{(\pi/9)(z + 30)^2 dz}{0 - (0.62)2\sqrt{2gz}}$$

$$= -\frac{(\pi/9)}{(0.62)2\sqrt{2g}}\int_{36}^{0}[z^{3/2} + 60z^{1/2} + 900z^{-1/2}]dz = -\frac{(\pi/9)}{1.24\sqrt{2g}}\left[\frac{z^{2.5}}{2.5} + \frac{60z^{1.5}}{1.5} + \frac{900z^{0.5}}{0.5}\right]_{36}^{0} = 791\text{ sec} \blacktriangleleft$$

12.2.4 *The cone-shaped tank in Fig X12.2.3 has the shape of the frustrum of a cone with a 0.2-m² orifice (C_d = 0.60) in the bottom. Given D_1 = 12 m, D_2 = 5.6 m, z_0 = 10.4 m, and that the water level outside the tank is constant at section 2, how long will it take the water level in the tank to drop from section 1 to section 2? (Note: The tank diameter = ky, and y = z + h_2, where z is the variable distance between surface levels.)*

See Fig. X12.2.3 with Solution 12.2.3.

SI

$$\frac{h_2}{Z_0} = \frac{D_2}{D_1 - D_2}; \quad h_2 = \frac{D_2 z_0}{D_1 - D_2} = \frac{5.6(10.4)}{12 - 5.6} = 9.1 \text{ m}$$

$$z = h_2 - y; \quad y = z + h_2 = z + 9.1; \quad k = d/y = (12 - 5.6)/10.4 = 0.615$$

$$A_s = (\pi/4)d^2 = (\pi/4)(0.615y)^2 = 0.297y^2 = 0.297(z + 9.1)^2 \text{ m}^2$$

$$Q_i = 0; \quad Q_o = C_d A\sqrt{2gz} = 0.60(0.2)\sqrt{2(9.81)}\,z^{1/2} = 0.531 z^{1/2}$$

$$\text{Eq. 12.2:} \quad t = \int \frac{A_s dz}{Q_i - Q_o} = \int_{10.4}^{0} \frac{0.297(z + 9.1)^2 dz}{0 - 0.531 z^{1/2}} = -0.560\int_{10.4}^{0} [z^{1.5} + 18.2z^{0.5} + 82.8z^{-0.5}]dz$$

$$= -0.560\left[\frac{z^{2.5}}{2.5} + 18.2\frac{z^{1.5}}{1.5} + 82.8\frac{z^{0.5}}{0.5}\right]_{10.4}^{0} = 605 \text{ s} \quad ◄$$

12.2.5 *Given the same tank and C_d as in Exer. 12.2.3, but now with the water surface outside constant at section 1 instead and the tank initially empty, how long will it take for the water level in the tank to rise from section 2 to section 1?*

See Fig. X12.2.3 with Solution 12.2.3. Exer. 12.2.3: D_1 = 44 ft, D_2 = 20 ft, z_0 = 36 ft, 2-ft² orifice (C_d = 0.62). (Note: The tank diameter = ky, and y = z + h_2, where z is the variable distance between surface levels.)

BG

$$\frac{h_2}{z_0} = \frac{D_2}{D_1 - D_2} \text{ so } h_2 = \frac{D_2 z_0}{D_1 - D_2} = \frac{20(36)}{44 - 20} = 30 \text{ ft}; \quad A_s = (\pi/4)(2y/3)^2 = (\pi/9)y^2 = (\pi/9)(66 - z)^2$$

$$z = h_1 - y; \quad y = h_1 - z = 30 + 36 - z = 66 - z; \quad k = d/y = (44 - 20)/36 = 2/3$$

$$Q_i = C_d A\sqrt{2gz} = 0.62(2)\sqrt{2gz}; \quad Q_o = 0. \qquad \text{Eq 12.2:} \quad t = \int \frac{A_s dz}{Q_i - Q_o} = \int_{36}^{0} \frac{(\pi/9)(66 - z)^2(-dz)}{(0.62)2(2gz)^{1/2}}$$

$$= -\frac{(\pi/9)}{(0.62)2\sqrt{2g}}\int_{36}^{0}[4356z^{-1/2} - 132z^{1/2} + z^{3/2}]dz = -\frac{(\pi/9)}{1.24\sqrt{2g}}\left[\frac{4356z^{0.5}}{0.5} - \frac{132z^{1.5}}{1.5} + \frac{z^{2.5}}{2.5}\right]_{36}^{0} = 1276 \text{ sec} \quad ◄$$

12.2.6 *Given the same tank and C_d as in Exer. 12.2.4, but now with the water surface outside constant at section 1 instead and the tank initially empty, how long will it take for the water level in the tank to rise from section 2 to section 1?*

See Fig. X12.2.3 with Solution 12.2.3. Exer. 12.2.4: D_1 = 12 m, D_2 = 5.6 m, z_0 = 10.4 m, 0.2-m² orifice (C_d = 0.60). (Note: The tank diameter = ky, and y = z + h_2, where z is the variable distance between surface levels.)

SI

$$\frac{h_2}{z_0} = \frac{D_2}{D_1 - D_2}; \quad h_2 = \frac{D_2 z_0}{D_1 - D_2} = \frac{5.6(10.4)}{12 - 5.6} = 9.1 \text{ m}$$

$$z = h_1 - y; \quad y = h_1 - z = 9.1 + 10.4 - z = 19.5 - z. \quad k = d/y = (12 - 5.6)/10.4 = 0.615$$

$$A_s = (\pi/4)d^2 = (\pi/4)(0.615y)^2 = 0.297y^2 = 0.297(19.5 - z)^2$$

$$Q_i = C_d A\sqrt{2gz} = 0.6(0.2)\sqrt{2(9.81)}\,z^{1/2} = 0.531\sqrt{z}; \quad Q_o = 0$$

/cont...

Eq 12.2: $t = \int \frac{A_s dz}{Q_i - Q_o} = \int_{10.4}^{0} \frac{0.297(19.5 - z)^2(-dz)}{0.531 z^{1/2} - 0} = -0.560\int_{10.4}^{0} [380 z^{-0.5} - 39 z^{0.5} + z^{1.5}]dz$

$$= -0.560\left[380\frac{z^{0.5}}{0.5} - 39\frac{z^{1.5}}{1.5} + \frac{z^{2.5}}{2.5}\right]_{10.4}^{0} = 962 \text{ s} \quad ◄$$

12.2.7 *The tank in Fig. X12.2.7 has vertical sides, b_1 = 12 ft, b_2 = 24 ft, z_0 = 14 ft, and its dimension normal to the plane of the paper is a = 6 ft. The vertical dividing plate has a submerged orifice 0.7 ft² in area (C_d = 0.65). How long will it take for the two water surfaces to equalize?*

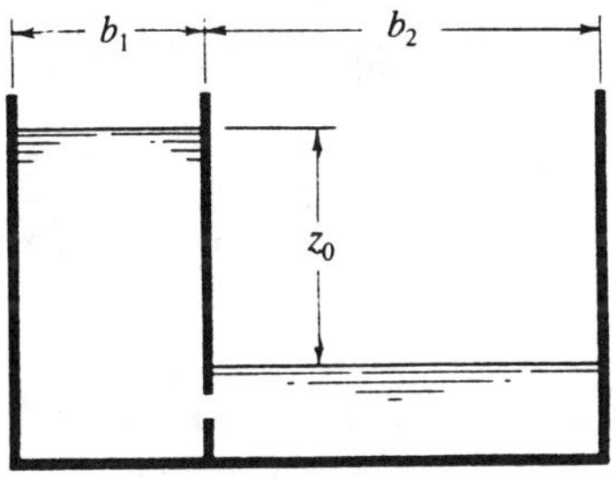

Figure X12.2.7

BG

$d\forall = A_1 dz_1 = A_2 dz_2$; $A_1/A_2 = b_1/b_2 = 12/24 = 1/2 = dz_2/dz_1$

$\therefore\ dz_1 = 2\,dz_2$; $dz = dz_1 + dz_2 = 3dz_2 = 1.5\,dz_1$; $dz_1 = (2/3)dz$

Eq 12.2: $t = \int \frac{A_1 dz_1}{C_d A\sqrt{2gz}} = \int \frac{b_1 a(2/3)dz}{C_d A\sqrt{2gz}}$

$$= \frac{(12)6(2/3)}{0.65(0.7)\sqrt{2 \times 32.2}} \int_0^{14} \frac{dz}{z^{1/2}} = 13.15\left[\frac{z^{0.5}}{0.5}\right]_0^{14} = 98.4 \text{ sec} \quad ◄$$

12.2.8 *The tank in Fig. X12.2.7 has vertical sides, b_1 = 3.6 m, b_2 = 7.2 m, z_0 = 4.2 m, and its dimension normal to the plane of the paper is a = 1.8 m. The vertical dividing plate has a submerged orifice 0.063 m² in area (C_d = 0.65). How long will it take for the two water surfaces to equalize?*

See Fig. X12.2.7 with Solution 12.2.7.

SI

$d\forall = A_1 dz_1 = A_2 dz_2$; $A_1/A_2 = b_1/b_2 = 3.6/7.2 = 1/2 = dz_2/dz_1$; $\therefore\ dz_1 = 2\,dz_2$

$dz = dz_1 + dz_2 = 3dz_2 = 1.5\,dz_1$; $dz_1 = (2/3)dz$

Eq 12.2: $t = \int \frac{A_1 dz_1}{C_d A\sqrt{2gz}} = \int \frac{b_1 a(2/3)dz}{C_d A\sqrt{2gz}} = \frac{(3.6)1.8(2/3)}{0.65(0.063)\sqrt{2 \times 9.81}} \int_0^{4.2} \frac{dz}{z^{1/2}} = 23.8\left[\frac{z^{0.5}}{0.5}\right]_0^{4.2} = 97.6 \text{ s} \quad ◄$

Sec. 12.2: Discharge With Varying Head -- *Problems*

12.1 *(a) Outflow from a reservoir with vertical sides is over a spillway [$Q = C_W L H^{3/2}$, Eq. (11.26)]. Initially there is a steady inflow so that the height of the water surface above the spillway level is z_1. If the inflow is suddenly cut off, prove that the time required for the water level to fall from z_1 to z_2 is $t = (2As/C_W L)(z_2^{-0.5} - z_1^{-0.5})$. (Note: z = H.) (b) How long will it take theoretically for the outflow to cease entirely? What factors make this theoretical answer unrealistic? (c) Given the crest of the overflow spillway is 100 ft long, the value of C_w in Eq. (11.26) is 3.45, and the constant water surface area is 700,000 ft². With no inflow, find the time required for the water surface to fall from 3 ft to 1 ft above the spillway level.*

BG

(*a*) By Eq 12.2 with $Q_i = 0$ and $Q_o = C_W L z^{3/2}$:

$$t = \int_{z_1}^{z_2} \frac{A_s dz}{-C_W L z^{3/2}} = \frac{A_s}{C_W L}\left[\frac{2}{z^{1/2}}\right]_{z_1}^{z_2} = \frac{2A_s}{C_W L}\left[\frac{1}{z_2^{1/2}} - \frac{1}{z_1^{1/2}}\right] \quad \text{Q.E.D.} \quad ◄$$

(*b*) As $z_2 \to 0$, $t \to \infty$, so theoretically the outflow would never cease. ◄ Factors which make this theoretical answer unrealistic include surface tension, surface ripples, evaporation, and leakage/seepage losses.

(*c*) Eq 12.25 with $H = z$: $Q_o = C_W L z^{3/2}$; substituting this into Eq 12.2 with $Q_i = 0$:

$$t = \int_{z_1}^{z_2} \frac{A_s dz}{-C_W L z^{3/2}} = \frac{2A_s}{C_W L}\left[\frac{1}{z_2^{1/2}} - \frac{1}{z_1^{1/2}}\right] = \frac{2(700{,}000)}{3.45(100)}\left[\frac{1}{1^{1/2}} - \frac{1}{3^{1/2}}\right] = 1715 \text{ sec} = 28.6 \text{ min} \quad ◄$$

12.2 *Repeat parts (a) and (b) of Prob. *12.1. (c) Given the crest of the overflow spillway is 30m long, the value of C_W in Eq. (11.26) is 1.91, and the constant water surface area is 65 000 m^2, with no inflow find the time required for the water surface to fall from 900 mm to 300 mm above the spillway level. [*Note this correction to the textbook.]*

Refer to parts (a) and (b) of Prob. 12.1.

SI

See parts (*a*) and (*b*) of Solution 12.1

Eq 12.25 with $H = z$: $Q_o = C_W L z^{3/2}$; substituting this into Eq 12.2 with $Q_i = 0$:

$$t = \int_{z_1}^{z_2} \frac{A_s dz}{-C_W L z^{3/2}} = \frac{2A_s}{C_W L}\left[\frac{1}{z_2^{1/2}} - \frac{1}{z_1^{1/2}}\right] = \frac{2(65\,000)}{1.91(30)}\left[\frac{1}{0.3^{1/2}} - \frac{1}{0.9^{1/2}}\right] = 1751 \text{ sec} = 29.2 \text{ min} \quad ◄$$

12.3 *Figure P12.3 shows a tank with vertical sides containing liquid with a surface area of A_s. The liquid discharges through an orifice under a head z which varies from the initial height h to zero as the tank empties down to the orifice level. (a) Neglecting friction losses, what is the cumulative kinetic energy of the jet during the time required for the liquid surface to drop from h to zero? (b) How does this kinetic-energy summation compare with the total energy of the mass of fluid initially in the tank above the orifice level?*

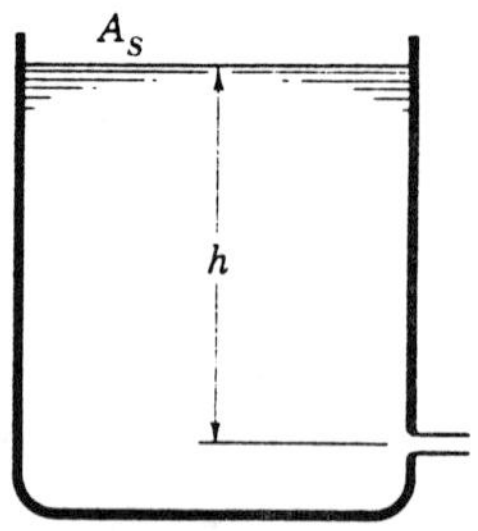

Figure P12.3

N

(*a*) $V = \sqrt{2gz}$; $Q_o = AV = A\sqrt{2gz}$; Eq 12.1: $A_s dz = -Q_o\, dt$

Sec 5.1: Cumulative K.E. $= \int(\Delta Wt \times V^2/2g)dt = \int \gamma(-Q_o dt)V^2/2g$

$$= -\gamma\int_h^0 A_s dz(2gz/2g) = \gamma A_s\int_0^h z dz = \gamma A_s(z^2/2)]_0^h = \gamma A_s h^2/2 \quad ◄$$

(*b*) Total potential energy in tank above orifice = weight × ave. height $= (\gamma A_s h)h/2 = \gamma A_s h^2/2$

This is the same as the cumulative K.E. ◄

12.4 *How long will it take for the tank in Prob. 12.3 to empty down to the orifice level if A_s is 8 ft^2, h is 16 ft, and the jet diameter is 5 in? Plot a graph of h versus time, using increments of 4 ft.*

See Fig. P12.3 with Solution 12.3. Prob. 12.3: A_s = surface area; head z on orifice varies from h to zero.

BG

$Q_i = 0$, $Q_o = A(2gz)^{1/2} = A(2g)^{1/2}z^{1/2}$

$$\text{Eq 12.2: } t = \int_{z_1}^{z_2}\frac{A_s dz}{0 - A(2g)^{1/2}z^{1/2}} = \frac{-A_s}{A(2g)^{1/2}}\int_{z_1}^{z_2} z^{-1/2}dz = \frac{-A_s}{A(2g)^{1/2}}[2z^{1/2}]_{z_1}^{z_2} = \frac{2A_s}{A\sqrt{2g}}\left[\sqrt{z_1} - \sqrt{z_2}\right]$$

$2A_s/(A\sqrt{2g}) = 2(8)/[0.1363\sqrt{2(32.2)}] = 14.62 \text{ s}\cdot\text{ft}^{-1/2}$

z_1 ft	$h = z_2$ ft	t sec	Σt sec
—	16	0	0
16	12	7.84	7.84
12	8	9.29	17.13
8	4	12.11	29.2
4	0	29.2	58.5
	▲		▲

Time to empty = 58.5 sec ◄

12.5 *How long will it take for the tank in Prob. 12.3 to empty down to the orifice level if A_s is 0.72 m², h is 5 m, and the jet diameter is 125 mm? Plot a graph of h versus time, using increments of 1 m.*

See Fig. P12.3 with Solution 12.3. Prob. 12.3: A_s = surface area; head z on orifice varies from h to zero.

SI

$Q_i = 0,\ \ Q_o = A(2gz)^{1/2} = A(2g)^{1/2}z^{1/2}$

Eq 12.2: $t = \int_{z_1}^{z_2} \frac{A_s dz}{0 - A(2g)^{1/2}z^{1/2}} = \frac{-A_s}{A(2g)^{1/2}} \int_{z_1}^{z_2} z^{-1/2}\, dz = \frac{-A_s}{A(2g)^{1/2}}\left[2z^{1/2}\right]_{z_1}^{z_2} = \frac{2A_s}{A\sqrt{2g}}\left[\sqrt{z_1} - \sqrt{z_2}\right]$

$2A_s/(A\sqrt{2g}) = 2(0.72)/[0.01227\sqrt{2(9.81)}) = 26.5\ \text{s}\cdot\text{m}^{-1/2}$

z_1 m	$h = z_2$ m	t sec	$\sum t$ sec
–	5	0	0
5	4	6.25	6.25
4	3	7.10	13.35
3	2	8.42	21.8
2	1	10.97	32.7
1	0	26.5	59.2
	▲		▲

Time to empty = 59.2 s ◄

12.6 *Water enters a reservoir at such a rate that the height z of water above the level of the spillway crest is 3 ft. The spillway ($Q = C_W LH^{3/2}$, $H = z$) is 100 ft long, and the value of C_W is 3.50. The area of the water surface for various water levels is as follows:*

z, ft:	*3.00*	*2.50*	*2.00*	*1.50*	*1.25*	*1.00*
A_s, ft²:	*850,000*	*800,000*	*700,000*	*550,000*	*510,000*	*470,000*

If the inflow is suddenly reduced to 150 cfs, what will be the water level for equilibrium? How long will it take, theoretically, for equilibrium to be attained? How long will it take for the level to drop from 3 to 1 ft above the spillway level?

BG

$Q_o = 3.50(100)z^{3/2}$; when $Q_o = Q_i = 150$ cfs, $z = 0.568$ ft ◄

and the time for equilibrium is theoretically infinite. ◄

From Eq 12.2: $t = \int \frac{A_s dz}{350z^{3/2} - 150}$ (z gets smaller)

z (ft)	Δz (ft)	A_s (ft²)	$\bar{A}_s$ (ft²)	$\bar{A}_s(\Delta z)$ (ft³)	$Q_o = 350z^{3/2}$ (cfs)	$\bar{Q}_o$ (cfs)	$\bar{Q}_o - 150$ (cfs)	$\frac{\bar{A}_s(\Delta z)}{\bar{Q}_o - 150}$ (sec)
3.0		850,000			1819			
	0.5		825,000	412,500		1601	1451	284
2.5		800,000			1383			
	0.5		750,000	375,000		1187	1037	362
2.0		700,00			990			
	0.5		625,000	312,500		816	667	469
1.5		550,000			643			
	0.25		530,000	132,500		566	416	318
1.25		510,000			489			
	0.25		490,000	122,500		420	270	454
1.00		470,000			350			
								Σ = 1888 sec

Total time for z to change from 3 ft to 1.0 ft is 1888 seconds = 31.4 minutes. ◄

12.7 *Work Prob. 12.6 using the same numbers but changing ft to m, ft^2 to m^2, and cfs to m^3/s.*
Refer to Prob. 12.6.

SI

From Eqs. 11.27: $C_W = 3.50(1.83/3.32) = 1.929$; so $Q_o = 1.929(100)z^{3/2} = 192.9\,z^{3/2}$

When $Q_o = Q_i = 150\text{ m}^3\text{/s},\ z = (150/194)^{2/3} = 0.846\text{ m}$ ◀ and theoretically $t_{\text{equilib}} = \infty$ ◀

From Eq. 12.2: $t = \int \dfrac{A_s\,dz}{192.9z^{3/2} - 150}$ (z gets smaller)

z (m)	Δz (m)	A_s (m^2)	$\bar{A}_s$ (m^2)	$\bar{A}_s(\Delta z)$ (m^3)	$Q_o = 192.9z^{3/2}$ (m^3/s)	$\bar{Q}_o$ (m^3/s)	$\bar{Q}_o - 150$ (m^3/s)	$\dfrac{\bar{A}_s(\Delta z)}{\bar{Q}_o - 150}$ (sec)
3		850 000			1002			
	0.5		825 000	412 500		883	733	563
2.5		800 000			763			
	0.5		750 000	375 000		654	504	744
2		700 000			547			
	0.5		625 000	312 500		460	300	1042
1.5		550 000			354			
	0.25		530 000	132 500		312	162.0	818
1.25		510 000			270			
	0.25		490 000	122 500		231	81.3	1507
1.00		470 000			192.9			
								Σ = 4670 sec

Total time for z to change from 3 m to 1.0 m is approximately 4670 s = 77.9 min. ◀

12.8 *A reservoir has an overflow spillway with a 40-ft-long crest, and a value of $C_W = 3.50$ [Eq. (11.26)]. For the range of water levels considered here, the area of the water surface is essentially constant at 600,000 ft^2. Initially, the water surface is 3 ft below the level of the spillway crest. If a flow of 500 cfs is suddenly discharged into this reservoir, what will be the height z of water in the reservoir (above spillway crest) for equilibrium? How much time will be required for this equilibrium height to be reached? How much time will be required for the water surface to reach 2 ft above the spillway level? (Note: This last part can be solved by integration after substituting x^3 for $z^{3/2}$ and consulting integral tables. However, it will be easier to solve it graphically, either by plotting and actually measuring the area under the curve, or by computing the latter by some method such as Simpson's rule.)*

BG

Eq. 11.26 with $H = z$: $Q_o = 3.50(40)z^{3/2} = 140z^{3/2}$. For equilibrium

$Q_i = Q_o = 500$ cfs, from which $z = (500/140)^{2/3} = 2.34$ ft ◀

The time required to reach this equilibrium height is theoretically infinite, but surface ripples and minor irregularities in flow cause practical equilibrium to be reached in a finite time. ◀ However, this does point out that true equilibrium is not quickly obtained. This should be borne in mind in experimental work.

Time for water level to reach spillway crest = hA/Q = 3(600,000/500) = 3600 sec. Time for water level to rise from crest to 2 ft above it is

Using Eq 12.2: $t = 600{,}000\displaystyle\int_0^2 \frac{dz}{500 - 140z^{3/2}}$. Letting $z^{3/2} = x^3$, then $z^{1/2} = x$, $z = x^2$, and $dz = 2x\,dx$

$\therefore\ t = 2(600{,}000)\displaystyle\int_0^{\sqrt{2}} \frac{x\,dx}{500 - 140x^3}$ (refer to a published table of integrals), using: $-1.529 = \left(\dfrac{500}{-140}\right)^{1/3}$

$$t = \frac{1{,}200{,}000}{6(-140)1.529}\left[\ln\frac{1.529^2 + 1.529x + x^2}{(x - 1.529)^2} + 2\sqrt{3}\tan^{-1}\frac{2x + 1.529}{-1.529\sqrt{3}}\right]_0^{\sqrt{2}}$$

$t = 935[(6.21 - 3.55) - (0 - 1.814)] = 4180$ sec

Total time = 3600 + 4180 = 7780 seconds = 129.7 minutes = 2.16 hours ◀

Sec. 12.3: Unsteady Flow of Incompressible Fluids in Pipes -- *Exercises*

12.3.1 *Verify that the neglect of the accelerative head was justified in Sample Problem 12.1. Find its magnitude at z = 5 ft from values of V and t at z = 4.5 and 5.5 ft.*

Sample Prob. 12.1: Pipe length L = 10 ft. Energy equation from water surface to jet, neglecting h_a is $z - V^2/2g = [(fL/D) + \Sigma k]V^2/2g = 1.42V^2/2g$ from which $V = 5.16z^{1/2}$ and the time for the surface to fall from z = 8 ft is $t = -35.5[(2/3)z^{3/2}]_8^z$.

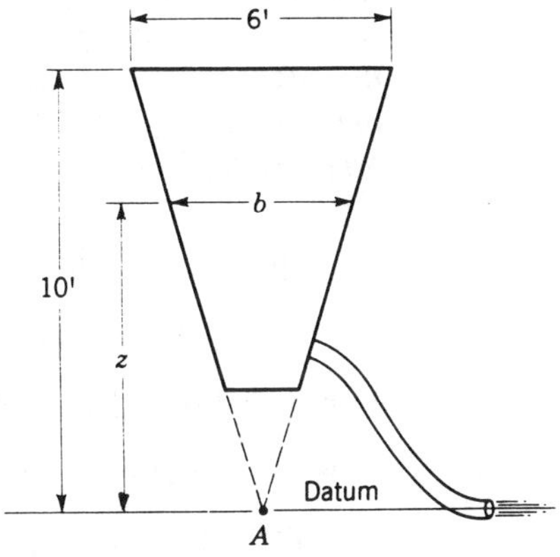

Figure S12.1

BG

Energy equation from water surface to jet, including h_a:

$$\text{Eq. 12.6}a\text{:}\quad z - \frac{V^2}{2g} = \left(\frac{fL}{D} + \Sigma k\right)\frac{V^2}{2g} + h_a\ ;\quad z = 2.42\frac{V^2}{2g} + h_a$$

$$\text{where } h_a = \frac{L}{g}\frac{dV}{dt}$$

Assuming the solution of Sample Prob. 12.1 to be approximately correct:

$$\text{At } z = 5.5\text{ ft:}\quad V = 5.16(5.5)^{1/2} = 12.10\text{ fps},\quad t = -35.5\left[\frac{2}{3}z^{3/2}\right]_8^{5.5} = 230\text{ sec}$$

$$\text{At } z = 4.5\text{ ft:}\quad V = 5.16(4.5)^{1/2} = 10.94\text{ fps},\quad t = -35.5\left[\frac{2}{3}z^{3/2}\right]_8^{4.5} = 310\text{ sec}$$

$$\therefore\ \text{at } z = 5\text{ ft},\ h_a = \frac{L}{g}\frac{dV}{dt} \approx \frac{L}{g}\frac{\Delta V}{\Delta t} = \frac{10}{32.2}\left(\frac{12.10 - 10.94}{230 - 310}\right) = -0.004\,52\text{ ft} \quad ◀$$

This is 0.09% of z = 5 ft. So, in comparison with other terms in the energy equation, h_a is negligible as assumed. ◀

12.3.2 *If the speed of the pump in Sample Prob. 12.2 is instead reduced instantaneously from 1650 to 1150 rpm, what is the deceleration of the flow rate immediately after the change in pump speed?*

Sample Prob. 12.2: Q = 1600 gpm at n = 1650 rpm, and $h \propto n^2$; D_2 = 6 in. From energy and continuity: $50 + h_p = 32.0(V_2^2/2g) + (822/g)dV_2/dt$ (a). At original n = 1650: V_2 = 18.16 fps, h_p = 113.8 ft.

BG

$$h_p' = h_p\left(\frac{n_2}{n_1}\right)^2 113.8\left(\frac{1150}{1650}\right)^2 = 55.3\text{ ft.}$$ Substituting this into Eq (*a*) of Sample Prob. 12.2:

$$50 + 55.3 = 32.0\frac{V_2^2}{2g} + \frac{822}{g}\frac{dV_2}{dt}\quad \text{where } V_2 = 18.16\text{ fps.}\quad \therefore\ \frac{dV_2}{dt} = \frac{105.3 - 163.8}{(822/32.2)} = -2.29\text{ fps/sec}$$

$$dQ/dt = A(dV/dt) = (\pi/4)(0.5)^2(-2.29) = -0.450\text{ cfs/sec} \quad ◀$$

12.9 *Attached to the tank in Fig. P12.9 is a flexible 1.5-in-diameter hose (f = 0.020) 250 ft long. The tank is hoisted in such a manner that h = 25 + 2t, where h is the head in feet and t is the time in seconds. (a) Find as accurately as you can the flow rate at t = 10 sec. (b) Suppose h were decreasing at the same rate. What would be the flow rate when h = 45 ft?*

BG

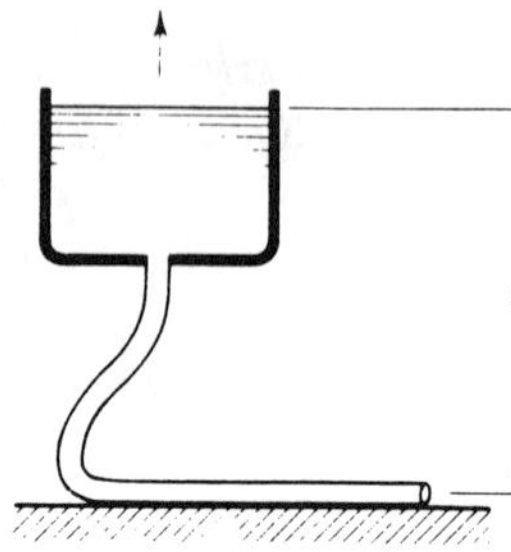

Figure P12.9

Find steady state velocity for $t = 10$ sec; then $h = 25 + 2(10) = 45$ ft.

Eq. 12.6*a*: $45 - V^2/2g = 0.020(250 \times 12/1.5)(V^2/2g)$

$V = 8.41$ fps at steady state

Now find effect of changing head:

Given $h = 25 + 2t$, $\therefore$ $dh/dt = 2.0$ ft/sec

Eq 12.6*a*: $h - 40(V^2/2g) = V^2/2g + (L/g)(dV/dt)$; $dh/dt = 41(2V/2g)(dV/dt) + (L/g)(d^2V/dt^2)$

Neglect the last term (of higher order); $\therefore$ $dh/dt = 2.0 = 41(V/g)(dV/dt)$

$dV/dt = 2g/(41\,V) = 0.1868$ fps/sec

(*a*) Substituting back into Eq 12.6*a*: $45 = 41(V^2/2g) + (250/32.2)(0.1868)$

$41(V^2/2g) = 45 - 1.451 = 43.5$; $V = 5.27$ fps, $Q = 0.1015$ cfs ◄

(and neglected higher order term = 0.0325 fps)

(*b*) For decreasing *h*, one sign (of dV/dt) is reversed: $41(V^2/2g) = 45 + 1.451 = 46.5$

$V = 8.54$ fps, $Q = 0.1048$ cfs ◄

12.10 *Attached to the tank in Fig. P12.9 is a flexible 40-mm-diameter hose (f = 0.024) 65 m long. The tank is hoisted in such a manner that h = 10 + 0.5t, where h is the head in meters and t is the time in seconds. (a) Find as accurately as you can the flow rate at t = 10 s. (b) Suppose h were decreasing at the same rate. What would be the flow rate when h = 15 m?*

See Fig. P12.9 with Solution 12.9.

SI

Find steady state velocity for $t = 10$ s; then $h = 10 + 0.5(10) = 15$ m

Eq. 12.6*a*: $15 - V^2/2g = 0.024(65/0.040)(V^2/2g)$; $V = 2.21$ m/s at steady state

Now find effect of changing head. Given $h = 10 + 0.5t$, $\therefore$ $dh/dt = 0.5$ m/s

Eq 12.6*a*: $h - 39(V^2/2g) = V^2/2g + (L/g)(dV/dt)$; $dh/dt = 40(2V/2g)(dV/dt) + (L/g)(d^2V/dt^2)$

Neglect the last term (of higher order);

$\therefore$ $0.5 = 40(V/g)(dV/dt)$; $dV/dt = 0.5g/(40\,V) = 0.0452$ m/s per s

(*a*) Substituting back into Eq 12.6*a*:

$15 = 40(V^2/2g) + (65/9.81)(0.0452)$; $40(V^2/2g) = 15 - 0.300 = 14.70$

$V = 2.69$ m/s, $Q = 0.003\,37$ m^3/s ◄

(*b*) For decreasing *h*, one sign (of dV/dt) is reversed: $40(V^2/2g) = 15 + 0.300 = 15.30$

$V = 2.74$ m/s, $Q = 0.003\,44$ m^3/s ◄

12.11 *Repeat Prob. 12.9 for the case of a 6-in-diameter hose with all other data remaining the same.*

See Fig. P12.9 with Solution 12.9. Prob. 12.9: The flexibile hose has f = 0.020, L = 250 ft. The tank is hoisted so that $h = 25 + 2t$, where h = ft, t = sec. (a) Find as accurately as possible Q when t = 10 sec. (b) Find Q when h = 45 ft if h were decreasing at the same rate.

BG

Find steady state velocity for t = 10 sec; then $h = 25 + 2t = 25 + 2(10) = 45$ ft

Eq. 12.6a: $45 - V^2/2g = 0.020(250/0.5)(V^2/2g)$; V = 16.23 fps at steady state

Now find effect of changing head

Given: $h = 25 + 2t$, $\therefore dh/dt = 2$ fps

Eq 12.6a: $h = 11(V^2/2g) + (L/g)(dV/dt)$

Differentiating with respect to t and neglecting the higher order term:

$dh/dt = 2.0 = 11(2V/2g)(dV/dt)$; $dV/dt = 2g/11V = 0.361$ fps/sec

(a) Substituting back into Eq. 12.6a: $45 = 11(V^2/2g) + (250/32.2)(0.361)$; $11(V^2/2g) = 45 - 2.80 = 42.2$

V = 15.72 fps, Q = 3.09 cfs ◄

(b) For decreasing h, one sign (of dV/dt) is reversed: $45 = 11(V^2/2g) + (250/32.2)(-0.361)$;

V = 16.73 fps, Q = 3.28 cfs ◄

12.12 *Repeat Prob. 12.10 for the case of a 150-mm-diameter hose with all other data remaining the same.*

See Fig. P12.10 with Solution 12.10. Prob. 12.10: The flexibile hose has f = 0.024, L = 65 m. The tank is hoisted so that $h = 10 + 0.5t$, where h = m, t = s. (a) Find as accurately as possible Q when t = 10 s. (b) Find Q when h = 15 m if h were decreasing at the same rate.

SI

Find steady state velocity for t = 10 sec: then $h = 70 + 0.5t = 15$ m

Eq. 12.6a: $15 - V^2/2g = 0.024(65/0.15)(V^2/2g)$; V = 5.08 m/s, at steady state

Now find effect of changing head. Given: $h = 10 + 0.5t$, $\therefore dh/dt = 0.5$

Eq 12.6a: $h = 11.4(V^2/2g) + (L/g)(dV/dt)$

Differentiating with respect to t and neglecting the higher order term:

$dh/dt = 0.5 = 11.4(2V/2g)(dV/dt)$; $dV/dt = 0.5g/(11.4V) = 0.0847$ m/s per s

(a) Substituting back into Eq. 12.6a: $15 = 11.4(V^2/2g) + (65/9.81)(0.0847)$; V = 4.98 m/s,

$Q = 0.0881\ m^3/s$ ◄

(b) For decreasing h, one sign (of dV/dt) is reversed: $15.0 = 11.4(V^2/2g) + (65/9.81)(-0.0847)$;

V = 5.18 m/s, $Q = 0.0915\ m^3/s$ ◄

12.13 *Work Sample Prob. 12.2 for the case where the pipe lengths are 400 and 1500 ft rather than 200 and 750 ft. All other data are the same.*

Sample Prob. 12.2: $Q = 1600$ gpm at $n = 1650$ rpm, suppose n is increased instantaneously to 2000 rpm, and $h \propto n^2$. $D_1 = 10$ in, $D_2 = 6$ in. At original $n = 1650$: $V_2 = 18.16$ fps, $h_p = 113.8$ ft. Find Q as a function of t.

BG

Subsituting 400 and 1500 for 200 and 750 in Sample Prob. 12.2:

Energy equation will now be: $50 + h_p = 62.9\dfrac{V_2^2}{2g} + \dfrac{1644}{g}\dfrac{dV_2}{dt}$

Original steady flow condition: $V = 18.16$ fps

$h_p = 62.9(V_2^2/2g) - 50 = 322 - 50 = 272$ ft

After pump speed is increased to 2000 rpm; $h_p' = 272(2000/1650)^2 = 400$ ft

Thus $50 + 400 = 62.9\dfrac{V_2^2}{2g} + \dfrac{1644}{g}\dfrac{dV_2}{dt}$

Expressing this equation in terms of Q: $450 = 62.9\dfrac{Q^2}{2gA_2^2} + \dfrac{1644}{gA_2}\dfrac{dQ}{dt} = 25.3Q^2 + 260\dfrac{dQ}{dt}$

Dividing all terms by 25.3 (the coef. of Q^2) yields:

$$dt = \frac{10.26dQ}{17.75 - Q^2} = \frac{10.26dQ}{4.21^2 - Q^2};\quad t = \int_0^t dt = \int_{3.57}^{Q}\frac{10.26Q}{4.21^2 - Q^2}$$

$$t = \frac{10.26}{8.43}\left[\ln\frac{4.21 + Q}{4.21 - Q}\right]_{3.57}^{Q} = 1.217\ln\frac{4.21 + Q}{4.21 - Q} - 3.03$$

$$\frac{t + 3.03}{1.217} = \ln\frac{4.21 + Q}{4.21 - Q};\quad e^{0.821t + 2.49} = \frac{4.21 + Q}{4.21 - Q};\quad Q = 4.22\frac{e^{0.821t + -2.49} - 1}{e^{0.821t + 2.49} + 1}$$ ◄

12.14 *A 1-in-diameter smooth brass pipe 1000 ft long drains an open 4-ft-diameter cylindrical tank that contains oil having $\rho = 1.8$ slugs/ft^3 and $\mu = 0.0006$ lb·s/ft^2. The pipe discharges at elevation 100 ft. Find the time required for the oil level to drop from elevation 120 to elevation 108 ft.*

BG

Check if flow is always laminar. When $V = V_{max}$, the elevation of oil in the tank is at 120 ft, then

$120 - f(L/D)(V^2/2g) = 100$; i.e., $20 = f(L/D)(V^2/2g)$

Assuming laminar flow, Eqs. 8.20 and 8.1: $f = 64/\mathbf{R}$, $\mathbf{R} = DV\rho/\mu$

Substituting: $20 = \dfrac{64\mu}{DV\rho}\left(\dfrac{L}{D}\right)\dfrac{V^2}{2g} = \dfrac{64(0.0006)}{(1/12)V(1.8)}\dfrac{1000}{(1/12)}\dfrac{V^2}{2(32.2)} = 47.7\,V$; $V = V_{max} = 0.419$ fps

$\mathbf{R}_{max} = (1/12)0.419(1.8)/0.0006 = 104.8$; $\mathbf{R}_{max} < 2000$, so the flow is always laminar.

In general, from the above: elev $- 100 = 47.7\,V$

$V = \dfrac{\text{elev} - 100}{47.7} = \dfrac{h}{47.7}$ (defining h = elev $-$ 100); $Q = AV = 0.00545(h/47.7)$

Eq 12.1: $Q\,dt = -A\,dh$, $\therefore$ $dt = -\dfrac{(\pi 4^2/4)dh}{0.00545(h/47.7)} = -110{,}000\,dh/h$

$$\int_0^t dt = -110{,}000\int_{20}^{8} dh/h = -110{,}000[\ln h]_{20}^{8}$$

$t = -110{,}000(2.08 - 3.00) = 100{,}800$ seconds $= 28.0$ hours ◄

12.15 *A 25-mm-diameter smooth brass pipe 300 m long drains an open 1.2-m-diameter cylindrical tank which contains oil having $\rho = 950\ kg/m^3$ and $\mu = 0.03\ N{\cdot}s/m^2$. The pipe discharges at elevation 30 m. Find the time required for the oil level to drop from elevation 36 to elevation 32.5 m.*

SI

Check if flow is always laminar. When $V = V_{max}$, the elevation of oil in the tank is at 36 m, then

$36 - f(L/D)V^2/2g = 30$; i.e. $6 = f(L/D)V^2/2g$

Assuming laminar flow, Eqs. 8.20 and 8.1: $f = 64/\mathbf{R}$; $\mathbf{R} = DV\rho/\mu$

$$\text{Substituting: } 6 = \frac{64\mu}{DV\rho}\frac{L}{D}\frac{V^2}{2g} = \frac{64(0.03)}{(0.025)V(950)}\frac{300}{(0.025)}\frac{V^2}{2(9.81)} = 49.4\,V$$

$V = V_{max} = 0.1213$ m/s; $\mathbf{R}_{max} = (0.025)0.1213(950)/0.03 = 96.1$

$\mathbf{R}_{max} < 2000$, so the flow is always laminar.

$$\text{In general: elev} - 30 = 49.4\,V;\quad V = \frac{\text{elev} - 30}{49.4} = \frac{h}{49.4}\quad (\text{where } h = \text{elev} - 30)$$

$Q = AV = (\pi/4)(0.025)^2(h/49.4) = 0.000491(h/49.4)$

$$\text{Eq 12.1: } Q\,dt = -A\,dh,\ \therefore\ dt = -\frac{(0.6)^2\pi\,dh}{0.000\,491(h/49.4)} = -113{,}900\frac{dh}{h}$$

$$\int_0^t dt = -113{,}900\int_6^{2.5} dh/h = -113{,}900[\ln h]_6^{2.5}$$

$t = -113{,}900(0.916 - 1.792) = 99\,700$ sec $= 27.7$ hours ◀

12.16 *Figure P12.16 depicts a large reservoir being drained with a pipe system ($f = 0.025$, $L = 2000$ ft, $D = 9$ in, $z_0 = 45$ ft). The flow rate is 3.5 cfs when the pump is initially rotating at 180 rpm. If the pump speed is increased instantaneously to 220 rpm, determine the flow rate as a function of time. Assume that the head h_p developed by the pump is proportional to the square of the rotative speed; that is $h_p \propto n^2$.*

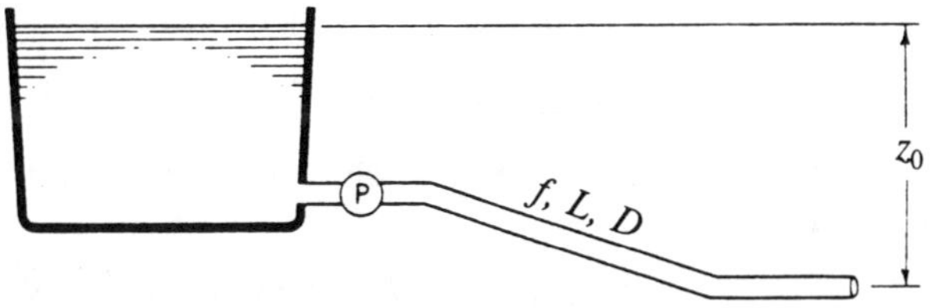

Figure P12.16

BG

Note to instructor: This problem is more challenging!

Initial condition (energy from water surface to jet) neglecting minor losses: $z_0 - h_{L_f} + h_p = V^2/2g$

$45 - 0.025(2000/0.75)V^2/2g + h_p = V^2/2g$, where $V = Q/A = Q/[\pi(9/12)^2/4] = 3.5/0.442 = 7.92$ fps

from which $h_p = 20.9$ ft; $h_p' = 20.9(220/180)^2 = 31.3$ ft

New conditions (per Eq. 12.6):

$45 - 66.7(V^2/2g) + 31.3 = V^2/2g + (L/g)(dV/dt)$

Eventually $dV/dt = 0$, when $V^2/2g = (45 + 31.3)/(66.7 + 1) = 1.127$ ft

$V = 8.52$ fps when steady flow is eventually achieved

In general: $76.3 - 67.7(V^2/2g) = (2000/32.2)(dV/dt)$; $dt = 59.1\,dV/(72.6 - V^2)$

Integrate left side from zero to t ; integrate right side from 7.92 to V_t. Select various V_t's and find the corresponding t's.

/cont...

$$\int dt = t = \frac{59.1}{17.04}\left[\ln\frac{8.52 + V}{8.52 - V}\right]_{7.92}^{V_t} = 3.47\ln\frac{8.52 + V_t}{8.52 - V_t} - 11.49\ ; \quad Q = AV_t = 0.442V_t$$

V_t (fps)	t (sec)	Q (cfs)
7.92	0	3.50
7.95	0.1700	3.51
8.00	0.498	3.53
8.15	1.707	3.60
8.40	5.64	3.71
8.45	7.50	3.73
8.52	∞	3.76
	▲	▲

12.17 *Figure P12.16 depicts a large reservoir being drained with a pipe system (f = 0.022, L = 825 m, D = 250 mm, z_0 = 14 m). The flow rate is 120 L/s when the pump is initially rotating at 200 rpm. If the pump speed is increased instantaneously to 250 rpm, determine the flow rate as a function of time. Assume that the head h_p developed by the pump is proportional to the square of the rotative speed; that is $h_p \propto n^2$.*
See Fig. P12.16 with Solution 12.16.

SI

Initial conditions (energy from water surface to jet) neglecting minor losses: $z - h_{L_f} + h_p = V^2/2g$

$14.00 - 0.022(825/0.250)V^2/2g + h_p = V^2/2g$, where $V = Q/A = 0.120/(\pi 0.250^2/4) = 2.44$ m/s

from which $h_p = 8.42$ m; $h_p' = 8.42(250/200)^2 = 13.15$ m

New conditions (per Eq. 12.6):

$$14 - 72.6(V^2/2g) + 13.15 = V^2/2g + (L/g)(dV/dt)$$

Eventually $dV/dt = 0$, when $V^2/2g = (14 + 13.15)/(72.6 + 1) = 0.369$ m

$V = 2.69$ m/s when steady flow is eventually achieved.

In general: $27.2 - 73.6\dfrac{V^2}{2g} = \dfrac{825}{9.81}\dfrac{dV}{dt}$; $dt = \dfrac{84.1 dV}{27.2 - 3.75V^2} = \dfrac{22.4 dV}{7.24 - V^2}$

Integrate left side from zero to t; integrate right side from 2.44 to V_t.

Select various V_t's and find the corresponding t's.

$$\int dt = t = \frac{22.4}{5.38}\left[\ln\frac{2.69 + V}{2.69 - V}\right]_{2.44}^{V_t} = 4.17\ln\frac{2.69 + V_t}{2.69 - V_t} - 12.75\ ; \quad Q = (\pi 0.250^2/4)V_t = 0.0491V_t$$

V_t (m/s)	t (s)	Q (m^3/s)
2.44	0	0.1200
2.50	1.021	0.1227
2.55	2.33	0.1252
2.60	4.20	0.1276
2.65	7.60	0.1301
2.69	∞	0.1321
	▲	▲

12.18 *(a) Repeat Prob. 12.16 with all the data the same except use an 18-in-diameter pipe rather than a 9-in pipe. (b) Repeat also for the case of an 8-in-diameter pipe.*

Prob. 12.16: The large reservoir in Fig. P12.16 (see with Solution 12.16) is being drained with a pipe system (f = 0.025, L = 2000 ft, z_0 = 45 ft). Initially Q = 3.5 cfs when the pump n = 180 rpm. If n is increased instantaneously to 220 rpm, determine the variation of Q with time. Assume $h_p \propto n^2$.

BG

(*a*) 18-inch-dia pipe:

Initial conditions (energy from water surface to jet): $45 - 0.025(2000/1.5)(V^2/2g) + h_p = V^2/2g$

$V = Q/A = 3.5/(\pi 1.5^2/4) = 1.981$ fps; $h_p = 34.3(V^2/2g) - 45 = -42.9$ ft

Thus a pump is not required; actually a head of 42.9 ft is available to drive a turbine when the flow rate is 3.5 cfs.

(*b*) 8-inch-dia pipe:

Initial conditions: $45 - 0.025[2000/(8/12)](V^2/2g) + h_p = V^2/2g$

where $V = Q/A = 3.5/[\pi(8/12)^2/4] = 3.5/0.349 = 10.03$ fps

$h_p = 76(V^2/2g) - 45 = 73.6$ ft; $h_p' = 73.6(220/180)^2 = 110.0$ ft

New conditions (per Eq. 12.6): $45 - 75(V^2/2g) + 110.0 = V^2/2g + (L/g)(dV/dt)$

Eventually $dV/dt = 0$, when $V^2/2g = (45 + 110.0)/(75 + 1) = 2.04$ ft

$V = 11.46$ fps when steady flow is eventually achieved

In general: $155.0 - 76(V^2/2g) = (2000/32.2)(dV/dt)$

$dt = 52.6\, dV/(131.4 - V^2)$; $\int_0^t dt = t = \int_{10.03}^{V_t} 52.6 dV/(131.4 - V^2)$

$$t = \frac{52.6}{22.9}\left[\ln\frac{11.46 + V}{11.46 - V}\right]_{10.03}^{V_t} = 2.30\ln\frac{11.46 + V_t}{11.46 - V_t} - 6.22\ ; \quad Q = AV_t = 0.349V_t$$

V_t (fps)	t (sec)	Q (cfs)
10.03	0	3.50
10.25	0.412	3.58
10.50	0.969	3.67
11.00	2.71	3.84
11.46	∞	4.00

12.19 *(a) Repeat Prob. 12.17 with all the data the same except use a 500-mm-diameter pipe rather than a 250-mm pipe. (b) Repeat also for the case of a 200-mm-diameter pipe.*

Prob. 12.17: The large reservoir in Fig. P12.17 (see with Solution 12.16) is being drained with a pipe system (f = 0.022, L = 825 m, z_0 = 14 m). Initially Q = 120 L/s when the pump n = 200 rpm. If n is increased instantaneously to 250 rpm, determine the variation of Q with time. Assume $h_p \propto n^2$.

SI

(*a*) 50-cm-dia pipe:

Initial conditions (energy from water surface to jet): $14.0 - 0.022(825/0.150)(V^2/2g) + h_p = V^2/2g$

where $V = Q/A = 0.120/(\pi 0.5^2/4) = 0.611$ m/s; $h_p = 37.3(V^2/2g) - 14.0 = -13.29$ m

Thus a pump is not required; actually a head of 13.29 m is available to drive a turbine when the flow rate is 0.120 m³/s.

/cont...

(*b*) 25-cm-dia pipe:

Initial conditions: $14.0 - 0.022(825/0.25)(V^2/2g) + h_p = V^2/2g$

where $V = Q/A = 0.12/[\pi(0.25/2)^2] = 0.12/0.0491 = 2.44$ m/s

$h_p = 73.6(V^2/2g) - 14.0 = 8.42$ m; $h_p' = 8.42(250/200)^2 = 13.15$ m

New conditions (per Eq. 12.6): $14.0 - 72.6(V^2/2g) + 34.6 = V^2/2g + (L/g)(dV/dt)$

Eventually $dV/dt = 0$, when $V^2/2g = (14.0 + 34.6)/(72.6 + 1) = 0.660$ m

$V = 3.60$ m/s when steady flow is eventually achieved

In general: $48.6 - 73.6(V^2/2g) = (825/9.81)(dV/dt)$

i.e., $dt = 22.4\,dV/(12.94 - V^2)$; $\int_0^t dt = t = \int_{2.44}^{V_t} 22.4\,dV/(12.94 - V^2)$

$$t = \frac{22.4}{7.20}\left[\ln\frac{3.60 + V}{3.60 - V}\right]_{2.44}^{V_t} = 3.12\ln\frac{3.60 + V_t}{3.60 - V_t} - 5.16\ ; \quad Q = AV_t = 0.0491V_t$$

V_t (m/s)	t (s)	Q (cfs)
2.44	0	0.1200
2.50	0.1818	0.1227
3.00	2.32	0.1473
3.50	10.50	0.1718
3.60	∞	0.1766
	▲	▲

Sec. 12.4: Establishment of Steady Flow -- *Exercises*

12.4.1 *A 4-in-diameter drain pipe (flush entrance, f = 0.020) 3000 ft long will freely discharge reservoir water at an elevation 100 ft below the reservoir water surface. Initially there is no flow, since there is a plug in the pipe outlet. What will be the steady flow velocity in this pipe, and how long after the plug is removed will the flow velocity be half the steady velocity? See Sec. 8.24 regarding the neglect of minor losses.*

BG

Sec. 8.24: $L/D = 3000/(4/12) = 9000 > 1000$, ∴ neglect minor losses.

Sec. 12.4: $K = fL/D + 1 + 0 = 0.020(9000) + 1 = 181$

Eq. 12.7: $V_s = \sqrt{2gH/K} = \sqrt{2(32.2)100/181} = 5.96$ fps ◄

Eq. 12.8 for $V = 0.5V_s$: $t = \frac{3000}{181(5.96)}\ln\left(\frac{1 + 0.5}{1 - 0.5}\right) = 3.05$ sec ◄

12.4.2 *A 150-mm-diameter drain pipe (flush entrance, f = 0.030) 500 m long will freely discharge reservoir water at an elevation 60 m below the reservoir water surface. Initially there is no flow, since there is a plug in the pipe outlet. What will be the steady flow velocity in this pipe, and how long after the plug is removed will the flow velocity be half the steady velocity? See Sec. 8.24 regarding the neglect of minor losses.*

SI

Sec. 8.24: $L/D = 500/0.15 = 3333 > 1000$, ∴ neglect minor losses.

Sec. 12.4: $K = fL/D + 1 + 0 = 0.030(3333) + 1 = 101$

Eq. 12.7: $V_s = \sqrt{2gH/K} = \sqrt{2(9.81)60/101} = 3.41$ m/s ◄

Eq. 12.8 for $V = 0.5V_s$: $t = \frac{500}{101(3.41)}\ln\left(\frac{1 + 0.5}{1 - 0.5}\right) = 1.593$ s ◄

12.4.3 *A 250-mm-diameter pipe (flush entrance, $f = 0.026$) 1000 m long can deliver water from a higher reservoir to a lower one (submerged discharge). The difference between the two water surface elevations is 75 m. After a valve near the downstream end is suddenly opened, what will be the steady flow velocity in this pipe, and how long after the valve is opened will the flow velocity be one-third and two-thirds of the steady velocity?*

SI

Sec. 8.24: $L/D = 1000/0.25 = 4000 > 1000$, $\therefore$ neglect minor losses.

Sec. 12.4: $K = fL/D + 0 + 0 = 0.026(4000) = 104$

Eq. 12.7: $V_s = \sqrt{2gH/K} = \sqrt{2(9.81)75/104} = 3.76$ m/s ◄

Eq. 12.8 for $V = (1/3)V_s = 1.254$ m/s: $t = \dfrac{1000}{104(3.76)} \ln\left(\dfrac{1+0.333}{1-0.333}\right) = 1.772$ s ◄

Eq. 12.8 for $V = (2/3)V_s = 2.51$ m/s: $t = \dfrac{1000}{104(3.76)} \ln\left(\dfrac{1+0.667}{1-0.667}\right) = 4.11$ s ◄

12.4.4 *A 4-in-diameter drain pipe (flush entrance, $f = 0.022$) 300 ft long will freely discharge reservoir water at an elevation 40 ft below the reservoir water surface. Initially there is no flow, since there is a plug in the pipe outlet. What will be the steady flow velocity in this pipe, and how long after the plug is removed will the flow velocity be 50% and 75% of the steady velocity?*

BG

Sec. 8.24: $L/D = 300/(4/12) = 900 < 1000$, $\therefore$ include minor losses.

Sec. 12.4: $K = fL/D + 1 + 0.5 = 0.022(900) + 1.5 = 21.3$

Eq. 12.7: $V_s = \sqrt{2gH/K} = \sqrt{2(32.2)40/21.3} = 11.00$ fps ◄

$t = \dfrac{300}{21.3(11.00)} \ln\left(\dfrac{1+0.5}{1-0.5}\right) = 1.407$ sec ◄

Eq. 12.8 for $V = 0.75V_s = 8.25$ fps: $t = \dfrac{300}{21.3(11.00)} \ln\left(\dfrac{1+0.75}{1-0.75}\right) = 2.49$ sec ◄

12.4.5 *A 6-in-diameter pipe (reentrant entrance, $f = 0.028$) 475 ft long can deliver water from a higher reservoir to a lower one (submerged discharge). The difference between the two water surface elevations is 25 ft. After a valve near the downstream end is suddenly opened, what will be the steady flow velocity in this pipe, and how long after the valve is opened will the flow velocity be 90% of the steady velocity?*

BG

Sec. 8.24: $L/D = 475/(6/12) = 950 < 1000$, $\therefore$ include minor losses.

Sec. 12.4: $K = fL/D + 0 + (0.5 + 1) = 0.028(950) + 1.5 = 28.1$

Eq. 12.7: $V_s = \sqrt{2gH/K} = \sqrt{2(32.2)25/28.1} = 7.57$ fps ◄

Eq. 12.8 for $V = 0.9V_s$: $t = \dfrac{475}{28.1(7.57)} \ln\left(\dfrac{1+0.9}{1-0.9}\right) = 6.58$ sec ◄

12.4.6 *A 150-mm-diameter pipe (reentrant entrance, $f = 0.023$) 140 m long can deliver water from a higher reservoir to a lower one (submerged discharge). The difference between the two water surface elevations is 12 m. After a valve near the downstream end is suddenly opened, what will be the steady flow velocity in this pipe, and how long after the valve is opened will the flow velocity be 90% of the steady velocity?*

SI

Sec. 8.24: $L/D = 140/0.15 = 933 < 1000$, $\therefore$ include minor losses.

Sec. 12.4: $K = fL/D + 0 + (0.5 + 1) = 0.023(933) + 1.5 = 23.0$

Eq. 12.7: $V_s = \sqrt{2gH/K} = \sqrt{2(9.81)12/23.0} = 3.20$ m/s ◄

Eq. 12.8 for $V = 0.9V_s$: $t = \dfrac{140}{23.0(3.20)} \ln\left(\dfrac{1+0.9}{1-0.9}\right) = 5.61$ s. ◄

Sec. 12.4: Establishment of Steady Flow -- *Problems*

12.20 *A reservoir is to be drained by a 6-in-diameter pipe (flush entrance, $f = 0.020$) 3500 ft long. The pipe outlet is at an elevation 110 ft below the reservoir water surface. Initially there is no flow since there is a plug at the pipe outlet. Plot Q versus t for time after the plug is removed.*

BG

Sec. 8.24: $L/D = 3500/0.50 = 7000 > 1000$, ∴ neglect minor losses, i.e., $\Sigma k \approx 0$

Sec. 12.4: $K = fL/D + 1 + 0 = 0.02(3500)/(6/12) + 1 = 141$

Eq. 12.7: $V_s = \sqrt{2gH/K} = \sqrt{2(32.2)110/141} = 7.09$ fps

Eq 12.8: $t = \dfrac{3500}{141(7.09)}\ln\dfrac{V_s + V}{V_s - V} = 3.50\ln\dfrac{7.09 + V}{7.09 - V}$; $Q = AV = (\pi/4)D^2V = (\pi/4)(6/12)^2V = 0.1963\,V$

V (fps)	t (sec)	Q (cfs)
0	0	0
1	0.995	0.1963
3	3.16	0.589
5	6.15	0.9817
6.0	8.71	1.178
7.09	∞	1.392
	▲	▲

12.21 *Repeat Prob. 12.20 for the case where the pipe length is 400 ft rather than 3500 ft.*
Prob. 12.20: A reservoir is to be drained by a 6-in-diameter pipe (flush entrance, $f = 0.020$). The pipe outlet is at an elevation 110 ft below the reservoir water surface. Initially there is no flow since there is a plug at the pipe outlet. Plot Q versus t for time after the plug is removed.

BG

Sec. 8.24: $L/D = 400/0.50 = 800 < 1000$, ∴ minor losses are not negligible ;

Sec. 8.18: $k_e = 0.5$ (flush entrance).

Sec. 12.4: $K = fL/D + 1 + \Sigma k = 0.02(400)/(6/12) + 1 + 0.5 = 17.5$

Eq. 12.7: $V_s = \sqrt{2gH/K} = \sqrt{2(32.2)110/17.5} = 20.1$ fps

Eq 12.8: $t = \dfrac{400}{17.5(20.1)}\ln\dfrac{V_s + V}{V_s - V} = 1.136\ln\dfrac{20.1 + V}{20.1 - V}$

Continuity: $Q = AV = (\pi/4)D^2V = (\pi/4)(6/12)^2V = 0.1963\,V$

V (fps)	t (sec)	Q (cfs)
0	0	0
5	0.577	0.982
10	1.239	1.963
15	2.188	2.95
18	3.283	3.53
20.1	∞	3.95
	▲	▲

12.22 *A reservoir is to be drained by a 200-mm-diameter pipe (f = 0.028) 700 m long. The elevation of the pipe outlet is 50 m below the reservoir water surface. Initially there is no flow because a valve at the pipe outlet is closed. Plot Q (L/s) versus t for time after the valve is suddenly opened.*

SI

Sec. 8.24: $L/D = 700/0.20 = 3500 > 1000$, $\therefore$ neglect minor losses, i.e., $\Sigma k \approx 0$

Sec. 12.4: $K = fL/D + 1 + 0 = 0.028(700)/0.2 + 1 = 99.0$

Eq. 12.7: $V_s = \sqrt{2gH/K} = \sqrt{2(9.81)50/99.0} = 3.15$ m/s

Eq 12.8: $t = \dfrac{700}{99.0(3.15)} \ln\dfrac{V_s + V}{V_s - V} = 2.25 \ln\dfrac{3.15 + V}{3.15 - V}$

Continuity: $Q = AV = (\pi/4)D^2V = (\pi/4)(0.2^2)V = 0.0314V$

V (m/s)	t (s)	Q (m^3/s)	Q (L/s)
0	0	0	0
1.0	1.478	0.0314	31.4
2.0	3.37	0.0628	62.8
3.0	8.37	0.0942	94.2
3.1	10.94	0.0974	97.4
3.15	∞	0.0989	98.9

▲ ▲

12.23 *A reservoir with its water surface at elevation 850 ft may be drained with a long pipe that discharges at elevation 50 ft. The pipe consists of 4000 ft of 4-in-diameter pipe (f = 0.020) followed by 500 ft of 2-in-diameter pipe (f = 0.020). Initially a valve at the discharge end is closed and the water in the pipe is at rest. The valve is then suddenly opened and as time ensues the flow rate gradually increases. Determine the time rate of change of the flow rate (cfs/sec) at the instant when the flow rate is 0.40 cfs.*

BG

$\dfrac{L_1}{D_1} = \dfrac{4000}{4/12} = 12{,}000$; $\dfrac{L_2}{D_2} = \dfrac{500}{2/12} = 3000$; minor losses ($\Sigma k$) are negligible in both lines, so $k_L = \dfrac{fL}{D}$

$$H_1 - H_2 = H = k_L\frac{V^2}{2g} + \frac{L_1}{g}\frac{dV_1}{dt} + \frac{L_2}{g}\frac{dV_2}{dt}; \qquad z_1 - \left(z_2 + \frac{V_2^2}{2g}\right) = \left(\frac{fL}{D}\right)_1\frac{V_1^2}{2g} + \left(\frac{fL}{D}\right)_2\frac{V_2^2}{2g} + \frac{L_1}{g}\frac{dV_1}{dt} + \frac{L_2}{g}\frac{dV_2}{dt}$$

$$850 - 50 = \frac{(0.02)4000}{4/12}\frac{V_1^2}{2g} + \left(\frac{(0.02)500}{2/12} + 1\right)\frac{V_2^2}{2g} + \frac{L_1}{g}\frac{dV_1}{dt} + \frac{L_2}{g}\frac{dV_2}{dt}$$

$$800 = 240\frac{V_1^2}{2g} + 61\frac{V_2^2}{2g} + \frac{L_1}{g}\frac{dV_1}{dt} + \frac{L_2}{g}\frac{dV_2}{dt}$$

Continuity: $A_1V_1 = A_2V_2$; $(\pi/4)(4/12)^2V_1 = (\pi/4)(2/12)^2V_2$; $16V_1 = 4V_2$; $V_2 = 4V_1$; $dV_2/dt = 4dV_1/dt$

$\therefore$ Putting all velocities in terms of V_1: $800 = 240\dfrac{V_1^2}{2g} + 61\dfrac{(4V_1)^2}{2g} + \dfrac{4000}{g}\dfrac{dV_1}{dt} + \dfrac{500}{g}4\dfrac{dV_1}{dt}$

$$800g = 120V_1^2 + 30.5(16)V_1^2 + \frac{dV_1}{dt}(4000 + 2000) = 608V_1^2 + 6000\frac{dV_1}{dt}$$

But $Q = A_1V_1 = \dfrac{\pi}{4}\left(\dfrac{4}{12}\right)^2V_1$; $\dfrac{\pi}{4}\left(\dfrac{4}{12}\right)^2\dfrac{dV_1}{dt} = \dfrac{dQ}{dt}$; $V_1 = Q\dfrac{4}{\pi}\left(\dfrac{12}{4}\right)^2$

$$\therefore\ 800g = 608Q^2\left(\frac{4}{\pi}\right)^2\left(\frac{12}{4}\right)^4 + 6000\frac{dQ}{dt}\frac{4}{\pi}\left(\frac{12}{4}\right)^2$$

When $Q = 0.4$ cfs: $6000\dfrac{dQ}{dt}\left(\dfrac{4}{\pi}\right)9 = 800(32.2) - 608(0.4)^2\left(\dfrac{4}{\pi}\right)^2 81 = 12{,}990$; $\dfrac{dQ}{dt} = 0.1889$ cfs/sec ◀

12.24 *The pipeline for draining a large reservoir consists of 200 ft of 6-in-diameter pipe (f = 0.030) followed by 500 ft of 10-in-diameter pipe (f = 0.020). The elevation of the outlet is 100 ft below the reservoir water surface. A valve at the outlet, initially closed, is quickly opened. Derive an equation similar to Eq. (12.8) that is applicable to this situation, and plot flow rate versus time. Neglect minor losses.*

BG

Eq 12.6*a*: $$h - f_1\frac{L_1}{D_1}\frac{V_1^2}{2g} - f_2\frac{L_2}{D_2}\frac{V_2^2}{2g} = \frac{V_2^2}{2g} + \frac{L_1}{g}\frac{dV_1}{dt} + \frac{L_2}{g}\frac{dV_2}{dt}$$

By continuity: $V_1 = V_2(D_2/D_1)^2;\ dV_1 = dV_2(D_2/D_1)^2$

Eliminate V_1 (to get Eq. 12.6*a* in terms of V_2 only):

$$h - \left[f_1\frac{L_1}{D_1}\left(\frac{D_2}{D_1}\right)^4 + f_2\frac{L_2}{D_2} + 1\right]\frac{V_2^2}{2g} = \frac{L_1(D_2/D_1)^2 dV_2}{g\,dt} + \frac{L_2}{g}\frac{dV_2}{dt}$$

For steady state condition when $dV_2/dt = 0$ and $V_2 = V_{2s}$, substituting known values yields steady-flow:

$$100 - \left[0.03\left(\frac{200}{0.5}\right)\left(\frac{10}{6}\right)^4 + 0.02\left(\frac{500}{10/12}\right) + 1\right]\frac{V_{2s}^2}{2g} = 0 \quad \text{or} \quad V_{2s} = 7.81 \text{ fps}$$

In general, subtracting the steady-state equation and rearranging:

$$dt = 2\left[\frac{L_1(D_2/D_1)^2 + L_2}{1 + f_1\frac{L_1}{D_1}\left(\frac{D_2}{D_1}\right)^4 + f_2\frac{L_2}{D_2}}\right]\frac{dV_2}{V_{2s}^2 - V_2^2}\ ; \quad \text{integrating,} \quad t = \left[\frac{L_1(D_2/D_1)^2 + L_2}{1 + f_1\frac{L_1}{D_1}\left(\frac{D_2}{D_1}\right)^4 + f_2\frac{L_2}{D_2}}\right]\frac{1}{V_{2s}}\ln\frac{V_{2s} + V_2}{V_{2s} - V_2}$$

$$t = \left[\frac{200(10/6)^2 + 500}{1 + 0.03(200/0.5)(10/6)^4 + 0.02\left(\frac{500}{10/12}\right)}\right]\frac{1}{7.81}\ln\frac{7.81 + V_2}{7.81 - V_2} = 1.280\ln\frac{7.81 + V_2}{7.81 - V_2} \quad ◄$$

Continuity: $Q = A_1V_1 = (\pi/4)D_1^2V_1 = (\pi/4)(6/12)^2V_1 = 0.1963\,V_1\ ; \quad V_1 = V_2(D_2/D_1)^2 = 2.78V_2$

V_1(fps)	V_2(fps)	t(sec)	Q(cfs)
0	0	0	0
8.33	3	1.037	1.636
13.89	5	1.942	2.73
19.44	7	3.72	3.82
20.8	7.5	4.99	4.09
21.5	7.75	7.12	4.23
21.7	7.81	∞	4.26
		▲	▲

12.25 *The pipeline for draining a large reservoir consists of 60 m of 150-mm-diameter pipe (f = 0.030) followed by 150 m of 250-mm-diameter pipe (f = 0.020). The elevation of the outlet is 30 m below the reservoir water surface. A valve at the outlet, initially closed, is quickly opened. Derive an equation similar to Eq. (12.8) that is applicable to this situation, and plot flow rate (L/s) versus time. Neglect minor losses.*

SI

Eq 12.6*a*: $$h - f_1\frac{L_1}{D_1}\frac{V_1^2}{2g} - f_2\frac{L_2}{D_2}\frac{V_2^2}{2g} = \frac{V_2^2}{2g} + \frac{L_1}{g}\frac{dV_1}{dt} + \frac{L_2}{g}\frac{dV_2}{dt}$$

By continuity: $V_1 = V_2(D_2/D_1)^2;\ dV_1 = dV_2(D_2/D_1)^2$

/cont..

Eliminate V_1 (to get Eq. 12.6*a* in terms of V_2 only):

$$h - \left[f_1\frac{L_1}{D_1}\left(\frac{D_2}{D_1}\right)^4 + f_2\frac{L_2}{D_2} + 1\right]\frac{V_2^2}{2g} = \frac{L_1(D_2/D_1)^2 dV_2}{g\,dt} + \frac{L_2}{g}\frac{dV_2}{dt}$$

For steady state condition when $dV_2/dt = 0$ and $V_2 = V_{2s}$, substituting known values yields steady-flow:

$$30 - \left[0.03\left(\frac{60}{0.150}\right)\left(\frac{250}{150}\right)^4 + 0.02\left(\frac{150}{0.250}\right) + 1\right]\frac{V_{2s}^2}{2g} = 0 \quad \text{or} \quad V_{2s} = 2.36 \text{ m/s}$$

In general, subtracting the steady-state equation and rearranging:

$$dt = 2\left[\frac{L_1(D_2/D_1)^2 + L_2}{1 + f_1\frac{L_1}{D_1}\left(\frac{D_2}{D_1}\right)^4 + f_2\frac{L_2}{D_2}}\right]\frac{dV_2}{V_{2s}^2 - V_2^2}\ ; \quad \text{integrating}, \quad t = \left[\frac{L_1(D_2/D_1)^2 + L_2}{1 + f_1\frac{L_1}{D_1}\left(\frac{D_2}{D_1}\right)^4 + f_2\frac{L_2}{D_2}}\right]\frac{1}{V_{2s}}\ln\frac{V_{2s} + V_2}{V_{2s} - V_2}$$

$$t = \left[\frac{60(250/150)^2 + 150}{1 + 0.03\left(\frac{60}{0.150}\right)\left(\frac{250}{150}\right)^4 + 0.02\left(\frac{150}{0.250}\right)}\right]\frac{1}{2.36}\ln\frac{2.36 + V_2}{2.36 - V_2} = 1.270\ln\frac{2.36 + V_2}{2.36 - V_2}$$ ◄

Continuity: $Q = A_1V_1 = (\pi/4)D_1^2V_1 = (\pi/4)(0.150)^2V_1 = 0.017\,67\,V_1 \text{ m}^3\text{/s}$

$$V_1 = V_2(D_2/D_1)^2 = V_2(0.250/0.150)^2 = 2.78V_2$$

V_1 (m/s)	V_2 (m/s)	t (s)	Q (m³/s)	Q (L/s)
0	0	0	0	0
2.78	1	2.94	0.0491	49.1
4.17	1.5	3.52	0.0736	73.6
5.56	2	4.63	0.0982	98.2
6.25	2.25	6.13	0.1104	110.4
6.39	2.3	6.89	0.1129	112.9
6.53	2.35	9.06	0.1154	115.4
6.56	2.36	∞	0.1159	115.9
		▲		▲

12.26 *Water in a large reservoir may be released through a 12-in-diameter pipe (f = 0.020) 350 ft long. Losses at the pipe entrance are negligible; a nozzle (C_v = 0.95) at the outlet end produces a 6-in-diameter jet. The jet discharges at an elevation 50 ft below the reservoir water surface. Initially, there is a tight-fitting plug in the nozzle, which is then removed. For this situation derive an equation similar to Eq. (12.8) and plot flow rate versus time.*

BG

Using Eq 11.14 for nozzle h_L, with Eq 12.6*a*: $h - k_L\frac{V^2}{2g} - \left[\frac{1}{C_v^2} - 1\right]\left[1 - \left(\frac{D_j}{D}\right)^4\right]\frac{V_j^2}{2g} = \frac{V_j^2}{2g} + \frac{L}{g}\frac{dV}{dt}$

Sec 12.4: $k_L = fL/D + \Sigma k = 0.020(350)/(12/12) + 0 = 7.0$

But $V_j = V(A/A_j) = V(D/D_j)^2$, so $h = C\frac{V^2}{2g} + \frac{L}{g}\frac{dV}{dt}$

where $C = k_L + \left(\frac{D}{D_j}\right)^4\left\{1 + \left[\frac{1}{C_v^2} - 1\right]\left[1 - \left(\frac{D_j}{D}\right)^4\right]\right\} = 7 + \left(\frac{12}{6}\right)^4\left\{1 + \left[\frac{1}{0.95^2} - 1\right]\left[1 - \left(\frac{6}{12}\right)^4\right]\right\} = 24.6$

For steady-state conditions when $dV/dt = 0$ and $V = V_s$:

$$h = CV_s^2/2g, \quad \text{or} \quad V_s = \sqrt{2gh/C} = \sqrt{2(32.2)50/24.6} = 11.44 \text{ fps}$$

Eliminating h between the above two equations and rearranging:

$$dt = \frac{2L}{C}\frac{dV}{V_s^2 - V^2} \quad \text{and integrating} \quad t = \frac{L}{CV_s}\ln\frac{V_s + V}{V_s - V}$$

so that $t = \dfrac{350}{24.6(11.44)}\ln\dfrac{V_s + V}{V_s - V} = 1.243\ln\dfrac{11.44 + V}{11.44 - V}$ ◄

Continuity: $Q = AV = (\pi/4)D^2V = (\pi/4)(12/12)^2V = 0.785\,V$ and $V_j = V/(D/D_j) = V(12/6)^2 = 4V$

V_j (fps)	V (fps)	t (sec)	Q (cfs)
0	0	0	0
8	2	0.500	1.57
16	4	0.954	3.14
24	6	1.70	4.71
32	8	2.35	6.28
40	10	2.82	7.85
45.7	11.44	∞	8.98
		▲	▲

12.27 *A large open tank containing oil (s = 0.85, μ = 0.0005 lb·s/ft²) is connected to a 3-in-diameter smooth pipe 4000 ft long. The elevation of the discharge point is 10 ft below the liquid surface in the tank. A closed valve at the discharge end of the pipe is opened suddenly. Plot the ensuing flow rate versus time.*

BG

Sec 8.22: $L/D = 4000/(3/12) = 16{,}000 > 1000$, ∴ neglect minor losses.

Assume laminar flow:

Eq 8.19: $h_L = 10 = 32(\mu/\gamma)(L/D^2)V_s \quad \therefore \; V_s = \dfrac{10(0.85 \times 62.4)(0.25)^2}{32(0.0005)(4000)} = 0.518$ fps

Eq 8.1: $\mathbf{R} = \dfrac{DV\rho}{\mu} = \dfrac{0.25(0.518)(0.85 \times 1.940)}{0.0005} = 427$

R < 2000, therefore flow is laminar, assumption was O.K.

Assuming the velocity head is negligible:

From Eqs. 8.19 and 12.6*a*: $h - 32\dfrac{\mu}{\gamma}\dfrac{L}{D^2}V = \dfrac{L}{g}\dfrac{dV}{dt}$, or $h - \alpha V = \beta\dfrac{dV}{dt}$,

where $\alpha = 32\dfrac{\mu}{\gamma}\dfrac{L}{D^2} = \dfrac{32(0.0005)4000}{(0.85 \times 62.4)(0.25)^2} = 19.31$ and $\beta = L/g = 4000/32.2 = 124.2$

Rearranging: $dt = \dfrac{\beta dV}{h - \alpha V} = \dfrac{dV}{(h/\beta) - (\alpha/\beta)V}$; integrating: $t = \dfrac{-1}{(\alpha/\beta)}\ln[(h/\beta) - (\alpha/\beta)V]_0^V$

from which $t = -6.43\ln[0.0805 - 0.1554V]_0^V = -6.43\ln\left(\dfrac{0.0805 - 0.1554V}{0.0805}\right)$ or $t = 6.43\ln(1 - 1.931V)$

Continuity: $Q = AV = (\pi/4)D^2V = (\pi/4)(2/12)^2V = 0.0218\,V$

/cont..

V (fps)	t (sec)	Q (cfs)
0	0	0
0.10	1.38	0.0049
0.20	3.14	0.0098
0.30	5.57	0.0147
0.40	9.52	0.0196
0.45	13.07	0.0221
0.518	∞	0.0254
	▲	▲

Check velocity head assumption: $\dfrac{V^2/2g}{h_L} = \dfrac{V^2/2g}{\alpha V} = \dfrac{V}{2g\alpha} = \dfrac{V}{2100}$

As V varies from 0 to 0.518 fps, $(V^2/2g)/h_L$ varies from 0 to 0.000 417

$\therefore$ $V^2/2g < 0.042\ \%$ of h_L, and velocity head is negligible (as assumed).

12.28 *A large open tank containing oil (s = 0.82, $\nu = 6.5 \times 10^{-5}$ m²/s) is connected to a 150-mm-diameter pipe 450 m long. The elevation of the discharge point is 2.2 m below the liquid surface in the tank. A closed valve at the discharge end of the pipe is opened suddenly. Plot the ensuing flow rate (L/s) as a function of time.*

SI

Sec 8.22: $L/D = 450/0.15 = 3000 > 1000$, $\therefore$ neglect minor losses.

Assuming laminar flow (Eq 8.19) for h_L, and assuming the velocity head is negligible:

Eqs. 12.6*a* and 8.19: $H - 32\nu\dfrac{L}{gD^2}V = \dfrac{L}{g}\dfrac{dV}{dt}$; $2.2 - 32(0.000\,065)\dfrac{450}{9.81(0.15)^2}V = \dfrac{450}{9.81}\dfrac{dV}{dt}$

i.e. $2.2 - 4.24V = 45.9dV/dt$. At steady state, $2.2 - 4.24V_s = 0$ so steady-flow velocity $V_s = 0.519$ m/s

Rearranging, $dt = \dfrac{45.9dV}{2.2 - 4.24V}$ and integrating: $t = -\dfrac{1}{1.928}[\ln(1 - 1.928V)]_0^V = -0.519\ln(1 - 1.928V)$

V (m/s)	t (s)	Q (m/s)	Q (L/s)
0	0	0	0
0.1	0.1111	1.77×10^{-3}	1.77
0.2	0.253	3.54×10^{-3}	3.54
0.3	0.448	5.30×10^{-3}	5.30
0.4	0.765	7.07×10^{-3}	7.07
0.5	1.721	8.84×10^{-3}	8.84
0.51	2.12	9.01×10^{-3}	9.01
0.519	∞	9.17×10^{-3}	9.17
	▲		▲

$\mathbf{R}_{max} = DV/\nu = 0.15(0.519)/(6.5 \times 10^{-5}) = 1198$. $\mathbf{R} < 2000$, so flow is laminar (as assumed).

Check velocity head assumption: max $\dfrac{V^2/2g}{h_L} = \dfrac{V^2/2g}{4.24V} = \dfrac{V}{2g(4.24)} = \dfrac{0.519}{2(9.81)4.24} = 0.006\,24$

$\therefore$ $V^2/2g < 0.63\%$ of h_L, and velocity head is negligible, as assumed. Both assumptions are valid, so answers are valid.

12.29 *A vertical pipe full of oil is allowed to drain by removing a plug from its lower end. (a) Obtain an equation for the varying flow velocity as a function of time t, the pipe diameter D, and the specific weight γ and viscosity μ of the oil. Assume that the head loss is given by the equation of established laminar flow and that surface-tension effects are negligible. (b) Obtain an equation for the distance x of the oil surface from the top of the pipe, in terms of the same variables. (c) Given that the pipe is 10 ft long, its diameter is 1 in, and the oil has specific gravity 0.88 and viscosity 0.004 lb·s/ft², find the time required to drain the pipe after the plug is removed. Check whether the laminar flow assumption is reasonable.*

BG

(*a*) Method 1: Energy equation from (1) to (2): $h + V^2/2g - h_L = V^2/2g + (h/g)(dV/dt)$

Given established laminar flow: Eq 8.19: $h_L = 32\dfrac{\mu}{\gamma}\dfrac{h}{D^2}V = khV$ where $k = \dfrac{32\mu}{\gamma D^2}$

Thus $h - khV = \left(\dfrac{h}{g}\right)\dfrac{dV}{dt}$, or $\dfrac{dV}{dt} = g(1 - kV)$; $dt = \dfrac{1}{g}\left[\dfrac{dV}{1 - kV}\right] = \dfrac{dV}{g[1 - (32u/\gamma D^2)V]}$

Method 2 (Newton's Law, $\Sigma F_z = ma$):

At any time t liquid column of height h remains in the pipe.

Gravity force = $\rho(Ah)g$ acting down

Friction force = $\tau_0(\pi D)h$ acting up, where by Eq. 2.10, $\tau_0 = \mu(du/dy)_{\text{wall}}$

For laminar flow: Substituting: $r = (r_0 - y)$ into laminar flow velocity profile Eq. 8.16:

$u = (V_c/r_0^2)(2yr_0 - y^2)$. Differentiating: $du/dy = 2(V_c/r_0^2)(r_0 - y)$ so when $y = 0$ then

$(du/dy)_{\text{wall}} = 2V_c/r_0 = 2(2V)/(D/2) = 8V/D$. Thus (Eq. 2.9) $\tau_0 = \mu(8V/D)$

Σ Forces = (mass)(accel); $\rho g\left(\dfrac{\pi D^2}{4}\right)h - \mu\dfrac{8V}{D}\pi Dh = \rho\left(\dfrac{\pi D^2}{4}h\right)\dfrac{dV}{dt}$

$\dfrac{dV}{dt} = g - kVg$ since $32\mu = k\gamma D^2$ and $\gamma = \rho g$. From which again $dt = \dfrac{1}{g}\left[\dfrac{dV}{1 - kV}\right]$

Integrating: $t = -(1/gk)[\ln(1 - kV)]_0^V = -(1/gk)[\ln(1 - kV) - \ln(1)]$

$= -(1/gk)\ln(1 - kV)$ or $V = (1 - e^{-gkt})/k$ ◄

(*b*) Writing $V = dx/dt$, where x is the distance of the oil surface from the top of the tank:

Rearranging: $kdx = (1 - e^{-gkt})dt$. Integrating: $kx = t + (1/gk)e^{-gkt} + C$

$x = 0$ when $t = 0$, $\therefore C = -1/gk$. Thus $x = t/k + (1/gk^2)(e^{-gkt} - 1)$ ◄

(*c*) $k = \dfrac{32\mu}{\gamma D^2} = \dfrac{32(0.004)}{(0.88 \times 62.4)(1/12)^2} = 0.336$ sec/ft

Note that for $t > 2$, $e^{-gkt} < 5 \times 10^{-10}$ (negligible), so that $V = dx/dt = 1/k$ and $t = kx + 1/gk$

For $x = 10$ ft: $t = 0.336(10) + 1/(32.2 \times 0.336) = 3.45$ sec ◄

$V_{\max} = V_{t\,=\,3.45\text{ sec}} = 1/k = 1/0.336 = 2.98$ fps

$\mathbf{R}_{\max} = DV_{\max}\rho/\mu = (1/12)2.98(0.88 \times 1.940)/0.004 = 106.0$; $\mathbf{R} < 2000$, so flow is laminar. ◄

12.30 *Repeat parts (a) and (b) of Prob. 12.29. (c) Given that the pipe is 2.8 m long, its diameter is 20 mm, and the oil has specific gravity 0.88 and viscosity 0.25 N·s/m², find the time required to drain the pipe after the plug is removed. Check whether the laminar flow assumption is reasonable.*

Refer to Prob. statement 12.29

SI

(a,b) Same as Solution 12.29(a) and (b).

$$k = \frac{32\mu}{\gamma D^2} = \frac{32(0.25)}{(0.88 \times 9810)(0.020)^2} = 2.32 \text{ s/m}$$

Note that for $t > 1.1$, $e^{-gkt} < 10^{-10}$ (negligible), so that $V = dx/dt = 1/k$ and $t = kx + 1/gk$

For $x = 2.8$ m: $t = 2.32(2.8) + 1/(9.81 \times 2.32) = 6.53$ s ◄

$V_{max} = V_{t\,=\,6.53\,s} = 1/k = 1/2.32 = 0.432$ m/s

$\mathbf{R}_{max} = DV_{max}\rho/\mu = = (0.020)0.432(0.88 \times 1000)/0.25 = 30.4$; $\mathbf{R} < 2000$, so flow is laminar. ◄

Sec. 12.5: Velocity of Pressure Wave in Pipes -- ***Exercises***

12.5.1 *Find the celerity of a pressure wave in open fresh water at 70°F. What will the celerity be if the water is contained in a cast iron pipe with 12.00 in inside diameter and 3/8-in wall thickness? Express the celerity in the pipe as a percentage of the celerity in open water.*

BG

Table A.1 at 70°F: $\rho = 1.936 \text{ slug/ft}^3$, $E_v = 320{,}000$ psi

Eq. 12.9: $c = \sqrt{E_v/\rho} = \sqrt{320{,}000(144)/1.936} = 4879$ fps ◄

Sec. 12.5 for cast iron: $E \approx 15{,}000{,}000$ psi

Eq. 12.10: $$c_p = \frac{4879}{\sqrt{1 + \left(\frac{12}{0.375}\right)\left(\frac{320{,}000}{15{,}000{,}000}\right)}} = 3761 \text{ fps}$$ ◄

Ratio = 3761/4879 = 0.771 = 77.1% ◄

12.5.2 *Find the celerity of a pressure wave in open fresh water at 20°C. What will the celerity be if the water is contained in a cast iron pipe with 489 mm inside diameter and 14.3 mm wall thickness? Express the celerity in the pipe as a percentage of the celerity in open water.*

SI

Table A.1 at 20°C: $\rho = 998.2 \text{ kg/m}^3$, $E_v = 2.18 \times 10^6 \text{ kN/m}^2$

Eq. 12.9: $c = \sqrt{E_v/\rho} = \sqrt{2.18 \times 10^6(1000)/998.2} = 1478$ m/s ◄

Sec. 12.5 for cast iron: $E \approx 103 \times 10^6 \text{ kN/m}^2$

Eq. 12.10: $$c_p = \frac{1478}{\sqrt{1 + \left(\frac{489}{14.3}\right)\left(\frac{2.18 \times 10^6}{103 \times 10^6}\right)}} = 1126 \text{ m/s}$$ ◄

Ratio = 1126/1478 = 0.762 = 76.2% ◄

12.5.3 *Find the celerity of a pressure wave in benzene (Appendix A, Table A.4) contained in an 8-in diameter steel pipe having a wall thickness of 0.32 in.*

BG

Table A.4 for benzene: $s = 0.88, \quad E_v = 150{,}000$ psi

Eq. 12.9: $c = \sqrt{E_v/\rho} = \sqrt{150{,}000 \times 144/(0.88 \times 1.940)} = 3557$ fps

Eq. 12.10: $c_p = \dfrac{3557}{\sqrt{1 + \left(\dfrac{8}{0.32}\right)\left(\dfrac{150{,}000}{30 \times 10^6}\right)}} = 3354$ fps ◄

12.5.4 *Find the celerity of a pressure wave in benzene (Appendix A, Table A.4) contained in an 200-mm diameter steel pipe having a wall thickness of 8.0 mm.*

SI

Table A.4 for benzene: $s = 0.88, \quad E_v = 1030 \times 10^6$ N/m^2

Eq. 12.9: $c = \sqrt{1030 \times 10^6/(0.88 \times 1000)} = 1082$ m/s

Eq. 12.10: $c_p = \dfrac{1082}{\sqrt{1 + \left(\dfrac{200}{8.0}\right)\left(\dfrac{1030 \times 10^6}{207 \times 10^9}\right)}} = 1020$ m/s ◄

Sec. 12.6: Water Hammer -- *Exercises*

12.6.1 *Given that the pipe in Fig. 12.5a is 1500 ft long, that the water is initially flowing through it at 2 fps, and that the pressure waves travel along it at 3750 fps, find the maximum water hammer pressure at the valve, and at points 500 ft and 1000 ft from the valve, (a) if the valve is completely closed instantaneously, and (b) if the valve is partially closed instantaneously, reducing the flow velocity to 0.6 fps.*

Fig. 12.5a: The valve is at the outlet end of the straight pipe.

BG

(*a*) Complete closure: $\Delta V = V_2 - V = 0 - 2 = -2$ fps

Eq. 12.12: $\Delta p = -\rho c_p \Delta V = -(62.4/32.2)(3750)(-2) = 14{,}530$ psf $= 100.9$ psi at all locations ◄

(*b*) Partial closure: $\Delta V = V_2 - V = 0.6 - 2 = -1.4$ fps

Eq. 12.12: $\Delta p = -\rho c_p \Delta V = -(62.4/32.2)3750(-1.4) = 10{,}170$ psf $= 70.7$ psi at all locations ◄

12.6.2 *Given that the pipe in Fig. 12.5a is 300 m long, that the water is initially flowing through it at 0.8 m/s, and that the pressure waves travel along it at 1000 m/s, find the maximum water hammer pressure at the valve, and at points 100 m and 200 m from the valve, (a) if the valve is completely close instantaneously, and (b) if the valve is partially closed instantaneously, reducing the flow velocity to 0.2 m/s.*

Fig. 12.5a: The valve is at the outlet end of the straight pipe.

SI

(*a*) Complete closure: $\Delta V = V_2 - V = 0 - 0.8 = -0.8$ m/s

Eq. 12.12: $\Delta p = -1000$ kg/m^3(1000 m/s)(−0.8 m/s) $= 800$ kN/m^2 $= 800$ kPa at all locations ◄

(*b*) Partial closure: $\Delta V = V_2 - V = 0.2 - 0.8 = -0.6$ m/s

Eq. 12.12: $\Delta p = -\rho c_p \Delta V = -1000(1000)(-0.6) = 600$ kN/m^2 $= 600$ kPa at all locations ◄

12.6.3 *Repeat Exer. 12.6.1 for valve closure times of 0.36 sec.*

Exer. 12.6.1: Straight pipe L = 1500 ft, initial V = 2 fps, c_p = 3750 fps. Find the maximum waterhammer Δp at the outlet valve and at 500 ft and 1000 ft from it (a) for complete closure, and (b) for partial closure which reduces V to 0.6 fps.

BG

Eq. 12.11: $T_r = 2(1500)/3750 = 0.8$ sec $> t_c = 0.36$ sec, so closure is "rapid."

Eq. 12.15: $x_0 = 3750(0.36)/2 = 675$ ft.

(*a*) Complete closure: $\Delta V = V_2 - V = 0 - 2 = -2$ fps

Eq. 12.12: $\Delta p = -\rho c_p \Delta V = -(62.4/32.2)(3750)(-2) = 14{,}530$ psf $= 100.9$ psi

At outlet valve, $x = L = 1500$ ft: $x > x_0$, ∴ have full $\Delta p = 100.9$ psi ◀

At $x_2 = L - 500 = 1000$ ft: $x_2 > x_0$, ∴ have full $\Delta p = 100.9$ psi ◀

At $x_1 = L - 1000 = 500$ ft: $x_1 < x_0$, ∴ have partial Δp

$\Delta p_{500} = (x/x_0)\Delta p = (500/675)100.9 = 74.8$ psi ◀

(*b*) Partial closure: $\Delta V = V_2 - V = 0.6 - 2.0 = -1.4$ fps

Eq. 12.12: $\Delta p = -\rho c_p \Delta V = -(62.4/32.2)3750(-1.4) = 10{,}170$ psf $= 70.7$ psi

At valve and x_2: have full $\Delta p = 70.7$ psi ◀

At x_1: $\Delta p_{500} = (500/675)70.65 = 52.3$ psi ◀

12.6.4 *Repeat Exer. 12.6.2 for valve closure times of 0.33 s.*

Exer. 12.6.2: Straight pipe L = 300 m, initial V = 0.8 m/s, c_p = 1000 m/s. Find the maximum waterhammer Δp at the outlet valve and at 100 m and 200 m from it (a) for complete closure, and (b) for partial closure which reduces V to 0.2 m/s.

SI

Eq. 12.11: $T_r = 2L/c_p = 2(300)/1000 = 0.6$ s > 0.33 s, so closure is "rapid."

Eq. 12.15: $x_0 = c_p t_c/2 = 1000(0.33)/2 = 165$ m

(*a*) Complete closure: $\Delta V = V_2 - V = 0 - 0.8 = -0.8$ m/s

Eq. 12.12: $\Delta p = -\rho c_p \Delta V = -(1000\ \text{kg/m}^3)(1000\ \text{m/s})(-0.8\ \text{m/s}) = 800\ \text{kN/m}^2 = 800$ kPa

At outlet valve, $x = L = 300$ m: $x > x_0$, ∴ have full $\Delta p = 800$ kPa ◀

At $x_2 = L - 100 = 200$ m: $x_2 > x_0$, ∴ have full $\Delta p = 800$ kPa ◀

At $x_1 = L - 200 = 100$ m: $x_1 < x_0$, ∴ have partial Δp

$\Delta p_{100} = (x/x_0)\Delta p = (100/165)800 = 485$ kPa ◀

(*b*) Partial closure: $\Delta V = V_2 - V = 0.2 - 0.8 = -0.6$ m/s

Eq. 12.12: $\Delta p = -\rho c_p \Delta V = -1000(1000)(-0.6) = 600\ \text{kN/m}^2 = 600$ kPa

At valve and x_2: full have $\Delta p = 600$ kPa ◀

At x_1: $\Delta p_{100} = (100/165)600 = 364$ kPa ◀

12.6.5 *Repeat Exer. 12.6.1 for valve closure times of 1.00 sec.*

Exer. 12.6.1: Straight pipe L = 1500 ft, initial V = 2 fps, c_p = 3750 fps. Find the maximum waterhammer Δp at the outlet valve and at 500 ft and 1000 ft from it (a) for complete closure, and (b) for partial closure which reduces V to 0.6 fps.

BG

Eq. 12.11: $T_r = 2(1500)/3750 = 0.8$ sec $< t_c = 1.0$ sec, so closure is "slow."

Eq. 12.15: $x_0 = c_p t_c/2 = 3750(1)/2 = 1875$ ft

(*a*) Complete closure: $\Delta V = V_2 - V = 0 - 2.0 = -2.0$ fps

Eq. 12.12: $\Delta p = -\rho c_p \Delta V = -(62.4/32.2)3750(-2) = 14{,}530$ psf $= 100.9$ psi

As $L < x_0$, at valve Eq. 12.16: $\Delta p' = (L/x_0)\Delta p = (1500/1875)100.9 = 80.7$ psi ◄

Assuming a straight line variation within x_0.

At $x_2 = L - 500 = 1000$ ft: $\Delta p = (1000/1500)80.7$ psi $= 53.8$ psi ◄

At $x_1 = L - 1000 = 500$ ft: $\Delta p = (500/1500)80.7$ psi $= 26.9$ psi ◄

(*b*) Partial closure: $\Delta V = V_2 - V = 0.6 - 2.0 = -1.4$ fps

At valve, Eq. 12.16: $\Delta p' = -\dfrac{2L\rho\Delta V}{t_c} = -\dfrac{2(1500)62.4(-1.4)}{(1.0)32.2} = 8140$ psf $= 56.5$ psi ◄

Assuming a straight line variation within x_0.

At $x_2 = 1000$ ft, $\Delta p = (1000/1500)56.5 = 37.7$ psi ◄

At $x_1 = 500$ ft, $\Delta p = (500/1500)56.5 = 18.84$ psi ◄

12.6.6 *Repeat Exer. 12.6.2 for valve closure times of 0.84 s.*

Exer. 12.6.2: Straight pipe L = 300 m, initial V = 0.8 m/s, c_p = 1000 m/s. Find the maximum waterhammer Δp at the outlet valve and at 100 m and 200 m from it (a) for complete closure, and (b) for partial closure which reduces V to 0.2 m/s.

SI

Eq. 12.11: $T_r = 2(300)/1000 = 0.6$ s $< t_c = 0.84$ s, so closure is "slow."

Eq. 12.15: $x_0 = c_p t_c/2 = 1000(0.84)/2 = 420$ m

(*a*) Complete closure: $\Delta V = V_2 - V = 0 - 0.8 = -0.8$ m/s

As $L < x_0$, at valve, Eq. 12.16:

$\Delta p' = -2L\rho\Delta V/t_c = -2(300)1000(-0.8)/0.84 = 571\,000$ N/m^2 = 571 kPa at valve ◄

Assuming a straight line variation within x_0.

At $x_2 = L - 100 = 200$ m: $\Delta p = (200/300)571 = 381$ kPa ◄

At $x_1 = L - 200 = 100$ m: $\Delta p = (100/300)571 = 190.5$ kPa ◄

(*b*) Partial closure: $\Delta V = V_2 - V = 0.2 - 0.8 = -0.6$ m/s

At valve, Eq. 12.16: $\Delta p' = -2L\rho\Delta V/t_c = -2(300)1000(-0.6)/0.84$ N/m^2 = 429 kPa ◄

Assuming a straight line variation within x_0.

At $x_2 = 200$ m: $\Delta p = (200/300)429 = 286$ kPa ◄

At $x_1 = 100$ m: $\Delta p = (100/300)429 = 142.9$ kPa ◄

12.31 *(a) What is the celerity of a pressure wave in a 6-ft-diameter water pipe with 0.65-in steel walls? (b) If the pipe is 4200 ft long, what is the time required for a pressure wave to make the round trip from the valve? (c) If the initial water velocity is 7 fps, what will be the rise in pressure at the valve if the time of closure is less than the time of a round trip? (d) If the valve is closed at such a rate that the velocity in the pipe decreases uniformly with respect to time and closure is completed in a time $t_c = 5L/C_p$, approximately what will be the increase in pressure head at the valve when the first pressure unloading wave reaches the valve?*

BG

(*a*) Sec. 12.5: For water c = 4720 fps and E_v = 300,000 psi, for steel E = 30 × 10⁶ psi.

$$\text{Eq 12.10:}\quad c_p = \frac{4720}{\sqrt{1 + \dfrac{6 \times 12}{0.65}\left(\dfrac{300{,}000}{30 \times 10^6}\right)}} = 3251 \text{ fps} \quad ◀$$

(*b*) Eq 12.11: $T_r = 2L/c_p = 2(4200)/3251 = 2.58$ sec ◀

(*c*) Eq 12.14: $\Delta p = (\gamma/g)c_p V = (62.4/32.2)(3251)7 = 44{,}100$ psf $= 306$ psi ◀

(*d*) If velocity changes uniformly from 7.0 fps to 0 in time $5L/c_p$, the change in V in $t = T_r = 2L/c_p$ is $\Delta V = V_2 - V = (2/5)(0 - 7.0) = -2.80$ fps

when, from Eq. 12.12: $\Delta p'/\gamma \approx -c_p \Delta V/g = -3251(-2.80)/32.2 = 283$ ft ◀

Alternatively, from Eq. 12.16 with $t_c = 5L/c_p$:

$$\frac{\Delta p'}{\gamma} = -\frac{2L\,\Delta V}{g t_c} = -\frac{2}{5}\frac{c_p \Delta V}{g} = -\frac{2(3251)(-7.0)}{5(32.2)} = 283 \text{ ft} \quad ◀$$

12.32 *(a) What is the celerity of a pressure wave in a 1.8-m diameter water pipe with 125-mm concrete walls? (b) If the pipe is 1250 m long, what is the time required for a pressure wave to make the round trip from the valve? (c) If the initial water velocity is 2.75 m/s, what will be the rise in pressure at the valve if the time of closure is less than the time of a round trip? (d) If the valve is closed at such a rate that the velocity in the pipe decreases uniformly with respect to time and closure is completed in a time $t_c = 5L/c_p$, approximately what will be the increase in pressure head at the valve when the first pressure unloading wave reaches the valve?*

SI

(*a*) Sec. 12.5: For water c = 1440 m/s and E_v = 2.07 × 10⁶ kPa, for concrete E = 207 × 10⁶ kPa.

$$\text{Eq 12.10:}\quad c_p = \frac{1440}{\sqrt{1 + \dfrac{1.8}{0.125}\left(\dfrac{2.07 \times 10^6}{20.7 \times 10^6}\right)}} = 922 \text{ m/s} \quad ◀$$

(*b*) Eq 12.11: $T_r = 2L/c_p = 2(1250)/922 = 2.71$ s ◀

(*c*) Eq 12.14: $\Delta p = (\gamma/g)c_p V = (9810/9.81)922(2.75) = 2.54 \times 10^6$ N/m² $= 2.54$ MPa ◀

(*d*) If the velocity changes uniformly from 2.75 m/s to 0 in time $5L/c_p$, the change in V in $t = T_r = 2L/c_p$ is $\Delta V = V_2 - V = (2/5)(0 - 2.75) = -1.100$ m/s

when, from Eq. 12.12: $\Delta p'/\gamma \approx -c_p \Delta V/g = -922(-1.100)/9.81 = 103.4$ m ◀

Alternatively, from Eq. 12.16 with $t_c = 5L/c_p$:

$$\frac{\Delta p'}{\gamma} = -\frac{2L\,\Delta V}{g t_c} = -\frac{2}{5}\frac{c_p \Delta V}{g} = -\frac{2(922)(-2.75)}{5(9.81)} = 103.4 \text{ m} \quad ◀$$

12.33 *Using Eqs. (12.12) and (12.16) and the data for Figs. 12.8 and 12.9 as given in footnote 3, compute the water-hammer pressure for each case and compare the answers with the actual measurements. Also, for the given data, compute f.*

Footnote 3: $L = 3060$ ft, $D = 2.06$ in, $c_p = 4371$ fps, $V = 1.11$ fps, $T_r = 1.40$ sec, $h_L = 5.1$ ft, $t_c = 1$ sec (rapid complete closure, Fig. 12.8) or 3 sec (slow complete closure, Fig. 12.9).

BG

For Fig. 12.8 (rapid complete closure):

Eq 12.14: $\Delta p = \rho c_p V = 1.940(4371)1.11 = 9410$ psf. Thus $\Delta p/\gamma = 9410/62.4 = 150.8$ ft ◄

This compares very well with the slightly more than 151 ft recorded in Fig 12.8 ◄

For Fig. 12.9 (slow complete closure):

Eq 12.16: $\Delta p' \approx -2L\rho\Delta V/t_c = -2(3060)1.940(-1.11)/3 = 4390$. Thus $\Delta p'/\gamma \approx 4390/62.4 = 70.4$ ft ◄

This compares fairly well with the 55 ft or so recorded in Fig 12.9. ◄

From conditions before valve closure, Eq 8.10: $5.1 = f\dfrac{3060}{(2.06/12)}\dfrac{1.11^2}{2(32.2)}$; $f = 0.01495$ ◄

12.34 *Water at 15°C is flowing through a 300-mm-diameter welded-steel pipe 2400 m long that drains a reservoir under a head of 50 m. The pipe wall has a thickness of 8.5 mm. (a) If a valve at the end of the pipe is closed in 10 s, approximately what water-hammer pressure will be developed? (b) If the steady-state flow is instantaneously reduced to one-half its original value, what water-hammer pressure would you expect?*

SI

Table A.1 at 15°C: $\nu = 1.139 \times 10^{-6}$ m/s². Table 8.1 for welded steel: $e = 0.046$ mm

Eq. 8.42a: $V = -2\sqrt{\dfrac{(2)9.81(0.3)50}{2400}}\log\left[\dfrac{0.046/300}{3.7} + \dfrac{2.51(1.139)}{10^6(2400)}\sqrt{\dfrac{2400}{2(9.81)0.3(50)}}\right] = 3.07$ m/s

Sec. 12.5: For water $c = 1440$ m/s and $E_v = 2.07 \times 10^6$ kPa, for steel $E = 207 \times 10^6$ kPa.

Eq. 12.10: $c_p = \dfrac{1440}{\sqrt{1 + \dfrac{300}{8.5}\left(\dfrac{2.07 \times 10^6}{207 \times 10^6}\right)}} = 1238$ m/s

(*a*) Eq. 12.11: $T_r = 2(2400)/1238 = 3.88$ s $\le t_c = 10$ s, so closure is "slow."

Eq 12.16: $\Delta p' \approx -2L\rho\Delta V/t_c = -2(2400)1000(-3.07)/10$ N/m² $= 1473$ kN/m² ◄

(*b*) Eq 12.12: $\Delta p = -\rho c_p \Delta V = -(1000)1238(0.5 - 1)3.07$ N/m² $= 1900$ kN/m² ◄

12.35 *For the situation described in Sample Prob. 12.5, find the water-hammer pressure at the valve if a flow of 90 cfs is reduced to 30 cfs in 3.2 sec. Under these conditions, what would be the maximum water-hammer pressures at points 500 and 1500 ft from the reservoir?*

Sample Prob. 12.5: Length of straight pipe to valve, $L = 2000$ ft, $D = 4$ ft, $c_p = 3200$ fps.

BG

$V_1 = Q/A = 90/(\pi 2^2) = 7.16$ fps, $V_2 = 30/(\pi 2^2) = 2.39$ fps. $\Delta V = 2.39 - 7.16 = -4.77$ fps over 3.2 sec

Eq 12.11: $T_r = 2L/c_p = 2(2000)/3200 = 1.250$ sec $< t_c = 3.2$ sec, so closure is "slow."

At valve, Eq 12.16: $\Delta p' \approx -2(2000)1.940(-4.77)/3.2 = 11{,}580$ psf $= 80.4$ psi ◄

500 ft from reservoir: $\Delta p' \approx (500/2000)80.4 = 20.1$ psi ◄

1500 ft from reservoir: $\Delta p' \approx (1500/2000)80.4 = 60.3$ psi ◄

12.36 *In Fig. P12.36 the total length of pipe is 10,000 ft, it diameter is 42 in, and its wall thickness is 0.89 in. Assume $E = 32{,}000{,}000$ psi and $E_v = 300{,}000$ psi. If the initial velocity for steady flow is 8 fps and the valve at G is partially closed so as to reduce the flow to half of the initial velocity in 3.6 sec, find (a) the maximum pressure rise due to the water hammer; (b) the location of the point of maximum total pressure.*

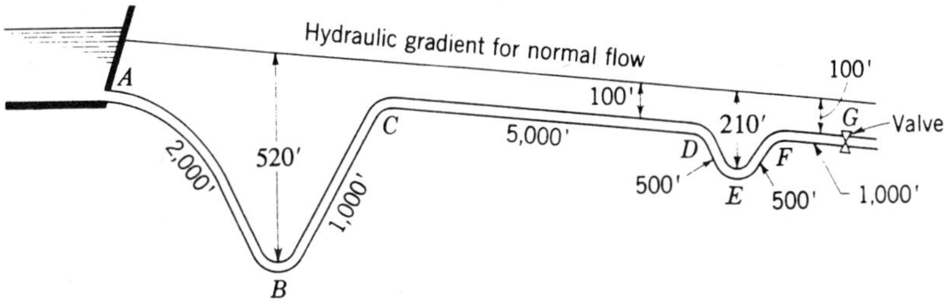

Figure P12.36

BG

(*a*) Sec. 12.5: $c = 4720$ fps. Eq 12.10: $c_p = \dfrac{4720}{\sqrt{1 + \dfrac{42}{0.89}\left(\dfrac{300{,}000}{32{,}000{,}000}\right)}} = 3930$ fps

Eq 12.12: $\Delta p = -\rho c_p \Delta V = -1.940(3930)(4 - 8) = 30{,}500$ psf $= 212$ psi ◄

(*b*) Eq. 12.11: $T_r = 2(10{,}000)/3930 = 5.09$ sec $> t_c = 3.6$ sec, so closure is "rapid"

Eq. 12.15: $x_0 = c_p t_c/2 = 3930(3.6)/2 = 7070$ ft

Thus water hammer pressure rise,

at B ($x = 2000$ ft $< x_0$): $\Delta p = 212(2000/7070) = 59.9$ psi

at E ($x = 8500$ ft $> x_0$): $\Delta p = 212(100\%) = 212$ psi

Total pressure = static pressure plus water hammer pressure = $\gamma h + \Delta p$

Total pressure at B: $520(62.4/144) + 59.9 = 285$ psi

at E: $210(62.4/144) + 212 = 303$ psi

∴ maximum total pressure occurs at point E ◄

/more...

12.37 *Refer to Fig. P12.36, but take all the dimensions given in feet to be in meters instead. This 10-km-long pipe has a diameter of 1.35 m and a wall thickness of 21 mm. Assume E = 205 GPa and E_v = 2 GPa. The initial steady flow velocity is 5 m/s. The valve at G is then partially closed so as to reduce the velocity to 1 m/s in 14 s. Find (a) the maximum pressure rise due to water hammer and (b) the location of the point of maximum total pressure.*

See Fig. P12.36 with Solution 12.36.

SI

(*a*) Sec. 12.5: c = 1440 m/s. Eq. 12.10: $c_p = \dfrac{1440}{\sqrt{1 + \dfrac{1350}{21}\left(\dfrac{2}{205}\right)}} = 1129$ m/s

Eq. 12.12: $\Delta p = -\rho c_p \Delta V = -1000(1129)(1-5) = 4520$ kN/m^2 (or kPa) ◄

(*b*) Eq. 12.11: $T_r = 2(10{,}000)/1129 = 17.72$ s $> t_c = 14$ s, so closure is "rapid."

Eq. 12.15: $x_0 = c_p t_c/2 = 1129(14)/2 = 7900$ m

Thus the water hammer pressure rise,

at B ($x = 2000$ m $< x_0$): $\Delta p = 4520(2000/7900) = 1143$ kN/m^2 (or kPa)

at E ($x = 8500$ m $> x_0$): $\Delta p = 4520(100\%) = 4520$ kN/m^2 (or kPa)

Total pressure = static pressure plus water hammer pressure = $\gamma h + \Delta p$

Total pressure at B = 9.81(520) + 1143 = 6240 kPa

at E = 9.81(210) + 4520 = 6580 kPa

∴ maximum total pressure occurs at point E ◄

Sec. 12.7: Surge Tanks -- ***Exercises***

12.7.1 *A 48-in-diameter steel pipe (f = 0.020, flush inlet) 4000 ft long delivers 275 cfs of water to a power plant. It is protected from water hammer by a 7-ft-diameter surge tank located at the downstream end just before the control valve. Without using trial and error, obtain a safe, preliminary estimate of the largest height (above the reservoir surface elevation) to which a surge would rise. Neglect all velocity heads and minor losses; in the surge tank (only) neglect fluid friction and inertial effects.*

BG

From Eq. 12.17: $K = fL/D + 0 + 0 = 0.020(4000/4) = 20$

From Eq. 12.24: $z_{max} < \dfrac{LA}{KA_s} = \dfrac{L}{K}\left(\dfrac{D}{D_s}\right)^2 = \dfrac{4000}{20}\left(\dfrac{4}{7}\right)^2 = 65.3$ ft ◄

12.7.2 *A 900-mm-diameter steel pipe (f = 0.018, flush inlet) 850 m long delivers 4 m^3/s of water to a power plant. It is protected from water hammer by a surge tank located at the downstream end just before the control valve. Without using trial and error, obtain a safe, preliminary estimate of the smallest surge tank diameter that would be required to prevent surges from exceeding 12 m above the reservoir water surface elevation. Neglect all velocity heads and minor losses; in the surge tank (only) neglect fluid friction and inertial effects.*

SI

From Eq. 12.17: $K = fL/D + 0 + 0 = (0.018)850/0.9 = 17.0$

From Eq. 12.24: $\left(\dfrac{D_s}{D}\right)^2 = \dfrac{A_s}{A} < \dfrac{L}{Kz_{max}} = \dfrac{850}{17.0(12)} = 4.17$ so $\dfrac{D_s}{D} < 2.04$

$D_s < 2.04D = 2.04(0.9) = 1.837$ m ◄

12.38 *Derive Eq. (12.20).*

N

Substitute into the original differential equation for K from Eq. 12.17 and for dz/dt from Eq. 12.19:

$$-z = K\frac{V^2}{2g} + \frac{L}{g}\frac{dV}{dz}\frac{AV}{A_s} \quad \text{or} \quad z + \frac{K}{2g}V^2 + \frac{LA}{2gA_s}\frac{2V\,dV}{dz} = 0$$

Define $N = K/2g$; $M = (L/2g)(A/A_s)$; and $V^2 = x$ so that $2V\,dV = dx$

That gives: $z + Nx + M\,dx/dz = 0$ or $dx/dz + (N/M)x + (1/M)z = 0$

Now define $\alpha = N/M$, $\beta = 1/M$; then $dx/dz + \alpha x + \beta z = 0$

Multiply by $e^{\alpha z}$: $dx/dz\, e^{\alpha z} + \alpha x e^{\alpha z} + \beta z e^{\alpha z} = 0$

i.e., $d/dz(xe^{\alpha z}) + e^{\alpha z}\beta z = 0$ or $d(xe^{\alpha z}) + e^{\alpha z}\beta z dz = 0$. Integrate: $\int d(xe^{\alpha z}) + \int e^{\alpha z}\beta z dz = 0$

i.e., $xe^{\alpha z} + \beta(e^{\alpha z}/\alpha^2)(\alpha z - 1) + C = 0$ or $x = (\beta/\alpha^2)(1 - \alpha z) - Ce^{-\alpha z}$

But $\dfrac{\beta}{\alpha^2} = \dfrac{M}{N^2} = \dfrac{M(2g)^2}{K^2} = \dfrac{LA(2g)}{A_s K^2}$, $\alpha = \dfrac{N}{M} = \dfrac{KA_s}{LA}$, and $x = V^2$

$$\therefore\ V^2 = \frac{LA(2g)}{A_s K^2}\left(1 - \frac{KA_s z}{LA}\right) - Ce^{-(KA_s/LA)z} = \frac{2g}{K}\left(\frac{LA}{KA_s} - z\right) - Ce^{-(KA_s/LA)z} \quad \text{(Eq. 12.20) Q.E.D.}$$ ◄

12.39 *A 1.5-m-diameter steel pipeline 1000 m long (flush inlet, $f = 0.020$) carries water at Q_0 (m^3/s) from a reservoir to a power plant. When the valve at the outlet end is closed instantaneously, water rises in the 2.5-m-diameter simple surge tank immediately adjacent to the valve. Determine the maximum allowable initial discharge Q_0 so that the resulting surge will not rise more than 12 m above the reservoir surface. In the surge tank only, neglect the velocity head, minor losses, fluid friction, and inertial effects.*

SI

Fig. 8.13: Flush inlet: $\Sigma k = 0.5$; $K = (0.02)1000/1.5 + 1 + 0.5 = 14.83$. $\dfrac{A_s}{A} = \left(\dfrac{D_s}{D}\right)^2 = 2.78$

$$\text{Eq. 12.25: } V_0^2 = \frac{-2(9.81)}{14.83}\left[12 + \frac{1000}{14.83(2.78)}\ln\left(1 - \frac{14.83}{1000}(2.78)12\right)\right] = 6.02;\ V_0 = 2.45 \text{ m/s}$$

so $Q_0 = AV_0 = (\pi/4)(1.5)^2 2.45 = 4.34\ m^3/s$ ◄

12.40 *Repeat Prob. 12.39 while also neglecting the velocity head and minor losses in the pipeline.*
Prob. 12.39: Pipeline $L = 1000$ m, $D = 1.5$ m, $f = 0.020$, flush inlet so $\Sigma k = 0.5$; $D_s = 2.5$, $t_c = 0$. Find the initial Q_0 so that $z_0 = 12$ m. In the surge tank only, neglect the velocity head, minor losses, fluid friction, and inertial effects.

SI

Sec. 12.7: Neglecting minor losses and $V^2/2g$ in the pipeline: $K = fL/D = 0.02(1000)/1.5 = 13.33$

$$\frac{A_s}{A} = \left(\frac{D_s}{D}\right)^2 = 2.78. \quad \text{Eq 12.25: } V_0^2 = \frac{-2(9.81)}{13.33}\left[12 + \frac{1000}{13.33(2.78)}\ln\left(1 - \frac{13.33}{1000}(2.78)12\right)\right] = 5.70$$

$V_0 = 2.39$ m/s so $Q_0 = AV_0 = (\pi/4)(1.5)^2 2.39 = 4.22\ m^3/s$ ◄

12.41 *Refer to Fig. 12.10. A 42-in-diameter steel pipe MN 3600 ft long (flush inlet, $f = 0.017$) supplies water to a small power plant. The discharge is 200 cfs, JN = 100 ft, and the elevations of J and the valve N are respectively *130 ft and *145 ft below the reservoir water surface. To protect against instaneous closure of the valve, what height would be required for the simple 6.5-ft-diameter surge tank if it is not to overflow? In the surge tank only, neglect the velocity head, minor losses, fluid friction, and inertial effects. [*Note these corrections to the textbook.]*

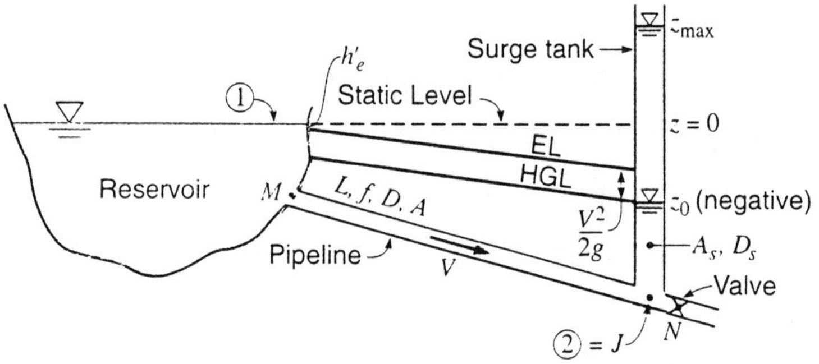

Figure 12.10

BG

Given: $Q = 200$ cfs, $f = 0.017$, $k_e = 0.5$ (Fig. 8.13)

$A_s/A = (D_s/D)^2 = 3.45$; $A = (\pi/4)(3.5)^2 = 9.62\ \text{ft}^2$; $L = MJ = 3600 - 100 = 3500$ ft

$V_0 = Q/A = 200/9.62 = 20.8$ fps; $V_0^2/2g = 20.8^2/(2 \times 32.2) = 6.71$ ft

From Eq. 12.17: $K = (fL/D) + 1 + \Sigma k = 0.017(3500/3.5) + 1 + 0.5 = 18.5$

$$\frac{KA_s}{LA} = \frac{18.5(3.45)}{3500} = \frac{1}{54.9} \quad \text{and} \quad KV_0^2/2g = 18.5(6.71) = 124.1\ \text{ft} \quad (= -z_0)$$

From Eq. 12.24: $z_{max} < LA/(KA_s) = 54.9$ ft

From Eq. 12.23: $1 - \dfrac{z_{max}}{54.9} - e^{-(1/54.9)(124.1 + z_{max})} = 0$

So by trial and error:

Trial z_{max}	Left side = error	
50	0.046 7	
53	−0.005 79	
52	0.011 71	
52.5	0.002 96	
52.7	0.000 543	
52.69	−0.000 368	Good enough

$z_{max} = 52.7$ ft. $\therefore$ height of surge tank $= 130 + 52.7 = 182.7$ ft ◄

/more...

12.42 **BG** *Repeat Prob. 12.41 while also neglecting the velocity head and minor losses in the pipeline.*
Refer to Prob. 12.41.

Given: $Q = 200$ cfs, $f = 0.017$, $k_e = 0.5$ (Fig. 8.13)

$A_s/A = (D_s/D)^2 = 3.45$; $A = (\pi/4)(3.5)^2 = 9.62$ ft^2; $L = MJ = 3600 - 100 = 3500$ ft

$V_0 = Q/A = 200/9.62 = 20.8$ fps; $V_0^2/2g = 20.8^2/(2 \times 32.2) = 6.71$ ft

From Eq. 12.17: $K = fL/D = 0.017(3500)/3.5 = 17$

$$\frac{KA_s}{LA} = \frac{17(3.45)}{3500} = \frac{1}{59.7} \quad \text{and} \quad KV_0^2/2g = 17(6.71) = 114.1 \text{ ft} \quad (= -z_0)$$

From Eq. 12.24: $z_{max} < LA/(KA_s) = 59.7$ ft

$$\text{From Eq. 12.23:} \quad 1 - \frac{z_{max}}{59.7} - e^{-(1/59.7)(114.1 + z_{max})} = 0$$

Solve for z_{max} by trial and error:

Trial z_{max}	Left side = error	
55	0.020 6	
60	−0.058 5	
57	−0.011 00	
56	0.047 9	
56.5	−0.003 10	
56.3	0.000 0588	
56.31	−0.000 0992	Close enough

$\therefore z_{max} = 56.3$ ft and height of surge tank = 130 + 56.3 = 186.3 ft ◀

12.43 **BG** *Repeat Prob. 12.41 for a surge tank diameter of 10 ft.*
Refer to Prob. 12.41.

Given: $Q = 200$ cfs, $f = 0.017$, $k_e = 0.5$ (Fig. 8.13)

$$\frac{A_s}{A} = \left(\frac{D_s}{D}\right)^2 = \frac{100}{12.25} = 8.16; \quad A = \pi\frac{3.5^2}{4} = 9.62 \text{ ft}^2; \quad L = MJ = 3600 - 100 = 3500 \text{ ft}$$

$V_0 = Q/A = 200/9.62 = 20.8$ fps; $V_0^2/2g = 20.8^2/(2 \times 32.2) = 6.71$ ft

From Eq. 12.17: $K = (fL/D) + 1 + \Sigma k = 0.017(3500)/3.5 + 1 + 0.5 = 18.5$

$$\frac{KA_s}{LA} = \frac{18.5(8.16)}{3500} = \frac{1}{23.2} \quad \text{and} \quad \frac{KV_0^2}{2g} = 18.5(6.71) = 124.1 \text{ ft} \quad (= -z_0)$$

From Eq. 12.24: $z_{max} < LA/(KA_s) = 23.2$ ft

From Eq. 12.23: $1 - z_{max}/23.2 - e^{-(1/23.2)(124.1 + z_{max})} = 0$

Trial z_{max}	Left side = error	
20	0.135	
23	0.005 83	
23.2	−0.002 78	
23.13	0.000 232	Close enough

$z_{max} = 23.1$ ft. $\therefore$ height of surge tank = 130 + 23.1 = 153.1 ft ◀

12.44 Using the data of Prob. 12.41, find the diameter of surge tank that will produce a surge requiring a tank height of 165 ft.

Refer to Prob. 12.41.

BG

Given: $Q = 200$ cfs, $f = 0.017$, $k_e = 0.5$ (Fig. 8.13)

$D = 42/12 = 3.5$ ft; $A = (\pi/4)(3.5)^2 = 9.62$ ft^2; $L = MJ = 3600 - 100 = 3500$ ft

$z_{max} = 165 - 130 = 35$ ft. $V_0 = Q/A = 200/9.62 = 20.8$ fps ; $V_0^2/2g = 20.8^2/(2 \times 32.2) = 6.71$ ft

From Eq. 12.17: $K = (fL/D) + 1 + \Sigma k = 0.017(3500/3.5) + 1 + 0.5 = 18.5$

For Eq. 12.23: $\dfrac{K}{LA} = \dfrac{18.5}{3500(9.62)} = \dfrac{1}{1820}$ and $(KV_0^2/2g + z_{max}) = 18.5(6.71) + 35 = 159.1$ ft

So Eq. 12.23: $1 - A_s(35)/1820 = e^{-A_s(159.1)/1820}$ or $1 - (A_s/52.0) = e^{-(A_s/11.44)}$

But $A_s = \pi D_s^2/4 = 0.785D_s^2$ so $1 - 0.785D_s^2/52.0 - e^{-0.785D_s^2/11.44} = 0$ or $1 - D_s^2/66.2 - e^{-D_s^2/14.56} = 0$

Per Eq. 12.24, here: $D_s^2/66.2 < 1$, i.e., $D_s < \sqrt{66.2} = 8.13$ ft

Trial D_s	Left side = error
8.0	0.021
8.1	-0.001 89
8.09	0.000 662
8.092	-0.000 034 15

Close enough $\therefore D_s = 8.09$ ft ◄

12.45 Using the data of Prob. 12.42, find the diameter of surge tank that will produce a surge requiring a tank height of 165 ft.

Refer to Prob. 12.41, while also neglecting the velocity head and minor losses in the pipeline.

BG

Given: $Q = 200$ cfs, $f = 0.017$

$D = 42/12 = 3.5$ ft; $A = (\pi/4)(3.5)^2 = 9.62$ ft^2; $L = MJ = 3600 - 100 = 3500$ ft

$z_{max} = 165 - 130 = 35$ ft. $V_0 = Q/A = 200/9.62 = 20.8$ fps ; $V_0^2/2g = 20.8^2/(2 \times 32.2) = 6.71$ ft

From Eq. 12.17: $K = fL/D = 0.017(3500/3.5) = 17$

For Eq. 12.23: $\dfrac{K}{LA} = \dfrac{17}{3500(9.62)} = \dfrac{1}{1981}$ and $(KV_0^2/2g + z_{max}) = 17(6.71) + 35 = 149.1$ ft

So Eq. 12.23: $1 - \dfrac{A_s(35)}{1981} = e^{-A_s(149.1)/1981}$ or $1 - (A_s/56.6) = e^{-(A_s/13.29)}$

But $A_s = \pi D_s^2/4 = 0.785D_s^2$ so $1 - 0.785D_s^2/56.6 - e^{-0.785D_s^2/13.28} = 0$ or $1 - D_s^2/72.1 - e^{-D_s^2/16.92} = 0$

Per. Eq. 12.24, here: $D_s^2/72.1 < 1$, i.e., $D_s < \sqrt{72.1} = 8.49$ ft

Trial D_s	Left side = error
8.4	0.005 36
8.45	-0.005 59
8.420	0.000 992
8.425	-0.000 102
8.4245	0.000 007 62

Close enough $\therefore D_s = 8.42$ ft ◄

☐ 12.46 *Refer to Fig. 12.10. A 1-m-diameter steel pipe MN 1070 m long (flush inlet, f = 0.016) supplies water to a small power plant. The discharge is 2.45 m³/s, JN = 20 m, and the elevations of J and the valve N are respectively 25 m and 30 m below the reservoir water surface. To protect against instantaneous closure of the valve, what height would be required for the simple 3.5-m-diameter surge tank if it is not to overflow? In the surge tank only, neglect the velocity head, minor losses, fluid friction, and inertial effects.*

See Fig. 12.10 with Solution 12.41.

SI

Given: $Q = 2.45\ \text{m}^2/\text{s}$, $f = 0.016$, $k_e = 0.5$ (Fig. 8.13)

$A_s/A = D_s^2/D^2 = 3.5^2/1 = 12.25$; $L = MJ = 1070 - 20 = 1050$ m

$A = \pi 1^2/4 = 0.785\ \text{m}^2$; $V_0 = Q/A = 2.45/0.785 = 3.12$ m/s ; $V_0^2/2g = 3.12^2/(2 \times 9.81) = 0.496$ m

From Eq. 12.17: $K = fL/D + 1 + \Sigma k = 0.016(1050/1) + 1 + 0.5 = 18.3$

For Eq. 12.23: $\dfrac{KA_s}{LA} = \dfrac{18.3(12.25)}{1050} = \dfrac{1}{4.68}$ and $KV_0^2/2g = 18.3(0.496) = 9.08$ m $(= -z_0)$

So Eq. 12.23: $1 - z_{max}/4.68 - e^{-(1/4.68)(9.08 + z_{max})} = 0$

Per Eq. 12.24, here: $z_{max}/4.68 < 1$, i.e. $z_{max} < 4.68$ m

Trial z_{max}	Left side = error	
4.4	0.004 31	
4.5	−0.015 9	
4.45	−0.005 77	
4.42	0.002 76	
4.423	−0.000 329	Close enough

$z_{max} = 4.42$ m. ∴ height of surge tank $= 25 + 4.42 = 29.42$ m ◄

☐ 12.47 *Repeat Prob. 12.46 while also neglecting the velocity head and minor losses in the pipeline.*

Refer to Prob. 12.46, and see Fig. 12.10 with Solution 12.41.

SI

Given: $Q = 2.45\ \text{m}^3/\text{s}$, $f = 0.016$

$A_s/A = D_s^2/D^2 = 3.5^2/1 = 12.25$; $L = MJ = 1070 - 20 = 1050$ m

$A = \pi 1^2/4 = 0.785\ \text{m}^2$; $V_0 = Q/A = 2.45/0.785 = 3.12$ m/s ; $V_0^2/2g = 3.12^2/(2 \times 9.81) = 0.496$ m

From Eq. 12.17: $K = fL/D = 0.016(1050/1) = 16.8$

For Eq. 12.23: $\dfrac{KA_s}{LA} = \dfrac{16.8(12.25)}{1050} = \dfrac{1}{5.10}$ and $KV_0^2/2g = 16.8(0.496) = 8.33$ m $(= -z_0)$

So Eq. 12.23: $1 - z_{max}/5.10 - e^{-(1/5.10)(8.33 + z_{max})} = 0$

Per Eq. 12.24, here: $z_{max}/5.10 < 1$, i.e. $z_{max} < 5.10$ m

Trial z_{max}	Left side = error	
5	−0.0533	
4.5	0.0371	
4.7	0.001 06	
4.8	−0.017	
4.71	−0.000 753	
4.705	0.000 15	
4.706	−0.000 0295	Close enough

$z_{max} = 4.71$ m. ∴ height of surge tank $= 25 + 4.71 = 29.7$ m ◄

Chapter 13

Steady Flow of Compressible Fluids

PROBLEM SELECTION GUIDE

Sec	Exer/Prob	Units	Difficulty	Length	Parts	Similar	Special features
13.1	***Thermodynamic Considerations***						
	X[1]13.1.1	BG	Easy	Short	1	13.1.2	
	13.1.2	SI	Easy	Short	1	13.1.1	
	13.1.3	BG	Medium	Medium	1	13.1.4	Uses $pv = RT$ (Sec. 2.7)
	13.1.4	SI	Medium	Medium	1	13.1.3	Uses $pv = RT$ (Sec. 2.7)
	13.1.5	SI	Easy	Short	1		
	P 13.1	SI	Medium	Long	1		Uses $pv = RT$, and a graphical approach
13.3	***Speed of Sound***						
	X 13.3.1	BG	Easy	Short	1	13.3.2	
	13.3.2	SI	Easy	Short	1	13.3.1	
13.5	***Stagnation Pressure***						
	X 13.5.1	BG	Medium	Medium	1	13.5.2	
	13.5.2	SI	Medium	Medium	1	13.5.1	
	13.5.3	BG	Medium	Medium	1	13.5.4	
	13.5.4	SI	Medium	Medium	1	13.5.3	
	P 13.2	BG	Medium	Medium	1		Optionally uses $pv = RT$ (Sec. 2.7)
13.6	***Isentropic Flow***						
	X 13.6.1	BG	Easy	Medium	2	13.6.2	
	13.6.2	SI	Easy	Medium	2	13.6.1	
	13.6.3	BG	Easy	Short	1	13.6.4	Check if $V > c$
	13.6.4	SI	Easy	Short	1	13.6.3	Check if $V > c$
	P 13.3	N	Medium	Medium	1		Derivation.
	13.4	BG	Medium	Medium	1		Optionally uses $pv = RT$ (Sec. 2.7)
13.7	***Effect of Area Variation on One-Dimensional Compressible Flow***						
	X 13.7.1	N	Easy	Short	1		Derivation.

/cont...

[1] For all Exercises (marked "X"), answers are given in Appendix D of the textbook.
* Note a correction made to early printings of the textbook.
† Answers are sensitive to values (β) which must be measured from a figure.
X = Exercise, P = (end-of-chapter) Problem, S = Sample Problem. ☐ = could use computing aids.

Sec	Exer/Prob	Units	Difficulty	Length	Parts	Similar	Special features
13.8	*Compressible Flow Through a Converging Nozzle*						
	X 13.8.1	N	Medium	Medium	1		Derivation
	13.8.2	N	Medium	Medium	1		Derivation. Differentiation.
	13.8.3	BG	Medium	Medium	1	13.8.4	
	13.8.4	SI	Medium	Medium	1	13.8.3	
	13.8.5	BG	Medium	Medium	1	13.8.6	
	13.8.6	SI	Medium	Medium	1	13.8.5	
	P 13.5	BG	Hard	Long	1	13.6-7	Uses $pv = RT$ (Sec. 2.7)
	13.6	SI	Hard	Long	1	13.5-7	Uses $pv = RT$ (Sec. 2.7)
	13.7	SI	Hard	Long	1	13.5-6	Uses $pv = RT$ (Sec. 2.7)
	13.8	BG	Hard	Long	1		Uses $pv = RT$ (Sec. 2.7)
	13.9	BG	Hard	Long	1		* □ Uses $pv^k = C$ and $pv = RT$. T&E
13.9	*Isentropic Flow Through a Converging-Diverging Nozzle*						
	X 13.9.1	BG	Medium	Short	1	13.9.2	Uses $pv^k = C$ and $pv = RT$ (Sec. 2.7)
	13.9.2	SI	Medium	Short	1	13.9.1	Uses $pv^k = C$ and $pv = RT$ (Sec. 2.7)
	13.9.3	BG	Medium	Medium	1	S13.5	Uses $pv = RT$ (Sec. 2.7)
	13.9.4	BG	Easy	Short	1		Uses $pv = RT$ (Sec. 2.7)
	13.9.5	BG	Easy	Short	2	13.9.6	
	13.9.6	SI	Easy	Short	2	13.9.5	
	P 13.10	BG	Hard	Long	1	13.11	□ Uses $pv = RT$ (Sec. 2.7). T & E.
	13.11	SI	Hard	Long	1	13.10	□ Uses $pv = RT$ (Sec. 2.7). T & E.
	13.12	BG	Medium	Medium	2	13.13	Uses $pv = RT$. A check is required.
	13.13	SI	Medium	Medium	2	13.12	Uses $pv = RT$. A check is required.
13.10	*One-Dimensional Shock Wave*						
	X 13.10.1	BG	Medium	Medium	2	13.10.2	
	13.10.2	SI	Medium	Medium	2	13.10.1	
	P 13.14	SI	Hard	Long	2		Uses $pv = RT$ (Sec. 2.7)
	13.15	BG	Hard	Long	1		Uses $pv = RT$ (Sec. 2.7)
13.11	*The Oblique Shock Wave*						
	X 13.11.1	BG	Easy	Short	1		†
	13.11.2	BG	Easy	Short	1	13.11.3	
	13.11.3	SI	Easy	Short	1	13.11.2	
13.12	*Isothermal Flow*						
	P 13.16	BG	Hard	Long	1	13.17	Integration; uses Sec. 2.7
	13.17	SI	Hard	Long	1	13.16	Integration; uses Sec. 2.7

/cont...

Sec	Exer/Prob	Units	Difficulty	Length	Parts	Similar	Special features
13.13	***Isothermal Flow in a Constant-Area Duct***						
	X 13.13.1	BG	Medium	Medium	4	13.13.2	Uses $pv = RT$ (Sec. 2.7)
	13.13.2	SI	Medium	Medium	4	13.13.1	Uses $pv = RT$ (Sec. 2.7)
	P 13.18	BG	Medium	Long	1	S13.7	Uses Sec. 2.7
	13.19	BG	Medium	Medium	1	13.20	□ Uses $pv = RT$; assume/confirm **R**.
	13.20	SI	Medium	Medium	1	13.19	□ Uses $pv = RT$; assume/confirm **R**.
	13.21	BG	Hard	Long	1		□ Uses Secs 2.7, 2.11, 8.11-13;
	"						assume and correct **R**.
13.14	***Adiabatic Flow in a Constant-Area Duct***						
	P 13.22	BG	Hard	V Long	3	S13.9	□ Uses Secs 2.7-8,7.4; long T&E, plots
	13.23	BG	Medium	Long	1	13.24	Uses Secs. 2.7 and 8.11-12.
	13.24	SI	Medium	Long	1	13.23	Uses Secs. 2.7 and 8.11-12.

Chapter 13
STEADY FLOW OF COMPRESSIBLE FLUIDS

Sec. 13.1: Thermodynamic Considerations -- *Exercises*

13.1.1 BG *Compute the change in enthalpy of 15 slugs of oxygen if its temperature is increased from 120°F to 155°F.*

Table A.5 for oxygen: $c_p = 5437$ ft·lb/(slug·°R)

From Eq. 13.1: $\Delta h = c_p(T_2 - T_1) = 5437(155 - 120) = 190{,}300$ ft·lb/slug

Thus change in enthalpy $= 190{,}300(15) = 2.85 \times 10^6$ ft·lb ◄

13.1.2 SI *Compute the change in enthalpy of 250 kg of oxygen if its temperature is increased from 50°C to 70°C.*

Table A.5 for oxygen: $c_p = 909$ N·m/(kg·K)

From Eq. 13.1: $\Delta h = c_p(T_2 - T_1) = 909(70 - 50) = 18\,180$ N·m/kg

Thus change in enthalpy $= 18\,180(250) = 4.55 \times 10^6$ N·m (or J) ◄

13.1.3 *Suppose the 15 slugs of oxygen of Exer. 13.1.1 were compressed isentropically to 80 percent of its original volume. Find the final temperature and pressure, the work required, and the change in enthalpy. Assume $T_1 = 120°F$ and $p_1 = 200$ psia.*

BG

pv^k = constant (isentropic, Sec. 13.1), and $pv = RT$ (Eq. 2.4)

Eliminating p: $(RT/v)v^k = \text{constant} = RTv^{k-1}$

Since also R = constant, $\therefore Tv^{k-1}$ = constant; $k = 1.40$ (Table A.5 for oxygen)

From which: $T_2 = T_1(v_1/v_2)^{k-1} = (120 + 460)(1/0.8)^{0.4} = 580(1.25)^{0.4} = 634.2°R = 174.2°F$ ◄

From Eq. 2.4: $R = \dfrac{p_1 v_1}{T_1} = \dfrac{p_2 v_2}{T_2}$; so $\dfrac{200 v_1}{580} = \dfrac{p_2(0.8 v_1)}{634}$; $p_2 = 273$ psia ◄

Table A.5 for oxygen: $c_v = 3883$, $c_p = 5437$ ft·lb/(slug·°R)

Work required = change of internal energy (as isentropic); using Eq. 13.2

$= \Delta i \times m = c_v(\Delta T) \times 15$ where $\Delta T = 174.2 - 120 = 54.2°F = 54.2°R$

$= 3883(54.2)15 = 3.15 \times 10^6$ ft·lb ◄

Using Eq. 13.11: change in enthalpy, $\Delta h = 5437(54.2)15 = 4.42 \times 10^6$ ft·lb increase ◄

13.1.4 *Suppose the 250 kg of oxygen of Exer. 13.1.2 were compressed isentropically to 80 percent of its original volume. Find the final temperature and pressure, the work required, and the change in enthalpy. Assume $T_1 = 50°C$ and $p_1 = 1400$ kN/m² abs.*

SI

pv^k = constant (isentropic, Sec. 13.1), and $pv = RT$ (Eq. 2.4)

Eliminating p: $(RT/v)v^k = \text{constant} = RTv^{k-1}$

Since also R = constant, $\therefore Tv^{k-1}$ = constant; $k = 1.40$ (Table A.5 for oxygen)

From which: $T_2 = T_1(v_1/v_2)^{k-1} = (50 + 273)(1/0.8)^{0.4} = 323(1.25)^{0.4} = 353.2\text{ K} = 80.2°C$ ◄

From Eq. 2.4: $R = \dfrac{p_1 v_1}{T_1} = \dfrac{p_2 v_2}{T_2}$; so $\dfrac{1400 v_1}{323} = \dfrac{p_2(0.8 v_1)}{353.2}$; $p_2 = 1913$ kPa abs ◄

/cont...

Table A.5 for oxygen: $c_v = 649$, $c_p = 909$ N·m/(kg·K)

Work required = change of internal energy (as isentropic); Using Eq. 13.2

$= \Delta i \times m = c_v(\Delta T) \times 250$ where $\Delta T = 80.2 - 50 = 30.2°C = 30.2$ K

$= 649(30.2)250 = 4.89 \times 10^6$ N·m $= 4.89 \times 10^6$ J ◄

Using Eq. 13.1: change in enthalpy, $\Delta h = 909(30.2)250 = 6.85 \times 10^6$ N·m $= 6.85 \times 10^6$ J increase ◄

13.1.5 *Using the data of Sample Prob. 13.2 compute $\Delta(p/\rho)$ and thus show that $\Delta h = \Delta i + \Delta(p/\rho)$.*
Sample Prob. 13.2: $\Delta h = 130\,100$ J/kg, $\Delta i = 92\,900$ J/kg, $T_2 = 423$ K.

SI

Table A.5 for air: $R = 287$ J/(kg·K)

$p_1/\rho_1 = RT_1 = 287(20 + 273) = 84\,100$ J/kg; $p_2/\rho_2 = RT_2 = 287(423) = 121\,400$ J/kg

$\Delta(p/\rho) = p_2/\rho_2 - p_1/\rho_1 = 121\,400 - 84\,100 = 37\,300$ J/kg ◄

$\Delta i + \Delta(p/\rho) = 92\,900 + 37\,300 = 130\,200$ which checks accurately ◄ with $\Delta h = 130100$.

Sec. 13.1: Thermodynamic Considerations -- *Problem*

13.1 *Using the data of Sample Prob. 13.2, determine the work done in compressing the air by finding the area under a pressure-vs-volume curve. Compute and tabulate volumes and pressures using volume increments that are 10 percent of the original volume.*
Sample Prob. 13.2: 15 kg of air are compressed isentropically. Originally $T_1 = 20°C$, $p_1 = 95$ kPa abs.

SI

Table A.5 for air: $R = 287$ N·m/(kg·K), $k = 1.40$

Using Eq. 2.4: $v_1 = RT_1/p_1 = 287(20 + 273)/95\,000 = 0.885$ m³/kg

After compression, from Sample Prob. 13.2: $T = T_1(v_1/v)^{k-1} = 293(v_1/v)^{0.4}$;

$v = v_1/(v_1/v)$; $\forall = mv = 15v$; $p = RT/v$

v_1/v	T, K	v, m³/kg	$\forall$, m³	p, N/m²
100/100	293	0.885	13.28	95 000
100/90	306	0.797	11.95	110 100
100/80	320	0.708	10.62	129 900
100/70	338	0.620	9.29	156 500
100/60	359	0.531	7.97	194 200
100/50	387	0.443	6.64	251 000
100/40	423	0.354	5.31	343 000
			▲	▲

Measure or calculate area A under p–$\forall$ curve.

Work done in compression = A

$= 1.397 \times 10^6$ N·m ◄

(here calculated by Weddle's Rule)

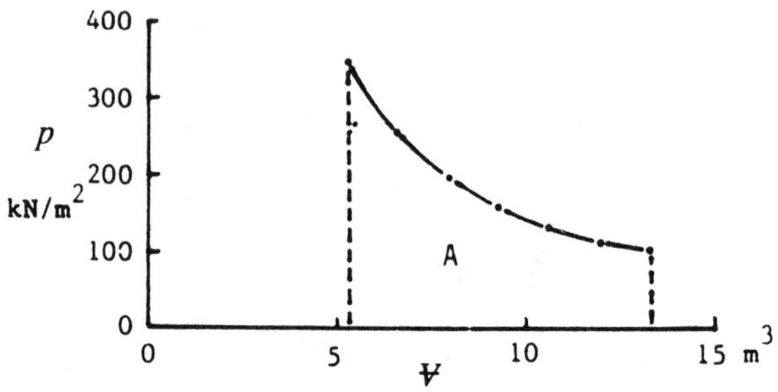

Sec. 13.3: Speed of Sound -- *Exercises*

13.3.1 *Use Sec. 13.3 to calculate the sonic velocity in air at sea level and at elevations 5000, 10,000, 20,000, and 30,000 ft. Assume standard atmosphere (Appendix A, Table A.3).*

BG

Eq. 13.15 (Sec. 13.3): $c = \sqrt{kRT}$; Table A.5: In all cases $k = 1.40$ and $R = 1715$ ft²/(sec²·°R)

Table A.3: Temps = 59°F, 41.2°F, 23.4°F, −12.3°F, and −47.8°F

Substitute for T = (460 + Temp) and compute the resulting c's.

Elev (ft)	Temp (°F)	c (ft/sec)
0	59.0	1116
5,000	41.2	1097
10,000	23.4	1077
20,000	−12.3	1037
30,000	−47.8	995

▲

13.3.2 *Repeat Exer. 13.3.1 for sea level, 2000 and 10 000 m, expressing the answers in SI units.*
Exer. 13.3.1: Use Sec. 13.3 to calculate the sonic velocity in air. Assume standard atmosphere (Appendix A, Table A.3).

SI

Eq. 13.15 (Sec. 13.3): $c = \sqrt{kRT}$; Table A.5: In all cases $k = 1.40$ and $R = 287$ m²/(s²·K)

Table A.3: Temps = 15.0, 2.0, and −49.9°C

Substitute for T = (273 + Temp °C) and compute the resulting c's.

Elev (m)	Temp (°C)	c (m/s)
0	15.0	340
2 000	2.0	332
10 000	−49.9	299

▲

Sec. 13.5: Stagnation Pressure -- *Exercises*

13.5.1 *Air flows past an object at 600 ft/sec. Determine the stagnation pressures and temperatures in the standard atmosphere at elevations of sea level, 5000 and 30,000 ft.*

BG

Table A.5 for air: $R = 1715$ ft²/(sec²·°R), $c_p = 6000$ ft²/(sec²·°R), $k = 1.40$

Eq. 13.15: $c = \sqrt{kRT} = \sqrt{1.40(1715)T}$; Eq. 13.10: $\mathbf{M} = V/c = 600/c$

From Eq. 13.25: $T_O = T + (V^2/2c_p) = T + [600^2/(2 \times 6000)] = T + 30$

Eq. 13.30: $p_O = p + (1/2)\rho V^2(1 + \mathbf{M}^2/4)/144 = p + (1/2)\rho(600^2)(1 + \mathbf{M}^2/4)/144$

$= p + 1250\rho(1 + \mathbf{M}^2/4)$

Elev (ft)	*T (°F)	*p (psi)	*ρ (slug/ft³)	c (ft/sec)	$\mathbf{M}$	T_O (°F)	p_O (psi)
0	59.0	14.70	0.002 38	1116	0.537	89.0	17.89
5000	41.2	12.23	0.002 05	1097	0.547	71.2	14.98
30,000	−47.8	4.37	0.000 891	995	0.603	−17.8	5.59

*From Table A.3 for standard atmosphere. ▲ ▲

13.5.2 *Air flows past an object at 200 m/s. Determine the stagnation pressures and temperatures in the standard atmosphere at elevations of sea level, 2000 and 10 000 m.*

SI

Table A.3 at 2000 m: $T = 2.00°\text{C}$, $p = 79.50$ kPa abs, $\rho = 1.007\ \text{kg/m}^3$

Table A.5 for air: $R = 287\ \text{m}^2/(\text{s}^2{\cdot}\text{K})$, $c_p = 1003\ \text{m}^2/(\text{s}^2{\cdot}\text{K})$, $k = 1.40$

Eq. 13.15: $c = \sqrt{kRT} = \sqrt{1.40(287)(273 + 2.00)} = 332$ m/s; Eq. 13.10: $\mathbf{M} = V/c = 200/332 = 0.602$

From Eq. 13.25: $T_O = T + \dfrac{V^2}{2c_p} = 2.00°\text{C} + \dfrac{200^2}{2(1003)} = 2.00°\text{C} + 19.94°\text{C} = 21.9°\text{C}$ ◀

Eq. 13.30: $p_O = 79.50 + (1/2)1.007(200^2)[1 + (1/4)0.602^2]/10^3 = 101.5$ kPa abs ◀

Similarly at 10 000 m, $T_O = -30.0°\text{C}$ ◀ $p_O = 35.7$ kPa abs ◀

and at sea level, $T_O = 34.9°\text{C}$ ◀ $p_O = 127.9$ kPa abs ◀

13.5.3 *Air at 250 psia is moving at 500 ft/sec in a high-pressure wind tunnel at a temperature of 100°F. Find the stagnation pressure and temperature. Note the magnitude of the sonic velocity for the 250-psia 100°F air.*

BG

Table A.5 for air: $R = 1715\ \text{ft}^2/(\text{sec}^2{\cdot}°\text{R})$, $c_p = 6000\ \text{ft}^2/(\text{sec}^2{\cdot}°\text{R})$, $k = 1.40$

Eq. 13.15: $c = \sqrt{kRT} = \sqrt{(1.40)1715(460 + 100)} = 1160$ ft/sec ◀

Eq. 13.10: $\mathbf{M} = V/c = 500/1160 = 0.431$; 500 ft/sec $= V < c$ so can use Eq. 13.27

From Eq. 13.27: $p_O = 250[1 + 0.5(1.40 - 1)0.431^2]^{1.4/0.4} = 284$ psia ◀

From Eq. 13.25: $T_O = T + V^2/(2g \times c_p) = (460 + 100) + 500^2/(2 \times 6000) = 581°\text{R} = 121°\text{F}$ ◀

13.5.4 *Air at 1750 kPa is moving at 150 m/s in a high-pressure wind tunnel at a temperature of 40°C. Find the stagnation pressure and temperature. Note the magnitude of the sonic velocity for the 1750-kPa 40°C air.*

SI

Table A.5 for air: $R = 287\ \text{m}^2/(\text{s}^2{\cdot}\text{K})$, $c_p = 1003\ \text{m}^2/(\text{s}^2{\cdot}\text{K})$, $k = 1.40$

Eq. 13.15: $c = \sqrt{kRT} = \sqrt{1.40(287)(273 + 40)} = 355$ m/s ◀

Eq. 13.10: $\mathbf{M} = V/c = 150/355 = 0.423$; 150 m/s $= V < c$ so can use Eq. 13.27

From Eq. 13.27: $p_O = 1750[1 + 0.5(1.40 - 1)0.423^2]^{1.4/0.4} = 1979$ kPa abs ◀

From Eq. 13.25: $T_O = T + V^2/(2g \times c_p) = (273 + 40) + 150^2/(2 \times 1003) = 324\text{K} = 51.2°\text{C}$ ◀

Sec. 13.5: Stagnation Pressure -- *Problem*

13.2 *Find the stagnation pressure and temperature in air flowing at 88 ft/sec if the static pressure and static temperature are 14.7 psia and 50°F respectively.*

BG

Table A.5 for air: $R = 1715\ \text{ft}^2/(\text{sec}^2{\cdot}°\text{R})$, $c_p = 6000\ \text{ft}^2/(\text{sec}^2{\cdot}°\text{R})$, $k = 1.40$

Eq. 13.15: $c = \sqrt{kRT} = \sqrt{1.40(1715)(460 + 50)} = 1107$ ft/sec

Eq. 13.10: $\mathbf{M} = V/c = 88/1107 = 0.0795$

From Eq. 13.25: $T_O = T + \dfrac{V^2}{2c_p} = (460 + 50) + \dfrac{88^2}{2(6000)} = 511°\text{R} = 50.6°\text{F}$ ◀

From Eq. 13.27: $p_O = 14.7[1 + 0.5(1.40 - 1)0.0795^2]^{1.4/0.4} = 14.77$ psia ◀

Alternatively, obtain $\rho = 0.002\,42\ \text{slug/ft}^3$ from Eq. 2.4, then find p_O from Eq. 13.30.

Sec. 13.6: Isentropic Flow -- ***Exercises***

13.6.1 *Air at a pressure of 150 psia and a temperature of 100°F expands in a suitable nozzle to 15 psia. (a) If the flow is frictionless and adiabatic and the initial velocity is negligible, find the final velocity by Eq. 13.35. (b) Find the final temperature at the end of the expansion through use of Eq. 13.23.*

BG

Table A.5 for air: $R = 1715$ ft²/(sec²·°R), $c_p = 6000$ ft²/(sec²·°R), k = 1.40

$$(a)\ \text{Eq. 13.35:}\quad \frac{V_2^2 - V_1^2}{2} = RT_1\frac{k}{k-1}\left[1 - (p_2/p_1)^{(k-1)/k}\right]\quad \text{where } V_1 \approx 0 \text{ (given)}$$

$$V_2^2/2 = 1715(100 + 460)(1.40/0.40)\left[1 - (15/150)^{0.4/1.4}\right];\quad V_2 = 1800 \text{ ft/sec}\ ◄$$

$$(b)\ \text{Eq. 13.23:}\quad V_2^2 - V_1^2 = 2c_p(T_1 - T_2)$$

$$1800^2 - 0 = 2(6000)(560 - T_2);\quad T_2 = 290°\text{R} = -170°\text{F}\ ◄$$

13.6.2 *Air at a pressure of 1000 kPa and a temperature of 40°C expands in a suitable nozzle to 100 kPa. (a) If the flow is frictionless and adiabatic and the initial velocity is negligible, find the final velocity by Eq. 13.35. (b) Find the final temperature at the end of the expansion through use of Eq. 13.23.*

SI

Table A.5 for air: $R = 287$ m²/(s²·K), $c_p = 1003$ m²/(s²·K), k = 1.40

$$(a)\ \text{Eq. 13.35:}\quad \frac{V_2^2 - V_1^2}{2} = RT_1\frac{k}{k-1}\left[1 - (p_2/p_1)^{(k-1)/k}\right]\quad \text{where } V_1 \approx 0 \text{ (given)}$$

$$V_2^2/2 = 287(40 + 273)(1.40/0.40)\left[1 - (100/1000)^{0.4/1.4}\right];\quad V_2 = 551 \text{ m/s}\ ◄$$

$$(b)\ \text{Eq. 13.23:}\quad V_2^2 - V_1^2 = 2c_p(T_1 - T_2)$$

$$551^2 - 0 = 2(1003)(313 - T_2);\quad T_2 = 161.9\ \text{K} = 111.1°\text{C}\ ◄$$

13.6.3 *Carbon dioxide flows isentropically. At a point in the flow the velocity is 50 ft/sec and the temperature is 125°F. At a second point on the same streamline the temperature is 80°F. What is the velocity at the second point? Check if your answer is valid.*

BG

Table A.5 for carbon dioxide: $R = 1123$ ft²/(sec²·°R), $k = 1.28$,

Flow is isentropic, therefore adiabatic.

$$\text{Eq. 13.24:}\quad V_2^2 - 50^2 = \frac{2(1.28)}{0.28}1123(460 + 125)\left(1 - \frac{460 + 80}{460 + 125}\right);\quad V_2 = 682 \text{ ft/sec}$$

Check to see if sonic velocity is exceeded. Eq. 13.15: $c = \sqrt{kRT} = \sqrt{1.28(1123)(460+80)} = 881$ ft/sec

Since 628 = $V_2 < c$ = 881, the flow is subsonic, and V_2 = 682 ft/sec is correct. ◄

13.6.4 *Carbon dioxide flows isentropically. At a point in the flow the velocity is 15 m/s and the temperature is 50°C. At a second point on the same streamline the temperature is 25°C. What is the velocity at the second point? Check if your answer is valid.*

SI

Table A.5 for carbon dioxide: $R = 188$ m²/(s²·K), $k = 1.28$

Flow is isentropic, therefore adiabatic.

$$\text{Eq. 13.24:}\quad V_2^2 - 15^2 = \frac{2(1.28)}{0.28}188(273 + 50)\left(1 - \frac{273 + 25}{273 + 50}\right);\quad V_2 = 208 \text{ m/s}$$

Check to see if sonic velocity is exceeded. Eq. 13.15: $c = \sqrt{kRT} = \sqrt{1.28(188)(273+25)} = 268$ m/s

Since 208 = $V_2 < c$ = 286, the flow is subsonic, and V_2 = 208 m/s is correct. ◄

Sec. 13.6: Isentropic Flow -- *Problems*

13.3 N *Derive Eq. 13.35 for isentropic flow by integrating the Euler equation.*

Euler Eq. 13.9: $dp/\rho + VdV = 0$ (1)

Sec. 13.1 for isentropic flow: pv^k = constant = $C = p/\rho^k$ because $v = 1/\rho$ (Eq. 2.2)

$p = C\rho^k;\ dp = Ck\rho^{k-1}d\rho$ (2)

Substitute (2) into (1) and integrate: $Ck\rho^{k-2}d\rho + VdV = 0$

$$\frac{Ck\rho^{k-1}}{k-1}\Bigg|_{\rho_1}^{\rho_2} + \frac{V^2}{2}\Bigg|_{V_1}^{V_2} = 0;\quad \frac{V_2^2 - V_1^2}{2} = \frac{Ck}{k-1}\left(\rho_1^{k-1} - \rho_2^{k-1}\right) = \frac{k}{k-1}C\rho_1^{k-1}\left[1 - \left(\frac{\rho_2}{\rho_1}\right)^{k-1}\right]$$

We note that $\dfrac{p}{\rho} = C\rho^{k-1}$ and $\left(\dfrac{p}{C}\right)^{(k-1)/k} = \rho^{k-1}$

So substituting: $$\frac{V_2^2 - V_1^2}{2} = \frac{p_1}{\rho_1}\frac{k}{k-1}\left[1 - \left(\frac{p_2}{p_1}\right)^{(k-1)/k}\right]$$ (Eq. 13.35) Q.E.D. ◄

13.4 BG *Carbon dioxide flows isentropically. At a point in the flow the velocity is 50 ft/sec, the temperature is 125°F, and the pressure is 20 psia. Determine the pressure and temperature on the nose of a streamlined object placed in the flow at that point.*

Table A.5 for carbon dioxide: $R = 1123\ \text{ft}^2/(\text{sec}^2\cdot°\text{R})$, $k = 1.28$, $c_p = 5132\ \text{ft}^2/(\text{sec}^2\cdot°\text{R})$

Eq. 13.15: $c = \sqrt{kRT} = \sqrt{1.28(1123)(125 + 460)} = 917$ ft/sec

Eq. 13.10: $\mathbf{M} = V/c = 50/917 = 0.0545$

From Eq. 13.27: $p_o = 20[1 + 0.5(1.28 - 1)0.0545^2]^{1.28/0.28} = 20.0$ psia ◄

Alternatively, obtain $\rho = 0.004\,38\ \text{slug/ft}^3$ from Eq. 2.4, then find p_o from Eq. 13.30.

From Eq. 13.25: $T_o = T + V^2/(2 \times c_p) = (125 + 460) + 50^2/(2 \times 5132) = 585°\text{R} = 125°\text{F}$ ◄

Temperature and pressure increases are negligible.

Sec. 13.7: Effect of Area Variation on One-Dimensional Compressible Flow -- *Exercise*

13.7.1 N *Show in detail the development of Eq. (13.39) from Eqs. (13.38) and (13.37).*

Eq. 13.37: $\dfrac{dA}{A} + \dfrac{d\rho}{\rho} + \dfrac{dV}{V} = 0$; Eq. 13.38: $c^2\dfrac{d\rho}{\rho} + VdV = 0$ or $\dfrac{d\rho}{\rho} = -\dfrac{VdV}{c^2}$

Substituting into Eq. 13.37: $\dfrac{dA}{A} = \dfrac{VdV}{c^2} - \dfrac{dV}{V} = \dfrac{dV}{V}\left[\dfrac{V^2}{c^2} - 1\right] = \dfrac{dV}{V}(\mathbf{M}^2 - 1)$ using Eq. 13.10

Rearranging: $$\frac{dV}{V} = \frac{1}{\mathbf{M}^2 - 1}\frac{dA}{A}$$ (Eq. 13.39) Q.E.D. ◄

Sec. 13.8: Compressible Flow Through a Converging Nozzle -- *Exercises*

13.8.1 *Start with Eq. 13.40 and derive Eq. 13.43.*

N

Eq. 13.40: $\dfrac{V_2^2}{2} = \dfrac{p_2}{\rho_2}\dfrac{k}{k-1}\left[(p_O/p_2)^{(k-1)/k} - 1\right]$; Eq. 13.5: $\dot m = (\rho A V)_2$

$\therefore\ \dot m^2 = \rho_2^2 A_2^2 V_2^2 = A_2^2 2 p_2 \rho_2 \dfrac{k}{k-1}\left[(p_O/p_2)^{(k-1)/k} - 1\right]$

Isentropic flow, Sec. 13.1: $pv^k = \text{const}$; Eq. 2.2: $v = 1/\rho$

Eliminating v: $p/\rho^k = \text{const}$, from which $\dfrac{p_O}{\rho_O^k} = \dfrac{p_2}{\rho_2^k}$, thus $\rho_2 = \left(\dfrac{p_2}{p_O}\right)^{1/k}\rho_O$

Substituting for ρ_2: $\dot m^2 = A_2^2 2 p_2 (p_2/p_O)^{1/k}\rho_O \dfrac{k}{k-1}\left[(p_O/p_2)^{(k-1)/k} - 1\right]$

and noting that $p_2(p_2/p_O)^{1/k} = p_O^{-1/k} p_2^{(k+1)/k}$, we divide outside the bracketed term by $(p_2/p_O)^{(k+1)/k}$ and multiply inside by the same quantity to get

$$\dot m^2 = A_2^2 2 p_O \rho_O \frac{k}{k-1}\left[\left(\frac{p_2}{p_O}\right)^{2/k} - \left(\frac{p_2}{p_O}\right)^{(k+1)/k}\right] \quad \text{i.e.,}\quad \dot m = A_2\sqrt{\frac{2k}{k-1}p_O\rho_O\left[\left(\frac{p_2}{p_O}\right)^{2/k} - \left(\frac{p_2}{p_O}\right)^{(k+1)/k}\right]}$$

(Eq. 13.43) Q.E.D. ◄

13.8.2 *Differentiate Eq. 13.43 with respect to p_2/p_O and set to zero to find the value of p_2/p_O for which $\dot m$ is a maximum. The answer should correspond to Eq. 13.42.*

N

Eq. 13.43: $\dot m = A_2\sqrt{\dfrac{2k}{k-1}p_O\rho_O\left[\left(\dfrac{p_2}{p_O}\right)^{2/k} - \left(\dfrac{p_2}{p_O}\right)^{(k+1)/k}\right]}$ so $\dfrac{\dot m^2}{A^2 \rho_O p_O}\dfrac{k-1}{2k} = \left(\dfrac{p_2}{p_O}\right)^{2/k} - \left(\dfrac{p_2}{p_O}\right)^{(k+1)/k}$

$$d/d\left(\frac{p_2}{p_O}\right) \Rightarrow 0 = \frac{2}{k}\left(\frac{p_2}{p_O}\right)^{(2/k)-1} - \frac{k+1}{k}\left(\frac{p_2}{p_O}\right)^{\left(\frac{k+1}{k}\right)-1} = \frac{2}{k}\left(\frac{p_2}{p_O}\right)^{(2-k)/k} - \frac{k+1}{k}\left(\frac{p_2}{p_O}\right)^{1/k}$$

$$\frac{k+1}{k}\left(\frac{p_2}{p_O}\right)^{1/k} = \frac{2}{k}\left(\frac{p_2}{p_O}\right)^{(2-k)/k};\quad \left(\frac{p_2}{p_O}\right)^{(1/k)-(2-k)/k} = \frac{2}{k}\left(\frac{k}{k+1}\right);\quad \left(\frac{p_2}{p_O}\right)^{(k-1)/k} = \frac{2}{k+1}$$

$$\frac{p_2{*}}{p_O} = \left(\frac{2}{k+1}\right)^{k/(k-1)}$$ (Eq. 13.42) Q.E.D. ◄

13.8.3 *Carbon dioxide within a tank at 40 psia and 80°F discharges through a convergent nozzle into a 14.2 psia atmosphere. Find the velocity, pressure, and temperature at the nozzle outlet. Assume isentropic conditions.*

BG

Table A.5 for CO_2: $R = 1123\ \text{ft}^2/(\text{sec}^2\cdot{}^\circ\text{R})$, $k = 1.28$; Eq. 13.42: $p_2^*/p_O = (2/2.28)^{1.28/0.28} = 0.549$

Sec. 13.8: If $p_b > 0.549p_O = 0.549(40) = 22.0$ psia, the flow is subsonic.

But (given) $p_b = 14.2$ psia < 22.0 psia, $\therefore$ the flow is sonic.

Sec. 13.8: For sonic flow, $p_2/p_O = p_2^*/p_O$. $\therefore$ The pressure at the nozzle outlet is 22.0 psia ◄

From Eq. 13.36: $T_2 = (80 + 460)(22.0/40)^{0.28/1.28} = 474°\text{R} = 13.68°\text{F}$ ◄

Eq. 13.24 with $V_1 = V_O =$ zero: $V_2^2 = 2(1.28/0.28)(1123)(540)[1 - (474/540)]$; $V_2 = 825$ ft/sec ◄

13.8.4 *Carbon dioxide within a tank at 280 kPa abs and 25°C discharges through a convergent nozzle into a 98 kPa abs atmosphere. Find the velocity, pressure, and temperature at the nozzle outlet. Assume isentropic conditions.*

SI

Table A.5 for carbon dioxide: $R = 188\ \text{m}^2/(\text{s}^2\cdot\text{K})$, $k = 1.28$

Eq. 13.42: $p_2^*/p_o = (2/2.28)^{1.28/0.28} = 0.549$

Sec. 13.8: If $p_b > 0.549p_o = 0.549(280) = 153.8\ \text{kN/m}^2$ abs, the flow is subsonic.

But (given) $p_b = 98\ \text{kN/m}^2$ abs < 153.8, $\therefore$ the flow is sonic.

Sec. 13.8: For sonic flow, $p_2/p_o = p_2{}^*/p_o$. $\therefore$ The pressure at the nozzle outlet is 153.8 kPa abs ◄

From Eq. 13.36: $T_2 = (25 + 273)(153.8/280)^{0.28/1.28} = 261\text{K} = -11.60°\text{C}$ ◄

Eq. 13.24 with $V_1 = V_o =$ zero: $V_2^2 = 2(1.28/0.28)(188)(298)[1 - (261/298)]$; $V_2 = 251$ m/s ◄

13.8.5 *In Exer. 13.8.3 if the pressure and temperature within the tank had been 20 psia and 100°F, what would have been the velocity, pressure, and temperature at the nozzle outlet? Assume isentropic conditions.*
Exer. 13.8.3: Carbon dioxide flows through a convergent nozzle, $p_b = 14.2$ psia.

BG

Table A.5 for carbon dioxide: $R = 1123\ \text{ft}^2/(\text{sec}^2\cdot°\text{R})$, $k = 1.28$

Eq. 13.42: $p_2^*/p_o = (2/2.28)^{1.28/0.28} = 0.549$

Sec. 13.8: If $p_b > 0.549p_o = 0.549(20) = 10.99$ psia, the flow is subsonic.

But (given) $p_b = 14.2$ psia > 10.99 psia, $\therefore$ the flow is subsonic.

Sec. 13.8: With subsonic flow $p_2 = p_b = 14.2$ psia ◄

From Eq. 13.36: $T_2 = (100 + 460)(14.2/20)^{0.28/1.28} = 520°\text{R} = 59.6°\text{F}$ ◄

Eq. 13.24 with $V_1 = V_o =$ zero: $V_2^2 = 2(1.28/0.28)(1123)(560)[1 - (520/560)]$; $V_2 = 644$ ft/sec ◄

13.8.6 *In Exer. 13.8.4 if the pressure and temperature within the tank had been 140 kPa abs and 40°C, what would have been the velocity, pressure, and temperature at the nozzle outlet? Assume isentropic conditions.*
Exer. 13.8.4: Carbon dioxide flows through a convergent nozzle, $p_b = 98$ kPa abs.

SI

Table A.5 for carbon dioxide: $R = 188\ \text{m}^2/(\text{s}^2\cdot\text{K})$, $k = 1.28$

Eq. 13.42: $p_2^*/p_o = (2/2.28)^{1.28/0.28} = 0.549$

Sec. 13.8: If $p_b > 0.549p_o = 0.549(140) = 76.9\ \text{kN/m}^2$ abs, the flow is subsonic.

But (given) $p_b = 98\ \text{kN/m}^2$ abs > 76.9, $\therefore$ the flow is subsonic.

Sec. 13.8: With subsonic flow $p_2 = p_b = 98\ \text{kN/m}^2$ abs ◄

From Eq. 13.36: $T_2 = (40 + 273)(98/140)^{0.28/1.28} = 290\text{K} = 16.51°\text{C}$ ◄

Eq. 13.24 with $V_1 = V_o = 0$: $V_2^2 = 2(1.28/0.28)(188)(313)[1 - (290/313)]$; $V_2 = 201$ m/s ◄

13.5 *Air flows at 150°F from a large tank through a 1.5-in-diameter converging nozzle. Within the tank the pressure is 85 psia. Calculate the flow rate for back pressures of 10, 30, 50, and 70 psia. Assume isentropic conditions. Plot ṁ as a function of p_b. Assume that the temperature within the tank is 150°F in all cases. Compute also the temperature at the nozzle outlet for each condition.*

BG

Table A.5 for air: $R = 1715$ ft²/(sec²·°R), $k = 1.40$

Eq. 13.33 or 13.42: $p_2^*/p_o = (2/2.4)^{1.4/0.4} = 0.528$

Sec. 13.8: If $p_b > 0.528(85) = 44.9$ psia the flow is subsonic and $p_2 = p_b$

If $p_b < 0.528(85) = 44.9$ psia the flow is sonic and $p_2 = 44.9$ psia.

For sonic flows (i.e., $p_b < 44.9$ psia) $\dot{m}_{max}$ is given by Eq. 13.45. $A_2 = (\pi/4)(1.5/12)^2 = 0.012\,27$ ft².

Eq. 13.45: $$\dot{m}_{max} = \frac{0.012\,27(85)144}{\sqrt{460 + 150}}\sqrt{\frac{1.4}{1715}\left(\frac{2}{2.4}\right)^{2.4/0.4}} = 0.1006 \text{ slug/sec} \quad ◄$$

For subsonic flows (i.e., $p_b > 44.9$ psia) $\dot{m}$ is given by Eq. 13.43

From Eq. 2.4: $\rho_o = p_o/RT_o = 85(144)/(1715 \times 610) = 0.011\,70$ slug/ft³

Eq. 13.43: $$\dot{m} = 0.012\,27\sqrt{\frac{2(1.4)}{0.4}85(144)0.011\,70\left[\left(\frac{p_2}{85}\right)^{2/1.4} - \left(\frac{p_2}{85}\right)^{2.4/1.4}\right]} \quad \text{where } p_2 = p_b$$

Temperatures, from Eq. 13.36: $T_2 = (460 + 150)(p_2/85)^{0.4/1.4}$ °R

Results:

p_o (psia)	T_o (°F)	p_b (psia)	p_2 (psia)	Flow Eq.	$\dot{m}$ or $\dot{m}_{max}$ (slug/sec)	T_2 (°F)
85	150	10	44.9	sonic 13.45	0.1006	48.3
85	150	30	44.9	sonic 13.45	0.1006	48.3
85	150	50	50	subsonic 13.43	0.0998	64.2
85	150	70	70	subsonic 13.43	0.0786	117.1
85	150	85	85	subsonic 13.43	0	150.0
					▲	▲

13.6 *Air flows at 65°C from a large tank through a 4-cm-diameter converging nozzle. Within the tank the pressure is 600 kPa abs. Calculate the flow rate for back pressures of 50, 200, 350, and 500 kPa abs. Assume isentropic conditions. Plot ṁ as a function of p_b. Assume that the temperature within the tank is 65°C in all cases. Compute also the temperature at the nozzle outlet for each condition.*

SI

Table A.5 for air: $R = 287$ m²/(s²·K), $k = 1.40$

Eq. 13.33 or 13.42: $p_2^*/p_o = (2/2.4)^{1.4/0.4} = 0.528$

Sec. 13.8: If $p_b > 0.528(600) = 317$ kPa abs the flow is subsonic and $p_2 = p_b$

If $p_b < 0.528(600) = 317$ kPa abs the flow is sonic and $p_2 = 317$ kPa abs.

For sonic flows (i.e., $p_b < 317$ kPa abs) $\dot{m}_{max}$ is given by Eq. 13.45. $A_2 = (\pi/4)0.04^2 = 0.001\,257$ m²

Eq. 13.45: $$\dot{m}_{max} = 0.001\,257\frac{600\,000}{\sqrt{273 + 65}}\sqrt{\frac{1.4}{287}\left(\frac{2}{2.4}\right)^{2.4/0.4}} = 1.658 \text{ kg/s} \quad ◄$$

/cont...

For subsonic flows (i.e., p_b > 317 kPa abs) $\dot{m}$ is given by Eq. 13.43

From Eq. 2.4: $\rho_o = p_o/RT_o = 600\,000/(287 \times 338) = 6.19$ kg/m^3

Eq. 13.43: $\dot{m} = 0.001\,257\sqrt{\frac{2(1.4)}{0.4}(600\,000)6.19\left[\left(\frac{p_2}{600}\right)^{2/1.4} - \left(\frac{p_2}{600}\right)^{2.4/1.4}\right]}$ where $p_2 = p_b$

Temperatures, from Eq. 13.36: $T_2 = (273 + 65)(p_2/600)^{0.4/1.4}$ K

Results:

p_o (kPa abs)	T_o (°C)	p_b (kPa abs)	p_2 (kPa abs)	Flow Eq.	$\dot{m}$ or $\dot{m}_{max}$ (kg/s)	T_2 (°C)
600	65	50	317	sonic 13.45	1.658	8.67
600	65	200	317	sonic 13.45	1.658	8.67
600	65	350	350	subsonic 13.43	1.647	16.76
600	65	500	500	subsonic 13.43	1.267	47.8
600	65	600	600	subsonic 13.43	0	65.0

▲ ▲

13.7 *Air flows at 25°C from a large tank through a 10-cm-diameter converging nozzle. Within the tank the pressure is 50 kPa abs. Calculate the flow rate for back pressures of 30, 20, and 10 kPa abs. Assume isentropic conditions. Plot $\dot{m}$ as a function of p_b. Assume that the temperature within the tank is 25°C in all cases. Compute also the temperature at the nozzle outlet for each condition.*

SI

Table A.5 for air: $R = 287$ m^2/(s$^2\cdot$K), $k = 1.40$

Eq. 13.33 or 13.42: $p_2^*/p_o = (2/2.4)^{1.4/0.4} = 0.528$

Sec. 13.8: If $p_b > 0.528(50) = 26.4$ kPa abs the flow is subsonic and $p_2 = p_b$

If $p_b < 0.528(50) = 26.4$ kPa abs the flow is sonic and $p_2 = 26.4$ kPa abs.

For sonic flows (i.e., p_b < 26.4 kPa abs) $\dot{m}_{max}$ is given by Eq. 13.45. $A_2 = (\pi/4)0.10^2 = 0.007\,85$ m^2

Eq. 13.45: $\dot{m}_{max} = 0.007\,85\,\frac{50\,000}{\sqrt{273+25}}\sqrt{\frac{1.4}{287}\left(\frac{2}{2.4}\right)^{2.4/0.4}} = 0.919$ kg/s

For subsonic flows (i.e., p_b > 26.4 kPa abs) $\dot{m}$ is given by Eq. 13.43.

From Eq. 2.4: $\rho_o = p_o/RT_o = 50\,000/(287 \times 298) = 0.585$ kg/m^3

Eq. 13.43: $\dot{m} = 0.007\,85\sqrt{\frac{2(1.4)}{0.4}50\,000(0.585)\left[\left(\frac{p_2}{50}\right)^{2/1.4} - \left(\frac{p_2}{50}\right)^{2.4/1.4}\right]}$

Results:

p_o (kPa, abs)	T_o (°C)	p_b (kPa, abs)	p_2 (kPa, abs)	Flow Eq.	$\dot{m}$ or $\dot{m}_{max}$ (kg/s)	T_2 (°C)
50	25	10	26.4	sonic 13.45	0.919	−24.6
50	25	20	26.4	sonic 13.45	0.919	−24.6
50	25	30	30	subsonic 13.43	0.909	−15.47

▲ ▲

13.8 *Air within a tank at 120°F flows isentropically through a 2-in-diameter convergent nozzle into a 14.2-psia atmosphere. Find the flow rate for air pressures within the tank of 5, 10, 20, 40, and 50 psia.*

BG

Table A.5 for air: $R = 1715 \text{ ft}^2/(\text{sec}^2\cdot{}^\circ\text{R})$, $k = 1.40$

Eq. 13.33 or 13.42: $p_2^*/p_o = (2/2.4)^{1.4/0.4} = 0.528$; $p_2 = 14.2$ psia (given)

Sec. 13.8: If $p_b/p_o > 0.528$, i.e., $p_o < 14.2/0.528 = 26.9$ psia, the flow is subsonic and $p_2 = p_b$,

$\dot{m}$ is given by Eq. 13.43.

Sec. 13.8: If $p_b/p_o \le 0.528$, i.e., $p_o \ge 14.2/0.528 = 26.9$ psia, the flow is sonic and $p_2 = 0.528p_o$

For sonic flow (i.e., $p_o \ge 26.9$ psia), $\dot{m}_{max}$ is given by Eq. 13.45. $A_2 = (\pi/4)(2/12)^2 = 0.0218 \text{ ft}^2$.

Eq. 13.45: $$\dot{m}_{max} = \frac{0.0218(144p_o)}{\sqrt{580}}\sqrt{\frac{1.4}{1715}\left(\frac{2}{2.4}\right)^{2.4/0.4}} = 0.002\,16p_o \text{ slug/sec} \quad ◄$$

For subsonic flows (i.e., $p_o < 26.9$ psia) $\dot{m}$ is given by Eq. 13.43.

From Eq. 2.4: $\rho_o = p_o/RT_o = 144p_o/(1715 \times 580)$

Eq. 13.43: $$\dot{m} = 0.0218\sqrt{2\left(\frac{1.4}{0.4}\right)\frac{(144p_o)^2}{1715(580)}\left[\left(\frac{14.2}{p_o}\right)^{2/1.4} - \left(\frac{14.2}{p_o}\right)^{2.4/1.4}\right]} \text{ slug/sec}$$

If $p_o < 14.2$ psia (negative gage pressure) reverse flow will occur. Results are:

p_o (psig)	p_o (psia)	T_o (°F)	p_b (psia)	p_2 (psia)	Flow Eq.	$\dot{m}$ or $\dot{m}_{max}$ (slug/sec)
0	14.2	120	14.2	14.2	–	0
5	19.2	120	14.2	14.2	subsonic 13.43	0.0371
10	24.2	120	14.2	14.2	subsonic 13.43	0.0518
12.7	26.9	120	14.2	14.2	sonic 13.45	0.0580
20	34.2	120	14.2	18.07	sonic 13.45	0.0738
40	54.2	120	14.2	28.6	sonic 13.45	0.1169

▲

13.9 *Refer to Sample Prob. 13.4. If the pressure in the tank is 5 psig, confirm *by calculation that $\dot{m} = 0.0374$ slug/sec, $p_2 =$ *13.50 psia, and $T_2 = 494°R$. [Note these corrections to the textbook.]*
Sample Prob. 13.4: $D = 2$ in, $T_o = 80°F = 540°R$, $p_b = 13.5$ psia, $p_2^*/p_o = 0.528$.

BG

Table A.5 for air: $R = 1715 \text{ ft}^2/(\text{sec}^2\cdot{}^\circ\text{R})$, $k = 1.40$

$p_o = 5$ psig; $p_o = 5 + p_{atm} = 5 + 13.5 = 18.50$ psia. $A_2 = (\pi/4)(2/12)^2 = 0.0218 \text{ ft}^2$.

From Eq. 2.4: $\rho_o = p_o/RT_o = 18.50(144)/(1715 \times 540) = 0.002\,88 \text{ slug/ft}^3$

$p_b/p_o = 13.5/18.5 = 0.730 > 0.528 = p_2{*}/p_o$

so flow is subsonic, ($p_2 = p_b$), and $\dot{m}$ is given by Eq. 13.43.

$$\dot{m} = 0.0218\sqrt{2\frac{1.4}{0.4}(18.5)144(0.002\,88)\left[\left(\frac{13.5}{18.5}\right)^{2/1.4} - \left(\frac{13.5}{18.5}\right)^{2.4/1.4}\right]} = 0.0374 \text{ slug/sec} \quad ◄$$

/cont...

(1) $V_2^2/2 = (p_2/\rho_2)[k/(k-1)]\left[(p_O/p_2)^{(k-1)/k} - 1\right]$ (Eq. 13.40)

(2) $\dot{m} = \rho_2 A_2 V_2$ (Eq. 13.5)

(3) $\rho_2 = p_2/RT_2$ (from Eq. 2.4)

(4) $p_O/\rho_O^k = p_2/\rho_2^k$ (from pv^k = const and $v = 1/\rho$, Sec. 13.1 and Eq. 2.2)

From (4): $p_2 = p_O\dfrac{\rho_2^k}{\rho_O^k} = (18.5 \times 144)\dfrac{\rho_2^k}{0.002\,88^{1.4}} = 9{,}620{,}000\rho_2^k$ (5)

From (2) $\rho_2 = \dot{m}/(A_2V_2) = 0.0374/(0.0218V_2) = 1.716/V_2$ (6)

Substitute for ρ_2 from (6) into (5): $p_2 = 9{,}620{,}000(1.716/V_2)^{1.4} = 20.5 \times 10^6/V_2^{1.4}$ (7)

Substitute into (1): $V_2^2 = 2\dfrac{20.5 \times 10^6(1.4/0.4)}{V_2^{1.4}(1.716/V_2)}\left[\left(\dfrac{18.5(144)V_2^{1.4}}{20.5 \times 10^6}\right)^{0.4/1.4} - 1\right]$

or $V_2^{2.4} = 83.6 \times 10^6[0.0776V_2^{0.4} - 1]$; by T and E $V_2 = 747$ ft/sec

From (2): $\rho_2 = \dot{m}/(A_2V_2) = 0.0374/(0.0218 \times 747) = 0.002\,30$ slug/ft^3

From (4): $p_2 = (18.5)0.002\,30^{1.4}/0.002\,88^{1.4} = 13.50$ psia ◄

From (3): $T_2 = p_2/(R\rho_2) = (13.50 \times 144)/(1715 \times 0.002\,30) = 494°$R ◄

Sec. 13.9: Isentropic Flow Through a Converging-Diverging Nozzle -- *Exercises*

13.9.1 *Air enters a converging-diverging nozzle at a pressure of 120 psia and a temperature of 90°F. Neglecting the entrance velocity and assuming a frictionless process, find the Mach number at the cross section where the pressure is 35 psia.*

BG

Table A.5 for air: $R = 1715$ ft^2/(sec$^2\cdot$°R), $k = 1.40$

From pv^k = const and $pv = RT$: $T_2 = T_O(p_2/p_O)^{(k-1)/k} = (460 + 90)(35/120)^{0.4/1.4} = 387°$R

Eq. 13.24 with $V_1 = 0$: $V_2^2 = 2(1.4/0.40)1715(460 + 90)[1 - (387/550)]$; $V_2 = 1400$ ft/sec

Eq. 13.15: $c = \sqrt{kRT_2} = \sqrt{1.4(1715)387} = 964$ ft/sec

Eq. 13.10: $\mathbf{M} = V/c = 1400/964 = 1.453$ ◄

13.9.2 *Air enters a converging-diverging nozzle at a pressure of 830 kPa abs and a temperature of 32°C. Neglecting the entrance velocity and assuming a frictionless process, find the Mach number at the cross section where the pressure is 240 kPa abs.*

SI

Table A.5 for air: $R = 287$ m^2/(s$^2\cdot$K), $k = 1.40$

From pv^k = const and $pv = RT$: $T_2 = T_O(p_2/p_O)^{(k-1)/k} = (273 + 32)(240/830)^{0.4/1.4} = 214$ K

Eq. 13.24 with $V_1 = 0$: $V_2^2 = 2(1.40/0.40)287(273 + 32)[1 - (214/305)]$; $V_2 = 428$ m/s

Eq. 13.15: $c = \sqrt{kRT} = \sqrt{1.4(287)214} = 293$ m/s

Eq. 13.10: $\mathbf{M} = V_2/c = 428/293 = 1.459$ ◄

13.9.3 *Work Sample Prob. 13.5 with all data the same except for the pressure within the tank, which is 100 rather than 50 psia.*

Sample Prob. 13.5: $T_O = 80°F = 540°R$, $D_2 = 2$ *in,* $A_2 = 0.0218$ ft^2, $p_3 = p_b = 13.5$ *psia. Find* D_3, $\dot{m}$ $(=0.0118$ *slug/sec),* V_2, V_3, T_2, T_3. *Assume isentropic flow.*

BG

Table A.5 for air: $R = 1715$ ft²/(sec²·°R), $k = 1.40$

Pressure at throat (where $V_2 = c$): $p_2 = p_2^* = 0.528p_O = 0.528(100) = 52.8$ psia

Eq. 13.35 (isentropic, given): $V_3^2 = 2(1.4/0.4)1715(540)[1 - (13.5/100)^{0.4/1.4}]$; $V_3 = 1681$ ft/sec ◀

Eq. 13.45: $\dot{m}_{max} = 0.224$ slug/sec (exactly twice that of Sample Prob. 13.5) ◀

Eq. 13.24 with $V_1 = 0$: $1681^2 = 2(1.4/0.4)(1715)540(1 - T_3/540)$; $T_3 = 305°R = -155.3°F$ ◀

From Eq. 2.4: $\rho_3 = p_3/(RT_3) = (13.5 \times 144)/(1715 \times 305) = 0.003\,72$ slug/ft³

p/ρ^k = constant, (isentropic, Sec. 13.1), ∴ $52.8/\rho_2^{1.4} = 13.5/0.003\,72^{1.4}$; $\rho_2 = 0.009\,86$ slug/ft³

From Eq. 13.5: $V_2 = \dot{m}/(\rho_2 A_2) = 0.224/(0.009\,84 \times 0.0218) = 1039$ ft/sec ◀

Eq. 13.24 with $V_1 = 0$: $1039^2 = 2(1.4/0.4)(1715)540(1 - T_2/540)$; $T_2 = 450°R = -10.00°F$ ◀

Finally, from Eq. 13.5: $A_3 = \dot{m}/(\rho_3 V_3) = 0.224/(0.003\,72 \times 1681) = 0.0358$ ft² $= 5.15$ in² $= \pi D_3^2/4$

so $D_3 = 2.56$ in ◀

13.9.4 *Air is to flow through a converging-diverging nozzle at 0.6 slug/sec. At the throat the pressure, temperature, and velocity are to be 20 psia, 100°F and 500 ft/sec, respectively. At outlet the velocity is to be 200 ft/sec. Determine the throat diameter. Assume isentropic flow.*

BG

Table A.5 for air: $R = 1715$ ft·lb/(slug·°R)

From Eq. 2.4: $\rho_2 = p_2/(RT_2) = 20(144)/(1715 \times 560) = 0.003\,00$ slug/ft³

Eq. 13.5: $\dot{m} = 0.6 = \rho_2 A_2 V_2 = 0.003\,00(A_2)500$

$A_2 = 0.400$ ft² $= 57.6$ in² $= \pi D_2^2/4$; $D_2 = 8.57$ in ◀

13.9.5 *Air in a tank at a pressure of 140 psia and 70°F flows out into the atmosphere through a 1.00-in-diameter converging nozzle. (a) Find the mass flow rate. (b) If a diverging section with an outlet diameter of 1.50 in were attached to the converging nozzle, what then would be the flow rate? Neglect friction.*

BG

Table A.5 for air: $R = 1715$ ft²/(sec²·°R), $k = 1.40$. Assume $p_b = p_{atm} = 14.70$ psia.

(*a*) Sec. 13.8: $p_b/p_O = 14.7/140 = 0.105 < 0.528$; so flow at the throat is sonic; use Eq. 13.45:

$$\dot{m}_{max} = \frac{\pi}{4}\left(\frac{1}{12}\right)^2 \frac{(140)144}{\sqrt{460 + 70}} \sqrt{\frac{1.4}{1715}\left(\frac{2}{2.4}\right)^{2.4/0.4}} = 0.0790 \text{ slug/sec}$$ ◀

(*b*) Sec. 13.9: With sonic flow (and no change in Δp), attaching a diverging nozzle will not change the flow rate. ◀

13.9.6 *Air in a tank at a pressure of 950 kPa abs and 20°C flows out into the atmosphere through a 2.50-cm-diameter converging nozzle. (a) Find the mass flow rate. (b) If a diverging section with an outlet diameter of 4 cm were attached to the converging nozzle, what then would be the flow rate? Neglect friction.*

SI

Table A.5 for air: $R = 287\ \text{m}^2/(\text{s}^2\cdot\text{K})$, $k = 1.40$. Assume $p_b = p_{atm} = 101$ kPa abs.

(*a*) Sec. 13.8: $p_b/p_o = 101/950 = 0.106 < 0.528$; so flow at the throat is sonic; use Eq. 13.45:

$$\dot{m}_{max} = \frac{\pi}{4}\left(\frac{2.50}{100}\right)^2 \frac{950\,000}{\sqrt{273 + 20}}\sqrt{\frac{1.4}{287}\left(\frac{2}{2.4}\right)^{2.4/0.4}} = 1.101\ \text{kg/s} \quad ◀$$

(*b*) Sec. 13.9: With sonic flow (and no change in Δp), attaching a diverging nozzle will not change the flow rate. ◀

Sec. 13.9: Isentropic Flow Through a Converging-Diverging Nozzle -- *Problems*

□ 13.10 *Air discharges from a large tank through a converging-diverging nozzle. The throat diameter is 3.0 in and the exit diameter is 4.0 in. Within the tank the air pressure and temperature are 40 psia and 150°F respectively. Calculate the flow rate for back pressures of 39, 38, 36, and 30 psia. Assume no friction.*

BG

Table A.5 for air: $R = 1715\ \text{ft}^2/(\text{sec}^2\cdot°\text{R})$, $k = 1.40$

From Eq. 2.4: $\rho_o = p_o/(RT_o) = 40(144)/(1715 \times 610) = 0.005\,51\ \text{slug/ft}^3$

If sonic flow occurs, Eq. 13.45: $\dot{m}_{max} = \frac{\pi}{4}\left(\frac{3}{12}\right)^2 \frac{40(144)}{\sqrt{460 + 150}}\sqrt{\frac{1.4}{1715}\left(\frac{2.0}{2.4}\right)^{2.4/0.4}} = 0.1893\ \text{slug/sec}$

For subsonic flow: Eq. 13.43: $\dot{m} = \frac{\pi}{4}\left(\frac{4}{12}\right)^2\sqrt{2\left(\frac{1.4}{0.4}\right)(40)144(0.005\,51)\left[\left(\frac{p_b}{40}\right)^{2/1.4} - \left(\frac{p_b}{40}\right)^{2.4/1.4}\right]}$

By T and E using Eq. 13.43, at threshold point with $\dot{m} = 0.1893$ slug/sec, $p_b = p_2 = 36.75$ psia.

p_b (psia)	40	39	38	36.75	36	30	
$\dot{m}$ (slug/sec), Eq. 13.43	0	0.1084	0.1512	0.1893	--	--	◀◀
$\dot{m}_{max}$ (slug/sec), Eq. 13.45	--	--	--	0.1893	0.1893	0.1893	◀◀

□ 13.11 *Air discharges from a large tank through a converging-diverging nozzle. The throat diameter is 7.5 cm and the exit diameter is 10 cm. Within the tank the air pressure and temperature are 290 kPa and 65°C, respectively. Calculate the flow rate for back pressures of 280, 270, 250, and 200 kPa abs. Assume no friction.*

SI

Table A.5 for air: $R = 287\ \text{m}^2/(\text{s}^2\cdot\text{K})$, $k = 1.40$. Eq. 2.4: $\rho_o = p_o/(RT_o) = 290\,000/(287 \times 338) = 2.99$

If sonic flow occurs, Eq. 13.45: $\dot{m}_{max} = \frac{\pi}{4}\left(\frac{7.5}{100}\right)^2 \frac{290\,000}{\sqrt{273 + 65}}\sqrt{\frac{1.4}{287}\left(\frac{2}{2.4}\right)^{2.4/0.4}} = 2.82\ \text{kg/s}$

For subsonic flow, Eq. 13.43: $\dot{m} = \pi 0.05^2\sqrt{2\left(\frac{1.4}{0.4}\right)290\,000 \times 2.99\left[\left(\frac{p_b}{290}\right)^{2/1.4} - \left(\frac{p_b}{290}\right)^{2.4/1.4}\right]}$

By T and E using Eq. 13.43, at threshold point with $\dot{m} = 2.82$ kg/s, $p_b = p_2 = 266$ kPa abs.

p_b (kPa abs)	290	280	270	266	250	200	
$\dot{m}$ (kg/sec), Eq. 13.43	0	1.885	2.61	2.82	--	--	◀◀
$\dot{m}_{max}$ (kg/sec), Eq. 13.45	--	--	--	2.82	2.82	2.82	◀◀

13.12 *Repeat Exer. 13.9.5 for the case where the air within the tank is at 20 psia. Assume all other data to be the same.*

Exer. 13.9.5: Converging nozzle with D_2 = 1.00 in, air T_O = 70°F = 530°R. (a) Find $\dot{m}$. (b) Find $\dot{m}$ after a diverging section with D_3 = 1.50 in is attached.

BG

Table A.5 for air: $R = 1715$ ft²/(sec²·°R), $k = 1.40$. Assume $p_b = p_{atm} = 14.70$ psia.

(*a*) Sec. 13.8: $p_b/p_O = 14.7/20 = 0.735 > 0.528$; so flow at the throat is subsonic, $p_2 = p_b$, $\dot{m}$ is given by Eq. 13.43.

From Eq. 2.4: $\rho_O = p_O/(RT_O) = (20 \times 144)/(1715 \times 530) = 0.003\,17$ slug/ft³

Eq. 13.43: $\dot{m} = \frac{\pi}{4}\left(\frac{1}{12}\right)^2\sqrt{2\frac{1.4}{0.4}(20 \times 144)0.003\,17\left[0.735^{2/1.4} - 0.735^{2.4/1.4}\right]} = 0.01015$ slug/sec ◀

(*b*) Sec. 13.9: Attaching a diverging nozzle with subsonic flow (when the air behaves like an incompressible fluid) will increase the flow rate. It only changes A_2 in Eq. 13.43.

Thus in Eq. 13.43: $\dot{m} = 0.01015(1.5/1.0)^2 = 0.0228$ slug/sec providing subsonic flow still occurs in the throat. Check this by computing $\dot{m}_{max}$, which is the maximum $\dot{m}$ if p_O is not increased.

Eq. 13.45: $\dot{m}_{max} = \frac{\pi}{4}(1.0)^2\frac{20}{\sqrt{460 + 70}}\sqrt{\frac{1.4}{1715}\left(\frac{2}{2.4}\right)^{2.4/0.4}} = 0.011\,28$ slug/sec

0.0228 slug/sec exceeds $\dot{m}_{max} = 0.011\,28$, ∴ with the 1.5-in-diameter outlet the flow at the throat has become sonic, and $\dot{m} = 0.011\,28$ slug/sec. ◀

13.13 *Repeat Exer. 13.9.6 for the case where the air within the tank is at 140 kPa abs. Assume all other data to be the same.*

Exer. 13.9.6: Converging nozzle with D_2 = 2.50 cm, air T_O = 20°C = 293 K. (a) Find $\dot{m}$. (b) Find $\dot{m}$ after a diverging section with D_3 = 4.00 cm is attached.

SI

Table A.5 for air: $R = 287$ m²/(s²·K), $k = 1.40$. Assume $p_b = p_{atm} = 101$ kPa abs.

(*a*) Sec 13.8: $p_b/p_O = 101/140 = 0.721 > 0.528$; so flow at the throat is subsonic, $p_2 = p_b$, $\dot{m}$ is given by Eq. 13.43.

From Eq. 2.4: $\rho_O = p_O/(RT_O) = 140\,000/(287 \times 293) = 1.665$ kg/m³

Eq. 13.43: $\dot{m} = \frac{\pi}{4}\left(\frac{2.50}{100}\right)^2\sqrt{2\frac{1.4}{0.4}(140\,000)1.665\left[0.721^{2/1.4} - 0.721^{2.4/1.4}\right]} = 0.1482$ kg/s ◀

(*b*) Sec. 13.9: Attaching a diverging nozzle with subsonic flow (when the air behaves like an incompressible fluid) will increase the flow rate. It only changes A_2 in Eq. 13.43.

Thus in Eq. 13.43: $\dot{m} = 0.1482(4/2.5)^2 = 0.379$ kg/s providing subsonic flow still occurs in the throat. Check this by computing $\dot{m}_{max}$, which is the maximum $\dot{m}$ if p_O is not increased.

Eq. 13.45: $\dot{m}_{max} = \frac{\pi}{4}\left(\frac{2.5}{100}\right)^2\frac{140\,000}{\sqrt{273 + 20}}\sqrt{\frac{1.4}{287}\left(\frac{2}{2.4}\right)^{2.4/0.4}} = 0.1623$ kg/s

0.379 kg/s exceeds $\dot{m}_{max} = 0.1623$ kg/s, ∴ with the 4-cm-diameter outlet the flow at the throat has become sonic, and $\dot{m} = 0.1623$ kg/s. ◀

Sec. 13.10: One Dimensional Shock Wave -- *Exercises*

13.10.1 *Just downstream of a normal shock wave the pressure, velocity, and temperature are 52 psia, 400 ft/sec and 120°F. Compute the Mach number upstream of the shock wave. Consider (a) air and (b) carbon dioxide as the working fluids.*

BG

(*a*) Table A.5 for air: $R = 1715\ \text{ft}^2/(\text{s}^2\cdot{}^\circ\text{R})$, $k = 1.40$

Eq. 13.15: $c_2 = \sqrt{1.4(1715)(460 + 120)} = 1180$ ft/sec; Eq. 13.10: $\mathbf{M}_2 = 400/1180 = 0.339$

Eq. 13.52: $\left(2 + 0.4\mathbf{M}_1^2\right)/\left(2(1.4)\mathbf{M}_1^2 - 0.4\right) = \mathbf{M}_2^2 = (0.339)^2$; $\mathbf{M}_1^2 = -26.1$ (impossible)

∴ This situation cannot happen with air. ◀

(*b*) Table A.5 for carbon dioxide: $R = 1123\ \text{ft}^2/(\text{s}^2\cdot{}^\circ\text{R})$, $k = 1.28$

Eq. 13.15: $c_2 = \sqrt{1.28(1123)(460 + 120)} = 913$ ft/sec; $\mathbf{M}_2 = 400/913 = 0.438$

Eq. 13.52: $\left(2 + 0.28\mathbf{M}_1^2\right)/\left(2(1.28)\mathbf{M}_1^2 - 0.28\right) = \mathbf{M}_2^2 = (0.438)^2$; $\mathbf{M}_1^2 = 9.72$; $\mathbf{M}_1 = 3.12$ ◀

13.10.2 *Just downstream of a normal shock wave the pressure, velocity, and temperature are 360 kPa abs, 110 m/s and 50°C. Compute the Mach number upstream of the shock wave. Consider (a) air and (b) carbon dioxide as the working fluids.*

SI

(*a*) Table A.5 for air: $R = 287\ \text{m}^2/(\text{s}^2\cdot\text{K})$, $k = 1.40$

Eq. 13.15: $c_2 = \sqrt{1.40(287)(273 + 50)} = 360$ m/s; Eq. 13.10: $\mathbf{M}_2 = 110/360 = 0.305$

From Eq. 13.52: $\mathbf{M}_1{}^2 = (2 + 0.4\mathbf{M}_2{}^2)/(2 \times 1.4\mathbf{M}_2{}^2 - 0.4) = -14.66$ (impossible)

∴ This situation cannot happen with air. ◀

(*b*) Table A.5 for carbon dioxide: $R = 188\ \text{m}^2/(\text{s}^2\cdot\text{K})$, $k = 1.28$

Eq. 13.15: $c_2 = \sqrt{1.28(188)(273 + 50)} = 279$ m/s; Eq. 13.10: $\mathbf{M}_2 = 110/279 = 0.395$

From Eq. 13.52: $\mathbf{M}_1^2 = (2 + 0.28\mathbf{M}_2^2)/(2 \times 1.28\mathbf{M}_2^2 - 0.28) = 17.24$; $\mathbf{M}_1 = 4.15$ ◀

Sec. 13.10: One Dimensional Shock Wave -- *Problems*

13.14 *Air discharges from a large tank through a converging-diverging nozzle with a 2.5-cm-diameter throat into the atmosphere. The gage pressure and temperature in the tank are 700 kPa and 40°C respectively, the barometric pressure is 995 millibars. (a) Find the nozzle tip diameter required for p_3 to be equal to the atmospheric pressure. For this case, what are the flow velocity, sonic velocity, and Mach number at the nozzle exit? (b) Determine the value of p_b that will cause the shock wave to be located at the nozzle exit.*

SI

Table A.5 for air: $R = 287\ \text{m}^2/(\text{s}^2\cdot\text{K})$, $c_p = 1003\ \text{m}^2/(\text{s}^2\cdot\text{K})$, $k = 1.40$

$p_{atm} = 995$ millibars $= 99.5$ kPa abs; $p_o = 700 + 99.5 = 799.5$ kPa abs; $T_o = 40 + 273 = 313$ K

(*a*) Assume sonic velocity at the throat:

Eq. 13.35 with $V_1 = 0$: $\dfrac{V_3^2}{2} = 287(313)\dfrac{1.4}{0.4}\left[1 - \left(\dfrac{99.5}{799.5}\right)^{0.4/1.4}\right]$; $V_3^2 = 282\ 000$; $V_3 = 531$ m/s ◀

and Eq. 13.45: $\dot{m}_{max} = \dfrac{(\pi \times 0.0125^2)799\ 500}{\sqrt{273 + 40}}\sqrt{\dfrac{1.4}{287}\left(\dfrac{2}{2.4}\right)^{2.4/0.4}} = 0.897$ kg/s

Eq. 13.23: $V_3^2 - V_1^2 = 2c_p(T_1 - T_3)$; $531^2 - 0 = 2(1003)(313 - T_3)$; $T_3 = 172.4$ K

From Eq. 2.4: $\rho_3 = \dfrac{p_3}{RT_3} = \dfrac{99\ 500}{287(172.4)} = 2.01\ \text{kg/m}^3$ /cont...

From Eq. 13.5: $A_3 = \dot{m}/(\rho_3 V_3) = 0.897/(2.01 \times 531) = 0.000\ 839\ \text{m}^2 = 8.39\ \text{cm}^2 = (\pi/4)D_3^2$

$D_3 = 3.27$ cm ◄

Eq. 13.15: $c_3 = \sqrt{kRT_3} = \sqrt{1.4(287)172.4} = 263$ m/s ◄

Eq. 13.10: $\mathbf{M}_3 = V_3/c_3 = 531/263 = 2.02$ ◄

Thus flow is supersonic at the nozzle exit, sonic velocity occurs at the throat, and the shock wave is located downstream of the nozzle tip.

(*b*) Across a shock wave, Eq. 13.49: $p_2/p_1 = \left[2k\mathbf{M}_1^2 - (k - 1)\right]/(k + 1)$

where Section (1) is just inside the nozzle tip and Section (2) is just outside the tip.

Relating this to Fig. 13.6 we have $\dfrac{p_b}{p_3} = \dfrac{p_2}{p_1} = \dfrac{2(1.4)2.02^2 - 0.4}{2.4} = 4.59$

$p_b = 4.59 p_3 = 4.59(99.5) = 456$ kPa abs ◄

13.15 *The pressure, velocity, and temperature just upstream of a normal shock wave in air are 10 psia, 2200 fps, and 23°F. Determine the pressure, velocity, and temperature just downstream of the wave.*

BG

Table A.5 for air: $R = 1715\ \text{ft}^2/(\text{sec}^2\cdot{}^\circ\text{R})$, $k = 1.40$

From Eq. 2.4: $\rho_1 = p_1/(RT_1) = (10)144/(1715 \times 483) = 0.001\ 738\ \text{slug/ft}^3$

Eq. 13.15: $c_1 = \sqrt{kRT_1} = \sqrt{1.40(1715)483} = 1077$ ft/sec

Eq. 13.10: $\mathbf{M}_1 = 2200/1077 = 2.04$

Eq. 13.49: $p_2 = p_1\left[2(1.4)2.04^2 - 0.4\right]/2.4 = 10(4.70) = 47.0$ psia ◄

From Eq. 13.47 and continuity: $p_1 - p_2 = \rho_1 V_1 (V_2 - V_1)$

i.e., $V_2 = V_1 + \dfrac{p_1 - p_2}{\rho_1 V_1} = 2200 + \dfrac{(10 - 47.0)144}{(0.001\ 738)2200} = 806$ ft/sec ◄

From Eqs. 13.48 and 2.2: $806^2 - 2200^2 = \dfrac{2(1.4)}{0.4}\left(\dfrac{10(144)}{0.001\ 738} - \dfrac{47.0(144)}{\rho_2}\right)$; $\rho_2 = 0.004\ 75\ \text{slug/ft}^3$

From Eq. 2.4: $T_2 = \dfrac{p_2}{R\rho_2} = \dfrac{47.0(144)}{1715(0.004\ 75)} = 832^\circ\text{R} = 372^\circ\text{F}$ ◄

Sec. 13.11: The Oblique Shock Wave -- *Exercises*

13.11.1 *Assuming the tip of the model in Fig. 13.9 to be a point source of infinitesimal disturbance, find the air velocity if the temperature is −60°F and k = 1.4. If the actual Mach number is 1.38, what is the percentage error involved in the preceding assumption?*

BG

Table A.5 for air: $R = 1715\ \text{ft}^2/(\text{s}^2\cdot{}^\circ\text{R})$, $k = 1.40$

Eq. 13.15: $c = \sqrt{kRT} = \sqrt{1.40(1715)(460 - 60)} = 980$ ft/sec

At $\mathbf{M} = 1.38$ (given), true $V = 1.38(980) = 1352$ ft/sec (from Eq. 13.10)

Fig. 13.9: $\beta = 49^\circ$; Eq. 13.53: $\mathbf{M} = V/c = 1/\sin\beta = 1/\sin 49^\circ = 1.325$

$\therefore\ V = c\mathbf{M} = (980)1.325 = 1299$ ft/sec ◄

% error is $(1.38 - 1.325)100/1.38 = 3.98\%$ (low) ◄

13.11.2 *A schlieren photograph of a bullet shows a Mach angle of 40°. The air is at a pressure of 14 psia and 50°F. Find the approximate speed of the bullet.*

BG

Table A.5 for air: $R = 1715\ \text{ft}^2/(\text{s}^2\cdot°\text{R})$, $k = 1.40$

Eq. 13.15: $c = \sqrt{kRT} = \sqrt{1.40(1715)(460 + 50)} = 1107$ ft/sec

Eq. 13.53: $\mathbf{M} = 1/\sin\beta = 1/\sin 40° = 1.556$; $\therefore\ V = c\mathbf{M} = 1107(1.556) = 1722$ ft/sec ◄

13.11.3 *A schlieren photograph of a bullet shows a Mach angle of 25°. The air is at a pressure of 9.5 kPa abs and 20°C. Find the approximate speed of the bullet.*

SI

Table A.5 for air: $R = 287\ \text{m}^2/(\text{s}^2\cdot\text{K})$, $k = 1.40$

Eq. 13.15: $c = \sqrt{kRT} = \sqrt{1.40(287)(273 + 20)} = 343$ m/s

Eq. 13.53: $\mathbf{M} = 1/\sin\beta = 1/\sin 25° = 2.37$; $\therefore\ V = c\mathbf{M} = 343(2.37) = 812$ m/s ◄

Sec. 13.12: Isothermal Flow -- *Problems*

13.16 *Air flows isothermally in a long pipe. At one section the pressure is 90 psia, the temperature is 80°F and the velocity is 100 fps. At a second section some distance from the first the pressure is 15 psia. Find the energy head loss due to friction, and determine the thermal energy that must have been added to or taken from the fluid between the two sections. The diameter of the pipe is constant.*

BG

Table A.5 for air: $R = 1715\ \text{ft}^2/(\text{sec}^2\cdot°\text{R})$

From Eq. 2.4 for isothermal flow: $\rho_1/\rho_2 = p_1/p_2 = 90/15 = 6$

But from continuity $\rho_1 A_1 V_1 = \rho_2 A_2 V_2$, thus as A is constant, $V_2 = 6V_1 = 6(100) = 600$ fps

From Eq. 13.54: $q_H = 600^2/2 - 100^2/2 = 175{,}000$ ft·lb/slug of air

$Q_H = 175{,}000/32.2 = 5430$ ft·lb/lb of air ◄

For head loss: From Sec. 13.13, ignoring z terms: $dp/\rho + d(V^2/2) = -(2\tau dx/\rho r) = -g\,dh_L$

Integrating: $gh_L = RT\ln(p_1/p_2) + (V_1^2 - V_2^2)/2 = 1715(460 + 80)\ln 6 + (100^2 - 600^2)/2$

$= 1{,}659{,}000 - 175{,}000 = 1{,}484{,}000$ ft·lb/slug

$h_L = 1{,}484{,}000/32.2 = 46{,}100$ ft·lb/lb of air ◄

Note: The total energy at section (2) is greater than that at section (1) because external heat is added. Thus h_L represents a degradation of mechanical energy, but not a net loss of energy, because it is converted to another form of energy, namely internal heat.

/more...

13.17 *Air flows isothermally in a long pipe. At one section the pressure is 600 kPa abs, the temperature is 25°C, and the velocity is 30 m/s. At a second section some distance from the first the pressure is 100 kPa abs. Find the energy head loss due to friction, and determine the thermal energy that must have been added to or taken from the fluid between the two sections. The diameter of the pipe is constant.*

SI

Table A.5 for air: $R = 287\ \text{m}^2/(\text{s}^2\cdot\text{K})$

From Eq. 2.4 for isothermal flow: $\rho_1/\rho_2 = p_1/p_2 = 600/100 = 6$

But from continuity $\rho_1 A_1 V_1 = \rho_2 A_2 V_2$, thus as A is constant, $V_2 = 6V_1 = 6(30) = 180$ m/s.

From Eq. 13.54: $q_H = (180^2 - 30^2)/2 = 15\,750$ N·m/kg of air

$Q_H = 15\,750/9.81 = 1606$ N·m/N of air ◄

For head loss: From Sec. 13.13, ignoring z terms: $dp/\rho + d(V^2/2) = -(2\tau dx/\rho r) = -g dh_L$

Integrating: $gh_L = RT\ln(p_1/p_2) + (V_1^2 - V_2^2)/2 = 287(273 + 25)\ln 6 + (30^2 - 180^2)/2$

$= 153\,200 - 15\,750 = 137\,500$ N·m/kg

$h_L = 137\,500/9.81 = 14\,020$ N·m/N of air ◄

Note: The total energy at section (2) is greater than that at section (1) because external heat is added. Thus h_L represents a degradation of mechanical energy, but not a net loss of energy, because it is converted to another form of energy, namely internal heat.

Sec. 13.13: Isothermal Flow in a Constant-Area Duct -- *Exercises*

13.13.1 *Air flows isothermally through a long horizontal pipe of uniform diameter. At a section where the pressure is 100 psia, the velocity is 120 ft/sec. Because of fluid friction the pressure at a distance point is 40 psia. (a) What is the increase in kinetic energy per slug of air? (b) What is the amount of thermal energy in Btu per slug of air that must be transferred in order to maintain the temperature constant? (c) Is this heat transferred to the air in the pipe or removed from it? (d) If the temperature of the air is 100°F and the diameter of the pipe is 3 in, find the total heat transferred in Btu per hour.*

BG

Table A.5 for air: $R = 1715\ \text{ft}^2/(\text{sec}^2\cdot°\text{R})$

Sec. 13.13: pV = const for isothermal flow in a constant-area pipe,

$\therefore\ V_2 = p_1V_1/p_2 = 100(120)/40 = 300$ ft/sec ; $V_1^2/2 = 120^2/2 = 7200$ ft·lb/slug, $V_2^2/2 = 45{,}000$ ft·lb/slug

(*a*) Eq. 13.54: Increase in K.E. of air = 45,000 − 7200 = 37,800 ft·lb/slug ◄

(*b*) Thermal energy/slug transferred = $q_H = \Delta KE = 37{,}800/778 = 48.6$ Btu/slug ◄

(*c*) KE increases, ∴ (Sec. 13.12) heat is transferred <u>to</u> the air in the pipe. ◄

(*d*) From Eq. 2.4: $\rho_1 = p_1/(RT_1) = (100 \times 144)/[1715(460 + 100)] = 0.014\,99\ \text{slug/ft}^3$

Eq. 13.5: $\dot{m} = \rho_1 V_1 A = 0.014\,99(120)(\pi/4)(3/12)^2 = 0.0883$ slug/sec

Heat transfer rate = (48.6 Btu/slug)(0.0833 slug/sec) = 4.29 Btu/sec = 15,450 Btu/hr ◄

13.13.2 *Air flows isothermally through a long horizontal pipe of uniform diameter. At a section where the pressure is 700 kPa abs, the velocity is 35 m/s. Because of fluid friction the pressure at a distance point is 280 kPa abs. (a) What is the increase in kinetic energy per kg of air? (b) What is the amount of thermal energy in J/kg air that must be transferred in order to maintain the temperature constant? (c) Is this heat transferred to the air in the pipe or removed from it? (d) If the temperature of the air is 40°C and the diameter of the pipe is 7.5 cm, find the total heat transferred in joules per hour.*

SI

Table A.5 for air: $R = 287\ \text{m}^2/(\text{s}^2\cdot\text{K})$. Sec. 13.13: pV = const for isothermal flow in a const-A pipe,

$\therefore\ V_2 = p_1V_1/p_2 = 700(35)/280 = 87.5$ m/s ; $V_1^2/2 = 35^2/2 = 613$ N·m/kg, $V_2^2/2 = 3830$ N·m/kg

(*a*) Eq. 13.54: Increase in K.E. of air = 3830 − 613 = 3220 N·m/kg ◀

(*b*) Thermal energy/kg transferred = $q_H = \Delta KE$ = 3220 J/kg ◀

(*c*) KE increases, ∴ (Sec. 13.12) heat is transferred <u>to</u> the air in the pipe. ◀

(*d*) From Eq. 2.4: $\rho_1 = p_1/(RT_1) = 700\ 000/[287(273 + 40)] = 7.79\ \text{kg/m}^3$

Eq. 13.5: $\dot{m} = \rho_1V_1A = 7.79(35)(\pi/4)(7.5/100)^2 = 1.205$ kg/s

Heat transfer rate = (3220 J/kg)(1.205 kg/s) = 3780 J/s = 13.95×10^6 J/hr ◀

Sec. 13.13: Isothermal Flow in a Constant-Area Duct -- *Problems*

13.18 *Refer to Sample Prob. 13.7. Neglecting the logarithm term in Eq. 13.59, find the pressure and velocities at sections 100, 300, and 800 ft downstream of the section where the pressure is 80 psia. Plot the pressure and velocity as a function of distance along the pipe.*

Sample Prob. 13.7: Air flows isothermally (65°F) at 3.0 slug/sec through a horizontal 10 in × 14 in duct. $R_h = 0.243$ *ft,* $f = 0.0083$, $\rho_1 = 0.012\ 79$ *slug/ft*^{3}, $V_1 = 242$ *fps.*

BG

Table A.5 for air: $R = 1715\ \text{ft}^2/(\text{sec}^2\cdot°\text{R})$. Substitute into Eq. 13.59 (neglecting the logarithm term):

$$(80\times 144)^2 - p_2^2 = \frac{(3.0)^2 1715(460+65)}{(10\times 14/144)^2}\left[0.0083\frac{L}{4(0.243)}\right];\quad p_2^2 = 1.327\times 10^8 - 73{,}200L$$

Isothermal, using Eq. 2.4: $p_1/\rho_1 = p_2/\rho_2$, ∴ $\rho_1/\rho_2 = p_1/p_2$

Continuity Eq. 13.5 with A = const: ρV = const, ∴ $V_2 = V_1(\rho_1/\rho_2) = V_1(p_1/p_2) = 242(80/p_2)$

Results:

L (ft)	p (psia)	V (ft/sec)
0	80.0	242
100	77.8	249
300	73.1	265
500	68.1	284
800	59.8	324

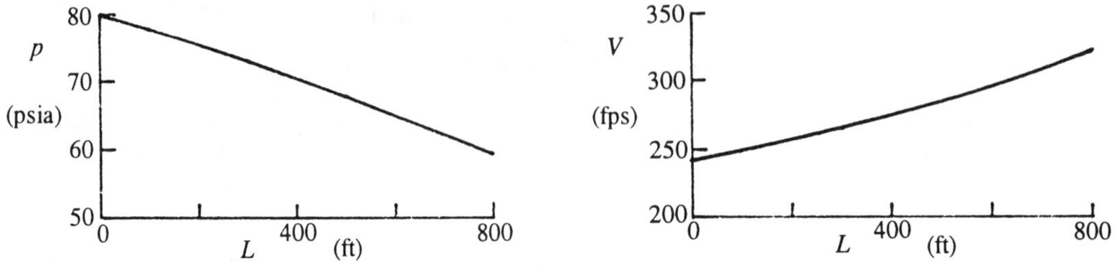

Note: Greater accuracy can be achieved by considering the logarithmic term.

13.19 *Carbon dioxide flows isothermally at 100°F through a horizontal 6-in-diameter pipe. At this temperature μ = 4.0 × 10^{-7} lb·sec/ft². The pressure changes from 150.0 to 140.0 psig in a 100-ft length of pipe. Determine the mass flow rate if the atmospheric pressure is 14.5 psia and e for the pipe is 0.002 ft.*

BG

Table A.5: $R = 1123\ \text{ft}^2/(\text{s}^2\cdot°\text{R})$; $A = (\pi/4)(6/12)^2 = 0.1963\ \text{ft}^2$

At section 1 (inlet), p_1 = 150.0 psig + 14.5 psi = 164.5 psia

From Eq. 2.4: $\rho_1 = (164.5 \times 144)/[1123(460+100)] = 0.0377\ \text{slug/ft}^3$;

$e/D = 0.002/0.5 = 0.004$; Assume $\mathbf{R} > 10^6$, Fig. 8.11 or Eq. 8.40 or 41: $f = 0.0284$

At section 2, p_2 = 140.0 psig + 14.5 psi = 154.5 psia.

Eq. 13.59: $$(164.5 \times 144)^2 - (154.5 \times 144)^2 = \frac{\dot{m}^2 1123(460 + 100)}{0.1963^2}\left[0.0284\left(\frac{100}{0.5}\right) + 2\ln\left(\frac{164.5}{154.5}\right)\right]$$

$5.61 \times 10^8 - 4.95 \times 10^8 = 1.632 \times 10^6\dot{m}^2(5.70 + 0.1254) = 95.1 \times 10^6$

$\therefore\ \dot{m}^2 = (66.1 \times 10^6)/(94.7 \times 10^6) = 0.699$; $\dot{m} = 0.836$ slug/sec (based on the above assumption for **R**)

From Eq. 13.5: $V_1 = \dot{m}/(\rho_1 A) = 0.834/(0.0377 \times 0.1963) = 113.0$ ft/sec

Eq. 13.58: $\mathbf{R} = [0.5(113.0)0.0377]/(4.0 \times 10^{-7}) = 5.32 \times 10^6$

Hence the assumption for **R** was O.K. and $\dot{m}$ = 0.836 slug/sec ◄

Alternative solutions: (1) Solve Eq. 13.59 and 8.40 simultaneously for f and $\dot{m}$ using Mathcad or similar mathematics software. (2) Eliminate f between these two equations, and solve for $\dot{m}$ using an equation solver on a programmable calculator (see Sec. 8.14, Type 2).

13.20 *Carbon dioxide flows isothermally at 40°C through a horizontal 15-cm-diameter pipe. At this temperature μ = 1.95 × 10^{-5} N·s/m². The pressure changes from 1000 to 930 kPa gage in a 30-m length of pipe. Determine the mass flow rate if the atmospheric pressure is 100 kPa and e for the pipe is 0.60 mm.*

SI

Table A.5: $R = 188\ \text{N·m/(kg·K)}$; $A = (\pi/4)(15/100)^2 = 0.017\,67\ \text{m}^2$

At section 1 (inlet), p_1 = 1000 kN/m² gage + 100 kN/m² = 1100 kN/m² abs

From Eq. 2.4: $\rho_1 = 1\,100\,000/[188(273 + 40)] = 18.69\ \text{kg/m}^3$

$e/D = 0.002/0.5 = 0.004$; Assume $\mathbf{R} > 10^6$, Fig. 8.11 or Eq. 8.40 or 41: $f = 0.0284$

At section 2, p_2 = 930 kN/m² + 100 kN/m² = 1030 kN/m² abs.

Eq. 13.59: $$(1\,100\,000)^2 - (1\,030\,000)^2 = \frac{\dot{m}^2 188(273 + 40)}{0.017\,67^2}\left(0.0284\frac{30}{0.15} + 2\ln\frac{1100}{1030}\right)$$

$1.210 \times 10^{12} - 1.061 \times 10^{12} = 1.884 \times 10^8\dot{m}^2(5.68 + 0.1315) = 1.095 \times 10^9\dot{m}^2$

from which $\dot{m}$ = 11.67 kg/s (based on the above assumption)

Eq. 13.5: $V_1 = \dot{m}/(\rho_1 A) = 11.67/(18.69 \times 0.017\,67) = 35.3$ m/s

Eq. 13.58: $\mathbf{R} = [0.15(35.3)18.69]/(1.95 \times 10^{-5}) = 5.08 \times 10^6$

Hence the assumption for **R** was O.K. and $\dot{m}$ = 11.67 kg/s ◄

Alternative solutions: (1) Solve Eq. 13.59 and 8.40 simultaneously for f and $\dot{m}$ using Mathcad or similar mathematics software. (2) Eliminate f between these two equations, and solve for $\dot{m}$ using an equation solver on a programmable calculator (see Sec. 8.14, Type 2).

13.21 *Methane gas is to be pumped through a 24-in-diameter welded-steel pipe connecting two compressor stations 25 mi apart. At the upstream station the pressure is not to exceed 60 psia, and at the downstream station it is to be at least 20 psia. Determine the maximum possible rate of flow (in cubic feet per day at 60°F and 1 atm). Assume isothermal flow at 60°F.*

BG

For methane: $R = 3100$ ft²/(sec²·°R) (Table A.5); $\mu = 2.40 \times 10^{-7}$ lb·sec/ft² (Fig. 2.3 at 60°F)

Absolute viscosity μ is independent of pressure over a wide range of pressures but kinematic viscosity ν varies with pressure (see end of Sec. 2.11).

From Eqs. 2.11 and 2.4: $\nu_1 = \dfrac{\mu}{\rho_1} = \dfrac{\mu}{p_1/RT_1} = \dfrac{2.40 \times 10^{-7}}{(60 \times 144)/(3100 \times 520)} = 4.48 \times 10^{-5}$ ft²/sec

Table 8.1 for welded steel: $e = 0.000\,15$ ft; $e/D = 0.000\,15/2 = 0.000\,075$

As a first trial, assume $\mathbf{R} > 10^7$, in which case from Fig. 8.11 or Eq. 8.40 or 41, $f = 0.0116$

Eq. 13.59: $(60 \times 144)^2 - (20 \times 144)^2 = \dfrac{\dot{m}^2(3100)520}{\pi^2}\left[0.0116\dfrac{25(5280)}{2} + 2\ln\left(\dfrac{60}{20}\right)\right]$

$66{,}400{,}000 = \dot{m}^2 163{,}300(766 + 2.20) = \dot{m}^2 1.254 \times 10^8$; $\dot{m}^2 = 0.529$; $\dot{m} = 0.727$ slug/sec

Eq. 13.58: $\mathbf{R} = \dot{m}D/(\mu A) = (0.727 \times 2)/(2.40 \times 10^{-7} \times \pi) = 1.930 \times 10^6$

Thus, from Fig. 8.11: $f = 0.0124$; The initially-assumed f was too low.

Substituting $f = 0.0124$ into Eq. 13.59 gives $\dot{m} = 0.704$ slug/sec

From Eq. 2.4: $\rho_1 = p_1/RT_1 = (60 \times 144)/(3100 \times 520) = 0.005\,36$ slug/ft³

From Eq. 13.5: $Q_1 = \dot{m}/\rho_1 = 0.705/0.005\,36 = 131.3$ ft³/sec $= 11.34 \times 10^6$ ft³/day at 60°F and 60 psia,

From Eqs. 2.4 and 13.5: $\dot{m} = \rho Q = pQ/RT$, i.e. here $pQ =$ const

∴ at 60°F and 14.7 psia: $Q = Q_1(p_1/p) = 11.34 \times 10^6(60/14.7) = 46.3 \times 10^6$ ft³/day ◄

Alternative solutions: (1) Solve Eq. 13.59 and 8.40 simultaneously for f and $\dot{m}$ using Mathcad or similar mathematics software. (2) Eliminate f between these two equations, and solve for $\dot{m}$ using an equation solver on a programmable calculator (see Sec. 8.14, Type 2).

/more...

13.22 *Refer to Sample Prob. 13.9. Find the distance along the pipe to (a) where $\rho_2 = 0.9\rho_1$; (b) where $\rho_2 = 0.7\rho_1$; (c) where subsonic adiabatic flow ends. Compute the corresponding values of p, V, T, and* **M** *and plot the first three as a function of distance along the pipe.*

Sample Prob. 13.9: Air flows adiabatically at 3.0 slug/sec through a 10 in × 14 in rectangular duct (e = 0 on Fig. 8.11). p_1 = 80 psia, T_1 = 65°F = 525°R. R = 1715 ft·lb/(slug·°R), ρ_1 = 0.012 79 slug/ft³, $\mathbf{R}_1 = 7.93 \times 10^6$, f = 0.0083, C = 6,360,000 ft²/sec².

BG

Follow the procedure given in Sample Prob. 13.9, as follows:

(1) $\rho_2 = \rho_1(\rho_2/\rho_1)$

(2) Find p_2 from Eq. 13.63.

(3) Find $\int_1^2 \rho dp$ from Eq. 13.64.

(4) Find L from Eq. 13.65; check (from T_2) if **R** and f changes.

(5) $V_2 = \dot{m}/(\rho_2 A)$

(6) $T_2 = p_2/(\rho_2 R)$

In addition, for **M**, Eq. 7.10: $\mathbf{M} = V/\sqrt{E_v/\rho}$; Sec. 2.8: $E_v = kp$; combining these, $\mathbf{M} = V\sqrt{\rho/(1.4p)}$

For part (*c*), find by T and E the value of ρ_2 which yields $\mathbf{M}_2 = 1$

Results:

	ρ_2/ρ_1	ρ_2 (slug/ft³)	p_2 (psia)	$\int_1^2 \rho dp$	L (ft)	V_2 (ft/sec)	T_2 (°F)	$\mathbf{M}_2$
	1.0	0.012 79	80.0	0	0	242	65.0	0.215
(*a*)	0.9	0.011 51	71.8	−14.26	326	268	63.8	0.239
	0.8*	0.010 23	63.6	−27.1	613	302	62.2	0.270
(*b*)	0.7	0.008 95	55.4	−38.4	861	345	59.9	0.309
(*c*)	0.2341	0.002 99	15.74	−72.2	1454**	1031	−18.7	1.000
			▲			▲	▲	▲

* From Sample Prob. 13.9
** With **R** = 9.22×10^6, f = 0.0082

The required plots look like the adiabatic pressure, velocity, and temperature curves in Fig 13.12. ◀
<u>Suggestion</u>: Solve using a spreadsheet.

/more...

13.23 *Air flows adiabatically at 3.0 slug/sec in a 12-in-diameter horizontal pipe. At a certain section the pressure is 150 psia and the temperature 140°F. Determine the distance along the pipe to the section where $\rho_2 = 0.80\rho_1$. Assume e/D = 0.0004.*

BG

Table A.5 for air: $R = 1715$ ft²/(sec²·°R), $k = 1.40$. $A = (\pi/4)1^2 = 0.785$ ft².

From Eq. 2.4: $\rho_1 = p_1/RT_1 = 150(144)/[1715(460 + 140)] = 0.0210$ slug/ft³

From Eq. 13.5: $V_1 = \dot{m}/(\rho_1 A_1) = 3.0/(0.021 \times 0.785) = 182.0$ ft/sec

Table A.2 for air at 140°F: $\mu = 4.14 \times 10^{-7}$ lb·sec/ft².

Eq. 13.58: $\mathbf{R}_1 = DV_1\rho_1/\mu_1 = 1(182.0)0.0210/(4.14 \times 10^{-7}) = 9.23 \times 10^6$; $e/D = 0.0004$ (given)

At this **R** and e/D the pipe behaves as wholly rough and $f = 0.0160$ (Fig. 8.11 or Eq. 8.40 or 41).

Eq. 13.62: $$C = \frac{\dot{m}^2}{\rho_1^2 A^2} + \frac{2k}{k-1}\frac{p_1}{\rho_1} = \frac{3.0^2}{(0.0210^2)0.785^2} + \frac{2(1.4)}{0.4}\frac{150(144)}{0.0210} = 7.24 \times 10^6$$

Eq. 13.63 with $\rho_2 = 0.8\rho_1$:

$$p_2 = \frac{0.4}{2.8}\left(7.24 \times 10^6(0.8)0.0210 - \frac{3.0^2}{(0.785^2)(0.8 \times 0.0210)}\right) = 17{,}240 \text{ lb/ft}^2 \text{ abs} = 119.7 \text{ psia}$$

Eq. 13.64: $\int_1^2 \rho\,dp = -82.5$; putting this into Eq. 13.65 yields $L = 679$ ft ◄

13.24 *Air flows adiabatically at 50 kg/s in a 30-cm-diameter horizontal pipe. At a certain section the pressure is 1000 kPa abs and the temperature 60°C. Determine the distance along the pipe to the section where $\rho_2 = 0.80\rho_1$. Assume e/D = 0.0004.*

SI

Table A.5 for air: $R = 287$ m²/(s²·K), $k = 1.40$. $A = (\pi/4)(30/100)^2 = 0.0707$ m².

From Eq. 2.4: $\rho_1 = p_1/RT_1 = (1\,000\,000)/[287(273 + 60)] = 10.46$ kg/m³

From Eq. 13.5: $V_1 = \dot{m}/(\rho_1 A_1) = 50/(10.46 \times 0.0707) = 67.6$ m/s

Table A.2 for air at 60°C: $\mu = 2.00 \times 10^{-5}$ N·s/m²

Eq. 13.58: $\mathbf{R}_1 = DV_1\rho_1/\mu_1 = 0.3(67.6)10.46/(2.00 \times 10^{-5}) = 10.61 \times 10^6$

At this **R** and $e/D = 0.0004$ (given) the pipe behaves as wholly rough

and $f = 0.0160$ (Fig. 8.11 or Eq. 8.40 or 41).

Eq. 13.62: $$C = \frac{\dot{m}^2}{\rho_1^2 A^2} + \frac{2k}{k-1}\frac{p_1}{\rho_1} = \frac{50^2}{10.46^2(0.0707)^2} + \frac{2(1.4)}{0.4}\frac{1\,000\,000}{10.46} = 674\,000 \text{ m}^2/\text{s}^2$$

Eq. 13.63 with $\rho_2 = 0.8\rho_1$:

$$p_2 = \frac{0.4}{2.8}\left(674\,000(0.8)10.46 - \frac{50^2}{0.0707^2(0.8 \times 10.46)}\right) = 797\,000 \text{ Pa abs}$$

Eq. 13.64: $\int_1^2 \rho\,dp = -1.912 \times 10^6$; putting this in Eq. 13.65 yields $L = 134.9$ m ◄

Chapter 14

Ideal Flow Mathematics

PROBLEM SELECTION GUIDE

Sec	Exer/Prob	Units	Difficulty	Length	Parts	Similar	Special features	
14.1	***Differential Equation of Continuity***							
	X[1]14.1.1	N	Easy	Medium	6		Differentiation	
	14.1.2	N	Easy	Medium	7		Differentiation	
	P 14.1	N	Easy	Short	1			
	14.2	N	Easy	V Short	1			
	14.3	N	Easy	Short	1		Derivations	
14.2	***Rotational and Irrotational Flow***							
	X 14.2.1	N	Easy	Medium	6		Differentiation	
	14.2.2	N	Easy	Medium	7		Differentiation	
14.3	***Circulation and Vorticity***							
	X 14.3.1	N	Easy	Medium	6	14.2.1	Differentiation	
	14.3.2	N	Easy	Medium	7	14.2.2	Differentiation	
14.4	***The Stream Function***							
	X 14.4.1	N	Medium	Short	1		Differentiation, integration	
	14.4.2	N	Medium	Medium	6		Differentiation, integration	
	14.4.3	N	Medium	Medium	7		Differentiation, integration	
	P 14.4	N	Easy	Short	3		Differentiation	Sketch
	14.5	N	Easy	Short	1		Differentiation	Sketch
	14.6	N	Medium	Medium	1		Differentiation	Plot
	14.7	N	Medium	Medium	2		Differentiation, integration.	Sketch
	14.8	N	Medium	Short	1		Differentiation	Sketch
14.5	***Basic Flow Fields***							
	X 14.5.1	SI	Easy	Short	2			
	14.5.2	SI	Medium	Medium	2		Differentiation (to check).	
	14.5.3	SI	Easy	Short	1		Differentiation	
	14.5.4	BG	Medium	Medium	1		Differentiation	
	P 14.9	N	Medium	Long	4		Differentiation	Plot
	14.10	SI	Medium	Long	1		Interpolation	Plot
	14.11	BG	Medium	Long	1		Differentiation	Plot
	14.12	N	Hard	Long	1		Differentiation	Sketch
	14.13	N	Hard	Long	5		Different'n, integration.	Plot & sketch

/cont...

[1] For all Exercises (marked "X"), answers are given in Appendix D of the textbook.

† Answers are sensitive to the accuracy of sketched flowfields and to scaling from them.

X = Exercise, P = (end-of-chapter) Problem, S = Sample Problem.

Sec	Exer/Prob	Units	Difficulty	Length	Parts	Similar	Special features	
14.6	*Velocity Potential*							
	X 14.6.1	N	Easy	Short	1		Differentiation	
	14.6.2	N	Medium	Short	1		Differentiation, integration	
	14.6.3	N	Medium	Medium	2	14.6.4	Differentiation, integration	
	14.6.4	N	Medium	Medium	2	14.6.3	Differentiation, integration	
14.7	*Orthogonality of Streamlines and Equipotential Lines*							
	X 14.7.1	N	Medium	Medium	6		Differentiation, integration	
	14.7.2	N	Medium	Medium	7		Differentiation, integration	
	P 14.14	BG	Hard	Medium	3	14.15	Differentiation	
	14.15	SI	Hard	Medium	3	14.14	Differentiation	
	14.16	BG	Medium	Short	1	14.17	Differentiation	
	14.17	SI	Medium	Short	1	14.16	Differentiation	
	14.18	BG	Medium	Medium	2	14.19	† Differentiation, integration	Sketch
	14.19	SI	Medium	Medium	2	14.18	† Differentiation, integration	Sketch
	14.20	N	Medium	Medium	1		Differentiation	Plot
	14.21	BG	Medium	Medium	1		Differentiation	Plot
	14.22	BG	Hard	Long	1		†	Sketch
	14.23	BG	Hard	Long	1	14.24	†	Sketch
	14.24	SI	Hard	Long	1	14.23	†	Sketch
	14.25	BG	Hard	Medium	1		Differentiation	Sketch
	14.26	N	Hard	Long	1		Proof, differentiation	Plot
	14.27	BG	Hard	Long	4		*	Plot
14.8	*Flow Through Porous Media*							
	X 14.8.1	BG	Easy	Short	1	14.8.2		
	14.8.2	SI	Easy	Short	1	14.8.1		
	14.8.3	BG	Easy	Short	1	14.8.4		
	14.8.4	SI	Easy	Short	1	14.8.3		

Chapter 14
IDEAL FLOW MATHEMATICS

Sec. 14.1: Differential Equation of Continuity -- *Exercises*

14.1.1 *Which of the following incompressible flows satisfy continuity?*

(a) $u = 2$
(b) $u = 2,\ v = 3$
(c) $u = 2 + 3x,\ v = 4$
(d) $u = 2y,\ v = 3x$
(e) $u = 2y,\ v = -3x$
(f) $u = 3xy,\ v = 1.5x^2$

N

(*a*) $u = 2,\ v = 0$; $\dfrac{\partial u}{\partial x} + \dfrac{\partial v}{\partial y} = 0 + 0 = 0$, so continuity is satisfied.

(*b*) $u = 2,\ v = 3$; $\dfrac{\partial u}{\partial x} + \dfrac{\partial v}{\partial y} = 0 + 0 = 0$, so continuity is satisfied.

(*c*) $u = 2 + 3x,\ v = 4$; $\dfrac{\partial u}{\partial x} + \dfrac{\partial v}{\partial y} = 3 + 0 \neq 0$, so continuity is not satisfied.

(*d*) $u = 2y,\ v = 3x$; $\dfrac{\partial u}{\partial x} + \dfrac{\partial v}{\partial y} = 0 + 0 = 0$, so continuity is satisfied.

(*e*) $u = 2y,\ v = -3x$; $\dfrac{\partial u}{\partial x} + \dfrac{\partial v}{\partial y} = 0 + 0 = 0$, so continuity is satisfied.

(*f*) $u = 3xy,\ v = 1.5x^2$; $\dfrac{\partial u}{\partial x} + \dfrac{\partial v}{\partial y} = 3y + 0 \neq 0$, so continuity is not satisfied.

Flows (*a*), (*b*), (*d*), and (*e*) satisfy continuity ◄

14.1.2 *Which of the following incompressible flows satisfy continuity?*

(a) $u = 3y,\ v = 0$
(b) $u = 3x,\ v = 3y$
(c) $u = 3x,\ v = -3y$
(d) $u = 4 + 2x,\ v = -6 - 2y$
(e) $u = 4xy + x^2,\ v = -2xy - 2y^2$
(f) $u = -2xy + 2x^2,\ v = 4xy - y^2$
(g) $u = -2xy - 2x^2 + 2y^2,\ v = 4xy - x^2 + y^2$

N

(*a*) $u = 3y,\ v = 0$; $\dfrac{\partial u}{\partial x} + \dfrac{\partial v}{\partial y} = 0 + 0 = 0$, so continuity is satisfied.

(*b*) $u = 3x,\ v = 3y$; $\dfrac{\partial u}{\partial x} + \dfrac{\partial v}{\partial y} = 3 + 3 \neq 0$, so continuity is not satisfied.

(*c*) $u = 3x,\ v = -3y$; $\dfrac{\partial u}{\partial x} + \dfrac{\partial v}{\partial y} = 3 - 3 = 0$, so continuity is satisfied.

(*d*) $u = 4 + 2x,\ v = -6 - 2y$; $\dfrac{\partial u}{\partial x} + \dfrac{\partial v}{\partial y} = 2 - 2 = 0$, so continuity is satisfied.

(*e*) $u = 4xy + x^2,\ v = -2xy - 2y^2$; $\dfrac{\partial u}{\partial x} + \dfrac{\partial v}{\partial y} = (4y + 2x) + (-2x - 4y) = 0$,
so continuity is satisfied.

(*f*) $u = -2xy + 2x^2,\ v = 4xy - y^2$; $\dfrac{\partial u}{\partial x} + \dfrac{\partial v}{\partial y} = (-2y + 4x) + (4x - 2y)$

$= 8x - 4y \neq 0$, so continuity is not satisfied

(*g*) $u = -2xy - 2x^2 + 2y^2,\ v = 4xy - x^2 + y^2$; $\dfrac{\partial u}{\partial x} + \dfrac{\partial v}{\partial y} = (-2y - 4x) + (4x + 2y) = 0$,
so continuity is satisfied

Flows (*a*), (*c*), (*d*), (*e*), and (*g*) satisfy continuity ◄

Sec. 14.1: Differential Equation of Continuity -- *Problems*

14.1 *Given a flow defined by $u = 3 + 2x$. If this flow satisfies continuity, what can be said about the density of the fluid?*

N

The flow satisfies continuity, so Eq. 14.1 holds. Flow in the y and z directions = 0, doesn't change with t,

so $\dfrac{\partial}{\partial y}(\rho u) = \dfrac{\partial}{\partial z}(\rho u) = \dfrac{\partial \rho}{\partial t} = 0$ and Eq. 14.1 becomes $\dfrac{\partial}{\partial x}(\rho u) = 0$, i.e., ρu = constant

i.e. $\rho = \dfrac{\text{constant}}{u} = \dfrac{\text{constant}}{3 + 2x}$; $\therefore$ the density ρ must decrease as x increases ◄

14.2 *Why are Eqs. (14.2) and (14.3) applicable to real fluids as well as ideal fluids?*

N

No mention is made of friction in their derivation, so the equations do not depend on the presence or absence of friction.

14.3 *Derive Eqs. (14.4) and (14.5)*

N

Flow in: $\rho v_r r\, d\theta + \rho v_t\, dr$

Flow out: $\left[\rho v_r + \dfrac{\partial(\rho v_r)}{\partial r}dr\right](r + dr)d\theta + \left[\rho v_t + \dfrac{\partial(\rho v_t)}{\partial(r\, d\theta)}r\, d\theta\right]dr$

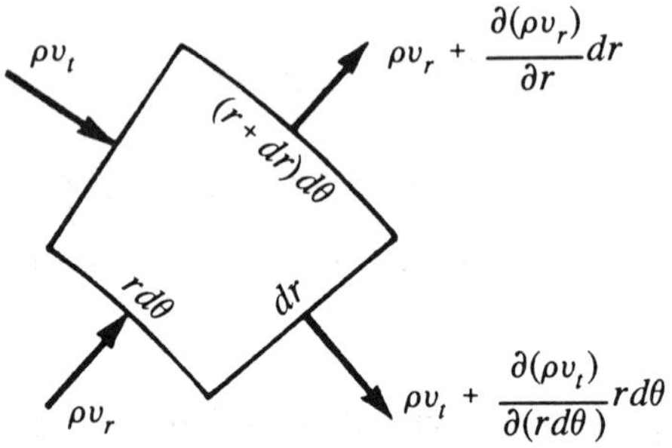

Continuity: Flow in − Flow out = 0

Substituting, collecting terms, and simplifying:

$$-\rho v_r\, dr\, d\theta - \frac{\partial(\rho v_r)}{\partial r}dr\, r\, d\theta - \frac{\partial(\rho v_t)}{\partial(r\, d\theta)}r\, d\theta\, dr = 0$$

$$\frac{\rho v_r}{r} + \frac{\partial(\rho v_r)}{\partial r} + \frac{\partial(\rho v_t)}{\partial(r\, d\theta)} = 0 \qquad \text{(14.4) for steady compressible flow}$$

And if ρ = constant, dividing through by ρ yields

$$\frac{v_r}{r} + \frac{\partial v_r}{\partial r} + \frac{\partial v_t}{\partial(r\, d\theta)} = 0 \qquad \text{(14.5) for steady incompressible flow}$$

Sec. 14.2: Rotational and Irrotational Flow -- *Exercises*

14.2.1 *Which of the flows of Exer. 14.1.1 are irrotational?*

N

(*a*) $u = 2,\ v = 0$; $\dfrac{\partial v}{\partial x} - \dfrac{\partial u}{\partial y} = 0 - 0 = 0$, irrotational

(*b*) $u = 2,\ v = 3$; $\dfrac{\partial v}{\partial x} - \dfrac{\partial u}{\partial y} = 0 - 0 = 0$, irrotational

(*c*) $u = 2 + 3x,\ v = 4$; $\dfrac{\partial v}{\partial x} - \dfrac{\partial u}{\partial y} = 0 - 0 = 0$, irrotational

(*d*) $u = 2y,\ v = 3x$; $\dfrac{\partial v}{\partial x} - \dfrac{\partial u}{\partial y} = 3 - 2 \neq 0$, rotational

(*e*) $u = 2y,\ v = -3x$; $\dfrac{\partial v}{\partial x} - \dfrac{\partial u}{\partial y} = -3 - 2 \neq 0$, rotational

(*f*) $u = 3xy,\ v = 1.5x^2$; $\dfrac{\partial v}{\partial x} - \dfrac{\partial u}{\partial y} = 3x - 3x = 0$, irrotational

Flows (*a*), (*b*), (*c*), and (*f*) are irrotational ◄

14.2.2 *Which of the flows of Exer. 14.1.2 are irrotational?*
N

(*a*) $u = 3y, \quad v = 0 \; ; \qquad \frac{\partial v}{\partial x} - \frac{\partial u}{\partial y} = 0 - 3 \neq 0,$ rotational

(*b*) $u = 3x, \quad v = 3y \; ; \qquad \frac{\partial v}{\partial x} - \frac{\partial u}{\partial y} = 0 - 0 = 0,$ irrotational

(*c*) $u = 3x, \quad v = -3y \; ; \qquad \frac{\partial v}{\partial x} - \frac{\partial u}{\partial y} = 0 - 0 = 0,$ irrotational

(*d*) $u = 4 + 2x, \quad v = -6 - 2y \; ; \qquad \frac{\partial v}{\partial x} - \frac{\partial u}{\partial y} = 0 - 0 = 0,$ irrotational

(*e*) $u = 4xy + x^2, \quad v = -2xy - 2y^2 \; ; \qquad \frac{\partial v}{\partial x} - \frac{\partial u}{\partial y} = -2y - 4x \neq 0,$ rotational

(*f*) $u = -2xy + 2x^2, \quad v = 4xy - y^2 \; ; \qquad \frac{\partial v}{\partial x} - \frac{\partial u}{\partial y} = 4y + 2x \neq 0,$ rotational

(*g*) $u = -2xy - 2x^2 + 2y^2, \quad v = 4xy - x^2 + y^2;$

$$\frac{\partial v}{\partial x} - \frac{\partial u}{\partial y} = (4y - 2x) - (-2x + 4y) = 0, \text{ irrotational}$$

Flows (*b*), (*c*), (*d*), and (*g*) are irrotational ◀

Sec. 14.3: Circulation and Vorticity -- *Exercises*

14.3.1 *Find the vorticity of each of the flows in Exer. 14.1.1.*
N

(*a*) $u = 2, \quad v = 0 \; ; \qquad \xi = \frac{\partial v}{\partial x} - \frac{\partial u}{\partial y} = 0 - 0 = 0$ ◀

(*b*) $u = 2, \quad v = 3 \; ; \qquad \xi = \frac{\partial v}{\partial x} - \frac{\partial u}{\partial y} = 0 - 0 = 0$ ◀

(*c*) $u = 2 + 3x, \quad v = 4 \; ; \qquad \xi = \frac{\partial v}{\partial x} - \frac{\partial u}{\partial y} = 0 - 0 = 0$ ◀

(*d*) $u = 2y, \quad v = 3x \; ; \qquad \xi = \frac{\partial v}{\partial x} - \frac{\partial u}{\partial y} = 3 - 2 = 1$ ◀

(*e*) $u = 2y, \quad v = -3x \; ; \qquad \xi = \frac{\partial v}{\partial x} - \frac{\partial u}{\partial y} = -3 - 2 = -5$ ◀

(*f*) $u = 3xy, \quad v = 1.5x^2 \; ; \qquad \xi = \frac{\partial v}{\partial x} - \frac{\partial u}{\partial y} = 3x - 3x = 0$ ◀

14.3.2 *Find the vorticity of each of the flows in Exer. 14.1.2.*
N

(*a*) $u = 3y, \quad v = 0 \; ; \qquad \xi = \frac{\partial v}{\partial x} - \frac{\partial u}{\partial y} = 0 - 3 = -3$ ◀

(*b*) $u = 3x, \quad v = 3y \; ; \qquad \xi = \frac{\partial v}{\partial x} - \frac{\partial u}{\partial y} = 0 - 0 = 0$ ◀

(*c*) $u = 3x, \quad v = -3y \; ; \qquad \xi = \frac{\partial v}{\partial x} - \frac{\partial u}{\partial y} = 0 - 0 = 0$ ◀

(*d*) $u = 4 + 2x, \quad v = -6 - 2y \; ; \qquad \xi = \frac{\partial v}{\partial x} - \frac{\partial u}{\partial y} = 0 - 0 = 0$ ◀

(*e*) $u = 4xy + x^2, \quad v = -2xy - 2y^2 \; ; \qquad \xi = \frac{\partial v}{\partial x} - \frac{\partial u}{\partial y} = -2y - 4x$ ◀

(*f*) $u = -2xy + 2x^2, \quad v = 4xy - y^2 \; ; \qquad \xi = \frac{\partial v}{\partial x} - \frac{\partial u}{\partial y} = 4y + 2x$ ◀

(*g*) $u = -2xy - 2x^2 + 2y^2, \quad v = 4xy - x^2 + y^2 \; ; \qquad \xi = \frac{\partial v}{\partial x} - \frac{\partial u}{\partial y} = (4y - 2x) - (-2x + 4y) = 0$ ◀

Sec. 14.4: The Stream Function -- *Exercises*

14.4.1 N

The flow of an incompressible fluid is defined by $u = 2$, $v = 8x$. Does a stream function exist for this flow? If so, determine the expression for the stream function.

$u = \partial\psi/\partial y = 2$; integrating: $\psi = 2y + f(x)$

$-v = \partial\psi/\partial x = -8x$; integrating: $\psi = -4x^2 + f(y)$

Comparing: $f(x) = -4x^2$ and $f(y) = 2y$

∴ A stream function does exist, it is $\psi = -4x^2 + 2y$ ◄

14.4.2 N

For each of the flows of Exer. 14.1.1, write an expression for the stream function if one exists.

Sec. 14.4: Stream functions exist only for those flows that satisfy continuity.

Per Solution 14.1.1, continuity Eq. 14.3, $\partial u/\partial x + \partial v/\partial y = 0$, is satisfied for flows (*a*), (*b*), (*d*) and (*e*).

Flows (*c*) and (*f*) do not satisfy continuity and so do not have a stream function.

By integrating Eq. 14.14 we get:

(*a*) $\psi = 2y$ ◄ (*d*) $\psi = -1.5x^2 + y^2$ ◄

(*b*) $\psi = 2y - 3x$ ◄ (*e*) $\psi = 1.5x^2 + y^2$ ◄

14.4.3 N

For each of the flows of Exer. 14.1.2, write an expression for the stream function if one exists.

Sec. 14.4: Stream functions exist only for those flows that satisfy continuity.

Per Solution 14.1.2, continuity Eq. 14.3, $\partial u/\partial x + \partial v/\partial y = 0$, is satisfied for flows (*a*), (*c*), (*d*), (*e*) and (*g*).

Flows (*b*) and (*f*) do not satisfy continuity and so do not have a stream function.

(*a*) $u = \partial\psi/\partial y = 3y$. Integrating: $\psi = (3/2)y^2 + f(x)$

$-v = \partial\psi/\partial x = 0$. Integrating: $\psi = f(y)$

Comparing: $f(x) = 0$, $f(y) = (3/2)y^2$. ∴ $\psi = (3/2)y^2$ ◄

(*c*) $u = \partial\psi/\partial y = 3x$. Integrating: $\psi = 3xy + f(x)$

$-v = \partial\psi/\partial x = 3y$. Integrating: $\psi = 3xy + f(y)$

Comparing: $f(x) = f(y) = 0$. ∴ $\psi = 3xy$ ◄

(*d*) $u = \partial\psi/\partial y = 4 + 2x$. Integrating: $\psi = 4y + 2xy + f(x)$

$-v = \partial\psi/\partial x = 6 + 2y$. Integrating: $\psi = 6x + 2xy + f(y)$

Comparing: $f(x) = 6x$, $f(y) = 4y$. ∴ $\psi = 6x + 2xy + 4y$ ◄

(*e*) $u = \dfrac{\partial\psi}{\partial y} = 4xy + x^2$. Integrating: $\psi = 2xy^2 + x^2y + f(x)$

$-v = \dfrac{\partial\psi}{\partial x} = 2xy + 2y^2$. Integrating: $\psi = x^2y + 2xy^2 + f(y)$

Comparing: $f(x) = f(y) = 0$. ∴ $\psi = x^2y + 2xy^2$ ◄

(*g*) $u = \dfrac{\partial\psi}{\partial y} = -2xy - 2x^2 + 2y^2$. Integrating: $\psi = -xy^2 - 2x^2y + (2/3)y^3 + f(x)$

$-v = \dfrac{\partial\psi}{\partial x} = -4xy + x^2 - y^2$. Integrating: $\psi = -2x^2y + (1/3)x^3 - xy^2 + f(y)$

Comparing: $f(x) = (1/3)x^3$, $f(y) = (2/3)y^3$. ∴ $\psi = (1/3)x^3 - 2x^2y - xy^2 + (2/3)y^3$ ◄

14.4 *Sketch streamlines (ψ = 0, 1, 2, 3) for the following flow fields, note the values of u and v, and verify that continuity is satisfied in all cases. (a) $\psi = 10y$, (b) $\psi = -20x$, (c) $\psi = 10y - 20x$.*

N

Eq. 14.16: (*a*) $u = \partial\psi/\partial y = 10$, $v = 0$; (*b*) $u = 0$, $v = -\partial\psi/\partial x = 20$; (*c*) $u = 10$, $v = 20$ ◄

In all cases $(\partial u/\partial x) + (\partial v/\partial y) = 0$, thus continuity (Eq. 14.3) is satisfied ◄

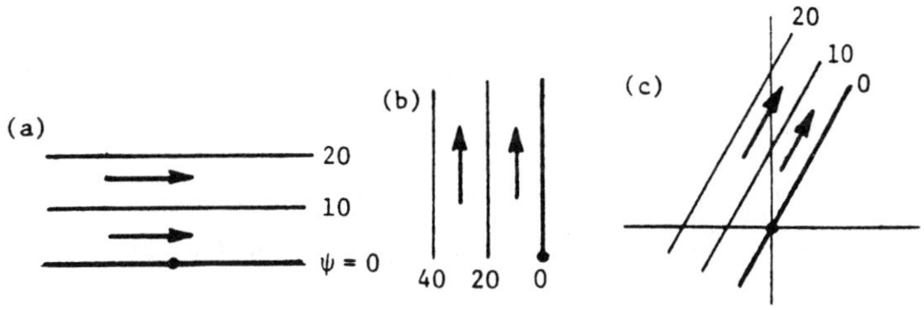

◄◄

14.5 *A flow field is described by the equation $\psi = 1.2xy$. Sketch the streamlines in one quadrant for ψ = 0, 1, 2, 3, 4.*

N

$\psi = 1.2xy$

Eq. 14.16: $u = \partial\psi/\partial y = 1.2x$

$v = -\partial\psi/\partial x = -1.2y$

Thus:

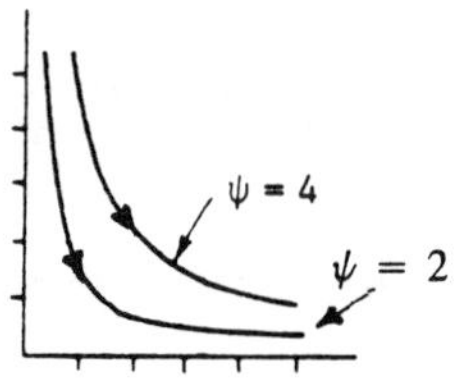

◄

14.6 *Plot the streamlines in the upper right-hand quadrant for the flow defined by $\psi = 1.5x^2 + y^2$ and determine the value of the velocity at $x = 4$, $y = 2$.*

N

Eq. 14.16: $u = \partial\psi/\partial y = 2y$, $v = -\partial\psi/\partial x = -3x$

At (4,2): $u = 4$, $v = -12$

Velocity $= \sqrt{4^2 + 12^2} = 12.65$ ◄

To plot streamlines, rewrite as $y = \sqrt{\psi - 1.5x^2}$.

Assume ψ, compute y's for different x's

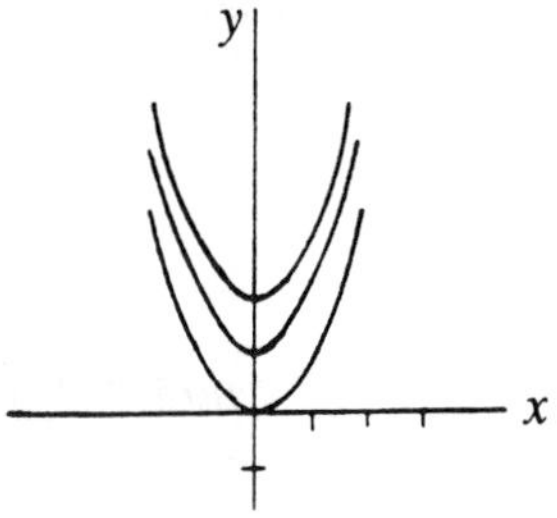

◄

14.7 *The components of the velocities of a certain flow system are*

$$u = -\frac{Q}{2\pi}\left(\frac{x}{x^2 + y^2}\right) + By + C \quad \text{and} \quad v = -A\left(\frac{y}{x^2 + y^2}\right) + Dx + E$$

(a) Calculate a value of A consistent with continuous flow. (b) Sketch the streamlines for this flow system, assuming B = C = D = E = 0.

N

(*a*) To satisfy continuity (Eq. 14.3 for 2-D flow): $\partial u/\partial x + \partial v/\partial y = 0$

$$\frac{\partial u}{\partial x} = -\frac{Q}{2\pi}\frac{(x^2 + y^2) - x(2x)}{(x^2 + y^2)^2} = -\frac{Q}{2\pi}\frac{(y^2 - x^2)}{(x^2 + y^2)^2}$$

$$\frac{\partial v}{\partial y} = -A\frac{(x^2 + y^2) - y(2y)}{(x^2 + y^2)^2} = -A\frac{(x^2 - y^2)}{(x^2 + y^2)^2}$$

/cont...

Substituting, $-Q/2\pi(y^2 - x^2) - A(x^2 - y^2) = 0$

thus $A = Q/2\pi$ ◄

(*b*) $u = -\dfrac{Q}{2\pi}\dfrac{x}{(x^2 + y^2)}$; $\upsilon = -\dfrac{Q}{2\pi}\dfrac{y}{(x^2 + y^2)}$

Eq. 14.14: $d\psi = -\upsilon dx + u dy = \dfrac{Q}{2\pi}\dfrac{y\,dx}{(x^2 + y^2)} - \dfrac{Q}{2\pi}\dfrac{x\,dy}{(x^2 + y^2)}$

Integrating, $\psi = \dfrac{Q}{2\pi}\tan^{-1}\left(\dfrac{x}{y}\right) - \tan^{-1}\left(\dfrac{y}{x}\right)$

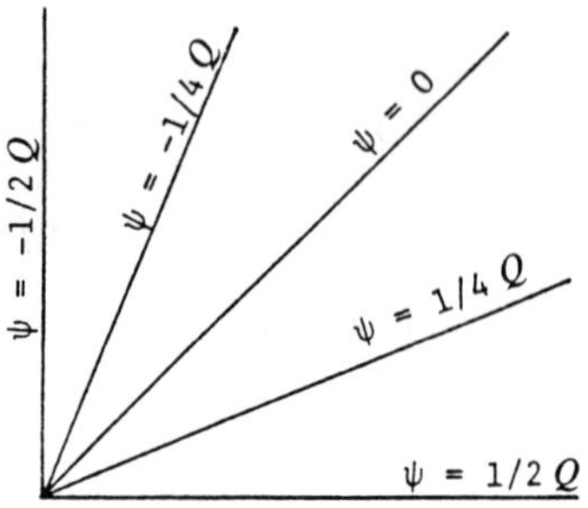

◄

14.8 *A flow field is described by $\psi = x^2 - y$. Sketch the streamlines for $\psi = 0$, 1, and 2. Derive an expression for the velocity at any point in the flow field and determine the vorticity of the flow.*

N

From the equation for ψ, the flow field is a family of parabolas symmetric about the *y*-axis. The streamline $\psi = 0$ passes through the origin. ◄

From Eq. 14.16: $u = \partial\psi/\partial y = -1$, $\upsilon = -\partial\psi/\partial x = -2x$

Thus $V = \sqrt{u^2 + \upsilon^2} = \sqrt{1 + 4x^2}$ ◄

Eq. 14.12: Vorticity $\xi = (\partial\upsilon/\partial x) - \partial u/\partial y) = -2$ ◄ (Since $\xi \neq 0$, the flow is rotational)

Sec. 14.5: Basic Flow Fields -- ***Exercises***

14.5.1 *A source discharging 25 m³/s per m is located at the origin and a uniform flow with a velocity of 5 m/s from left to right is superimposed on the source flow. Determine the stream function of the flow (a) in polar and (b) in rectangular coordinates.*

SI

Per Section 14.5: $\psi_1 = 25\dfrac{\theta \text{ rad}}{2\pi} = \dfrac{25}{2\pi}\arctan\left(\dfrac{y}{x}\right)$; $\psi_2 = 5y = 5r\sin\theta$

(*a*) $\psi_{\text{polar}} = \dfrac{12.5\,\theta}{\pi} + 5r\sin\theta$ ◄ (*b*) $\psi_{\text{rect}} = \dfrac{12.5}{\pi}\arctan\left(\dfrac{y}{x}\right) + 5y$ ◄

14.5.2 *For the flow of Exer. 14.5.1, (a) find the location of the stagnation points and (b) find the velocity at x = 3 m, y = 4 m.*

Exer. 14.5.1: A source (25 m³/s per m) discharging at the origin is superimposed on a uniform flow (V = 5 m/s) from left to right.

SI

(*a*) Stagnation point is to the left of origin where source velocity = 5 m/s

i.e., where $V = 25/(2\pi r) = 5$; $r = 25/(10\pi) = 0.796$ m

Stagnation point is at $x = -0.796$ m, $y = 0$ (or $\theta = 180°$, $r = 0.796$ m) ◄

(*b*) At the point (3 m, 4 m):

From rectilinear flow: $u_1 = 5$ m/s, $\upsilon_1 = 0$

From source flow: $\upsilon_r = \dfrac{Q}{2\pi r} = \dfrac{25}{2\pi(5)} = 0.796$ m/s ;

$u_2 = (3/5)\upsilon_r = (3/5)(0.796) = 0.477$ m/s

/cont...

and $v_2 = (4/5)v_r = (4/5)(0.796) = 0.637$ m/s

$u = u_1 + u_2 = 5.48$ m/s ; $V = \sqrt{u^2 + v^2} = \sqrt{5.48^2 + 0.637^2} = 5.51$ m/s ◀

Check: (using Eq. 14.16 and $x = 3$ m, $y = 4$ m):

$$u = \frac{\partial\psi}{\partial y} = 5 + \frac{12.5}{\pi}\left(\frac{1}{1 + y^2/x^2}\right)\left(\frac{1}{x}\right) = 5.48 \text{ m/s}$$

$$v = -\frac{\partial\psi}{\partial x} = -\frac{12.5}{\pi}\left(\frac{1}{1 + y^2/x^2}\right)\left(-\frac{y}{x^2}\right) = 0.637 \text{ m/s}$$

14.5.3 *Refer to Exer. 14.5.1. Use Bernoulli's theorem to find the difference in pressure head between point A (−14 m, 0) and point B (0, 2 m).*

Exer. 14.5.1: A source (25 m³/s per m) discharging at the origin is superimposed on a uniform flow (V = 5 m/s) from left to right.

SI

Differentiate ψ_{rect} (Exer. 14.5.1) per Eq 14.16:

$$u = \frac{\partial\psi}{\partial y} = \frac{12.5}{\pi}\frac{x}{(x^2 + y^2)} + 5 \; ; \qquad v = -\frac{\partial\psi}{\partial x} = \frac{12.5}{\pi}\frac{y}{(x^2 + y^2)}$$

At point A (−14 m, 0): $u = 4.72$ m/s, $v = 0$; $V_A = 4.72$ m/s

At point B (0, 2 m): $u = 5$ m/s, $v = 1.989$ m/s ; $V_B = \sqrt{5^2 + 1.989^2} = 5.38$ m/s

From Bernoulli's theorem (Eq. 4.10):

$$p_A/\gamma - p_B/\gamma - V_B^2/(2g) - V_A^2/(2g) = \frac{5.38^2 - 4.72^2}{2(9.81)} = 0.342 \text{ m}$$ ◀

14.5.4 *A source discharging 13 cfs/ft is at (−1, 0) and a sink taking in 13 cfs/ft is at (+1, 0). If a uniform flow with velocity 8 fps from left to right is superimposed on the source-sink combination, what is the length of the resulting closed body contour?*

BG

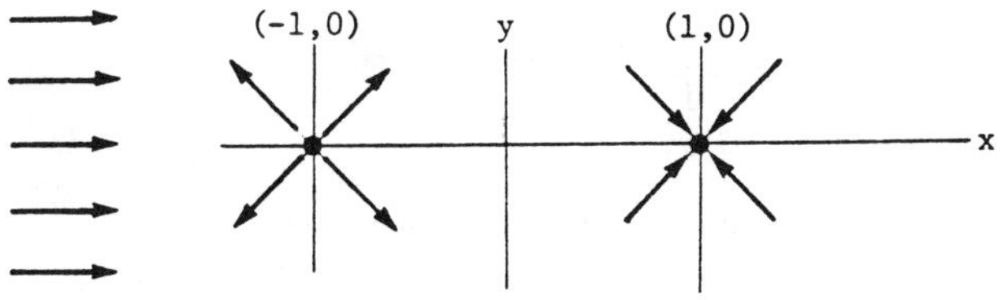

Differentiate Eq. 14.18 with respect to y (per Eq. 14.16):

$$u = \frac{\partial\psi}{\partial y} = U + \frac{q}{2\pi}\left[\frac{x + a}{(x + a)^2 + y^2} - \frac{x - a}{(x - a)^2 + y^2}\right]$$

Along the x-axis ($y = 0$), with $U = 8$ fps and $q = 13$ cfs/ft,

$$u = 8 + \frac{13}{2\pi(x + 1)} - \frac{13}{2\pi(x - 1)} = 0 \quad \text{at ends of body.}$$

Solving for x: $x = \pm 1.232$ ft. ∴ Length of body $= 2x = 2.46$ ft ◀

14.9 *A source of strength 8π is located at (2, 0). Another source of strength 16π is located at $(-3, 0)$. For the combined flow field produced by these two sources: (a) find the location of the stagnation point; (b) plot the $\psi = 0$, $\psi = 4\pi$, $\psi = 8\pi$ lines; (c) find the values of ψ at (0, 2) and at $(3, -1)$; (d) find the velocity at $(-2, 5)$.*

N

(*a*) In general (Sec. 14.5), $\psi = \frac{q_1\theta_1}{2\pi} + \frac{q_2\theta_2}{2\pi} = \frac{8\pi}{2\pi}\arctan\frac{y}{x-2} + \frac{16\pi}{2\pi}\arctan\frac{y}{x+3}$

Per Eq. 14.16 (differentiating and simplifying): $u = 4\frac{(x-2)}{(x-2)^2+y^2} + 8\frac{(x+3)}{(x+3)^2+y^2}$

Due to symmetry, stagnation point is on x-axis ($y = 0$). If $y = 0$, $u = \frac{4}{x-2} + \frac{8}{x+3}$

At stagnation point, $u = 0 = \frac{4}{x-2} + \frac{8}{x+3}$; thus $x = 1/3$ ◀

(*b*) ψ's at various points may be found by substituting coordinates of the point into the general expression for ψ (above). Lines of equal u may then be plotted by interpolation. ◀

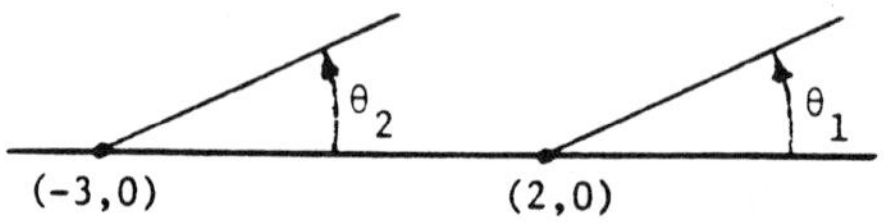

(*c*) At (0, 2): $\psi = 4\arctan\frac{2}{-2} + 8\arctan\frac{2}{3} = 4(135°/57.3°) + 8(33.7°/57.3°) = 14.13$ ◀

At $(3, -1)$: $\psi = 4\arctan\frac{-1}{1} + 8\arctan\frac{-1}{6} = 4(315°/57.3°) + 8(350.5°/57.3°) = 70.9$ ◀

(*d*) Differentiating ψ per Eq. 14.16 and simplifying: $u = \frac{\partial\psi}{\partial y} = 4\left[\frac{(x-2)}{(x-2)^2+y^2}\right] + 8\left[\frac{(x+3)}{(x+3)^2+y^2}\right]$

At $(-2, 5)$: $u = 4\left[\frac{-4}{16+25}\right] + 8\left[\frac{1}{1+25}\right] = -0.0826$

Similarly, $\upsilon = -\frac{\partial\psi}{\partial x} = 4\left[\frac{y}{(x-2)^2+y^2}\right] + 8\left[\frac{y}{(x+3)^2+y^2}\right]$

At $(-2, 5)$: $\upsilon = 4\left[\frac{5}{16+25}\right] + 8\left[\frac{5}{1+25}\right] = 2.03$

Adding vectorially, $V = \sqrt{u^2+\upsilon^2} = 2.03$, upwards at an angle of $2°\,18'$ from the vertical ◀

/more...

14.10 *Using the method described in Sample Prob. 14.4, plot the boundary of the body and a set of streamlines for a steady two-dimensional flow past a body such as that of Fig. 4.12, for b = 15 m using a scale of 1 cm = 2 m.*

SI

For Fig. 4.12, $\psi = Uy + (q/2\pi)\tan^{-1}[y/(x - a)]$

Placing the source at the origin ($a = 0$), with $q = bU = 15U$

$\psi = U[y + (15/2\pi)\tan^{-1}(y/x)]$

Note from Fig. 4.12: Stagnation point S is on the x-axis at $x = -b/2\pi = -15/2\pi = -2.39$ m

x m	y m	y/x	$\tan^{-1}(y/x)$ degrees	$\tan^{-1}(y/x)$ radians $= A$	$\psi/U = y + 15A/2\pi$
−15	0	0	0	0	0
−15	2.5	−0.167	−9.4	−0.165	2.11
−15	5	−0.333	−18.4	−0.322	4.23
−15	7.5	−0.500	−26.6	−0.464	6.39
−15	10	−0.667	−33.7	−0.588	8.60
−5	2.5	−0.5	−26.6	−0.464	1.39
−5	5	−1.0	−45.0	−0.785	3.13
−5	7.5	−1.5	−56.3	−0.983	5.15
−5	10	−2.0	−63.4	−1.107	7.36
5	0	0	−180.0	−3.142	−7.50
5	2.5	0.5	−153.4	−2.678	−3.89
5	5	1.0	−135.0	−2.356	−0.63
5	7.5	1.5	−123.7	−2.159	2.35
5	10	2.0	−116.6	−2.034	5.14
etc.					

Note: Need $-180° \le A \le 0$ for consistency $[\tan(A - \pi) = \tan A]$.

Plot ψ/U values and interpolate for flow lines:

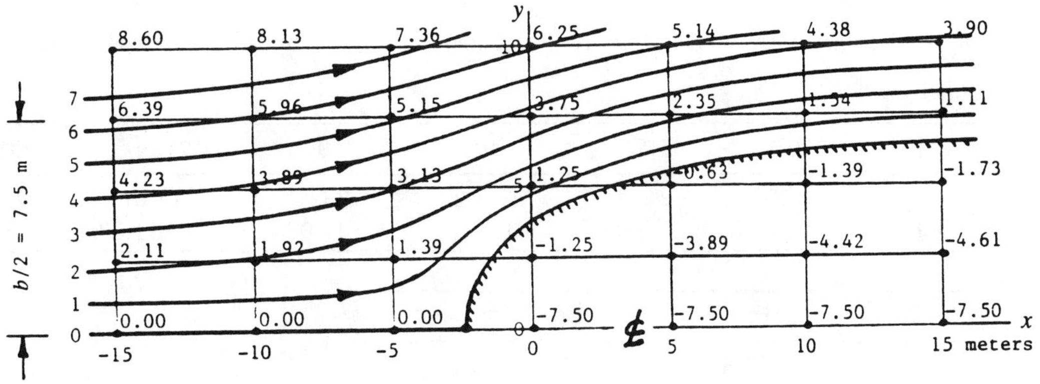

14.11 *Combine the uniform flow defined by $u = 16$ fps with the doublet $2qa = m$, where $q = 10$ cfs/ft and $a = 2$ in. Sketch the streamlines for $\psi = -3, -2, -1, -½, 0, ½, 1, 2, 3$ cfs/ft. Use the scale of 1 in = 1 in.*

BG

From Eq. 14.18: $\psi = 16y + 1.592\left(\tan^{-1}\dfrac{y}{x+2} - \tan^{-1}\dfrac{y}{x-2}\right)$

Substitute various values of x and y and compute the corresponding u's. Plot the points and sketch the streamlines.

To find location of $\psi = 0$ on y-axis ($x = 0$):

$0 = 16y + 1.592\left(\tan^{-1}\dfrac{y}{2} - \tan^{-1}\dfrac{y}{-2}\right)$; by trial $y = 0.283$ in ◀

To find location of stagnation points:

$\psi = 0$, note that $\tan\theta = \theta$ for small angles. $\psi = 0 = 16y + 1.592\left(\dfrac{y}{x+2} - \dfrac{y}{x-2}\right)$

$16y(x^2 - 4) + 1.592y(x - 2) - 1.592y(x + 2) = 0$; $16(x^2 - 4) + 1.592(-4) = 0$

$x^2 - 4 = 6.37/16 = 0.398$; $x^2 = 4.398$; $x = \pm 2.10$ in ◀

The streamline pattern is similar to Fig. S14.4. ◀

14.12 *A flow is defined by the stream function $\psi = 15r\sin\theta - 30\ln r - (20/r)\sin\theta$. Sketch this flow field. Calculate the velocities at $r = 3$ for $\theta = 0, 45, 90, 150, 210$, and $315°$.*

N

Given: $\psi = 15r\sin\theta - 30\ln r - (20/r)\sin\theta$

Polar form of Eq. 14.16:

$\upsilon_t = -\partial\psi/\partial r = -15\sin\theta + 30/r - (20/r^2)\sin\theta$; $\upsilon_r = \partial\psi/r\partial\theta = 15\cos\theta - (20/r^2)\cos\theta$

For case where $r = 3$,

$\upsilon_t = -15\sin\theta + 10 - 2.22\sin\theta = 10 - 17.22\sin\theta$; $\upsilon_r = 15\cos\theta - 2.22\cos\theta = 12.78\cos\theta$

For $r = 3$:

θ	0°	45°	90°	150°	210°	315°	
υ_t	10	−2.17	−7.22	1.389	18.61	22.2	◀◀
υ_r	12.78	9.04	0	−11.07	−11.07	9.04	◀◀

θ	$\psi(r = 2)$	$\psi(r = 4)$
0°	−20.8	−41.6
45°	−6.65	−2.70
90°	−0.794	13.41
135°	−6.65	−2.70
180°	−20.8	−41.6
225°	−34.9	−80.5
270°	−40.8	−96.6
315°	−34.9	−80.5

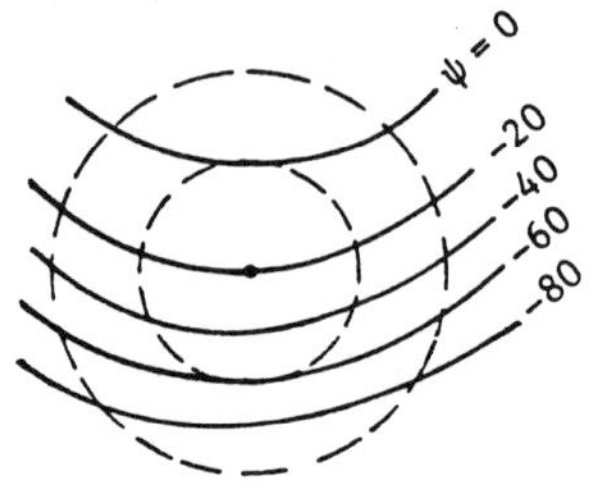

◀

14.13 *Given is the two-dimensional flow described by $u = x^2 + 2x - 4y$, $\upsilon = -2xy - 2y$. (a) Does this satisfy continuity? (b) Compute the vorticity. (c) Plot the velocity vectors for $0 < x < 5$ and $0 < y < 4$ and sketch the general flow pattern. (d) Find the location of all stagnation points in the entire flow field. (e) Find the expression for the stream function.*

N

Note: This problem is based on Sections 14.1 and 14.3–5.

(*a*) From Eq. 14.3: $\dfrac{\partial u}{\partial x} + \dfrac{\partial \upsilon}{\partial y} = 2x + 2 - 2x - 2 = 0$; Continuity is satisfied. ◀

(*b*) From Eq. 14.12: $\xi = \dfrac{\partial \upsilon}{\partial x} - \dfrac{\partial u}{\partial y} = -2y - 4 \neq 0$ ◀ The flow is rotational

(*c*)

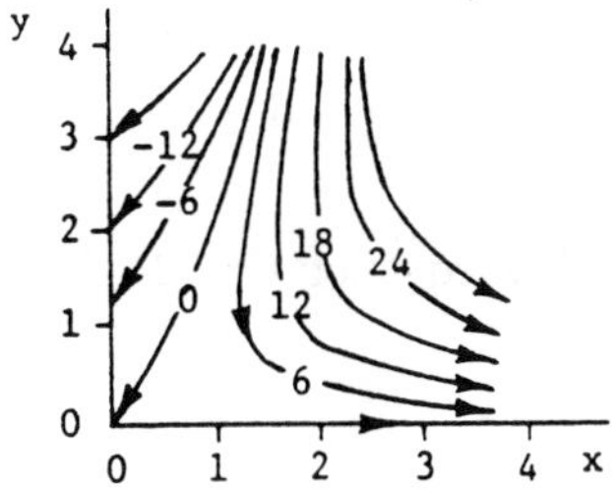

◀

(*d*) Stagnation points occur where both $u = 0$ and $\upsilon = 0$.

$\upsilon = 0 = -2xy - 2y = -2y(x + 2)$; this is true when $y = 0$ or $x = -1$.

$u = 0 = x^2 + 2x - 4y$; If $x = -1$, $y = -1/4$; If $y = 0$, $x = 0$ or -2.

Thus there are three stagnation points: (0, 0), (−2, 0), (−1, −1/4). ◀

(*e*) Eq. 4.16: $u = \dfrac{\partial \psi}{\partial y}$ $\psi = \int u \partial y = \int (x^2 + 2x - 4y)\partial y = x^2y + 2xy - 2y^2 + f(x)$

$\upsilon = -\dfrac{\partial \psi}{\partial x}$ $\psi = -\int \upsilon \partial x = +\int (2xy + 2y)\partial x = x^2y + 2xy + f(y)$

So $\psi = x^2y + 2xy - 2y^2$ ◀ Check: $\dfrac{\partial \psi}{\partial y} = u = x^2 + 2x - 4y$; $\dfrac{\partial \psi}{\partial x} = -\upsilon = 2xy + 2y$

Sec. 14.6: Velocity Potential -- *Exercises*

14.6.1 *If $\phi = y + 2x^2$ is the velocity potential function for a two-dimensional flow, is it irrotational? Does it satisfy the Laplace equation? If not, suggest why.*

N

Eq. 14.23: $\dfrac{\partial \phi}{\partial x} = 4x = -u$; $\dfrac{\partial \phi}{\partial y} = 1 = -\upsilon$ does not satisfy Laplace Eq. 14.25 ◀

Eq. 14.12: $\xi = \dfrac{\partial \upsilon}{\partial x} - \dfrac{\partial u}{\partial y} = 0 - 0 = 0$; Sec.14.6: Yes, it is a potential flow, is irrotational. ◀

Continuity: $\dfrac{\partial \upsilon}{\partial x} + \dfrac{\partial \upsilon}{\partial y} = 4 + 0 = 4$, non-zero, so continuity (Eq. 14.3) is not satisfied. This is the reason why the Laplace equation is not satisfied. ◀

14.6.2 *Given the stream function $\psi = 10x - 7y$, is this a potential flow? If it is, determine the velocity potential function. Does it satisfy the Laplace equation?*

N

Eq. 14.16: $u = \dfrac{\partial\psi}{\partial y} = -7, \quad v = -\dfrac{\partial\psi}{\partial x} = -10$; Eq. 14.12: $\xi = \dfrac{\partial v}{\partial x} - \dfrac{\partial u}{\partial y} = 0$

Yes, this <u>is</u> potential flow, because $\xi = 0$. ◀

From Eq. 14.23: $u = -\dfrac{\partial\phi}{\partial x} = -7, \quad v = -\dfrac{\partial\phi}{\partial y} = -10$

Eq. 14.22: $d\phi = 7dx + 10dy$; integrating: $\phi = 7x + 10y$ ◀

By differentiation: $\dfrac{\partial^2\phi}{\partial x^2} = 0, \quad \dfrac{\partial^2\phi}{\partial y^2} = 0$

So yes, it <u>does</u> satisfy the Laplace equation (Eq. 14.25) in ϕ. ◀

14.6.3 *For the following stream functions, determine if a potential function exists, find the potential function if it does, and determine if the Laplace equation in ϕ is satisfied: (a) $\psi = 3xy + 2x$; (b) $\psi = 3xy + 2x^2$.*

N

(*a*) $\psi = 3xy + 2x$. Eq. 14.16: $u = \dfrac{\partial\psi}{\partial y} = 3x$; $v = -\dfrac{\partial\psi}{\partial x} = -3y - 2$

$\xi = \dfrac{\partial v}{\partial x} - \dfrac{\partial u}{\partial y} = 0 - 0 = 0$ ∴ Yes, it <u>is</u> potential flow, a potential function exists ◀

Eq. 14.23: $\dfrac{\partial\phi}{\partial x} = -u = -3x$; $\dfrac{\partial\phi}{\partial y} = -v = 3y + 2$

$\dfrac{\partial^2\phi}{\partial x^2} = -3$; $\dfrac{\partial^2\phi}{\partial y^2} = 3$. ∴ $\dfrac{\partial^2\phi}{\partial x^2} + \dfrac{\partial^2\phi}{\partial y^2} = 0$, the Laplace equation <u>is</u> satisfied. ◀

Integrating $\dfrac{\partial\phi}{\partial x}$ and $\dfrac{\partial\phi}{\partial y}$: $\phi = \dfrac{-3x^2}{2} + f(y)$; $\phi = 2y + \dfrac{3y^2}{2} + f(x)$

Comparing: $f(y) = 2y + \dfrac{3y^2}{2}$; $f(x) = \dfrac{-3x^2}{2}$. ∴ $\phi = 2y + \dfrac{3y^2}{2} - \dfrac{3x^2}{2} + C$ ◀

(*b*) $\psi = 3xy + 2x^2$. Eq. 14.16: $u = \dfrac{\partial\psi}{\partial y} = 3x$; $v = -\dfrac{\partial\psi}{\partial x} = -3y - 4x$

$\xi = \dfrac{\partial v}{\partial x} - \dfrac{\partial u}{\partial y} = -4 + 0 = -4$, no, it is <u>not</u> potential flow, a potential function does not exist ◀

Eq. 14.23: $\dfrac{\partial\phi}{\partial x} = -u = -3x$; $\dfrac{\partial\phi}{\partial y} = -v = 3y + 4x$

$\dfrac{\partial^2\phi}{\partial x^2} = -3$; $\dfrac{\partial^2\phi}{\partial y^2} = 3$, ∴ yes, the Laplace Eq. 14.25 <u>is</u> satisfied. ◀

14.6.4 *For the following stream functions, determine if a potential function exists, find the potential function if it does, and determine if the Laplace equation in ϕ is satisfied: (a) $\psi = 6xy$; (b) $\psi = x\sin y$*

N

(*a*) $\psi = 6xy$. Eq. 14.16: $u = \dfrac{\partial\psi}{\partial y} = 6x$; $v = -\dfrac{\partial\psi}{\partial x} = -6y$

$\xi = \dfrac{\partial v}{\partial x} - \dfrac{\partial u}{\partial y} = 0 - 0 = 0$; ∴ Yes, it is potential flow, a potential function exists ◀

Eq. 14.23: $\dfrac{\partial\phi}{\partial x} = -u = -6x$; $\dfrac{\partial\phi}{\partial y} = -v = 6y$

Integrating: $\phi = -3x^2 + f(y)$; $\phi = 3y^2 + f(x)$. Comparing: $\phi = 3(y^2 - x^2)$ ◀

$\dfrac{\partial^2\phi}{\partial x^2} = -6$; $\dfrac{\partial^2\phi}{\partial y^2} = 6$ ∴ the Laplace Eq. 14.25 is satisfied ◀

(*b*) $\psi = x\sin y$. Eq. 14.16: $u = \dfrac{\partial\psi}{\partial y} = x\cos y$; $v = -\dfrac{\partial\psi}{\partial x} = -\sin y$

$\xi = \dfrac{\partial v}{\partial x} - \dfrac{\partial u}{\partial y} = 0 - (-x\sin y) \neq 0$ ∴ not potential flow, a potential function does not exist ◀

Eq. 14.23: $\dfrac{\partial\phi}{\partial x} = -u = -x\cos y$; $\dfrac{\partial\phi}{\partial y} = -v = \sin y$

$\dfrac{\partial^2\phi}{\partial x^2} = -\cos y$; $\dfrac{\partial^2\phi}{\partial y^2} = \cos y$ ∴ yes, the Laplace Eq. 14.25 is satisfied ◀

Sec. 14.7: Orthogonality of Streamlines and Equipotential Lines -- *Exercises*

14.7.1 *Which of the flows in Exer. 14.1.1 can be described by a flow net? Write expressions for the stream functions and the potential functions.*

N

By evaluating $\partial u/\partial x + \partial v/\partial y$ (to determine if continuity is satisfied, as in Exer. 14.1.1) and $\partial v/\partial x - \partial u/\partial y$ (to determine if the flow is irrotational, as in Exer. 14.2.1), we find that only flows (*a*) and (*b*) can be represented by a flow net because they satisfy both continuity and irrotationality. ◀

Integrating Eqs. 14.16 and 14.23 we obtain:

(*a*) $\psi = 2y$, $\phi = -2x$ ◀

(*b*) $\psi = 2y - 3x$, $\phi = -3y - 2x$ ◀

14.7.2 *Which of the flows in Exer. 14.1.2 can be described by a flow net? Write expressions for the stream functions and the potential functions.*

N

By evaluating $\partial u/\partial x + \partial v/\partial y$ (to determine if continuity is satisfied, as in Exer. 14.1.1) and $\partial v/\partial x - \partial u/\partial y$ (to determine if the flow is irrotational, as in Exer. 14.2.1), we find that only flows (*c*), (*d*), and (*g*) can be represented by a flow net because they satisfy both continuity and irrotationality. ◀

Integrating Eqs. 14.16 and 14.23 we obtain:

(*c*) $\psi = 3xy$, $\phi = -1.5x^2 + 1.5y^2$ ◀

(*d*) $\psi = 4y + 6x + 2xy$, $\phi = 6y - 4x - x^2 + y^2$ ◀

(*g*) $\psi = x^3/3 - 2x^2y - xy^2 + 2y^3/3$, $\phi = (2/3)x^3 + x^2y - 2xy^2 - (1/3)y^3$ ◀

14.14 *The flow around the body of Fig. 4.12 may be considered as that due to the sum of two velocity potentials, $\phi_1 = -Ux$, representing an undisturbed flow of velocity U in the x direction, and $\phi_2 = -S \ln r$, representing the radial flow from a source located inside the body behind the stagnation point. To relate U and S, it is observed that the total flow $2\pi S$ from the source (which is hydrodynamically equivalent to the body itself) must be equal to the flow of the main stream which is not passing through the body of width b, or $2\pi S = Ub$. This gives*

$$\phi_2 = -\frac{Ub}{2\pi} \ln r$$

(a) The distance from the stagnation point to the source is determined by setting the radial velocity from the source, $v_r = -\partial\phi/\partial r$, equal and opposite to the undisturbed velocity U. Prove that this establishes the source at a distance $b/2\pi$ behind the stagnation point. The absolute velocity at any point of the field may be determined by the vector sum of the components U and v_r.

There follows an ingenious method of plotting the boundary of such a streamlined body, as shown in Fig. P14.14. Suppose that the streamlines in the undisturbed flow are spaced a distance a apart, where $b/2a = n$, an integer. Next divide the upper half of the source into n radial sectors, each of angle α, that is, $n\alpha = \pi$. Then the undisturbed flow between the x axis and the first streamline is associated with the source flow in the first sector from the stagnation point. Thus the intersection of the first streamline with the first line must be a point on the boundary of the body, through which there can be no flow. Similarly, the intersection of the horizontal line at 2a with the radial line at 2α forms another point, and so on. Further streamlines can be plotted by connecting successive intersections of the original horizontal lines with the radial lines, recognizing that the same flow must exist between any adjacent pair of streamlines. Thus the intersection of a horizontal line ea above the axis with a radial line at fα from the stagnation point must lie on a streamline which is (e − f)a distant from the axis in the undisturbed region, where e and f are integers.

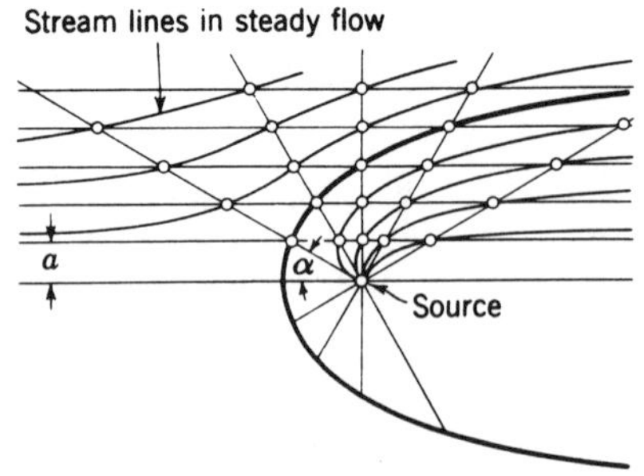

Figure P14.14

(b) Assume a value of U = 20 fps and a two-dimensional flow past a streamlined body for which b = 36 ft. Compute the distance from the source to the stagnation point and to the surface of the body at a radius of 90° to the axis. What is the value of the source velocity at the latter point?
(c) What is the magnitude of the velocity of the fluid along the surface at the 90° point? (Compare with the results of Sample Prob. 4.2.) What is its direction relative to the axis?

BG

(*a*) Given: Source potential, $\phi_2 = (Ub/2\pi)\ln r$; ∴ (Eq. 14.24) $v_r = -\partial\phi/\partial r = Ub/2\pi r$

Setting $v_r = U$ for stagnation condition (on tip of nose) and solving for r, the desired distance behind the nose, $r = b/2\pi$ Q.E.D. ◄

(*b*) For $b = 36$ ft, distance to stagnation point $= -36/2\pi = -5.73$ ft ◄

Since $b/2a = n = 6 = 36/2a$ = the number of radial sectors from half of source,

streamline spacing $= a = 36/12 = 3$ ft

From Fig. P14.14: At 90° $= 3\alpha$, $r = 3a = 3(3) = 9$ ft ◄

Source velocity at 90°, $r = 3a$: $v_r = \dfrac{Ub}{2\pi r} = \dfrac{20(36)}{2\pi(9)} = 12.73$ fps ◄

(*c*) At 90°, $V = \sqrt{U^2 + v_r^2} = \sqrt{20^2 + 12.73^2} = 23.7$ fps ◄

Thus, on the surface at 90°, $V/U = 23.7/20 = 1.185$

In Sample Prob. 4.2 at 90° (≈ 1.6 ft from the stagnation point), $V/U \approx 12/10 = 1.2$

Thus the two agree closely ◄

Direction relative to axis is $\tan^{-1}(12.73/20) = 32.5°$ ◄

14.15 SI

Solve Prob. 14.14 in SI units; for parts (b) and (c) use $U = 5$ m/s and $b = 12$ m.

(*a*) Given: Source potential, $\phi_2 = (Ub/2\pi)\ln r$; $\therefore$ (Eq. 14.24) $v_r = -\partial\phi/\partial r = Ub/2\pi r$

Setting $v_r = U$ for stagnation condition (on tip of nose) and solving for r, the desired distance behind the nose, $r = b/2\pi$ Q.E.D. ◄

(*b*) For $b = 12$ m, distance to stagnation point $= -12/2\pi = -1.910$ m ◄

Since $b/2a = 12/2a = n = 6$ = the number of radial sectors from half of source,

streamline spacing $= a = 12/12 = 1$ m

Distance (source to surface at 90°) $= r = 3a = 3(1\text{ m}) = 3$ m ◄

Source velocity at 90°, $r = 3a$: $v_r = \dfrac{Ub}{2\pi r} = \dfrac{5(12)}{2\pi(3)} = 3.18$ m/s ◄

(*c*) At 90°, $r = 3a$: $V = \sqrt{U^2 + v_r^2} = \sqrt{5^2 + 3.18^2} = 5.93$ m/s ◄

Thus, on the surface at 90°, $V/U = 5.93/5 = 1.185$

In Sample Prob. 4.2 at 90° ($\approx$ 1.6 ft from the stagnation point), $V/U \approx 12/10 = 1.2$

Thus the two agree closely ◄

Direction relative to axis is $\tan^{-1}(3.18/5) = 32.5°$ ◄

14.16 BG

Find the distance to the surface of the body and the two velocities called for in Prob. 14.14 for an angle of 30°. (Compare with Sample Prob. 4.2.)

At 30°, $r = a/\sin 30° = 3/0.5 = 6.00$ ft ◄

$$v_r = \frac{\partial\phi}{\partial r} = \frac{Ub}{2\pi r} = \frac{20(36)}{2\pi(6.0)} = 19.10 \text{ fps} \quad ◄$$

$V^2 = U^2 + v_r^2 - 2Uv_r\cos\alpha = 20^2 + 19.10^2 - 2(20)19.10\cos 30° = 400 + 365 - 662 = 103$

$V = 10.16$ fps ◄ (Comparison with Sample Prob. 4.2 is good.)

14.17 SI

Find the distance to the surface of the body and the two velocities called for in Prob. 14.15 for an angle of 30°. (See Sample Prob. 4.2.)

At 30°, $r = a/\sin 30° = 1/0.5 = 2$ m ◄

$$v_r = \frac{\partial\phi}{\partial r} = \frac{Ub}{2\pi r} = \frac{5(12)}{2\pi(2)} = 4.77 \text{ m/s} \quad ◄$$

$V^2 = U^2 + v_r^2 - 2Uv_r\cos\alpha = 5^2 + 4.77^2 - 2(5)4.77\cos 30° = 6.45$; $V = 2.54$ m/s ◄

/more...

14.18 *An ideal fluid flows in a two-dimensional 90° bend. The inner and outer radii of the bend are 0.6 and 1.6 ft. (a) Sketch the flow net and estimate the velocity at the inner and outer walls of the bend if the velocity in the 1.0-ft-wide straight section is 8 fps. (b) Develop an analytic expression for the stream function, in this case noting that* $\upsilon_t = -\dfrac{\partial\psi}{\partial r}$ *and* $\upsilon_r = \dfrac{\partial\psi}{r\partial\theta}$. *Determine the inner and outer velocities accurately.*

BG

(*a*) Velocity estimates:

Sec. 14.7: $V = -\dfrac{\partial\phi}{\partial s}$

$\therefore\ -\Delta\phi \approx V\Delta s \approx 8\Delta s_0$ so $V \approx 12\,\Delta s_0/\Delta s$

From sketch: $V_{outside} \approx 8(4\text{ mm}/7\text{ mm}) = 4.57$ fps ◄

$V_{inside} \approx 8(4\text{ mm}/2\text{ mm}) = 16$ fps ◄

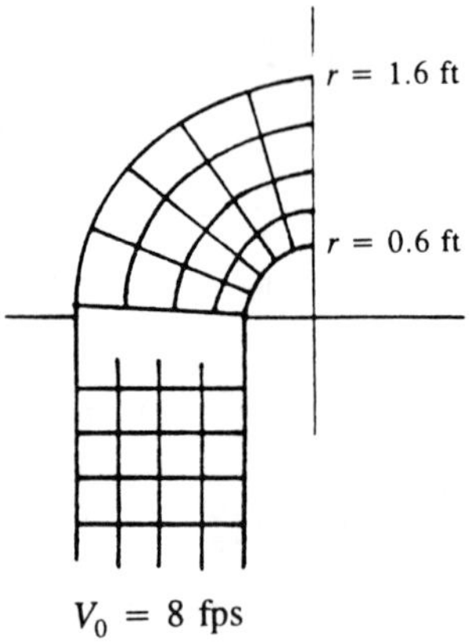

◄

(*b*) Accurate velocity determinations:

$\upsilon_t(\Delta r)$ = constant, Δr is proportional to r

$\upsilon_t(kr)$ = constant, therefore $\upsilon_t = C/r$. Also $\upsilon_r = 0$

$\psi = \psi(r, \theta)$, so by chain rule of differentiation

$$d\psi = \frac{\partial\psi}{r\partial\theta}rd\theta + \frac{\partial\psi}{\partial r}dr = \upsilon_r rd\theta - \upsilon_t dr = 0 - \upsilon_t dr$$

If $\psi_{inside} = 0$, $\psi_{outside} = \psi_{inside} + dq = 0 + (1.0\text{ ft})(8\text{ fps}) = 8$ cfs/ft

Integrating: $\int_0^8 d\psi = -\int_{0.6}^{1.6}\upsilon_t dr = -C\int_{0.6}^{1.6}dr/r$

$\psi = 8 = -C[\ln r]_{0.6}^{1.6} = -C\ln(1.6/0.6) = -C\ln 2.67$ and $C = -8.16$

Thus the analytical expression is: $\psi = 8.16\ln r$ ◄

$\upsilon_t = C/r = -8.16/r$; $(\upsilon_t)_{r=0.6} = 13.59$ fps ◄

$(\upsilon_t)_{r=1.6} = 5.10$ fps ◄

14.19 *An ideal fluid flows in a two-dimensional 90° bend. The inner and outer radii of the bend are 0.2 and 0.6 m. (a) Sketch the flow net and estimate the velocity at the inner and outer walls of the bend if the velocity in the 0.4-m-wide straight section is 2.5 m/s. (b) Develop an analytic expression for the stream function, in this case noting that* $\upsilon_t = -\dfrac{\partial\psi}{\partial r}$ *and* $\upsilon_r = \dfrac{\partial\psi}{r\partial\theta}$. *Determine the inner and outer velocities accurately.*

SI

(*a*) Velocity estimates: Sec 14.7: $V = -\dfrac{\partial\phi}{\partial s}$ $\therefore\ -\Delta\phi \approx V\Delta s \approx 2.5\,\Delta s_0$; $V \approx 2.5\,\Delta s_0/\Delta s$

From sketch (with Solution 14.18): $V_{outside} \approx 2.5(4\text{ mm}/7\text{ mm}) = 1.429$ m/s ◄

$V_{inside} \approx 2.5(4\text{ mm}/2\text{ mm}) = 5$ m/s ◄

(*b*) Accurate velocity determinations:

$\upsilon_t(\Delta r)$ = constant, Δr is proportional to r

$\upsilon_t(kr)$ = constant, therefore $\upsilon_t = C/r$. Also $\upsilon_r = 0$

$\psi = \psi(r, \theta)$, so by chain rule of differentiation

$$d\psi = \frac{\partial\psi}{r\partial\theta}rd\theta + \frac{\partial\psi}{\partial r}dr = \upsilon_r rd\theta - \upsilon_t dr = 0 - \upsilon_t dr$$

/cont...

If $\psi_{inside} = 0, \quad \psi_{outside} = \psi_{inside} + dq = 0 + (0.4 \text{ m})(2.5 \text{ m/s}) = 1 \text{ m}^3\text{/s per m}$

Integrating: $\int_0^{2.5} d\psi = \int_{0.2}^{0.6} v_t \, dr = -C\int_{0.2}^{0.6} dr/r$

$\psi = 1 = -C[\ln r]_{0.2}^{0.6} = -C\ln(0.6/0.2) = -C\ln 3 \; ; \qquad C = -0.910$ ◄

Thus the analytical expression is $\psi = 0.910 \ln r$ ◄

$v_t = C/r = -0.910/r \; ; \quad (v_t)_{r=0.2} = 4.55$ m/s ◄ $(v_t)_{r=0.6} = 1.517$ m/s ◄

14.20 *For the two-dimensional flow of a frictionless incompressible fluid against a flat plate normal to the initial velocity, the stream function is given by $\psi = -2axy$, while its conjugate function, the velocity potential, is*

$$\phi = a(x^2 - y^2)$$

where a is a constant and the flow is symmetrical about the yz plane (Fig. 14.1). By direct differentiation, demonstrate that these functions satisfy Eq. (14.25). Using a scale of 1 in = 1 unit of distance, plot the streamlines given by $\psi = \pm 2a, \pm 4a, \pm 6a, \pm 8a$, and the equipotential lines given by $\phi = 0, \pm 2a, \pm 4a, \pm 6a, \pm 8a$. Observe that this flow net also gives the ideal flow around an inside square corner. Compare your results with Sample Prob. 14.5 and note the effect of changing the sign of ψ and ϕ.

N

Differentiating: $\partial\psi/\partial y = -2ax; \; \partial\phi/\partial x = 2ax; \; -\partial\psi/\partial x = 2ay; \; \partial\phi/\partial y = a(-2y);$

$\partial^2\phi/\partial x^2 = 2a; \; \partial^2\phi/\partial y^2 = -2a;$ Thus Eq. 14.25 is satisfied ◄

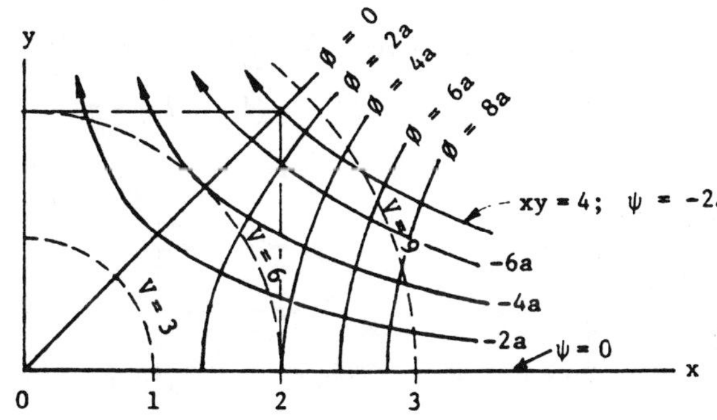

◄

14.21 *In Prob. 14.20 determine the velocity components u and v, and demonstrate that they satisfy the differential equations for continuity and irrotational flow. In which direction is the flow? Prove that the absolute velocity is given by V = 2ar, where r is the radius to the point from the origin. Now assume that the linear scale is 1 in = 1 ft. Determine the constant a such that the flow net of Prob. 14.20 will represent a flow of 3 ft²/s between any two adjacent streamlines. What are the dimensions of a? Draw curves of equal velocity for values of 3, 6, 9, 12 fps. How does the velocity vary along the surface of the plate?*

BG

Given: $\psi = -2axy$, then: $u = \partial\psi/\partial y = -2ax; \; v = -\partial\psi/\partial x = 2ay$

Differentiating further: $(\partial u/\partial x) + (\partial v/\partial y) = -2a + 2a = 0$, so continuity (Sec. 14.4) is satisfied ◄

$(\partial v/\partial x) - (\partial u/\partial y) = 0 - 0 = 0$, so (from Eq. 14.6) flow is irrotational ◄

$V = \sqrt{u^2 + v^2} = \sqrt{4a^2x^2 + 4a^2y^2} = 2a\sqrt{x^2 + y^2} = 2ar$ ◄

Flow between two streamlines in flow net of Prob. 14.20 is given by $\psi_2 - \psi_1 = 2a$

$\therefore \; 2a = 3$ ft²/sec (given); $a = 3/2 \text{ sec}^{-1}$ ◄

Dimensions of "a" are most clearly seen from $V = 2ar; \; a = V/2r = (L/T)/L = T^{-1}$ ◄

$V \propto r$, so the velocity varies linearly along the plate. ◄

14.22 *A cylindrical drum with a 2 ft radius is securely held in position in an open channel of rectangular section. The channel is 10 ft wide, and the flow rate is 240 cfs. Water flows beneath the drum as shown in Fig. P14.22. Sketch the flow net, and determine from flow net measurements the pressure at the points indicated along the wetted drum surface. Neglect fluid friction. Sketch the pressure distribution, and by numerical integration determine an approximate value of the horizontal thrust on the cylinder. (Compare with Exer. 6.4.1.)*

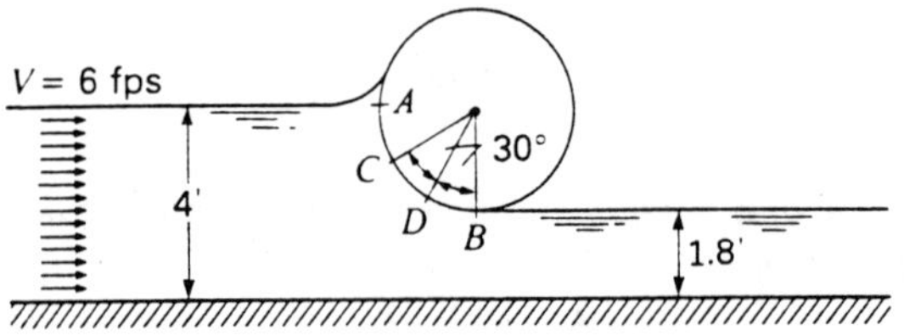

Figure P14.22

BG

Sketch the streamlines by eye and use continuity to determine the velocities. A high degree of accuracy is not expected.

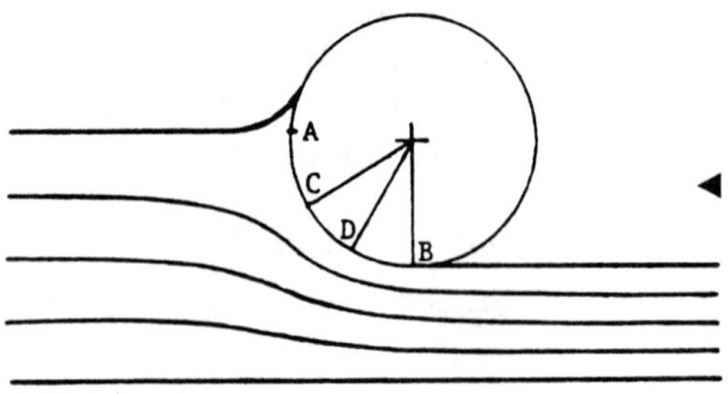

◄

Energy Eq. from upstream to point on drum:

At A, $p/\gamma = p_s/\gamma = V^2/2g = 6^2/2g = 0.559$ ft ◄

At B, in contact with the atmosphere ($p_{atm} = 0$ psig):

$p/\gamma = 0$ ◄

At C, from flow net, $V \approx 7$ fps. $4 + 6^2/2g \approx (4 - 1.0) + p/\gamma + 7^2/2g$; $p/\gamma \approx 0.798$ ft. ◄

At D, from flow net, $V \approx 11.5$ fps. $4 + 6^2/2g \approx (4 - 1.73) + p/\gamma + 11.5^2/2g$; $p/\gamma \approx 0.235$ ft ◄

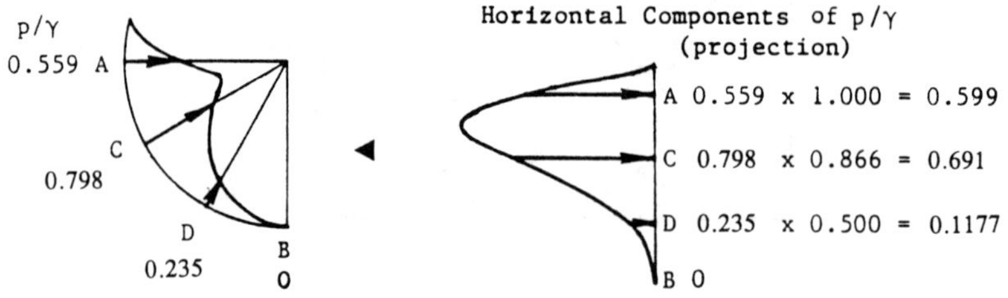

◄

Area under projected p/γ diagram $\approx$ average horizontal component = Area/2.2 = 0.44 ft = 0.19 psi

Horizontal thrust $\approx 0.19(2.2 \times 12)(15 \times 12) = 547$ lb ◄

Note to instructor: See Exer. 6.4.1 for the application of impulse-momentum to this problem.

14.23 *Refer to Sample Prob. 6.1. Sketch a flow net. Using the given dimensions in BG units through application of Bernoulli's principle, determine the approximate pressure distribution along the channel bottom and around the curved structure. By numerical integration estimate the magnitude of the horizontal and vertical components of the force of the water on the structure.*

Sample Prob. 6.1: Q = 481 cfs = 48.1 cfs/ft.

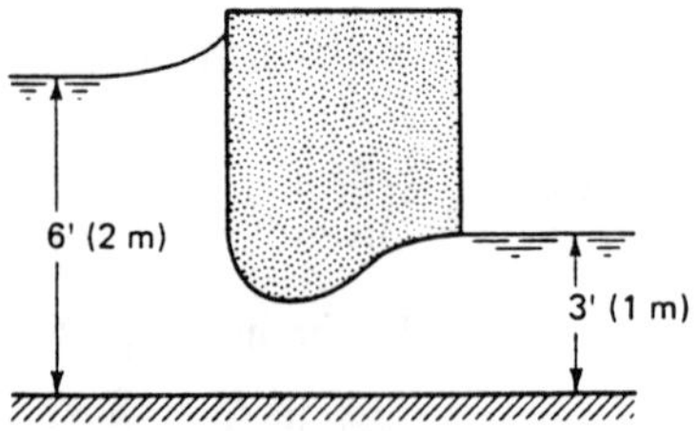

Figure S6.1

BG

$V_A = 48.1/6.0 = 8.02$ ft ; $V_B = 48.1/1 \approx 16$ fps

$V_C = 48.1/2 \approx 24$ fps ; $V_D = 48.1/3.0 = 16.03$ fps

Energy equation between points:

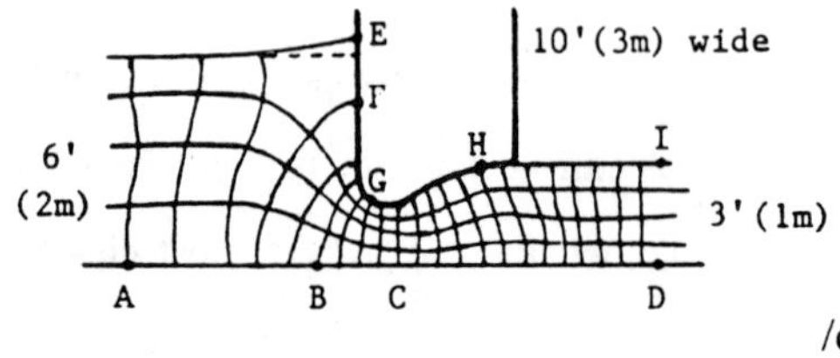

◄

$$6 + \frac{V_A^2}{2g} = \frac{p_B}{\gamma} + \frac{V_B^2}{2g} = \frac{p_C}{\gamma} + \frac{V_C^2}{2g} = \frac{p_D}{\gamma} + \frac{V_D^2}{2g}$$

/cont...

$$6 + 1.00 \approx \frac{p_B}{\gamma} + \frac{16^2}{64.4} \approx \frac{p_C}{\gamma} + \frac{24^2}{64.4} = \frac{p_D}{\gamma} + \frac{16.03^2}{64.4}$$

from which $p_B/\gamma \approx 3.02$ ft, $p_C/\gamma \approx -1.944$ ft, $p_D/\gamma = 3.01$ ft ◄◄

Selecting arbitrary points F, G, and H: $V_E = 0$, $V_F \approx 48.1/6 = 8$ fps

$V_G = 48.1/1.6 \approx 30$ fps, $V_H \approx 48.1/2.8 = 17$ fps, $V_I = 16.03$ fps

Scaling off the sketch: $z_F \approx 4$ ft, $z_G \approx 1.7$ ft, $z_H \approx 2.8$ ft; $z_I = 3.0$ ft (given);

$z_E = 7.0$ ft (from above calculations).

Energy: $$7 = 4 + \frac{p_F}{\gamma} + \frac{V_F^2}{2g} = 1.7 + \frac{p_G}{\gamma} + \frac{V_G^2}{2g} = 2.8 + \frac{p_H}{\gamma} + \frac{V_H^2}{2g} = 3.0 + \frac{p_I}{\gamma} + \frac{V_I^2}{2g}$$

Substituting the approximate values of V gives:

$$\frac{p_E}{\gamma} = 0, \quad \frac{p_F}{\gamma} \approx 2 \text{ ft}, \quad \frac{p_G}{\gamma} \approx -8.7 \text{ ft}, \quad \frac{p_H}{\gamma} \approx -0.5 \text{ ft}, \quad \frac{p_I}{\gamma} = 0$$ ◄◄

With careful detail a reasonably accurate estimate of the horizontal and vertical components of the force on the structure can be obtained. ◄

(From Sample Prob. 6.1, by impulse-momentum analysis, $F_x = 936$ lb.)

14.24 *Work Prob. 14.23 using the dimensions as given in SI units in Sample Prob. 6.1.*
Refer to Fig. S6.1 with Prob. 14.23. Sample Prob. 6.1: $Q = 15.34$ m³/s $= 5.11$ m³/s per meter.

SI

Refer to sketch for above solution to Prob 14.23.

$V_A = 5.11/2.0 = 2.56$ m/s ; $V_B \approx 5.11/1.0 = 5.11$ m/s

$V_C \approx 5.11/0.67 = 7.63$ m/s ; $V_D = 5.11/1.0 = 5.11$ m/s

Energy equation between points: $$2 + \frac{V_A^2}{2g} = \frac{p_B}{\gamma} + \frac{V_B^2}{2g} = \frac{p_C}{\gamma} + \frac{V_C^2}{2g} = \frac{p_D}{\gamma} + \frac{V_D^2}{2g}$$

from which $p_B/\gamma \approx 1.0$ m, $p_C/\gamma \approx -0.637$ m, $p_D/\gamma = 1.0$ m ◄◄

Selecting arbitrary points F, G, and H: $V_E = 0$, $V_F \approx 5.11/2 = 2.56$ m/s

$V_G \approx 5.11/0.5 = 10.23$ m/s, $V_H \approx 5.11/0.95 = 5.38$ m/s, $V_I = 5.11$ m/s

Scaling off sketch: $z_F \approx 1.3$ m, $z_G \approx 0.6$ m, $z_H \approx 0.95$ m ; $z_I = 1$ m (given);

$z_E = 2.33$ m (from above calculations).

$$\therefore\ 2.33 = 1.3 + \frac{p_F}{\gamma} + \frac{V_F^2}{2g} = 0.6 + \frac{p_G}{\gamma} + \frac{V_G^2}{2g} = 0.95 + \frac{p_H}{\gamma} + \frac{V_H^2}{2g} = 1.0 + \frac{p_I}{\gamma} + \frac{V_I^2}{2g}$$

Substituting the approximate values of V gives:

$$\frac{p_E}{\gamma} = 0, \quad \frac{p_F}{\gamma} \approx 0.7 \text{ m}, \quad \frac{p_G}{\gamma} \approx -3.6 \text{ m}, \quad \frac{p_H}{\gamma} \approx -0.094 \text{ m}, \quad \frac{p_I}{\gamma} = 0$$ ◄◄

With careful detail a reasonably accurate estimate of the horizontal and vertical components of the force on the structure can be obtained. ◄

(From Sample Prob. 6.1, by impulse-momentum analysis, $F_x = 4.91$ kN)

14.25 *The three-dimensional counterpart of the flow in Probs. 14.20 and 14.21 is that of flow along the y axis approaching the plate in the xz plane. As the flow must be symmetrical about the y axis, the traces of stream and equipotential surfaces in the xy plane will be representative of those in all planes containing the y axis. The velocity potential is now given by $\phi = -a(0.5x^2 - y^2)$, and the stream function by $\psi = -ax^2y$. Notice that these functions no longer satisfy Eq. (14.25). Why not? Again plot streamlines and equipotential lines for the values given in Prob. 14.20. The velocities u and υ may still be determined by Eq. (14.23). Prove that the absolute velocity for this case is given by $V = a\sqrt{x^2+4y^2}$. With the value of a = 1.5sec^{-1} found in Prob. 14.21, draw curves of equal velocity for values of 3, 6, 9, 12 fps. How does the velocity vary along the surface? What is the total flow between any adjacent stream surfaces?*

BG

Given: $\phi = -a(0.5x^2 - y^2)$; $\psi = -ax^2y$. Eq. 14.23: $u = -\partial\phi/\partial x = ax$; $\upsilon = -\partial\phi/\partial y = -2ay$

Adding vectorially: $V = \sqrt{u^2 + \upsilon^2} = a\sqrt{x^2 + 4y^2}$, Q.E.D. ◄

Substituting for $a = 1.5$ sec^{-1} and $V = 3$ fps: $3 = 1.5\sqrt{x^2 + 4y^2}$

or $x^2 + 4y^2 = 2^2 = 4$; from which $x = 2\sqrt{1 - y^2}$

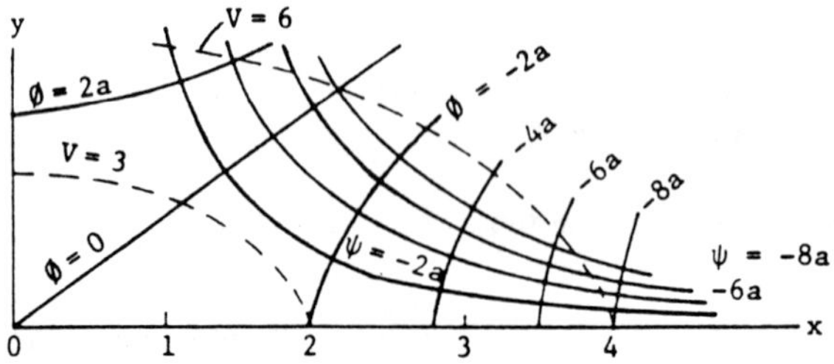

◄

On the surface ($y = 0$), the velocity varies linearly with the distance from the y-axis. ◄

Flow between two stream surfaces is $Q = 2\pi(\Delta\psi) = 2\pi(2a) = 2\pi(2)(3/2) = 6\pi$ cfs ◄

Note: Although Eq. 14.25 is not satisfied as it stands, the Laplace equation is still valid for cylindrical

coordinates: $$\frac{\partial^2\phi}{\partial r^2} + \frac{1}{r}\frac{\partial\phi}{\partial r} + \frac{1}{r^2}\frac{\partial^2\phi}{\partial\theta^2} + \frac{\partial^2\phi}{\partial z^2} = 0$$

In this example, "x" is the radial direction while "y" is the z-direction. Also $\partial\phi/\partial\theta = 0$

/more...

14.26 *For the two-dimensional flow around any angle α, the velocity potential and stream function are given in polar coordinates as $\phi = -ar^{\pi/\alpha}\cos(\pi\theta/\alpha)$ and $\psi = -ar^{\pi/\alpha}\sin(\pi\theta/\alpha)$, respectively. Prove that the functions given in Prob. 14.20 are a specialization of these expressions for $\alpha = \pi/2$. Take the case of $\alpha = 3\pi/2$, and plot streamlines and equipotential lines for the values given in Prob. 14.20. Compare the velocity at the corner with that at the corner in Prob. 14.20.*

N

Given: $\phi = -ar^{\pi/\alpha}\cos(\pi\theta/\alpha)$; $\psi = -ar^{\pi/\alpha}\sin(\pi\theta/\alpha)$

Substituting for $\alpha = \pi/2$: $\phi = -ar^2\cos(2\pi\theta/\pi)$; $\psi = -ar^2\sin 2\theta$.

For rectangular coordinates, $x = r\cos\theta$, $y = r\sin\theta$; $r = \sqrt{x^2 + y^2}$

Thus: $\phi = -ar^2\cos 2\theta = -ar^2(2\cos^2\theta - 1) = -ar^2[2(x^2/r^2) - 1]$

$= -ar^2(2x^2 - r^2)/r^2 = -a[2x^2 - (x^2 + y^2)] = -a(x^2 - y^2)$ Q.E.D. (cf Prob. 14.20) ◄

$\psi = -ar^2\sin 2\theta = -ar^2(2\sin\theta\cos\theta) = -ar^2[2(y/r)(x/r)] = -2axy$ Q.E.D. (cf Prob. 14.20) ◄

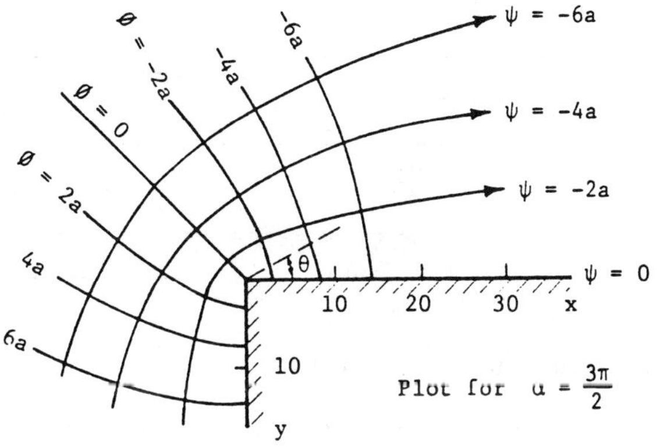

◄

For $\alpha = 3\pi/2$: At corner, for $\phi = -ar^{2/3}\cos(2\theta/3)$

Eq. 14.24: $$\upsilon_t = -\frac{1}{r}\frac{\partial\phi}{\partial\theta} = -\frac{1}{r}(-ar^{2/3})\left(-\frac{2}{3}\sin\frac{2}{3}\theta\right) = -\frac{2}{3}\frac{a}{r^{1/3}}\sin\frac{2}{3}\theta$$

This reveals that as $r \to 0$, $\upsilon_t \to \infty$ ◄ while for inside corner, as $r \to 0$, $\upsilon_t \to 0$ ◄

In either case, $\upsilon_r = 0$ along a line bisecting the total angle θ ◄

14.27 *Superimpose a point source (Q = 100 cfs) on a rectilinear flow field (U = 20 fps). Plot the body contour at θ = 30, 60, 90, 120, 150, 180° using a scale of 1 in = 1 ft. Compute the velocities along the body contour at these points. Determine the pressures at these points assuming ρ = 1.94 slug/ft^3 with zero pressure in the undisturbed rectilinear flow field. What is the velocity and pressure in the combined flow field at the following points? Hint: Refer to Prob. 14.14.*

(a) $\theta = 45°$, $r = 4.0$ ft
(b) $\theta = 90°$, $r = 2.0$ ft
(c) $\theta = 90°$, $r = 4.0$ ft
(d) $\theta = 135°$, $r = 2.0$ ft

BG

Given: $U = 20$ fps, $\upsilon_r = 100/2\pi r$ fps

Distance from source to stagnation point: $U = \upsilon_r$; $r = 5/2\pi = 0.796$ ft

Refer to Prob. 14.14; for upper half let $n = 6$, thus $\theta = 30°$

$q = Ua = 100/2n$; thus $a = 100/(2nU) = 0.417$ ft

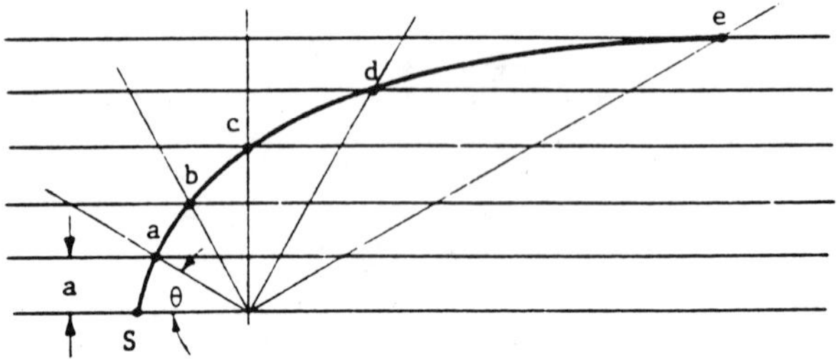

Scale off radial distances to points a, b, c, d, e and f

$\upsilon_x = U - \upsilon_r \cos\theta$; $\upsilon_y = \upsilon_r \sin\alpha$. Vector addition: $\upsilon = \sqrt{\upsilon_x^2 + \upsilon_y^2}$

Point	r (ft)	θ	υ_r	υ_x	υ_y	υ (fps)
S	0.80	0°	20.0	0	0	0
a	0.81	30°	19.64	2.99	9.90	10.34
b	0.96	60°	16.57	11.71	14.46	18.71
c	1.24	90°	12.83	20.0	12.83	23.8
d	1.90	120°	8.38	24.2	7.26	25.3
e	4.16	150°	3.83	23.3	1.915	23.4

▲

$p_0 + \rho V_0^2/2 = p + \rho V^2/2$. $0 + 1.940(20^2/2) = 388$ lb/ft^2 $= p + \rho V^2/2$; $\therefore\ p = 388 - \rho V^2/2$

Hence the pressures at 0°, 30°, etc. are: 388, 284, 48.4, −161, −232, −143 psf, respectively ◀

At other points in the flow field:

(*a*) $\upsilon_x = [20 - 100/(2\pi \times 4)]\cos 45°$; $\upsilon_y = [100/(2\pi \times 4)]\sin 45°$

$\upsilon_x = 11.33$ fps; $\upsilon_y = 2.81$ fps; $\upsilon = 11.67$ fps ◀

$p = \rho 20^2/2 - \rho V^2/2 = 256$ psf ◀

(*b*) Similarly $\upsilon_x = 20.0$ fps, $\upsilon_y = 7.95$ fps, $\upsilon = 21.5$ fps, $p = -60.3$ psf ◀

(*c*) $\upsilon_x = 20.0$ fps, $\upsilon_y = 4.98$ fps, $\upsilon = 20.6$ fps, $p = -23.6$ psf ◀

(*d*) This point is inside the bounds of the body. Therefore, it is not in the flow field. ◀

Sec. 14.8: Flow Through Porous Media -- *Exercises*

14.8.1 *The estimated average depth of saturated flow in a sloping water table aquifer (porosity 0.25, K = 5 ft/day) is 11.8 ft. The distance between 1-ft water table contours is 34 ft. What is the flow rate through a 1-ft width of aquifer, and what is the average pore velocity?*

BG

$V = -K(dh/ds) = -5(-1/34) = 0.1471$ ft/day

$q = AV = 1(11.8)0.1471 = 1.735$ ft^3/d per ft width ◄

$V_p = V/n = 0.1471/0.25 = 0.588$ ft/day ◄

14.8.2 *The estimated average depth of saturated flow in a sloping water table aquifer (porosity 0.23, K = 1.6 m/d) is 3.8 m. The distance between 1-m water table contours is 37 m. What is the flow rate through a 1-m width of aquifer, and what is the average pore velocity?*

SI

$V = -K(dh/ds) = -1.6(-1/37) = 0.0432$ m/d

$q = AV = 1(3.8)0.0432 = 0.1643$ m^3/d per m width ◄

$V_p = V/n = 0.432/0.23 = 0.1880$ m/d ◄

14.8.3 *From dye tests between two wells, ground water is estimated to travel at 0.034 ft/day. If the corresponding water table slope is 0.0018, and the average porosity of the aquifer is 0.15, estimate its hydraulic conductivity.*

BG

From Eq. 14.28: $V = nV_p = 0.15(0.034) = 0.00510$ ft/day

Eq. 14.27: $0.00510 = K(0.0018)$; $K = 2.83$ ft/day ◄

14.8.4 *From dye tests between two wells, ground water is estimated to travel at 0.0094 m/d. If the corresponding water table slope is 0.0016, and the average porosity of the aquifer is 0.18, estimate its hydraulic conductivity.*

SI

From Eq. 14.28: $V = nV_p = 0.18(0.0094) = 0.001692$ m/d

Eq. 14.27: $0.001692 = K(0.0016)$; $K = 1.058$ m/d ◄

Chapter 15

Hydraulic Machinery -- Pumps

PROBLEM SELECTION GUIDE

Sec	Exer/Prob	Units	Difficulty	Length	Parts	Similar	Special features
15.2	***Head Developed by a Pump***						
	P 15.1	SI	Easy	Short	1		
15.3	***Pump Efficiency***						
	P 15.2	BG	Easy	Short	1		
	15.3	BG	Easy	Short	2		
	15.4	BG	Easy	Short	1		
	15.5	BG	V Easy	V Short	1		
	15.6	SI	V Easy	V Short	1		
15.4	***Similarity Laws for Pumps***						
	X[1]15.4.1	SI	Easy	Short	1	S15.2	
	15.4.2	BG	Easy	Short	2	15.4.3	
	15.4.3	SI	Easy	Short	2	15.4.2	
	15.4.4	BG	Easy	Short	1		*
	P 15.7	BG	Easy	Short	2		
	15.8	SI	Easy	Short	1		
	15.9	SI	Easy	Short	1		
	15.10	N	V Easy	V Short	1		
15.6	***Performance Characteristics at Different Speeds and Sizes***						
	X 15.6.1	BG	Easy	Short	1		
	15.6.2	BG	Easy	Short	1		
	15.6.3	BG	Easy	Medium	1		Plot graph
	P 15.11	BG	Easy	Short	1		
	15.12	BG	Easy	Medium	1		Plot graph
	15.13	BG	Easy	Medium	1		Plot graph
15.7	***Operating Point of a Pump***						
	X 15.7.1	BG	Medium	Long	3	S15.4,15.7.2	† Plot curves
	15.7.2	BG	Medium	Long	3	S15.4,15.7.1	* † Plot curves
	P 15.14	BG	Medium	Medium	1		
	15.15	BG	Medium	Long	3	S15.4,15.16	† Plot curves
	15.16	BG	Hard	Long	3	S15.4,15.15	† Minor losses. Plot curves
	15.17	BG	Medium	Long	2	15.18	† Plot curve
	15.18	BG	Medium	Long	2	15.17	† Plot curve
	15.19	SI	Medium	Long	1		† Plot graph

/cont...

[1] For all Exercises (marked "X"), answers are given in Appendix D of the textbook.
* Note a correction made to early printings of the textbook.
† Answers are sensitive to values (ϕ_e, σ_c, etc.) read from a figure or plotted graph.
X = Exercise, P = (end-of-chapter) Problem, S = Sample Problem.

Sec	Exer/Prob	Units	Difficulty	Length	Parts	Similar	Special features
15.8	***Specific Speed of Pumps***						
	X 15.8.1	N	Medium	Medium	1		Derivation
	15.8.2	BG	Easy	V Short	1	S15.5	
	15.8.3	BG	Easy	Short	1	S15.5	
	15.8.4	BG	Easy	V Short	1		
	15.8.5	SI	Easy	V Short	1		
	P 15.20	BG	Easy	Short	1		
	15.21	BG	Easy	Medium	1		
	15.22	BG	Easy	Medium	1		
	15.23	BG	Easy	Short	1		
	15.24	SI	Easy	Short	1		
	15.25	SI	Easy	Medium	3		
15.9	***Peripheral-Velocity Factor***						
	X 15.9.1	BG	Easy	Short	1	S15.7	†
	15.9.2	BG	Easy	V Short	1	S15.7	†
	15.9.3	B	Easy	Medium	1		†
	P 15.26	BG	Easy	Short	1	15.27	†
	15.27	B	Easy	Short	1	15.26	†
	15.28	B	Easy	Short	1		†
	15.29	B	Easy	Short	1		†
	15.30	BG	Hard	Long	1		† Trial and Error
	15.31	BG	Medium	Medium	1		* †
	15.32	BG	Easy	Short	1		†
	15.33	BG	Easy	Short	1		†
	15.34	BG	Easy	V Short	1		†
	15.35	SI	Medium	V Short	1		
15.10	***Cavitation in Pumps***						
	X 15.10.1	BG	Medium	Short	1	15.10.2	†
	15.10.2	BG	Medium	Short	1	15.10.1	†
	15.10.3	SI	Medium	Short	1	S15.8, P15.37, P15.44	
	P 15.36	SI	Medium	V Short	1		
	15.37	BG	Medium	Short	1	S15.8, X15.10.3, 15.44	
	15.38	BG	Medium	Short	1	15.39	
	15.39	BG	Medium	Medium	1	15.38	
	15.40	BG	Medium	Short	1		
	15.41	SI	Medium	Medium	1		†
	15.42	SI	Hard	Long	1	15.43	†
	15.43	SI	Hard	Long	1	15.42	†
	15.44	SI	Medium	Short	1	S15.8, X15.10.3, 15.37	
	15.45	BG	Medium	Medium	1		†

/cont...

Sec	Exer/Prob	Units	Difficulty	Length	Parts	Similar	Special features
15.12	***Selection of Pumps***						
	X 15.12.1	BG	Medium	Medium	2	S15.10	†
	15.12.2	BG	Medium	Medium	2	S15.10	†
	15.12.3	SI	Hard	Long	1		†
	P 15.46	BG	Medium	Long	4	15.47	†
	15.47	BG	Medium	Long	4	15.46	†
	15.48	BG	Medium	Medium	3		†
	15.49	BG	Hard	Medium	1		†
	15.50	BG	Medium	Medium	1	15.51	† Other solutions are possible.
	15.51	BG	Medium	Long	1	15.50	† Other solutions are possible.
	15.52	BG	Medium	Medium	1		Must estimate *f* (Secs. 8.12-13)
	15.53	SI	Hard	Medium	1		Partly unanswerable.
15.13	***Pumps Operating in Series and Parallel***						
	X 15.13.1	BG	Medium	Long	3	S15.13	† Plot curves
	15.13.2	BG	Medium	Medium	2	S15.13	†
	P 15.54	BG	Medium	Medium	1		† Other solutions are possible.
	15.55	BG	Medium	Medium	1		† Other solutions are possible.
	15.56	BG	Hard	Long	2		† Plot curves
	15.57	BG	Medium	Medium	1		
Ch.15	***Miscellaneous***						
	P 15.58	BG	Medium	Long	1		† Plot curves, graph
	15.59	BG	Easy	Short	1		†
	15.60	B	Easy	Medium	1		
	15.61	BG	Easy	Short	1		†

Chapter 15
HYDRAULIC MACHINERY – PUMPS

Sec. 15.2: Head Developed by a Pump -- *Problem*

15.1 *The diameter of the discharge side of a pump is 100 mm, and that of the intake pipe is 120 mm. The pressure gage at discharge reads 240 kPa and the pressure gage at intake reads 60 kPa. If Q = 45 L/s of water and the pump efficiency is 0.82, find the power delivered to the pump by the drive shaft. The intake of the pump is 350 mm below the discharge side. Neglect the effects of pre-rotation in the entry pipe and assume smooth flow at discharge from the pump.*

SI

$V_s = Q/A_s = (0.045\ \text{m}^3/\text{s})/(\pi 0.120^2/4) = 3.98\ \text{m/s};\quad V_d = V_s(120/100)^2 = 5.73\ \text{m/s}$

Eq. 15.1: $h = \left(\dfrac{p_d}{\gamma} + \dfrac{V_d^2}{2g} + z_d\right) - \left(\dfrac{p_s}{\gamma} + \dfrac{V_s^2}{2g} + z_s\right)$

$$h = \left(\frac{240\ \text{kN/m}^2}{9.81\ \text{kN/m}^3} + \frac{5.73^2}{2(9.81)} + 0.35\right) - \left(\frac{60\ \text{kN/m}^2}{9.81\ \text{kN/m}^3} + \frac{3.98^2}{2(9.81)} + 0\right) = 19.56\ \text{m}$$

Eq. 15.2: $P = \gamma Qh/\eta = 9.81(0.045)19.56/0.82 = 10.53\ \text{kW}$ ◄

Sec. 15.3: Pump Efficiency -- *Problems*

15.2 *The diameter of the discharge side of a pump is 6 in, and that of the intake pipe is 8 in. The pressure gage at discharge reads 30 psi, and the vacuum gage at intake reads 10 inHg. If Q = 3.0 cfs of water and the brake horsepower is 35.0, find the efficiency of the pump. The intake and discharge are at the same elevation.*

BG

$V_d = Q/A_d = 3.0/(\pi 0.5^2/4) = 15.29\ \text{fps};\quad V_s = V_d(6/8)^2 = 8.60\ \text{fps}$

10 inHg vacuum is equivalent to $-33.9(10.0/29.9) = -11.34$ ft of water

Eq. 15.1: $h = \left(\dfrac{p_d}{\gamma} + \dfrac{V_d^2}{2g}\right) - \left(\dfrac{p_s}{\gamma} + \dfrac{V_s^2}{2g}\right) = \left(\dfrac{30(144)}{62.4} + \dfrac{15.29^2}{64.4}\right) - \left(-11.34^2 + \dfrac{8.60^2}{64.4}\right)$

$= 69.2 + 3.6 + 11.34 - 1.15 = 83.0\ \text{ft}$

Eq. 15.2: $\eta = (\gamma Qh/550)/35.0 = 62.4(3.0)83.0/(550 \times 35) = 0.81 = 81\%$ ◄

15.3 *When operating at its BEP, the pump of Fig. 15.6 develops 60 ft of head at a capacity of 10,500 gpm. (a) Calculate the horsepower delivered to the water by the pump. (b) If the brake horsepower is 192 hp, what is the efficiency of the pump at its BEP? Does your answer to (b) agree with Fig. 15.6?*

Fig. 15.6: At Q = 10,500 gpm (the BEP), pump efficiency = 83%.

BG

(*a*) Water Power $= \gamma Qh/550 = 62.4(10{,}500/449)60/550 = 159.2$ hp ◄

(*b*) $\eta = 159.2/192 = 0.83 = 83\%$ ◄ Yes, answer agrees with Fig. 15.6 ◄

15.4 *The pump of Fig. 15.6 delivers 26 ft of head at a capacity of 15,000 gpm. The pump efficiency under these conditions of operation is 44%. Calculate the water horsepower and the brake horsepower, and check to see if your answers agree with Fig. 15.6.*

Fig. 15.6: At Q = 15,000 gpm, water horsepower = 93, brake horsepower = 230.

BG

Water horsepower $= \gamma Qh/550 = 62.4(15{,}000/449)26/550 = 98.5$ hp ◄

Brake horsepower = Water horsepower$/\eta = 98.5/0.44 = 224$ hp ◄

These check closely with Fig. 15.6. ◄

15.5 *A centrifugal pump delivers 40 gpm of gasoline (s = 0.72) when developing a net head of 15 ft. If the shaft power is 0.126 hp, what is the efficiency of the pump?*

BG

Eq. 15.2: $\eta = \gamma Qh/\text{shaft power} = 0.72(62.4)(40/449)15/(550 \times 0.126) = 0.87$ ◄

15.6 *A pump delivers 1000 L/s of water at a head of 8.5 m. If the efficiency of the pump is 68%, what is the shaft power?*

SI

Eq. 15.2: Shaft power $= \gamma Qh/\eta = 9.81(1.0)8.5/0.68 = 122.6$ kW ◄

Sec. 15.4: Similarity Laws for Pumps -- ***Exercises***

15.4.1 *A model centrifugal pump has a scale ratio of 1:15. The model when tested at 3600 rpm, delivered 0.10 m^3/s of water at a head of 40 m with an efficiency of 80 percent. Assuming the prototype has an efficiency of 88 percent, what will be its speed, capacity, and power requirement at a head of 50 m?*

SI

Eq. 15.5: $h_m = K_h D_m^2 n_m^2$; $40 = K_h(D_p/15)^2 3600^2$; $K_h = 40(225)/[D_p^2 3600^2] = 0.000\,694/D_p^2$

Eq. 15.5: $h_p = K_h D_p^2 n^2 = 50 = (0.000\,694/D_p^2)D_p^2 n_p^2$; $n_p^2 = 72{,}046$; $n_p = 268$ rpm ◄

From Eq. 15.4: $Q \propto D^3 n$; $Q_p = Q_m(15)^3(268/3600) = 251Q_m$; $Q_p = 251(0.10) = 25.1\ \text{m}^3/\text{s}$ ◄

From Eq. 15.2: $P_p = \gamma Q_p h_p/\eta = 9.81(25.1)50/0.88 = 14\,000$ kW ◄

15.4.2 *A 19½-in-diameter centrifugal-pump impeller discharges 20 cfs at a head of 100 ft when running at 1200 rpm. (a) If its efficiency is 85 percent, what is the brake horsepower, i.e., what is the horsepower input to the shaft of the pump? (b) If the same pump were run at 1500 rpm, what would be h, Q, and the brake horsepower for homologous conditions?*

BG

(*a*) Eq. 15.2: $bhp = \gamma Qh/(550\eta) = 62.4(20)100/[500(0.85)] = 267$ bhp ◄

(*b*) Eq. 15.5: $h = 100(1500/1200)^2 = 156.3$ ft ◄

Eq. 15.4: $Q = 20(1500/1200) = 25$ cfs ◄

Eq. 15.6: $bhp = 267(1500/1200)^3 = 521$ bhp ◄

15.4.3 *An axial-flow pump delivers 300 L/s at a head of 6.0 m when rotating at 1800 rpm. (a) If its efficiency is 80 percent, how many kilowatts of power must the shaft deliver to the pump? (b) If this same pump were operated at 1500 rpm, what would be h, Q, and the power delivered by the shaft for homologous conditions?*

SI

(*a*) Eq. 15.2: Power $= \gamma Qh/\eta = 9.81(0.3)6/0.8 = 22.1$ kW ◄

(*b*) Eqs. 15.4, 15.5, and 15.6: $Q \propto n$, $h \propto n^2$, $hp \propto n^3$

$\therefore Q = 300(1800/1500) = 360$ L/s ◄

$h = 6(1800/1500)^2 = 8.64$ m ◄

$P = 22.1(1800/1500)^3 = 38.1$ kW ◄

15.4.4 *In Sample Prob. 15.2 the model with a scale ratio of *1:10 has a maximum efficiency of 84%. Using Eq. (15.7), calculate the maximum efficiency of the prototype pump and compare the result with the data given in the sample problem. Note that Eq. (15.7) is empirical, and will provide only approximate values. [*Note this correction to the textbook.]*

Sample Prob. 15.2: Assume a prototype pump efficiency of 90%.

BG

From Eq. 15.7: $1 - \eta_p = (1 - \eta_m)(D_m/D_p)^{1/5} = (1 - 0.84)(1/10)^{1/5} = 0.1010$

$\eta_p = 1 - 0.1010 = 0.899 = 89.9\%$ ◄

This compares very well with the prototype efficiency of 90% assumed in Sample Prob. 15.2 ◄

Sec. 15.4: Similarity Laws for Pumps -- *Problems*

15.7 *A centrifugal pump with an 18-in-diameter impeller is rated at 25 cfs under a head of 100 ft when running at 1200 rpm. (a) If its efficiency is 85%, what is the brake horsepower? (b) If the same pump is run at 1800 rpm, what would be its rating in terms of h, Q, and the brake horsepower? Assume the efficiency does not change with the change in speed.*

BG

(*a*) Eq. 15.2: Brake hp $= \gamma Qh/(550\eta) = 62.4(25)100/(550 \times 0.85) = 333.7$ ◄

(*b*) Eq. 15.4: $Q \propto n$, $Q = 25(1800/1200) = 37.5$ cfs ◄

Eq. 15.5: $h \propto n^2$, $h = 100(1800/1200)^2 = 225$ ft ◄

Eq. 15.2: Brake hp $= 62.4(37.5)2.25/(550 \times 0.85) = 1126$

or Eq. 15.6: $P \propto n^3$, $P = 333.7(1800/1200)^3 = 1126$ hp ◄

15.8 *A centrifugal pump with a 300-mm-diameter impeller is rated at 40 L/s against a head of 25 m when rotating at 1750 rpm. What would be the rating of a pump of identical geometric shape with a 200-mm-diameter impeller? Assume pump efficiencies and pump speeds are identical.*

SI

Eq. 15.4: $Q \propto D^3$, $Q = 40(200/300)^3 = 11.85$ L/s ◄

Eq. 15.5: $h \propto D^2$, $h = 25(200/300)^2 = 11.11$ m ◄

Pump rating: 11.85 L/s against a head of 11.11 m ◄

15.9 *At its optimum point of operation a given centrifugal pump delivers 3.2 m³/s of water against a head of 25 m when rotating at 1450 rpm. (a) If the efficiency is 82%, what is the brake power of the drive shaft? (b) If a homologous pump with an impeller diameter one half as large is rotating at 1200 rpm, what would be the discharge, head, and shaft power at its point of optimum efficiency? Assume both pumps operate at the same efficiency.*

SI

Eq. 15.4: $Q \propto nD^3$, $Q = 3.2(1200/1450)(1/2)^3 = 0.33$ m³/s ◄

Eq. 15.5: $h \propto n^2D^2$, $h = 25(1200/1450)^2(1/2)^2 = 4.28$ m ◄

Eq. 15.2: Brake power $= \gamma Qh/\eta = 9.81(0.33)4.28/0.82 = 16.9$ kW ◄

15.10 *Using Eq. (15.7), find the approximate efficiency of the smaller pump of Problem 15.9.*

Prob. 15.9: $D_m = 0.5D_p$, $\eta_p = 0.82$

N

Eq. 15.7: $(1 - \eta_p)/(1 - \eta_m) = (D_m/D_p)^{1/5}$; $(1 - 0.82)/(1 - \eta_m) = (1/2)^{1/5} = 0.87$

From which $\eta_m = 0.793 = 79.3\%$ ◄

Sec. 15.6: Performance Characteristics at Different Speeds and Sizes -- *Exercises*

15.6.1 *Refer to Fig. 15.8. Transfer the point (h = 70 ft and Q = 7500 gpm) to a point on Curve 3 if n_1 = 1450 rpm and n_3 = 1160 rpm. Check your result with Fig. 15.8.*

BG

Eqs. 15.4 and 15.5: $Q \propto n$, $h \propto n^2$

$Q_3/Q_1 = n_3/n_1$; $Q_3 = Q_1(n_3/n_1) = 7500(1160/1450) = 6000$ gpm ◄

$h_3/h_1 = (n_3/n_1)^2$; $h_3 = h_1(n_3/n_1)^2 = 70(1160/1450)^2 = 44.8$ ft ◄

These check with curve 3 of Fig. 15.8. ◄

15.6.2 *Refer to Fig. 15.9. Transfer the point (h = 70 ft and Q = 7500 gpm) to a point on the D_2 curve if D_1 = 16.7 in and D_2 = 17.5 in. Check your result with Fig. 15.9.*

BG

Eqs. 15.4 and 15.5: $Q \propto D^3, \quad h \propto D^2$

$Q_2 = Q_1(D_2/D_1)^3 = 7500(17.5/16.7)^3 = 8630$ gpm ◀

$h_2 = h_1(D_2/D_1)^2 = 70(17.5/16.7)^2 = 76.9$ ft ◀ These check with the D_2 curve ◀

15.6.3 *Plot h versus Q for the pump of Fig. 15.7 for the case where n = 600 rpm.*
Fig. 15.7: For Q = 0, 75, 122 cfs respectively h = 40, 23, 16 ft, etc.

BG

Eqs. 15.4 and 15.5: $Q \propto n, \quad h \propto n^2$; $n_2/n_1 = 600/450 = 1.333$; $(n_2/n_1)^2 = 1.778$

Shutoff head = 40(1.778) = 71.1 ft, Q = 0 cfs

BEP: h = 16(1.778) = 28.4 ft ◀ Q = 122(1.333) = 162.6 cfs

etc.: 75 cfs at 23 ft becomes Q = 75(1.333) = 100 cfs ◀ at h = 23(1.778) = 40.9 ft ◀

Sec. 15.6: Performance Characteristics at Different Speeds and Sizes -- *Problems*

15.11 *At its BEP, a given pump develops 30 ft of head when operating at 720 rpm. The flow rate is 4.2 cfs. Find the head and flow rate for a homologous pump operating at 600 rpm if its diameter is 90% that of the given pump.*

BG

Eq. 15.4: $Q \propto nD^3$, $Q = 4.2(600/720)0.9^3 = 2.55$ cfs ◀

Eq. 15.5: $h \propto n^2D^2$, $h = 30(600/720)^2 0.9^2 = 16.88$ ft ◀

15.12 *Plot h versus Q for the pump of Fig. 15.6 for the case where n = 1200 rpm.*
Fig. 15.6 (n = 1450 rpm): For Q = 0, 6000, 10,500, 15,000 gpm respectively, h = 80, 72, 60, 26 ft, etc.

BG

Eq. 15.4: $Q \propto n$, $Q' = Q(1200/1450) = 0.828Q$; Eq. 15.5: $h \propto n^2$, $h' = h(1200/1450)^2 = 0.685h$

Q = 0, h = 80 ft becomes 0 gpm, 54.8 ft ◀

Q = 6000 gpm, h = 72 ft becomes 4970 gpm, 49.3 ft ◀

Q = 10,500 gpm, h = 60 ft becomes 8690 gpm, 41.1 ft ◀

Q = 15,000 gpm, h = 26 ft becomes 12,420 gpm, 17.8 ft ◀

etc.

15.13 *Plot h versus Q for the pump of Fig. 15.7 for the case where n = 600 rpm.*
Fig. 15.7: For Q = 0, 50, 122 cfs respectively h = 40, 26, 16 ft, etc.

BG

Eq. 15.4: $Q \propto n$, $Q' = Q(600/450) = 1.333Q$; Eq. 15.5: $h \propto n^2$, $h' = h(600/450)^2 = 1.777h$

0 cfs, 40 ft becomes 0 cfs, 71.1 ft ◀

50 cfs, 26 ft becomes 66.7 cfs, 46.2 ft ◀

122 cfs, 16 ft becomes 162.6 cfs, 28.4 ft ◀ etc.

Sec. 15.7: Operating Point of a Pump -- ***Exercises***

15.7.1 *Repeat Sample Prob. 15.4 for the case where the length of the 18-inch diameter pipe is 1000 ft.*
Sample Prob. 15.4: $f = 0.032$, $L = 500$ ft, $D = 18$ in, *resulting in* $h = \Delta z + 0.0530Q^2$.

BG

Refer to the Solution to Sample Problem 15.4.

The 1000 ft pipe is twice as long as the 500 ft pipe of Sample Prob. 15.4.

Hence, $h = \Delta z + 2(0.0530)Q^2 = \Delta z + 0.1061Q^2$ where Q is in cfs. Coordinates of system curves:

Q (cfs)	Q (gpm)	(*a*) $h = h_L$ (ft)	(*b*) $20 + h_L$	(*c*) $65 + h_L$
0	0	0	20	65
10	4890	10.6	30.6	75.6
20	9780	42.4	62.4	107.4
30	14,670	95.6	115.6	160.6

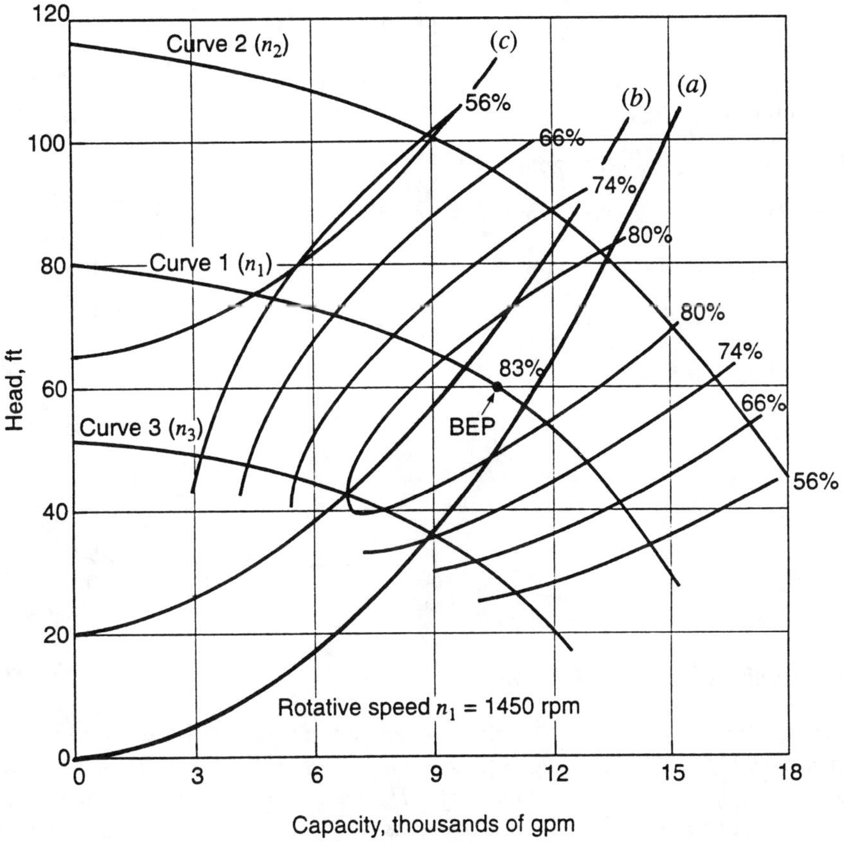

Figure 5.8, modified

By plotting the system curves on Fig. 15.8 we find the points of intersection with Curve 1:

(*a*) $h \approx 56$ ft and $Q \approx 11{,}300$ gpm ◄

(*b*) $h \approx 63$ ft and $Q \approx 9700$ gpm ◄

(*c*) $h \approx 76$ ft and $Q \approx 4800$ gpm ◄

From the same figure, approx. pump efficiencies are: 81%, 81%, and 54% respectively ◄

15.7.2 *Repeat Sample Prob. 15.4 for the case where the pipe diameter is *12 inches rather than 18 inches, all other data remaining the same. [*Note this correction to the textbook.]*

Sample Prob. 15.4: $f = 0.032$, $L = 500$ ft, $D = 18$ in, resulting in $h = \Delta z + 0.0530Q^2$.

BG

Refer to the Solution to Sample Problem 15.4.

Note: Pipe friction $h_L = f(L/D)V^2/2g$; $V = Q/A = Q/(\pi D^2/4)$; $V \propto 1/D^2$, $V^2 \propto 1/D^4$

Hence pipe friction $h_L \propto 1/D^5$; $\dfrac{\text{Pipe friction } h_L \text{ for } D = 12 \text{ in}}{\text{Pipe friction } h_L \text{ for } D = 18 \text{ in}} = \left(\dfrac{18}{12}\right)^5 = 7.59$

Hence, for the 12-in-diameter pipe $h = \Delta z + 7.59(0.0530)Q^2 = \Delta z + 0.403Q^2$

Coordinates of system curves:

Q (cfs)	Q (gpm)	(a) $h = h_L$ (ft)	(b) $20 + h_L$	(c) $65 + h_L$
0	0	0	20	65
5	2445	10.1	30.1	75.1
10	4890	40.3	60.3	105.3
15	7335	90.7	110.7	155.7

By plotting the system curves on Fig. 15.8 we find the points of intersection with Curve 1.

(a) $h \approx 72$ ft and $Q \approx 6500$ gpm ◄

(b) $h \approx 74$ ft and $Q \approx 5500$ gpm ◄

(c) $h \approx 77$ ft and $Q \approx 2500$ gpm ◄

Approx. pump efficiencies are: 66%, 60%, and 40% respectively ◄

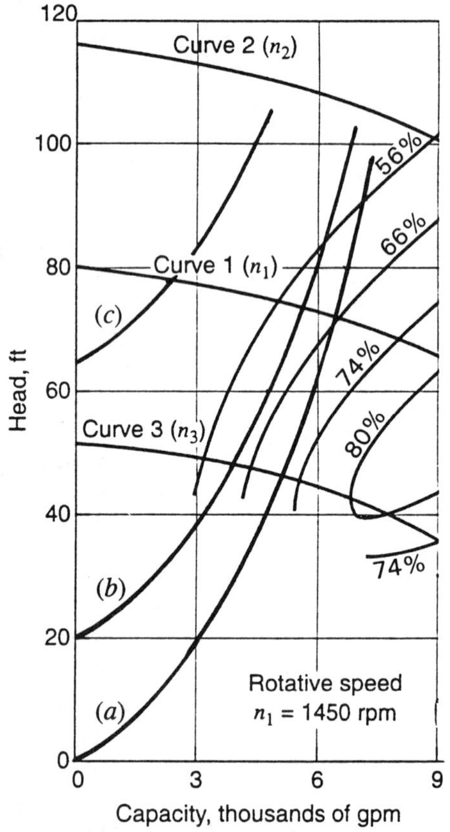

Figure 15.8, modified

Sec. 15.7: Operating Point of a Pump -- *Problems*

15.14 *A centrifugal pump is installed to deliver water from a reservoir of water surface elevation zero to another of elevation 300 ft. The 12-in-diameter suction pipe ($f = 0.020$) is 100 ft long and the 10-in-diameter discharge pipe ($f = 0.026$) is 5000 ft long. The pump characteristic at 1200 rpm is defined by $h_p = 375 - 24Q^2$, where the pump head h_p is in ft and Q is in cfs. Calculate the rate at which this pump will deliver water, assuming the setting is low enough to avoid cavitation.*

BG

Energy equation neglecting minor loss at discharge:

$0 - 0.02(100/1)V_{12}^2/2g + h_p - 0.026[5000/(10/12)]V_{10}^2/2g = 306$

$V_{12} = Q/[\pi(6/12)^2] = 1.273Q$, $V_{10} = Q/[\pi(5/12)^2] = 1.833Q$

From the above: $h_p = 300 + 8.19Q^2$ (system characteristic)

Given: $h_p = 375 - 24Q^2$ (pump characteristic)

Equating the h's: $300 + 8.19Q^2 = 375 - 24Q^2$

$32.19Q^2 = 75$; $Q = 1.526$ cfs $= 1.526(449$ gpm$) = 685$ gpm ◄

15.15 *A pump whose characteristics are shown on Fig. 15.7 when operating at 450 rpm is used to pump water from reservoir A to reservoir B through a 24-in-diameter pipe (f = 0.028) of length 170 ft. Neglecting minor losses, find the flow rate and pump efficiency for the following conditions; (a) reservoir water surface elevations identical; (b) water surface in reservoir B is 10 ft higher than that in reservoir A; (c) reservoir water surface elevation in reservoir B is 10 ft lower that in reservoir A.*

BG

Pump characteristic curve is given in Fig. 15.7 (see below).

Friction head loss: $h_L = f(L/D)(V^2/2g) = f(L/D)Q^2/[(\pi r^2)^2 2g] = 0.028(170/2.0)Q^2/[(\pi 1^2)^2(2)32.2]$

$h_L = 0.003\,75Q^2$ (Q in cfs)

System characteristic curves: $\Delta z + h_p - h_L = 0;\quad h_p = -\Delta z + h_L$

$(a)\ h_p = h_L\ ;\quad (b)\ h_p = -10 + h_L\ ;\quad (c)\ h_p = 10 + h_L$

Coordinates of system curves:

Q (cfs)	(*a*) h_L (ft)	(*b*) $-10 + h_L$	(*c*) $10 + h_L$
50	9.4	−0.6	19.4
70	18.4	8.4	29.4
90	30.5	20.5	40.5
100	37.7	27.7	47.7*

*For plotting purposes only

By plotting the curves we find h and Q, the points of intersection:

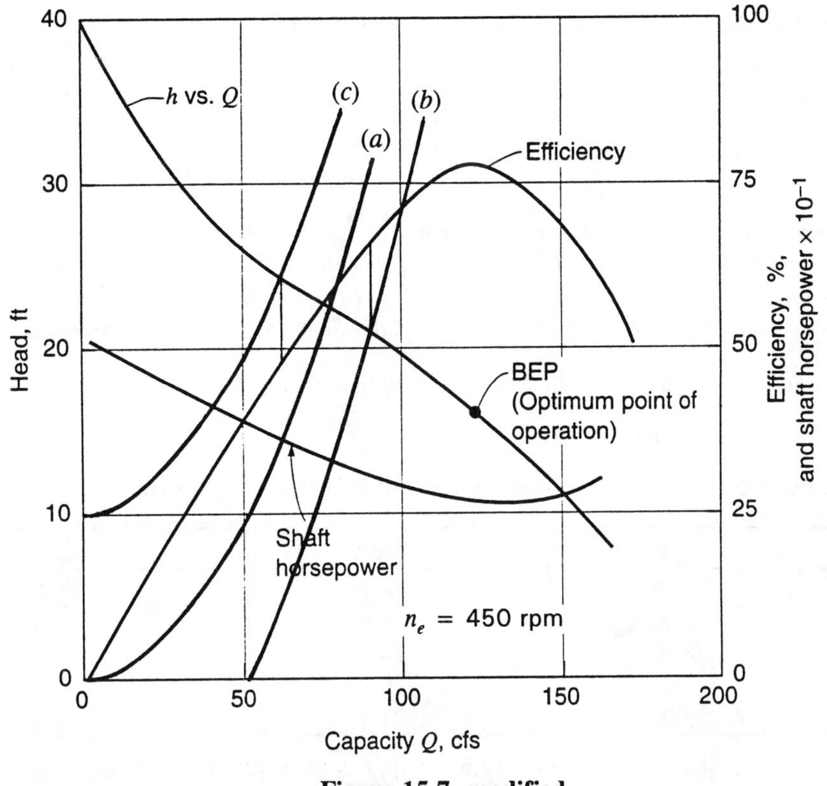

Figure 15.7, modified

(*a*) $h \approx 22$ ft and $Q \approx 76$ cfs, $\eta \approx 58\%$ ◀

(*b*) $h \approx 20$ ft and $Q \approx 92$ cfs, $\eta \approx 63\%$ ◀

(*c*) $h \approx 24$ ft and $Q \approx 65$ cfs, $\eta \approx 50\%$ ◀

15.16 *Repeat Prob. 15.15 using the same data except consider minor losses: square entrance and submerged discharge. Neglect bend losses.*

Prob. 15.15: A pump with the characteristics of Fig. 15.7 (see with Solution 15.15) pumps water from reservoir A to reservoir B through a 170-ft-long pipeline ($f = 0.028$, $D = 24$ in). Find Q and the pump η for (a) reservoir water surface elevations identical, (b) water surface B 10 ft higher than A; (c) water surface B 10 ft lower than A.

BG

The pump characteristic curve is given in Fig. 15.7 (see previous problem, above).

Friction head loss: $h_L = f(L/D)(V^2/2g) = f(L/D)Q^2/[(\pi r^2)^2 2g] = 0.028(170/2.0)Q^2/[(\pi 1^2)^2(2)32.2]$

$(h_L)_{\text{fric}} = 0.003\,75Q^2$ (Q in cfs). Consider minor losses: $0.5(V^2/2g) + 1.0(V^2/2g) = 1.5(V^2/2g)$

$V = Q/A = Q/(\pi 1^2) = Q/\pi$; $1.5(V^2/2g) = 1.5Q^2/[(\pi)^2(2)32.2] = 0.002\,36Q^2$

Total head loss $= 0.003\,75Q^2 + 0.002\,36Q^2 = 0.006\,11Q^2$

System charateristic curves: (a) $h = h_L$; (b) $h = -10 + h_L$; (c) $h = 10 + h_L$

Coordinates of system curves:

Q (cfs)	(a) h_L (ft)	(b) $-10 + h_L$	(c) $10 + h_L$
30	5.5	−4.5	15.5
50	15.3	5.3	25.3
70	29.9	19.9	39.9
90	49.5	39.5	59.5

By plotting the curves we find h and Q, the points of intersection:

(a) $h \approx 24$ ft and $Q \approx 60$ cfs, $\eta \approx 48\%$ ◀
(b) $h \approx 23$ ft and $Q \approx 74$ cfs, $\eta \approx 57\%$ ◀
(c) $h \approx 26$ ft and $Q \approx 50$ cfs, $\eta \approx 39\%$ ◀

15.17 *A pump whose characteristics are shown in Fig. 15.8, operating at 1450 rpm, is placed in a 24-in-diameter pipe ($f = 0.032$) of length 5000 ft. The discharge end of the pipe is 30.0 ft lower than the water surface elevation in the reservoir. The water discharges freely into the atmosphere. (a) Find the flow rate. (b) What would be the flow rate if the pump had not been installed in the pipeline?*

BG

(a) Energy equation: $30 + h_p - 0.032(5000/2)(V^2/2g) = 0$ (1)

$V = Q/A = Q/(\pi 1^2) = Q/\pi$ (2)

Substitute (2) into (1) to obtain system characteristic: $h_p = 0.1259Q^2 - 30$

Q (gpm)	Q (cfs)	h_p (ft)
3000	6.68	−24.4
6000	13.36	−7.5
9000	20.0	20.4
12,000	26.3	59.8
15,000	33.4	110.6

Plot the system curve on Fig. 15.8 (see Fig. with Solution 15.18) and find the point of intersection:

$h \approx 53$ ft, $Q \approx 11{,}600$ gpm, $\eta \approx 80\%$ ◀

(b) If there was no pump: $30 - 0.1260Q^2 = 0$ or $Q^2 = 30/0.1260 = 238$

$Q = \sqrt{238} = 15.43$ cfs $= 6930$ gpm ◀

15.18 *Repeat Problem 15.17 with all data the same except operate the pump at 1200 rpm rather than 1450 rpm. Note the pump efficiency under this mode of operation.*

Prob. 15.17: A pump with the characteristics of Fig. 15.8 ($n = n_1 = 1450$ rpm) pumps water from a reservoir through a 5000-ft-long pipeline ($f = 0.032$, $D = 24$ in) to discharge freely into the atmosphere 30 ft lower than the reservoir surface. (a) Find Q. (b) Find Q if the pump had not been installed.

BG

Sec. 15.6: Characteristic performance curves for the same pump operating at different speeds are related approximately by Eqs. 15.4 and 15.5, as shown in Fig. 15.8.

At shutoff ($Q = 0$), $h = 80$ ft at 1450 rpm

Eq. 15.5: $h \propto n^2$

With $n = 1200$ rpm, $h = 80(1200/1450)^2 = 54.8$ ft

In Fig. 15.8 when $Q = 0$ we see the n_3 curve has an h of approx 55 ft. Hence the n_3 curve may be taken to represent the pump characteristic for operation at 1200 rpm.

Energy equation:

(*a*) $30 + h_p - 0.032(5000/2)(V^2/2g) = 0$ (1)

$V = Q/A = Q/(\pi 1^2) = Q/\pi$ (2)

Substitute (2) into (1) to obtain system characteristic: $h_p = 0.1259Q^2 - 30$

Q (gpm)	Q (cfs)	h_p (ft)
3000	6.68	−24.4
6000	13.36	−7.5
9000	20.0	20.4
12,000	26.3	59.8
15,000	33.4	110.6

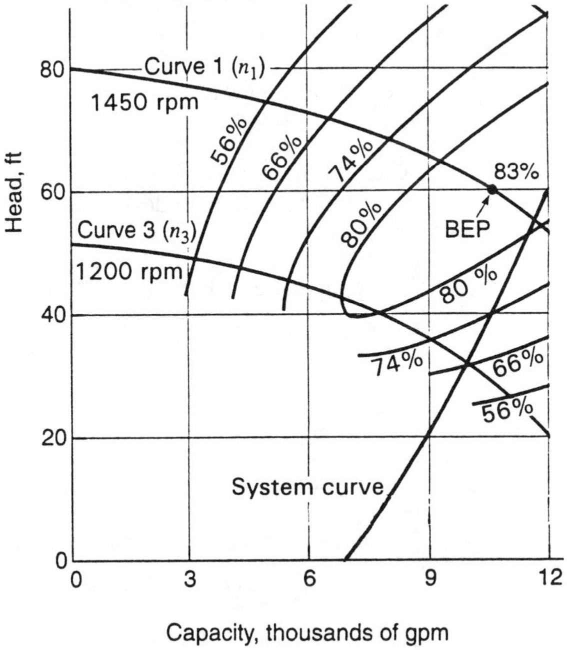

Figure 15.8, modified

The n_3 curve and plotted system curve (see figure) intersect at $h \approx 31$ ft and $Q \approx 9900$ gpm at which point $\eta \approx 66\%$ ◄

(*b*) If there was no pump: $30 - 0.1260Q^2 = 0$ or $Q^2 = 30/0.1260 = 238$

$Q = \sqrt{238} = 15.43$ cfs $= 6930$ gpm ◄

/more...

15.19 *The characteristic curve for a pump operating at 1200 rpm is given by the following:*

Head, m	*Q, L/s*
20	*0*
15	*165*
10	*250*

This pump is installed in a pipeline to augment the flow from reservoir A to reservoir B. When the water surface elevation in reservoir A exceeds that in reservoir B by 15 m, the flow rate is 150 L/s. What will be the flow rate when the elevation difference is 3 m? Assume the system curve is parabolic.

SI

Plot the pump characteristic curve from the given data (see figure).

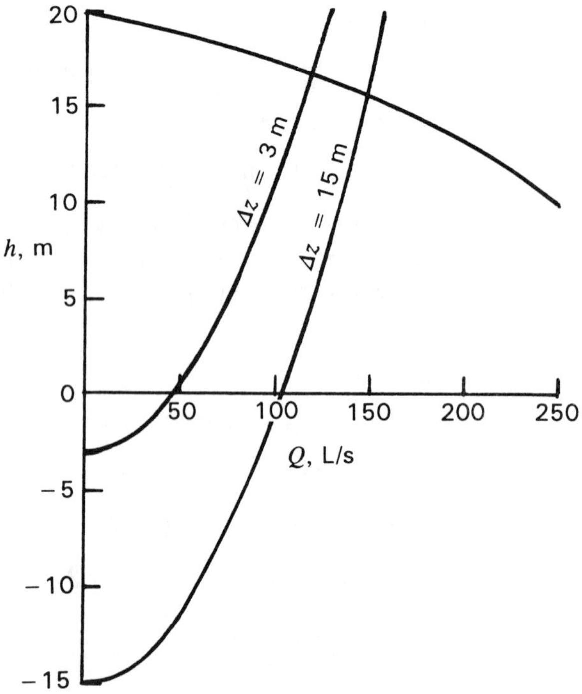

Note (from the pump characteristic curve) that when $Q = 150$ L/s the head developed by the pump is 15.5 m.

Energy equation when $\Delta z = 15$ m:

$$15 + h_p - h_L = 0, \quad \text{where} \quad h_L = kQ^2$$

i.e., $\quad 15 + 15.5 - k(150)^2 = 0 \qquad (Q \text{ in L/s})$

$\therefore\ k = 30.5/150^2 = 0.001\,356$ and

$$\Delta z + h_p - 0.001\,356Q^2 = 0$$

When $\Delta z = 3$ m the system curve is:

$$3 + h_p - kQ^2 = 0 \quad \text{or} \quad h_p = -3 + 0.001\,356Q^2$$

Assume values of Q, calculate values of h_p, and plot on the diagram to find the point of intersection.

Q (L/s)	$0.001\,355Q^2$	h_p (m)
0	0	−3.0
50	3.39	0.4
100	13.55	10.6
120	19.5	16.5
130	22.9	19.9

Point of intersection is at $h \approx 17$ m, $\quad Q \approx 122$ L/s ◄

Sec. 15.8: Specific Speed of Pumps -- *Exercises*

15.8.1 *Derive the expression for the specific speed of a pump by eliminating D from Eqs. (15.4) and (15.5) in such a way that n becomes a term in the numerator.*

N

Eq. 15.4: $Q = K_Q n D^3$; Eq. 15.5: $h = K_h n^2 D^2$

To obtain an expression in the form of Eq. 15.8 we must eliminate the D from these two equations.

Raise Eq. 15.4 to the 1/2 power: $\quad Q^{1/2} = K_Q^{1/2} n^{1/2} D^{3/2} \qquad (a)$

Raise Eq. 15.5 to the 3/4 power: $\quad h^{3/4} = K_h^{3/4} n^{3/2} D^{3/2} \qquad (b)$

Divide Eq. (*a*) by Eq. (*b*): $\quad \dfrac{Q^{1/2}}{h^{3/4}} = \dfrac{K_Q^{1/2} n^{1/2} D^{3/2}}{K_h^{3/2} D^{3/2}} = \dfrac{K_Q^{1/2} n^{1/2}}{K_h^{3/4} n^{3/2}}; \quad \dfrac{K_Q^{1/2}}{K_h^{3/4}} = \dfrac{nQ^{1/2}}{h^{3/4}} = N_s$ ◄

15.8.2 *What is the specific speed of the pump whose performance characteristics are given in Fig. 15.7? What type of pump is this?*

Fig. 15.7: BEP is at Q = 122 cfs and h = 16 ft (n_e = 450 rpm).

BG

Eq. 15.8*a*: $N_s = n_e\sqrt{\text{gpm}}/h^{3/4} = 450\sqrt{122(449)}/16^{3/4} = 13{,}165$ ◄

Fig. 15.11 for N_s = 13,165: This is an axial-flow pump. ◄

15.8.3 *What are the specific speeds of the pumps on the Colorado River Aqueduct at the Hayfield, Gene and Iron Mountain Plants? (See Sec. 15.14).*

Sec. 15.14: The Hayfield pumps (n_e = 450 rpm) each deliver 200 cfs at h = 444 ft; the Gene pumps (n_e = 400 rpm) each deliver 200 cfs at h = 310 ft; the Iron Mountain pumps (n_e = 300 rpm) each deliver 200 cfs at h = 146 ft.

BG

Eq. 15.8*a*: $N_s = n_e\sqrt{\text{gpm}}/h^{3/4}$

Hayfield: $N_s = 450\sqrt{200(449)}/444^{3/4} = 1395$ ◄

Gene: $N_s = 400\sqrt{200(449)}/310^{3/4} = 1622$ ◄

Iron Mountain: $N_s = 300\sqrt{200(449)}/146^{3/4} = 2140$ ◄

Sec. 15.8: These are all radial-flow pumps. The corresponding values of $(N_s)_{\text{cfs}}$ are 65.8, 76.6 and 101 respectively. ◄

15.8.4 *Find the specific speed of a 10-stage pump that develops a total head of 600 ft at a capacity of 1600 gpm when operating at maximum efficiency at a rotative speed of 900 rpm.*

BG

h = 600/10 = 60 ft per stage ; Eq. 15.8*a*: $N_s = n_e\sqrt{\text{gpm}}/h^{3/4} = 900\sqrt{1600}/60^{3/4} = 1670$ ◄

Fig. 15.11 for N_s = 1670: This is a radial-flow pump.

15.8.5 *A pump is to discharge 0.8 m^3/s at a head of 40 m when running at 300 rpm. What type of pump will be required?*

SI

Eq. 15.8*b*: $(N_s)_{\text{SI}} = \omega\sqrt{Q}/(gh)^{3/4}$ where $\omega = 2\pi n/60$ (ω rad/s, n rev/min)

$(N_s)_{\text{SI}} = 2\pi(300/60)\sqrt{0.8}/(9.81\times 40)^{3/4} = 0.318$

Sec. 15.8 and Fig. 15.11: This is a radial-flow centrifugal pump. ◄

Sec. 15.8: Specific Speed of Pumps -- *Problems*

15.20 *All dimensions of a model pump are one-third as large as the corresponding dimensions of a prototype pump. The model pump operating at 1800 rpm delivers 3.0 cfs at a head of 35 ft. Find the speed and capacity of the prototype if it is to develop a head of 35 ft. Find the specific speed of the model pump and the prototype. They should be the same. Assume the efficiency of the model and prototype are the same.*

BG

Use Eqs. 15.4 and 15.5: $Q \propto nD^3$ and $h \propto n^2D^2$

Eq. 15.5: $h_p/h_m = 1.0 = n_p^2D_p^2/(n_m^2D_m^2) = n_p^2/(1800)^2(3D_m/D_m)^2$; n_p = 600 rpm ◄

Eq. 15.4: $Q_p/Q_m = n_pD_p^3/(n_mD_m^3) = (600/1800)(3D_m/D_m)^3 = 9$; $Q_p = 9Q_m = 27.0$ cfs ◄

Eq. 15.8*a*: $(N_s)_m = 1800\sqrt{3(449)}/(35)^{3/4} = 4590$ ◄

$(N_s)_p = 600\sqrt{27(449)}(35)^{3/4} = 4590$ ◄

15.21 *All dimensions of a model pump are one-third as large as the corresponding dimensions of a prototype pump. If the model pump operating at 1800 rpm delivers 3.0 cfs at a head of 35 ft, find the speed and head of the prototype when it delivers 3.0 cfs. Find also the specific speed of the model pump and the prototype. They should be the same. Assume the efficiency of the model and prototype are the same.*

BG

(*a*) Use Eqs. 15.4 and 15.5: $Q \propto nD^3$ and $h \propto n^2D^2$

Eq. 15.4: $Q_p/Q_m = n_pD_p^3/(n_mD_m^3) = 1 = (n_p/n_m)(3D_m/D_m)^3$; $n_p = n_m/27 = 1800/27 = 66.7$ rpm ◄

Eq. 15.5: $h_p/h_m = n_p^2D_p^2/(n_m^2D_m^2) = (66.7/1800)^2(3D_m/D_m)^2$

$h_p = h_m(1/27)^2(9/1) = h_m/81 = 35/81 = 0.432$ ft ◄

(*b*) Eq. 15.8*a*: $(N_s)_m = 1800\sqrt{3(449)}/(35)^{3/4} = 4590$ ◄

$(N_s)_p = 66.7\sqrt{3(449)}/(0.432)^{3/4} = 4590$ ◄

15.22 *All dimensions of a model pump are one-third as large as the corresponding dimensions of the prototype pump. If the model operating at 1800 rpm delivers 3.0 cfs at a head of 35 ft, find the speed and capacity of the prototype if it is to develop a head of 140 ft. Assume the efficiency of the prototype and the model are the same. Find also the specific speed of the model and the prototype.*

BG

Use Eqs. 15.4 and 15.5: $Q \propto nD^3$ and $h \propto n^2D^2$

Eq. 15.5: $h_p/h_m = n_p^2D_p^2/(n_m^2D_m^2) = (140/35) = (n_p^2/n_m^2)(3D_m/D_m)^2$

$n_p = n_m(4/9)^{1/2} = 1800(2/3) = 1200$ rpm ◄

Eq. 15.4: $Q_p/Q_m = n_pD_p^3/(n_mD_m^3) = (1200/1800)(3D_m/D_m)^3$; $Q_p = Q_m(2/3)(27/1) = 18Q_m = 54.0$ cfs ◄

Eq. 15.8*a*: $(N_s)_m = 1800\sqrt{3(449)}/(35)^{3/4} = 4590$ ◄

$(N_s)_p = 1200\sqrt{27(449)}/(140)^{3/4} = 4590$ ◄

15.23 *A centrifugal pump is required to deliver 1600 gpm against a head of 350 ft at a rotative speed 1150 rpm. Would a two-stage pump be more efficient than a single-stage?*

BG

One-stage pump, $h = 350$ ft ; $N_s = n_e\sqrt{\text{gpm}}/h^{3/4} = 1150\sqrt{1600}/(350)^{3/4} = 569$

Two-stage pump, $h = 350/2 = 175$ ft ; $N_s = 1150\sqrt{1600}/(175)^{3/4} = 956$

Three-stage pump, $h = 350/3 = 116.7$; $N_s = 1150\sqrt{1600}/(116.7)^{3/4} = 1295$

Fig. 15.11 indicates that pumps with $N_s = 569$ have a rather low efficiency. Therefore yes, it is better to use a two-stage (or three-stage) pump. ◄

15.24 *A pump is to deliver 2.0 m³/s against a head of 160 m when operating at 300 rpm. What type of pump will be required?*

SI

300 rpm $= 2\pi(300)/60 = 31.4$ rad/s $= \omega_e$

Eq. 15.9*b*: $(N_s)_{SI} = \omega_e\sqrt{Q}/(gh)^{3/4} = 31.4\sqrt{2}/(9.81 \times 160)^{3/4} = 0.178$

Fig. 15.11 indicates that this is a radial-flow pump with a low specific speed (outside the range of most radial-flow centrifugal pumps). It is not very desirable because of low efficiency. ◄

15.25 *At its optimum point of operation, a given centrifugal pump with an impeller diameter of 500 mm delivers 3.2 m^3/s of water against a head of 25 m when rotating at 1450 rpm. (a) If its efficiency is 82%, what is the brake power of the drive shaft? (b) If a homologous pump with a diameter of 800 mm is rotating at 1200 rpm, what would be the discharge, head and shaft power? Assume both pumps operate at the same efficiency. (c) Compute the specific speed of both pumps.*

SI

(*a*) Brake power $= \gamma Qh/\eta = 9.81(3.2)25/0.82 = 957$ kW ◄

(*b*) Eq. 15.4: $Q \propto nD^3$, $Q = 3.2(1200/1450)(800/500)^3 = 10.85$ m^3/s ◄

Eq. 15.5: $h \propto n^2D^2$, $h = 25(1200/1450)^2(800/500)^2 = 43.8$ m ◄

Eq. 15.6: $P \propto n^3D^5$, $P = 957(1200/1450)^3(800/500)^5 = 5687$ kW ◄

(*c*) For $n = 1450$ rpm, $\omega = 2\pi n/60 = 151.8$ rad/s ; $(N_s)_{SI} = 151.8(3.2)^{1/2}/(9.81 \times 25)^{3/4} = 4.38$ ◄

For $n = 1200$ rpm, $\omega = 2\pi(1200)/60 = 125.6$ rad/s

$(N_s)_{SI} = 125.6(10.85)^{1/2}/(9.81 \times 43.8)^{3/4} = 4.38$ ◄ This is an axial-flow pump.

Sec. 15.9: Peripheral-Velocity Factor -- *Exercises*

15.9.1 *Estimate the diameter of the impeller of the pump whose operating characteristics are shown in Fig. 15.7.*

BG

BEP is $Q = 122$ cfs at $h = 16$ ft ; Eq. 15.8*a*: $N_s = 450\sqrt{122(449)}/16^{3/4} = 13{,}165$

Fig. 15.11 for $N_s = 13{,}165$: $\phi_e \approx 2.3$

Eq. 15.10: $D = 153.3\phi_e\sqrt{h}/n_e \approx 153.3(2.3)\sqrt{16}/450 = 3.1$ ft $= 38$ in ◄

15.9.2 *Estimate the diameter of the impeller of a pump whose BEP is defined by Q = 5000 gpm at a head of 82 ft when operating at 3000 rpm.*

BG

Eq. 15.8*a*: $N_s = 3000\sqrt{5000}/82^{3/4} = 7785$; Fig. 15.11 for $N_s = 7785$: $\phi_e \approx 1.8$

Eq. 15.10: $D \approx 153.3(1.8)\sqrt{82}/3000 = 0.83$ ft $= 10$ in ◄

15.9.3 *At maximum efficiency, a four-stage pump delivers 400 L/s against a head of 300 m at a rotative speed ω of 125.6 rad/sec. Estimate the diameter of the impellers if all four impellers are identical. Calculate $(N_s)_{SI}$, convert to N_s, and then obtain an estimated value of D.*

B

Four-stage pump: $h = 320/4 = 80$ m

Eq. 15.8*b*: $(N_s)_{SI} = \omega_e\sqrt{Q}/(gh)^{3/4} = 125.6\sqrt{0.4}/(9.81 \times 80)^{3/4} = 0.54$

From Sec. 15.8: $N_s = 0.54/0.000\,368 = 1467$. Fig. 15.11 for $N_s = 1467$: $\phi_e \approx 1.0$; $n_e = \omega_e(60/2\pi)$

Eq. 15.10: $D \approx 153.3(1.0)\sqrt{80(3.28)}/[125.6(60/6.28)] = 2.07$ ft $= 24.8$ in ◄

$D \approx 2.07(0.3048) = 0.630$ m $= 630$ mm ◄

Sec. 15.9: Peripheral-Velocity Factor -- *Problems*

15.26 *A three-stage pump is rated at 7500 gpm against a head of 600 ft at a speed of 1200 rpm. The three impellers are identical. Find the N_s of this pump and the approximate diameter of the impellers.*

BG

$h = 600/3 = 200$ ft/stage. Eq. 15.8*a*: $N_s = n_e\sqrt{\text{gpm}}/h^{3/4} = 1200\sqrt{7500}/200^{3/4} = 1954$ ◄

From Fig. 15.11 for $N_s = 1954$: $\phi_e \approx 1.05$

Eq. 15.10: $D = 153.3\phi_e\sqrt{h}/n \approx 153.3(1.05)\sqrt{200}/1200 = 1.90$ ft $= 22.8$ in ◄

15.27 *A four-stage pump is designed to deliver 65 L/s against a head of 120 m when operating at 1450 rpm. The four impellers are identical. Find the specific speed of this pump. What type of pump is it? Estimate the diameter of the impellers.*

B

$h = 120/4 = 30$ m/stage

Eq. 15.8*b*: $(N_s)_{SI} = \omega_e\sqrt{Q}/(gh)^{3/4} = [2\pi(1450)/60]\sqrt{0.065}/(9.81 \times 30)^{3/4} = 0.545$

From Sec. 15.8: $N_s = (N_s)_{SI}/0.000\,368 = 0.545/0.000\,368 = 1481$

Fig. 15.11 for $N_s = 1481$: $\phi_e \approx 1.02$

Eq. 15.10: $D = 153.3P_e\sqrt{h}/n \approx 153.3(1.02)\sqrt{30(3.28)}/1450 = 1.07$ ft $= 12.8$ in $= 326$ mm ◄

This is a radial flow pump ◄

15.28 *Estimate the diameter of the impeller of the model pump depicted in Exer. 15.4.1.*
Exer. 15.4.1: The pump (n_e = 3600 rpm) delivers 0.10 m³/s at h = 40 m.

B

Eq. 15.8*b*: $(N_s)_{SI} = \omega_e\sqrt{Q}/(gh)^{3/4} = 2\pi(3600/60)\sqrt{0.10}/(9.81 \times 40)^{3/4} = 1.35$

Fig. 15.11 for $(N_s)_{SI} = 1.35$: $\phi_e \approx 1.3$

Eq. 15.10: $D \approx 153.3(1.3)\sqrt{40(3.28)}/3600 = 0.64$ ft $= 8$ in $= 190$ mm ◄

15.29 *Approximately what would be the diameter of the impeller of a pump that delivers at its BEP 150 L/s of water at a head of 45 m when operating at 1750 rpm.*

B

Eq. 15.8*b*: $(N_s)_{SI} = \omega_e\sqrt{Q}/(gh)^{3/4} = 2\pi(1750/60)\sqrt{0.150}/(9.81 \times 45)^{3/4} = 0.737$

From Sec. 15.8: $N_s = 0.737/0.000\,368 = 2000$; Fig. 15.11 for $N_s = 2000$: $\phi_e \approx 1.10$

Eq. 15.10: $D \approx 153.3(1.10)\sqrt{45(3.28)}/1450 = 1.41$ ft $= 17.0$ in $= 430$ mm ◄

Sec. 15.8 and Fig. 15.11 for $N_s = 2000$: This is a radial-flow centrifugal pump. ◄

15.30 *Approximately how much head will be developed by a pump when operating at its point of optimum efficiency if it delivers 6500 gpm when running at 1200 rpm? The diameter of the impeller is 15.0 in.*

BG

This problem must be solved by trial using Eq. 15.8*a*, Eq. 15.10 and Fig. 15.11.

Eq. 15.10 assuming $\phi_e = 1.0$: $D = 153.3(1.0)\sqrt{h}/1200 = 15/12$ so $\sqrt{h} = 9.78$ ft, $h = 95.7$ ft

Eq. 15.8*a*: $N_s = 1200\sqrt{6500}/95.7^{3/4} = 3160$; Fig. 15.11 for $N_s = 3160$: $\phi_e \approx 1.25$

Try $\phi_e = 1.2$: $D = 153.3(1.2)\sqrt{h}/1200 = 15/12$, so $\sqrt{h} = 8.15$, $h = 66.5$ ft

Eq. 15.8*a*: $N_s = 1200\sqrt{6500}/66.5^{3/4} = 4150$; Fig. 15.11 for $N_s = 4150$: $\phi_e \approx 1.4$

Try $\phi_e = 1.5$: $D = 153.3(1.5)\sqrt{h}/1200 = 15/12$ so $\sqrt{h} = 6.52$, $h = 42.6$ ft

Eq. 15.8*a*: $N_s = 1200\sqrt{6500}/42.6^{3/4} = 5790$; Fig. 15.11 for $N_s = 5790$: $\phi_e \approx 1.52$ (close!)

Hence $h \approx 43$ ft ◄

15.31 *Calculate the values of ϕ_e for the pumps of Sample Prob. 15.2 and compare these values with those given in Fig. 15.11. Perform these calculations for both the model and the prototype, *given the model diameter is 7.4 in. [*Note this correction to the textbook.]*

Sample Prob. 15.2: Model pump delivers 3.0 cfs at 3600 rpm when h = 125 ft; prototype delivers 300 cfs at 360 rpm when h = 125 ft.

BG

Model: Eq. 15.8*a*: $N_s = 3600\sqrt{3(449)}/125^{3/4} = 3535$; Fig. 15.11 for $N_s = 3535$: $\phi_e \approx 1.3$

Eq. 15.10: $D \approx 7.4/12 \approx 153.3\phi_e\sqrt{125}/3600$ so $\phi_e \approx 1.29$, good check ◄

Prototype: Eq. 15.8*a*: $N_s = 360\sqrt{300(449)}/125^{3/4} = 3535$; Fig. 15.11 for $N_s = 3535$: $\phi_e \approx 1.3$

Eq. 15.10: $D \approx 10(7.4/12) \approx 153.3\phi_e\sqrt{125}/360$ so $\phi_e \approx 1.29$, good check ◄

15.32 *At maximum efficiency, the pump at Rocky River delivers 280 cfs at a head of 238 ft when operating at 327 rpm. Calculate the specific speed of this pump and provide an estimate of the diameter of the impeller.*

BG

Eq. 15.8*a*: $N_s = 327\sqrt{280(449)}/238^{3/4} = 1913$

Fig. 15.11, for $N_s = 1913$: $\phi_e \approx 1.04$; Eq. 15.10: $D \approx 153.3(1.04)\sqrt{238}/327 = 7.5$ ft ◄

15.33 *At the Grand Coulee project on the Columbia River there are several identical pumps, each with an impeller diameter of 167⅜ inches. The rotative speed of these pumps is 200 rpm and the maximum efficiency is 90.8% when discharging 1250 cfs at a head of 344 ft. Find the specific speed and calculate ϕ_e. How does the calculated value of ϕ_e compare with the value shown in Fig. 15.11?*

BG

Eq. 15.8*a*: $N_s = 200\sqrt{1250(449)}/344^{3/4} = 1877$

Eq. 15.10: $D = 153.3\phi_e\sqrt{344}/200 = 167.4/12$ so $\phi_e = 0.98$ ◄

Fig. 15.11 for $N_s = 1877$: $\phi_e \approx 1.0$ Good check

15.34 *Approximately what head would you expect a pump (impeller diameter = 10.0 in) to develop at 1200 rpm if its specific speed is 3500?*

BG

Fig. 15.11 for $N_s = 3500$: $\phi_e \approx 1.3$

Eq. 15.10: $D = 153.3\phi_e\sqrt{h}/n$; $10/12 \approx 153.3(1.3)\sqrt{h}/1200$, so $h \approx 25$ ft ◄

15.35 *A given axial-flow pump delivers 300 L/s at a head of 6 m when operating at 1800 rpm. If this pump were operating on the moon at the same speed, what head would it develop when delivering water at 300 L/s? Assume the gravitational constant of the moon is one-sixth that of the earth. Assume cavitation does not occur.*

SI

Eq. 15.3 indicates that if n and D are constant, gh = constant. Hence, since g of moon is $(1/6)g$ of earth, h on moon = $6h$ on earth.

So $h_{moon} = 6h_{earth} = 6 \times 6 = 36$ m ◄

Sec. 15.10: Cavitation in Pumps -- ***Exercises***

15.10.1 *A mixed-flow pump, located at sea level, with a specific speed of 6000 is to be used to pump 80°F water from a reservoir. The head to be developed is 40 ft. What is the greatest elevation above the reservoir water surface that the pump can be placed such that cavitation will not occur in the pump? Assume the head loss from the reservoir to the pump is 1.5 ft.*

BG

Fig. 15.12 for N_s = 6000: $\sigma_c \approx 0.63$; Table A.1 at 80°F: p_v/γ = 1.17 ft and γ = 62.22 lb/ft^3

Eq. 15.14: $(z_s)_{max} \approx (p_0)_{abs}/\gamma - p_v/\gamma - \sigma_c h - h_L$ = 14.7(144)/62.22 - 1.17 - 0.63(40) - 1.5

$(z_s)_{max} \approx$ 34.02 - 1.17 - 25.20 - 1.50 = 6.2 ft above the reservoir water surface ◄

An approximate answer (10 ft) can be found through use of Fig. 15.13.

15.10.2 *Solve Exer. 15.10.1 for the case where the pump is located at elevation 10,000 ft with a water temperature of 50°F. Assume all other data remain the same.*

Exer. 15.10.1: The pump (N_s = 6000) delivering water from a reservoir is to develop h = 40 ft, when the friction h_L = 1.50 ft. Find $(z_s)_{max}$ above the reservoir without cavitation in the pump.

BG

Fig. 15.12 for N_s = 6000: $\sigma_c \approx 0.63$; Table A.1 at 50°F: p_v/γ = 0.41 ft and γ = 62.41 lb/ft

Table A.3 at elevation 10,000 ft: p_{atm} = 10.11 psia

Eq. 15.14: $(z_s)_{max} \approx$ 10.11(144)/62.41 - 0.41 - 0.63(40) - 1.5 = 23.33 - 0.41 - 25.20 - 1.50

$(z_s)_{max} \approx$ -3.8 ft

To safeguard against cavitation this pump must be placed about 3.8 ft or more below the elevation of the reservoir water surface. ◄

15.10.3 *A pump with a critical value of σ = 0.20 is to pump against a head of 20 m. The barometric pressure is 98.5 kPa abs, and the vapor pressure is 5.2 kPa abs. Friction losses from the reservoir to the pump are 0.5 m. Find the maximum allowable height of the pump relative to the water surface in the reservoir.*

SI

Eq. 15.14: $(z_s)_{max} = (p_0)_{abs}/\gamma - p_v/\gamma - \sigma_c h - h_L$ = (98.5 - 5.2)/9.81 - 0.2(20) - 0.5

$(z_s)_{max}$ = 9.51 - 4.00 - 0.50 = 5.01 m

To prevent cavitation, the pump should be set less than 5.01 m above the reservoir water surface. ◄

Sec. 15.10: Cavitation in Pumps -- ***Problems***

15.36 *Is the pump in Prob. 15.35 likely to cavitate? Why?*

Prob. 15.35: An axial-flow pump (n = 1800 rpm) which on earth delivers 300 L/s at h = 6 m is to deliver the same flow on the moon.

BG

Yes, the pump is quite likely to cavitate because the atmospheric pressure on the moon is very low. ◄

15.37 *A pump with a critical value of σ_c of 0.20 is to pump against a head of 200 ft. The barometric pressure is 14.3 psia, and the vapor pressure of the water is 0.8 psia. Assume the friction losses in the intake piping are 4 ft. Find the maximum allowable height of the pump relative to the water surface at intake to assure that cavitation will not occur.*

BG

Eq. 15.14: $(z_s)_{max} = (p_0)_{abs}/\gamma - p_v/\gamma - \sigma_c h - h_L$

$(z_s)_{max}$ = (14.3 - 0.8)144/62.4 - 0.2(200) - 4 = 31.15 - 40 - 4 = -12.85 ft

The pump must be placed at least 12.85 ft below the reservoir water surface. ◄

15.38 *A boiler feed pump delivers water at 200°F, which it draws from an open hot well with a friction loss of 2 ft in the intake pipe. The barometric pressure is 29 inHg, and the value of σ_c for the pump is 0.16. What must be the elevation of the water surface in the hot well relative to that of the pump intake? The total pumping head is 140 ft.*

BG

29 inHg = (29/12)13.55 = 32.75 ft of water = p_{atm}/γ ; Table A.1 at 200°F: $p_v/\gamma = 26.59$ ft

Eq. 15.14: $(z_s)_{max} = 32.75 - 26.59 - 0.16(140) - 2 = 4.16 - 22.4 = -18.24$ ft

The pump must be placed at least 18.24 ft below the hot well water surface to prevent cavitation. ◀

15.39 *Repeat Prob. 15.38 with all data identical except for the total pumping head. Consider three cases as follows: total pumping heads of 40 ft, 100 ft, and 180 ft.*

Prob. 15.38: A pump ($\sigma_c = 0.16$) draws 200°F water through intake piping with friction $h_L = 2$ ft. If $p_0 = 29$ inHg, find $(z_s)_{max}$ for no cavitation in the pump.

BG

29 inHg = (29/12)13.55 = 32.75 ft of water = p_{atm}/γ ; Table A.1 at 200°F: $p_v/\gamma = 26.59$ ft

Eq. 15.14: $(z_s)_{max} = 32.75 - 26.59 - 0.16h - 2 = 4.16 - 0.16h$ ft

h (ft)	Required distance below water surface (ft)
40	2.24
100	11.84
180	24.6

▲

15.40 *Suppose a pump was pumping water at a head of 30 ft with a water temperature of 100°F and a barometric pressure of 14.2 psia. At intake to the pump the pressure is 17 inHg vacuum and the pipeline velocity is 16 ft/sec. What are the values of NPSH and σ?*

BG

Table A.1 at 100°F: $\gamma = 62.00$ lb/ft^3 and $p_v/\gamma = 2.19$ ft

$(p_s)_{abs}/\gamma = p_{atm}/\gamma + p_{gage}/\gamma = 14.2(144)/62.00 + (-17 \text{ inHg})33.9/29.9 = 33.0 - 19.3 = 13.7$ ft

Eq. 15.11: $NPSH = (p_s)_{abs}/\gamma + V_s^2/2g - p_v/\gamma = 13.7 + 16^2/(2 \times 32.2) - 2.19 = 15.49$ ft ◀

Eq. 15.11: $\sigma = NPSH/h = 15.49/30 = 0.516$ ◀

15.41 *A centrifugal pump with $(N_s)_{SI} = 1.48$ is to pump 30 L/s of gasoline ($s = 0.82$, vapor pressure = 32 kPa) from an open tank. The pump impeller rotates at 157 rad/s. Assuming the free surface in the tank is at elevation 100.0 m and the barometric pressure is 101 kPa abs, what is the highest elevation at which the pump centerline may be placed without cavitation problems? The suction pipe ($f = 0.022$) is 40 m long and has a diameter of 120 mm. Assume square entrance conditions where the pipe takes off from the tank. Express the answer in terms of h.*

SI

Fig. 15.12 for $(N_s)_{SI} = 1.48$: $\sigma_c \approx 0.40$

$h_L = 0.5V^2/2g + 0.022(40/0.120)V^2/2g = 7.83V^2/2g$; $V = Q/A = 0.030/[\pi(0.120)^2/4] = 2.65$ m/s

$\therefore h_L = 7.83(2.65)^2/(2 \times 9.81) = 2.81$ m

Eq. 15.14: $(z_s)_{max} = (p_s)_{abs}/\gamma - p_v/\gamma - \sigma_c h - h_L \approx 101/(0.82 \times 9.81) - 32/(0.82 \times 9.81) - 0.40h - 2.81$

$(z_s)_{max} \approx 12.56 - 3.98 - 0.40h - 2.81 = 5.77 - 0.40h$

Max elev. (meters) = $100.0 + (z_s)_{max} \approx 105.77 - 0.40h$ ◀

15.42 *At rated capacity a four-stage centrifugal pump rotates at 1200 rpm and delivers 360 L/s of water against a head of 180 m. What diameter of impellers should be used? The suction and discharge flanges have inside diameters of 300 mm and 320 mm respectively, and both are located 0.7 m below the pump centerline. At what gage pressure at the suction flange may cavitation be expected? Water temperature is 34°C and barometric pressure is 98 kPa abs.*

SI

Eq. 15.8*b*: $(N_s)_{SI} = \omega_e\sqrt{Q}/(gh)^{3/4} = 2\pi(1200/60)\sqrt{0.360}/(9.81 \times 180/4)^{3/4} = 0.782$

Fig. 15.11 for $(N_s)_{SI} = 0.782$: $\phi_e \approx 1.1$; $h = 180/4 = 45$ m per stage

Eq. 15.10: $D = 153.3\phi_e\sqrt{h}/n_e \approx 153.3(1.1)\sqrt{45(3.28)}/1200 = 1.7 \text{ ft} = 21 \text{ in} = 520 \text{ mm}$ ◄

Eq. 15.14: $(z_s)_{max} = (p_0)_{abs}/\gamma - p_v/\gamma - \sigma_c h - h_L$

where we let $(z_s)_{max} = -0.7$ m for the condition of incipient cavitation

Table A.1 at 34°C: $\gamma = 9.75 \text{ kN/m}^3$ and $p_v/\gamma = 0.56$ m

$(p_0)_{abs}/\gamma = p_{atm}/\gamma = 98/9.75 = 10.05$ m ; Fig. 15.12 for $(N_s)_{SI} = 0.782$: $\sigma_c \approx 0.16$

Hence Eq. 15.14 can be expressed as: $-0.7 \approx 10.05 - 0.56 - 0.16(180/4) - h_L$; $h_L \approx 3$ m

Energy equation from surface of reservoir to suction side of pump, with reservoir water surface as datum:

$0 - h_L = +z_s + p_s/\gamma + V_s^2/2g$ where $z_s = -0.7$ m and $V_s = Q/A_s$

$V_s = 0.36/(\pi 0.3^2/4) = 5.10$ m/s; $V_s^2/2g = 5.10^2/(2 \times 9.81) = 1.326$ m

Hence, energy equation may be written as: $0 - 3 \approx -0.7 + p_s/\gamma + 1.33$

from which $p_s/\gamma \approx -3 - 1.33 + 0.7 = -3.63$ m ; $p_s \approx -3.63(9.75) = -35.4 \text{ kN/m}^2$

Inside cover: 760 mmHg is equivalent to 101.32 kN/m^2.

Hence the pressure at suction $\approx -35.4(760/101.32)$ = Vacuum of 210 mmHg ◄

15.43 *Rework Prob. 15.42 for the case of a three-stage pump. Assume speed, flow rate, and total head are unchanged.*

Prob. 15.42: At rated capacity a four-stage centrifugal pump rotates at 1200 rpm and delivers 360 L/s of water against a head of 180 m. What diameter of impellers should be used? The suction and discharge flanges have inside diameters of 300 mm and 320 mm respectively, and both are located 0.7 m below the pump centerline. At what gage pressure at the suction flange may cavitation be expected? Water temperature is 34°C and barometric pressure is 98 kPa abs.

SI

Eq. 15.8*b*: $(N_s)_{SI} = \omega_e\sqrt{Q}/(gh)^{3/4} = 2\pi(1200/60)\sqrt{0.360}/(9.81 \times 180/3)^{3/4} = 0.631$

Fig. 15.11 for $(N_s)_{SI} = 0.631$: $\phi_e \approx 1.02$; $h = 180/3 = 60$ m

Eq. 15.10: $D = 153.3\phi_e\sqrt{h}/n_e \approx 153.3(1.02)\sqrt{60(3.28)}/1200 = 1.83 \text{ ft} = 21.9 \text{ in} = 560 \text{ mm}$ ◄

Fig. 15.12 for $(N_s)_{SI} = 0.631$: $\sigma_c \approx 0.10$. Table A.1 at 34°C: $\gamma = 9.75 \text{ kN/m}^3$ and $p_v/\gamma = 0.56$ m

Eq. 15.14 with $(z_s)_{max} = -0.7$ m for the condition of incipient cavitation:

$-0.7 \approx 10.05 - 0.56 - 0.10(180/3) - h_L$ from which $h_L \approx 10.05 - 0.56 - 6.0 + 0.7 = 4.2$ m

Energy equation from surface of reservoir to suction side of pump, with reservoir water surface as datum:

$0 - h_L = +z_s + p_s/\gamma + V_s^2/2g$ where $z_s = -0.7$ and $V_x = Q/A_s$

$V_s = 0.36/(\pi 0.3^2/4) = 5.10$ m/s ; $V_s^2/2g = 5.10^2/(2 \times 9.81) = 1.326$ m

$p_s/\gamma = -z_s - V_s^2/2g - h_L \approx -(-0.7) - 1.3 - 4.2 = -4.8$ m ; $p_s \approx -4.8(9.75) = -47 \text{ kN/m}^2$

Using the inside cover: The pressure at suction $\approx -47(760/101.32)$ = Vacuum of 350 mmHg ◄

15.44 *A pump with a critical value of $\sigma_c = 0.18$ is to pump against a head of 80 m. The barometric pressure is 98.5 kPa abs, and the vapor pressure is 5.4 kPa abs. Assume the friction losses in the intake are 1.2 m. Find the maximum allowable elevation of the pump relative to the water surface at intake.*

SI

Eq. 15.14: $(z_s)_{max} = (p_s)_{abs}/\gamma - p_v/\gamma - \sigma_c h - h_L = (98.5 - 5.4)/9.81 - 0.18(80) - 1.2$

$(z_s)_{max} = 9.49 - 14.40 - 1.20 = -6.11$ m

The pump must be set at least 6.11 m below the elevation of the reservior water surface. ◀

15.45 *Determine the specific speed of a centrifugal pump that is rated at 3500 gpm under a head of 70 ft at 1750 rpm? What would be the head and capacity of this pump if operated at 1160 rpm? For each rotative speed note the maximum tolerable suction lift as recommended in Fig. 15.13.*

BG

Eq. 15.8*a*: $N_s = 1750\sqrt{3500}/70^{3/4} = 4280$ ◀

Fig. 15.13 for $N_s = 4280$ (1750 rpm): the maximum tolerable suction lift is about 3.0 ft. ◀

Changing the rotative speed will not change N_s as N_s is defined at the (rated) BEP.

From Eq. 15.5 with D unchanged: $h_2 = h_1(n_2D_2/n_1D_1)^2 = 70(1160/1750)^2 = 30.8$ ft ◀

From Eq. 15.4 with D unchanged: $Q_2 = Q_1(n_2D_2^3/n_1D_1^3) = 3500(1160/1750) = 2320$ gpm ◀

Fig. 15.13 for $N_s = 4280$ and $h = 30.8$ ft (1160 rpm): the tolerable suction lift is about 21 ft. ◀

Note: The lower speed results in a lower flow rate which permits a higher tolerable lift.

Sec. 15.12: Selection of Pumps -- *Exercises*

15.12.1 *Repeat Sample Prob. 15.10 for the case where the reservoir water surface elevations are 120 ft and 160 ft respectively, with all other data remaining the same.*

Sample Prob. 15.10: Water is pumped at 20 cfs from reservoir A to reservoir B (total friction h_L = 18.88 ft). The intake piping friction h_L = 2.83 ft, and p_v/γ = 0.41 ft. Find (a) N_s of the most suitable pump with n = 600 rpm, and (b) whether the pump is safe from cavitation at the surface elevation of reservoir A.

BG

(*a*) $h = (160 - 120) + 18.88 = 58.9$ ft ; Eq. 15.8*a*: $N_s = 600\sqrt{20(449)}/58.9^{3/4} = 2670$ ◀

(*b*) Eq. 15.14: $(z_s)_{max} = (p_0)_{abs}/\gamma - p_v/\gamma - \sigma_h - h_L$

From Table A.3 at elevation 140 ft: $(p_0)_{abs}/\gamma = 14.63(144)/62.4 = 33.8$ ft

Fig. 15.12 for $N_s = 2670$: $\sigma_c \approx 0.21$; $\therefore$ $\sigma_c h \approx 0.21(58.9) = 12$ ft

$(z_s)_{max} \approx 33.8 - 0.41 - 12 - 2.83 = 18$ ft

The pump can be placed as much as about 18 ft above the water surface elevation in reservoir A.

$\therefore$ with $z_s = 0$, yes, the pump will be safe against cavitation. ◀

15.12.2 *Repeat Sample Prob. 15.10 for the case where the flow rate is 50 cfs rather than 20 cfs, with all other data remaining the same.*

Sample Prob. 15.10: Water is pumped from reservoir A (surface elevation 5910 ft) to reservoir B (6060 ft) through a 24-in-diameter pipeline (f = 0.03) 2000 ft long. The pump is 300 ft from reservoir A, at the same elevation as its surface. $(p_0)_{abs}/\gamma$ = 27.3 ft and p_v/γ = 0.41 ft. Find (a) N_s of the most suitable pump with n = 600 rpm, and (b) whether the pump is safe from cavitation.

BG

(*a*) $V = Q/A = 50/\pi(1)^2 = 15.92$ ft/sec

$h = (6060 - 5910) + 0.03(2000/2)15.92^2/(2 \times 32.2) = 268$ ft

Eq. 15.8*a*: $N_s = 600\sqrt{50(449)}/268^{3/4} = 1357$ ◀

(*b*) Eq. 15.14: $(z_s)_{max} = (p_0)_{abs}/\gamma - p_v/\gamma - \sigma_c h - h_L$

Fig. 15.12 for $N_S = 1357$: $\sigma_c \approx 0.08$, $\therefore$ $\sigma_c h \approx 0.08(268) \approx 21$ ft

Eq. 8.10: $h_L = 0.03(300/2)15.92^2/(2 \times 32.2) = 17.7$ ft

$(z_s)_{max} = 27.3 - 0.41 - 21 - 17.7 = -12$ ft

This pump must be placed at least approx. 12 ft below the water surface elevation in reservoir A.

$\therefore$ with $z_s = 0$, no, the pump will not be safe against cavitation. ◀

15.12.3 *Water flows at 80°C through a 100-mm-diameter pipe (f = 0.024) of length 300 m from a large pressurized tank (p = 200 kPa, water surface elevation 60.0 m) to discharge into the atmosphere at elevation 68.0 m. If the flow rate is to be 20 L/s, find the head that a pump, rotating at 1200 rpm placed in the pipeline, must develop and determine its specific speed. Assume the pressure in the tank remains constant and neglect minor losses. This pump is placed 6.0 m above the water level in the tank. Will cavitation occur if the head loss to the pump is 0.8 m?*

SI

Table A.1 at 80°C: $\gamma = 9.53$ kN/m³ ; $V = Q/A = (20 \times 10^{-3}\ \text{m}^3/\text{s})/(\pi 0.05^2\ \text{m}^2) = 2.55$ m/s

Energy, from tank atmosphere to outlet: $z_1 + p_1/\gamma - f(L/D)V^2/2g + h_p = z_2$

$60.0\ \text{m} + 200\ \text{kN/m}^2/(9.53\ \text{kN/m}^3) - 0.024(300/0.10)2.55^2/[2(9.81)] + h_p = 68.0$ m

$h_p = 8.0 - 21.0 + 23.9 = 10.9$ m ◀

Eq. 15.8*b*: $(N_s)_{SI} = \omega\sqrt{Q}/(gh)^{3/4}$ where $\omega = 2\pi n/60 = 6.28(1200)/60 = 125.6$ rad/sec

$(N_s)_{SI} = 125.6\sqrt{20 \times 10^{-3}}/(9.81 \times 10.9)^{3/4} = 0.534$

Eq. 15.14: $(z_s)_{max} = (p_0)_{abs}/\gamma - p_v/\gamma - \sigma_c h - h_L$

Table A.3: $(p_0)_{abs} = 101.33$ kN/m² ; Table A.1: $\gamma = 9.53$ kN/m³, $p_v/\gamma = 4.96$ m

Fig. 15.12 for $(N_s)_{SI} = 0.534$: $\sigma_c \approx 0.09$

Thus: $(z_s)_{max} \approx (101.33/9.53) - 4.96 - 0.09(10.9) - 0.8 = 10.63 - 4.96 - 0.98 - 0.8 = 3.9$ m

$z_s = 6.0$ m is above $(z_s)_{max} \approx 3.9$ m, so yes, the pump will cavitate. ◀

15.46 *Water at 60°F flows by gravity between two reservoirs. The water surface in Reservoir A is at elevation 5010 ft and that in Reservoir B is at elevation 4980 ft. The water flows through a 24-in-diameter pipe (f = 0.026) of length 8000 ft. (a) What is the flow rate? Neglect minor losses. (b) If one wishes to double the flow rate using a pump operating at 1800 rpm, what specific speed pump would you recommend? (c) Approximately what would be the diameter of the impeller of the pump? (d) If the pump were set at the upper end of the pipeline, at what elevation must the pump be set to safeguard against cavitation if the head loss from Reservoir A to the pump is 1.5 ft?*

BG

(*a*) Energy equation from water surface A to water surface B: $5010 - 0.026(8000/2)V^2/2g = 4980$

$104V^2/2g = 5010 - 4980 = 30; \quad V^2/2g = 30/104 = 0.2885$

$V^2 = 2(32.2)0.288 = 18.58; \quad V = 4.31$ ft/sec, $Q = \pi(2^2/4)4.31 = 13.53$ cfs ◄

(*b*) Double the flow rate: $5010 - 0.026(8000/2)(2 \times 4.31)^2/2g + h_p = 4980$

$30 - 104(8.62)^2/(2 \times 32.2) + h_p = 0; \quad h_p = 120.0 - 30 = 90.0$ ft

$Q = AV = \pi(2^2/4)8.62 = 27.1$ cfs ; Eq. 15.8*a*: $N_s = 1800\sqrt{27.1(449)}/90.0^{3/4} = 6795$

Recommend specific speed of 6800 ◄

(*c*) Fig. 15.11 for $N_s = 6800$: $\phi_e \approx 1.75$

Eq. 15.10: $D \approx 153.3(1.75)\sqrt{90.0}/1800 = 1.41$ ft $= 16.9$ in ◄

(*d*) Eq. 15.14: $(z_s)_{max} = (p_0)_{abs}/\gamma - p_v/\gamma - \sigma_c h - h_L$

where (Table A.3 at 4995 ft) $(p_0)_{abs}/\gamma = p_{atm}/\gamma = 12.23(144)/62.4 = 28.2$ ft

Table A.1 at 60°F: $p_v/\gamma = 0.59$ ft ; Fig. 15.12 for $N_s = 6800$: $\sigma_c \approx 0.74$

$(z_s)_{max} \approx 28.2 - 0.6 - 0.74(90.0) - 1.5 = -40$ ft

Must set pump at least about 40 ft below the elevation of the reservoir water surface. ◄

15.47 *Repeat Prob. 15.46 for the case where the pump is to operate at 600 rpm rather than 1800 rpm, with all other data remaining the same.*

Prob. 15.46: Water at 60°F flows by gravity from reservoir A (surface elevation 5010 ft) to reservoir B (4980 ft) through a pipeline (f = 0.026, L = 8000 ft, D = 24 in). (a) Find Q, neglecting minor losses. (b) What N_s would you use for a pump (n = 1800 rpm) to double the Q? (c) Find the approximate pump impeller diameter. (d) With the pump at rreservoir A, and a 1.5-ft intake h_L, find $(z_s)_{max}$ above the surface of reservoir A for no cavitation in the pump.

BG

(*a*) Energy equation from water surface A to water surface B: $5010 - 0.026(8000/2)V^2/2g = 4980$

$104V^2/2g = 5010 - 4980 = 30; \quad V^2/2g = 30/104 = 0.2885$

$V^2 = 2(32.2)0.288 = 18.58; \quad V = 4.31$ ft/sec, $Q = \pi(2^2/4)4.31 = 13.53$ cfs ◄

(*b*) Double the flow rate: $5010 - 0.026(8000/2)(2 \times 4.31)^2/2g + h_p = 4980$

$30 - 104(8.62)^2/(2 \times 32.2) + h_p = 0; \quad h_p = 120.0 - 30 = 90.0$ ft

$Q = AV = \pi((2^2/4)8.62 = 27.1$ cfs ; Eq. 15.8*a*: $N_s = 600\sqrt{27.1(449)}/90.0^{3/4} = 2264$

Recommend specific speed of 2260 ◄

(*c*) Fig. 15.11 for $N_s = 2260$: $\phi_e \approx 1.1$

Eq. 15.10: $D \approx 153.3(1.1)\sqrt{90.0}/600 = 2.7$ ft $= 32$ in ◄

/cont...

(d) Eq. 15.14: $(z_s)_{max} = (p_0)_{abs}/\gamma - p_v/\gamma - \sigma_c h - h_L$

where (Table A.3 at 4995 ft) $(p_0)_{abs}/\gamma = p_{atm}/\gamma = 12.23(144)/62.4 = 28.2$ ft

Table A.1 at 60°F: $p_v/\gamma = 0.59$ ft ; Fig. 15.12 for $N_s = 2260$: $\sigma_c \approx 0.185$

Eq. 15.14: $(z_s)_{max} \approx 28.2 - 0.6 - 0.185(90.0) - 1.5 = 9.5$ ft

The pump can be set as much as about 9.5 ft above the reservoir water surface. ◀

15.48 *Water at 170°F is pumped through a 15-in-diameter pipe ($f = 0.028$) of length 2000 ft from a large pressurized tank ($p = 30$ psi) whose water surface is at elevation 200 ft to a point of discharge in the atmosphere at elevation 242 ft. (a) If the flow rate is 5800 gpm, find the head that a pump, placed in the pipeline and rotating at 1200 rpm, must develop and determine the specific speed of the pump. Assume the pressure in the tank remains constant and neglect minor losses. (b) Approximately what will be the diameter of the impeller? (c) At what elevation must the pump be set to safeguard against cavitation if the head loss from the lower reservoir to the pump is 1.6 ft?*

BG

Table A.1 at 170°F: $\gamma = 60.8$ lb/ft^3, $p_v = 5.99$ psia

(a) $V = Q/A = (5800/449)/[\pi(15/12)^2/4] = 10.53$ fps ; $V^2/2g = (10.53)^2/(2 \times 32.2) = 1.721$ ft

Energy eq. from tank atmos. to outlet: $200 + 30(144)/60.8 - 0.028[(2000)/(15/12)]1.721 + h_p = 242$

$200 + 71.1 - 77.1 + h_p = 242$; $h_p = 48.0$ ft ◀

Eq. 15.8a: $N_s = n_e\sqrt{\text{gpm}}/h^{3/4} = 1200\sqrt{5800}/(48.0)^{3/4} = 5010$ ◀

(b) Fig. 15.11 for $N_s = 5010$: $\phi_e \approx 1.5$

Eq. 15.10: $D = 153.3\phi_e\sqrt{h}/n \approx 153.3(1.5)\sqrt{48.0}/1200 = 1.3$ ft $= 16$ in ◀

(c) Eq. 15.14: $(z_s)_{max} = p_0/\gamma - p_v/\gamma - \sigma_c h - h_L$ where $p_0 = p_{atm} + p_{tank}$

Fig. 15.14 for $N_s = 5010$: $\sigma_c \approx 0.5$

$(z_s)_{max} \approx (14.7 + 30.0)144/60.8 - 5.99(144)/60.8 - 0.5(48.0) - 1.6 = 66$ ft

The pump could be placed as much as about 66 ft above the water level in the lower tank. ◀

15.49 *A pump manufacturer is asked to provide a pump that will deliver 84,500 gpm against a head of 225 ft. Several speeds of operation are considered, namely 225 rpm, 600 rpm, 1200 rpm, and 1800 rpm. Determine the specific speeds of pumps that will operate at these speeds and estimate the diameter of each of their impellers. What factors might one consider in making a choice among them?*

BG

Eq. 15.8a: $N_s = n_e\sqrt{\text{gpm}}/h^{3/4} = n_e\sqrt{84{,}500}/225^{3/4} = 4.3n_e$

Eq. 15.10: $D = 153.3\phi_e\sqrt{h}/n = 153.8\phi_e\sqrt{225}/n = 2300\phi_e/n$ where ϕ_e depends on N_s (Fig. 15.11)

n_e	N_s	ϕ_e	D (ft)
225	968	≈0.95	≈9.7
600	2580	≈1.15	≈4.41
1200	5160	≈1.55	≈2.97
1800	7740	≈1.80	≈2.30
	▲		▲

Low speed pumps tend to have large inpellers. High speed pumps are more likely to have problems with cavitation.

15.50 *You are asked to select a single-stage or multi-stage pump to deliver 10 cfs against a head of 300 ft when operating at 450 rpm. Specify the number of stages, the specific speed, and the approximate diameter of the impeller.*

BG

Assume a single-stage pump. Eq. 15.8*a*: $N_s = 450\sqrt{10(449)}/300^{3/4} = 418$

N_s is too low, so select a multi-stage pump.

Stages	h	N_s	ϕ_e*	D**
2	150	703	≈0.90	≈3.8 ft
3	100	953	≈0.95	≈3.2 ft
4	75	1183	≈0.98	≈2.9 ft
6	50	1603	≈1.01	≈2.43 ft

*From Fig. 15.11; **From Eq. 15.10

Select a multi-stage pump, say 3-stage with $N_s = 953$ having an impeller diameter of about 3.2 ft. ◄

A 4-stage pump ($N_s = 1183$, $D \approx 2.9$ ft) has higher efficiency (Fig. 15.11) but costs more than 3 stages.

15.51 *You are asked to select a single-stage or multi-stage pump to deliver 300 cfs against a head of 10 ft when operating at 450 rpm. Specify the number of stages, the specific speed and the approximate diameter of the impeller.*

BG

Assume a single-stage pump. Eq. 15.8*a*: $N_s = 450\sqrt{300(449)}/10^{3/4} = 29{,}370$

Impossible, there is no pump with $N_s = 29{,}370$ (Fig. 15.11).

Select for example, 10 pumps acting in parallel. Each pump must deliver 300/10 = 30 cfs

Eq. 15.8*a*: $N_s = 450\sqrt{30(449)}/10^{3/4} = 9280$, a mixed flow pump (Fig. 15.11).

Fig. 15.11 for $N_s = 9280$: $\phi_e \approx 2.0$; Eq. 15.10: $D = 153.2\phi_e\sqrt{10}/450 \approx 153.3(2.0)\sqrt{10}/450 = 2.1$ ft

Hence, 10 mixed-flow pumps ($N_s = 9280$, impeller dia ≈ 2.1 ft) in parallel is a possible solution. ◄
As an alternative use 6 pumps operating in parallel, each pump to deliver 300/6 = 50 cfs.

$N_s = 450\sqrt{50(449)}/10^{3/4} = 11{,}990$, an axial-flow pump (Fig. 15.11).

Fig. 15.11 for $N_s = 11{,}990$: $\phi_e \approx 2.2$; $D \approx 153.3(2.2)\sqrt{50}/450 = 5.3$ ft

An alternative would be to use 6 axial-flow pumps ($N_s = 11{,}990$) with impeller diameter ≈ 5.3 ft. ◄
Many alternatives are possible. A change in speed of operation adds alternatives. To decide which is the better choice an economy study is necessary. Such a study would include investment, maintenance and operating costs, etc. Also, the selected pumps must be determined to be free from cavitation.

15.52 *A submersible centrifugal pump in a water well is to be set 120 ft below the land surface. The steel discharge pipe from the pump to the surface has a diameter of 6.0 in. The water surface in the well at maximum drawdown is 106 ft below the land surface. How many stages (N_s = 2650) would you recommend for the pump if the discharge from the well at maximum drawdown is 800 gpm and the pump operates at 1450 rpm?*

BG

Eq. 15.8*a*: $N_s = n_e\sqrt{\text{gpm}}/h^{3/4}$ or $h = \left(n_e\sqrt{\text{gpm}}/N_s\right)^{4/3} = \left[1450(800)^{1/2}/2650\right]^{4/3} = 38.5$ ft per stage

But $h = 120 + f(L/D)V^2/2g$ where $V = Q/A = (800/449)/(\pi 0.25^2) = 9.08$ ft/sec

f will have a value of about 0.015 (Table 8.1 and Fig. 8.11). Let $f = 0.02$ to be on safe side

$\therefore\ h = 106 + 0.02(120/0.5)9.08^2/(2 \times 32.2) = 112.1$ ft

112.1/38.5 = 2.91 stages. Use a three-stage pump. ◄

15.53 *A pump is required to deliver 830 L/s against a head of 230 m. If the minimum desirable specific speed from an efficiency standpoint is $(N_s)_{SI}$ = 0.35 and the motor speed in 1500 rpm, how many stages would you recommend? Also, what is the maximum permissible suction lift?*

SI

Eq. 15.8*b*: $(N_s)_{SI} = \omega_e\sqrt{Q}/(gh)^{3/4}$; $0.35 = 2\pi(1500/60)\sqrt{0.830}/(9.81h)^{3/4} = 25.8/h^{3/4}$; $h = 308.6$ m

308.6 m > 230 m ; Hence one stage is adequate.

$(N_s)_{SI} = \omega_e\sqrt{Q}/(gh)^{3/4} = 2\pi(1500/60)\sqrt{0.830}/(9.81 \times 230)^{3/4} = 0.437$

Use a single-stage pump having $(N_s)_{SI} = 0.437$ ◀

The maximum permissible suction lift cannot be calculated because the following data are missing: elevation where the pump is to be used, temperature of the water to be pumped, and head loss from the source of the water to the pump (see Equation 15.14). ◀

Sec. 15.13: Pumps Operating in Series and Parallel -- *Exercises*

15.13.1 *Suppose the pumps of Sample Prob. 15.13 were operated at 1500 rpm. What then would have been the flow rates for (a) a single pump; (b) two pumps in series; (c) two pumps in parallel. All other data to remain the same.*

Prob. 15.13: Water is pumped from reservoir A to reservoir B, which has a 35-ft higher water surface. The loss in the pipeline is given by $h_L = 20Q^2$ (h_L in ft, Q in 100's of gpm). Two available pumps (n = 1800 rpm) each have the following characteristics:

h, ft:	*100*	*90*	*80*	*60*	*40*	*20*
Q, gpm:	*0*	*110*	*180*	*250*	*300*	*340*

BG

With changes in speed the pump characteristics will change in accordance with Eqs. 15.4 and 15.5:

$Q \propto n$ and $h \propto n^2$. Hence $Q' = Q(1500/1800) = 0.833Q$ and $h' = h(1500/1800)^2 = 0.694h$

The pump characteristics shown in Sample Problem 15.13 thus become for 1500 rpm:

Head (ft)	Flow (gpm)
83.3	0
75.0	76.3
66.6	124.7
50.0	173.5
33.3	203.2
16.7	236.0

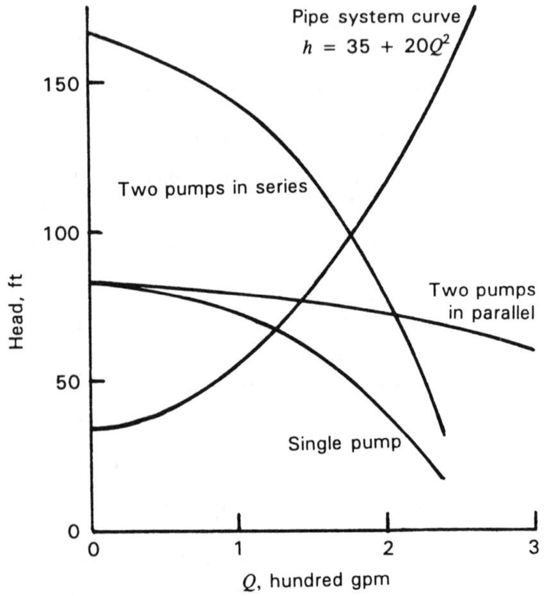

From plot, the points of intersection are:

(*a*) Single pump: $Q \approx 124$ gpm ◀

(*b*) Two pumps in series: $Q \approx 172$ gpm ◀

(*c*) Two pumps in parallel: $Q \approx 142$ gpm ◀

15.13.2 *Two pumps whose characteristics are given in Sample Prob. 15.13 are to be used in parallel. They must develop a head $h = 35 + 20Q^2$ as in the sample problem. (a) What is the flow rate with both pumps operating at 1800 rpm? (b) The speed of one of the pumps is reduced until it no longer delivers water. At approximately what speed will this happen?*

Sample Prob. 15.13: The two pumps (n = 1800 rpm) each have the following characteristics:

h, ft:	100	90	80	60	40	20
Q, gpm:	0	110	180	250	300	340

BG

(*a*) From Fig. S15.13 (intersection of curves for system and two pumps in parallel):

$Q = 170$ gpm ◀

(*b*) Slow down one pump until its shutoff head drops to the point where the single pump curve intersects the pipe system curve. This point of intersection (Fig. S15.13) is

$h \approx 84$ ft, $Q \approx 156$ gpm.

The reduced speed to give a shutoff head of 84 ft is given by Eq. 15.5:

$h \propto n^2$, so $84/100 = (n/1800)^2$

$n = 1650$ rpm ◀

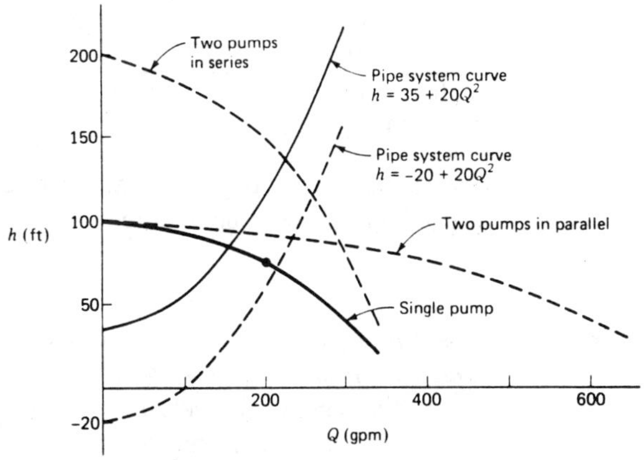

Figure S15.13

Sec. 15.13: Pumps Operating in Series and Parallel -- *Problems*

15.54 *In planning a pumping station an engineer decided to use two identical pumps operating in parallel. These pumps, when both are operating, must deliver a total of 200 cfs against a head of 40 ft. Calculate the specific speed of each of these pumps and estimate the diameter of the impellers.*

BG

Pumps in parallel -- each pump must deliver 100 cfs against a head of 40 ft.

Eq. 15.8*a*: $N_s = n_e\sqrt{\text{gpm}}/h^{3/4} = n_e\sqrt{100(449)}/40^{3/4} = 13.32n_e$

Analyze for different speeds of operation:

n_e (rpm)	N_s	ϕ_e*	D**
450	5994	≈1.65	≈3.55 ft
600	7992	≈1.85	≈2.99 ft
900	11,988	≈2.20	≈2.37 ft
*From Fig. 15.11	▲		▲
**From Eq. 15.10			

The above represent possible alternatives. Lower rotation speeds will result in lower values of N_s, but larger impellers will be required.

15.55 *In planning a pumping station, an engineer decided to use two identical pumps operating in series. These pumps, when both are operating, must deliver 200 cfs against a total head of 40 ft. Calculate the specific speed of each of these pumps and estimate the diameter of the impellers.*

BG

Pumps in series -- each pump must deliver 200 cfs against a head of 20 ft.

Eq. 15.8*a*: $N_s = n_e\sqrt{200(449)}/20^{3/4} = 31.7n_e$

n_e (rpm)	N_s	ϕ_e*	D**
300	9550	≈2.00	≈4.57 ft
450	14,265	≈2.35	≈3.58 ft
*From Fig. 15.11	▲		▲
**From Eq. 15.10			

The above represents possible alternatives. Lower rotation speeds will result in lower values of N_s, but larger impellers will be required.

15.56 *A very long pipe connects two reservoirs whose water levels are the same. Identical pumps A and B are connected to the pipe in parallel. The operating characteristics of each of the pumps are shown in Fig. 15.6. The pumps are operated at 1450 rpm. When only one pump is operating the flow rate was found to be 13,200 gpm. (a) How much flow will there be when both pumps operate? (b) Find also the rate of delivery (with one and two pumps) if the pump speeds are reduced to 1200 rpm.*

BG

Sec. 8.24: The pipe is very long, so minor losses are negligible.

Fig. 15.6: Single pump (n = 1450 rpm) at Q = 13,200 gpm: h = 44 ft.

Since the two water levels are the same $h = h_L$, and the system curve is $h = kQ^2$.

Thus (with Q in thousands of gpm):

$$44 = k(13.2)^2$$

from which $k = 44/(13.2)^2 = 0.253$

Hence $h = 0.253Q^2$

Q	h
6	9.0
9	20.2
12	36.0
15	56.2
18	81.0

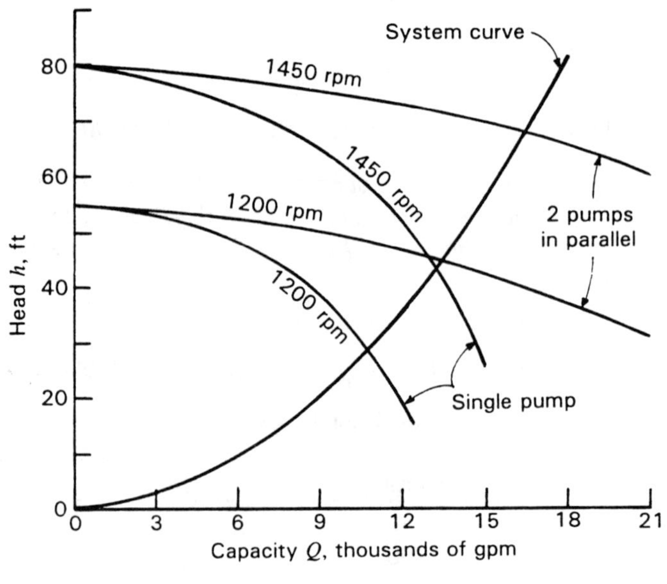

(a) From graph for two pumps in parallel (n = 1450 rpm): Curves intersect at $h \approx 68$ ft,

$Q \approx 16{,}500$ gpm ◄

(b) Eq. 15.4: $Q \propto nD^3$. So for same pump (D), $Q' = (n'/n)Q = (1200/1450)Q = 0.828Q$

Eq. 15.5: $h \propto n^2D^2$. So for same pump (D), $h' = (n'/n)^2h = 0.828^2h = 0.685h$

So corresponding points are (plotted on graph above):

/cont...

1450 rpm		1200 rpm	
Q (1000 gpm)	h	Q (1000 gpm)	h
0	80 ft	0	54.8 ft
6	74 ft	4.96	50.7 ft
10.5	60 ft	8.69	41.1 ft
13.2	43 ft	10.92	29.5 ft
15.0	25 ft	12.41	17.1 ft

Single pump (n = 1200 rpm): Curves intersect at Q = 10,700 gpm, h = 28.5 ft ◀

Two pumps in parallel (n = 1200 rpm): Curves intersect at Q = 13,500 gpm, h = 45 ft ◀

15.57 *Select the specific speed of the pump or pumps required to convey 1.0 cfs of water against a head of 115 ft. The pump rotative speed is 1750 rpm. Consider the following cases: a single pump; two pumps in parallel; three pumps in parallel; two pumps in series; three pumps in series.*

BG

Inside cover: Q = 1 cfs = 448.8 gpm. Eq. 15.8 for single pump: $N_s = 1750\sqrt{448.8}/115^{3/4} = 1055$ ◀

Pumps in parallel: h = 115 ft, $Q = 448.8/no.$ where $no.$ = number of pumps in parallel

2 pumps in parallel: $N_s = 1750\sqrt{448.8/2}/115^{3/4} = 740$ ◀
3 pumps in parallel: $N_s = 1750\sqrt{448.8/3}/115^{3/4} = 610$ ◀

This value of N_s is very low, having a low efficiency (Fig. 15.11).

Pumps in series: Q = 448.8 gpm, $h = 115/no.$ where $no.$ = number of pumps in series

2 pumps in series: $N_s = 1750\sqrt{448.8}/(115/2)^{3/4} = 1781$ ◀
3 pumps in series: $N_s = 1750\sqrt{448.8}/(115/3)^{3/4} = 2399$ ◀

Chapter 15: Miscellaneous -- *Problems*

15.58 *The pump of Fig. 15.6 is placed in an 8-in-diameter pipe (f = 0.030) 150 ft long that is used to lift water from one pond to another. The difference in water surface elevation between the two ponds varies from 20 ft to 50 ft. Plot a curve showing delivery rate versus water-surface elevation difference.*

BG

$h = \Delta z + f(L/D)V^2/2g$ where

$V = Q/A = Q/[(\pi/4)(8/12)^2] = 2.865Q$

$h_L = 0.030[150/(8/12)](2.865Q)^2/(2 \times 32.2)$

$h_L = 0.860Q^2$ where Q is expressed in cfs

$h_L = 0.860(449Q/1000)^2 = 0.1734Q^2$

where Q is expressed in thousands of gpm

Q (1000 gpm)	h_L (ft)
1	0.173
5	4.34
10	17.3
15	38.9

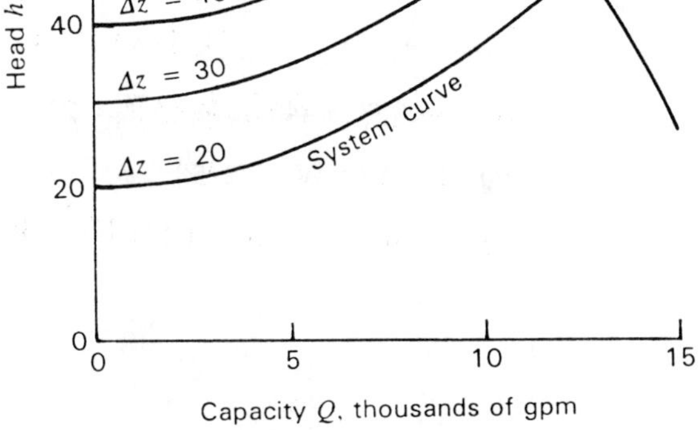

Plot system curves, $h = \Delta z + h_L$, for Δz = 20, 30, 40, 50 ft (see graph). From graph, these curves intersect with the pump's head-capacity curve at:

Δz (ft)	Q (gpm)
20	≈12,600
30	≈11,600
40	≈10,500
50	≈9,200

The required curve is:

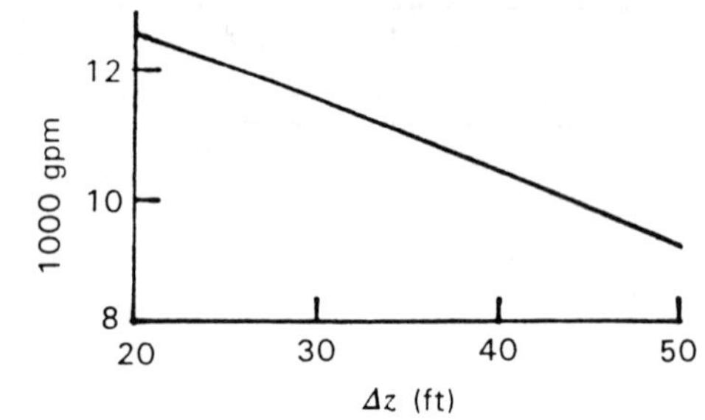

15.59 *Under normal operating conditions a centrifugal pump with an impeller diameter of 20.0 in delivers water at a head of 170 ft with an efficiency of 70% at 1800 rpm. Compute the peripheral velocity of the impeller, the specific speed, and the flow rate.*

BG

Peripheral velocity $= \omega r = 2\pi(n/60)10/12 = 2\pi(1800/60)10/12 = 157.1$ ft/sec ◀

Eq. 15.10: $D = 20/12 = 153.3\phi_e\sqrt{h}/n_e = 153.3\phi_e\sqrt{170}/1800$; $\phi_e = 1.501$

Fig. 15.11 for $\phi_e = 1.5$: $N_s \approx 4700$ ◀

Eq. 15.8*a*: $N_s \approx 4700 \approx n_e\sqrt{\text{gpm}}/h^{3/4}$; $\sqrt{\text{gpm}} \approx 4700(170)^{3/4}/1800 = 123$; gpm ≈ 15,000 gpm ◀

15.60 *A centrifugal pump driven by an electric motor lifts water a total height of 200 ft. The pump efficiency is 78% and the motor efficiency is 88%. The lift is through 1220 ft of 6-in-diameter pipe and the pumping rate is 350 gpm. If $f = 0.022$ and power costs 70 mils/kWh, what is the cost of pumping a million gallons of water? (A mil is one-thousandth of a dollar = 0.10 cent).*

B

$h = \Delta z + f(L/D)V^2/2g$ where $V = Q/A = (350/448.8)/(\pi 0.25^2) = 3.97$ ft/sec

$\therefore\ h = 200 + 0.022[1220/(6/12)](3.97)^2/(2 \times 32.2) = 213.1$ ft

Power used $= \gamma Qh/[0.78(0.88)550] = 27.48$ hp $= 0.746(27.48)$ kW $= 20.49$ kW

$$\text{Time to pump} = \frac{1{,}000{,}000\ \text{gal}}{350\ \text{gal/min}}\left(\frac{\text{hr}}{60\ \text{min}}\right) = 47.6\ \text{hr}$$

Cost = (\$0.070/kWh)(20.49 kW)(47.6 hr) = \$68.29 ◀

15.61 *What is the specific speed of the four-stage pumps mentioned in the last paragraph of Sec. 15.14? Approximately what is the diameter of the impellers?*

Sec. 15.14: The pumps (n = 600 rpm) each deliver 315 cfs against a total head of 1970 ft.

BG

Eq. 15.8*a*: $N_s = n_e\sqrt{\text{gpm}}/h^{3/4}$ where $h = 1970/4 = 492.5$ ft per stage

$N_s = 600\sqrt{315(448.8)}/492.5^{3/4} = 2158$ ◀

Fig. 15.11 for $N_s = 2158$: $\phi_e \approx 1.13$

Eq. 15.10: $D \approx 153.3(1.13)\sqrt{492.5}/600 = 6.4$ ft $= 77$ in ◀

Chapter 16

Hydraulic Machinery -- Turbines

PROBLEM SELECTION GUIDE

Sec	Exer/Prob	Units	Difficulty	Length	Parts	Similar	Special features
16.3	*Action of the Impulse Turbine*						
	X[1] 16.3.1	BG	Medium	Short	1		Uses Sec. 6.7
	16.3.2	BG	Easy	Short	3	S16.1	Uses Sec. 6.8
	16.3.3	SI	Medium	Short	3		Uses Sec. 6.8
	P 16.1	BG	Medium	Long	3	S16.1	
	16.2	BG	Medium	Long	3	S16.1	
	16.3	SI	Medium	Medium	3		Uses Sec. 6.8
	16.4	BG	Medium	Medium	4	16.5	Uses Sec. 6.8
	16.5	BG	Medium	Medium	4	16.4	Uses Sec. 6.8
	16.6	BG	Medium	Long	3	S16.1	Uses Sec. 6.8
16.4	*Head on an Impulse Turbine and Efficiency*						
	X 16.4.1	BG	Easy	Medium	3	S16.4	
	16.4.2	BG	Medium	Long	1	S16.4	
	16.4.3	BG	Easy	Short	1		
	P 16.7	BG	Easy	Long	1	S16.4	
	16.8	BG	Easy	V Short	1		
	16.9	BG	Easy	Short	2		Uses Sec. 6.8
	16.10	SI	Hard	Long	1		
	16.11	BG	Medium	Long	1		Uses Sec. 6.8
	16.12	SI	Medium	Medium	1		* T & E for f using Sec. 8.12
	16.13	BG	Medium	Long	3		
16.5	*Nozzles for Impulse Turbines*						
	X 16.5.1	BG	Medium	Medium	1	S16.5, 16.5.2	Trials for D_j
	16.5.2	BG	Medium	Medium	1	S16.5, 16.5.1	Trials for D_j
	P 16.14	BG	Medium	Medium	2		
	16.15	BG	Hard	Medium	1		* Uses calculus
16.7	*Action of the Reaction Turbine*						
	X 16.7.1	SI	Easy	V Short	1		
	16.7.2	BG	Easy	V Short	1		

/cont...

[1] For all Exercises (marked "X"), answers are given in Appendix D of the textbook.
* Note a correction made to early printings of the textbook.
† Answers are sensitive to values (ϕ_e, σ_c, η, h_{max}) read from a figure.
X = Exercise, P = (end-of-chapter) Problem, S = Sample Problem.

Sec	Exer/Prob	Units	Difficulty	Length	Parts	Similar	Special features
16.8	*Draft Tubes and Effective Head on Reaction Turbines*						
	X 16.8.1	BG	Medium	Medium	1	S16.7, P16.18	Uses Sec. 8.21
	16.8.2	BG	V Easy	V Short	1	16.8.3	
	16.8.3	SI	V Easy	V Short	1	16.8.2	*
	P 16.16	SI	V Easy	V Short	1		
	16.17	SI	Medium	Medium	1		Trial and Error for *f* using Sec. 8.12
	16.18	SI	Medium	Medium	1	S16.7, X16.8.1	Uses Sec. 8.21
	16.19	BG	Medium	Medium	3		
16.9	*Efficiency of Turbines*						
	P 16.20	N	V Easy	V Short	1	16.24	
16.10	*Similarity Laws for Reaction Turbines*						
	X 16.10.1	SI	Medium	Medium	3	P16.21	*
	16.10.2	BG	Easy	V Short	2		
	16.10.3	BG	Medium	Medium	1		
	P 16.21	BG	Easy	Medium	2	X16.10.1	
	16.22	BG	Easy	Short	1		
	16.23	SI	Easy	V Short	1	16.25	
	16.24	BG	V Easy	V Short	1	16.20	
	16.25	BG	Easy	V Short	1	16.23	
16.11	*Peripheral Velocity Factor and Specific Speed of Turbines*						
	X 16.11.1	BG	Easy	Short	1	16.11.2-3	†
	16.11.2	BG	Easy	Short	1	16.11.1-3	†
	16.11.3	BG	Easy	Short	1	16.11.1-2	†
	P 16.26	BG	Medium	Medium	4		* †
	16.27	BG	Easy	Short	1		† Partly unsolvable
	16.28	B	Easy	Short	1	16.29	† Unit conversions
	16.29	BG	Easy	Short	1	16.28	†
	16.30	B	Easy	Short	1	16.31	† Assume η from Fig 16.13; unit conv's
	16.31	BG	Easy	Short	1	16.30	†
	16.32	BG	Easy	Short	2	16.33-34	†
	16.33	BG	Easy	Short	2	16.32-34	†
	16.34	BG	Easy	Short	2	16.32-33	†
16.12	*Cavitation in Turbines*						
	X 16.12.1	BG	Easy	Short	1		* †
	16.12.2	BG	Easy	Medium	2		† Interpolation
	16.12.3	BG	Easy	Short	1		* †

/cont...

Sec	Exer/Prob	Units	Difficulty	Length	Parts	Similar	Special features
	P 16.35	BG	Easy	Short	1	16.36	†
	16.36	BG	Easy	Short	1	16.35	†
	16.37	BG	Easy	Medium	4	16.38	† Assume $T°$
	16.38	BG	Easy	Medium	4	16.37	† Interpolation. Assume $T°$
	16.39	BG	Easy	Short	1	16.38	† Assume $T°$
	16.40	BG	Easy	Short	2		†
16.13	***Selection of Turbines***						
	X 16.13.1	BG	Medium	Medium	2	16.13.2	†
	16.13.2	BG	Medium	Medium	2	16.13.1	†
	16.13.3	BG	Easy	Short	1		†
	16.13.4	BG	Easy	Short	2		
	P 16.41	BG	Hard	Long	2	16.52-53	†
	16.42	BG	Medium	Medium	1		†
	16.43	BG	Medium	Medium	1		
	16.44	BG	Easy	Medium	2		*
	16.45	BG	Easy	Medium	1		†
	16.46	BG	Medium	Medium	1		†
	16.47	BG	Medium	Long	2		†
	16.48	BG	Easy	Medium	1		†
	16.49	B	Medium	Long	1		† Unit conversions
	16.50	BG	Hard	Long	2	16.51	
	16.51	BG	Medium	Long	1	16.50	†
	16.52	B	Medium	Medium	1	16.41*a*,53	† Unit conversions
	16.53	BG	Medium	Medium	1	16.41*a*,52	†

Chapter 16
HYDRAULIC MACHINERY – TURBINES

Sec. 16.3: Action of the Impulse Turbine -- *Exercises*

16.3.1 *In Sample Prob. 16.1 subtract the horsepower of the absolute velocity of the water at bucket exit from the horsepower of the jet at bucket entrance. What does the result represent? Why?*

Sample Prob. 16.1: $V_1 = 200$ *fps,* $u = 125.6$ *fps,* $\upsilon_2 = \upsilon_1 = 74.4$ *fps,* $Q = 4.36$ *cfs.*

BG

Eq. 6.17: $V_{2x} = u + \upsilon_2 \cos\beta_2 = 125.6 + 74.4\cos 160° = 55.7$ fps

Eq. 6.18: $V_{2y} = \upsilon_2 \sin\beta_2 = 74.4 \sin 160° = 25.4$ fps. $V_2 = \sqrt{55.7^2 + 25.4^2} = 61.2$ fps

$$\text{HP at entrance} = \frac{\gamma Q(V_1^2/2g)}{550} = \frac{62.4(4.36)}{550}\left(\frac{200^2}{2 \times 32.2}\right) = 307$$

$$\text{HP at exit} = \frac{\gamma Q(V_2^2/2g)}{550} = \frac{62.4(4.36)}{550}\left(\frac{61.2^2}{2 \times 32.2}\right) = 28.8\ ; \quad \Delta\,\text{HP} = 307 - 28.8 = 278\ \text{hp}$$ ◄

This represents the HP transferred to the turbine runner. ◄

This is so because we assumed $\upsilon_2 = \upsilon_1$ ◄

16.3.2 *Repeat Sample Prob. 16.1 for the case where* $\upsilon_2 = 0.8\upsilon_1$ *and calculate the percent error made by assuming* $\upsilon_2 = \upsilon_1$ *in Sample Prob. 16.1.*

Sample Prob. 16.1: $u = 125.6$ *fps,* $\upsilon_2 = \upsilon_1 = 74.4$ *fps,* $Q = 4.36$ *cfs,* $D = 10.0$ *ft, HP to runner = 278.6. Find (a) F on buckets, (b) T on runner, (c) HP transferred to runner.*

BG

$\upsilon_2 = 0.8\upsilon_1 = 0.8(74.4) = 59.5$ fps ; $\Delta\upsilon_x = \upsilon_{2x} - \upsilon_{1x} = 59.5\cos 160° - 74.4 = -130.3$ fps

(*a*) Eq. 6.24: $F = 1.940(4.36)130.3 = 1102$ lb ◄

(*b*) $T = Fr = 1102(5) = 5510$ ft·lb ◄

(*c*) Eqs. 5.32 and 5.35: HP transfered to runner $= Fu/550 = 1102(125.6)/550 = 251.7$ hp ◄

(*d*) Error $= (278.6 - 251.7)/251.7 = 0.1069 = 10.69\%$ ◄

16.3.3 *Consider an impulse turbine with a pitch diameter of 2.0 m and a bucket angle* β_2 *of 165°. If the jet velocity is 80 m/s, the jet diameter 112 mm, and the rotative speed 375 rpm, find (a) the force on the buckets, (b) the torque on the runner, and (c) the horsepower transferred to the runner. Assume that the velocity relative to the bucket does not change (i.e. assume* $\upsilon_2 = \upsilon_1$*).*

SI

$\omega = 375(2\pi/60) = 39.3$ rad/s ; $u = \omega r = 39.3(1\ \text{m}) = 39.3$ m/s

$Q = A_1V_1 = \pi(0.112/2)^2 80 = 0.788\ \text{m}^3/\text{s}$; $V_1 = 80 = \upsilon_1 + u = \upsilon_1 + 39.3$; $\upsilon_1 = 40.7$ m/s

$\Delta\upsilon_x = \upsilon_{2x} - \upsilon_{1x} = 40.7\cos 165° - 40.7 = -80.1$ m/s

(*a*) Eq. 6.24: $F = 1000(\text{kg/m}^3)0.788(\text{m}^3/\text{s})80.1\ \text{m/s} = 63\,100\ \text{kg·m/s}^2 = 63\,100$ N ◄

(*b*) $T = Fr = 63\,100(1) = 63\,100$ N·m ◄

(*c*) Eq. 5.35: $P = T\omega = 63\,100(39.3) = 2480\,000$ N·m/s $= 2480$ kW ◄

16.1 *Repeat Sample Prob. 16.1 for rotative speeds of 80, 160, 300, and 360 rpm, with all other data remaining the same. Also find the hydraulic efficiency of the flow through the buckets for each speed. Plot horsepower as ordinate versus rotative speed as abscissa to find the approximate speed at which maximum power will be developed. At maximum power, approximately what is the value of* u/V_1*?*

Sample Prob. 16.1: Impulse turbine, pitch $D = 10$ *ft,* $\beta_2 = 160°$, $V_i = V_j = 200$ *fps,* $Q = 4.36$ *cfs,* $v_2 = v_1$. *Find (a) F on buckets, (b) T on runner, (c) HP transferred to runner.*

BG

Horsepower of jet (= hydraulic horsepower input to turbine buckets)

$$= \gamma Q(V_j^2/2g)/550 = 62.4(4.36)(200^2/64.4)/550 = 307.2 \text{ hp}$$

$u = \omega r = n(2\pi/60)5 = 0.524n$; $v_2 = v_1 = V_1 - u = 200 - u$; $T = Fr = 5F$

n (rpm)	u (fps) $0.524n$	$v_2 = v_1$ $200 - u$	$v_2 \cos 160°$	Δv_x (fps) $v_2 \cos 160° - v_1$	F (lb) $\rho Q(\Delta v_x)$	T (ft·lb) $5F$	HP $Fu/550$	η' (%) HP/307.2
80	41.9	158.1	−148.5	−306.6	2593	12,965	197.5	64.3
160	83.8	116.2	−109.2	−225.4	1906	9,530	290.0	94.4
240	125.6	74.4	−69.9	−144.3	1220	6,100	278.6	90.7
300	157.0	43.0	−40.4	−83.4	705	3,525	201.2	65.5
360	188.4	11.6	−10.9	−22.5	190	950	65.0	21.1
▲					▲	▲	▲	▲

Maximum power and maximum efficiency of flow through the turbine occur at about 190 rpm. ◄

$u = \omega r \approx 190(2\pi/60)5 = 99$ ft/sec ; $u/V_1 \approx 99/200 = 0.5$ ◄

16.2 *Same as Prob. 16.1, except also consider hydraulic friction in the buckets by letting* $v_2 = 0.85\, v_1$.

Prob. 16.1: Impulse turbine, pitch $D = 10$ *ft,* $\beta_2 = 160°$, $V_i = V_j = 200$ *fps,* $Q = 4.36$ *cfs. Find (a) F on buckets, (b) T on runner, (c) HP transferred to runner, each for rotative speeds of 80, 160, 300, and 360 rpm. Also find the hydraulic efficiency of the flow through the buckets for each speed. Plot horsepower as ordinate versus rotative speed as abscissa to find the approximate speed at which maximum power will be developed. At maximum power, approximately what is the value of* u/V_1*?*

BG

Horsepower of jet (= hydraulic horsepower input to turbine buckets)

$$= \gamma Q(V_j^2/2g)/550 = 62.4(4.36)(200^2/64.4)/550 = 307.2 \text{ hp}$$

$u = \omega r = n(2\pi/60)5 = 0.524n$; $v_1 = V_1 - u = 200 - u$; $v_2 = 0.85 v_1$; $v_{2x} = v_2 \cos 160°$

$\Delta v_x = v_2 \cos 160° - v_1$; $F = \rho Q \Delta v_x$; $T = Fr = 5F$

n rpm	u (fps) $0.524n$	v_1 (fps) $200 - u$	v_2 (fps) $0.85 v_1$	Δv_x $v_2 \cos 160° - v_1$	F (lb) $\rho Q(\Delta v_x)$	T (ft·lb) $5F$	HP $Fu/550$	η' HP/307.2
80	41.9	158.1	126.2	−284.3	2405	12,025	183.2	59.6%
160	83.8	116.2	92.8	−209.0	1768	8,840	269.4	87.7%
240	125.6	74.4	59.4	−133.8	1132	5,660	258.5	84.2%
300	157.0	43.0	34.3	−77.3	654	3,270	186.7	60.8%
360	188.4	11.6	9.3	−20.9	177	885	60.6	19.7%
▲					▲	▲	▲	▲

Maximum power and maximum efficiency of flow through the turbine occur at about 180 rpm. ◄

$u = \omega r \approx 180(2\pi/60)5 = 94.2$ ft/sec ; $u/V_1 \approx 94.2/200 = 0.47$ ◄

16.3 *Consider an impulse wheel with a pitch diameter of 2.65 m and a bucket angle of 168°. If the jet velocity is 55.0 m/s, the jet diameter is 100 mm, and the rotative speed is 300 rpm, find (a) the force on the buckets, (b) the torque on the runner, and (c) the power transferred to the runner. Assume* $v_2 = 0.85v_1$.

SI

$u = \omega r = 300(2\pi/60)2.65/2 = 41.6$ m/s ; $V_1 = 55$ m/s $= v_1 + u = v_1 + 41.6$; $v_1 = 13.4$ m/s

$v_2 = 0.85v_1 = 11.4$ m/s, $v_{2x} = v_2 \cos 168° = -11.15$ m/s

Hence $\Delta v_x = v_{2x} - v_{1x} = -11.15 - 13.4 = -24.55$ m/s

Note also, $V_{2x} = u + v_2 \cos 168° = 41.6 - 11.15 = 30.45$ m/s

Thus $\Delta V_x = V_{2x} - V_{1x} = 30.45 - 55 = -24.55$ m/s. Hence $\Delta V_x = \Delta v_x = -24.55$ m/s

(*a*) Eq. 6.24: $F = \rho Q(\Delta v_x) = \rho Q(\Delta V_x)$ where $Q = A_j V_j$

$F = 1000 \text{ kg/m}^3[\pi(0.1/2)^2]55 \text{ m}^3\text{/s } (24.55 \text{ m/s}) = 10\,580 \text{ N}$ ◀

(*b*) $T = Fr = 10\,580 \text{ N}(2.65/2) \text{ m} = 14\,020 \text{ N·m}$ ◀

(*c*) $P = Fu = 10\,580 \text{ N}(41.6 \text{ m/s}) = 440 \text{ kW}$ ◀

or $P = T\omega = 14\,020(300)(2\pi/60) \text{ N·m/s} = 440 \text{ kW}$ ◀

16.4 *An impulse turbine with a pitch diameter of 4.0 ft and a bucket angle of 160° is rotating at 514.7 rpm. The nozzle produces a jet with a velocity of 240 ft/sec and diameter of 4 in. Find (a) the force on the buckets, (b) the torque on the runner, (c) the horsepower transferred to the runner, and (d) the efficiency* η *of the turbine under these conditions of operation. Assume the head loss through the nozzle can be expressed as* $0.05V_j^2/2g$ *and assume the velocity relative to the bucket does not change (i.e., assume* $v_2 = v_1$*). Neglect bearing friction and windage.*

BG

$Q = A_j V_j = \pi(2/12)^2 240 = 20.93$ cfs ; $h = 1.05V_j^2/2g = 1.05(240^2/64.4) = 939$ ft

$u = \omega r = 514.7(2\pi/60)2 = 107.7$ fps

$V_1 = V_j = 240$ fps $= v_1 + u = v_1 + 107.7$; thus $v_1 = 240 - 107.7 = 132.3$ fps $= v_2$

$\Delta v_x = v_{2x} - v_{1x} = v_1 \cos 160° - v_1 = -124.3 - 132.3 = -256.6$ fps

Note also $V_{2x} = u + v_2 \cos 160° = 107.7 - 124.3 = -16.6$ fps

$\Delta V_x = V_{2x} - V_{1x} = -16.6 - 240 = -256.6$ fps. So $\Delta v_x = \Delta V_x$

(*a*) Eq. 6.24: $F = \rho Q(\Delta V_x) = 1.940(20.93)256.6 = 10{,}420$ lb ◀

(*b*) $T = Fr = 10{,}420(2.0) = 20{,}820$ ft·lb ◀

(*c*) Eq. 5.35: Shaft horsepower $= T\omega/550 = 20{,}820(2\pi/60)514.7/550 = 2039$ hp ◀

(*d*) Eq. 5.34: $\eta = \dfrac{\text{shaft horsepower}}{\gamma Qh/550} = \dfrac{2039}{62.4(20.93)939/550} = 0.915$ ◀

16.5 *Work Prob. 16.4 where the speed of rotation is 300 rpm, all other data remaining the same.*

Prob. 16.4: Impulse turbine, pitch D = 4.0 ft, $\beta_2 = 160°$, *n = 514.7 rpm,* $D_j = 4$ *in,* $V_j = 240$ *fps. Find (a) F on buckets, (b) T on runner, (c) HP transferred to runner, (d)* η *of turbine. Assume nozzle* $h_L = 0.05V_j^2/2g$, $v_2 = v_1$, *and neglect bearing friction and windage.*

BG

$Q = A_jV_j = \pi(2/12)^2 240 = 20.93$ cfs ; $h = 1.05V_j^2/2g = 1.05(240^2/64.4) = 939$ ft

$u = \omega r = 300(2\pi/60)2 = 62.8$ fps

$V_1 = V_j = 240$ fps $= v_1 + u = v_1 + 62.8$; thus $v_1 = 240 - 62.8 = 177.2$ fps $= v_2$

$\Delta v_x = v_{2x} - v_{1x} = v_2 \cos 160° - v_1 = -166.5 - 177.2 = -343.7$ fps

(*a*) Eq. 6.24: $F = \rho Q(\Delta v_x) = 1.940(20.93)343.7 = 13{,}956$ lb ◀

(*b*) $T = Fr = 13{,}956(2) = 27{,}910$ ft·lb ◀

(*c*) Eq. 5.35: Shaft horsepower $= T\omega/550 = 27{,}910(2\pi/60)300/550 = 1593$ hp ◀

(*d*) Eq. 5.34: $\eta = \dfrac{\text{shaft horsepower}}{\gamma Qh/550} = \dfrac{1593}{62.4(20.93)939/550} = \dfrac{1593}{2230} = 0.714$ ◀

16.6 *(a) Solve Sample Prob. 16.1 for blade angles of 160°, 165°, and 170° for the case of an impulse turbine whose pitch diameter is 8 ft rather than 10 ft. All other data are to remain the same. (b) Repeat the above for the case where* $v_2 = 0.80\,v_1$.

Sample Prob. 16.1: Impulse turbine, $V_1 = V_j = 200$ *fps, Q = 4.36 cfs,* $v_2 = v_1$, *n = 240 rpm. Find (a) F on buckets, (b) T on runner, (c) HP transferred to runner.*

BG

(*a*) $u = \omega r = 240(2\pi/60)4 = 100.5$ ft/sec

$V_1 = 200$ ft/sec $= v_1 + u = v_1 + 100.5$ ft/sec, $v_1 = 99.5$ ft/sec $= v_2$

$\Delta v_x = v_2 \cos\beta_2 - v_1 = 99.5\cos\beta_2 - 99.5 = 99.5(\cos\beta_2 - 1)$

Eq. 6.24: $F = \rho Q(\Delta v_x) = 1.940(4.36)\Delta v_x = 8.46\Delta v_x$

$T = Fr = 4F$. Eq. 5.35: $HP = Fu/550 = F(100.5)/550 = 0.1828F$

β_2	$\cos\beta_2$	Δv_x fps	F lb	T lb·ft	HP
160°	−0.940	−193.0	1632	6528	298
165°	−0.966	−195.6	1654	6616	302
170°	−0.985	−197.5	1670	6680	305
			▲	▲	▲

Note: As β_2 gets larger the HP gets larger. However as β_2 approaches 170° the discharge from a bucket may interfere with the following bucket. Hence $\beta_2 \approx 165°$ is a good design.

(*b*) $v_2 = 0.80v_1 = 0.80(99.5) = 79.6$ ft/sec ; $\Delta v_x = v_2\cos\beta_2 - v_1 = 79.6\cos\beta_2 - 99.5$

β_2	$\cos\beta_2$	$v_2\cos\beta_2$ fps	Δv_x fps	F lb	T lb·ft	HP
160°	−0.940	74.8	174.3	1474	5896	269
165°	−0.966	76.9	176.4	1492	5968	273
170°	−0.985	78.4	177.9	1505	6020	275
				▲	▲	▲

Sec. 16.4: Head on an Impulse Turbine and Efficiency -- *Exercises*

16.4.1 *Refer to Sample Prob. 16.4. (a) Find the head h" delivered to the buckets of the turbine if the absolute velocity at discharge from the buckets is 60 ft/sec. Assume a bucket loss coefficient of 0.30. (b) Calculate the hydraulic efficiency of the flow through the turbine buckets. (c) What is the overall efficiency of the turbine?*

Sample Prob. 16.4: Impulse turbine, V_j = 377 fps, bucket u = 166.5 fps, net h = 2314 ft, k_n = 0.05.

BG

(*a*) $V_1 = 377 = u + v_1 = 166.5 + v_1$; $v_1 = 211$ fps

Eq. 16.6 with $\left(\frac{1}{C_v^2} - 1\right)\left[1 - \left(\frac{A_j}{A_B}\right)^2\right]$ replaced by k_n, the nozzle loss coefficient:

$$2314 = 0.05(377^2)/(2 \times 32.2) + 0.30(211^2)/(2 \times 32.2) + 60^2/(2 \times 32.2) + h'' = 110 + 206 + 56 + h''$$

$h'' = 1942$ ft ◀

(*b*) Hydraulic effeciency of flow through the buckets =

$$\frac{\text{energy head transfered to buckets}}{\text{energy head entering buckets}} = \frac{h''}{V_j^2/2g} = \frac{1942(2)32.2}{377^2} = 0.88 = 88\% \quad ◀$$

(*c*) Eq. 5.34: Overall turbine $\eta = \frac{\text{Shaft power}}{\text{Input power}} = \frac{\gamma Q h''}{\gamma Q h} = \frac{h''}{h} = \frac{1942}{2314} = 0.84 = 84\%$ ◀

16.4.2 *Repeat Sample Prob. 16.4 for the case of a 4-in-diameter jet with all other data remaining the same.*

Sample Prob. 16.4: Delivery pipe, L = 10,000 ft, f = 0.025, D = 4.0 ft, delivers water to a nozzle at an elevation 2420 ft below the reservoir water surface, nozzle loss = $0.05V_j^2/2g$. The jet drives an impulse wheel, D_w = 106 in, β_2 = 165°, connected to a 20-pole generator in a 60-cycle system. Find the pipe V, V_j, Q, u, u/V_1, net head h, F on bucket, T on wheel, shaft HP, HP at base of nozzle, overall turbine η. Assume $v_2 = v_1$, neglect bearing friction and windage.

BG

$L/D = 10{,}000/4 = 2500 > 1000$ so neglect minor loss at pipe entrance.

Energy equation from reservoir surface to jet: $2420 - 0.025(10{,}000/4)V^2/2g - 0.05V_j/2g = V_j^2/2g$

Continuity: $V_j = (48/4)^2V = 144V$; ∴ $2420 - 62.5V^2/2g - 0.05(144V)^2/2g = (144V)^2/2g$

i.e., $2420 = (62.5 + 1037 + 20{,}736)V^2/2g = 21{,}835V^2/2g$

$V = 2.67$ fps ◀ $V_j = 144(2.67) = 385$ fps ◀ $Q = AV = \pi 2^2(2.67) = 33.5$ cfs ◀

Sec. 16.1: $N = 7200/n$ so $n = 7200/N = 7200/20 = 360$ rpm

$\omega = 2\pi n/60 = 2\pi(360)/60 = 37.7$ rad/sec ; $u = \omega r = 37.7(106/2)/12 = 166.5$ fps ◀

$u/V_1 = 166.5/385 = 0.432$ ◀

Eq. 16.5: Net head, $h = z_1 - h_L = 2420 - 62.5(2.67)^2/(2 \times 32.2) = 2413$ ft ◀

Eq. 16.2: $F = 1.940(33.5)(385 - 166.5)(1 - \cos 165°) = 27{,}900$ lb ◀

$T = Fr = 27{,}900(106/2)/12 = 123{,}300$ ft·lb ◀

Eqs. 5.32 and 5.35: $P = T\omega/550 = 123{,}300(37.7)/550 = 8450$ hp ◀

Power at head (inlet, base) of nozzle = power input to turbine (including nozzle) =

$P = \gamma Q h/550 = 62.4(33.5)2413/550 = 9170$ hp ◀

Eq. 5.34: Overall turbine $\eta = \frac{\text{Shaft power}}{\text{Input power}} = \frac{8450}{9170} = 0.921 = 92.1\%$ ◀

16.4.3 *Refer to Sample Prob. 16.1. Calculate the hydraulic efficiency of the flow through the turbine buckets. Note, this is different than the hydraulic efficiency η' of the turbine because the nozzle is considered an integral part of the turbine.*

Sample Prob. 16.1: $V_j = V_1 = 200$ fps, $Q = 4.36$ cfs, P transferred to runner = 278.6 hp.

BG

Power extracted by runner = power transferred to runner = 278.6 hp (given).

$HP_{jet} = \gamma Q(V_j^2/2g)/550 = 62.4(4.36)[200^2/(2 \times 32.2)]/550 = 307.2$ hp

Hydraulic efficiency of flow through the turbine buckets =

$$\eta_b = \frac{P \text{ extracted by runner}}{P \text{ of jet}} = \frac{278.6}{307.2} = 0.906 = 90.6\% \quad ◄$$

Note: Here, with $v_2 = v_1$ as the water flows through the buckets, there is no head loss due to fluid friction. However, the bucket hydraulic efficiency is less than 100% because the velocity head at discharge, $V_2^2/2g$ (i.e., the residual velocity head), does not act on the buckets and is effectively lost.

Sec 16.4: Head on an Impulse Turbine and Efficiency -- *Problems*

16.7 *Repeat Sample Prob. 16.4 for the case of a 6-in-diameter jet, with all other data remaining the same.*

Sample Prob. 16.4: Delivery pipe, L = 10,000 ft, f = 0.025, D = 4.0 ft, delivers water to a nozzle at an elevation 2420 ft below the reservoir water surface, nozzle loss = $0.05V_j^2/2g$. The jet drives an impulse wheel, $D_w = 106$ in, $\beta_2 = 165°$, connected to a 20-pole generator in a 60-cycle system. Find the pipe V, V_j, Q, u, u/V_1, net head h, F on bucket, T on wheel, shaft HP, HP at base of nozzle, overall turbine η. Assume $v_2 = v_1$, neglect bearing friction and windage.

BG

$L/D = 10{,}000/4 = 2500 > 1000$ so neglect minor loss at pipe entrance.

Energy equation from surface of reservoir to jet: $2420 - 0.025(10{,}000/4)V^2/2g - 0.05V^2/2g = V_j^2/2g$

Continuity: $V_j = V(48/6)^2 = 64V$. $\therefore$ $2420 - 62.5V^2/2g - 0.05(64V)^2/2g = (64V)^2/2g$

$2420 = (62.5 + 204.8 + 4096)V^2/2g = 4363V^2/2g$; $V = \sqrt{2(32.2)2420/4363} = 5.98$ fps ◄

$V_j = 64V = 64(5.98) = 383$ fps ◄ $Q = AV = \pi 2^2(5.98) = 75.1$ cfs ◄

Sec. 16.1: $N = 7200/n$ so $n = 7200/N = 7200/20 = 3600$ rpm

$\omega = 2\pi n/60 = 2\pi 360/60 = 37.7$ rad/sec

$u = \omega r = 37.7(106/2)/12 = 166.5$ fps ◄ $u/V_1 = 166.5/383 = 0.435$ ◄

Eq. 16.5: Net head, $h = z_1 - h_L = 2420 - f(L/D)V^2/2g = 2420 - 0.025(10{,}000/4)(5.98)^2/64.4$

$= 2385$ ft ◄

Eq. 16.2: $F = 1.940(75.1)(383 - 166.5)(1 - \cos 165°) = 62{,}000$ lb ◄

Torque on wheel $T = Fr = 62{,}000(106/2)/12 = 273{,}800$ ft·lb ◄

Eqs. 5.32 and 5.35: Shaft power $P = T\omega/550 = 273{,}800(37.7)/550 = 18{,}770$ hp ◄

Power at base (entrance) of nozzle = power input to turbine (including nozzle) =

$P = \gamma Qh/550 = 62.4(75.1)2385/550 = 20{,}320$ hp ◄

Eq. 5.34: Overall turbine efficiency $\eta = \dfrac{\text{Shaft power}}{\text{Input power}} = \dfrac{18{,}770}{20{,}320} = 0.924 = 92.4\%$ ◄

Note: The turbine efficiency is still high. However, the power developed (18,770 hp) using this nozzle with a 6-in-diameter jet is much less than the power (31,960 hp, Sample Prob. 16.4) with the nozzle that produces the 8-in-diameter jet. See Sample Problem 16.5 for an explanation.

16.8 *A torque of 157,600 ft·lb is transmitted to a turbine that is developing a brake horsepower of 6000. How many poles does the generator have if it is producing 50-cycle electricity?*

BG

Eq. 5.35: bhp = 6000 = $T\omega/550$ = $157{,}600(2\pi/60)n/550$; n = 200 rpm

Sec. 16.1: Number of poles = $6000/n$ = 6000/200 = 30 ◄

16.9 *A 1.5-in-diameter water jet with a velocity of 120 ft/sec impinges on a bucket of a stationary impulse wheel that has a pitch diameter of 8.0 ft. If the bucket blade angle is 165°, what torque must be applied to the turbine shaft to prevent it from rotating? Consider two cases: (a) $v_2 = v_1$; (b) $v_2 = 0.9v_1$.*

BG

Eq. 6.24: $F = \rho Q(\Delta v_x)$; $Q = A_jV_j = [\pi(1.5/12)^2/4]120 = 1.472$ cfs

(*a*) $\Delta v_x = 120\cos 165° - 120 = -115.9 - 120 = -235.9$ fps

$T = Fr = \rho Q(\Delta v_x)r = 1.940(1.472)235.9(4) = 2695$ ft·lb ◄

(*b*) $\Delta v_x = 0.9(120)\cos 165° - 120 = -104.3 - 120 = -224.3$ fps

$T = \rho Q(\Delta v_x)r = 1.940(1.472)(224.3)4 = 2562$ ft·lb ◄

16.10 *Water is delivered from a reservoir through a 1200-mm-diameter pipe (f = 0.022) 3200 m long to a nozzle that emits a 200-mm-diameter jet. The jet impinges on the buckets of an impulse turbine, pitch diameter 2.5 m, which drives a 16-pole generator that produces 50-Hz electricity. The surface of the reservoir providing the water is at an elevation 800 m higher than the nozzle. Assume that the head loss through the nozzle can be expressed as $0.05V_j^2/2g$ where V_j is the jet velocity. Find the following: velocity of flow in the pipe, jet velocity, flow rate, speed of the buckets, value of u/V_1, net head, force F on the bucket, torque on the wheel, shaft power, power at the head of the nozzle, and overall turbine efficiency. Assume $\beta_2 = 165°$ and $v_2 = v_1$; neglect bearing friction and windage.*

SI

Energy equation from reservoir surface to jet: $800 - (0.022)(3200/1.2)V^2/2g - 0.05V_j^2/2g = V_j^2/2g$

Continuity: $V_j = (1200/200)^2V = 36V$. $\therefore$ $800 - 58.7V^2/2g = 1.05(36V)^2/2g = 1360V^2/2g$

$800 = 1420V^2/2g$; $V = 3.325$ m/s ◄ $V_j = 36V = 36(3.325) = 119.7$ m/s ◄

$Q = AV = \pi(0.200/2)^2 119.7 = 3.76$ m^3/s ◄ $N = 6000/n$, $n = 6000/16 = 375$ rpm ◄

$\omega = 2\pi n/60 = 2\pi(375/60) = 39.25$ rad/sec ; $u = \omega r = 39.25(2.500/2) = 49.0$ m/s ◄

Hence $u/V_1 = 49.0/119.7 = 0.41$ ◄

Net head (Fig. 16.1): $h = z_1 - h_L = 800 - 0.022(3200/1.2)3.325^2/[2(9.81)] = 767$ m

Eq. 16.2: $F = \rho Q(V_1 - u)(1 - \cos\beta_2) = 1000(3.76)(119.7 - 49.0)(1 + 0.966) = 523\,000$ N ◄

Torque on wheel: $T = Fr = 523{,}000(2.5/2) = 654\,000$ N·m ◄

Shaft power = $T\omega$ = 654 000(39.25)/1000 = 25 500 kW ◄

Power at base (entrance) of nozzle = power input to turbine (including nozzle) =

$P = \gamma Qh = 9.81(3.76)767 = 28\,300$ kW ◄

Eq. 5.34: Overall efficiency of turbine: $\eta = T\omega/\gamma Qh = 25\,500/28\,300 = 0.90$ ◄

16.11 *The diameter of a Pelton wheel is 6 ft. The velocity of the 4-in-diameter jet is 400 ft/sec and the bucket angle is 165°. Assuming a k value of 0.21, for bucket speeds of 100, 200, 300 and 400 ft/sec find the torque in ft·lb, the horsepower, and the hydraulic efficiency of the buckets. Neglect bearing friction and windage.*

BG

Flow through buckets: $v_1^2/2g - h_L = v_2^2/2g$ where $h_L = k(v_2^2/2g)$; $v_1^2/2g - 0.21 v_2^2/2g = v_2^2/2g$

$v_1^2 = (0.21 + 1)v_2^2$ from which $v_2 = 0.909 v_1$

$\Delta v = v_2 \cos 165° - v_1 = 0.909 v_1 \cos 165° - v_1 = -1.879 v_1$; $\Delta V = (u + v_2 \cos 165°) - V_1$

It can be shown that $\Delta v = \Delta V$. $Q = AV = (\pi/4)(4/12)^2 400 = 34.9$ cfs

Eq. 6.24: $F = \rho Q(\Delta V) = 1.940(34.9)\Delta V = 67.7 \Delta V = 67.7 \Delta v$; $T = Fr = 3F = 3(67.7)\Delta v = 203 \Delta v$

Shaft HP $= Fu/550 = 67.7(\Delta v)u/550$; Jet HP $= \gamma Q(V_j^2/2g)/550 = 62.4(34.9)[400^2/(2 \times 32.2)]/550 = 9840$

Bucket η = Shaft HP/Jet HP = $[67.7(\Delta v)u/550]/9840$

μ (ft/sec)	v_1 (ft/sec)	Δv (ft/sec)	T (ft·lb)	shaft hp	bucket η
100	300	563.7	114,300	6940	0.71
200	200	375.8	76,300	9250	0.94
300	100	187.9	38,100	6940	0.71
400	0	0	0	0	0
			▲	▲	▲

Note: This assumes that all water issuing from the jet acts on the wheel.

16.12 *An impulse turbine is supplied by a 1200-mm-diameter steel *pipe (e = 0.10 mm) 500 m long. The jet diameter of the single nozzle is 400 mm and its loss coefficient is 0.05. When the water level in the reservoir is 150 m above the nozzle of the turbine, the brake power is measured to be 7900 kW. What is the turbine efficiency? [*Note this correction to the textbook.]*

SI

Energy equation: $150 - f(L/D)V^2/2g - 0.05 V_j^2/2g = V_j^2/2g$

$e/D = 0.10/1200 = 0.000\,083$; Fig. 8.11: Assume $f = 0.013$ (depends on Reynolds number)

Continuity: $V_j(400)^2 = V(1200)^2$; $V_j = (1200/400)^2 V = 9V$

$\therefore$ $150 - 0.013(500/1.2)V^2/2g = 1.05(9V)^2/2g = 85.1V^2/2g$; $V = 5.71$ m/s

Check f: $\mathbf{R} = DV/\nu = 1.2(5.71)/(1 \times 10^{-6}) = 6.85 \times 10^6$; Fig. 8.11 shows $f = 0.012$ is closer

This changes V very little, $V = 5.72$ m/s

Net head at base of nozzle: $h = 150 - 0.012(500/1.2)5.72^2/(2 \times 9.81) = 141.7$ m

Power at base (entrance) of nozzle = power input to turbine (including nozzle) =

$P = \gamma Q h = 9.81(6.47)141.7 = 8990$ kW

Turbine η = brake P/delivered P = 7900/8990 = 0.879 ◄

16.13 *Using the data of Prob. 16.11, find values of (a) head loss in bucket friction, (b) velocity head at discharge, and (c) total head loss for bucket speeds of 180, 190, 200, and 210 fps. At what bucket speed is the discharge loss a minimum? At what bucket speed is the total head loss a minimum?*

Prob. 16.11: Pelton wheel, V_1 = 400 fps, β_2 = 165°, k = 0.21.

BG

For u = 180 fps:

(*a*) $v_1 = V_1 - u = 400 - 180 = 220$ fps

Sec. 16.4: $v_1^2/2g - kv_2^2/2g = v_2^2/2g$, $\therefore$ $v_2 = v_1/\sqrt{k+1} = 220/\sqrt{0.21+1} = 200$ fps

Bucket friction $h_L = kv_2^2/2g = 0.21(200^2/64.4) = 130.4$ ft ◄

(*b*) $V_{2x} = u + v_2\cos 165° = 180 + 200(-0.966) = -13.19$ fps ; $V_{2y} = 200\sin 165° = 51.8$ fps

$V_2 = \sqrt{V_{2x}^2 + V_{2y}^2} = \sqrt{(-13.19)^2 + (51.8)^2} = 53.4$ fps ; $V_2^2/2g = (53.4)^2/(2\times 32.2) = 44.3$ ft ◄

(*c*) Total head loss = 130.4 + 44.3 = 174.7 ft ◄

u fps	v_2 fps	(*a*) bucket h_L ft	V_{2x} fps	V_{2y} fps	V_2 fps	(*b*) $V_2^2/2g$ ft	(*c*) Total loss ft
180	200.0	130.4	−13.2	51.8	53.4	44.3	174.7
190	191.1	119.1	5.4	49.5	49.8	38.5	157.6
200	182.0	108.0	24.2	47.1	53.7	44.8	152.8
210	172.9	97.5	43.0	44.8	62.0	59.7	157.2
		▲				▲	▲

The discharge loss (*b*) is a minimum at a bucket speed of slightly less than 190 fps. ◄

The total head loss (*c*) is a minimum at a bucket speed of slightly more than 200 fps. ◄

Sec. 16.5: Nozzles for Impulse Turbines -- *Exercises*

16.5.1 *Repeat Sample Prob. 16.5 for the case where the pipe length is 10,000 ft. All other data are to remain the same.*

Sample Prob. 16.5: Impulse turbine, pipe D = 6 in, $h_e' = 0$, f = 0.020, nozzle $h_L = 0.04V_j^2/2g$. Reservoir surface elev. = 500 ft, nozzle elev. = 300 ft. Find D_j for maximum P_j.

BG

Energy Eq: $500 - 0.020(10{,}000/0.5)V^2/2g - 0.04V_j^2/2g = 300 + V_j^2/2g$ (1)

Continuity: $D^2V = D_j^2V_j$; $V = (D_j/0.5)^2V_j = 4D_j^2V_j$ (2)

Combining (1) and (2) gives $200 = 400(4D_j^2V_j)^2/2g + 1.04V_j^2/2g = (6400D_j^4 + 1.04)V_j^2/2g$

Assume values of D_j (as in Sample Prob. 16.5) and find corresponding values for V_j.

Then find corresponding $Q = A_jV_j = (\pi/4)D_j^2V_j$ and from Eq. 16.9: $P_j = \gamma Q(V_j^2/2g)/550$

D_j in	D_j ft	V_j fps	A_j ft²	Q cfs	P_j hp	
0.5	0.0416	110.3	0.001 36	0.150	3.21	
1.0	0.0833	97.7	0.005 45	0.53	8.95	(max)
2.0	0.167	46.3	0.021 9	1.014	3.83	

Thus the maximum power (about 8.95 hp) occurs with a jet diameter of about 1.0 inch. ◄

An alternative mode of solution would be to use differential calculus, see Prob. 16.15.

16.5.2 *Repeat Sample Prob. 16.5 for the case where the pipe diameter is 12 in. All other data are to remain the same.*

Sample Prob. 16.5: Impulse turbine, pipe L = 1000 ft, $h_e' = 0$, $f = 0.020$, *nozzle* $h_L = 0.04V_j^2/2g$. *Reservoir surface elev. = 500 ft, nozzle elev. = 300 ft. Find* D_j *for maximum* P_j.

BG

Energy Eq: $500 - 0.020(1000/1.0)V^2/2g - 0.04V_j^2/2g = 300 + V_j^2/2g$ (1)

Continuity: $D^2V = D_j^2V_j$; $V = (D_j/1)^2V_j = D_j^2V_j$ (2)

Combining (1) and (2) gives $200 = 20(D_j^2V_j)^2/2g + 1.04V_j^2/2g = (20D_j^4 + 1.04)V_j^2/2g$

Assume values of D_j (as in Sample Prob. 16.5) and find corresponding values for V_j.

Then find corresponding $Q = A_jV_j = (\pi/4)D_j^2V_j$ and from Eq. 16.9: $P_j = \gamma Q(V_j^2/2g)/550$

D_j in	D_j ft	V_j fps	A_j ft^2	Q cfs	P_j hp	
3.0	0.250	107.3	0.0491	5.27	106.9	
4.0	0.333	100.1	0.0855	8.58	152.4	
5.0	0.417	88.5	0.1365	12.08	166.7	(max)
6.0	0.500	75.0	0.1963	14.72	145.9	
8.0	0.667	50.7	0.352	17.74	79.4	

Thus the maximum power (about 167 hp) occurs with a jet diameter of almost 5.0 in. ◀

An alternative mode of solution would be to use differential calculus, see Prob. 16.15.

Sec 16.5: Nozzles for Impulse Turbines -- *Problems*

16.14 *A 36-in pipeline (f = 0.020) 10,000 ft long connects a reservoir whose water surface elevation is 1800 ft to a nozzle at elevation 1000 ft. The jet from the nozzle is used to drive an impulse wheel. (a) If the head loss through the nozzle is expressible as* $0.04V_j^2/2g$, *determine the horsepower that will be developed by jets having diameters of 8 in, 10 in, and 12 in. (b) On the basis of the answers to (a), approximately what would be the horsepower and diameter of the jet that gives the maximum horsepower?*

BG

Energy: $1800 - 0.020(10{,}000/3)V^2/2g - 0.04V_j^2/2g = 1000 + V_j^2/2g$ (1)

Continuity: $D^2V = D_j^2V_j$; $V = (D_j/3)^2V_j = D_j^2V_j/9$ (2)

Combining (1) and (2) gives $800 = 66.7(D_j^2/9)^2V_j^2/2g + 1.04V_j^2/2g = (0.823D_j^4 + 1.04)V_j^2/2g$

For the given values of D_j find corresponding values of V_j.

Next find corresponding $Q = A_jV_j = (\pi/4)D_j^2V_j$. Then from Eq. 16.9: $P_j = \gamma Q(V_j^2/2g)/550$

D_j in	D_j ft	V_j fps	A_j ft^2	Q cfs	$V_j^2/2g$ ft	P_j hp	
8	0.667	207	0.349	72.2	665	5450	
10	0.833	189.3	0.545	103.2	556	6510	(max)
12	1.000	166.2	0.785	130.5	429	6350	

So the maximum jet power of about 6510 hp occurs with a jet diameter slightly larger than 10 inches. ◀

An alternative mode of solution would be to use differential calculus, see Problem 16.15.

16.15 *Using the data of Prob. 16.14, express the jet power analytically in terms of D_j and V_j, differentiate, and equate to zero to find the value of D_j for which the jet power is a maximum. Evaluate the value of the maximum jet power. Hint: Express power as P = f(x), where x = *D_j^2. [*Note this correction to the textbook.]*

Prob. 16.14: Impulse turbine, pipe D = 3 ft, L = 10,000 ft, f = 0.020, nozzle $h_L = 0.04V_j^2/2g$. Reservoir surface elev. = 1800 ft, nozzle elev. = 1000 ft.

BG

Energy: $1800 - 0.020(10{,}000/3)V^2/2g - 0.04V_j^2/2g = 1000 + V_j^2/2g$ (1)

Continuity: $D^2V = D_j^2V_j$; $V = (D_j/3)^2V_j = D_j^2V_j/9$ (2)

Combining (1) and (2) and rearranging gives: $V_j^2 = 2g(800)/(1.04 + 0.823D_j^4)$

Power of jet: $P = \gamma Qh = \gamma A_jV_j(V_j^2/2g) = 62.4(\pi D_j^2/4)V_j^2/2g = 0.761D_j^2V_j^3$ ft·lb/sec

Let $x = D_j^2$ and $y = V_j^2 = \dfrac{2g(800)}{1.04 + 0.823x^2}$

Then $P = 0.761xy^{3/2}$ and $\dfrac{dy}{dx} = \dfrac{-2g(800)2(0.823)x}{(1.04 + 0.823x^2)^2} = \dfrac{-2(0.823)xy}{1.04 + 0.823x^2}$

$$\frac{dP}{dx} = 0.761y^{3/2} + 0.761x\left(\frac{3}{2}\right)y^{1/2}\left[\frac{-2(0.823)xy}{1.04 + 0.823x^2}\right] = 0$$

i.e. $1 = 3(0.823)x^2/(1.04 + 0.823x^2)$; $1.04 + 0.823x^2 = 3(0.823)x^2$

$1.04 = 2(0.823)x^2$; $x^2 = 0.632$; $D_j = \sqrt{x} = 0.632^{1/4} = 0.892$ ft $= 10.70$ in ◀

Then $y = 2g(800)/[1.04 + 0.823(0.632)] = 33{,}000$; $V_j = \sqrt{y} = 181.7$ fps

$P = 0.761xy^{3/2} = 0.761\sqrt{0.632}(33{,}000)^{3/2} = 3.63 \times 10^6$ ft·lb/sec = 6600 hp ◀

Sec. 16.7: Action of the Reaction Turbine -- *Exercises*

16.7.1 *A Kaplan turbine operating at 327.3 rpm develops a shaft power of 300 kW when the flow is 5.5 m³/s. Find the torque transmitted from the flowing water to the shaft.*

SI

Eq. 16.10 (and 5.35): Shaft power $P = T\omega = 300$ kW = 300 000 N·m/s

$\omega = 327.3(2\pi/60) = 34.3$ rad/sec ; $T = 300\,000/\omega = 300\,000/34.3 = 8750$ N·m ◀

16.7.2 *A torque of 7500 ft·lb is transferred from the flowing water to a reaction turbine whose shaft power is 285.5 horsepower. How many poles must the generator have to develop 50-cycle electricity?*

BG

Eq. 16.10 and (5.35): Shaft power $= T\omega$; 285.5(550) ft·lb/sec = $7500(2\pi n/60)$; $n = 200$ rpm

Sec. 16.1: Number of poles, $N = 6000/n = 6000/200 = 30$ ◀

/more...

Sec. 16.8: Draft Tubes and Effective Head on Reaction Turbines -- *Exercises*

16.8.1 *Repeat Sample Prob. 16.7 for the case where the flow rate is 90 cfs with all other data remaining the same.*
Sample Prob. 16.7: Original draft tube = pipe, L = 40 ft, f = 0.03, constant D = 3 ft. If replaced by a diverging tube, D increasing uniformly from 3 ft to 8 ft over 40 ft length, how much additional effective head will be developed? Neglect effects of the bend and the tailrace velocity.

BG

Original (straight) draft tube: $V = Q/A = 90/(\pi 1.5^2) = 12.73$ ft/sec

Head loss due to draft tube: $h_L = f(L/D)V^2/2g = 0.03(40/3)12.73^2/(2 \times 32.2) = 1.007$ ft

Head loss at discharge: $h_{Lx} = V_2^2/2g - V_C^2/2g = 12.73^2/(2 \times 32.2) - 0 = 2.52$ ft

Thus original total head loss $= 1.007 + 2.52 = 3.52$ ft

Replacement (diverging) draft tube:

Head loss due to draft tube: Eq. 8.59: $h_L = k'(V_1 - V_2)^2/2g$ where $V_1 = 12.73$ ft/sec

Continuity: $V_2 = (D_1/D_2)^2 V_1 = (3/8)^2 12.73 = 1.790$ ft/sec

Cone angle (Fig. 8.18): $\tan(\alpha/2) = (8 - 3)/(2 \times 40) = 0.0625$, $\alpha = 7.15°$

Fig. 8.19*a* (solid curve) for $\alpha = 7.15°$: $k' = 0.133$

Thus $h_L = 0.133(12.73 - 1.790)^2/(2 \times 32.2) = 0.247$ ft

Head loss at discharge: $h_{Lx} = 1.790^2/(2 \times 32.2) - 0 = 0.0498$ ft

New total head loss $= 0.247 + 0.0498 = 0.297$ ft

Reduction in total head loss = original − new $= 3.52 - 0.297 = 3.23$ ft

∴ new draft tube will provide 3.23 ft of additional net head. ◄

16.8.2 *A Francis turbine operating at 300 rpm develops a shaft horsepower of 1125 hp when the flow is 120 cfs. What is the net head on the runner if the turbine efficiency is 85%?*

BG

Eq. 16.8: Shaft power $= \gamma Q h \eta$; $1125 = \gamma Q h \eta/550 = 62.4(120)h(0.85)/550$; $h = 97.2$ ft ◄

16.8.3 *A Kaplan turbine operating at 327.3 rpm develops a shaft power of 300 kW when the flow is 5.5 m³/s. What is the *net head on the runner if the turbine efficiency is 92%? [*Note this correction to the textbook.]*

SI

Eq. 16.8: Shaft power $= \gamma Q h \eta$; $300 = 9810(5.5)h(0.92)/1000$; $h = 6.04$ m ◄

Sec. 16.8: Draft Tubes and Effective Head on Reaction Turbines -- *Problems*

16.16 *In the test of a Francis turbine, a pressure of 140 kPa was measured at the flange at the entrance to a spiral turbine-case where the diameter is 600 mm. Neglecting the small velocity head in the tailrace, find the net head on the turbine if the flow rate was 2.5 m³/s and the flange was 3.0 m above the tailrace.*

SI

At entrance: $V = 2.5/(\pi 0.3^2) = 8.84$ m/s

Eq. 16.12: $h = z + p/\gamma + V^2/2g - V_C^2/2g = 3.0 + 140(1000)/9810 + 8.84^2/(2 \times 9.81) - 0$

$= 3.0 + 14.27 + 3.98 = 21.3$ m ◄

16.17 *A reaction turbine is supplied with water through a 1500-mm-diameter pipe (e = 0.10 mm) that is 50 m long. The water surface in the reservoir is 27 m above the draft-tube inlet that is 4.1 m above the water level in the tailrace. If the turbine efficiency is 92% and the discharge is 12 m³/s, what is the power output of the turbine in kilowatts? By how much must the discharge be increased to increase power production by 500 kW? Assume the velocity in the tailrace can be neglected.*

SI

$V = Q/A = 12/(\pi 1.5^2/4) = 6.79$ m/s ; Table A.1 for water at 20°C: $\nu = 1.00 \times 10^{-6}$ m²/s

$\mathbf{R} = 6.79(1.5)/(1 \times 10^{-6}) = 1.019 \times 10^7$; $e/D = 0.10/1500 = 0.000\,0667$, $f = 0.0114$ (Fig. 8.11)

Net $h = 27 + 4.1 - 0.0114(50/1.5)6.79^2/(2 \times 9.81) = 30.2$ m

$P = \gamma Q h \eta = 9810(12)30.2(0.92) = 3270\,000$ W $= 3270$ kW ◀

$(3270 + 500)$ kW $= 9810(\pi 1.5^2/4)V[27 + 4.1 - f(50/1.5)V^2/(2 \times 9.81)]0.92/1000$

By trials: $V = 7.36$ m/s, $\mathbf{R} = 1.104 \times 10^7$, $f = 0.0113$

$Q = AV = (\pi 1.5^2/4)7.36 = 13.00$ m³/s ; $\Delta Q = 13.00 - 12.0 = 1.001$ m³/s ◀

16.18 *A draft tube leading from the discharge side of a turbine to the submerged discharge in the tailrace as in Fig. 16.11 consists of a pipe (f = 0.025) of constant diameter 0.8 m and length 10.0 m. The flow rate is 9.2 m³/s. If this draft tube were to be replaced by a diverging tube 12.0 m long whose diameter changed uniformly from 0.8 m to 2.5 m over the 12.0 m length, how much additional head would be developed by the replacement draft tube? Neglect head loss due to curvature in the bend.*

SI

Original (straight) draft tube, $L = 10$ m: $V = Q/A = 9.2/(\pi 0.4^2) = 18.30$ m/s

Assume the tailrace velocity V_C is negligible.

$h_L = h_{Lf} + h_d' = 0.025(10/0.8)V^2/2g + V^2/2g = (0.313 + 1)(18.30)^2/(2 \times 9.81) = 22.4$ m

Replacement (diverging) draft tube, $L = 12$ m: $V_1 = 18.30$ m/s; $V_2 = Q/A_2 = 9.2/(\pi 1.25^2) = 1.874$ m/s

Cone angle (Fig. 8.18): $\tan(\alpha/2) = (2.5 - 0.8)/(2 \times 12) = 0.0708$, $\alpha = 8.10°$

Fig. 8.19*a* (solid curve) for $\alpha = 8.10°$: $k' = 0.144$

Eq. 8.59: $h' = k'(V_1 - V_2)^2/2g = 0.144(18.30 - 1.874)^2/[2(9.81)] = 1.981$ m

$h_d' = V_2^2/2g = 1.874^2/[2(9.81)] = 0.1790$ m. New total head loss $= 1.981 + 0.1790 = 2.16$ m

Additional head that would be developed = original − new = $22.4 - 2.16 = 20.2$ m ◀

16.19 *At its upper end a draft tube has a diameter of 24.5 in where it joins the discharge side of the turbine at a point 11.0 ft above the surface of the water in the tailrace. The discharge end of the draft tube has a diameter of 42 in and the velocity in the tailrace is negligible. The total head loss in the draft tube is $0.15 V_2^2/2g$ plus the submerged discharge loss of $V_3^2/2g$, where subscripts 2 and 3 refer to the upper and lower ends of the draft tube, respectively. (a) When the flow is 38.8 cfs, what is the pressure at the upper end of the draft tube? (b) Suppose the draft tube were of uniform diameter; what then would be the pressure at the upper end of the tube? (c) How much head is saved by the diverging tube? Assume the draft tube has a length of 18 ft and f = 0.020.*

BG

(*a*) Upper end: $V_2 = Q/A_2 = 38.8/[\pi(24.5/12)^2/4] = 11.85$ fps ; $V_2^2/2g = 11.85^2/(2 \times 32.2) = 2.18$ ft

Lower end: $V_3 = Q/A_3 = 38.8(4)/[\pi(42/12)^2] = 4.03$ fps ; $V_3^2/2g = 4.03^2/(2 \times 32.2) = 0.253$ ft

From 2 to tailwater (given): $\Sigma h_L = 0.15V_2^2/2g + V_3^2/2g = 0.15(2.18) + 0.253 = 0.580$ ft

Energy equation from 2 to tailwater: $z_2 + p_2/\gamma + V_2^2/2g - \Sigma h_L = 0$

$11.0 + p_2/\gamma + 2.18 - 0.580 = 0$; $p_2/\gamma = -12.60$ ft ; $p_2 = -12.60(62.4/144) = -5.46$ psi ◀

/cont...

or $p_2 = -5.46(29.92/14.70) = 11.11$ inHg vacuum ◄

(*b*) Straight tube: $h_{Lf} = 0.020[18/(24.5/12)]2.18 = 0.385$ ft

From 2 to tailwater: $\Sigma h_L = h_{Lf} + h_d' = 0.385 + V_2^2/2g = 0.385 + 2.18 = 2.57$ ft

Energy Eq. from 2 to tailwater: $11.0 + p_2/\gamma + 2.18 - 2.57 = 0$

$p_2/\gamma = -10.62$ ft ; $p_2 = -10.62(62.4/144) = -4.60$ psi ◄

or (from inside covers) $p_2 = -4.60(29.92/14.70) = 9.36$ inHg vacuum ◄

(*c*) Head saved by diverging tube $= \Sigma h_L(b) - \Sigma h_L(a) = 2.57 - 0.580 = 1.986$ ft ◄

Sec. 16.9: Efficiency of Turbines -- *Problems*

16.20 *A model turbine (one-twentieth of prototype size) has a maximum hydraulic efficiency of 86.2%. Estimate the efficiency of the prototype utilizing the Moody step-up formula.*

N

Eq. 16.15: $(1 - 0.862)/(1 - \eta_p) = (20/1)^{1/5} = 1.821$; $\eta_p = 0.924 = 92.4\%$ ◄

Sec. 16.10: Similarity Laws for Reaction Turbines -- *Exercises*

16.10.1 *At a maximum efficiency a turbine runs at 150 rpm, discharges 10.0 m³/s, and develops a shaft power of 2600 kW under a net head of 30 m. (a) What is its efficiency? (b) What would be the revolutions per minute, the flow rate, and brake power of this same turbine under a net head of *48 m for homologous conditions? (c) How many poles would be required for 60-cycle electricity if the net head is *48 m? [*Note these corrections to the textbook.]*

SI

(*a*) Eq. 16.14: $\eta = \dfrac{\text{shaft power}}{\text{water power}} = \dfrac{2600}{\gamma Q h} = \dfrac{2600(1000)}{9810(10)30} = 0.883 = 88.3\%$

(*b*) Eq. 16.16: $\dfrac{n_2}{n_1} = \left(\dfrac{h_2}{h_1}\right)^{1/2}\left(\dfrac{D_1}{D_2}\right) = \left(\dfrac{48}{30}\right)^{1/2}\left(\dfrac{D}{D}\right) = 1.265$

$n_2 = 1.265 n_1 = 1.263(150) = 189.7$ rpm ◄

Eq. 16.17: $\dfrac{Q_2}{Q_1} = \left(\dfrac{h_2}{h_1}\right)^{1/2}\left(\dfrac{D_2}{D_1}\right)^2 = \left(\dfrac{48}{30}\right)^{1/2}\left(\dfrac{D}{D}\right)^2 = 1.265$; $Q_2 = 1.265(10) = 12.65$ m³/s ◄

Eq. 16.18: $\dfrac{P_2}{P_1} = \left(\dfrac{h_2}{h_1}\right)^{3/2}\left(\dfrac{D_2}{D_1}\right)^2 = \left(\dfrac{48}{30}\right)^{3/2} = 2.02$; $P_2 = 2.02(2600) = 5260$ kW ◄

(*c*) Sec. 16.1: Number of poles $= 7200/n_2 = 7200/189.7 = 37.9$. Use 38 poles ◄

16.10.2 *At a maximum efficiency of 93% a turbine runs at 300 rpm and discharges 310 cfs under a net head of 64 ft. This turbine drives a generator that is developing 50-cycle electricity. (a) How many poles are required for this generator? (b) What is the brake horsepower?*

BG

(*a*) Sec. 16.11: Number of poles $= 6000/300 = 20$ ◄

(*b*) From Eqs. 16.14, 5.32: $P = \eta\gamma Q h/550 = (0.93)62.4(310)64/550 = 2093$ hp ◄

16.10.3 *If the turbine of Exer. 16.10.2 is operated under a head of 48 ft, find the speed, flow rate and brake power for homologous conditions. Be sure to use an even number of poles. Hint: It may not be possible to have perfectly homologous conditions.*

Exer. 16.10.2: Turbine (50-cycle generator), n = 300 rpm, Q = 310 cfs, η = 93%, net h = 64 ft, P = 2093 bhp.

BG

Eq. 16.16: $n_2/n_1 = (h_2/h_1)^{1/2}(D_1/D_2) = (48/64)^{1/2}(1) = 0.866$; $n_2 = n_1(0.866) = 300(0.866) = 260$ rpm

Sec. 16.1 for 50 cycles: $N_2 = 6000/n_2 = 6000/260 = 23.1$ poles

N must be even so use 22 or 24 poles. These will result in conditions that are very close to being homologous to the original conditions.

For a given turbine, i.e. D = constant, with 22 poles: $n_2 = 6000/22 = 273$ rpm ◄

Eq. 16.16: $n \propto h^{1/2}$ so $h \propto n^2$; $h = 64(273/300)^2 = 52.9$ ft

Eq. 16.17: $Q \propto h^{1/2}$ so $Q \propto n$; $Q = 310(273/300) = 282$ cfs ◄

Eq. 16.18: $P \propto h^{3/2}$ so $P \propto n^3$; $P = 2093(273/300)^3 = 1573$ hp ◄

Likewise for 24-pole generator: n = 250 rpm, h = 44.4 ft, Q = 258 cfs, P = 1209 hp ◄◄

Sec. 16.10: Similarity Laws for Reaction Turbines -- *Problems*

16.21 *A turbine runs at 150 rpm, discharges 200 cfs, and develops 1600 bhp under a net head of 81 ft. (a) What is its efficiency? (b) What would be the revolutions per minute, Q, and brake horsepower of the same turbine under a net head of 162 ft for homologous conditions?*

BG

(*a*) Eq. 16.14: $\eta = T\omega/(\gamma Q h) = 1600(550)/[62.4(200)81] = 0.87 = 87\%$ ◄

(*b*) For homologous conditions with a fixed D:

Eq. 16.16: $n \propto h^{1/2}$; $n = 150(162/81)^{1/2} = 212$ rpm ◄

Eq. 16.17: $Q \propto h^{1/2}$; $Q = 200(162/81)^{1/2} = 283$ cfs ◄

Eq. 16.18: $P \propto h^{3/2}$; $P = 1600(162/81)^{3/2} = 4530$ bhp ◄

16.22 *If a turbine homologous to that of Prob. 16.21 has a runner of twice the diameter, what would be the revolutions per minute, Q, and brake horsepower under the same head of 81 ft?*

Prob. 16.21: n = 150 rpm, Q = 200 cfs, P = 1600 bhp, net h = 81 ft.

BG

For homologous conditions with a fixed h:

Eq. 16.16: $n \propto 1/D$; $n = 150(1/2) = 75$ rpm ◄

Eq. 16.17: $Q \propto D^2$; $Q = 200(2)^2 = 800$ cfs ◄

Eq. 16.18: $P \propto D^2$; $P = 1600(2)^2 = 6400$ bhp ◄

16.23 *A 2.6-m-diameter reaction turbine is to be operated at 200 rpm under a net head of 30 m. A 1:10 model of this turbine is built and tested in the laboratory. If the model is operated at 600 rpm, under what net head should it be tested to simulate normal operating conditions in the prototype?*

SI

Given: $D_m/D_p = 1/10$, and $n_m/n_p = 600/200 = 3$

Eq. 16.16: $n \propto h^{1/2}/D$ or $h \propto n^2D^2$; $h_m = h_p(3)^2(1/10)^2 = 30(0.09) = 2.70$ ◄

16.24 *A small Francis turbine (N_s = 30, D = 2 ft) is tested and found to have an efficiency of 0.893 when operating under optimum conditions. Approximately what would be the maximum efficiency of a homologous runner (N_s = 30) with a diameter of 6 ft?*

BG

Eq. 16.15: $(1 - 0.893)/(1 - \eta_p) = (6/2)^{1/5} = 1.246$; $\eta_p = 0.914$ ◄

16.25 *A 12-ft-diameter reaction turbine is to be operated at 100 rpm under a net head of 96 ft. A 1:8 model of this turbine is built and tested in the laboratory. If the model is operated at 450 rpm, under what net head should it be tested to simulate normal operating conditions?*

BG

Given: $D_m/D_p = 1/8$, $n_m/n_p = 450/100 = 4.5$

Eq. 16.16: $n \propto h^{1/2}/D$ or $h \propto n^2D^2$; $h_m = h_p(4.5)^2(1/8)^2 = 96(0.316) = 30.4$ ft ◄

Sec. 16.11: Peripheral Velocity Factor and Specific Speed of Turbines -- *Exercises*

16.11.1 *Find the specific speed of an impulse wheel operating at a maximum efficiency of 92.6% under a net head of 2314 ft at 360 rpm if the flow rate from the nozzle is 131.5 cfs. Estimate the diameter of the runner. Note: This is the turbine of Sample Prob. 16.4.*

BG

From Eq. 16.8: bhp = (0.926)62.4(131.5)2314/550 = 31,970

Eq. 16.21: $N_s = 360\sqrt{31,970}/2314^{5/4} = 4.01$ ◄

Fig. 16.14 for N_s = 4.1: $\phi_e \approx 0.43$; Eq. 16.20: $D \approx 153.3(0.43)\sqrt{2314}/360 = 8.8$ ft ◄

16.11.2 *Find the specific speed of a turbine that operates at a maximum efficiency of 90% at 150 rpm under a net head of 81 ft with a flow rate of 200 cfs. Estimate the diameter of the runner. What type of turbine is this?*

BG

From Eq. 16.8: bhp = (0.90)62.4(200)81/550 = 1654

Eq. 16.21: $N_s = 150\sqrt{1654}/81^{5/4} = 25.1$ ◄ Fig. 14.14 for N_s = 25.1: Francis turbine ◄

Fig. 16.14 for N_s = 25.1: $\phi_e \approx 0.70$; Eq. 16.20: $D \approx 153.3(0.70)\sqrt{81}/150 = 6.4$ ft ◄

16.11.3 *Find the specific speed of a turbine that operates at a maximum efficiency of 83% at 450 rpm under a net head of 20 ft with a flow rate of 100 cfs. Estimate the diameter of the runner. What type of turbine is this?*

BG

Eq. 16.8: bhp = (0.83)62.4(100)20/550 = 188.3

Eq. 16.21: $N_s = 450\sqrt{188.3}/20^{5/4} = 146.0$ ◄ Fig. 16.14 for N_s = 146: Propeller turbine ◄

Fig. 16.14 for N_s = 146: $\phi_e \approx 1.78$; Eq. 16.20: $D \approx 153.3(1.78)\sqrt{20}/450 = 2.71$ ft ◄

16.26 *A 1:8 model of a 12-ft-diameter turbine is operated at 600 rpm under a net head of 54.0 ft. Under this mode of operation the bhp and Q of the model were observed to be 332 hp and 62 cfs, respectively. From the above data compute (a) the specific speed of the model and the value of *ϕ_e, (b) the efficiency and shaft torque of the model, (c) the efficiency of the prototype, and (d) the flow rate and horsepower of the prototype if it is operated at 450 rpm under a net head of 200 ft. [*Note this correction to the textbook.]*

BG

(*a*) $(N_s)_m = n_e\sqrt{\text{bhp}}/h^{5/4} = 600\sqrt{332}/54^{5/4} = 74.7$ ◄ Fig. 16.14 for $N_s = 74.7$: $\phi_e \approx 0.74$ ◄

(*b*) $\text{bhp} = T\omega/550 = T(2\pi 600/60)/550 = 332$; $T = 2910$ lb·ft ◄

Eq. 16.8: $\eta = T\omega/(\gamma Qh) = 332(550)/[62.4(62)54] = 0.874$ ◄

(*c*) Eq. 16.15: $(1 - 0.874)/(1 - \eta_p) = (8/1)^{1/5} = 1.516$; $\eta_p = 0.917$ ◄

(*d*) Eq. 16.16 for model: $600 = (K_n)_m(54)^{1/2}/(12/8)$; $(K_n)_m = 122.5$

For prototype: $450 = (K_n)_p 200^{1/2}/12$; $(K_n)_p = 450(12)/200^{1/2} = 382$

Thus $K_m \neq K_p$ and homologous conditions do not occur with the given data. So predictions can not be made using similarity laws. ◄

16.27 *Calculate the specific speed of the turbine whose runner is shown in Fig. 16.2 and estimate the runner diameter.*

Fig. 16.2: Net h = 2200 ft, n = 250 rpm, D = 162 in.

BG

Eq. 16.21: $N_s = n_e\sqrt{bhp}/h^{5/4}$; Cannot solve this equation, as brake horsepower is not given. ◄

However, Eq. 16.20: $D = 162/12 = 153.3\phi_e\sqrt{2200}/250$: $\phi_e = 0.469$

From Fig. 16.14 with $\phi_e = 0.469$ it must be an impulse wheel, which agrees with Fig. 16.2

16.28 *Find the specific speed of the turbine shown in Fig. 16.7 and estimate the runner diameter.*

Fig. 16.7: P = 128 000 kW, n = 150 rpm, h = 87.6 m.

B

$h = (87.6\text{ m})3.28 = 287$ ft; $P = (128\,000\text{ kW})1.341 = 171{,}700$ hp

Eq. 16.21: $N_s = n_e\sqrt{\text{bhp}}/h^{5/4} = 150\sqrt{171{,}700}/287^{5/4} = 52.5$ ◄

Fig. 16.14 with $N_s = 52.5$: This is a Francis turbine, $\phi_e \approx 0.72$

Eq. 16.20: $D = 153.3\phi_e\sqrt{h}/n \approx 153.3(0.72)\sqrt{287}/150 = 12.5$ ft $= 3.80$ m ◄

This checks with Fig. 16.7 (see the man in the photo).

16.29 *Calculate the specific speed of the turbine shown in Fig. 16.9 and estimate the runner diameter.*

Fig. 16.9: P = 42,000 hp, n = 94.7 rpm, h = 52 ft.

BG

Eq. 16.21: $N_s = n_e\sqrt{\text{bhp}}/h^{5/4} = 94.7\sqrt{42{,}000}/52^{5/4} = 139.0$ ◄

Fig. 16.14 with $N_s = 139$: This is a propeller turbine, $\phi_e \approx 1.71$

Eq. 16.20: $D = 153.3\phi_e\sqrt{h}/n \approx 153.3(1.71)\sqrt{52}/94.7 = 20.0$ ft ◄

This checks with Fig. 16.9 (see the man in the photo).

16.30 *Find the specific speed of the propeller turbine mentioned in Exer. 16.7.1. Express the answer in BG units.*
Exer. 16.7.1: Kaplan turbine, n = 327.3 rpm, shaft P = 300 kW, Q = 5.5 m³/s.

B

Fig. 16.13 for Kaplan turbine: Assume η = 0.92

bhp = (300 kW)1.341 = 402.3 hp ; Q = (5.5 m^3/s)35.31 = 194.2 cfs

From Eq. 16.8: bhp = $\eta\gamma Qh/550$ = (0.92)62.4(194.2)h/550 = 402.3; h = 19.84 ft

Eq. 16.21: $N_s = 327.3\sqrt{402.3}/19.84^{5/4} = 156.7$ ◀

16.31 *Find the specific speed of a turbine that runs at a maximum efficiency of 90% at 300 rpm under a net head of 81 ft with a flow rate of 50 cfs. Estimate the runner diameter.*

BG

From Eq. 16.8: bhp = $\eta\gamma Qh/550$ = 0.90(62.4)50(81)/550 = 413.5

Eq. 16.21: $N_s = n_e\sqrt{\text{bhp}}/h^{5/4} = 300\sqrt{413.5}/81^{5/4} = 25.1$ ◀

Fig. 16.14 with N_s = 25.1: $\phi_e \approx 0.70$; Eq. 16.20: $D \approx 153.3(0.70)\sqrt{81}/300 = 3.2$ ft = 39 in ◀

16.32 *In Sec. 16.15 a double-overhung impulse turbine at Dixence, Switzerland is mentioned. Each wheel of the two-wheel unit develops 25,000 hp at 500 rpm under a head of 5330 ft. (a) What is the specific speed of these wheels? (b) Estimate their diameter and compare your answer with their actual diameter of 10.89 ft.*

BG

(*a*) Eq. 16.21: $N_s = n_e\sqrt{\text{bhp}}/5330^{5/4} = 500\sqrt{25{,}000}/5330^{5/4} = 1.736$ ◀

(*b*) Fig. 16.14 with N_s = 1.736: $\phi_e \approx 0.46$; Eq. 16.20: $D \approx 153.3(0.46)\sqrt{5330}/500 = 10.3$ ft ◀

This compares well with the actual diameter of 10.89 ft (5.4% difference). ◀

16.33 *In Sec. 16.15 a Francis turbine on the Susquehanna River is mentioned. This turbine develops 54,000 hp at 81.8 rpm under a head of 89 ft. (a) What is the specific speed of this turbine? (b) Estimate the runner diameter and compare your answer with the actual diameter of 18 ft.*

BG

(*a*) Eq. 16.21: $N_s = 81.8\sqrt{54{,}000}/89^{5/4} = 69.5$ ◀

(*b*) Fig. 16.14 with N_s = 69.5: $\phi_e \approx 0.74$; Eq. 16.20: $D \approx 153.3(0.74)\sqrt{89}/81.8 = 13.1$ ft ◀

This does not compare well with the actual diameter of 18 ft (27.2% difference). ◀

16.34 *A Kaplan turbine at Rock River, Illinois develops 800 hp at 80 rpm under a head of 7 ft. (a) What is the specific speed of this turbine? (b) Estimate the runner diameter and compare your answer with the actual diameter of 136 in.*

BG

(*a*) Eq. 16.21: $N_s = 80\sqrt{800}/7^{5/4} = 198.7$ ◀

(*b*) Fig. 16.14 with N_s = 198: $\phi_e \approx 2.39$; Eq. 16.20: $D \approx 153.3(2.39)\sqrt{7}/80 = 12.1$ ft ◀

This $D \approx 12.1$ ft = 145 in compares well with the actual D = 136 in (6.9% difference). ◀

Sec. 16.12: Cavitation in Turbines -- *Exercises*

16.12.1 *Using Eq. (16.24) and Fig. 16.17 for σ_c calculate the maximum permissible head under which a Francis turbine (N_s = *40) can operate if it is set 12 ft above tailwater. The installation is at sea level and the water temperature is 80°F. How does your answer compare to Fig. 16.16? [*Note this correction to the textbook.]*

BG

Fig. 16.17 for N_s = 40: $\sigma_c \approx 0.132$; Table A.1 for water at 80°F: p_v/γ = 1.17 ft abs

Inside cover for standard atmosphere: p/γ = 33.91 ft water abs.

Eq. 16.24: $z_B = p_{atm}/\gamma - p_v/\gamma - \sigma_c h$; $12 \approx 33.91 - 1.17 - 0.132h$; $h \approx 157$ ft ◄

Fig. 16.16 for N_s = 40 and 12 ft draft head: max $h \approx 205$ ft.

The above result (157 ft) compares fairly well with Fig. 16.16 ◄

16.12.2 *(a) Find the maximum permissible head under which an axial-flow turbine (N_s = 160) can operate if it is set 5 ft below tailwater. The installation is at elevation 3150 ft and the water temperature is 65°F. (b) What would be the maximum permissible head if the turbine had been set 5 ft above tailwater?*

BG

Table A.3 at 3150 ft, by interpolation: p_{atm} = 13.14 psia

Table A.1 for water at 65°F, by interpolation: p_v/γ = 0.715 ft abs

Fig. 16.17 for N_s = 160: $\sigma_c \approx 0.815$

Eq. 16.24: $z_B \approx 13.14(144)/62.4 - 0.715 - 0.815h = 29.6 - 0.815h$

(*a*) For z_B = −5 ft: $-5 \approx 29.6 - 0.815h$; $h \approx 42.5$ ft ◄

(*b*) For z_B = +5 ft: $5 \approx 29.6 - 0.815h$; $h \approx 30.2$ ft ◄

16.12.3 *Repeat Exer. 16.12.1 for the case where the turbine is at elevation 10,000 ft rather than at sea level. In this case one cannot compare the result with the curves presented in Fig. 16.16, because of the change in p_{atm}.*

*Exer. 16.12.1: N_s = *40, water T = 80°F, z_B = 12 ft. Find the maximum permissible head. [*Note this correction to the textbook.]*

BG

Fig. 16.17 for N_s = 40: $\sigma \approx 0.132$

Table A.3 at elev 10,000 ft: p_{atm} = 10.11 psia ; Table A.1 for water at 80°F: p_v/γ = 1.17 ft abs

Eq. 16.24: $z_B = p_{atm}/\gamma - p_v/\gamma - \sigma_c h$; $12 \approx 10.11(144/62.4) - 1.17 - 0.132h$; $h \approx 77$ ft ◄

Sec. 16.12: Cavitation in Turbines -- *Problems*

16.35 *Consider the case of a Francis turbine having a specific speed of 40 that is set 10 ft above tailwater elevation. Perform calculations using Fig. 16.17 and Eq. (16.22) to find the maximum permissible head under which this turbine should operate in order to be safe from cavitation. Check your calculated result against the information shown on Fig. 16.16. Do they agree?*

BG

Assume sea-level conditions with T = 80°F ; Table A.3 at sea level: p_{atm} = 14.70 psia

Table A.1 for water at 80°F: p_v/γ = 1.17 ft abs ; Fig. 16.17 for N_s = 40: $\sigma_c \approx 0.132$

Eq. 16.23: $h \approx [14.70(144/62.4) - 1.17 - 10]/0.132 = 172$ ft ◄

Fig. 16.16 for N_s = 40 and z_B = 10 ft: max $h \approx 230$ ft.

The above result (172 ft) compares fairly well with Fig. 16.16 ◄

16.36 *Consider the case of a propeller turbine having a specific speed of 150 that is set 5 ft below tailwater. Perform calculations using Fig. 16.17 and Eq. (16.22) to find the maximum permissible head under which this turbine should operate in order to be safe from cavitation. Check your calculated result against the information shown in Fig. 16.16. Do they agree?*

BG

Assume sea-level conditions with T = 80°F ; Table A.3 at sea level: p_{atm} = 14.70 psia

Table A.1 for water at 80°F: p_v/γ = 1.17 ft abs ; Fig. 16.17 for N_s = 150: $\sigma_c \approx 0.71$

Eq. 16.23: $h \approx [14.70(144/62.4) - 1.17 + 5]/0.71 = 53$ ft ◄

Fig. 16.16 for N_s = 150 and $z_B = -5$ ft: max $h \approx 56$ ft. The above (53 ft) checks well with this. ◄

16.37 *At its maximum efficiency of 93% a turbine delivers 3000 hp to the shaft under a head of 72 ft when operating at 300 rpm. Find the following: (a) the flow rate through the turbine; (b) the specific speed of the turbine, (c) the approximate diameter of the turbine runner; (d) the elevation at which the turbine should be set to be safe against cavitation if it is installed at sea level.*

BG

(*a*) From Eq. 16.8: HP to shaft = $\gamma Q h \eta/550 = 62.4Q72(0.93)/550 = 3000$ hp ; $Q = 395$ cfs ◄

(*b*) Eq. 16.21: $N_s = n_e\sqrt{\text{bhp}}/h^{5/4} = 300\sqrt{3000}/72^{5/4} = 78.3$ ◄

(*c*) Fig. 16.14 for N_s = 78.3: $\phi_e \approx 0.74$; Eq. 16.20: $D \approx 153.3(0.74)\sqrt{72}/300 = 3.2$ ft ◄

(*d*) Fig. 16.17 for N_s = 78.3: $\sigma_c \approx 0.38$; Table A.3 at sea level: p_{atm} = 14.70 psia

Assume T = 80°F. Table A.1 for water at 80°F: p_v/γ = 1.17 ft abs

Eq. 16.24: $z_B = p_{atm}/\gamma - p_v/\gamma - \sigma_c h \approx 14.70(144/62.4) - 1.17 - 0.38(72) = 5.4$ ft ◄

Best to limit draft head to zero, though turbine should operate satisfactorily with a draft head of 5 ft.

16.38 *A propeller turbine operates at a maximum efficiency of 92% under a head of 30 ft at 450 rpm and develops a shaft power of 620 hp. Find (a) the flow rate through the turbine, (b) the specific speed of the turbine, (c) the approximate diameter of the turbine runner, and (d) how far above trailwater elevation the turbine can be set and still be safe from cavitation. Assume the turbine is at an elevation of 100 ft.*

BG

(*a*) From Eq. 16.8: HP to shaft = $\gamma Q h \eta/550 = 62.4Q30(0.92)/550 = 620$ hp ; $Q = 198.0$ cfs ◄

(*b*) Eq. 16.21: $N_s = n_e\sqrt{\text{bhp}}/h^{5/4} = 450\sqrt{620}/30^{5/4} = 159.6$ ◄

(*c*) Fig. 16.14 for N_s = 160: $\phi_e \approx 1.94$; Eq. 16.20: $D \approx 153.3(1.94)\sqrt{30}/450 = 3.62$ ft = 43.4 in ◄

(*d*) Fig. 16.17 for N_s = 160: $\sigma_c \approx 0.81$; Table A.3 at 100 ft by interpolation: p_{atm} = 14.65 psia

Assume T = 80°F. Table A.1 for water at 80°F: p_v/γ = 1.17 ft abs

Eq. 16.24: $z_B = p_{atm}/\gamma - p_v/\gamma - \sigma_c h \approx 14.65(144/62.4) - 1.17 - 0.81(30) = 8.3$ ft

The turbine should be set no more than 8.3 ft above tailwater. ◄

16.39 *Repeat part (d) of Prob. 16.38 for the case where the turbine is installed at elevation 5000 ft.*
A propeller turbine at maximum η has n = 450 rpm, P = 620 bhp, h = 30 ft. (d) Find z_B.

BG

Eq. 16.21: $N_s = n_e\sqrt{\text{bhp}}/h^{5/4} = 450\sqrt{620}/30^{5/4} = 159.6$

(*d*) Fig. 16.17 for N_s = 160: $\sigma_c \approx 0.81$; Table A.3 at 5000 ft: p_{atm} = 12.23 psia

Assume temperature = 50°F. Table A.1 for water at 50°F: p_v/γ = 0.41 ft abs

Eq. 16.24: $z_B = p_{atm}/\gamma - p_v/\gamma - \sigma_c h \approx 12.23(144/62.4) - 0.41 - 0.81(30) = 3.5$ ft

The turbine should be set no more than 3.5 ft above tailwater. ◄

16.40 *A turbine whose specific speed is 80 is to operate under a head of 50 ft at an elevation where the atmospheric pressure is 12.8 psia. The water temperature is 50°F. (a) If this turbine is set 6 ft above tailwater, will it be safe from cavitation? (b) What is the highest permissible elevation of this turbine with respect to tailwater when operating under a head of 60 ft?*

BG

(*a*) Fig. 16.17 for N_s = 80: $\sigma_c \approx 0.395$; Table A.1 for water at 50°F: $p_v/\gamma = 0.41$ ft abs

Eq. 16.24: $z_B = p_{atm}/\gamma - p_v/\gamma - \sigma_c h \approx 12.8(144/62.4) - 0.41 - 0.395(50) = 9.4$ ft

The turbine can be set as high as 9.4 ft above tailwater.

So at 6 ft above tailwater the turbine will be safe from cavitation. ◄

(*b*) For h = 60 ft: $z_B \approx 29.54 - 0.41 - 0.395(60) = 5.4$ ft

The turbine can be set as much as 5.4 ft above tailwater. ◄

Sec. 16.13: Selection of Turbines -- *Exercises*

16.13.1 *(a) If Q = 40 cfs and h = 70 ft, would it be possible to use a single impulse wheel with a specific speed less than 7.0 at a site where 60-Hz current is to be generated? Assume a turbine efficiency of 90% and that the generator is to have no more than 96 poles. (b) If a single impulse wheel can be used under these conditions, calculate the approximate diameter of the turbine runner for the range of possible speeds of operation.*

BG

(*a*) bhp = (0.90)62.4(40)70/550 = 286 hp

For a (max) 96-pole generator, min n = 7200/96 = 75 rpm,

min $N_s = n\sqrt{\text{bhp}}/h^{5/4} = 75\sqrt{286}/70^{5/4} = 6.25$

Fig. 16.16 for N_s = 6.25 (with 80°F water at sea level): for no cavitation, max $h \approx$ **950** ft.

Since $N_s = 6.25 < 7$ and $h = 70$ ft < 950 ft, Yes, a single impulse wheel with $N_s < 7$ would be possible. ◄

(*b*) Fig. 16.14 for N_s = 6.25: $\phi_e \approx 0.40$; Eq. 16.20: $D \approx 153.3(0.40)\sqrt{70}/75 = 6.8$ ft

Consider alternative = min number of poles (max N_s).

If $N_s = 7.0 = n\sqrt{\text{bhp}}/h^{5/4} = n\sqrt{286}/(70)^{5/4}$, n = 83.8 rpm and N = 7200/83.8 = 85.9 poles

So use 86 poles. Then n = 7200/86 = 83.7 rpm and $N_s = 83.7\sqrt{286}/70^{5/4} = 6.99$

Fig. 16.14 for N_s = 7: $\phi_e \approx 0.395$; $D \approx 153.3(0.395)\sqrt{70}/83.7 = 6.1$ ft

Resulting ranges:

N_s	N poles	n rpm	D ft
6.25	96 (max)	75 (min)	6.8
6.99 (max)	86	83.7 (max)	6.1
▲		▲	▲

16.13.2 *(a) If Q = 40 cfs and h = 70 ft, would it be possible to use a single Francis turbine with a specific speed of at least 20 at a site where 60-Hz current is to be generated? Assume a turbine efficiency of 90% and assume the generator is to have no less than 12 poles. (b) If a single Francis turbine can satisfy these conditions, calculate the approximate diameter of the turbine runner for the range of possible speeds of operation.*

BG

(*a*) From Eq. 16.4: bhp = (0.90)62.4(40)70/550 = 286 hp

For a (min) 12-pole generator, max n = 7200/12 = 600 rpm,

max $N_s = n\sqrt{\text{bhp}}/h^{5/4} = 600\sqrt{286}/70^{5/4} = 50.1$

Fig. 16.16 for N_s = 50.1 (with 80°F water at sea level): for no cavitation, max $h \approx$ 130 ft (with draft head $z_B \le 15$ ft).

Since $N_s = 50.1 > 20$ and $N_s = 50.1 < 100$ = max for a Francis turbine, and h = 70 ft < 130 ft, yes a single Francis turbine with $N_s > 20$ would be possible. ◀

(*b*) Fig. 16.14 for N_s = 50.1: $\phi_e \approx 0.72$; Eq. 16.20: $D \approx 153.3(0.72)\sqrt{70}/600$ = 1.54 ft

Consider alternative = max number of poles (min N_s)

If $N_s = 20 = n\sqrt{286}/(70)^{5/4}$, n = 239 rpm and N = 7200/239 = 30.1 poles

So use 30 poles; then n = 7200/30 = 240 rpm and $N_s = 240\sqrt{286}/(70)^{5/4} = 20.05$

Fig. 16.14 for N_s = 20: $\phi_e \approx 0.70$; $D \approx 153.3(0.70)\sqrt{70}/240$ = 3.7 ft

Resulting ranges:

N_s	N poles	n rpm	D ft
50.1	12 (min)	600 (max)	1.54
20.05 (min)	30	240 (min)	3.7
▲		▲	▲

16.13.3 *Specify the specific speed and the approximate runner diameter at a site where Q = 2350 cfs and h = 480 ft. Use a 40-pole generator for 60-cycle current. Assume a turbine efficiency of 90%. These are the specifications for some of the turbines at Hoover Dam.*

BG

From Eq. 16.14: bhp = (0.90)62.4(2350)400/550 = 115,200

n = 7200/40 = 180 rpm ; Eq. 16.21: $N_s = n\sqrt{\text{bhp}}/h^{5/4} = 180\sqrt{115{,}200}/480^{5/4} = 27.2$ ◀

Fig. 16.14 for N_s = 27.2: $\phi_e \approx 0.70$; Eq. 16.20: $D \approx 153.3(0.70)\sqrt{480}/180$ = 13.1 ft ◀

16.13.4 *(a) What is the maximum head for which a propeller-type turbine can be used if the flowrate is 40 cfs? Assume a turbine efficiency of 90% with specific speeds for propeller-type turbines varying from 100 to 200. Also, assume a 12-pole generator with 50-Hz current. (b) Find the maximum head for which a 20-pole generator could have been used with the same turbine efficiency and 50-Hz current. Neglect the possibility of cavitation.*

BG

From Eq. 16.14: bhp = (0.90)62.4(40)h/550 = 4.08h

(*a*) n = 6000/12 = 500 rpm ; $N_s = n\sqrt{\text{bhp}}/h^{5/4} = 500\sqrt{4.08h}/h^{5/4} = 1010/h^{3/4}$, so $h^{3/4} = 1010/N_s$

h will be largest when N_s is smallest, so use N_s = 100 ; $h = (1010/100)^{4/3}$ = 21.8 ft ◀

(*b*) n = 6000/20 = 300 rpm ; $N_s = n\sqrt{\text{bhp}}/h^{5/4} = 300\sqrt{4.08h}/h^{5/4} = 606/h^{3/4}$, so $h^{3/4} = 606/N_s$

For min N_s = 100, max $h = (606/100)^{4/3}$ = 11.06 ft ◀

Sec. 16.13: Selection of Turbines -- *Problems*

16.41 *(a) A turbine is to be installed at a point where the available head is 175 ft and the available flow will average 1000 cfs. What type of turbine would you recommend? Specify the operating speed and the number of generator poles for 60-cycle electricity if a turbine with the highest tolerable specific speed to safeguard against cavitation is selected. Assume a draft head of 10 ft and 90% turbine efficiency. Approximately what size of turbine runner is required? (b) For the same conditions, select a set of two identical turbines to be operated in parallel. Specify the speed and size of the units.*

BG

(*a*) Assume sea-level conditions at 59°F, for single turbine

Table A.3 at sea level: p_{atm} = 14.70 psia

Table A.1 at 59°F by interpolation: γ = 62.37 pcf; p_v/γ = 0.57 ft

Eq. 16.22: σ_c = [14.70(144/62.37) − 0.57 − 10]/175 = 0.1335

From Fig. 16.17 for σ_c = 0.1335: $N_s \approx 41$, machine = Francis turbine ◀

From Eq. 16.8: $P = \gamma Q h \eta/550$ = 62.37(1000)175(0.90)/550 = 17,860 hp ◀

Eq. 16.21: $N_s = n_e\sqrt{\text{bhp}}/h^{5/4}$ so $41 \approx n_e\sqrt{17{,}860}/175^{5/4}$, $n_e \approx 195$ rpm

$N = 7200/n_e \approx 7200/195.3 = 36.9$ poles. So use N = 36 poles ◀

Then N_e = 7200/36 = 200 rpm ◀ and $N_s = 200\sqrt{17{,}860}/175^{5/4} = 42.0$

Fig. 16.14 for N_s = 42: $\phi_e \approx 0.71$; Eq. 16.20: $D \approx 153.3(0.71)\sqrt{175}/200 = 7.2$ ft ◀

(*b*) Use two turbines in parallel

Q = 1000/2 = 500 cfs per turbine, h = 175 ft, N_s = 41 from (*a*) above.

For each turbine, P = 17,860/2 = 8930 hp

Eq. 16.21: $N_s = n_e\sqrt{\text{bhp}}/h^{5/4}$; $41 = n_e\sqrt{8930}/175^{5/4}$ so n_e = 276 rpm and N = 7200/276 = 26.1

Choose 26 poles, so n_e = 7200/26 = 277 rpm ◀ and $N_s = 277\sqrt{8930}/175^{5/4} = 41.1$

Fig. 16.14 for N_s = 41.1: $\phi_e \approx 0.71$; Eq. 16.20: $D \approx 153.3(0.71)\sqrt{175}/277 = 5.2$ ft ◀

16.42 *What is the least number of identical turbines that can be used at a powerhouse where the available head is 1200 ft and Q = 1650 cfs? Assume turbine efficiency is 90% and speed of operation is 138.5 rpm. Specify the size and specific speed of the units.*

BG

From Eq. 16.8: Shaft power = bhp = $\gamma Q h \eta/550$ = 62.4(1650)1200(0.90)/550 = 202,000

Fig. 16.16 for h = 1200 ft: To be safe from cavitation the specific speed must be less than 5.4.

Let x = no. of identical units: Eq. 16.21: $N_s = n_e\sqrt{\text{bhp}}/h^{5/4} = 138.5\sqrt{202{,}000/x}/1200^{5/4}$

If x = 1, N_s = 8.81 (too large, cavitation). If x = 2, N_s = 6.23 (too large, cavitation).

If x = 3, N_s = 5.09, OK, no cavitation (and maximum efficiency of impulse wheel is at $N_s \approx 5$, see Sec. 16.9).

Thus, use three identical impulse turbines with N_s = 5.1 ◀

Fig. 16.14 for N_s = 5.09: $\phi_e \approx 0.43$; Eq. 16.20: $D \approx 153.3(0.43)\sqrt{1200}/138.5 = 16.5$ ft ◀

16.43 *A six-jet impulse turbine operating at 300 rpm develops 60,000 hp under a net head of 1060 ft. The runner has a diameter of 6.0 ft. How large a homologous runner would be needed for a single-jet machine operating under the same head and developing the same horsepower?*

BG

Eq. 16.21: $N_s = n_e\sqrt{\text{bhp}}/h^{5/4} = 300\sqrt{(60{,}000/6)}/(1060)^{5/4} = 4.96$

Single nozzle $N_s = 4.96 = n_e\sqrt{60{,}000}/1060^{5/4}$; $n = 122.5$ rpm

Assume 60-cycle electricity. $N = 7200/122.4 = 58.8$ poles. Use 58 poles, so $n = 7200/58 = 124.1$ rpm

Eq. 16.16: $n \propto h^{1/2}/D$, $h = 1060$ ft = const, hence $D \propto 1/n$

$D_1 = D_6(n_6/n_1) = 6.0(300/124.1) = 14.50$ ft ◀

Note: Fig. 16.16 (for $N_s = 4.96$, $h = 1060$ ft) indicates no problem with cavitation.

16.44 *An impulse turbine ($N_s \approx$ *4) develops 100,000 hp under a head of 2000 ft. (a) For 60-cycle electricity calculate the turbine speed (rpm), wheel diameter (ft), and number of poles in the generator. (b) Solve the problem for a six-nozzle unit using the same N_s, bhp, and head. In both instances assume *$\phi_e = 0.45$. *Neglect cavitation. [*Note these corrections to the textbook.]*

BG

(*a*) Eq. 16.21: $N_s = 4 = n_e\sqrt{100{,}000}/2000^{5/4}$; $n_e = 169.2$ rpm, $N = 7200/169.2 = 42.6$ poles

Use 42 poles ◀ $n = 7200/42 = 171.4$ rpm ◀ $N_s = 171.4\sqrt{100{,}000}/2000^{5/4} = 4.05$

Eq. 16.20: $D = 153.3(0.45)\sqrt{2000}/171.4 = 18.00$ ft ◀

(*b*) Eq. 16.21: $N_s = 4 = n_e\sqrt{(100{,}000/6)}/2000^{5/4}$; $n = 414.4$ rpm

$N = 7200/414.4 = 17.37$, use 18 poles ◀ $n = 7200/18 = 400$ rpm ◀

$N_s = 400\sqrt{(100{,}000/6)}/2000^{5/4} = 3.86$; $D = 153.3(0.45)\sqrt{2000}/400 = 7.71$ ft ◀

Note: Fig. 16.16 indicates (for $N_s = 4$, $h = 2000$ ft) no problem with cavitation.

16.45 *A multinozzle impulse turbine is to be designed to develop 60,000 hp at 300 rpm under a head of 1200 ft. How many nozzles should this turbine have? Specify the approximate wheel diameter for this design. How large would a single-jet machine have to be to satisfy these requirements?*

BG

For multi-nozzle turbine with x nozzles, Eq. 16.21: $N_s = n_e\sqrt{\text{bhp}/x}/h^{5/4} = 300\sqrt{(60{,}000/x)}/1200^{5/4}$

$\therefore N_s = 10.4, 7.35, 6.0, 5.2, 4.65, 4.25$, for x values of 1, 2, 3, 4, 5, 6.

Fig. 16.16 indicates that for a head of 1200 ft, N_s should be less than about 5.4.

Footnote page 742 and Fig. 16.14 indicate maximum efficiency occurs at $N_s \approx 5$.

Hence use 4 or 5 nozzles. ◀ (5 will be safer against cavitation.)

For 4 or 5 nozzles, $N_s \approx 5$; Fig. 16.14 for $N_s = 5$: $\phi_e \approx 0.43$

Eq. 16.20: $D \approx 153.3(0.43)\sqrt{1200}/300 = 7.6$ ft ◀

For single nozzle, $N_s = 10.4$; per footnote p. 742 the N_s of impulse wheels ranges from 2 to 8, so a single-nozzle machine will not work. (Also, if it would work it would experience cavitation, as noted above.) ◀

16.46 *It is desired to develop 15,000 bhp under a head of 1000 ft. Make any necessary assumptions and estimate the diameter of the wheel required and the rotative speed.*

BG

Fig. 16.16 indicates that an impulse wheel with $N_s < 6$ or a Francis turbine with $N_s < 23$ (zero draft head) would be safe against cavitation. Hence, try (*i*) an impulse wheel with $N_s \approx 5$ (see footnote page 742) or (*ii*) a Francis turbine with $N_s = 20$.

Impulse wheel: Eq. 16.21: $N_s = 5 = n_e\sqrt{\text{bhp}}/h^{5/4} = n_e\sqrt{15{,}000}/1000^{5/4}$; thus $n_e = 229.6$ rpm

Assume 60-cycle electricity: $N = 7200/229.6 = 31.4$; Use 32 poles, $n = 7200/32 = 225$ rpm ◄

Fig. 16.14 for $N_s = 5$: $\phi_e \approx 0.43$. Eq. 16.20: $D \approx 153.3(0.43)\sqrt{1000}/225 = 9.3$ ft ◄

Francis turbine: $N_s = 20 = n_e\sqrt{15{,}000}/1000^{5/4}$; $n_e = 918$ rpm

$N = 7200/918 = 7.8$ (use 8 poles). But most generators have at least 12 poles (see page 721).

If an 8-poles generator is not available, cannot use a Francis turbine.

Another alternative: Use a double-overhung impulse wheel (Fig. 16.3), 7500 hp/wheel.

$N_s = 5 = n_e\sqrt{(15{,}000/2)}/h^{5/4}$; $n_e = 324.7$ rpm

$N = 7200/324.7 = 22.2$ poles, use 22 poles. Then $n = 7200/22 = 327.3$ rpm ◄

Eq. 16.20: $D = 153.3\phi_e\sqrt{h}/n \approx 153.3(0.43)\sqrt{1000}/327.3 = 6.7$ ft ◄

Advantage of double-overhung installation: It equalizes the thrust on the bearings.

16.47 *A single hydraulic turbine is to be selected for a power site with a net head of 100 ft. The turbine is to produce 25,000 hp at maximum efficiency. What speed (rpm) and diameter should this turbine have if (a) a Francis turbine is selected; (b) a propeller turbine is selected? What are the highest "settings" (above or below tailwater) that should be recommended for each of these machines for them to run cavitation-free at their points of maximum efficiency?*

BG

Fig. 16.16 indicates two possibilities:

(*a*) With $h = 100$ ft a Francis turbine can be used with $N_s \leq 70$. If $N_s = 70$, use a draft head of zero, if $N_s = 60$, use a draft head of up to 10 ft, i.e., can set the turbine as much as 10 ft above tailwater.

(*b*) With $h = 100$ ft a propeller turbine can be used with $N_s = 100$ ft and a draft head of -8 ft (i.e. with runner 8 ft or more below water elevation at discharge).

Francis turbine with $N_s = 70$:

Eq. 16.21: $N_s = 70 = n_e\sqrt{\text{bhp}}/h^{5/4} = n_e\sqrt{25{,}000}/100^{5/4}$; $n_e = 140$ rpm

Assume 60-Hz electricity, $N = 7200/140 = 51.4$, use 52 poles. $n_e = 7200/52 = 138.5$ rpm ◄

$N_s = 138.5\sqrt{25{,}000}/100^{5/4} = 69.2$

Fig. 16.14 when $N_s = 69.2$: $\phi_e \approx 0.74$; Eq. 16.20: $D \approx 153.3(0.74)\sqrt{100}/138.5 = 8.2$ ft ◄

Recommend zero draft head for $N_s = 70$.

Francis turbine with $N_s = 60$:

Following the same procedure, $n_e = 120$ rpm, $N = 60$ poles, $\phi_e \approx 0.72$, and $D \approx 9.2$ ft. This turbine can be set as much as 10 ft above water level at discharge.

/cont...

Propeller turbine with $N_s = 100$:

Eq. 16.21: $N_s = 100 = n_e\sqrt{25{,}000}/100^{5/4}$; $n_e = 200$ rpm

$N = 7200/200 = 36$, use 36 poles, $n_e = 200$ rpm

Fig. 16.14 for $N_s = 100$: $\phi_e \approx 1.27$; Eq. 16.20: $D \approx 153.3(1.27)\sqrt{100}/200 = 9.7$ ft

This turbine needs a draft head of −8 ft. Hence set the runner at least 8 ft below tailwater.

Resulting selections:

	N_s	N poles	max z_B, ft	n, rpm	D, ft
Francis turbine	69.2	52	0	138.5	8.2
Francis turbine	60	60	10	120	9.2
Propeller turbine	100	36	−8	200	9.7

▲ ▲ ▲

16.48 *For 50-cycle electricity how many poles would you recommend for a generator that is connected to a turbine operating under a design head of 3000 ft with a flow of 80 cfs? Assume turbine efficiencies as given in Fig. 16.14 and be sure the turbine is free of cavitation.*

BG

Fig. 16.16 for $h = 3000$ ft: use impulse wheel with $N_s < 2.9$ to avoid cavitation.

Fig. 16.14 for $N_s = 2.9$: $\eta_{max} \approx 0.87$; From Eq. 16.8: bhp $\approx 62.4(80)3000(0.87)/550 = 23{,}700$ hp

Eq. 16.21: $N_s = 2.9 = n_e\sqrt{23{,}700}/3000^{5/4}$; $n_e = 418$ rpm

Sec. 16.1, for 50-cycle: $N = 6000/418 = 14.34$

Try 14 poles: $n = 6000/14 = 429$ rpm, $N_s = 2.9(429/418) = 2.97$ (not OK, > 2.9)

Try 16 poles: $n = 6000/16 = 375$ rpm, $N_s = 2.9(375/418) = 2.60$ (OK, < 2.9). Use 16 poles. ◄

16.49 *Specify the type, speed, and size of a single turbine to be installed at a site with an effective head of 48 m, a maximum draft head of 2 m, and a flow rate of 5 m³/s. How would your recommendation change if the available flow was 50 m³/s?*

B

(*a*) $h = 48$ m $= 157.5$ ft, $z_B = 2$ m $= 6.56$ ft, $Q = 5$ m³/s $= 176.6$ cfs

Fig. 16.16 for $h = 157.5$ ft, $z_B = 6.56$ ft: To be safe from cavitation, select Francis turbine with $N_s \le 53$

Fig. 16.14 for $N_s \approx 53$: $\eta = 0.950$

From Eq. 16.8: bhp $= \gamma Q h \eta/550 = 62.4(176.6)157.5(0.950)/550 = 3000$

Eq. 16.21: $N_s = n_e\sqrt{3000}/157.5^{5/4} \le 53$; $n_e \le 540$ rpm

Assume 60-cycle electricity, $N \ge 7200/540 = 13.33$, Use 14 poles

$n = 7200/14 = 514$ rpm ; $N_s = 514\sqrt{3000}/157.5^{5/4} = 50.5$

Fig. 16.14 for $N_s = 50.5$: $\phi_e \approx 0.72$; Eq. 16.20: $D \approx 153.3(0.72)\sqrt{157.5}/514 = 2.7$ ft

To be safer from cavitation (per Fig. 16.16), use a lower value of N_s such as $N_s = 40$, for example.

Fig. 16.14 for $N_s = 40$: $\eta = 0.948$. ∴ bhp $= 2990$ hp

$N_s = n_e\sqrt{2990}/157.5^{5/4} = 40$; $n_e = 408$ rpm , $N = 7200/408 = 17.65$

Use 18 poles: $n = 7200/18 = 400$ rpm , $N_s = 40(400/408) = 39.2$

Fig. 16.14 for $N_s \approx 40$: $\phi_e \approx 0.71$; Eq. 16.20: $D \approx 153.3(0.71)\sqrt{157.5}/400 = 3.4$ ft

/cont...

Possible solutions include:

N poles	n rpm	$(N_s)_{BG}$	D ft	D m	Turbine
14 (min)	514	50.5 (max)	2.7	0.82	Francis
16	450	43.0	3.2	0.98	Francis
18	400	39.2	3.4	1.04	Francis
20	360	34.4	4.0	1.20	Francis
24	300	28.6	7.2	2.19	Francis
	▲		▲	▲	▲

(*b*) Case where $Q = 50\ \mathrm{m^3/s} = 1766$ cfs

bhp = 30,000; $N_s = n_e\sqrt{30{,}000}/157.5^{5/4} \le 53$; $n_e \le 170.8$ rpm

With 60-cycle electricity, $N \ge 7200/170.8 = 42.2$, use 44 poles

$n = 7200/44 = 163.6$ rpm ; $N_s = 163.6\sqrt{30{,}000}/157.5^{5/4} = 50.8$

$D \approx 153.3(0.72)\sqrt{157.5}/163.6 = 8.5$ ft

Some possible solutions are:

N poles	n rpm	$(N_s)_{BG}$	D ft	D m	Turbine	
44 (min)	136.6	50.8 (max)	8.5	2.58	Francis	Close to cavitation
46	156.5	47.1	9.2	2.80	Francis	Best
52	138.5	41.7	10.4	3.17	Francis	Perhaps best
64	112.5	33.9	12.8	3.90	Francis	D is too large
	▲		▲	▲		

16.50 *It is desired to develop 300,000 hp under a head of 49 ft and to operate at 60 rpm. (a) If turbines with a specific speed of approximately 150 are to be used, how many units would be required? (b) If Francis turbines with a specific speed of 80 were to be used, how many units would be required?*

BG

Assume 60-cycle electricity

(*a*) Fig. 16.14 for $N_s \approx 150$: machine is a propeller turbine.

Eq. 16.21: $N_s = n_e\sqrt{\mathrm{bhp}}/h^{5/4} = 60\sqrt{300{,}000}/49^{5/4} = 253 > 150$

Hence, it is impossible to satisfy the given conditions with a single propeller turbine.

Try two identical turbines: $N_s = 60\sqrt{(300{,}000/2)}/49^{5/4} = 179.2 > 150$

Try three: $N_s = 60\sqrt{(300{,}000/3)}/49^{5/4} = 146.4 \approx 150$

So use three identical propeller turbines with $N_s = 146.4$. ◄

(*b*) Fig. 16.14 for $N_s \approx 80$: machine is a Francis turbine.

Eq. 16.21: $N_s = n_e\sqrt{\mathrm{bhp}}/h^{5/4} = 60\sqrt{300{,}000}/49^{5/4} = 253.5 > 80$

Try ten identical turbines: $N_s = 60\sqrt{30{,}000}/49^{5/4} = 80.2$

So use ten identical Francis turbines with $N_s = 80.2$ ◄

16.51 *Select two, four, and six identical turbines for an installation where $h = 400$ ft and total $Q = 300$ cfs. Develop 60-cycle electricity using either 36- or 72-pole generators. Be sure your selection is free of cavitation. Assume the turbine efficiency is 90%.*

BG

From Eq. 16.8: Total bhp $= \gamma Qh\eta/550 = 62.4(300)400(0.90)/550 = 12{,}250$ hp

2 units with 36 poles ($n = 7200/36 = 200$ rpm):

Eq. 16.21: $N_s = n_e\sqrt{\text{bhp}}/h^{5/4} = 200\sqrt{(12{,}250/2)}/400^{5/4} = 8.75$

Footnote on p. 748: N_s for impulse wheels are normally ≤ 8, $\therefore$ 8.75 is too high.

Fig. 16.14 for $N_s = 8.75$: The efficiency would be low.

Fig. 16.16 for $N_s = 8.75$: max $h = 300$ ft. $\therefore$ these machines with $h = 400$ ft will not be safe against cavitation.

4 units with 36 poles ($n = 200$ rpm):

$N_s = 200\sqrt{(12{,}250/4)}/400^{5/4} = 6.19$, OK; safe against cavitation (Fig. 16.16).

Fig. 16.14 for $N_s = 6.19$: $\phi_e \approx 0.42$; Eq. 16.20: $D \approx 153.3(0.42)\sqrt{400}/200 = 6.4$ ft ◄

6 units with 36 poles ($n = 200$ rpm):

$N_s = 200\sqrt{2042}/400^{5/4} = 5.05$, OK, cavitation safe; $\phi_e \approx 0.43$; $D \approx 6.6$ ft.

Note: The units for the 4-unit and 6-unit installations are of similar size, but their specific speeds are different.

2 units with 72 poles ($n = 7200/72 = 100$ rpm):

$N_s = 8.75/2 = 4.38$, OK, safe against cavitation (Fig. 16.16).

Fig. 16.14 for $N_s = 4.38$: $\phi_e \approx 0.43$; Eq. 16.20: $D \approx 153.3(0.43)\sqrt{400}/100 = 13.2$ ft

4 units with 72 poles ($n = 100$ rpm):

$N_s = 6.19/2 = 3.09$ (cavitation safe); $\phi_e \approx 0.43$; $D \approx 13.2$ ft

6 units with 72 poles ($n = 100$ rpm):

$N_s = 5.05/2 = 2.53$ (cavitation safe), on the low side, but OK; $\phi_e \approx 0.44$; $D \approx 13.5$ ft

Selections:

N poles	Units	N_s	Cavitation	*D* ft	Machine
36	2	8.75	Not safe	–	–
36	4	6.19	Safe	6.4	Impulse
36	6	5.05	Safe	6.6	Impulse
72	2	4.38	Safe	13.2	Impulse
72	4	3.09	Safe	13.2	Impulse
72	6	2.53	Safe	13.5	Impulse (low η?)

| | | ▲ | ▲ | ▲ | ▲ |

/more...

16.52 *A turbine is to be installed at a point where the net available head is 35 m and the available flow will average 23 m^3/s. What type of turbine would you recommend? Specify the operating speed and number of generator poles for 60-cycle electricity if a turbine with the highest tolerable specific speed to safeguard against cavitation is selected. Assume a draft head of 3 m and 90% turbine efficiency. Approximately what size of turbine runner is required?*

B

Assume sea-level conditions at 15°C. Table A.3 at sea level: p_{atm} = 101.3 kPa abs = 101.3 kN/m^2 abs

Table A.1 for water at 15°C: $\gamma = 9.80\ kN/m^3$, $p_v/\gamma = 0.17$ m abs

Eq. 16.24: $z_B = p_{atm}/\gamma - p_v/\gamma - \sigma_c h$; $3 = (101.3/9.80) - 0.17 - \sigma_c(35)$; $\sigma_c = 0.205$

Fig. 16.17 for $\sigma_c = 0.205$: $(N_s)_{BG} \le 54$ which is for a Francis turbine ◄

(Note that N_s is closely confirmed by Fig. 16.16.)

From Eq. 16.8: $P = \gamma Qh\eta = 9.80(23)35(0.90) = 7100$ kW

Eq. 16.21: $(N_s)_{BG} = n_e\sqrt{\text{bhp}}/h^{5/4}$ so $54 \ge n_e\sqrt{7100(1.341)}/(35 \times 3.28)^{5/4}$; $n_e \le 208$ rpm

$N \ge 7200/208 = 34.6$, ∴ choose N = 36 poles ◄ so $n_e = 7200/36 = 200$ rpm ◄

$(N_s)_{BG} = 54(200/208) = 51.9$; Fig. 16.14 for $N_s = 51.9$: $\phi_e \approx 0.72$

Eq. 16.20: $D \approx 153.3(0.72)\sqrt{35(3.28)}/200 = 5.9$ ft = 1.80 m ◄

16.53 *A turbine is to be installed where the net available head is 185 ft, and the available flow will average 900 cfs. What type of turbine would you recommend? Specify the operating speed and number of generator poles for 60-cycle electricity if a turbine with the highest tolerable specific speed that will safeguard against cavitation is selected. Assume the turbine is set 5 ft above tailwater. Assume the turbine efficiency is 90%. Approximately what size of runner is required?*

BG

Assume sea-level conditions at 59°F. Table A.3 at sea level: p_{atm} = 14.70 psia

Table A.1 for water at 59°F by interpolation: $\gamma = 62.4$ pcf, $p_v/\gamma = 0.57$ ft abs

Eq. 16.24: $5 = 14.70(144/62.4) - 0.57 - \sigma_c(185)$; $\sigma_c = 0.153$

Fig. 16.17 for $\sigma_c = 0.153$: $N_s \le 45$ which is for a Francis turbine ◄

(Note: Fig. 16.16 indicates that a 10 ft draft head is tolerable for a turbine with $N_s = 45$.)

$P = \gamma Qh\eta/550 = 62.4(900)185(0.90)/550 = 17{,}000$ hp

Eq. 16.21: $N_s = n_e\sqrt{17{,}000}/185^{5/4} \le 45$ from which $n_e \le 235$ rpm

$N = 7200/n_e \ge 7200/235 = 30.6$, ∴ choose N = 32 poles ◄ so $n_e = 7200/32 = 225$ rpm ◄

$N_s = 54(225/235) = 51.6$; Fig. 16.14 for $N_s = 51.6$: $\phi_e \approx 0.72$

Eq. 16.20: $D \approx 153.3(0.72)\sqrt{185}/225 = 6.7$ ft = 80 in ◄

Appendix A
SAMPLE SOLUTIONS BY MATHCAD

CONTENTS

General Comments on Mathcad

Two particular features of Mathcad are of special value and interest for application to the solution of problems in fluid mechanics. The first of these is the ability to solve an implicit equation (such as Eq. 8.40), even if it is nonlinear and/or transcendental. The second is the ability to solve n simultaneous such equations in n unknowns. Note, however, that these capabilities were **not** included in the *student* edition of Mathcad version 6 (1995), as noted in Appendix B.3 of the textbook, but they **are** included in the version 7 (1997) student edition.

The advantage of these equation solvers is that they avoid the need to use manual trial and error procedures (illustrated in Sample Prob. 3.5 of the textbook) which can be quite tedious; the solvers in fact have the trial and error programmed into them. Most problems involving trial and error[1] can be solved directly using Mathcad; in the Problem Selection Guides the ratings of the length and difficulty of such problems are for the *manual* solution methods.

In the textbook *Fluid Mechanics with Engineering Applications*, ninth edition, equation solvers are discussed in Secs. B.1 and B.3, and are mentioned at various locations throughout, with notable example applications in Secs. 8.14, 8.24, and 10.3. As indicated in these example applications, Mathcad solution procedures are quite standard. The additional examples which follow are provided to illustrate the comparative simplicity and clarity of Mathcad applications, and to help people start using Mathcad on fluid mechanics problems. Startup tips are provided on the next two pages. Copies of the examples and the tips could be made available to students at a library or distributed as course handouts.

---

[1] This does *not* include those requiring a value to be read from a figure. Trial and error problems are identified in the Problem Selection Guides.

Tips on Mathcad 7, relevant to the following example applications to fluid mechanics {Mathcad 6 equivalents in braces, when different}

1. ***Inputting, editing, executing:*** When Mathcad is in "automatic mode" (the default; see "auto" in the message line) you may have to wait for computations being made for each change as you edit. To disable this, i.e., switch to manual mode, click the light bulb on the toolbar, or remove the checkmark (click) beside **Automatic Mode** in the **Math** menu. Then you can update results, on the presently visible screen only (!), by pressing [F9], or by clicking on the equals button, or by choosing **Calculate** from the **Math** menu. But to reprocess the *entire* worksheet, visible *and* not visible, choose **Calculate Worksheet** from the **Math** menu or return to "auto" mode. Be sure to "Calculate Worksheet" before printing it. Individual equations can be made inactive/active by choosing **Properties {Toggle Equation}** from the **Format {Math}** menu, then clicking on the "Equations" tab {this and the following are not needed in Ver. 6}, and under "Calculation Options" putting a check next to "Disable Equation"; a small square (▫) following the equation indicates that it is inactive/disabled.
2. ***Variables:*** Variable names are case sensitive in Mathcad. See item 4 regarding subscripts.
3. ***Angles:*** Mathcad expects angles to be input as radians. Thus, when $\theta = 30°$, write $\cos(\pi/6)$. Mathcad will convert degrees to radians when input as follows: cos(30*deg) (* displays as ·).
4. ***Subscripts:*** Mathcad uses three kinds of subscripts. In text, subscripts may be specified by choosing **Text {Change Font}** from the **Format {Text}** menu, and using the "Effects" option. Variables may be given *literal* subscripts or *array* subscripts. Literal subscripts, such as in V_{max} or D_1, do not refer to an array or vector element and are instead just part of the variable name; create them by typing a period just before the subscript. These variables might just as easily be entered as Vmax or D1, and in fact this has been done in the following examples and in the textbook. Array subscripts, which vary numerically to indicate positions in arrays or vectors, are created by typing a left bracket ([ ) instead of a period before the subscript.
5. ***Ranges:*** For i := 0 .. 2 type *i*:0;2
6. ***Mathematical operations:*** Addition and subtraction are as usual. Multiplication is typed * and displayed as · Division is typed / and displayed as a vertical fraction. Powers are typed with a ^ and displayed as raised exponents.
7. ***Equals signs:*** Mathcad uses three different kinds. These are:
   - Colon equals or "definition symbol" (typed as a colon and displayed on screen as :=), used to assign a value or expression to a variable or function, eg., **u := 3.2** or **f(x) := x^3 + 10** (these are called "assignment statements");
   - Bold (or Boolean) equals (typed as [Ctrl]= and displayed on screen as), used to specify that two sides of an equation must have the same numerical value; and
   - "Regular" equals (typed/displayed as =), used to require the display of results, eg., **x** = .
8. ***Display of results:*** Type the expression wanted followed by = . For example **i**= or **f(x)**= . If the wanted expression contains range variables, the results will appear as a table.
9. ***Precision, Trailing Zeros:*** Click on an empty space and choose **Number {Numerical Format}** from the **Format {Math}** menu. For "Displayed Precision" select the number of *decimal places* to be shown (globally). Check the "Trailing Zeros" box. Use the "current region" {"LOCAL"} option to specify different precision for selected results.

/cont...

10. ***Solve Block:*** This is the area of the worksheet used to designate and enclose simultaneous equations (not more than 50) to be solved. Its top is defined by the word **Given** (in black lettering, not in blue "text") and its bottom is defined by any equation that involves the **Find** function. Besides equations, the Solve Block may include inequalities, eg., for constraints, and it may include text. Each unknown appearing in the Solve Block must be assigned a guessed value *above* (before) the Solve Block; if there are multiple roots, Mathcad will find the root closest to the guess. Be sure that the number of equations and unknowns are equal, and that you use the correct type of equals signs. With more equations, Mathcad becomes more sensitive to the reasonableness of the initial guesses; if no solution is found, try varying the guesses. Constraints that prevent unreasonable trials will also help.

11. ***Vectors:*** Most of the following examples make use of column vectors (single column arrays). Create such vectors by choosing **Matrix {Matrices}** from the **Insert {Math}** menu. Enter 1 column and the number of rows, "Create," click on each placeholder and type in the contents (or move around by tabbing). By default, Mathcad arrays begin at element zero; to change this to 1 type **ORIGIN~1** anywhere on the worksheet (~ is displayed as ≡), or reset it using the **Built-In Variables** tab under the **Options** command in the **Math** menu.

12. ***Text:*** This is displayed in blue and is ignored in mathematical processing. Two types can be selected from the **Text** menu. Choose **Create Text Paragraph** for page-width text (reports!); choose **Create Text Region** for small insertions. Text can be dragged, edited, and spellchecked. To insert an existing equation into text choose **Copy** then **Paste** from the **Edit** menu; to insert a new equation into text, click on the location, choose **Math Region {Embed Math}** from the **Insert {Text}** menu, type in the equation in the usual manner, and click on any text {press [Return]}. Make such equations "inactive" as per item 1.

13. ***Fonts:*** To set a different font for *default text* (blue; Book Antiqua was used in the following examples), click on the default (blue) text, choose **Style {Change Defaults⇒Font}** from the **Format {Text}** menu, click on the "Font" button, make the selection, and click on OK. To set a different font for *active variables* (and commands; black, Arial was used in the following examples), click on any (black) variable and choose **Equation {Font Tag}** from the **Format {Math}** menu, click on "Modify" {"Properties,"} make selections, and click on OK. To make *numbers* (results) match, click on a number and then repeat as for variables (it will now display "Constants" instead of "Variables").

14. ***Format:*** After selecting a region (enclosing rectangle), the contents (math and/or text) may be moved by dragging. Contents may be deleted or copied by using **Cut**, or **Copy** and **Paste**, from the **Edit** menu.

15. ***Graphs:*** Use ranges (item 5) with increments, eg., type **x:1,1.2;5** (where the increment is 0.2, displayed as **x:=1,1.2 .. 5**) to compute values for plotting. You may need to use array subscripts (item 4). Choose **Graph⇒X-Y Plot {Create X-Y Plot}** from the **Insert {Graphics}** menu to create a graph.

16. ***Help:*** Besides familiar online **Help**, there is Context-Sensitive Help which displays a help message relevant to the last click position of the mouse pointer when you press [F1] {[Shift][F1]} (useful for error messages), there are ToolTips which are displayed when the mouse pointer hovers over a palette or toolbar for a few seconds (also see the message line at the bottom of the window), and there are QuickSheets (from **Help⇒Resource Center {Help}**) which provide mathematical shortcuts for over 300 common procedures.

FLUID STATICS   3.8 Center of Pressure (p. 71)

Sample Problem 3.5   This gate is 2 feet wide perpendicular to the sketch. It is pivoted at hinge H. The gate weighs 500 lb. Its center of gravity is 1.2 ft to the right of and 0.9 ft above H. For what values of water depth x above H will the gate remain closed. Neglect friction at the pivot and neglect the thickness of the gate.

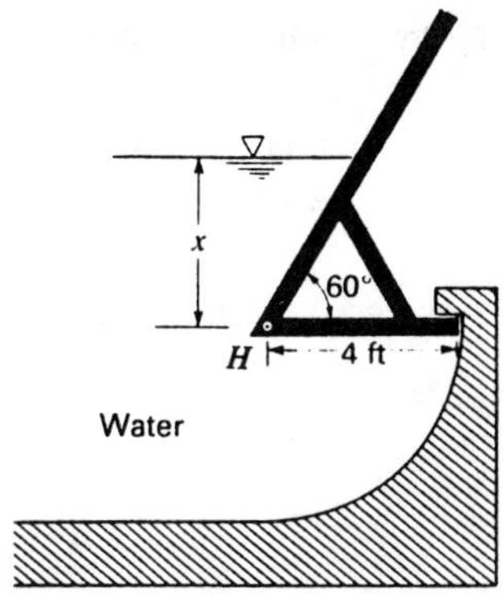

Solution 1

The sketch shows the forces acting on the gate. Because the pressure distribution on the sloping face OH is triangular, F_s acts at $OH/3$ from H; F_v acts at the mid-point of HE. Let the distances from H of the forces be represented by d with subscripts. We shall take moments about H when $R_E = 0$ (as the gate is about to open).

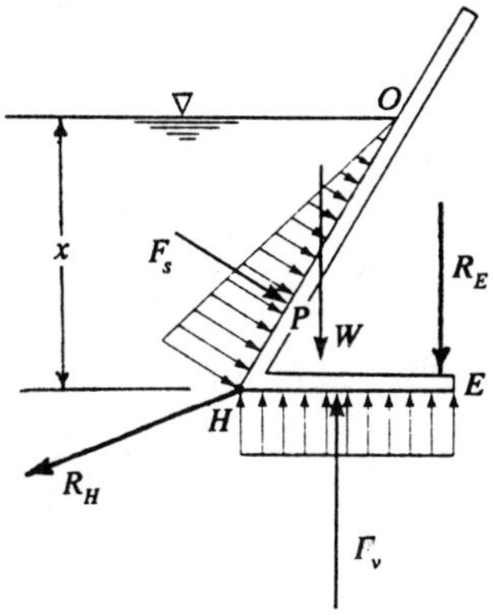

Known values: $W := 500$ $dw := 1.2$ $\gamma := 62.4$

$Av := 4 \cdot 2$ $dv := \frac{4}{2}$

Guessed values: $Fs := 30$ $OH := 1$ $As := 1$ $ds := 0.3$ $hcs = 0.3$

$Fv := 300$ $x := 1$ $hcv ;= 0.6$

Given

Clockwise moments: $W \cdot dw + Fs \cdot ds - Fv \cdot dv = 0$

where $Fs = \gamma \cdot hcs \cdot As$ $hcs = \frac{x}{2}$ $ds = \frac{OH}{3}$ $OH = \frac{x}{\cos(30 \cdot deg)}$ $As = OH \cdot 2$

$Fv = \gamma \cdot hcv \cdot Av$ $hcv = x$

with constraints $x > 0$ $x < 1$

$$\begin{bmatrix} Fs \\ Fv \\ As \\ OH \\ ds \\ hcs \\ hcv \\ x \end{bmatrix} := \text{Find}(Fs, Fv, As, OH, ds, hcs, hcv, x)$$

Results: $OH = 0.7011$ $Fs = 26.5637$ $hcs = 0.3036$ $As = 1.4022$ $ds = 0.2337$

$x = 0.6072$ $Fv = 303.1040$ $hcv = 0.6072$

Note: When the above calculations are repeated for $x > 1$, the following results are obtained:

OH = 6.551	Fs = 2319.2	hcs = 2.8367	As = 13.102	ds = 2.1837
x = 5.6733	Fv = 2832.1	hcv = 5.6733		

Answer: From these two sets of results we find the gate will remain closed for $0.607 < x < 5.67$ ft.

Solution 2

Known values: $W := 500$ $dw := 1.2$ $\gamma := 62.4$ $Av := 4 \cdot 2$ $dv := \frac{4}{2}$

Equations to solve:

Clockwise moments: $W \cdot dw + Fs \cdot ds - Fv \cdot dv = 0$

where $Fs = \gamma \cdot hcs \cdot As$ $hcs = \frac{x}{2}$ $ds = \frac{OH}{3}$ $OH = \frac{x}{\cos(30 \cdot deg)}$ $As = OH \cdot 2$

$Fv = \gamma \cdot hcv \cdot Av$ $hcv = x$

Eliminating unknowns, these are equivalent to: $a \cdot x^3 + b \cdot x^2 + c \cdot x + d = 0$

where the coefficients are: $a := 1$ $b := 0$ $c := -2.25 \cdot Av \cdot dv$ $d := \frac{2.25}{\gamma} \cdot W \cdot dw$

Solutions:

$$r := \begin{pmatrix} d \\ c \\ b \\ a \end{pmatrix} \qquad \text{polyroots}(r) = \begin{pmatrix} -6.2805 \\ 0.6072 \\ 5.6733 \end{pmatrix}$$

Note: The negative root (depth) is meaningless.

Answer: From the moment equation we see the gate will remain open when $0.607 < x < 5.67$ ft.

Problem An 18-in-diameter welded-steel pipe (mean roughness element size $e = 0.00015$ ft) carries water flowing at a Reynolds number $\mathbf{R} = 5 \times 10^4$. Find the friction factor f using the Colebrook equation.

Solution

$e := 0.00015$ $D := \frac{18}{12}$ $R := 5 \cdot 10^4$

Guessed values:

$f := 0.03$

Given

Colebrook equation (8.40):

$$\frac{1}{\sqrt{f}} = -2 \cdot \log\left(\frac{e}{3.7 \cdot D} + \frac{2.51}{R \cdot \sqrt{f}}\right)$$

with constraints $f > 0$ $f < 1$

$\text{Find}(f) = 0.02125$

PIPE FLOW **8.14 Single-Pipe Flow** (p. 299)

Sample Problem 8.8*b* A galvanized iron pipe (e = 0.0005 ft) 18,000 ft long is to carry ethyl alcohol (ν = 2.3 x 10^{-5} ft^2/sec) at a rate of 135 gpm. If the friction head loss is to be 215 ft, determine the pipe size theoretically required.

Solution

Known Values: $g := 32.2$ $L := 18000$ $e := 0.0005$ $\nu := 2.3 \cdot 10^{-5}$ $Q := \frac{135}{449}$ $h := 215$

Guessed values: $f := 0.03$ $V := 6$ $R := 9000$ $D := 0.9$

Given

Eq. (8.10): $h = \frac{f \cdot L \cdot V^2}{D \cdot 2 \cdot g}$ Eq. (8.40): $\frac{1}{\sqrt{f}} = -2 \cdot \log\left(\frac{e}{3.7 \cdot D} + \frac{2.51}{R \cdot \sqrt{f}}\right)$

where $Q = V \cdot \frac{\pi}{4} \cdot D^2$ $R = \frac{D \cdot V}{\nu}$

$$\begin{pmatrix} D \\ V \\ R \\ f \end{pmatrix} := \mathrm{Find}(D, V, R, f)$$

Results: $D = 0.3441$ $V = 3.2341$ $R = 4.8378 \cdot 10^4$ $f = 0.0253$

Sample Problem 8.11 A pump is located 15 ft above the surface of a liquid ($\gamma = 52\ \text{lb/ft}^3$) in a closed tank. The pressure in the space above the liquid surface is 5 psi. The suction line to the pump (flush inlet, $k_e = 0.5$) is 50 ft of 6-in-diameter pipe ($f = 0.025$). The discharge line from the pump is 200 ft of 8-in-diameter pipe ($f = 0.030$). The pipe discharges in a submerged fashion to an open tank whose free liquid surface is 10 ft lower than the liquid surface in the pressure tank. If the pump puts 2.0 hp into the liquid, determine the flow rate. Assume that the Reynolds numbers are high enough to assure turbulent flow.

Solution

Known values: $p := 5 \cdot 144$ $\gamma := 52$ $ke := 0.5$ $fs := 0.025$ $Ls := 50$ $Ds := \frac{6}{12}$

$Pwr := 2$ $g := 32.2$ $\Delta z := -10$ $fd := 0.030$ $Ld := 200$ $Dd := \frac{8}{12}$

Guessed values: $hp := 5$ $Vs := 10$ $Vd := 10$ $Q := 5$

Given

Energy:

$$\frac{p}{\gamma} - ke \cdot \frac{Vs^2}{2 \cdot g} - fs \cdot \frac{Ls}{Ds} \cdot \frac{Vs^2}{2 \cdot g} + hp - fd \cdot \frac{Ld}{Dd} \cdot \frac{Vd^2}{2 \cdot g} - \frac{Vd^2}{2 \cdot g} = \Delta z$$

where

$$Pwr = \frac{\gamma \cdot Q \cdot hp}{550} \qquad Q = \frac{\pi}{4} \cdot Ds^2 \cdot Vs \qquad Q = \frac{\pi}{4} \cdot Dd^2 \cdot Vd$$

with constraint

$$Q > 0$$

$$\begin{pmatrix} hp \\ Vs \\ Vd \\ Q \end{pmatrix} := \text{Find}(hp, Vs, Vd, Q)$$

Results: $hp = 6.0917$ $Vs = 17.6856$ $Vd = 9.9482$ $Q = 3.4726$

Sample Problem 8.12*b* Given that, in Fig. 8.27, pipe 1 is 6000 ft of 15 in diameter, pipe 2 is 1500 ft of 10 in diameter, and pipe 3 is 4500 ft of 8 in diameter, all asphalt-dipped cast iron (e = 0.0004 ft). The elevations of the water surfaces in reservoirs A and C are 250 ft and 160 ft respectively, and the discharge Q_2 of 60°F water ($\nu = 1.217 \times 10^{-5}$ ft^2/sec) into reservoir B is 3.3 cfs. Find the surface elevation of reservoir B.

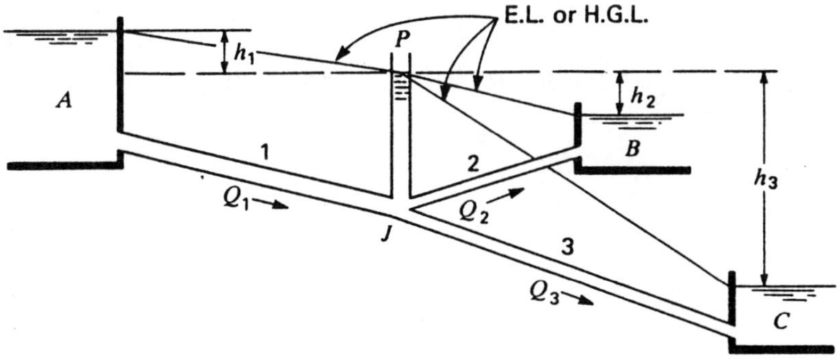

Figure 8.27 Branching pipes.

Solution

Known values:

$L1 := 6000 \quad D1 := \frac{15}{12} \quad e1 := 0.0004 \quad ElevA := 250 \quad g := 32.2$

$L2 := 1500 \quad D2 := \frac{10}{12} \quad e2 := 0.0004 \quad Q2 := 3.3$

$L3 := 4500 \quad D3 := \frac{8}{12} \quad e3 := 0.0004 \quad ElevC := 160 \quad \nu := 1.217 \cdot 10^{-5}$

Guessed values:

$f1 := 0.03 \quad V1 := 10 \quad R1 := 800000 \quad Q1 := 10 \quad h1 := 10$

$f2 := 0.03 \quad V2 := 10 \quad R2 := 800000 \quad h2 := 10$

$f3 := 0.03 \quad V3 := 10 \quad R3 := 800000 \quad Q3 := 10 \quad h3 := 10$

Given

$$\frac{1}{\sqrt{f1}} = -2 \cdot \log\left(\frac{e1}{3.7 \cdot D1} + \frac{2.51}{R1 \cdot \sqrt{f1}}\right) \quad R1 = \frac{D1 \cdot V1}{\nu} \quad h1 = \frac{f1 \cdot L1 \cdot (V1)^2}{D1 \cdot 2 \cdot g} \quad Q1 = \frac{V1 \cdot \pi \cdot (D1)^2}{4}$$

$$\frac{1}{\sqrt{f2}} = -2 \cdot \log\left(\frac{e2}{3.7 \cdot D2} + \frac{2.51}{R2 \cdot \sqrt{f2}}\right) \quad R2 = \frac{D2 \cdot V2}{\nu} \quad h2 = \frac{f2 \cdot L2 \cdot (V2)^2}{D2 \cdot 2 \cdot g} \quad Q2 = \frac{V2 \cdot \pi \cdot (D2)^2}{4}$$

$$\frac{1}{\sqrt{f3}} = -2 \cdot \log\left(\frac{e3}{3.7 \cdot D3} + \frac{2.51}{R3 \cdot \sqrt{f3}}\right) \quad R3 = \frac{D3 \cdot V3}{\nu} \quad h3 = \frac{f3 \cdot L3 \cdot (V3)^2}{D3 \cdot 2 \cdot g.} \quad Q3 = \frac{V3 \cdot \pi \cdot (D3)^2}{4}$$

$$h1 + h3 = ElevA - ElevC \qquad Q1 = Q2 + Q3$$

with constraints: $V1 > 0 \quad V2 > 0 \quad V3 > 0$

$$\begin{bmatrix} Q1 \\ Q3 \\ V1 \\ V2 \\ V3 \\ R1 \\ R2 \\ R3 \\ f1 \\ f2 \\ f3 \\ h1 \\ h2 \\ h3 \end{bmatrix} := \mathrm{Find}(Q1, Q3, V1, V2, V3, R1, R2, R3, f1, f2, f3, h1, h2, h3)$$

Results: $Q1 = 5.3368$ $V1 = 4.3488$ $R1 = 4.4667 \cdot 10^5$ $f1 = 0.0166$ $h1 = 23.3415$

$V2 = 6.0504$ $R2 = 4.1430 \cdot 10^5$ $f2 = 0.0177$ $h2 = 18.1287$

$Q3 = 2.0368$ $V3 = 5.8349$ $R3 = 3.1963 \cdot 10^5$ $f3 = 0.0187$ $h3 = 66.6585$

$ElevB := ElevA - h1 - h2$ Answer: $ElevB = 208.5298$

PIPE FLOW **8.28 Pipes in Parallel** (p. 341)

Problem Parallel pipes 1, 2, and 3 are as follows: 750 m of 5 cm, 1250 m of 10 cm, and 650 m of 15 cm; all have absolute roughness $e = 0.00015$ m. When the total flow of water ($\nu = 1.2 \times 10^{-6}\ m^2/s$) is 45 L/s, find the common head loss across the pipes and the flow in each pipe.

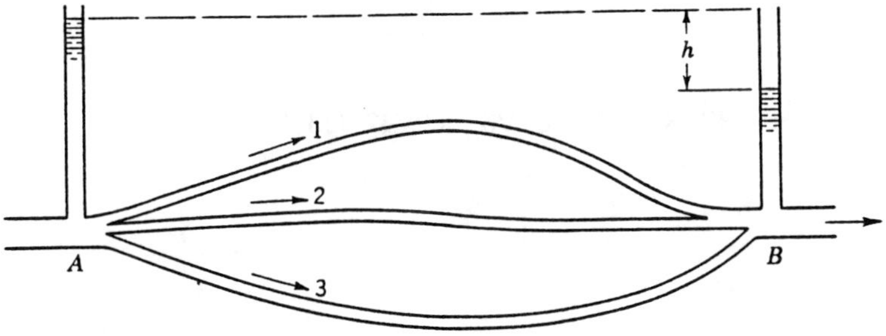

Figure 8.30 Pipes in parallel.

Solution

Known values:

$L1 := 750$ $\quad D1 := 0.05$ $\quad e := 0.00015$

$L2 := 1250$ $\quad D2 := 0.10$ $\quad \nu := 1.20 \cdot 10^{-6}$ $\quad g := 9.81$

$L3 := 650$ $\quad D3 := 0.15$ $\quad Q := 0.045$

Guessed values:

$f1 := 0.03$ $\quad R1 := 20000$ $\quad V1 := 1.0$ $\quad Q1 := 0.002$

$f2 := 0.025$ $\quad R2 := 100000$ $\quad V2 := 1.5$ $\quad Q2 := 0.008$

$f3 := 0.02$ $\quad R3 := 200000$ $\quad V3 := 2.0$ $\quad Q3 := 0.035$ $\quad hL := 30$

Given

Eq. (8.72): $\quad Q = Q1 + Q2 + Q3$

where, for each pipe:

Eq. (8.10)	Eq. (8.40)	Eq. (8.1)	Continuity
$hL = f1 \cdot \frac{L1}{D1} \cdot \frac{V1^2}{2 \cdot g}$	$\frac{1}{\sqrt{f1}} = -2 \cdot \log\left(\frac{e}{3.7 \cdot D1} + \frac{2.51}{R1 \cdot \sqrt{f1}}\right)$	$R1 = \frac{D1 \cdot V1}{\nu}$	$Q1 = \frac{\pi}{4} \cdot D1^2 \cdot V1$
$hL = f2 \cdot \frac{L2}{D2} \cdot \frac{V2^2}{2 \cdot g}$	$\frac{1}{\sqrt{f2}} = -2 \cdot \log\left(\frac{e}{3.7 \cdot D2} + \frac{2.51}{R2 \cdot \sqrt{f2}}\right)$	$R2 = \frac{D2 \cdot V2}{\nu}$	$Q2 = \frac{\pi}{4} \cdot D2^2 \cdot V2$
$hL = f3 \cdot \frac{L3}{D3} \cdot \frac{V3^2}{2 \cdot g}$	$\frac{1}{\sqrt{f3}} = -2 \cdot \log\left(\frac{e}{3.7 \cdot D3} + \frac{2.51}{R3 \cdot \sqrt{f3}}\right)$	$R3 = \frac{D3 \cdot V3}{\nu}$	$Q3 = \frac{\pi}{4} \cdot D3^2 \cdot V3$

with constraints: $\quad V1 > 0 \quad V2 > 0 \quad V3 > 0 \quad f1 > 0 \quad f2 > 0 \quad f3 > 0$

$$\begin{pmatrix} f1 \\ f2 \\ f3 \\ R1 \\ R2 \\ R3 \\ V1 \\ V2 \\ V3 \\ Q1 \\ Q2 \\ Q3 \\ hL \end{pmatrix} := \mathrm{Find}(f1, f2, f3, R1, R2, R3, V1, V2, V3, Q1, Q2, Q3, hL)$$

Results:

f1 = 0.0293	R1 = $3.7093 \cdot 10^4$	V1 = 0.8902	Q1 = 0.0017	hL = 17.7813
f2 = 0.0239	R2 = $9.0000 \cdot 10^4$	V2 = 1.0800	Q2 = 0.0085	
f3 = 0.0208	R3 = $2.4595 \cdot 10^5$	V3 = 1.9676	Q3 = 0.0348	

Problem 8.119 Solve the pipe network shown in Fig. P8.119 to find the flow in each pipe. Use Hazen-Williams Eq. (8.49*b*) in the form $h_L = KQ^n$ [Eq. (8.66), Sec. 8.26], with $C_{HW} = 120$. For initial flows, assume only values of 30, 15, and 0 L/s.

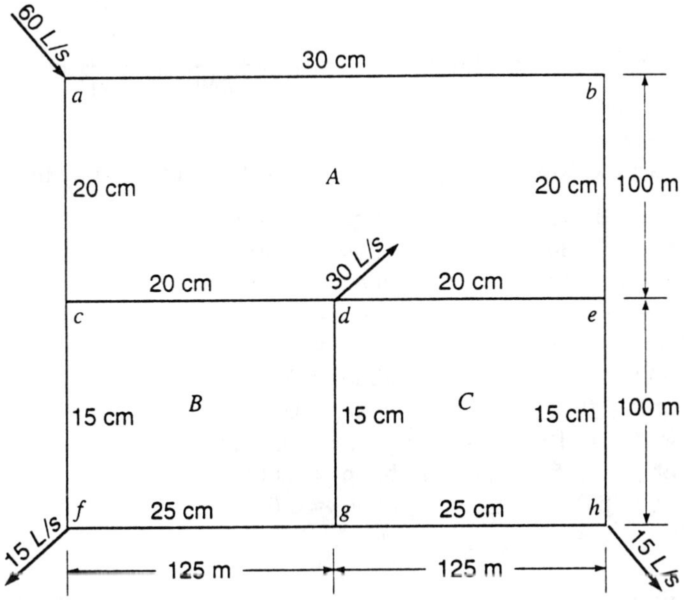

Figure P8.119

Solution

Known values:

$Chw = 120 \qquad n = 1.852$

$Lab = 250 \qquad Dab = 0.3 \qquad Kab = \dfrac{10.675}{Chw^{1.852} \cdot Dab^{4.87}}$

$Lbe = 100 \qquad Dbe = 0.2 \qquad Kbe = \dfrac{10.675}{Chw^{1.852} \cdot Dbe^{4.87}}$

$Led = 125 \qquad Ded = 0.2 \qquad Ked = \dfrac{10.675}{Chw^{1.852} \cdot Ded^{4.87}}$

$Ldc = 125 \qquad Ddc = 0.2 \qquad Kdc = \dfrac{10.675}{Chw^{1.852} \cdot Ddc^{4.87}}$

$Lca = 100 \qquad Dca = 0.2 \qquad Kca = \dfrac{10.675}{Chw^{1.852} \cdot Dca^{4.87}}$

$Ldg = 100 \qquad Ddg = 0.15 \qquad Kdg = \dfrac{10.675}{Chw^{1.852} \cdot Ddg^{4.87}}$

$Lgf = 125 \qquad Dgf = 0.25 \qquad Kgf = \dfrac{10.675}{Chw^{1.852} \cdot Dgf^{4.87}}$

$Lfc := 100 \qquad Dfc := 0.15 \qquad Kfc := \frac{10.675}{Chw^{1.852} \cdot Dfc^{4.87}}$

$Leh := 100 \qquad Deh := 0.15 \qquad Keh := \frac{10.675}{Chw^{1.852} \cdot Deh^{4.87}}$

$Lhg := 125 \qquad Dhg := 0.25 \qquad Khg := \frac{10.675}{Chw^{1.852} \cdot Dhg^{4.87}}$

Guessed values (assume flows are positive in the direction indicated by the 'subscripts'):

Qab := 0.030	hLab := 0.0005
Qbe := 0.030	hLbe := 0.0050
Qed := 0.015	hLed := 0.0020
Qdc := -0.015	hLdc := -0.0020
Qca := -0.030	hLca := -0.0050
Qdg := 0.0	hLdg := 0.0
Qgf := 0.0	hLgf := 0.0
Qfc := -0.015	hLfc := -0.0050
Qeh := 0.015	hLeh := 0.0050
Qhg := 0.00	hLhg := 0.0

Given

Condition 1 (continuity at node)		Condition 2 (Eq. 8.77)
b:	Qab=Qbe	$hLab = Kab \cdot Qab \cdot \lvert Qab^{n-1} \rvert$
e:	Qbe=Qed + Qeh	$hLbe = Kbe \cdot Qbe \cdot \lvert Qbe^{n-1} \rvert$
d:	Qed=Qdc + Qdg + 0.030	$hLed = Ked \cdot Qed \cdot \lvert Qed^{n-1} \rvert$
c:	Qdc + Qfc=Qca	$hLdc = Kdc \cdot Qdc \cdot \lvert Qdc^{n-1} \rvert$
g:	Qdg + Qhg=Qgf	$hLca = Kca \cdot Qca \cdot \lvert Qca^{n-1} \rvert$
f:	Qgf=Qfc + 0.015	$hLdg = Kdg \cdot Qdg \cdot \lvert Qdg^{n-1} \rvert$
h:	Qeh=Qhg + 0.015	$hLgf = Kgf \cdot Qgf \cdot \lvert Qgf^{n-1} \rvert$
		$hLfc = Kfc \cdot Qfc \cdot \lvert Qfc^{n-1} \rvert$
		$hLeh = Keh \cdot Qeh \cdot \lvert Qeh^{n-1} \rvert$
		$hLhg = Khg \cdot Qhg \cdot \lvert Qhg^{n-1} \rvert$

Condition 3 (Σlosses around loop)

A: hLab + hLbe + hLed + hLdc + hLca=0

B: hLdg + hLgf + hLfc − hLdc=0

C: hLeh + hLhg − hLdg − hLed=0

/continued...

$$\begin{bmatrix} Qab \\ Qbe \\ Qed \\ Qdc \\ Qca \\ Qdg \\ Qgf \\ Qfc \\ Qeh \\ Qhg \\ hLab \\ hLbe \\ hLed \\ hLdc \\ hLca \\ hLdg \\ hLgf \\ hLfc \\ hLeh \\ hLhg \end{bmatrix} := Find(Qab, Qbe, Qed, Qdc, Qca, Qdg, Qgf, Qfc, Qeh, Qhg, hLab, hLbe, hLed, hLdc, hLca, hLdg, hLgf, hLfc, hLeh, hLhg)$$

Results:

Kab = 0.5298387 Kbe = 3.8168768 Ked = 3.8168768 Kdc = 3.8168768 Kca = 3.8168768

Qab = 0.0292775 Qbe = 0.0292775 Qed = 0.0179628 Qdc = −0.0190265 Qca = −0.0307225

hLab = 0.0007659 hLbe = 0.0055173 hLed = 0.0022326 hLdc = −0.0024836 hLca = −0.0060322

Kdg = 15.4938656 Kgf = 1.2875271 Kfc = 15.4938656 Keh = 15.4938656 Khg = 1.2875271

Qdg = 0.0069893 Qgf = 0.0033041 Qfc = −0.0116959 Qeh = 0.0113148 Qhg = −0.0036852

hLdg = 0.0015778 hLgf = 0.0000327 hLfc = −0.0040941 hLeh = 0.0038504 hLhg = −0.0000401

Note: If *e* is given for the pipes instead of C_{HW}, four equations per pipe are required for head loss (as for the previous problem) instead of the one empirical equation used here, resulting in a total of 50 instead of 20 equations.

Sample Problem 10.1c This channel has a bed slope S of 0.0006 and a Manning's n of 0.016. Find the uniform flow depth y when the flow rate Q is 225 cfs.

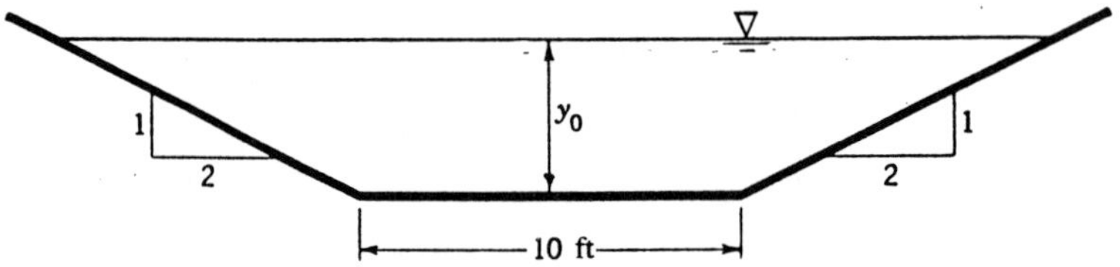

Figure S10.1

Solution

Known values (c = 1.486 for ft, 1.0 for m; side slope m = run/rise):

$Q := 225 \quad c := 1.486 \quad b := 10 \quad m := 2 \quad n := 0.016 \quad S := 0.0006$

Guessed values: $y := 1 \quad A := 10 \quad P := 10 \quad Rh := 1$

Given

Manning's Eq. (10.8): $Q = \frac{c}{n} \cdot A \cdot Rh^{\frac{2}{3}} \cdot S^{\frac{1}{2}}$

where $A = (b + m \cdot y) \cdot y \qquad P = b + 2 \cdot y \cdot \sqrt{1 + m^2} \qquad Rh = \frac{A}{P}$

$$\begin{pmatrix} y \\ A \\ P \\ Rh \end{pmatrix} := \text{Find}(y, A, P, Rh)$$

Results: $y = 3.4063 \quad A = 57.2684 \quad P = 25.2334 \quad Rh = 2.2695$

Sample Problem 10.6*a* Water in a rectangular channel flows 0.5 m deep at a velocity of 6 m/s. Find the alternate depth.

Solution 1

Known values:

$V1 := 6 \qquad y1 := 0.5 \qquad g := 9.81$

Guessed values:

$E := 2 \qquad q := 3 \qquad V2 := 0.5 \qquad y2 := 5$

Given

Specific energy: $E = y1 + \frac{V1^2}{2 \cdot g} \qquad E = y2 + \frac{V2^2}{2 \cdot g}$

Continuity per unit width: $q = V1 \cdot y1 \qquad q = V2 \cdot y2$

with constraint $y2 > y1$

$$\begin{pmatrix} E \\ q \\ V2 \\ y2 \end{pmatrix} := \text{Find}(E, q, V2, y2)$$

Results: $E = 2.3349 \qquad q = 3.0000 \qquad V2 = 1.3370 \qquad y2 = 2.2437$

Note: The constraint y2 < y1 (with negative guessed values for V2 and y2) yields y2 = -0.409 ft, which is a meaningless solution (root) for flow depth.

Solution 2

Known values: $E := 2.3349$ $q := 3$ $g := 9.81$

Equation to solve

$$E = y + \frac{q^2}{2 \cdot g \cdot y^2}$$

is equivalent to:

$$a \cdot y^3 + b \cdot y^2 + c \cdot y + d = 0$$

where coefficients are: $a := 1$ $b := -E$ $c := 0$ $d := \frac{q^2}{2 \cdot g}$

Solutions for y:

$$r := \begin{pmatrix} d \\ c \\ b \\ a \end{pmatrix} \qquad \text{polyroots}(r) = \begin{pmatrix} -0.4089 \\ 0.5000 \\ 2.2438 \end{pmatrix}$$

Check:

$i := 0..2$

$y_i := \text{polyroots}(r)_i$

y_i	$y_i + \frac{q^2}{2 \cdot g \cdot (y_i)^2}$	$y_i + \frac{q^2}{2 \cdot g \cdot (y_i)^2} - E$
-0.4089	2.3349	0
0.5000	2.3349	0
2.2438	2.3349	0

Sample Problem 10.12*b* Refer to Fig. 10.32. A rectangular channel changes in width from 4 ft to 6 ft. Measurements indicate that y_1 = 2.50 ft and Q = 50 cfs. Determine the depth y_2 by considering the head loss to be given as shown in the figure.

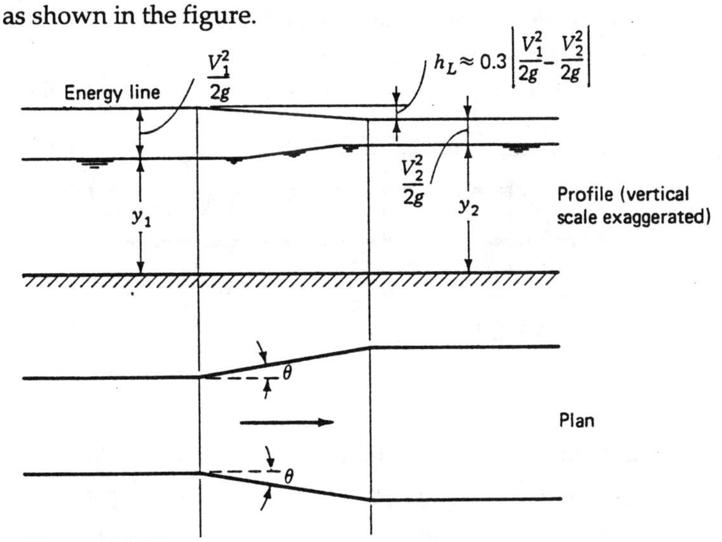

Figure 10.32
Simple open-channel wedge transition for decelerating flow.

Solution

Known values: g := 32.2 Q := 50 b1 := 4 b2 := 6 y1 := 2.5

Guessed values: E1 := 3 E2 := 2.9 hL := 0.1 V1 := 5 V2 := 3 y2 := 2

Given

$$E2 = E1 - hL \qquad hL = 0.3\cdot\left(\frac{V1^2}{2\cdot g} - \frac{V2^2}{2\cdot g}\right)$$

where $E1 = y1 + \frac{V1^2}{2\cdot g}$ $E2 = y2 + \frac{V2^2}{2\cdot g}$ $Q = b1\cdot y1\cdot V1$ $Q = b2\cdot y2\cdot V2$

with constraints $hL > 0$ $y2 > 0$

$$\begin{bmatrix} E1 \\ E2 \\ hL \\ V1 \\ V2 \\ y2 \end{bmatrix} := \mathrm{Find}(E1, E2, hL, V1, V2, y2)$$

Results: E1 = 2.8882 V1 = 5.0000 hL = 0.0709

E2 = 2.8173 V2 = 3.1264 y2 = 2.6655

Note: Substitution to eliminate unknowns from these equations will yield a cubic equation in y1.

Sample Problem 10.13 A culvert under a road must be 30 m long, have a slope of 0.003, and carry 4.3 m^3/s. If the maximum permissible headwater level is 3.6 m above the culvert invert, what size of corrugated-pipe culvert (n = 0.025) would you select? Neglect velocity of approach. Assume a square-edged inlet with k_e = 0.5 and C_d = 0.65. The outlet will discharge freely.

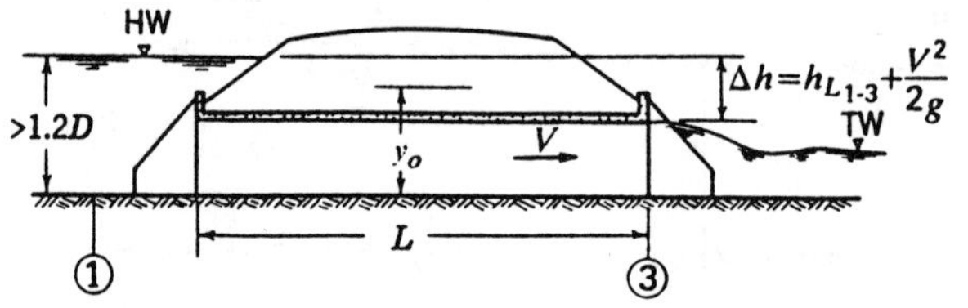

Figure 10.33*b*

Solution

Known values:	L := 30	n := 0.025	S0 := 0.003	g := 9.81
	ke := 0.5	Q := 4.3	y1 := 3.6	c := 1.0
Guessed values:	d0 := 2	A0 := 3	Rh0 := 1	he := 0.5
	D := 1	A := 1	Rh := 0.3	hf := 2
	Dh := 3	hL := 3	S := 0.05	hv := 1
	y3 := 1	V := 5		

where d0 is the maximum diameter that will just flow full with normal (uniform) flow.

Given

Equation (10.8*b*) for normal flow: $Q = \frac{c}{n} \cdot A0 \cdot Rh0^{\frac{2}{3}} \cdot S0^{\frac{1}{2}}$

where $A0 = \frac{\pi}{4} \cdot d0^2$ $Rh0 = \frac{d0}{4}$

Assuming the barrel flows full (Fig. 10.33*b*):

Energy (Fig. 10.33*b*): $Dh = hL + \frac{V^2}{2 \cdot g}$

where $hL = y1 - y3 + S0 \cdot L$ $y3 = D$ $Q = A \cdot V$ $A = \frac{\pi}{4} \cdot D^2$

Energy losses (Eq. 10.54): $Dh = he + hf + hv$

where $he = ke \cdot \frac{V^2}{2 \cdot g}$ $hf = S \cdot L$ $V = \frac{c}{n} \cdot Rh^{\frac{2}{3}} \cdot S^{\frac{1}{2}}$ $Rh = \frac{D}{4}$ $hv = \frac{V^2}{2 \cdot g}$

with constraints $d0 > 0$ $D > 0$ $hL > 0$ $S > 0$

$$\begin{bmatrix} d0 \\ Rh0 \\ A0 \\ D \\ Rh \\ A \\ Dh \\ hL \\ S \\ he \\ hf \\ hv \\ y3 \\ V \end{bmatrix} := \text{Find}(d0, Rh0, A0, D, Rh, A, Dh, hL, S, he, hf, hv, y3, V)$$

Results:

d0 = 1.9938	Rh0 = 0.4984	A0 = 3.1220	he = 0.5085
D = 1.1071	Rh = 0.2768	A = 0.9626	hf = 2.0744
Dh = 3.6000	hL = 2.5829	S = 0.0691	hv = 1.0171
y3 = 1.1071		V = 4.4671	

Comment: As $D < d0$ the culvert must run full, and so assumed case (*b*) of Fig. 10.33 does occur.

Sample Problem 12.6 You need to design a surge tank for a 1.5-m.diameter steel pipeline (f = 0.018) 1200 m long that supplies water to a small power plant. The surge tank will be connected to the pipeline at a point where the center line is 35 m below the water surface in the reservoir. For a discharge of 12 m^3/s, what is the smallest surge tank diameter that will prevent surges from exceeding 18.5 m above the reservoir surface? Neglect inertial effects and fluid friction in the surge tank only. The pipeline inlet is square-edged.

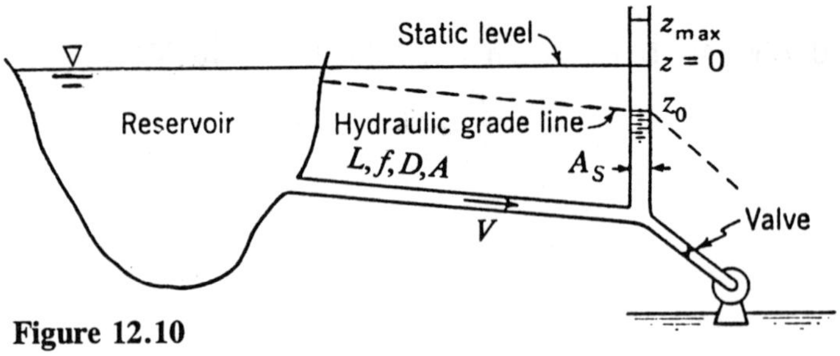

Figure 12.10

Solution

Known values: $g := 9.81$ $f := 0.018$ $L := 800$ $D := 1.5$ $\Sigma k := 0.5$ $Q0 := 12$ $zmax := 18.5$

Guessed values: $A := 2$ $K := 10$ $V0 := 10$ $z0 := -10$ $As := 10$ $Ds := 3$

Given

Initial head loss: $-z0 = K \cdot \frac{V0^2}{2 \cdot g}$ $K = \frac{f \cdot L}{D} + 1 + \Sigma k$

First surge:

$$1 - \frac{K \cdot As}{L \cdot A} \cdot zmax = e^{\frac{-K \cdot As}{L \cdot A} \cdot \left(K \cdot \frac{V0^2}{2 \cdot g} + zmax \right)}$$

where $A = \frac{\pi}{4} \cdot D^2$ $As = \frac{\pi}{4} \cdot Ds^2$ $Q0 = A \cdot V0$

with constraints $D > 0$ $Ds > 0$ $V0 > 0$ $z0 < 0$ $\frac{K \cdot As}{L \cdot A} \cdot zmax < 1$

$$\begin{bmatrix} A \\ K \\ V0 \\ z0 \\ As \\ Ds \end{bmatrix} := Find(A, K, V0, z0, As, Ds)$$

Results: $K = 11.100$ $V0 = 6.791$ $z0 = -26.088$

$A = 1.767$ $As = 6.059$ $Ds = 2.778$

Appendix B
SAMPLE KEYSTROKES FOR HP 48G CALCULATOR

Contents

A. To key in and save a single equation (eg, the Colebrook Eq.)

1. *Turn on calculator (it will resume where last turned off):*
   ON, ON
2. *Clear stack (screen) of unwanted numbers [required here; optional anytime]:*
   ↰, CLEAR
3. *Activate the Equation Writer:*
   ↰, EQUATION
4. *Key in the Colebrook Eq. (8.40):*
   (can use backspace anytime to correct, but it is slow; can also edit later)
   1, ÷, √, α, F, ▸, ▸, ↰, =, −, 2, ×, ↱, LOG, α, E, ÷, 3.7, ×, α, D, ▸,
   +, 2.51, ÷, α, R, ×, √, α, F, ▸, ▸, ▸
   [It helps to use unique variable names in the entire memory, eg, R *for Reynolds number,* RH *for hydraulic radius.]*
5. *Store this equation in {HOME} memory, name it COLE:*
   ENTER, α, C, α, O, α, L, α, E, STO
6. *Repeat for other equations [eg, for continuity Eq. named CNTY (not CONT or CON which are "reserved"), for Reynolds No. Eq. named REYN, and for Eq. (8.42b) named PIPE].*
   [Note: Calculator uses log functions to evaluate superscripts, so D×D is preferred to D^2.]

B. To display, edit, or "rearrange" saved equations

1. *Enter memory, display equations and variables, etc., in the {HOME} directory:*
   ↱, MEMORY
2. *Select the equation name (eg, COLE) where the edited equation is to be stored:*
   ▾ or ▴ as needed to highlight chosen equation, A (= EDIT), B (= CHOOS)
3. *Select the equation (eg, COLE) to be edited:*
   ▾ or ▴ as needed to highlight chosen equation
4. *Place chosen equation in command line for editing:*
   F (= OK), A (= EDIT)
5. *Move pointer to right of character (R) to be edited:*
   ▾ or ▸ etc., as needed
6. *Delete R:*
   ← (= backspace)
7. *Insert P:*
   α, P
8. *Place edited equation in memory:*
   ENTER, ENTER
   (or press E (= CANCL), E (= CANCL) *to abort editing without implementing any changes)*

/cont...

C. To purge a saved equation from memory

1. *Highlight the equation to be purged:*
   ↱, MEMORY, ▾ or ▴ as needed
2. *Purge the equation pointed to:*
   NXT, B (= PURG)

D. To solve with a single, pre-saved equation

1. *Display the types of equations available:*
   ON, ON, ↱, SOLVE
2. *Select the type of equation to be solved:*
   "Solve equation.." should be highlighted; ENTER
3. *Select one equation to be the "current" equation:*
   B (= CHOOS), ▾ or ▴ as needed to highlight the desired equation then: ENTER
4. *Enter a value (or estimated value) of 200 (say) for the variable in the first position:*
   2, 0, 0, ENTER
5. *Repeat Step D4 to enter all other variables (can move to other variables by ▸, ▴, etc.)*
6. *Solve for a chosen variable:*
   Highlight the desired variable by ▾ or ▴, then: F (= SOLVE)
7. *To display all the significant figures [optional]:*
   E (= INFO)
8. *To return to the "Solve Equation" display:*
   F (= OK)
9. *To return to {HOME}:*
   ON, ON

E. To solve using a set of (not linked) saved equations (eg, with PIPE, CNTY, REYN, COLE)

1. *Execute Steps D1-7 using the chosen equation (eg, PIPE) to solve for the desired variable (eg, H or Q or D)*
2. *Move to the next desired equation (eg, CNTY).*
   F (= OK), ▾ or ▴ as needed to highlight "EQ:"
   Note: The same variable values will be transferred to this equation.
3. *Execute Steps D3-7 to solve for the next unknown (eg, V).*
4. *Repeat Steps E2-3 to solve for R (in REYN) and for F (in COLE).*

/cont...

F. To solve using a set of linked, saved equations (eg, with PIPE, CNTY, REYN, COLE)
[Alternate to Procedure E; uses SOLVR]

1. Exit any program that you are presently in:
ON, MTH

2. Display the names of equations, etc., "variables" in memory:
VAR, NXT as needed

3. Place equations on the stack in the order they will be used:
Press appropriate white menu keys (A-F) for: PIPE, CNTY, REYN, COLE

4. Set the pointer to the first of these equations (on the highest stack level):
▲, ▲, ▲, ▲

5. Gather the four equations into a "list":
F (= →LIST), ENTER

6. Store the list in "EQ" as the current "equation":
↰, SOLVE, A (= ROOT), ↰, C (= EQ)

7. Enter the SOLVR environment (variables for the first equation = PIPE appear):
A (= SOLVR)

8. Enter a value (eg, 0.301) for the first variable, Q:
0.301, A (= Q)

9. Repeat Step F8 for all other variables (enter an estimate for the unknown), using NXT to complete the list if necessary.

10. [Optional] To display the value of a variable:
↱, appropriate white menu key (A-F)

11. To solve for the unknown (estimated) variable (eg, H or Q or D):
↰, white menu key for the unknown variable

12. Move to the next desired equation in the list (eg, CNTY):
NXT as needed, F (= NXEQ)
Note: Common variables will retain the same values.

13. Enter an estimated value (eg, 5) for the unknown variable, V (in B position, say):
5, B (= V)

14. Solve for V:
↰, B (= V)

15. Repeat Steps F12-14 to solve for R (in REYN) and for F (in COLE).

16. To leave the SOLVR environment:
MTH

G. Conclusion

Once installed in a calculator, procedures **E** or **F** with the equations just described will *very conveniently* solve *any* turbulent flow problem in a single pipe without minor losses. The equations for velocity and Reynolds number are necessary to confirm that the flow is turbulent. The equation for f is not necessary, it is for interest only.

Other flow problems can be equally conveniently solved on a programmable calculator, if they can be similarly reduced to a set of equations starting with a single master equation like Eq. (8.42*b*), so that each equation in turn will have only one unknown.

Appendix C

CORRECTIONS TO THE SECOND PRINTING[1] OF THE TEXTBOOK

Chapter 1
Page 9, "Acceleration" line of table, column 2: Change LT^{-1} to LT^{-2} .
Chapter 3
Page 52, bottom line: Change "kilonewtons per square meter" to " kilopascals (kilonewtons per square meter)".
Page 53, top line: Change "kilopascals (kilonewtons/cubic meter)" to "kilonewtons/cubic meter".
Page 84, Exer. 3.9.8, line 2: Change $k/N/m^2$ to kN/m^2 .
Page 96, sample problem, line of first ***ANS***: Change the minus sign before 0.660 to a plus sign.
Chapter 4
Page 133, Prob. 4.15, line 3: Change m/s to m^3/s .
Chapter 5
Page 181, Prob. 5.4, line 2: Change "drops" to "rises", change "even though" to "when".
Page 181, Fig. P5.7: Change "1 in" to "2 in".
Chapter 6
Page 206, Fig. 6.6, bottom right blue triangle: The v_1 arrowhead should point to the right (not to the left).
Page 231, Prob. 6.2, line 1: Insert "In" before "Sample".
Page 234, Prob. 6.19, line 1: Change 12 to 19 .
Page 238, Prob. 6.41, line 2: Change "height" to "the height z".
Page 238, Prob. 6.42, line 2: Change "at 15" to "of 15".
Chapter 8
Page 285, Equation (8.29): Change 5.73 to 5.76 .
Page 337, top equation: Change 4.727 to $4.727L$.
Page 347, Eq. 8.78: Change Kn/Q to KnQ (remove the /).
Page 356, Prob. 8.71, line 2: Change 55 to 5.5 .
Chapter 9
Page 384, 2nd line below Eq. 9.33: Change 25 to 24 .
Chapter 10
Page 440, 2nd new paragraph, line 2: Change "usually" to "often" .
Page 444, Exer. 10.8.3, line 1: After "wide" insert ", smooth," .
Page 449, Exer. 10.9.2, line 2: Insert "at" after "surface" .
Page 458, Sample Prob. 10.8, line 5: Change "jump" to "hump" .
Page 461, Exer. 10.13.4, line 3: Change 0.15 to 0.12 .
Page 492, Fig. 10.33(*b*): Move 3 in circle to center it under extension line at right end of dimension L.
Page 496, last 4 lines of Sample Prob: Delete last term on each line (corresponding to $V^2/2g$) (4 times).
Page 497, Eq. on line 4: Change 0.5 to 1.5 .
Page 497, Line 5: Change 1.107 to 1.196 .
Page 497, Line 11: Change 1.107 to 1.196 .
Page 506, Prob. 10.61, line 3: Change 0.00019 to 0.0019 .

/cont...

[1] The second printing of the companion textbook, *Fluid Mechanics with Engineering Applications*, ninth edition, is identifiable by 2 3 4 5 6 7 8 9 0 DOC DOC 9 0 8 7 6 (with the initial 1 missing) on the line above the ISBN on the copyright page.

Chapter 11

Page 536, Eq. 11.16: Remove right-most equals sign (should be multiplied).
Page 537, upper half of Fig. 11.19: Along **R**-axis, change "of approach" to "at throat".
Page 553, line 4: Add sentence: "The results are called Francis formulas."
Page 556, Exer. 11.11.4, line 3: Change h to H .
Page 567, Prob. 11.17, line 3: Change 75 to 7.5 .
Page 571, Prob. 11.45, line 8: Change $(k\sqrt{y})$ to $(k/\sqrt{y})$.

Chapter 12

Page 593, Footnote 3, line 3: Change $\Delta\rho$ to Δp (i.e., rho to 'pee').
Page 595, Eq. 12.16: After last =, insert − (minus).
Page 602, Prob. 12.2, line 1: Change 12.2 to 12.1 .

Chapter 13

Page 629, Sample Prob. 13.4, line 3 below "Solution": Change $p_b p_O$ to p_b/p_O
Page 630, text below Fig. S13.4, line 5: Change ρ_1 to ρ_O (sub ital cap 'Oh').
Page 630, text below Fig. S13.4, line 6: Change 746 to 747 , change 13.53 to 13.50 .
Page 633, line below first centered equation: Change ft^2/sec to ft^2/sec^2 .
Page 635, line below Eq. 13.46: Change $\rho_1 A_1 V_2$ to $\rho_1 A_1 V_1$.
Page 646, Eq. 13.64: Change lower limit of integral from 0 to 1 .
Page 652, Prob. 13.9, line 2: Insert "by calculator" before "that", change 13.53 to 13.50 .

Chapter 14

Page 673, line above Eq. 14.28: Change "flow area" to "area of the media".

Chapter 15

Page 688, Exer. 15.4.4, line 1: Change 1:15 to 1:10 .
Page 690, lower part of Fig. 15.6: Change "n = 1450" to "n_e = 1450" .
Page 691, lower part of Fig. 15.7: Change "n = 450" to "n_e = 450" .
Page 695, line 10 below "Solution": Change h to h_L (3 times).
Page 695, Exer. 15.7.2, line 1: Change 21 to 12 .
Page 710, first "word" of bottom line: Change h_L to h .
Page 716, Prob 15.31, end of line 3: Change to "prototype, given the model diameter is 7.4 in."

Chapter 16

Page 739, 2nd line above Eq. 16.13: Change "head h' that is effectively" to "***effective head h'*** ".

Page 739, Eq. 16.13: Change $\dfrac{(V_2 - V_C)^2}{2g}$ to $\left[\dfrac{V_2^2}{2g} - \dfrac{V_C^2}{2g}\right]$.

Page 739, Sample Prob. 16.7, line 6: Change "net" to "effective".

Page 740, line 3: Change $\dfrac{(V - V_C)^2}{2g} = \dfrac{(18.4 - 0)^2}{64.4}$ to $\dfrac{V_2^2}{2g} - \dfrac{V_C^2}{2g} = \dfrac{18.4^2}{2(32.2)} - 0$.

Page 740, line 13: Change $\dfrac{(V_2 - V_C)^2}{2g} = \dfrac{(2.6 - 0)^2}{64.4}$ to $\dfrac{V_2^2}{2g} - \dfrac{V_C^2}{2g} = \dfrac{2.6^2}{2(32.2)} - 0$.

Page 740, lines 15, 17, 18: Change "net" to "effective" (3 times).
Page 741, Exer. 16.8.3, line 2: Insert "net" before "head".
Page 746, lines 1 and 2 (Exer. 16.10.1): Change 50 to 48 (twice).
Page 758, Prob. 16.12, line 1: Change "penstock" to "pipe".
Page 758, Prob. 16.15, line 4: Change $D_j^2 V_j^3$ to D_j^2 .
Page 760, Prob. 16.26, line 4: Change ϕ to ϕ_e .
Page 761, Prob. 16.44, line 1: Change 5 to 4 .
Page 761, Prob. 16.44, line 5: Change ϕ to ϕ_e ; after "0.45." add "Neglect cavitation." .

Appendix B

Page 775, 3rd new paragraph, lines 3 and 4: Change "current student" to "version 6 *student*" .

/cont...

Appendix D: Answers to Exercises

Page 785, Exer. 2.11.8: (*a*) change 5.92 to 3.44 , (*b*) change 7.89 to 4.59 .
Page 786, Exer. 3.8.8(*b*): Change 2496 to 2500 , change 4.27 to 4.33 .
Page 786, Exer. 3.8.19(*a*): Change 1.8 to 1.80 .
Page 787, Exer. 4.10.2: Change 0.225 to 0.23 .
Page 787, Exer. 4.10.3: (*a*) change 13.86 to 13.9 , (*b*) change 10.37 to 10.4 .
Page 787, Exer. 4.10.4(*a*): Change 4.81 to 4.8 .
Page 787, Exer. 4.12.11: Change "57.0" to "Approx. 57", change 28.6 to 29 , change 63.8 to 64 .
Page 789, Exer. 6.4.4: Change lb to lb/ft .
Page 789, Exer. 6.6.2: (*a*) change 135.7 to 135.9 , (*c*) change 160.9 to 161.1 .
Page 789, Exer. 6.6.4: (*a*) change 125.1 to 125.2 , (*b*) change 63.7 to 63.8 , (*c*) change 140.3 to 140.5 .
Page 789, Exer. 6.6.6: Change 95.4 to 95.5 .
Page 789, Exer. 6.6.8(*b*): Change 5.52 to 5.53 .
Page 789, Exer. 6.6.9(*b*): Change 25.5 to 25.0 .
Page 789, Exer. 6.14.2: Change 15.69 to 15.70 .
Page 791, Exer. 8.25.4: Change 30.2 to 30.1 .
Page 791, Exer. 8.25.5: Change 20.4 to 20.3 .
Page 791, Exer. 9.3.3: Change 3450 to 3470 .
Page 791, Exer. 9.4.1: (*a*) change 0.0313 to 0.0337 , (*b*) change 0.0219 to 0.0236 ,
(*c*) change 0.019 36 to 0.0209 .
Page 791, Exer. 9.4.2: Change the terminal period to "; 2.8-4.5 times greater."
Page 791, Exer. 9.5.3: (*a*) change 0.1940 to 0.214 , (*b*) change 434 to 415 .
Page 791, Exer. 9.8.1: Change 2.76 to 8.47 .
Page 791, Exer. 9.10.1: Change −23.5 to −23.4 .
Page 791, Exer. 9.13.1: (*a*) change 117.0 to 116.7 , (*b*) change 13.40 to 13.37 ,
(*c*) change 139.8 to 140.0 .
Page 791, Exer. 9.13.2: Change 2.72 to 2.58 .
Page 791, Exer. 9.13.3: Change 2.72 to 2.58 .
Page 791, Exer. 9.14.1: (*a*) change 487 to −451 , (*b*) change 466 to −471 .
Page 791, Exer. 9.14.2: (*a*) change 97.5 to −96.0 , (*b*) change 90.6 to −102.9 .
Page 791, Exer. 10.3.1: Change "83.0, 309, 704, 1299" to "82.7, 308, 703, 1298".
Page 791, Exer. 10.3.2: Change 1285 to 970 .
Page 792, Exer. 10.3.5: (*a*) change 0.000 646 to 0.000 648 , (*b*) change 3.41 to 3.42 .
Page 792, Exer. 10.3.7: Change 0.017 73 to 0.017 83 .
Page 792, Exer. 10.3.10: Change 146.7 to 146.6 .
Page 792, Exer. 10.4.1: Change 13.57 to 13.55 .
Page 792, Exer. 10.4.2: Change 6.31 to 6.30 .
Page 792, Exer. 10.6.2: (*a*) change 98.8 to 98.5 , (*b*) change 85.1 to 84.9 .
Page 792, Exer. 10.7.1: Change "227" to "Approx. 237", change "3.68" to "approx. 3.55".
Page 792, Exer. 10.7.2: (*a*) change "33.6" to "Approx. 33.0", (*b*) change "36" to "approx. 35.4".
Page 792, Exer. 10.11.1: Change 0.003 04 to 0.003 05 .
Page 792, Exer. 10.13.4: Change 0.149 to 0.070 .
Page 792, Exer. 10.15.1: Change 1401 to 1377 .
Page 792, Exer. 10.15.2: Change 326 to 640 .
Page 792, Exer. 10.15.3: Change 0.009 90 to 0.009 88 .
Page 792, Exer. 10.15.4: Change 0.0114 to 0.011 41 .
Page 792, Exer. 10.17.7: Change 0.101 to 0.1009 .
Page 792, Exer. 10.17.9: Change 4.25 to 4.26 .
Page 792, Exer. 10.18.2(*b*): Change 2.82 to 2.81 .
Page 792, Exer. 10.22.2: Replace "1.518 m." by "(*a*) 1.518 m; (*b*) expanding."
Page 792, Exer. 10.23.1: Change 1.359 to 1.409 .

/cont...

Page 793, Exer. 11.1.4: Change 11 937 to 11 940 .
Page 793, Exer. 11.1.5: Change 0.0676 to 0.386 .
Page 793, Exer. 11.1.6: Change 1.059 to 20.4 .
Page 793, Exer. 11.1.7: Change 0.000 019 84 to 0.000 0227 .
Page 793, immediately below **11.1.9** insert the following:
11.1.10 0.0806 lb·sec/ft^2; 0.0647
Page 793, Exer. 11.3.2: Change 14.83 to 14.84 .
Page 793, Exer. 11.3.5: Change 1.155 to 1.156 .
Page 793, Exer. 11.6.5: Change −18 to −18.00 .
Page 793, Exer. 11.6.6: Change 10.51 to 10.90 .
Page 793, Exer. 11.7.1: Change 0.314 to 0.316 .
Page 793, Exer. 11.7.4: Change 4.04 to 4.84 .
Page 793, Exer. 11.8.1: (*a*) change 0.000 155 to 0.000 1548 , (*b*) change 0.014 67 to 0.014 65 , (*c*) change 1.467 to 1.465 .
Page 793, Exer. 11.9.1(*b*): Change 0.0382 to 0.0383 .
Page 793, Exer. 11.10.1: Change 14.96 to 169.4 .
Page 793, Exer. 11.10.3: Change 5.55 to 5.34 .
Page 793, Exer. 11.10.5: Change 6.27 to 5.87 .
Page 793, Exer. 11.10.7(*a*): Change 1029 to 1030 .
Page 793, Exer. 11.11.1: Change 23.7 to 23.6 .
Page 793, Exer. 11.11.2: Change 0.733 to 0.729 .
Page 793, Exer. 11.11.3(*a*): Change 22.8 to 22.7 .
Page 793, Exer. 11.11.4(*a*): Change 0.711 to 0.708 .
Page 793, Exer. 11.11.5(*b*): Change 137 to 137.5 .
Page 793, Exer. 11.12.1: Change 18.93 to 18.92 .
Page 793, Exer. 11.12.3: Change 2.458 to 2.46 .
Page 793, Exer. 11.14.2: Change 0.580 to 0.581 .
Page 794, Exer. 12.2.2: Change 15.6 to 15.64 .
Page 794, Exer. 12.2.7: Change 91.1 to 98.4 .
Page 794, Exer. 12.3.1: Change −0.0045 to −0.004 52 .
Page 794, Exer. 12.6.3(*a*): Change 74.7 to 74.8 .
Page 794, Exer. 12.7.2: Change 1.761 to 1.837 .
Page 794, Exer. 13.1.5: Change 37 200 to 37 300 , change 129 600 to 130 200 .
Page 794, Exer. 13.5.1: Change 14.99 to 14.98 , change 5.58 to 5.59 .
Page 794, Exer. 13.5.2: Change 106.1 to 127.9 , change 35.6 to 35.7 , change −29.9 to −30.0 .
Page 794, Exer. 13.5.3: Change 1159 to 1160 .
Page 794, Exer. 13.8.3: Change 823 to 825 , change 13.80 to 13.68 .
Page 794, Exer. 13.8.4: Change 153.7 to 153.8 , change −11.6 to −11.60 .
Page 794, Exer. 13.8.5: Change 645 to 644 , change 59.5 to 59.6 .
Page 794, Exer. 13.8.6: Change 196.2 to 201 , change 17.6 to 16.51 .
Page 794, Exer. 13.9.1: Change 1.451 to 1.453 .
Page 794, Exer. 13.9.2: Change 1.461 to 1.459 .
Page 794, Exer. 13.9.3: Change 1042 to 1039 , change −12 to −10.00 , change 1680 to 1681 , change −155 to −155.3 .
Page 795, Exer. 13.9.4: Change 8.56 to 8.57 .
Page 795, Exer. 13.9.5: Change "0.0789" to "and (*b*) 0.0790".
Page 795, Exer. 13.9.6: After "(*a*)" insert " and (*b*)".
Page 795, Exer. 13.10.1: Before "(*b*)" insert "(*a*) impossible situation; ", after "(*b*)" change 3.20 to 3.12 .
Page 795, Exer. 13.10.2: Before "(*b*)" insert "(*a*) impossible situation; ".
Page 795, Exer. 13.11.1: Change 1277 to 1299 , change 3.99 to 3.98 .

/cont...

Appendix D: Answers to Exercises -- cont.

Page 795, Exer. 14.4.3: (*c*) change $6xy$ to $3xy$, (*d*) change $4xy$ to $2xy$.
Page 795, Exer. 14.6.3(*a*): Change $2y-$ to $2y+$.
Page 795, Exer. 14.7.2: (*c*) change $6xy$ to $3xy$, (*d*) change $4xy$ to $2xy$, (*g*) change y^3/x to $y^3/3$.
Page 795, Exer. 14.8.4: Change 1.128 to 1.058 .
Page 795, Exer. 15.4.4: Change 90.7 to 89.9 .
Page 795, Exer. 15.7.1: Change "z + 0.1062" to "Δz + 0.1061", change "= 56" to "≈ 56", change "= 11,300" to "≈ 11,300", change "= 63" to "≈ 63", change "= 9700" to "≈ 9700", change "= 76" to "≈ 76", change "= 4800" to "≈ 4800", change "Pump" to "Approx. pump".
Page 795, Exer. 15.7.2: Change "z" to "Δz", change "= 72" to "≈ 72", change "= 6500" to "≈ 6500", change "= 74" to "≈ 74", change "= 5500" to "≈ 5500", change "= 77" to "≈ 77", change "= 2500" to "≈ 2500", change "Pump" to "Approx. pump", delete "about" .
Page 796, Exer. 15.9.1: Change "37.6" to "Approx. 38".
Page 796, Exer. 15.9.2: Change "10.0" to "Approx. 10".
Page 796, Exer. 15.9.3: Before "24.8" insert "Approx. ".
Page 796, Exer. 15.10.1: Change "6.1" to "Approx. 6.2".
Page 796, Exer. 15.10.2: Change "3.9" to "Approx. 3.8".
Page 796, Exer. 15.10.3: Change 5.0 to 5.01 .
Page 796, Exer. 15.12.1: Change 2674 to 2670 , change "18.3" to "about 18".
Page 796, Exer. 15.12.2: Change "12.3" to "about 12".
Page 796, Exer. 15.12.3: Change "than 3.9" to "than about 3.9".
Page 796, Exer. 15.13.1: Before "124" insert "Approx. ".
Page 796, Exer. 15.13.2: Before "1650" insert "Approx. 170 gpm; ".
Page 796, Exer. 16.3.1: Before "Horsepower" insert "278 hp. ", after "runner" insert a comma.
Page 796, Exer. 16.3.3: Change 63 000 to 63 100 (twice), change 247 to 2480 .
Page 796, Exer. 16.5.1: Before "1.0" insert "About ", after "with" insert "about".
Page 796, Exer. 16.5.2: Change 168 to 167 .
Page 796, Exer. 16.7.1: Change 8.76 to 8750 .
Page 796, Exer. 16.8.1: Change 3.2 to 3.23 .
Page 796, Exer. 16.10.1: Change 88 to 88.3 , change 12.63 to 12.65 .
Page 796, Exer. 16.10.3: Change "52.9 ft" to "273 rpm", change 281.8 to 282 , change 1572 to 1573 , change "44.4 ft" to "250 rpm", change 258.3 to 258 .
Page 796, Exer. 16.11.1: Change "4.0, 9.2 ft." to "4.01, approx. 8.8 ft.".
Page 796, Exer. 16.11.2: Change "6.4" to "approx. 6.4".
Page 796, Exer. 16.11.3: Change 146 to 146.0 , change "3.0" to "approx. 2.71".
Page 796, Exer. 16.12.1: Change "207" to "approx. 157".
Page 796, Exer. 16.12.2: Change "40.6" to "Approx. 42.5", change "30.6" to "approx. 30.2".
Page 796, Exer. 16.12.3: Change "32.5 ft, 20.7 ft." to "Approx. 77 ft.".
Page 796, Exer. 16.13.1: Change second 7.0 to 6.99 , change "7.7 ft" to "approx. 6.8 ft", change "6.9 ft." to "approx. 6.1 ft."
Page 796, Exer. 16.13.2: Change 50 to 50.1 (twice), change second 20 to 20.1 , change "3.74 ft" to "approx. 3.7 ft", change "1.60 ft." to "approx. 1.54 ft."
Page 796, Exer. 16.13.3: Change "13 ft." to "approx. 13.1 ft.".
Page 796, Exer. 16.13.4: Change 11.0 to 11.06 .

Back matter

Back inside cover, Flow rate Q, column 3 under "By": Change 33.314 7 to 35.314 7 .